Encyclopedia of Molecular Cell Biology
and Molecular Medicine

Edited by Robert A. Meyers

Volume 8
Mass Spectrometry-based Methods of
Proteome Analysis to Mucoviscidosis
(Cystic Fibrosis), Molecular Cell Biology of

Encyclopedia of Molecular Cell Biology and Molecular Medicine

Editorial Board

Encyclopedia of Molecular Cell Biology and Molecular Medicine

Edited by Robert A. Meyers

Second Edition

Volume 8
Mass Spectrometry-based Methods of Proteome Analysis to
Mucoviscidosis (Cystic Fibrosis), Molecular Cell Biology of

WILEY-
VCH

WILEY-VCH Verlag GmbH & Co. KGaA

Editor:

Dr. Robert A. Meyers
President, Ramtech Limited
7 Harbor Point Drive, 7A
Mill Valley, CA 94941
USA

Library of Congress Card No.: applied for

**British Library Cataloguing-in-Publication
Data**: A catalogue record for this book is
available from the British Library.

**Bibliographic information published by
Die Deutsche Bibliothek**
Die Deutsche Bibliothek lists this publication
in the Deutsche Nationalbibliografie; detailed
bibliographic data is available in the internet
at http://dnb.ddb.de.

©WILEY-VCH Verlag GmbH & Co. KGaA
Weinheim, 2005

Printed in the Federal Republic of Germany.
Printed on acid-free paper.

Composition: Laserwords Private Ltd,
Chennai, India
Printing: Druckhaus Darmstadt GmbH,
Darmstadt
Bookbinding: Litges & Dopf Buchbinderei
GmbH, Heppenheim
ISBN-13: 978-3-527-30550-6
ISBN-10: 3-527-30550-5

Preface

The *Encyclopedia of Molecular Cell Biology and Molecular Medicine*, which is the successor and second edition of the *Encyclopedia of Molecular Biology and Molecular Medicine* (VCH Publishers, Weinheim), covers the molecular and cellular basis of life at a university and professional researcher level. The first edition, published in 1996–97, was very successful and is being used in libraries around the world. This second edition will almost double the first edition in length and will comprise the most detailed treatment of both molecular cell biology and molecular medicine available today. The Board Members and I believe that there is a serious need for this publication, even in view of the vast amount of information available on the World Wide Web and in text books and monographs. We feel that there is no substitute for our tightly organized and integrated approach to selection of articles and authors and implementation of peer review standards for providing an authoritative single-source reference for undergraduate and graduate students, faculty, librarians, and researchers in industry and government.

Our purpose is to provide a comprehensive foundation for the expanding number of molecular biologists, cell biologists, pharmacologists, biophysicists, biotechnologists, biochemists, and physicians, as well as for those entering molecular cell biology and molecular medicine from majors or careers in physics, chemistry, mathematics, computer science, and engineering. For example, there is an unprecedented demand for physicists, chemists, and computer scientists who will work with biologists to define the genome, proteome, and interactome through experimental and computational biology.

The Board Members and I first divided the entire study of molecular cell biology and molecular medicine into primary topical categories and further defined each of these into subtopics. The following is a summary of the topics and subtopics:

- *Nucleic Acids:* amplification, disease genetics overview, DNA structure, evolution, general genetics, nucleic acid processes, oligonucleotides, RNA structure, RNA replication and transcription.
- *Structure Determination Technologies Applicable to Biomolecules:* chromatography, labeling, large structures, mapping, mass spectrometry, microscopy, magnetic resonance, sequencing, spectroscopy, X-ray diffraction.
- *Biochemistry:* carbohydrates, chirality, energetics, enzymes, biochemical genetics, inorganics, lipids, mechanisms, metabolism, neurology, vitamins.

Encyclopedia of Molecular Cell Biology and Molecular Medicine, 2nd Edition. Volume 8
Edited by Robert A. Meyers.
Copyright © 2005 Wiley-VCH Verlag GmbH & Co. KGaA, Weinheim
ISBN: 3-527-30550-5

- *Proteins, Peptides, and Amino Acids:* analysis, enzymes, folding, mechanisms, modeling, peptides, structural genomics (proteomics), structure, types.
- *Biomolecular Interactions:* cell properties, charge transfer, immunology, recognition, senses.
- *Cell Biology:* developmental cell biology, diseases, dynamics, fertilization, immunology, organelles and structures, senses, structural biology, techniques.
- *Molecular Cell Biology of Specific Organisms:* algae, amoeba, birds, fish, insects, mammals, microbes, nematodes, parasites, plants, viruses, yeasts.
- *Molecular Cell Biology of Specific Organs or Systems:* excretory, lymphatic, muscular, nervous, reproductive, skin.
- *Molecular Cell Biology of Specific Diseases:* cancer, circulatory, endocrinal, environmental stress, immune, infectious, neurological, radiational.
- *Pharmacology:* chemistry, disease therapy, gene therapy, general molecular medicine, synthesis, toxicology.
- *Biotechnology:* applications, diagnostics, gene-altered animals, bacteria and fungi, laboratory techniques, legal, materials, process engineering, nanotechnology, production of classes or specific molecules, sensors, vaccine production.

We then selected some 400 article titles and author or author teams to cover the above topics. Each article is designed as a self-contained treatment which begins with a keyword section including definitions, to assist the scientist or student who is unfamiliar with the specific subject area. The Encyclopedia includes more than 3000 key words, each defined within the context of the particular scientific field covered by the article. In addition to these definitions, the glossary of basic terms found at the back of each volume, defines the most commonly used terms in molecular cell biology. These definitions, along with the reference materials (the genetic code, the common amino acids, and the structures of the deoxyribonucleotides) printed at the back of each volume, should allow most readers to understand articles in the Encyclopedia without referring to a dictionary, textbook, or other reference work. There is, of course, a detailed subject index in Volume 16 as well as a cumulative table of contents and list of authors, as well as a list of scientists who assisted in the development of this Encyclopedia.

Each article begins with a concise definition of the subject and its importance, followed by the body of the article and extensive references for further reading. The references are divided into secondary references (books and review articles) and primary research papers. Each subject is presented on a first-principle basis, including detailed figures, tables and drawings. Because of the self-contained nature of each article, some articles on related topics overlap. Extensive cross-referencing is provided to help the reader expand his or her range of inquiry.

The articles contained in the Encyclopedia include core articles, which summarize broad areas, directing the reader to satellite articles that present additional detail and depth for each subject. The core article Brain Development is a typical example. This 45-page article spans neural induction, early patterning, differentiation, and wiring at a molecular through to cellular and tissue level. It is directly supported, and cross-referenced, by a number of molecular neurobiology satellite articles, for example, Behavior Genes, and further supported by other core presentations, for example,

Developmental Cell Biology; Genetics, Molecular Basis of, and their satellite articles. Another example is the core article on Genetic Variation and Molecular Evolution by Werner Arber. It is supported by a number of satellite articles supporting the evolutionary relatedness of genetic information, for example, Genetic Analysis of Populations.

Approximately 250 article titles from the first edition are retained, but rewritten, half by new authors and half by returning authors. Approximately 80 articles on cell biology and 70 molecular biology articles have been added covering areas that have become prominent since preparation of the first edition. Thus, we have compiled a totally updated single source treatment of the molecular and cellular basis of life.

Finally, I wish to thank the following Wiley-VCH staff for their outstanding support of this project: Andreas Sendtko, who provided project and personnel supervision from the earliest phases, and Prisca-Maryla Henheik and Renate Dötzer, who served as the managing editors.

November 2003

Robert A. Meyers
Editor-in-Chief

Editor-in-Chief

Robert A. Meyers

Dr. Meyers earned his Ph.D. in organic chemistry from the University of California Los Angeles, was a post-doctoral fellow at California Institute of Technology and manager of chemical processes for TRW Inc. He has published in *Science*, written or edited 12 scientific books and his research has been reviewed in the *New York Times* and the *Wall Street Journal*. He is one of the most prolific science editors in the world having originated, organized and served as Editor-in-Chief of three editions of the *Encyclopedia of Physical Science and Technology*, the *Encyclopedia of Analytical Chemistry* and two editions of the present *Encyclopedia of Molecular Cell Biology and Molecular Medicine*.

Encyclopedia of Molecular Cell Biology and Molecular Medicine, 2nd Edition. Volume 8
Edited by Robert A. Meyers.
Copyright © 2005 Wiley-VCH Verlag GmbH & Co. KGaA, Weinheim
ISBN: 3-527-30550-5

Editorial Board

Contents

Encyclopedia of Molecular Cell Biology and Molecular Medicine, 2nd Edition. Volume 8
Edited by Robert A. Meyers.
Copyright © 2005 Wiley-VCH Verlag GmbH & Co. KGaA, Weinheim
ISBN: 3-527-30550-5

List of Contributors

Abebe Akalu
New York University School of Medicine,
New York,
USA

Charles L. Asbury
Department of Biological Sciences,
Stanford University, Stanford, CA,
USA

and

Present Address: Department of Physiology
and Biophysics,
University of Washington,
Seattle, WA,
USA

Yoshinobu Baba
National Institute of Advanced Industrial
Science and Technology,
Takamatsu,
Japan

University of Tokushima,
Shomachi,
Tokushima,
Japan

and

Nagoya University,
Nagoya,
Japan

Rumiana Bakalova
National Institute of Advanced
Industrial Science and Technology,
Takamatsu,
Japan

Roland Benz
Biocenter of the University of Würzburg,
Würzburg,
Germany

Steven M. Block
Department of Biological Sciences,
Stanford University,
Stanford, CA,
USA

and

Department of Applied Physics,
Stanford University,
Stanford, CA,
USA

Peter C. Brooks
New York University School of Medicine,
New York,
USA

Jean-Marc Cavaillon
Institut Pasteur,
Paris,
France

Encyclopedia of Molecular Cell Biology and Molecular Medicine, 2nd Edition. Volume 8
Edited by Robert A. Meyers.
Copyright © 2005 Wiley-VCH Verlag GmbH & Co. KGaA, Weinheim
ISBN: 3-527-30550-5

Michael J. Clague
University of Liverpool,
Liverpool,
UK

Robert J. Cousins
University of Florida,
Gainesville, FL,
USA

Gerd Döring
Universitätsklinikum Tübingen,
Tübingen,
Germany

James Eberwine
University of Pennsylvania,
Philadelphia, PA,
USA

Walther R. Ellis
Utah State University,
Logan, UT,
USA

Caroline Engvall
Uppsala University,
Uppsala,
Sweden

Ashraf Ewis
National Institute of Advanced Industrial
Science and Technology,
Takamatsu,
Japan

and

El-Minia University,
El-Minia,
Egypt

James G. Ferry
The Pennsylvania State University,
University Park,
PA 16801,
USA

Paul Robert Fisher
La Trobe University,
Victoria,
Australia

David J. Grainger
Department of Medicine,
University of Cambridge,
Cambridge,
UK

Georg Haase
Institut de Neurobiologie de la
Méditerranée,
Marseille,
France

Zhaozhang Han
Argonne National Laboratory,
Argonne, IL,
USA

Ece Karatan
Argonne National Laboratory,
Argonne, IL,
USA

Brian Kay
Argonne National Laboratory,
Argonne, IL,
USA

Per Lundahl
Uppsala University,
Uppsala,
Sweden

Michael P. Murphy
MRC Dunn Human Nutrition Unit,
Cambridge,
UK

Keiichi Namba
Osaka University,
Osaka,
Japan

and

Dynamic NanoMachine Project,
Suita, Osaka,
Japan

Jeremy K. Nicholson
Department of Biological Chemistry,
Imperial College,
London,
UK

W. Olszewska
Imperial College,
Paddington, London,
UK

Peter J. M. Openshaw
Imperial College,
Paddington, London,
UK

Madeline E. Rasche
University of Florida,
Gainesville, FL,
USA

Felix Ratjen
University of Essen,
Germany

Anireooy S.N. Reddy
Colorado State University,
Fort Collins,
CO

Meredith F. Ross
MRC Dunn Human Nutrition Unit,
Cambridge,
UK

Emily Rozak
University of Pennsylvania,
Philadelphia, PA,
USA

Jeffrey H. Schwartz
University of Pittsburgh,
Pittsburgh, PA,
USA

Michael Sendtner
Institute of Clinical Neurobiology,
Würzburg,
Germany

Jennifer Spaethling
University of Pennsylvania,
Philadelphia, PA,
USA

David J. Triggle
State University of New York,
Buffalo, NY,
USA

Sylvie Urbé
University of Liverpool,
Liverpool,
UK

Michael P. Washburn
Stowers Institute for Medical Research,
Kansas City, MO,
USA

Koji Yonekura
University of California,
San Francisco, CA,
USA

and

Osaka University,
Osaka,
Japan

and

Dynamic NanoMachine Project,
Suita, Osaka,
Japan

Boris L. Zybailov
Stowers Institute for Medical Research,
Kansas City, MO,
USA

Color Plates

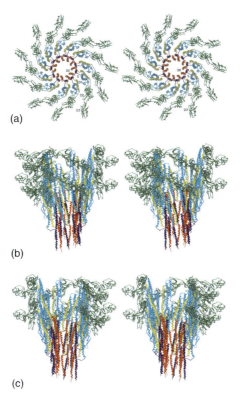

(a)

(b)

(c)

Fig. 7 (p. 400) Ribbon diagram of the Cα backbone of the filament model in stereo view. (a) End-on view from the distal end of the filament. Eleven subunits are displayed. (b) Side view from outside the filament. Three protofilaments on the far side have been removed for clarity. (c) Side view from inside the filament. Three protofilaments on the near side have been removed. Top and bottom of the side view images correspond to the distal and proximal ends of the filament respectively. The chain is color coded as in Fig. 6(a). Permission from Yonekura, K., Maki-Yonekura, S., Namba, K. (2003) Complete atomic model of the bacterial flagellar filament by electron cryomicroscopy, *Nature* **424**, 643–650.

Encyclopedia of Molecular Cell Biology and Molecular Medicine, 2nd Edition. Volume 8
Edited by Robert A. Meyers.
Copyright © 2005 Wiley-VCH Verlag GmbH & Co. KGaA, Weinheim
ISBN: 3-527-30550-5

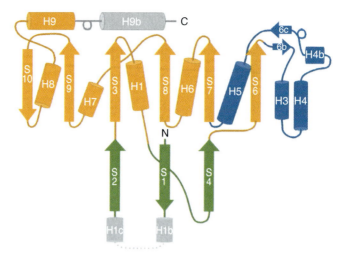

Fig. 2 (p. 628) Topology diagram of mNBD1. The F1-type ATP-binding core subdomain is shown in gold, the ABC α-subdomain in cyan, and the ABC ß-subdomain in green. Regions of mNBD1 that are different from previous ABC structures are shown in gray. Circles indicate the positions of 3_{10} helices. (From Lewis, H.A. et al. (2004) Structure of nucleotide-binding domain 1 of the cystic fibrosis transmembrane conductance regulator. *EMBO J.*, **23**, 282–293).

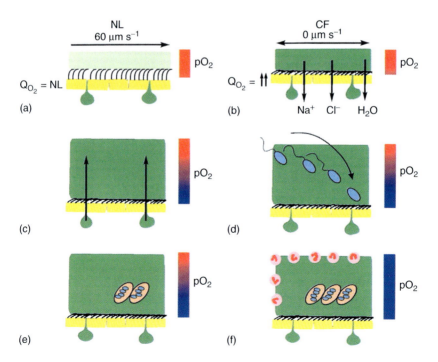

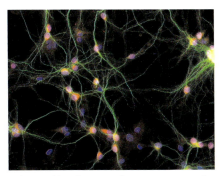

Fig. 2 (p. 504)   Immunohistochemical localization of Map2 protein to the somatodendritic region of neurons – A rabbit polyclonal antibody raised against Map2 was applied to neurons of rat hippocampal neurons in primary culture. This first antibody was localized with a goat antirabbit secondary antibody that was fluorescein labeled (Green color). Green staining was observed in the fine hairlike structures that radiate from the cell soma. The nucleus is stained with DAPI. Glial cells in the primary cell culture system have a strong DAPI staining but no Map2 staining consistent with the absence of Map2 protein form glial cells.

Fig. 3 (p. 631)   Schematic model of the pathogenic events hypothesized to lead to chronic *P. aeruginosa* infection in airways of CF patients. (a) On normal airway epithelia, a thin mucus layer (light green) resides atop the PCL (clear). The presence of the low-viscosity PCL facilitates efficient mucociliary clearance (denoted by vector). A normal rate of epithelial O_2 consumption (Q_{O2}; left) produces no O_2 gradients within this thin ASL (denoted by red bar). (b–f) CF airway epithelia. (b) Excessive CF volume depletion (denoted by vertical arrows) removes the PCL, mucus becomes adherent to epithelial surfaces, and mucus transport slows/stops (bidirectional vector). The raised O_2 consumption (left) associated with accelerated CF ion transport does not generate gradients in thin films of ASL. (c) Persistent mucus hypersecretion (denoted as mucus secretory gland/goblet cell units; dark green) with time increases the height of luminal mucus masses/plugs. The raised CF epithelial Q_{O2} generates steep hypoxic gradients (blue color in bar) in thickened mucus masses. (d) *P. aeruginosa* bacteria deposited on mucus surfaces penetrate actively and/or passively (due to mucus turbulence) into hypoxic zones within the mucus masses. (e) *P. aeruginosa* adapts to hypoxic niches within mucus masses with increased alginate formation and the creation of macrocolonies. (f) Macrocolonies resist secondary defenses, including neutrophils, setting the stage for chronic infection. The presence of increased macrocolony density and, to a lesser extent neutrophils, render the now mucopurulent mass hypoxic (blue bar) (From Worlitzsch, D., Tarran, R., Ulrich, M. et al(2002) Effects of reduced mucus oxygen concentration in airway Pseudomonas infections of cystic fibrosis patients, *J. Clin. Invest.* **109**, 317–325).

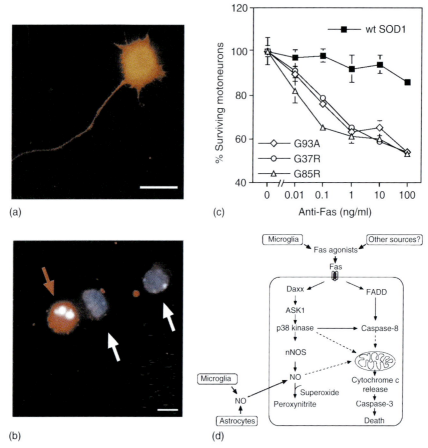

(a)

(b)

(c)

(d)

Fig. 1 (p. 546) Isolated motor neurons can be used to study mutant SOD1-linked motor neuron death. (a) Human superoxide SOD1 is expressed in cultured embryonic motor neurons from transgenic SOD1 G93A mice. (b) Treatment of SOD1 G85R motor neuron cultures with agonistic antibodies to the cell surface receptor Fas leads to increased apoptosis, as detected by staining with DAPI (in blue) and immunolabeling for activated caspase 3 (in red). An apoptotic motor neuron with strong caspase-3 activation and nuclear condensation c (red arrow) can be distinguished from healthy motor neurons displaying only weak caspase-3 activation and normal chromatin structure (white arrows). Scale bars: 25 μm. (c) Motor neurons from transgenic mutant SOD1 G93A, G85R, and G37R mice show higher susceptibility to Fas-triggered cell death than motor neurons from mice expressing wild-type SOD1. (d) Model of Fas-triggered motor neuron death. Cell death involves the classical FADD/caspase-8 pathway and a parallel pathway leading from Daxx, ASK1, and p38 activation to transcriptional upregulation of neuronal NO synthase (nNOS) and NO production. The presence of mutant SOD1 sensitizes motor neurons to Fas agonists and NO. Potential sources for these cell death triggers are astrocytes and microglia. See Fig. 2.

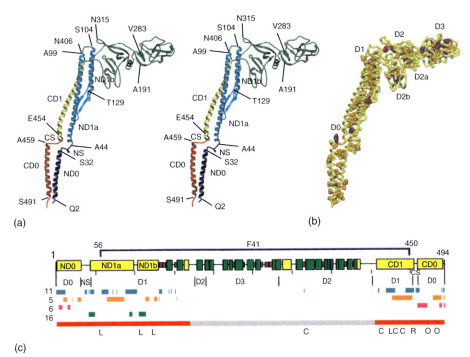

(a)

(b)

(c)

Fig. 6 (p. 399)   The Cα backbone trace, hydrophobic side-chain distribution, and structural information of flagellin. (a) Stereo diagram of the Cα backbone. The chain is color coded as follows: residues 1 to 44, blue; 44 to 179, cyan; 179 to 406, green; 406 to 454, yellow; 454 to 494, red. (b) Distribution of hydrophobic side chains, mainly showing hydrophobic cores that define domains D0, D1, D2a, D2b, and D3. Side-chain atoms are color-coded: Ala and Met, yellow; Leu, Ile, and Val, orange; Phe, Tyr, and Pro, purple (carbon) and red (oxygen). (c) Position and region of various structural features in the amino acid sequence of flagellin. Shown are, from top to bottom, the atomic model of a major fragment of flagellin called F41 in blue; the secondary structure distribution with α-helix in yellow, β-structure in green, and β-turn in purple; tic mark at every 50th residue in blue; domains D0, D1, D2, and D3, and spoke regions NS and CS; the subunit contact regions along the 11-start in cyan, along the 5-start in orange, along the 6-start in pink, and along the 16-start in green; the well-conserved amino acid sequence in red and variable region in violet; point mutations that produce the filament of different supercoils. Letters at the bottom indicate the morphology of mutant filaments: L (F53V, D107E, R124A, R124S, G426A), L-type straight; R (A449V), R-type straight; C (D313Y, A414V, A427V, N433D) curly; O (Q472L, Q481L, Q481S) coiled. Permission from Yonekura, K., Maki-Yonekura, S., Namba, K. (2003) Complete atomic model of the bacterial flagellar filament by electron cryomicroscopy, *Nature* **424**, 643–650.

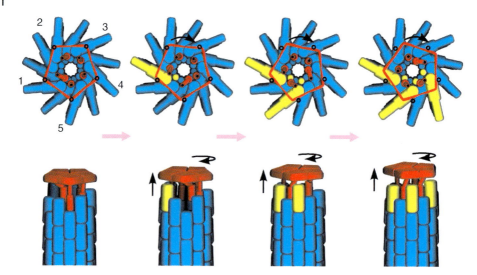

Fig. 10 (p. 407) Schematic diagram depicting the rotary cap mechanism promoting the flagellin assembly. This describes the rotation and axial translation of the cap plate and accompanied rearrangement of the legs upon every incorporation of a flagellin subunit (from left to right). Top view in the upper panel and an oblique view in the lower panel. In the upper panel, the cap plate is made transparent to show the different ways of leg domain binding, where black dots indicate the five positions of leg domain attachment to the plate. The plate and the black dots have strict fivefold symmetry, while the leg domains do not. In the lower panel, the outer domain of flagellin is removed for clarity. Subunits in yellow are newly incorporated flagellin molecules. Five open circles in the upper panel indicate the initial positions of the cap plate vertices as a reference for the cap rotation. The flagellin assembly proceeds along the 1-start helix, which is in the counterclockwise direction when viewed from the top, approximately at every $65.5°$ ($360 \times 2/11$). This is also the angle of rotation after which the next binding site appears. However, because the legs of the cap are located every $72°$ ($360/5$), a $6.5°$ clockwise rotation with permutation of the leg conformations is sufficient to make the appropriate interactions between the leg and flagellin subunits. Numbers indicate the directions of views in Fig. 9(a). Animation at http://www.npn.jst.go.jp/yone.html. Permission from Yonekura, K., Maki, S., Morgan, D.G., DeRosier, D.J., Vonderviszt, F., Imada, K., Namba, K. (2000) The bacterial flagellar cap as the rotary promoter of flagellin self-assembly, *Science* **290**, 2148–2152.

Fig. 13 (p. 315) Proposed pathways regulating cyclin-dependent kinases during pseudohyphal differentiation in *Saccharomyces cerevisiae*. From the nutritional signals that elicit filamentation, the pathways proceed via activation of a MAP kinase phosphorylation cascade, cAMP-activated protein kinase (PKA), Swe1 tyrosine protein kinase (a Cln/CDK inhibitor), or the stress-induced transcription factor Xbp1. Ste11 is the yeast equivalent of mammalian MAPKKK (MAPKK kinase), Ste7 the equivalent of MAPKK (MAPK kinase) and Kss1 is a MAP kinase (MAPK). Flo8 and Tek1/Ste12 are transcription factors. Unphosphorylated Kss1 inhibits Ste12 and its phosphorylation relieves this inhibition. Flo8 and Tek1/Ste12 are transcription factors. The pathways activating Swe1 and Xbp1 are uncertain and both may be coupled to the MAP kinase pathway. A variety of other proteins whose activities influence CDK activities also have corresponding effects on filamentation. Green ellipses indicate proteins whose activity stimulates apical growth, while red ellipses indicate proteins whose activity inhibits it. Arrowheads indicate stimulatory interactions and barred ends indicate inhibitory interactions. Blue arrows stand for interactions at the transcriptional level (gene induction or repression) and red arrows stand for interactions at the protein level (activation or inhibition).

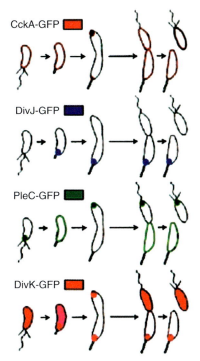

Fig. 4 (p. 298) Localization of some morphogenetic signaling proteins during the *Caulobacter* cell cycle. CckA and PleC are membrane-bound and cycle as shown between being dispersed throughout the membrane and concentrated at the indicated poles. DivJ, when present, is always concentrated at the stalked (ST) pole. DivK cycles between a diffuse cytoplasmic localization in swarmer cells and concentration at the indicated poles in predivisional and stalked cells. The length of the arrows indicates the time spent between each stage of differentiation. From Fig. 2 of Jacobs-Wagner, C. (2004) Regulatory proteins with a sense of direction: cell cycle signalling network in *Caulobacter, Mol. Microbiol.* **51**, 7–13.

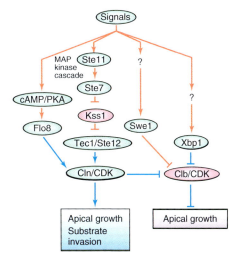

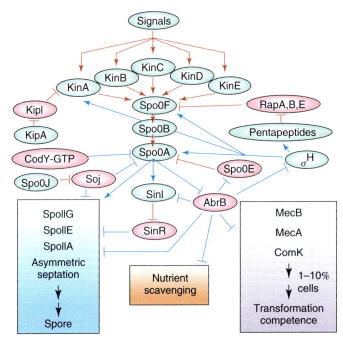

Fig. 7 (p. 302) Regulation of the phosphorelay and the initiation of sporulation in *Bacillus subtilis*. During the growth phase, AbrB represses genes involved in the three stationary phase survival strategies (boxed) – sporulation, nutrient scavenging, and transformation. At the onset of stationary phase, multiple signals result in activity of the phosphorelay, producing intermediate levels of phosphorylated Spo0A. These levels of Spo0A~P are sufficient to support the nutrient scavenging and, eventually, in some cells the transformation strategy for survival. Under more extreme circumstances of nutritional stress and high density, several positive feedbacks can combine with the lifting of several negative restraints to elicit production of high levels of Spo0A~P. At these high levels, the expression of sporulation-specific genes is induced and the cell progresses down the pathway of endospore formation. Green ellipses indicate proteins whose activity stimulates the phosphorelay, while red ellipses indicate proteins whose activity inhibits it. Arrowheads indicate stimulatory interactions and barred ends indicate inhibitory interactions. Blue arrows stand for interactions at the transcriptional level (gene induction or repression) and red arrows stand for interactions at the protein level (activation or inhibition). MekR, MekB, and ComK are AbrB-repressible proteins involved in the development of transformation competence. Other proteins are described in the text.

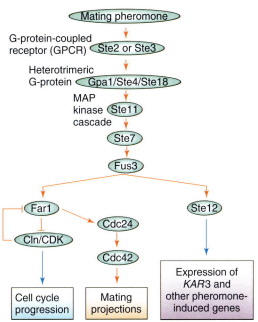

Fig. 16 (p. 318) Signal transduction pathways controlling gametogenesis in *Saccharomyces*. Binding of the pheromone to its cognate receptor (Ste2 = α factor receptor, Ste3 = a factor receptor) causes a conformational change and dissociation of the heterotrimeric G-protein (Gpa1 = α subunit, Ste4 = β subunit, Ste18 = γ subunit). The Ste4/Ste18 ($\beta\gamma$) heterodimer activates Ste20 a serine/threonine protein kinase, which initiates the MAP kinase phosphorylation cascade. Ste11 (MAPKKK), Ste7 (MAPKK) and Fus3 (MAPK) are each in their turn phosphorylated by the preceding kinase and thereby activated. At the end of the phosphorylation cascade, Fus3 phosphorylates and activates the Cln/CDK inhibitor Far1 as well as the transcription factor Ste12. As described in the text, the former causes cell cycle arrest in G1 and formation of mating projections, while the latter causes pheromone-inducible gene expression. Arrowheads indicate stimulatory interactions and barred ends indicate inhibitory interactions. Blue arrows stand for interactions at the transcriptional level (gene induction or repression) and red arrows stand for interactions at the protein level (activation or inhibition).

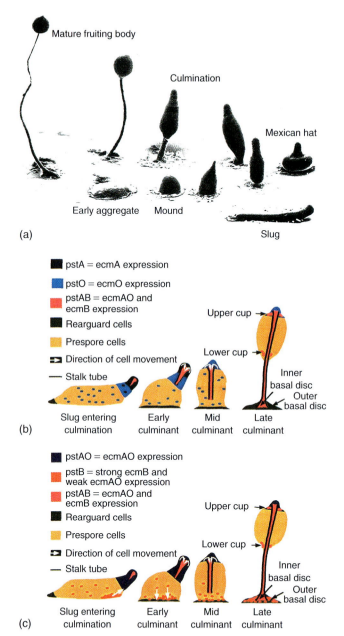

Fig. 18 (p. 322)(*Continued on page xxxiii*) Multicellular development in *Dictyostelium*. Panel A. A montage of scanning electron micrographs of stages in the *Dictyostelium* life cycle. Successive developmental stages are shown proceeding anticlockwise from the early aggregate formed by chemotactic aggregation of starving cells. The mature fruiting body is approximately 2 mm high. The original image was kindly provided by M.J. Grimson and R.L. Blanton, Biological Sciences Electron

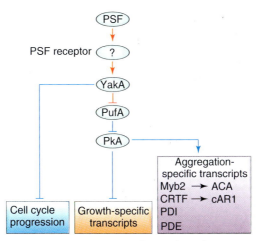

Fig. 19 (p. 325) The Prestarvation Factor (PSF) signaling pathway for initiating *Dictyostelium* development. PSF binds to an unknown receptor presumably belonging to the GPCR (G-protein-coupled Receptor) superfamily and elicits a signaling cascade that activates the protein kinase YakA, which, in turn, phosphorylates and inhibits the RNA-binding protein PufA. In its active, nonphosphorylated form, PufA binds to the 3′ end of cAMP-dependent protein kinase (PKA) mRNA preventing its translation. Once made and activated, PKA induces downstream genes directly or indirectly by phosphorylation of target proteins, ultimately regulating specific transcription factors such as Myb2 or CRTF (cAMP responsive transcription factor). Growth phase-specific transcripts are repressed and aggregation-specific transcripts are induced. These include those required for synthesizing and secreting cAMP such as adenylyl cyclase A (ACA) and for sensing and responding to extracellular cAMP signals, such as the cAMP receptor (cAR1). This establishes an autoactivatory feedback loop for induction of early development. YakA also phosphorylates other targets to arrest cell cycle progression at the phase shift point PS where cells exit from the cell cycle and enter differentiation. Arrowheads indicate stimulatory interactions and barred ends indicate inhibitory interactions. Blue arrows stand for interactions affecting protein expression (at the transcriptional or posttranscriptional level), and red arrows stand for interactions affecting protein activity (activation or inhibition).

(*Continued from page xxxii*) Microscopy Laboratory, Texas Tech University. Panel B. Diagrammatic representation of culmination where, for the sake of clarity, the band of pstB cells that will form the outer basal disc (see Panel C) is not shown. The *ecmA* promoter can be divided into two parts, a proximal part (the *ecmA* region) that directs expression predominantly in the cells within the tip (ie. in the pstA cells), and a distal part (the *ecmO* region) that directs expression in cells in the back of the prestalk region (the pstO cells) and in a subset (pstO:ALC cells) of the anterior-like cells (ALC). The whole *ecmA* promoter (the *ecmAO* promoter) directs expression in all these cell subtypes and has been termed the *pstAO population*. From Plate 6 of Maeda et al.(1997) *Dictyostelium discoideum – A Model Organism for Cell and Developmental Biology*. Universal Academy Press Inc., Tokyo, Japan. Image kindly provided by Prof. J.G. Williams, University of Dundee. Panel C. PstB cell behavior at culmination. The pstB cells are defined by selective staining with neutral red and because they express the *ecmB* gene at a high level relative to the *ecmA* gene. They have a complex movement pattern during slug migration. In this representation, for the sake of simplicity, separate pstA and pstO populations are not shown, but the behavior of the entire pstAO population is represented. From Plate 7 of Maeda et al. (1997). Image kindly provided by Prof. J.G. Williams, University of Dundee.

Fig. 1 (p. 240)   A modern-day Volta experiment.

Encyclopedia of Molecular Cell Biology and Molecular Medicine

Second Edition

Mass Spectrometry-based Methods of Proteome Analysis

Boris L. Zybailov and Michael P. Washburn
Stowers Institute for Medical Research, Kansas City, MO, USA

Encyclopedia of Molecular Cell Biology and Molecular Medicine, 2nd Edition. Volume 8
Edited by Robert A. Meyers.
Copyright © 2005 Wiley-VCH Verlag GmbH & Co. KGaA, Weinheim
ISBN: 3-527-30550-5

Keywords

2D Page
Two-dimensional polyacrylamide gel electrophoresis; a technology that separates intact proteins according to their pI in the first dimension and molecular weight in the second.

MS/MS
A mass spectrometry technique in which precursor ions are selectively fragmented, thus allowing detailed structural information to be obtained.

MudPIT
A multidimensional protein identification technology; a proteomic method based on liquid chromatography of peptide mixtures with subsequent identification by tandem mass spectrometry.

Quantitative Proteomics
A group of methods that allow large-scale quantitative assessment of protein expression and identification.

Proteomics aims to identify, characterize, and map gene functions at the protein level for whole cells or organisms. A typical experimental scheme for a large-scale proteomics inquiry involves fractionation of a complex protein mixture by electrophoretic or chromatographic means followed by subsequent identification of the components in the individual fractions by mass spectrometry. Owing to continuous and rapid improvement in instrument sensitivity, throughput capacity, software versatility, and techniques of statistical validation of the data, mass spectrometry–based approaches are becoming mainstream methods in a proteome analysis.

1
Principles and Instrumentation

1.1
Proteome and Proteomics

The term *proteome* was first introduced in mid-1990s to name the functional complement of a genome. By analogy with *genomics*, the term *proteomics* refers to studies of a gene's function at the protein level. Both these terms have a large-scale flavor to them. Indeed, studies that fall under the category "proteomics" frequently deal with large-scale analyses of proteins on the level of the whole organism, tissue, cell, or subcellular compartments.

Examples of typical biological questions addressed by modern proteomics experiments include but are not limited to establishing "news of difference" between healthy and pathological states, monitoring global gene expression during growth and development, establishing cellular localizations of a particular subset of an expressed genome, and identifying networks of protein–protein interactions. In fact, a recent review classifies proteomics studies according to the type of the addressed biological questions into "profiling proteomics," "functional proteomics," and "structural proteomics." Profiling proteomics amounts to large-scale identification of the proteins in a cell or tissue present at a certain physiological conditions. Functional proteomics refers to the studies of functional characteristics of proteins – posttranslational modifications,

protein–protein interactions, and cellular localizations. Structural proteomics includes studies of protein tertiary structure, typically made by a combination of X-ray, NMR, and computational techniques. Adopting this classification, this chapter primarily focuses on the profiling and functional proteomics.

1.2
Need for Large-scale Analyses of Gene Products at the Protein Level

Rapid success of the various genome-sequencing programs is one of the major factors that led to the development of large-scale proteomics methods. In fact, a search of the genome-derived sequence databases is usually an intrinsic part of mass spectrometry–based proteome analysis. While of tremendous value, genomic information is, in principle, insufficient for understanding the complex processes of cellular function. Indeed, in addition to the projection of the information from genes into proteins, cells of the living organisms must metabolically extract useful information from the environment. Therefore, the total informational content necessarily increases in a proteome compared to the corresponding genome.

An immediate consequence of this increase in the informational content is that all of the following – protein isoforms, protein posttranslational modifications, protein conformational states, protein abundance levels, as well as dynamical changes of these properties – have their own important functional implications in the living cells. In certain cases, genomic sequence still can be used to assess some of the functional properties of the corresponding proteins. For example, the codon adaptation index (CAI) and codon bias can be used to predict the expression level of genes. Another important case is a global sequence-based prediction of protein–protein interactions, an example of which is discussed in Sect. 3 of this chapter.

Also, to some degree, the functional properties of proteins can be assessed by analysis of the mRNA transcripts. Specifically, mRNA expression profiles are frequently used to estimate the corresponding protein levels. Most common methods that are used to measure global mRNA expression are cDNA microarray and serial analysis of gene expression (SAGE). However, in many cases, the correlation between mRNA and protein is insufficient to quantitatively predict protein expression. Potential reasons for these discrepancies in mRNA and protein levels include the differences in half-lives of proteins and mRNAs and posttranslational mechanisms that control the rates of translation. Thus, neither DNA nor mRNA sequence information is sufficient for understanding of the cellular functions. This fact provides motivation to improve the old and develop new technologies for the large-scale analyses of gene products at the protein level.

1.3
General Problems in Proteome Analyses

In a given organism, there are several times more different proteins than there are genes. Estimates for humans give such numbers as about 30 000 different genes and more than 100 000 different proteins. Factors such as posttranslational modifications, isoforms, and expression levels increase the complexity and interconnectedness of a proteome compared to a genome. Problems associated with modern proteomics methods include (1) difficulties in the detection of low-abundant proteins; (2) difficulties in obtaining quantitative information; (3) biases in a proteome coverage; (4) difficulties in characterization of posttranslational modifications; (5) difficulties in data analysis and interpretation in the large-scale experiments; (6) poor reproducibility.

1. Protein levels as low as several dozens copies per cell can be of functional significance. This is true for certain receptors, signaling proteins, and regulatory proteins. However, detection of the low-abundance proteins is often an issue in proteomics experiments. The ability of a proteomics method to identify low-abundant proteins in the presence of high-abundant ones is characterized by the dynamic range of a method. The dynamic range of a proteomics method is defined as the ratio of concentration of the most abundant to the concentration of the least abundant proteins identified by a method. Proteomics methods based on two-dimensional polyacrylamide gel-electrophoresis (2D PAGE) separations typically have a lower dynamic range than the methods based on affinity separation or multidimensional liquid chromatography.

2. Quantitative information on the protein abundance in different physiological contexts is the goal of comparative proteomics studies. 2D PAGE methods are good at solving these types of problems, when combined with scintillation

counting or fluorescence imaging spectroscopy. In the MS-based methods, quantitation is achieved by differential labeling (discussed in detail in Sect. 2).

3. Biases in proteome coverage can be inherent to a particular method, or be related to the design of an experimental scheme. For example, 2D PAGE methods are biased against highly hydrophobic proteins. At the same time, there is no inherent bias against hydrophobic proteins in MS identification. However, sample enrichment or fractionation steps necessarily preceding MS analysis are often biased against one or the other class of proteins.

4. Information on types and degrees of posttranslational modifications and protein isoforms is a necessary component of comprehensive functional description of a proteome. Unfortunately, it is difficult to design a proteomics method that would simultaneously give both good proteome coverage and identify all posttranslational modifications. Methods, such as 2D PAGE, which separate intact proteins, are good at the detection of different posttranslational modifications and isoforms present at the same time. However, owing to large diversity in protein masses and shapes, biases in proteome coverage are inherent in the intact protein methods.

5. During a large-scale proteome analysis, thousands of data points are generated. Hence, there are inherent problems related to the data reduction and extraction of the useful information. Additionally, the obtained information needs to be presented in an accessible format to the scientific community. These issues often require significant computational support and software development.

6. Most of the high-throughput, large-scale methods suffer from poor reproducibility. This is especially true in the case of 2D PAGE and MS-based schemes with several fractionation and separation steps involved. Possible ways to improve reproducibility include reduction in the number of separation and fractionation steps, adhering to standard protocols, and use of automation whenever possible.

Most of these challenges, however, are merely technical limitations. Modern technologies continuously progress toward higher sensitivity, higher range of proteome coverage, and higher reproducibility. Also, even if it may be impossible to have 100% proteome coverage with a single method, proper design of a proteomics program that takes advantage of several different methods can achieve impressive results.

1.4
MS Instrumentation

Mass spectrometry (MS) is a platform technology for all proteomics. It is the mass spectrometer that is used to identify the protein present in samples. Furthermore, there are a variety of mass spectrometry approaches that can be used to achieve this end. Figure 1 illustrates the general concept of MS analysis in proteomics. Generally, MS analysis involves creation of gaseous ions from an analyte followed by separation of the produced ions according to their mass-to-charge ratio (m/z). Instruments that perform analysis of this type are called *mass spectrometers*. Principle components of a mass spectrometer are an ion source, where ionization takes place; a mass analyzer, where the m/z is measured; and a detector, where amount

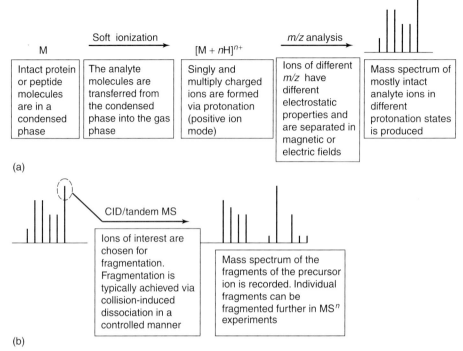

Fig. 1 Principle scheme of single MS (a) and MS/MS (b) analyses. (a) Intact proteins or peptides are transferred from the condensed phase into the gas phase and are ionized through capture (positive mode) or loss (negative mode) of protons. In high-throughput protein identification experiments, positive mode of ionization is used. Negative mode of ionization can be used for analysis of carbohydrates and nucleic acids. Ions of different m/z are discriminated by magnetic or electric fields. (b) Ions whose structure needs to be determined are chosen for fragmentation via CID reactions in the MS/MS experiment. The obtained fragmentation patterns can be searched against protein databases for identification.

of the ions corresponding to a particular m/z is recorded. Large macromolecules, such as proteins, are nonvolatile, and owing to a lack of proper ionization techniques for a long time, MS was limited to the analysis of smaller molecules. Finally, in the 1980s, protein ionization methods were developed, thus making MS-proteomics feasible.

1.4.1 Ionization Methods

The two most commonly used methods of protein ionization are electrospray ionization (ESI) and matrix-assisted laser desorption ionization (MALDI). In both of these techniques, ionization occurs through uptake (positive mode) or loss (negative mode) of one or several protons. In large-scale proteomics studies, the positive mode is typically used. The negative mode of ionization finds its uses in analyses of carbohydrates and nucleic acids. In any of the available ionization methods, there is no definite relationship between the amount of ions formed and the amount of the analyte. This fact makes mass spectrometers inherently nonquantitative devices.

ESI produces mostly multiply charged ions by creating a fine spray of charged droplets in a strong electric field. With the application of a dry gas, the droplets evaporate, and the electrostatic repulsion causes transfer of the analyte ions into the gas phase (Fig. 2a). The multiple charging allows analysis of very large molecules with analyzers that have relatively small m/z range. Another advantage of the multiple charging is that more accurate molecular weight can be obtained from analysis of the distribution of multiply charged peaks. ESI is used in a wide range of proteomics applications but is limited by susceptibility to high salt concentrations and to contaminants in the sample.

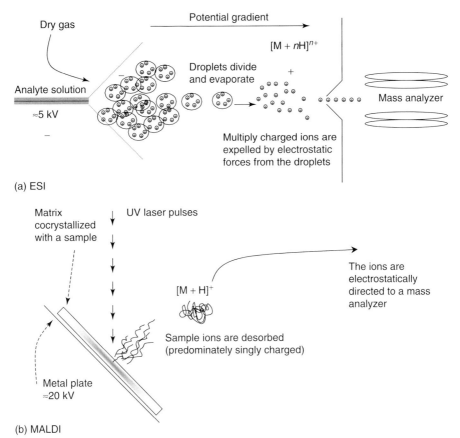

(a) ESI

(b) MALDI

Fig. 2 Conceptual schemes of (a) Electrospray ionization (ESI) and (b) Matrix-assisted laser desorption ionization (MALDI). (a) During ESI, a fine spray of charged droplets is created. The droplets evaporate and the multiply-charged ions are expelled by electrostatic forces. ESI is frequently used for analysis of complex peptide mixtures via coupling to multidimensional liquid chromatography. (b) During MALDI, the analyte molecules are ejected by laser pulses from the sample cocrystallized with a matrix. During MALDI, mostly singly charged ions are produced. Because the analyte molecules are ejected in bundles, MALDI method is ideally suited for coupling to TOF mass analyzers. MALDI-TOF instruments are frequently used to analyze individual spots on 2D PAGE.

MALDI produces predominately singly charged ions by aiming laser pulses at a sample cocrystallized with a molecular matrix on a metal plate under high voltage ($\sim$20 kV). Typical matrices used in MALDI-assisted protein analysis include α-Cyano-4-hydroxycinnamic acid and 2-(4-Hydroxyphenylazo)-benzoic acid. The laser is tuned to the absorption maximum of the matrix. The sample ions are preformed in the condensed phase in sufficient quantities. The matrix absorbs some of the laser pulse energy, thus minimizing sample damage. The sample ions and matrix molecules gain enough kinetic energy and are ejected into a gas phase (Fig. 2b). Table 1 outlines and compares principle characteristics of these ionization methods.

Importantly, in both the MALDI and ESI sources, ion fragmentation occurs rarely and the analyte molecules remain largely intact. This nondestructiveness is what makes these methods so attractive for characterization of biomolecules. However, in certain applications (e.g. peptide sequencing), it may be necessary to fragment molecular ions (preferably in a predictable manner) in order to extract additional information. This is typically achieved via collision-induced dissociation (CID). Such types of MS analysis are called *tandem MS* (MS/MS) and often are denoted by MS^n, where n is the number of generation of fragment ions analyzed (Fig.1).

1.4.2 Mass Analyzers

Mass analyzers separate ions according to their m/z in electric or magnetic fields. Most common types of mass analyzers used in proteomics research are quadrupole (QD), ion trap (IT), time-of-flight (TOF), and Fourier transform ion cyclotron resonance (FTICR). Often in modern instruments, several analyzers of the same or different types are combined together to achieve maximum performance.

Tab. 1 Comparison of ESI and MALDI sources.

	ESI	*MALDI*
Principle of action	Uses electric field to produce sprays of fine droplets; as the droplets evaporate, ions are formed	Uses laser pulses to desorb and ionize analyte molecules cocrystallized with a matrix on a metal surface
Ions formed	Multiply charged (The larger the analyte molecule, the more likely it acquires multiple charges)	Singly charged
Mass range	>100 kDa	>100 kDa
Resolution	~2500 (with IT/QD mass analyzers)	~10 000 (with TOF mass analyzers and ion reflectors)
Typical application	10^{-15} mole LC/MS of peptide mixtures; tandem MS; protein identification by comparing experimental MS/MS spectra with theoretical MS/MS spectra produced from protein databases	10^{-15} mole Analysis of spots on 2D PAGE; determination of molecular weight; protein identification by "peptide mass fingerprinting"

QD mass analyzers are frequently used in conjunction with ESI source. They offer moderate resolution (up to 2500 mass units) and moderate sensitivity. QD consists of four parallel rods with a hyperbolic cross section. Diagonally opposite rods are connected to radio frequency and direct-current voltage sources thus establishing the quadrupole field. Ions that are produced at an ion source are electrostatically accelerated into this quadrupole field. Mass-selection is achieved by proportional changes of amplitudes of radio frequency and direct current in a way that for any given pair of these amplitudes there, only the ions of specific m/z reach the detector. QDs are known for their tolerance of high pressures (up to 10^{-4} torr), which makes them attractive to use with ESI for liquid chromatography (LC)-coupled applications.

IT mass analyzers work by confining ions to a small volume via radio-frequently oscillating electric fields. ITs offer moderate sensitivity at a relatively low monetary cost and are good for MS/MS applications. In the latter case, ITs are often used cojointly with QDs. Mass accuracy of regular ITs is rather low, since only limited amount of ions can be accumulated in a small volume. Mass accuracy, as well as resolution can be significantly improved by using two-dimensional ITs (sometimes also called *linear*), which increase the ion storage volume.

TOF mass analyzers measure the time traveled by an ion from a source to a detector. The longer the flight path, the better the resolving power. However, longer flight paths also increase the scan time. In commercial TOF analyzers, compromise is achieved with the length of the flight path on the order of several meters. The resolving power of a simple TOF analyzer is poor – several to ten

times less than that of QD. Significant improvement in the resolving power of TOF is offered by additions of one or several ion reflectors. Upon reflection, the velocity distribution of ions at particular m/z narrows, thus increasing resolution and sensitivity. As a result, modern TOF and double-TOF spectrometers can achieve resolving power of 10 000 and more. Measurement of the ion's TOF can be achieved only if the analyte ions are presented to TOF analyzer in discrete bundles. Because of this need for discrete ion bundles, TOF analyzers are particularly suited to the pulsed nature of MALDI. In fact, MALDI-TOF is one of the most common types of mass spectrometers used in proteomics research.

FTICR mass analyzers are based on the resonance absorption of energy by ions that precess in a magnetic field. The recorded array of the precession time-curves is Fourier-transformed to obtain the component frequencies of the different ions. Next, the component frequencies are related to the ion's m/z. Because frequency is a parameter that is easy to measure with high precision and accuracy, FTICR has the highest resolving power amongst MS analyzers – up to 10^6 and more. With FTICR, it is also particularly easy to do MS^n experiments. Unfortunately, current FTICR instruments are cumbersome, expensive, and not readily available.

1.4.3 MS/MS

In some applications, it is necessary to fragment molecular ions produced by ESI or MALDI further to obtain additional structural information. Figure 3 illustrates a possible way to do this with a triple–QD ESI mass spectrometer. To obtain a tandem spectrum, the first quadrupole scans across a set m/z range and selects ions of interest. In the second quadrupole, CID

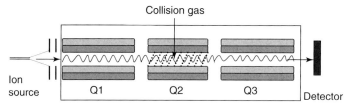

Fig. 3 A triple quadrupole mass spectrometer with MS/MS capability. Q1 selects ion to be fragmented and allows the selected ion to pass into Q2. The ions in Q2 are fragmented via collisions with inert gas (typically argon). Q3 analyzes the fragments generated in Q2. MS^n ($n \leq 4$) experiments can be performed with this setup as well. To achieve this, fragmentation must be increased at the level of ionization. In modern instruments, quadrupole-ion trap mass analyzers are used for MS^n ($n > 8$) experiments.

reaction takes place – ions that were selected by the first quadrupole undergo collisions with argon gas and fragment. Finally, the third quadrupole analyzes the resulted fragments. Importantly, the fragmentation via CID occurs in a predictable manner – protein and peptide ions break mostly at peptide bonds – which makes the large-scale sequencing easier.

1.5
Methods of Sample Fractionation

The large complexity of protein mixtures from biological samples poses additional challenges for proteomics experiments. Depending on the scope of a particular proteomics task, it is desirable to introduce sample enrichment and fractionation steps prior to MS identification. The scope of a particular task can range either from an analysis of a subset of a proteome – for example, analysis of proteins specific to a particular organelle and proteins modified in a certain way – to a simultaneous analysis of all the proteins in the proteome. In both of these cases, a proper choice of separation strategy determines the overall throughput and sensitivity of the proteomics experiment. While it may be

desirable to fractionate protein mixture down to individual proteins (aim of 2D PAGE separation), significant gain in the throughput can be achieved when mixtures of proteins are introduced to the mass spectrometer.

MS analysis of intact proteins is impractical in most of the high-throughput tasks with the current instrumentation because of the large range in protein masses. To simplify the measurements, in most of the MS analyses, proteins are enzymatically digested (either individually purified or in mixtures) to produce peptide fragments. The resulted peptide mixtures are often fractionated further. Fractionation of the peptide mixtures is usually achieved via in-liquid chromatography methods.

Isolation of a particular subcellular compartment or an organelle is usually achieved by a combination of centrifugation and solubilization steps. Proteins modified posttranslationally in a specific way (e.g. phosphoproteins) can be isolated by chemical- or immunoaffinity methods (see examples in Sect. 3). Large-scale isolation of protein complexes can be done by the epitope tagging and subsequent immunoaffinity purification (see examples in Sect. 3). Another promising affinity method is that of protein chips – a method that uses large arrays of antibodies, or other binding factors to isolate proteins of interests.

1.5.1 2D-electrophoretic Separation of Complex Protein Mixtures

2D PAGE separation methods were introduced in mid-1970s and were extensively used for analysis of complex protein mixtures. The principle of 2D PAGE separation is illustrated in Fig. 4. 2D PAGE separates proteins in the first dimension by their pI, and in the second, by their molecular weight. Prior to the development of MS-identification techniques suitable for macromolecules, identification of individual proteins in the gel spots was difficult. Typically, the identification of individual

proteins in the gel spots was done by immunostaining methods or by N-terminal degradation sequencing. Nowadays, in most laboratories, analysis of 2D PAGE spots is performed with MS. Typically, the individual spots are excised, digested, and analyzed by MALDI-MS. Protein identification is achieved by peptide mapping – comparison of the observed peptide peak patterns with the predicted digest fragments of proteins in a database. However, if some of the spots on the 2D PAGE are overlapping, the method of peptide mapping can give incorrect results.

As a separation technique, 2D PAGE offers high resolution, and is able to distinguish between different protein isoforms and also between different posttranslational modification

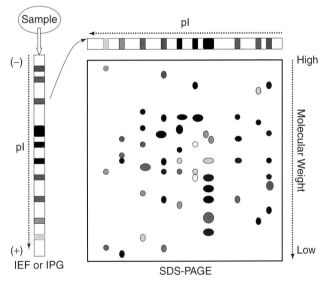

Fig. 4 Principle of the two-dimensional electrophoretic separation of intact proteins. The sample is loaded onto IEF to IPG strip, where proteins are separated according to their isoelectric points (pI). Next, the strip is loaded on SDS-PAGE, where the proteins are separated according to their molecular weights (MW). Visualization of protein spots is achieved via chemical staining methods. Individual spots can be further excised and analyzed by mass spectrometry.

states. However, 2D PAGE is a denaturing technique; hence, it is not suitable for direct analysis of protein complexes and protein–protein interactions. Also, 2D PAGE is known to discriminate against low-abundance and membrane proteins. Other limitations of 2D PAGE include biases against proteins with pIs outside the 2–10 range and biases against heavy proteins (>100 kDa). The dynamic range of 2D PAGE is also limited by resolution of the spot visualization methods. Recently, a number of ways to overcome some of these limitations of 2D PAGE have been developed.

Significant improvement to 2D PAGE was the development of immobilized pH gradients (IPGs), which increased both reproducibility and the load capacity of 2D PAGE. Sample prefractionation into discrete isoelectrical fractions and subsequent analysis by several narrow-range IPGs is yet a further improvement to this technology, which both increases the method's dynamic range and somewhat reduces the biases in proteome coverage. Despite these technological improvements, 2D PAGE remains a time-consuming and labor-intensive technique. Alternative methods of protein separations employing LC exist and can potentially overcome some of the 2D PAGE's limitations in the large-scale proteome analysis.

1.5.2 Liquid-phase Separation Methods

An ESI source allows MS characterization of proteins and peptides that are present in solutions and therefore is ideally suited for online coupling with liquid-phase separations. Liquid-phase separation can be achieved by using high performance liquid chromatography (HPLC), capillary isoelectric focusing (CIEF), capillary electrophoresis (CE), or by other methods. When two or more distinct liquid phases are used in a separation, with each of the phases relying on a unique independent physical property of the analyte, this liquid-phase separation is called *multidimensional*. Independent physical properties correspond to "dimensions" and these can be size, charge, hydrophobicity, or affinity to a particular substrate. The use of several independent dimensions significantly increases resolving power of a separation.

In-liquid separation methods can be used both for the intact proteins and for the peptide mixtures. One promising technique for characterization of global extracts of intact proteins is CIEF-FTICR mass spectrometry. In this method, the CIEF separates intact proteins by pI and subsequent analysis by FTCIR produces a two-dimensional display similar to the one obtained in 2D PAGE. Jensen et al. reported resolution of 400 to 1000 proteins in the mass range of 2–100 kDa from global protein extracts of *Escherichia coli* and *Deinococcus radiodurans*. While of good resolving power and sensitivity, this technique needs further improvement, perhaps by addition of protein fragmentation so that the resolved proteins could be identified.

Wall et al. developed a method that can potentially obtain both intact protein molecular weights as well as sequence information for a large portion of a proteome. In this method, proteins were separated by pI using isoelectric focusing (IEF) in the first dimension and by hydrophobicity using nonporous reversed-phase HPLC in the second dimension (IEF-NP RP HPLC). Next, the fractions that were eluted from HPLC were analyzed in two parallel experiments by (1) MALDI-TOF-MS and (2) ESI-TOF-MS. Analysis of human erythroleukemia cell lysate by this method resulted in resolution and identification of several hundreds of unique proteins in a mass range from 5 to 85 kDa. Having both intact protein molecular weights as well as sequence information alleviates characterization of protein posttranslational modifications and isoforms. Thus, Wall et al. reported detection of posttranslational modifications of cytosolic actin, heat shock 90 beta, HINT and α-enolase.

The intact protein approaches are usually limited to proteins of smaller sizes, mostly because the resolving power of mass analyzers drops with increase in the ions masses. Also, biases toward one or the other types of intact proteins are unavoidable with any chromatographic separation method. A conceptually different approach, in which purification of intact proteins is avoided, involves proteolytic digestion of protein mixtures. Proteolytic digestion transforms protein mixtures into more uniform (in terms of mass and chromatographic properties) mixtures of peptides. Because of the increased uniformity, analysis of peptide mixtures essentially eliminates biases in proteome coverage. Unfortunately, the absence of information on intact proteins makes characterization of posttranslational modifications more difficult.

Chemical or enzymatic cleavage of proteins into peptides followed by multidimensional liquid-phase separation is typically used in a global proteome survey type of experiments. Generally, proteins from whole cells or tissue homogenates are first enzymatically digested and then the resulting peptide mixtures are separated by electrophoresis or liquid chromatography in a microcapillary format. A microcapillary column is attached directly to the ionization source so that the eluting fractions are ionized and introduced to a mass analyzer. While fragmentation of the peptide ions is optional with MALDI-MS/peptide

mapping analysis of 2D PAGE spots, it is required with the liquid-phase/MS analysis of complex peptide mixtures. The sequence information obtained by MS/MS for individual peptide ions is used to match peptides to the corresponding proteins in a database.

Multidimensional protein identification technology (MudPIT) is a good example of a liquid-phase/MS method. In MudPIT, a biphasic SCX/RP microcapillary column is used to fractionate a complex peptide mixture (Fig. 5). The peptide fractions are eluted from the column by series of HPLC gradients into the tandem mass spectrometer. The obtained MS/MS spectra are used to search a sequence database via SEQUEST algorithm. MudPIT was used to detect and identify ~1500 proteins from *Saccaromyces cerevisiae* proteome, ~2500 proteins from the *Oryza sativa* proteome and ~2500 proteins from the *Plasmodium falciparum* proteome.

1.5.3 Affinity Methods (Epitope Tagging)

Affinity-purification methods are frequently employed in those proteomics programs that analyze specific subsets of a proteome, such as phosphorylated or glycosylated proteins. Also, affinity methods are often used in large-scale studies of protein–protein interactions and protein complexes. Affinity methods are based on targeted interactions between proteins and antibodies or between proteins and other protein-specific molecules, immobilized in a column or in the form of an array. For example, to isolate phosphorylated proteins from the rest of the proteome, one can use an immunoaffinity column prepared with antibodies specific to the phosphorylated amino acids.

Epitope tagging is used in large-scale analyses of protein–protein interactions and protein complexes (Fig. 6). In this

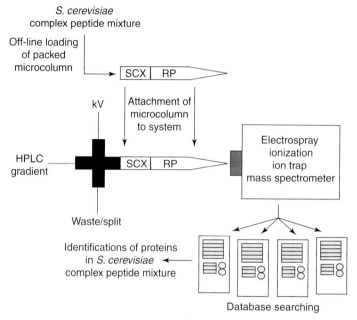

Fig. 5 Multidimensional protein identification technology (MudPIT).
Complex peptide mixtures are loaded onto a biphasic microcapillary
column packed with strong cation exchange (SCX) and reverse-phase (RP)
materials. Peptides directly elute into the tandem mass spectrometer
because a voltage (kV) supply is directly interfaced with the microcapillary
column. Peptides are first displaced from the SCX to the RP by a salt
gradient and are eluted off the RP into the MS/MS. The tandem mass
spectra generated are correlated to theoretical mass spectra generated
from protein or DNA databases by the SEQUEST algorithm. Figure is
reproduced from Washburn, M.P., Wolters, D., Yates, J.R. III (2001)
Large-scale analysis of the yeast proteome by multidimensional protein
identification technology, *Nat. Biotechnol.* **19**, 242–247 by permission of
Nature Publishing Group.

strategy, proteins to be isolated are fused
with a motif recognizable by a specific
antibody. This fusion is typically done by
incorporating the epitope sequence into
C-terminal of genes that encode proteins
of interest. Then the fused proteins are
expressed and purified along with their interaction partners by the immunoaffinity
chromatography using antibodies specific
to this particular epitope. If a particular
protein is of low abundance, it can be overexpressed to increase the overall recovery.

1.6
Protein Identification by MS

Accurate identification of proteins is the
essential requirement of proteomics studies. High complexity of protein mixtures
derived from tissues, whole cells, or subcellular compartments adds an additional
requirement of high throughput in the
identification of proteins. In this section,
we review common strategies that match
MS data with proteins present in a
biological sample.

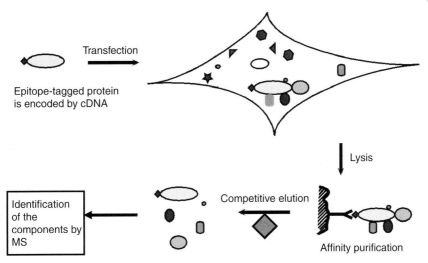

Fig. 6 Principle of affinity purification with epitope tagging. Bait proteins are expressed as fusion with a motif recognizable by a certain antibody (epitope). The cells are lysed, and proteins are purified by immunoaffinity chromatography. Next, the bound fraction is competitively eluted from the column, and the pulled proteins are analyzed by LC/MS or 1D PAGE/MS.

Typically, protein identification amounts to the deduction of the sequence-specific information from MS data followed by searches of the sequence databases. Databases that MS data can be matched against are protein, expressed sequence tag (EST), and genomic databases. Because currently there are no suitable ways to fragment intact proteins inside a mass spectrometer, high-throughput sequencing by MS is possible only if intact proteins are fragmented into peptides first. Depending on a particular type of proteomics scheme and MS instrumentation used, protein sequence information can be obtained by (1) peptide mass fingerprinting; (2) accurate mass tags (AMTs); (3) peptide fragmentation in MS/MS; (4) sequence tags; or by combinations of (1) through (4).

Peptide mass fingerprinting method is usually adopted in those cases in which individual proteins are separated during purification steps, such as in 2D PAGE.

In the case of 2D PAGE, individual spots are picked, digested, and masses of the peptides are recorded. Next, the experimentally obtained peptide masses are matched against theoretical peptide libraries generated from protein sequence databases. MS/MS data can also be obtained along with mass fingerprinting, and nowadays this is a common practice.

With MALDI-TOF instrumentation – which is typically used with the peptide mass fingerprinting type of analysis – several peptide masses are needed to unambiguously identify a protein. Using more accurate instrumentation, such as FTICR-MS, it is possible to identify proteins based on the mass of a single peptide, without MS/MS data. In this case, a peptide that uniquely corresponds to a protein is called *accurate mass tag* (AMT). Conrads et al. evaluated utility of AMTs for identification of proteins from *S. cerevisiae* and *Caenorhabditis elegans*. The authors demonstrated that up to 85% of the predicted tryptic peptides from these two organisms could be used as AMTs at mass accuracies typical of FTICR-MS instruments (~1 ppm). The authors also discussed utility of AMTs with highly accurate mass measurements in detection of phosphorylated proteins. They argued that because mass defect of P is larger than that of H, C, and O, the average mass of phosphopeptides is slightly lower than the mass of unmodified peptides of the same nominal weight; thus enabling the identification of phosphorylated peptides if the mass measurement accuracy is sufficient.

Whenever it is possible to match a single peptide to a protein, it is no longer necessary to purify samples down to individual proteins. Instead, fractions containing mixtures of proteins can be digested with trypsin followed by analysis of the resulted peptide fragments. For example, the above-discussed AMT method can be used for analysis of complex protein mixtures, because it matches single peptide to a unique protein. However, the requirement of high mass accuracy makes uses of AMT limited. Additionally, FTICR instrumentation is cumbersome and is not a widely distributed technology today. Far more superior in terms of proteome coverage per monetary cost are methods that employ peptide sequence analysis by MS/MS. The key fact that enables high-throughput peptide sequencing by MS/MS is predictability of the peptide precursor ion fragmentation in the CID reactions.

Figure 7 shows the adopted nomenclature for the fragment ions series. Ions that form through dissociation of the peptide bonds are the most abundant ones if moderate collision energies are used (30–50 V). The ions that retain their charge on the N-terminal part after fragmentation of the precursor are called y ions, and the

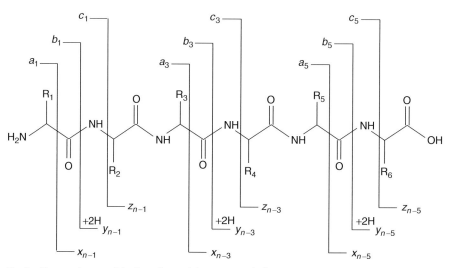

Fig. 7 Nomenclature of the ions formed during peptide fragmentation. In a typical MS/MS experiment via CID reactions, y and b ions are predominant. These two types are used for computer-assisted peptide identification. Through the loss of CO fragment, a ions can be produced from b ions. While not important for identification, a ion series can be used for independent result validation. Other types of ions are not produced in CID reactions.

ions that retain their charge on the C-terminal part after fragmentation of the precursor are called *b* ions. The subscript to the right of *b* and *y* symbols equals the number of amino acid residues in the corresponding fragment. The difference in *m/z* values between consecutive ions within a given series corresponds to the difference in the sequences of the two fragments. Because the consecutive ions within a series represent peptide fragments that differ in exactly one amino acid, and each amino acid residue has a unique nominal weight (except I and L), the pattern of *m/z* values of *y* and *b* ions corresponds to the amino acid sequence of the precursor peptide. Unfortunately, some expected peaks could be missing from the MS/MS. Additionally, experimental spectra can be complicated by unwanted fragmentation. Also, it is not always possible to unambiguously determine from which series a particular ion fragment comes. For these reasons, manual interpretation of MS/MS spectra can be tedious and ineffective. As a consequence, *de novo* sequencing is rarely done in high-throughput proteomics experiments. Instead of direct interpretation of the MS/MS spectra, computer programs that find the best matches to the spectra from a database are used.

Typical example of an algorithm that finds best matches to MS spectra is SE-QUEST software package developed by Eng et al. Analysis of MS data by SE-QUEST starts with reduction of tandem spectra complexity. Only a certain number of most abundant ions are considered and the rest are discarded. Also, the unfragmented precursor ion is removed from the spectra in order to prevent its misidentification as a fragment. Next, SEQUEST selects sequences from a database. First, SEQUEST creates list of peptides that have

masses at or near the mass of the precursor ion. Second, SEQUEST generates virtual MS/MS spectrum for each of the candidate and compares them to the observed spectrum. As the result of this comparison, a cross correlation score is produced (Xcorr). Another score, DeltaCN, reflects the difference in correlation of the second ranked match from the first one. The higher this score is, the more likely it is that the first match is the correct one.

SEQUEST also has the ability to detect posttranslational modifications. In this case, SEQUEST looks for increased masses of amino acids. For example, cysteines are typically carboxymethylated by iodoacetomide prior to enzymatic digestion. In this case, masses of all cysteine residues are considered to increase by 57 Da. To look for amino acids that can be either modified or not modified, the database size is increased by allowing different weights to represent the same amino acid. Unfortunately, only known posttranslational modification can be detected by SEQUEST – the types of modifications that are being looked at need to be explicitly specified in input parameters.

Another useful technique for peptide identification is sequence tag. This technique is implemented in algorithms such as GutenTag. In GutenTag, a partial sequence is inferred directly from the tandem spectrum, and then the database is searched for matches that include this partial sequence and that match the masses on C- and N-sides of the fragment. Importantly, GutenTag method is error-tolerant, and allows detection of unknown posttranslational modifications. A recent variation of this method, MultiTag also allows protein identification from organisms with unsequenced genomes.

2
Quantitative Methods of Proteome Analysis Using MS

Quantitative proteome analysis aims at large-scale identification of differences in protein expression. Two-dimensional difference gel electrophoresis (2D-DIGE) is one of the techniques that can achieve that aim. Recent example of application of this technique is the report by Friedman et al. The authors used 2D-DIGE in combination with MALDI-TOF to analyze the proteome of human colon cancer. However, while acknowledging the importance of the gel-based methods in modern quantitative proteomics, we believe that methods based on multidimensional chromatography have greater potential, and therefore, we limit our further discussion to quantitation within the LC/LC/MS/MS paradigm.

Even though absolute quantification (i.e. clear relationship between the peak intensity and amount of the analyte) is a challenge for modern mass spectrometers, methods of finding relative abundances exist. These methods are based on labeling through mass modifications of the whole proteome or some of its subsets. Labeling through mass modification is particularly suited for quantitative analysis in differential expression profiling, where two states under study are differentially labeled via "light" or "heavy" mass modifications. Figure 8 schematically depicts a quantitative proteomics approach. Mass modification can be introduced at different steps in the sample preparation – during growth, after growth, during digestion, or after digestion. Labeling at the early stages of sample preparation minimizes losses of the analytes. Instrumentation for quantitative studies is essentially the same as instrumentation for qualitative studies. Nevertheless, quantitative analysis is more laborious and typically achieves lower proteome coverage. Also, quantitative methods in proteomics are still at the stage of development and are not as broadly applied as qualitative methods.

2.1
Metabolic Labeling

Widely used in structural biology to prepare samples for analysis by nuclear magnetic resonance (NMR) analysis, metabolic labeling methods have become useful quantitative proteomics tools. Metabolic labeling is a mass modification method that is done very early in the experiment – at the stage of cell growth. Figure 8 illustrates the principle behind this strategy. The goal is to compare protein abundances in cells grown at different conditions. To do this, the growth media from one of the conditions is enriched in stable low-abundant isotopes. The enrichment can be done either by labeling all amino acids by ^{15}N, or by supplementation with a single labeled amino acid. The processed and digested cell extracts from the two different conditions are combined in a one-to-one ratio and analyzed by liquid-phase/MS/MS methods. In the resultant MS spectra, peaks corresponding to the same peptide from different conditions are offset according to the degree of labeling. The ratio between the two peaks corresponds to the difference in abundances.

The first demonstrations of metabolic labeling in quantitative proteomics used gel electrophoresis and spot excision as the protein isolation method. For example, Oda et al. identified proteins that were altered in expression between two strains of *S. cerevisiae* grown in ^{14}N or ^{15}N media, and determined the phosphorylation levels of a specific protein in the same

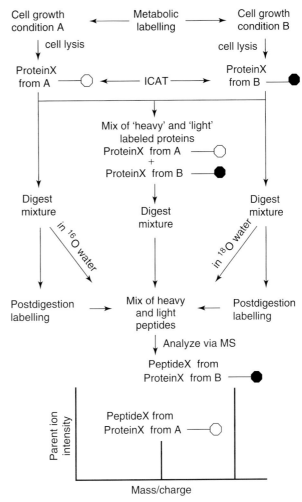

Fig. 8 Quantitative proteomics approach. When carrying out a quantitative proteomic analysis, the key is for the same peptide from two unique growth conditions to have unique masses when being analyzed by a mass spectrometer. "Heavy" and "light" peptides may be generated at many points in a sample preparation pathway. Metabolic labeling introduces a label during the growth of the organism and is therefore the earliest point of introduction of "heavy" and "light" labels. Metabolic labeling is followed by ICAT, digestion in ^{16}O and ^{18}O water, and lastly postdigestion labeling. Only after a label has been introduced can the samples be mixed and further processed. Figure and figure legend are reproduced from Washburn, M.P., Ulaszek, R., Deciu, C., Schieltz, D.M., Yates, J.R. III (2002) Analysis of quantitative proteomic data generated via multidimensional protein identification technology, *Anal. Chem.* **74**, 1650–1657 by permission of Analytical Chemistry.

sample. Complete metabolic labeling has been demonstrated using chromatography based proteomics methods on *D. radiodurans*, mouse B16 cells, and on *S. cerevisiae*.

In addition, isotopically enriched single amino acids may be used for the selective metabolic labeling of a cell type for a quantitative proteomic analysis. In *S. cerevisiae*, Jiang et al. have described the single amino acid isotopic enrichment of *S. cerevisiae* with D_{10}-Leu, and Berger et al. described the comparative analysis of *S. cerevisiae* cultured in media containing either ^{13}C-Lys or unlabeled lysine. Stable isotope labeling by amino acids in cell culture (SILAC) was explored as an alternative to ^{15}N labeling. Ong et al. studied mammalian cells grown either with deuterated or nondeuterated leucine. Importantly, parameters such as cell morphology, doubling time, and ability to differentiate did not change in deuterated sample compared to nondeuterated one. The authors used this technique to study changes in protein expression induced by muscle cell differentiation. The authors reported that glyceraldehyde-3-phosphate dehydrogenase, fibronectin, and pyruvate kinase M2 were upregulated. Another approach to metabolic labeling is the rare isotope depletion of the growth media from one of the conditions under study. If the rare isotopes are removed, then one expects m/z distributions to shift to the lighter values. With conventional instrumentation, this could work for large proteins. However, high-resolution instrumentation like FTCIR must be used to analyze peptide mixtures.

2.2
Isotope Coded Affinity Tags

Metabolic labeling with stable isotopes while analytically advantageous to other methods is available only in cases when studied cells or organisms are cultivable. For this reason, metabolic labeling is not suitable for diagnostic and clinical applications. When cultivation or controlling growth is difficult, different strategies need to be used for quantitative proteome analysis. One of the possible approaches is the method that uses cysteine-specific reagents – isotope coded affinity tags (ICATs) – to differentially label proteomics samples. ICATs have three functional elements: cysteine-specific reactive group, isotopically labeled linker, and affinity group (Fig. 9). For the comparative analysis, the cysteine residues in samples are separately labeled with either labeled or unlabeled ICAT. The derivatized samples are then combined in a one-to-one ratio and digested with a protease resulting in both labeled and unlabeled peptide fragments. The labeled peptide fragments are then purified by affinity chromatography, fractionated by reversed-phase chromatography and analyzed by tandem mass spectrometry, which provides both qualitative analysis and the relative abundance of the peptide isoforms in the samples. The MS analysis is analogous to metabolic labeling case – ratio of "heavy" to "light" peptide ions correlates with their relative abundance. Also, the ICAT approach has the obvious conceptual limitation – only peptides that contain cysteine can be detected. It has been shown that in yeast only ~10% of all tryptic peptides contain cysteine. Therefore, full sequence coverage may not be possible even for the most abundant proteins.

Significant improvement to the ICAT method is the cleavable ICAT (cICAT), which is presently supplied by Applied Biosystems in a kit format. The cICAT reagent has four essential structural elements. The first is a protein reactive group

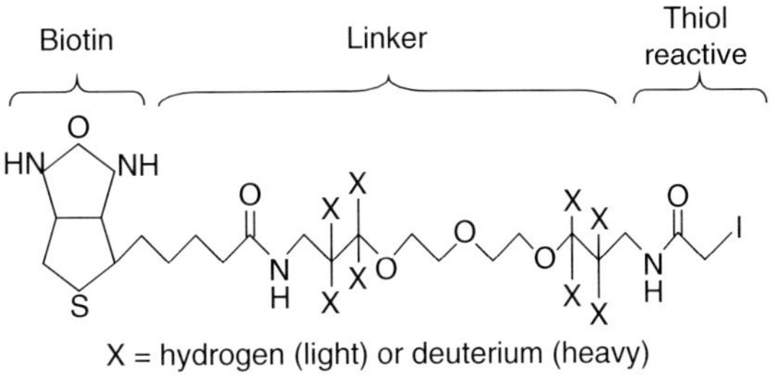

Fig. 9 Structure of isotope coded affinity tags. The ICAT reagent consists of a biotin group linked to a cysteine reactive group. The linker may be deuterated eight times or protonated at each site allowing for the generation of D_0- or D_8-ICAT. The differential masses of the linker group allow for the use of ICAT in a quantitative proteomic scheme. The figure and figure legend have been reproduced from Hunter, T.C., Washburn, M.P. (2003) The integration of chromatography and peptide mass modification for quantitative proteomics, *J. Liq Chromat. Rel. Technol.* **26**, 2285–2301 by permission of Marcel Dekker, Inc.

Biotin   Linker   Thiol reactive

X = hydrogen (light) or deuterium (heavy)

(iodoacetamide) that covalently links the isotope-coded affinity tag to the protein through alkylation of cysteines. The second structural element is the biotin affinity tag that allows enrichment of the tagged peptides. The third is an isotopically labeled linker ($C_{10}H_{17}N_3O_3$). Nine carbon atoms of the linker can be either ^{12}C or ^{13}C giving light and heavy version of the tag respectively. The light and heavy molecules have the same chromatographic properties, but differ in mass (9 Da). Once the sample is subjected to mass spectrometric analysis, the ratio of intensities between heavy and light peptides provides a relative quantitation of the proteins in the original sample. The fourth structural element of cICAT is an acid cleavage site that allows removal of part of the tag prior to MS analysis. After avidin-affinity purification of the cICAT-labeled peptides, biotin portion of the label and part of the linker can be removed by adding trifluoroacetic acid. This reduces the overall mass of the tag on the peptides and improves the overall peptide fragmentation efficiency.

The ICAT methodology has been successfully applied to a variety of biological questions including studies of several cell types, organelles, and different classes of proteins. Quantitative proteomic analysis via ICAT has been coupled with cDNA array analysis to investigate the galactose utilization pathway in *S. cerevisiae* and to investigate the mRNA and protein expression changes brought about by culturing *S. cerevisiae* in either galactose or ethanol. In addition, ICAT detected changes in protein expression of peripheral and integral membrane proteins by analyzing the effect on 12-phorbol 13-myristate acetate on the microsomes of HL-60 cells. Of all non-gel-based quantitative proteomic strategies, ICAT is the most mature as demonstrated by the successful use of ICAT in biologically driven analyses.

2.3
^{18}O Labeling

The global modification of all proteolytic peptides in a mixture may be carried out via the labeling of carboxyl groups that occurs through incorporation of ^{18}O from $H^{18}O$ during proteolytic hydrolysis (Fig. 10). In order to introduce a 4-Da mass shift into the C-terminus of a peptide, proteins may be digested in the presence of $H_2^{18}O$. Proteases like trypsin will carry out this

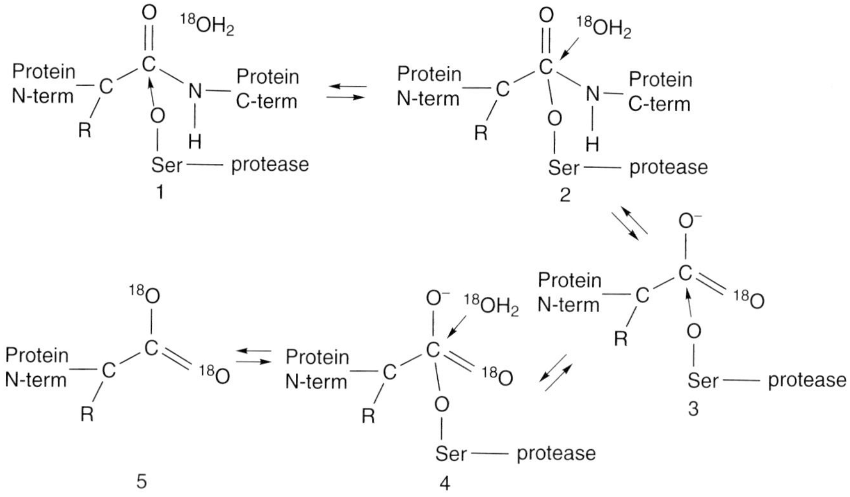

Fig. 10 C-terminal digestion modification with ^{18}O. ^{18}O may be incorporated into the C-terminus of a peptide during digestion with enzymes such as trypsin, endoproteinase Lys-C, and endoproteinase Glu-C. A simplified version of this reaction scheme is shown. (1) To begin, the peptide needs to be digested in $^{18}OH_2$ in order to then incorporate ^{18}O. The serine in the active site of the proteases listed attacks the carbonyl carbon in a peptide bond. (2) Next, $^{18}OH_2$ attacks the protein–protease intermediate also at the carbonyl carbon displacing the NH group on the peptide bond. (3) As a result, a peptide with a single ^{18}O has been generated. (4) A repeat of steps (1) and (2) is needed to drive the reaction to completion as shown in (5) where two atoms of ^{18}O have now been incorporated into the peptide C-terminus. Labeling of one sample with ^{18}O by digesting in $^{18}OH_2$ and mixing this with the other sample digested in ^{18}O depleted water allows for the determination of the relative abundance of peptides from a mixture. Figure and figure legend is reproduced from Hunter and Washburn (2003) by permission of Marcel Dekker, Inc.

reaction during the process of enzymatic cleavage. By mixing a sample with proteins digested in the presence of ^{18}O and the absence of ^{18}O, a pairwise comparison may be made to determine the relative abundance of peptides in a sample.

2.4
Postdigestion Labeling

There are several alternatives to the residue-specific modification of cysteine, which include methods for differential modification of lysine and O-phosphorylated serine residues. The phosphoprotein isotope coded affinity tag (PhIAT) method has been shown to be capable of

enriching and identifying mixtures of low-abundance phosphopeptides. The PhIAT method uses a chemical modification of phosphorylated serine and threonine residues to cysteine before introduction of a standard ICAT reagent. The mass-coded abundance tag (MCAT) approach uses a residue-specific modification lysine residues by O-methylisourea to introduce a differential tag. In addition, 2-methoxy-4,5-dihydro-1H-imidizole has also been used to modify lysine residues for the purpose of introducing a differential mass tag.

The N/C termini of peptides after digestion may also be labeled through a variety of means. A C-terminal modification method is methyl esterification of carboxyl groups using either methanolic HCl or the deuterated analog. In this case, multiple sites in a peptide may be modified with labels introduced at aspartic acids, glutamic acids, and C-termini. The N-terminal labeling of tryptic peptides with *N*-hydroxysuccinimide or 1-Nicotinoyloxy-succinimide esters and their stable isotope analogs is another approach that can be potentially used for quantitative proteomic analyses. In fact, coupling ^{18}O labeling and N-terminal labeling methods for protein expression profiling produced more comprehensive results than when either method was used alone.

2.5
Global mRNA and Protein Expression Analyses

An emerging application of quantitative proteomics approaches includes the large-scale analysis of protein expression correlated to large-scale mRNA expression analyses. In three independent comparisons of mRNA and protein levels in *Saccharomyces cerevisiae*, overall partial positive Spearman rank correlation coefficients ranging between 0.21, 0.45, and 0.57 were obtained. These studies employed ICAT, chromatography, and mass spectrometry, ^{15}N labeling and MudPIT or chemiluminescence of SDS-PAGE approaches to determine protein expression levels in cells grown under different conditions in each study. In all likelihood, this pattern of partial positive correlation between mRNA and protein expression levels could be expected to persist under a variety of conditions in *S. cerevisiae*. When these approaches begin to be globally applied to other organisms, it will be interesting to see if this trend persists.

3
Specific Examples of Applications

3.1
Global Proteome Sampling

The goal of global proteome sampling is the simultaneous identification of proteins in a cell or tissue at a given condition. Data obtained in such experiments can be further used to answer more specific biological questions, such as difference between healthy and pathological states. Typically, the proteins are identified by mass spectrometry and are grouped into functional categories.

3.1.1 Global Proteome Sampling Based on 2D Page
Global proteome analysis by 2D PAGE method is difficult because each spot needs to be picked and identified individually, thus increasing time and cost of the analysis. Also, as we discussed earlier, 2D PAGE separation suffers from biases – certain classes of proteins, such as hydrophobic or those of high molecular weight are difficult to detect. Apart from these limitations, 2D PAGE approach has an important advantage over other methods – it easily resolves protein isoforms.

In their analysis of *Haemophilus influenzae* proteome, Langen et al. used several techniques to maximize the 2D PAGE performance. To increase the proteome coverage, they used immobilized pH gradient strips covering several pH regions. Also, to visualize low-copy-number proteins, the authors performed a series of protein extractions, such as heparin chromatography, chromatofocusing, and hydrophobic interaction chromatography. In order to detect cell-envelope-bound proteins, the authors used immobilized pH gradient strips in combination with a two-detergent system with a cationic detergent in the first and an anionic detergent in the second dimensions. The isolated proteins were identified by MALDI/MS and peptide fingerprinting. As a result, 502 unique proteins were identified (about 30% of all ORFs)

Analysis of the mouse brain proteome performed by Klose et al. provides a good illustration of what 2D PAGE can do for the global proteome sampling type of experiments. By using 2D PAGE, the authors performed comparative analysis of the two distantly related mouse strains, *Mus musculus* C57Bl/6 (B6) and *Mus spretus* (SPR). About 8700 proteins from the cytosolic fraction of brain proteome were compared between the two species. By analyzing 2D PAGE of B6 and SPR strains, as well as of F_1 (B6 × SPR) and B_1 (F_1 × SPR) hybrids, the authors detected 1324 species-specific polymorphisms. Among these, 466 proteins were identified by MALDI-TOF/MS using peptide mass fingerprinting. To detect the polymorphisms, the authors considered variations in electrophoretic mobility, spot intensity, and the number of different isoforms corresponding to one protein. Additionally, through the analysis of F_1 and B_1 generations, the authors established which polymorphisms were genetically dominant. The key feature that enabled this comprehensive study was the high quality of the 2D PAGE. To analyze the mouse brain proteome, the authors used the large-gel 2D PAGE, which employs IEF gel incubation, and large (46 × 30 cm) format. Implemented in this way, the 2D PAGE gives both high resolution and high sensitivity – more than 10 000 protein spots from mouse tissues can be visualized simultaneously.

3.1.2 Global Proteome Sampling Based on Multidimensional LC

When it is necessary to catalog proteins present in a cell or an organism in a given environmental context, the multidimensional LC separation of peptide mixtures followed by MS/MS is the most convenient method to use. While it is not as good at determining protein isoforms and posttranslational modifications as 2D PAGE, the biases in proteome coverage are greatly reduced.

Florens et al. performed proteomics studies of the life cycle of the human pathogen *Plasmodium falciparum* (malaria) life cycle. The authors identified 2415 proteins and assigned them to functional groups at four stages of the cycles (sporozoites, merozoites, trophozoites, and gametocytes). The sporozoite is the form in which *P. falciparum* is injected by a mosquito. The merozoite is the form that invades erythrocytes. The trophozoite is the form that multiplies in the erythrocytes. The gametocyte is the sexual stage of malaria parasite life cycle. The analysis was performed by MudPIT. The authors found that about 50% of sporozoite proteins were unique to that stage. In sporozoites, about 25% were shared with any other stage. Trophozoites, merozoites, and gametocytes had 20 to 30% unique proteins and they had 40 to 60% of their proteins shared. Only 6% of all identified proteins were shared between all four stages, which were mainly histones, ribosomal proteins, and transcription factors. Out of the 2415 identified proteins, 51% were previously annotated as hypothetical.

Koller et al. used both 2D PAGE and MudPIT to analyze *Oryza sativa* (rice) proteome. The analyses were performed on the protein extracts from leaf, root, and seed tissue. The goal of this study was to determine tissue-specific expression of proteins. 2D PAGE separation followed by MS/MS yielded 556 unique protein identifications, comprising 348 proteins from leaf, 199 from root, and 152 from seed. MudPIT analysis resulted in significantly larger coverage: 2363 total proteins, with 867 from leaf, 1292 from root, and 822 from seed. A total of 165 proteins were uniquely detected by 2D PAGE, whereas 1972 proteins were uniquely detected by MudPIT. Next, the authors searched the nonredundant protein database by BLAST and grouped the identified proteins into functional categories. The largest category (32.8%) included proteins that had no homology to the predicted proteins. Proteins classified as involved in metabolism comprised 20.8% of all identified proteins. Out of the 2528 detected proteins, 189 were shared among all three tissues. These included housekeeping proteins that are involved in transcription, mRNA biosynthesis, translation, and protein degradation. However, most of the proteins had tissue-specific expression: 622 specific to leaf, 862 specific to root, and 512 specific to seeds.

To characterize the proteome of *S. cerevisiae* mitochondria, Sickmann et al. combined four separation methods: IEF-incubated 2D PAGE; digestion with four different proteases, followed by multidimensional LC/MS/MS; SDS/PAGE combined with multidimensional LC/MS/MS; treatment of mitochondria with trypsin, followed by SDS/PAGE or HPLC and MS/MS. The authors identified a total of 750 mitochondrial proteins. When classified into functional categories, 24.9% of all the identified proteins were of unknown function, 24.9% were involved in genome maintenance and gene expression, and 14.1% were involved in energy metabolism. The rest of the identified proteins were involved in metabolism, transport, and cell rescue.

3.1.3 MS-assisted Disease Diagnosis from Serum Samples

Proteomics technologies recently emerged as a useful tool in clinical disease diagnosis. For example, Petricoin et al. used surface-enhanced laser desorption ionization time-of-flight (SELDI-TOF) and artificial-intelligence-based informatics algorithms to discriminate between control group and ovarian cancer patients. First, the authors generated a preliminary training set of mass spectra derived from 50 unaffected women and 50 women with ovarian cancer. Next, they used an iterative searching algorithm to find the best discriminatory pattern amongst these MS data. As a result, the algorithm correctly identified all cancer cases in the masked set. Additionally, out of 66 cases of malignant disease, only 3 were recognized as cancer (false-positives). Thus, the study by Petricoin et al. demonstrated good sensitivity and predictive power of MS-based proteomics when applied to clinical disease diagnosis. For further information on this subject, we refer the interested reader to the comprehensive review by Rosenblatt et al.

3.2 Analysis of Protein Modifications by Mass Spectrometry

In living organisms, protein activity is regulated mainly by covalent modifications, which occur either co- or post-translationally. Identification of types of modifications and their locations is often a necessary requirement for an understanding of the regulation and function of a given protein. There are hundreds of known protein modifications. Among these, phosphorylation is, perhaps, the most important and widespread – about one-third of all proteins from mammalian genomes are thought to be phosphorylated. Another functionally important modification is glycosylation – glycosylated proteins are ubiquitous components of cellular surfaces where their oligosaccharide groups participate in a wide range of cell–cell recognition events. Comprehensive analysis of glycosylated proteins is more challenging than analysis of other modifications, mainly because the structure of oligosaccharide varies.

Other commonly occurring modifications that are involved in protein regulation and function are disulfide bonds, acetylations, and ubiquitinations. Some of these and other modifications are listed in Table 2. Changes in a protein length, either as a result of alternative splicing or protein truncations, also may be considered as protein modifications. Generally, it is difficult to identify protein truncations by methods that deal with protein/peptide mixtures, and often purification down to individual proteins is required in such cases (e.g. by 2D PAGE).

Currently available methods of large-scale analysis of modified proteins can be grouped into two major classes: those that use sample enrichment or chemical treatment prior to MS, and those that rely on MS data alone. Enrichment methods include affinity purification, chemical tagging followed by affinity purification, and immunoprecipitation. MS methods of detection and identification of modified peptides include neutral loss scan, precursor ion scan, postsource decay, and others.

Sometimes it is of special interest to obtain information on several types of modifications at once. If that is the case, computer programs such as GutenTag (see Sect. 2) can be used to analyze MS/MS data. Additionally, if mass changes introduced by modifications are known, the search for modified proteins can be done by SEQUEST with input parameters modified in accordance with the mass changes. However, generally, it is difficult to analyze modified peptides in the background of nonmodified peptides. Therefore, when it is clear what type of modification needs to be analyzed, the fractionation steps that enrich that particular modification need to be introduced into the experimental scheme.

Tab. 2 Common protein modification.

Modification	Monoisotopic/average mass change
Phosphorylation[a]	+79.9663/79.9799
Acetylation	+42.0106/42.0373
N-acetylglucosamine (GlcNAc)[a]	+203.0794/203.1950
Disulfide bond	−2.01565/2.0159
Methylation	+14.0157/14.0269
Hydroxylation	+15.9949/15.9994
Oxidation of methionine	+15.9949/15.9994
Ubiquitination of lysines	+114.0429/114.1040[b]

[a]Occurs on tyrosine, serine, threonine. Widespread throughout the proteome. Functions include protein regulation, signal transduction.
[b]Mass change is due to Gly–Gly residue, which is left on ubiquitinated lysines after trypsin digestion.

Below we discuss several illustrative examples of analysis of phosphorylated, glycosylated, and ubiquitinated proteins from recent literature. In addition, tools and techniques used in analysis of phosphorylation are generally applicable to analysis of many other modifications such as methylation and acetylation, while analysis of glycosylation poses additional analytical and instrumental challenges due to variability in structures of oligosaccharide groups. Ubiquitination, while not as frequent as phospho- and glyco modifications, is important for protein degradation in proteasomes.

3.2.1 Phosphorylated Proteins

Main tools in large-scale identification of phosphoproteins are enrichment by immobilized metal affinity chromatography (IMAC), chemical tagging, and immunoprecipitation by phosphor-specific antibodies. IMAC technology is based on methods developed by Andersson et al. and relies on interaction of phosphate group with immobilized Fe^{3+} ions. Ficarro et al. used IMAC combined with LC/MS/MS to characterize the phosphorylated portion of the yeast proteome. The authors showed that carboxylic acid interfered with IMAC purification, and needed to be protected. The protection was achieved by esterification with methanol in the presence of HCl. Phosphorylated tryptic peptides were identified via SEQUEST. From the whole cell lysate, Ficarro et al. detected more than 1000 phosphopeptides. From these, 383 sites of phosphorylation were determined. A potential improvement to this analysis would be the use of other proteinases in parallel with trypsin to increase the sequence coverage.

An enrichment technique that could also complement IMAC is immunoprecipitation with antibodies that bind to any protein that contain phosphorylated residues. While antibodies exist for phosphorylated serine, threonine, and tyrosine, only the anti-phosphotyrosine antibody binds strongly enough to allow enrichment. Pandey et al. used phosphotyrosine immunoprecipitation to study phosphorylation in HeLa cells in response to epidermal growth factor (EGF). The phosphopeptides were immunoprecipitated from untreated and EGF-treated cell lysates and resolved by electrophoresis. Individual gel bands were excised and studied by MALDI-MS and ESI-MS/MS. As a result, the authors identified Vav-2 as a substrate of EGF-receptor.

A report by Salomon et al. also gives a nice demonstration of analysis of phosphorylation in human cells. The authors used phosphotyrosine immunoprecipitation along with methyl esterification and IMAC combined with multidimensional LC/MS to assess tyrosine phosphorylation that occurs over time in myelogenous leukemia cells in response to treatment. The authors reported identification of 64 unique tyrosine phosphorylation sites in 32 proteins.

In another report by Ficarro et al., the authors used anti-phosphotyrosine immunoblots to study capacitation of human sperm. Capacitation is a cAMP-dependent process that is necessary for fertilization. The authors performed a comparative analysis of capacitated versus noncapacitated sperm. First, they separated sperm proteins by 2D PAGE followed by western blotting with anti-phosphotyrosine antibodies. In the next step, they excised and digested spots that exhibited phosphorylation, followed by IMAC to enrich for phosphopeptides with subsequent MS/MS analysis. As a result, the authors pinpointed several proteins that undergo phosphorylation upon capacitation of the sperm. Additionally, the authors used differential isotopic labeling to quantify the phosphorylation. The labeling was achieved at the stage of protecting the carboxy groups prior to IMAC, by treatment of the peptide mixtures from capacitated and noncapacitated digests with CD_3OD/DCl and CH_3OH/HCl respectively. These two peptide mixtures were further combined in one-to-one ratio and analyzed by IMAC/LC/MS/MS. As a result of this quantitation, the authors found 20 unique peptides that exhibited different phosphorylation levels between capacitated and noncapacitated sperm.

Metabolic labeling strategy was first described to quantitate changes in phosphorylation. In this approach, cells from two batches that have potentially different levels of phosphorylated proteins are metabolically labeled with N^{14} and N^{15}. Next, the cells are lysed; the target proteins are purified, digested, and analyzed by MS. Changes in peak intensities that correspond to modified and unmodified peptides from the two conditions provide quantitation of phosphorylation.

Another way to quantitate phosphorylation levels is to use modified ICAT strategy. In this approach, the phosphate groups in phosphopeptides derived from two different conditions are chemically replaced with either labeled or unlabeled tags. The tagging involves the following steps: (1) beta-elimination of the phosphate groups; (2) addition of 1,2-ethanedithiol containing either four hydrogens (EDT-D_0) or four deuteriums (EDT-D_4); (3) biotinylation of the EDT group using (+)-biotinyl-iodoacetamidyl-3,6-dioxaoctanediamine. The tagged peptides are further affinity purified by avidin column and analyzed by LC/MS.

3.2.2 Glycosylated Proteins

The importance of protein glycosylation, especially during cell–cell communication in multicellular organisms, is often acknowledged by using the terms *glycobiology* and *glycomics*. Owing to a wide range of possible polysaccharide structures, analysis of glycosylation is not as straightforward as analysis of other posttranslational modification. Currently, there are no satisfactory methods for global, proteome-wide analysis of all glycoprotein forms. It is possible, however, to characterize glycoproteins with glycogroups of constant structure. It is also possible to globally map glycosylation sites.

As an example, consider a broad research question such as to identify and characterize glycosylated proteins in a given biological system. In such a case, hypothetical analysis could include the following steps: (1) proteolytic digestion; (2) enrichment for glycopeptides; (3) identification of glycopeptides by MS and MS/MS; (4) structure determination of polysaccharide groups by MS^n. Alternatively, one could also separate proteins by 2D PAGE, and use glycospecific staining methods to identify spots of interest. Also, during the MS part of analyses, it could be useful to separate constant glyco structures from the variable ones, as well as N-linked from O-linked ones. One of the problems in analysis of glycopeptides is that glycogroups are very labile. Because of this lability, peptide fragmentation in CID reactions is reduced, thus making sequencing by MS/MS more difficult. An alternative way is to chemically (e.g. with beta-elimination of O-linked oligosaccharides) or enzymatically (e.g. with *N*-glycosidase F) remove glycogroups prior to analysis and to use chemical tags.

In eukaryotes, the most widespread constant type of glycosylation is O-linked N-acetylglucosamine (*O*-GlcNAc), which is found on many nuclear and cytoplasmic proteins. Glycosylation of serine and threonine residues by *O*-GlcNAc is believed to compete with and complement phosphorylation in mediating protein–protein interactions. The proteins that are glycosylated by *O*-GlcNAc include RNA polymerase II, transcription factors, chromatin-associated proteins, nuclear pore proteins, protooncogenes, tumor suppressors, and proteins involved in translation.

Because *O*-GlcNAc and phospho groups modify essentially the same amino acids, it is of special interest to establish methods that can characterize these modifications simultaneously. In this pursuit, Wells et al. developed a method based on beta-elimination followed by Michael addition (BEMAD) of dithiothreitol (DTT) or biotine pentylamine (BAP). This method relies on the fact that *O*-GlcNAc groups are more prone to elimination than phosphate groups. With the right conditions of elimination, *O*-GlcNAc can be tagged selectively. The DTT and BAP tags also allow enrichment by affinity chromatography and are stable in MS/MS fragmentation, thus allowing identification of the modified sites. First, the authors tested their method on synthetic peptides, and then they performed analysis of several biological samples: Synapsin I from rat brain and nuclear pore complex (NPC). In Synapsin I, three novel *O*-GlcNAc sites, as well as three previously known sites were mapped, thus validating the method. In the nuclear pore complex, BEMAD also mapped novel *O*-GlcNAc sites in Lamin B receptor and Nup155. Using BEMAD along with modification-specific antibodies and enzymes, the authors were able to distinguish between *O*-GlcNAc- and phosphopeptides.

During CID reactions, oligosaccharide moieties fragment mainly at glycosidic bonds. This fact potentially allows discerning of a primary structure of an oligosaccharide in MS^n experiments. While technology for a large-scale structural analysis of glycoforms is not developed yet, oligosaccharide structure determination is certainly possible for individually isolated glycopeptides. Notable examples include characterization of lipooligosaccharides from *Haemophilus influenzae* and *Neisseria gonorrhoeae*.

3.2.3 Ubiquitinated Proteins

Degradation of proteins in living organisms is a complex, highly regulated process, which plays important roles in many cellular pathways. The first step in protein degradation is the attachment of ubiquitin moieties to lysines of the substrate. The second step is proteolysis of the tagged protein by proteasome. Given the importance of this posttranslational modification, it is surprising that the report by Peng et al. is perhaps the only paper in the literature that addresses large-scale analysis of protein ubiquitination.

Peng et al. described a systematic approach based on LC/LC/MS/MS to analyze protein ubiquitination in yeast. In this pursuit, Peng et al. expressed His-tagged ubiquitin in *S. cerevisiae* cells followed by purification over Ni-chelating resin. Denaturing conditions were used at the enrichment step in order to minimize copurification of proteins that are not ubiquitinated, but form complexes with ubiquitin. The enriched ubiquitinated proteins were digested with trypsin and analyzed by SCX/RP/MS/MS followed by identification by SEQUEST. The tryptic digestion of ubiquitinated peptides results in glycine–glycine fragments at the sites of modification. The corresponding increase in mass by 114 Da of the modified lysine residues allows localization of the sites of ubiquitination. As a result, the authors identified 110 ubiquitination sites present in 72 ubiquitinated proteins. However, some of the known ubiquitinations were not detected. This is probably due to the fact that the method is biased toward more abundant species. Indeed, precise localization of a posttranslational modification via SEQUEST requires a high sequence coverage. As a consequence, in their analysis, Peng et al. identified 1075 proteins totally after Ni-resin purification, but were able to confirm ubiquitination in only 10% of them.

3.3
Analysis of Protein–Protein Interactions by Mass Spectrometry

In a cell, proteins exert their functions through interactions with other proteins. Proteomics methods developed in the past few years can be applied directly to the analysis of the protein–protein interactions and protein complexes. Protein–protein interactions can be studied either on the level of individual protein complexes, or on the level of the whole proteome. Examples of different types of analyses include the possibility of predicting interactions from amino acid sequence, focused mass spectrometric analyses of individual multiprotein complexes: yeast ribosome, SAGA-like complex (SLIK), Pol II preinitiation complex (PIC), nuclear pore complex (NPC), and proteome-wide analyses of protein–protein interactions. In addition, quantitative proteomic methods may be used to analyze the dynamics of protein complexes.

3.3.1 Computational Methods of Protein–Protein Interaction Prediction
To some degree, protein–protein interactions can be deduced indirectly from the amino acid sequences. In this pursuit, Bock et al. created a learning algorithm that was trained to recognize and predict protein–protein interactions. The training was achieved on the experimentally known interactions from a variety of organisms. As an outcome of the training, the decision function was constructed, which was statistically evaluated using unseen test data. As a result, on average, about 80% of the interactions were predicted accurately from the unseen datasets. Obviously, computational methods alone do not give exhaustive description and the obtained results need to be validated experimentally. Nevertheless, the interaction datasets obtained *in silico* provide useful reference points for experimental types of analyses.

3.3.2 Yeast 2-hybrid Arrays
One of the most common approaches to analyze protein–protein interactions is the yeast-2-hybrid (Y2H) screen, which is a selection method that detects protein–protein interactions in the yeast nucleus. This is a well-developed technique that can be easily optimized for a high-throughput analysis. The Y2H screen was successfully applied to mapping of a large-scale protein interaction network in the yeast *S. cerevisiae*. The Y2H has a good resolution and can detect weak and transient interactions. However, the Y2H method detects only binary interactions between proteins and may not be used to study transcriptional activators. Another drawback of the Y2H is that interactions occur in the nucleus, so that interactions for many proteins take place out of their native environment.

3.3.3 Direct Analysis of Large Protein Complexes; Composition of the Yeast Ribosome

Link et al. employed multidimensional liquid chromatography coupled with MS/MS to study composition of the yeast ribosome. The authors named their method *direct analysis of large protein complexes*, and demonstrated its wide applicability. In this study, the ribosomes were purified by sucrose gradients and then denatured. Next, the ribosomal RNA was removed, the ribosomes were enzymatically digested, and the digests were loaded on a 2D separation column, consisting of the SCX and the RP dimensions. After the separation, peptides were eluted directly into the mass spectrometer for the MS/MS analysis. Finally, SEQUEST algorithm was used to search nucleotide databases and to match the peptide fragmentation patterns. The authors demonstrated the high-throughput capacity of this approach – more than 100 proteins could be identified in a single experiment. As a result of this study, new protein components of yeast and human 40 S subunit were discovered.

3.3.4 Analysis of Multisubunit Protein Complexes Involved in Ubiquitin-dependent Proteolysis by Mass Spectrometry

In a number of reports, Deshaies et al. (and references therein) used MS-based strategies to characterize the composition of various protein complexes involved in proteolysis. The authors used sequential epitope tagging, affinity purification, and mass spectrometry (SEAM) to study regulation and function of SCF ubiquitin ligases. In the application of this method, SCF subunits Skp1 and Cul1 were C-terminally tagged with Myc epitope. Next, the cells expressing the tagged proteins were lysed, and the soluble fractions were affinity purified, digested, and analyzed by mass spectrometry. As a result, a total of 16 Skp1- and Cul1-interacting proteins were detected. Several of these proteins were not previously known, including Hrt1, Rav1, and Rav2. These new proteins were further subjected to SEAM and this led to identification of the new complex Rav1/Rav2/Skp1. Subsequently, it was found that Rav1/Rav2/Skp1 complex interacts with V_1 component of V-ATPase, the vacuolar membrane ATPase. In different studies by biochemical methods, the authors further determined that Rav1/Rav2/Skp1 regulates the assembly of V-ATPase from V_1 and V_0 domains.

The same group used multidimensional protein identification technology (MudPIT) to identify proteins that interact with the 26 S proteasome. As it was introduced in the previous sections of this chapter, MudPIT allows analysis of immunoprecipitated fractions without a gel separation step. Instead, the immunoprecipitated fractions are digested, and the peptide mixtures are separated in two dimensions (SCX and RP) followed by MS analysis. Using MudPIT, the authors identified every known subunit within affinity-purified 26 S proteasome, as well as one subunit that was not previously known. Additionally, a set of proteins potentially interacting with the proteasome (PIPs) was found. By immunoblotting methods, six of these PIPs were further confirmed to associate with the proteasome.

3.3.5 Proteomics of the Nuclear Pore Complex

MS-based protein identification is a useful tool that can efficiently determine the composition of a given multiprotein complex. However, this method by itself does not necessarily provide information on the complex's spatial architecture. Nor does it directly answer questions about the multiprotein complex function. Therefore, when such questions arise, the mass spectrometric tools need to be complemented with other techniques, for example, immunoblotting, immunofluorescence, electron microscopy and so on. Several studies of the nuclear pore complexes (NPCs) that are discussed here give a good illustration of these types of integrative strategies.

In the early 1990s, studies by three-dimensional cryoelectron microscopy revealed the basic shape and architecture of NPCs in *Xenopus* nucleus. It was determined that NPCs are proteinaceous structures situated in the double membrane of the nuclear envelope. Estimated sizes of NPCs vary from ~125 MDa (*Xenopus*) to 66 MDa (*Saccharomyces*). NPCs have an eightfold rotational symmetry with the rotational axis normal to the nuclear envelope membrane and a twofold mirror symmetry with the symmetry plane parallel to the nuclear envelope membrane. Current research efforts are aimed at understanding the mechanism of the biological function of the NPCs – nucleotransport. To understand the NPC's function, it is useful to catalog all the protein components of NPCs in different organisms. Ideally, this should lead to testable hypotheses on how these components contribute to the overall structure and function of NPCs. Biological problems of this kind can be addressed by proteomics methods as was elegantly demonstrated for yeast and mammalian NPCs.

In the yeast study, Rout et al. prepared highly enriched fraction of NPC proteins, followed by separation on ceramic hydroxyapatite HPLC, which gave efficient recovery of the loaded proteins. Reverse-phase TFA-HPLC was used in parallel for resolution of the low molecular mass proteins from the NPC fraction. The next step in the separation involved SDS-PAGE with

visualization of protein bands by copper staining. Subsequently, peptide mixtures were prepared from bands of interest via in-gel trypsin digestion followed by analysis by MALDI-MS. The peptide mass matching method was used to search nonredundant protein sequence database. Previously known nucleoporins were genomically tagged with a protein A epitope (ProtA), which allowed further immunolocalization by fluorescence microscopy. As a result of this study, the authors identified 29 nucleoporins and 11 transport factors and NPC-associated proteins. The authors also determined stoichiometry and position of each of the nucleoporins found within the NPC by quantitative immunoblotting and by immunoelectron microscopy. In the immunoelectron microscopy analysis, the ProtA-tagged proteins were labeled using gold-conjugated antibody, which aided visualization. On the basis of the deciphered architecture of the NPC, Rout et al. proposed a model of nucleotransport called a *Brownian affinity gating model*. The core idea of the proposed model is that translocation through the NPC occurs via diffusion: diffusive movements of the filamentous nucleoporins on the cytoplasmic face of the NPC exclude macromolecules that do not bind to them, but when the binding does occur (and that happens when a cargo molecule is associated with its transport factor), the residence time of the cargo at the NPC gate increases, which in turn facilitates the diffusion of the cargo into the nucleus.

In the mammalian study, Cronshaw et al. enriched NPCs fractions from rat liver nuclei by sequential solubilization. At each step of the enrichment, the authors used electron microscopy, SDS-PAGE, and immunoblotting to confirm that NPCs remained intact. After the enrichment, NPCs were treated with detergent, which produced a solution of monomeric nucle-oporins. The individual proteins in this nucleoporin mixture were separated by C4 reverse-phase chromatography followed by SDS-PAGE. Protein identification was per-formed using single-MS and MS/MS. In addition to previously known 23 nucleo-porins, Cronshaw et al. identified six novel nucleoporins, and also four proteins con-taining WD repeats. One of these four WD-containing proteins was ALADIN, the gene mutated in Allgrove syndrome.

Spatial organization of a protein complex can be assessed by the cross-linking method. In this method, a protein complex is affinity-purified and then treated with a cross-linking agent, a chemical that introduces new covalent bonds between neighbor proteins. New bands, that appear on SDS-PAGE because of this treatment can be excised and identified by mass spectrometry. As a result, if a pair of the proteins gets cross-linked, this usually means that the two proteins are located close to each other within the complex. A good example of this strategy is the study by Rappsilber et al. in which the authors used cross-linking/MS method to deduce the spatial composition of the six-member subcomplex Nup84p of the yeast NPC. The authors emphasized generic applicability of this approach. One of the significant challenges in application of this method, however, is the choice of the proper cross-linking reaction condition – usually a number of different cross-linkers has to be screened, before the right degree of cross-linking is obtained. Additionally, interaction between subunits that are hidden deep inside a complex may be inaccessible to cross-linkers. Because of these and other difficulties, the cross-linking method is limited to complexes of small sizes and is difficult to apply on the broader scale. Nevertheless, the

cross-linking method may prove to be useful in the studies of conformational or compositional changes of individual protein complexes, when the specific structural states can be "frozen" through interaction with the cross-linking agents.

3.3.6 High-throughput Analyses of Protein–Protein Interactions

In most cases, MS-based analytical sch-emes similar to those employed in the studies of individual complexes can be redesigned for use at the proteome-wide scale. A report by Gavin et al. is one of the first examples of the MS-based proteome-wide analyses of the protein complexes. For the large-scale isolation of the protein complexes from yeast Gavin et al. used tandem affinity purification (TAP), as first introduced by Rigaut et al. In the TAP method, a gene-specific fu-sion cassette – which contains calmodulin-binding domain, a specific protease cleav-age site, and ProtA domain – is introduced at the C- or N-terminal of yeast's ORFs of interest. Then, assuming that expression of fusion proteins is maintained close to the natural level, the first affinity purifi-cation is performed. In this step, fusion proteins along with their interaction part-ners (so-called protein assemblies) are iso-lated from cell extract by affinity selection on IgG matrix. Next, the bound proteins are released by addition of the protease. Finally, the second affinity purification is done, which involves incubation with calmodulin beads in the presence of cal-cium. The advantage of tandem affinity pu-rification when compared to standard epi-tope tagging approaches is that it removes most of the nonspecific interactions. Out of 1548 yeast strains generated by Gavin et al., 1167 expressed the fusion proteins at detectable levels. After the purification of the protein complexes by TAP, the authors subjected the complexes to electrophoretic separation followed by trypsin digestion, and subsequent analysis by MALDI-MS. Overall, by MS analysis, Gavin et al. identified 1440 gene products (~25% of the genome) from various organelles. However, identification of membrane proteins in this study has proven to be difficult – only 40 membrane proteins were purified successfully out of total 293 membrane proteins detected. The authors then proceeded with grouping of the identified proteins into complexes. This was done by the analysis of overlaps in composition of the pulled-down assemblies from 589 different bait proteins. The authors reported a total of 245 purifications that corresponded to 98 known complexes from the yeast protein database (YPD). Another 242 purifications out of the 589 were assembled into 134 new complexes. The authors were able to identify proteins as low-abundant as 15 copies per cell, thus showing high sensitivity of the TAP method. However, reproducibility was rather poor – the authors estimated that probability of finding the same protein from the same bait in two purifications is about 70%. Another weakness of the TAP method comes from possibility of interference of the TAP tag with complexes assembly and protein function. In fact, Gavin et al. found that when the essential genes were TAP-tagged; in about 20% of these cases, nonviable strains were obtained. Also, the authors reported significant bias against proteins with molecular weight below 15 KDa.

Another notable report of a high-throughput protein complex identification is the study by Ho et al. In this case, the bait proteins contained Flag epitope tag and were overexpressed from *GAL1* or *tet* promoters. Next, the protein assemblies were isolated in one-step immunoaffinity purification followed by resolution on SDS-PAGE, digestion and MS, and MS/MS analyses. The authors called this method "high-throughput mass spectrometric protein complex identification" (HMS-PCI). The immunoaffinity purification of complexes assembled around overexpressed baits should, in theory, generate more false-positives than TAP method would generate, because of the nonphysiological concentrations of the baits. On the other hand, weak and transient interactions that would not be detected in the TAP method could be captured by HMS-PCI. In fact, Ho et al. were able to assess certain regulatory and signaling pathways, by studying complexes pulled-down with phosphatases and kinases used as baits.

As of today none of the methods of mapping of protein–protein interactions within a proteome is comprehensive enough to provide full coverage. Hence, it is useful to compare datasets obtained with different approaches. In their article, Christian von Mering et al. evaluated all available interaction datasets obtained in yeast. These included the data from MS-based studies discussed above, as well as data from Y2H, correlated mRNA expression, synthetic lethality, and *in silico* predictions. The evaluation was done through comparisons with the reference dataset (MIPS and YPD). As a result, percentage coverage (fraction of the reference covered) and accuracy (fraction of data confirmed by the reference) were estimated for every method. According to the authors' analysis, TAP method provides both higher coverage and higher accuracy than either Y2H or HMS-PCI. Also, the analysis shows that HMS-PCI is the least accurate method amongst the three.

In a series of reports, Bader and Hogue developed algorithmic approaches for finding molecular complexes from datasets obtained in different interaction studies. By analyzing combined data from TAP and HMS-PCI studies, they found a novel nucleolar complex of 148 proteins that included 39 proteins with unknown function. Further, they described a graph-theoretic clustering algorithm *molecular complex detection* (MCODE) that allows detection of dense regions (potential complexes) within the interaction networks. Importantly, the authors showed that MCODE algorithm is not affected by a high rate of false-positives in datasets from the high-throughput experiments.

To summarize, none of the current high-throughput experimental schemes provide sufficient coverage and accuracy. Therefore, integrative approaches that take advantage of different methods are necessary. Additionally, all of the discussed cases dealt with cells in a certain growth conditions. It is of special interest, however, to study dynamics within protein interaction networks in response to environmental stimuli, in progression through the cell cycle, or in pathological states. Some of these questions can be addressed by quantitative proteomics techniques.

3.3.7 Quantitative Proteomics Methods in the Studies of Protein Complexes

Methods of quantitative proteomics can be used to study dynamical changes in abundance, composition, and activities of multiprotein complexes. A good example of such a study is the work by Ranish et al. in which composition of a large RNA polymerase preinitiation complex (PIC) was assessed by the isotopically coded affinity tags (ICATs) method. This and other differential labeling methods are discussed in detail in the previous section of this chapter. The ICAT approach as employed by Ranish et al., consists of the four major steps summarized below:

1. The samples from two different conditions are labeled with either "light" or "heavy" tags.
2. The labeled samples are combined together and enzymatically digested.
3. The mixture is fractionated by SCX chromatography, then the labeled peptides are isolated by avidin-affinity chromatography followed by separation by the reversed-phase microcapillary chromatography.
4. The labeled peptides that are eluting from the reversed phase column are analyzed by ESI-MS/MS.

In MS analysis, the peak intensity ratios of the differentially labeled peptides on the ion chromatogram are related to the relative abundances of the corresponding proteins in the two different environments. The peptide identities are established in the MS/MS spectra and the corresponding proteins are identified by SEQUEST. With this quantitative ICAT method, the authors were able to distinguish between components of the PIC and the copurifying background proteins, some of which had higher abundances. Thus, the authors demonstrated the high analytical power of this approach. In their analysis, Ranish et al. identified a total of 326 proteins, 42% of which participate in the Pol II-mediated transcription. Also, the authors used the ICAT method to monitor changes in the PIC composition in the presence or absence of TBP. TBP is a transcription factor that binds to the TATA element and is required for the functional PIC assembly. According to Ranish et al., most of the Pol II components are increased in abundance by a factor of at least 1.9 upon addition of TBP, and several Pol II components showed no increase in abundance. In addition, potentially new component of the PIC was discovered. A limitation of this approach, as was noted by the authors, is that using only the cysteine specific tags leaves out tryptic peptides that do not contain cysteines. In this respect, strategies that use metabolic labeling or N-terminal labeling may be more promising.

As an example, in their recent paper, Blagoev et al. used stable isotopic amino acids in cell culture (SILAC) to study EGF signaling. The control and EGF-stimulated HeLa cell populations were labeled with ^{12}C-arginine and ^{13}C-arginine respectively via metabolic incorporation. Combined cell lysates from these two conditions were affinity-purified with SH2 domain of GST-SH2 fusion protein used as bait. SH2 domain specifically binds phosphorylated EGF receptor. Protein complexes obtained in this purification were digested with trypsin and the peptide mixtures were analyzed by MS. As a result, the authors identified 228 proteins, 28 of which were enriched upon EGF simulation.

Bibliography

Books and Reviews

Aebersold, R., Mann, M. (2003) Mass spectrometry-based proteomics, *Nature* **422**, 198–207.

Bakhtiar, R., Nelson, R.W. (2000) Electrospray ionization and matrix-assisted laser desorption ionization mass spectrometry. Emerging technologies in biomedical sciences, *Biochem. Pharmacol.* **59**, 891–905.

Chalmers, M.J., Gaskell, S.J. (2000) Advances in mass spectrometry for proteome analysis, *Curr. Opin. Biotechnol.* **11**, 384–390.

Choudhary, J.S., Blackstock, W.P., Creasy, D.M., Cottrell, J.S. (2001) Matching peptide mass spectra to EST and genomic DNA databases, *Trends Biotechnol.* **19**, S17–S22.

Figeys, D. (2003) Proteomics in 2002: a year of technical development and wide-ranging applications, *Anal. Chem.* **75**, 2891–2905.

Koonin, E.V. (2001) Computational genomics, *Curr. Biol.* **11**, R155–R158.

McLachlin, D.T., Chait, B.T. (2001) Analysis of phosphorylated proteins and peptides by mass spectrometry, *Curr. Opin. Chem. Biol.* **5**, 591–602.

Noordewier, M.O., Warren, P.V. (2001) Gene expression microarrays and the integration of biological knowledge, *Trends Biotechnol.* **19**, 412–415.

Siuzdak, G. (1994) The emergence of mass spectrometry in biochemical research, *Proc. Natl. Acad. Sci. U.S.A.* **91**, 11290–11297.

Watson, J.T. (1997) *Introduction to Mass Spectrometry*, Lippincott-Raven, Philadelphia, NY.

Wilkins, M.R., Appel, K.D., Hochstrasser, D.F. (Eds.) (1997) *Proteome Research: New Frontiers in Functional Genomics (Principles and Practice)*, Springer-Verlag, New York.

Primary Literature

Adam, S.A. (2001) The nuclear pore complex, *Genome Biol.* **2**, REVIEWS0007.1–0007.6.

Aebersold, R. (2003) Quantitative proteome analysis: methods and applications, *J. Infect. Dis.* **187**(()Suppl. 2), S315–S320.

Aebersold, R., Mann, M. (2003) Mass spectrometry-based proteomics, *Nature* **422**, 198–207.

Akey, C., Radermacher, M. (1993) Architecture of the Xenopus nuclear pore complex revealed by three-dimensional cryo-electron microscopy, *J. Cell. Biol.* **122**, 1–19.

Andersson, L., Porath, J. (1986) Isolation of phosphoproteins by immobilized metal (Fe^{3+}) affinity chromatography, *Anal. Biochem.* **154**, 250–254.

Ardekani, A.M., Liotta, L.A., Petricoin, E.F. III (2002) Clinical potential of proteomics in the diagnosis of ovarian cancer, *Expert Rev. Mol. Diagn.* **2**, 312–320.

Bacher, G., Korner, R., Atrih, A., Foster, S.J., Roepstorff, P., Allmaier, G. (2001) Negative and positive ion matrix-assisted laser desorption/ionization time-of-flight mass spectrometry and positive ion nano-electrospray ionization quadrupole ion trap mass spectrometry of peptidoglycan fragments isolated from various Bacillus species, *J. Mass Spectrom.* **36**, 124–139.

Bader, G.D., Hogue, C.W. (2002) Analyzing yeast protein-protein interaction data obtained from different sources, *Nat. Biotechnol.* **20**, 991–997.

Bader, G.D., Hogue, C.W. (2003) An automated method for finding molecular complexes in large protein interaction networks, *BMC Bioinform.* **4**, 2.

Bakhtiar, R., Nelson, R.W. (2000) Electrospray ionization and matrix-assisted laser desorption ionization mass spectrometry. Emerging technologies in biomedical sciences, *Biochem. Pharmacol.* **59**, 891–905.

Berger, S.J., Lee, S.W., Anderson, G.A., Pasa-Tolic, L., Tolic, N., Shen, Y., Zhao, R., Smith, R.D. (2002) High-throughput global peptide proteomic analysis by combining stable isotope amino acid labeling and data-dependent multiplexed-MS/MS, *Anal. Chem.* **74**, 4994–5000.

Bock, J.R., Gough, D.A. (2001) Predicting protein–protein interactions from primary structure, *Bioinformatics* **17**, 455–460.

Bock, J.R., Gough, D.A. (2003) Whole-proteome interaction mining, *Bioinformatics* **19**, 125–134.

Cagney, G., Emili, A. (2002) De novo peptide sequencing and quantitative profiling of complex protein mixtures using mass-coded abundance tagging, *Nat. Biotechnol.* **20**, 163–170.

Cagney, G., Uetz, P., Fields, S. (2000) High-throughput screening for protein-protein interactions using two-hybrid assay, *Methods Enzymol.* **328**, 3–14.

Carr, S.A., Huddleston, M.J., Bean, M.F. (1993) Selective identification and differentiation of N- and O-linked oligosaccharides in glycoproteins by liquid chromatography-mass spectrometry, *Protein Sci.* **2**, 183–196.

Chalmers, M.J., Gaskell, S.J. (2000) Advances in mass spectrometry for proteome analysis, *Curr. Opin. Biotechnol.* **11**, 384–390.

Choudhary, J.S., Blackstock, W.P., Creasy, D.M., Cottrell, J.S. (2001) Matching peptide mass spectra to EST and genomic DNA databases, *Trends Biotechnol.* **19**, S17–S22.

Comer, F.I., Hart, G.W. (1999) O-GlcNAc and the control of gene expression, *Biochim. Biophys. Acta* **1473**, 161–171.

Comisarow, M.B., Marshall, A.G. (1996) The early development of Fourier transform ion cyclotron resonance (FT-ICR) spectroscopy, *J. Mass Spectrom.* **31**, 581–585.

Conrads, T.P., Alving, K., Veenstra, T.D., Belov, M.E., Anderson, G.A., Anderson, D.J., Lipton, M.S., Pasa-Tolic, L., Udseth, H.R., Chrisler, W.B., Thrall, B.D., Smith, R.D. (2001) Quantitative analysis of bacterial and mammalian proteomes using a combination of

cysteine affinity tags and 15N-metabolic labeling, *Anal. Chem.* **73**, 2132–2139.

Conrads, T.P., Anderson, G.A., Veenstra, T.D., Pasa-Tolic, L., Smith, R.D. (2000) Utility of accurate mass tags for proteome-wide protein identification, *Anal. Chem.* **72**, 3349–3354.

Conrads, T.P., Issaq, H.J., Veenstra, T.D. (2002) New tools for quantitative phosphoproteome analysis, *Biochem. Biophys. Res. Commun.* **290**, 885–890.

Corthals, G.L., Wasinger, V.C., Hochstrasser, D.F., Sanchez, J.C. (2000) The dynamic range of protein expression: a challenge for proteomic research, *Electrophoresis* **21**, 1104–1115.

Cronshaw, J.M., Krutchinsky, A.N., Zhang, W., Chait, B.T., Matunis, M.J. (2002) Proteomic analysis of the mammalian nuclear pore complex, *J. Cell. Biol.* **158**, 915–927.

Deshaies, R.J., Seol, J.H., McDonald, W.H., Cope, G., Lyapina, S., Shevchenko, A., Verma, R., Yates, J.R. III (2002) Charting the protein complexome in yeast by mass spectrometry, *Mol. Cell. Proteomics* **1**, 3–10.

Dohmen, R.J., Madura, K., Bartel, B., Varshavsky, A. (1991) The N-end rule is mediated by the UBC2(RAD6) ubiquitin-conjugating enzyme, *Proc. Natl. Acad. Sci. U.S.A.* **88**, 7351–7355.

Dotsch, V., Wagner, G. (1998) New approaches to structure determination by NMR spectroscopy, *Curr. Opin. Struct. Biol.* **8**, 619–623.

Dreger, M. (2003) Subcellular proteomics, *Mass Spectrom. Rev.* **22**, 27–56.

Eng, J., McCormack, A., Yates, J.R. (1994) An approach to correlate tandem mass spectral data of peptides with amino acid sequence in a protein database, *J. Am. Soc. Mass Spectrom.* **5**, 976–989.

Fenn, J.B., Mann, M., Meng, C.K., Wong, S.F., Whitehouse, C.M. (1989) Electrospray ionization for mass spectrometry of large biomolecules, *Science* **246**, 64–71.

Fey, S.J., Larsen, P.M. (2001) 2D or not 2D. Two-dimensional gel electrophoresis, *Curr. Opin. Chem. Biol.* **5**, 26–33.

Ficarro, S., Chertihin, O., Westbrook, V.A., White, F., Jayes, F., Kalab, P., Marto, J.A., Shabanowitz, J., Herr, J.C., Hunt, D.F., Visconti, P.E. (2003) Phosphoproteome analysis of capacitated human sperm. Evidence of tyrosine phosphorylation of a kinase-anchoring protein 3 and valosin-containing protein/p97 during capacitation, *J. Biol. Chem.* **278**, 11579–11589.

Fields, S., Song, O. (1989) A novel genetic system to detect protein-protein interactions, *Nature* **340**, 245–246.

Figeys, D. (2003) Proteomics in 2002: a year of technical development and wide-ranging applications, *Anal. Chem.* **75**, 2891–2905.

Florens, L., Washburn, M.P., Raine, J.D., Anthony, R.M., Grainger, M., Haynes, J.D., Moch, J.K., Muster, N., Sacci, J.B., Tabb, D.L., Witney, A.A., Wolters, D., Wu, Y., Gardner, M.J., Holder, A.A., Sinden, R.E., Yates, J.R., Carucci, D.J. (2002) A proteomic view of the plasmodium falciparum life cycle, *Nature* **419**, 520–526.

Fountoulakis, M., Takacs, M.F., Berndt, P., Langen, H., Takacs, B. (1999) Enrichment of low abundance proteins of escherichia coli by hydroxyapatite chromatography, *Electrophoresis* **20**, 2181–2195.

Fountoulakis, M., Takacs, M.F., Takacs, B. (1999) Enrichment of low-copy-number gene products by hydrophobic interaction chromatography, *J. Chromatogr., A* **833**, 157–168.

Friedman, D.B., Hill, S., Keller, J.W., Merchant, N.B., Levy, S.E., Coffey, R.J., Caprioli, R.M. (2004) Proteome analysis of human colon cancer by two-dimensional difference gel electrophoresis and mass spectrometry, *Proteomics* **4**, 793–811

Futcher, B., Latter, G.I., Monardo, P., McLaughlin, C.S., Garrels, J.I. (1999) A sampling of the yeast proteome, *Mol. Cell. Biol.* **19**, 7357–7368.

Gaucher, S.P., Cancilla, M.T., Phillips, N.J., Gibson, B.W., Leary, J.A. (2000) Mass spectral characterization of lipooligosaccharides from haemophilus influenzae 2019, *Biochemistry* **39**, 12406–12414.

Ghaemmaghami, S., Huh, W.K., Bower, K., Howson, R.W., Belle, A., Dephoure, N., O'Shea, E.K., Weissman, J.S. (2003) Global analysis of protein expression in yeast, *Nature* **425**, 737–741.

Glickman, M.H., Ciechanover, A. (2002) The ubiquitin-proteasome proteolytic pathway: destruction for the sake of construction, *Physiol. Rev.* **82**, 373–428.

Goff, S.A., Ricke, D., Lan, T.H., Presting, G., Wang, R., Dunn, M., Glazebrook, J., Sessions, A., Oeller, P., Varma, H., Hadley, D., Hutchison, D., Martin, C., Katagiri, F., Lange, B.M., Moughamer, T., Xia, Y., Budworth, P.,

Zhong, J., Miguel, T., Paszkowski, U., Zhang, S., Colbert, M., Sun, W.L., Chen, L., Cooper, B., Park, S., Wood, T.C., Mao, L., Quail, P., Wing, R., Dean, R., Yu, Y., Zharkikh, A., Shen, R., Sahasrabudhe, S., Thomas, A., Cannings, R., Gutin, A., Pruss, D., Reid, J., Tavtigian, S., Mitchell, J., Eldredge, G., Scholl, T., Miller, R.M., Bhatnagar, S., Adey, N., Rubano, T., Tusneem, N., Robinson, R., Feldhaus, J., Macalma, T., Oliphant, A., Briggs, S. (2002) A draft sequence of the rice genome (Oryza sativa L. ssp. japonica), *Science* **296**, 92–100.

Goodlett, D.R., Keller, A., Watts, J.D., Newitt, R., Yi, E.C., Purvine, S., Eng, J.K., von Haller, P., Aebersold, R., Kolker, E. (2001) Differential stable isotope labeling of peptides for quantitation and de novo sequence derivation, *Rapid. Commun. Mass Spectrom.* **15**, 1214–1221.

Gorg, A., Obermaier, C., Boguth, G., Harder, A., Scheibe, B., Wildgruber, R., Weiss, W. (2000) The current state of two-dimensional electrophoresis with immobilized pH gradients, *Electrophoresis* **21**, 1037–1053.

Goshe, M.B., Conrads, T.P., Panisko, E.A., Angell, N.H., Veenstra, T.D., Smith, R.D. (2001) Phosphoprotein isotope-coded affinity tag approach for isolating and quantitating phosphopeptides in proteome-wide analyses, *Anal. Chem.* **73**, 2578–2586.

Goshe, M.B., Veenstra, T.D., Panisko, E.A., Conrads, T.P., Angell, N.H., Smith, R.D. (2002) Phosphoprotein isotope-coded affinity tags: application to the enrichment and identification of low-abundance phosphoproteins, *Anal. Chem.* **74**, 607–616.

Goto, N.K., Kay, L.E. (2000) New developments in isotope labeling strategies for protein solution NMR spectroscopy, *Curr. Opin. Struct. Biol.* **10**, 585–592.

Greenbaum, D., Colangelo, C., Williams, K., Gerstein, M. (2003) Comparing protein abundance and mRNA expression levels on a genomic scale, *Genome Biol.* **4**, 117.

Greenbaum, D., Jansen, R., Gerstein, M. (2002) Analysis of mRNA expression and protein abundance data: an approach for the comparison of the enrichment of features in the cellular population of proteins and transcripts, *Bioinformatics* **18**, 585–596.

Greis, K.D., Hayes, B.K., Comer, F.I., Kirk, M., Barnes, S., Lowary, T.L., Hart, G.W. (1996) Selective detection and site-analysis of

O-GlcNAc-modified glycopeptides by beta-elimination and tandem electrospray mass spectrometry, *Anal. Biochem.* **234**, 38–49.

Griffin, T.J., Gygi, S.P., Ideker, T., Rist, B., Eng, J., Hood, L., Aebersold, R. (2002) Complementary profiling of gene expression at the transcriptome and proteome levels in saccharomyces cerevisiae, *Mol. Cell. Proteomics* **1**, 323–333.

Gronborg, M., Kristiansen, T.Z., Stensballe, A., Andersen, J.S., Ohara, O., Mann, M., Jensen, O.N., Pandey, A. (2002) A mass spectrometry-based proteomic approach for identification of serine/threonine-phosphorylated proteins by enrichment with phospho-specific antibodies: identification of a novel protein, frigg, as a protein kinase a substrate, *Mol. Cell. Proteomics* **1**, 517–527.

Gygi, S.P., Corthals, G.L., Zhang, Y., Rochon, Y., Aebersold, R. (2000) Evaluation of two-dimensional gel electrophoresis-based proteome analysis technology, *Proc. Natl. Acad. Sci. U.S.A.* **97**, 9390–9395.

Gygi, S.P., Rist, B., Gerber, S.A., Turecek, F., Gelb, M.H., Aebersold, R. (1999) Quantitative analysis of complex protein mixtures using isotope-coded affinity tags, *Nat. Biotechnol.* **17**, 994–999.

Gygi, S.P., Rochon, Y., Franza, B.R., Aebersold, R. (1999) Correlation between protein and mRNA abundance in yeast, *Mol. Cell. Biol.* **19**, 1720–1730.

Ha, K.S. (2001) Principle and biological applications of protein chip system, *Exp. Mol. Med.* **33**, 127–132.

Haab, B.B., Dunham, M.J., Brown, P.O. (2001) Protein microarrays for highly parallel detection and quantitation of specific proteins and antibodies in complex solutions, *Genome Biol.* **2**, RESEARCH0004.1–0004.13.

Han, D.K., Eng, J., Zhou, H., Aebersold, R. (2001) Quantitative profiling of differentiation-induced microsomal proteins using isotope-coded affinity tags and mass spectrometry, *Nat. Biotechnol.* **19**, 946–951.

Hansen, A.P., Petros, A.M., Mazar, A.P., Pederson, T.M., Rueter, A., Fesik, S.W. (1992) A practical method for uniform isotopic labeling of recombinant proteins in mammalian cells, *Biochemistry* **31**, 12713–12718.

Hansen, K.C., Schmitt-Ulms, G., Chalkley, R.J., Hirsch, J., Baldwin, M.A., Burlingame, A.L. (2003) Mass spectrometric analysis of protein mixtures at low levels using cleavable 13C-isotope-coded affinity tag and multidimensional chromatography, *Mol. Cell. Proteomics* **2**, 299–314.

Hart, G.W., Haltiwanger, R.S., Holt, G.D., Kelly, W.G. (1989) Nucleoplasmic and cytoplasmic glycoproteins, *Ciba. Found. Symp.* **145**, 102–112, discussion 112–108.

Herbert, B. (1999) Advances in protein solubilization for two-dimensional electrophoresis, *Electrophoresis* **20**, 660–663.

Herbert, B., Righetti, P.G. (2000) A turning point in proteome analysis: sample prefractionation via multicompartment electrolyzers with isoelectric membranes, *Electrophoresis* **21**, 3639–3648.

Ideker, T., Thorsson, V., Ranish, J.A., Christmas, R., Buhler, J., Eng, J.K., Bumgarner, R., Goodlett, D.R., Aebersold, R., Hood, L. (2001) Integrated genomic and proteomic analyses of a systematically perturbed metabolic network, *Science* **292**, 929–934.

Imam-Sghiouar, N., Laude-Lemaire, I., Labas, V., Pflieger, D., Le Caer, J.P., Caron, M., Nabias, D.K., Joubert-Caron, R. (2002) Subproteomics analysis of phosphorylated proteins: application to the study of B-lymphoblasts from a patient with Scott syndrome, *Proteomics* **2**, 828–838.

Ito, T., Chiba, T., Ozawa, R., Yoshida, M., Hattori, M., Sakaki, Y. (2001) A comprehensive two-hybrid analysis to explore the yeast protein interactome, *Proc. Natl. Acad. Sci. U.S.A.* **98**, 4569–4574.

Jansen, R., Bussemaker, H.J., Gerstein, M. (2003) Revisiting the codon adaptation index from a whole-genome perspective: analyzing the relationship between gene expression and codon occurrence in yeast using a variety of models, *Nucleic Acids Res.* **31**, 2242–2251.

Jensen, P.K., Pasa-Tolic, L., Anderson, G.A., Horner, J.A., Lipton, M.S., Bruce, J.E., Smith, R.D. (1999) Probing proteomes using capillary isoelectric focusing-electrospray ionization Fourier transform ion cyclotron resonance mass spectrometry, *Anal. Chem.* **71**, 2076–2084.

Jensen, P.K., Pasa-Tolic, L., Peden, K.K., Martinovic, S., Lipton, M.S., Anderson, G.A., Tolic, N., Wong, K.K., Smith, R.D. (2000) Mass spectrometric detection for capillary isoelectric focusing separations of complex protein mixtures, *Electrophoresis* **21**, 1372–1380.

Ji, J., Chakraborty, A., Geng, M., Zhang, X., Amini, A., Bina, M., Regnier, F. (2000)

Strategy for qualitative and quantitative analysis in proteomics based on signature peptides, *J. Chromatogr., B. Biomed. Sci. Appl.* **745**, 197–210.

Jiang, H., English, A.M. (2002) Quantitative analysis of the yeast proteome by incorporation of isitopically labeled leucine, *J. Proteome Res.* **1**, 345–350.

Karas, M., Hillenkamp, F. (1988) Laser desorption ionization of proteins with molecular masses exceeding 10,000 daltons, *Anal. Chem.* **60**, 2299–2301.

Klose, J. (1999) Large-gel 2-D electrophoresis, *Methods Mol. Biol.* **112**, 147–172.

Klose, J., Kobalz, U. (1995) Two-dimensional electrophoresis of proteins: an updated protocol and implications for a functional analysis of the genome, *Electrophoresis* **16**, 1034–1059.

Koller, A., Washburn, M.P., Lange, B.M., Andon, N.L., Deciu, C., Haynes, P.A., Hays, L., Schieltz, D., Ulaszek, R., Wei, J., Wolters, D., Yates, J.R., III (2002) Proteomic survey of metabolic pathways in rice, *Proc. Natl. Acad. Sci. U.S.A.* **99**, 11969–11974.

Koonin, E.V. (2001) Computational genomics, *Curr. Biol.* **11**, R155–R158.

Krishna, R., Wold, F. (1993) Post-Translational Modification of Proteins, *Adv. Enzymol. Relat. Areas Mol. Biol.* **67**, 265–298.

Lahm, H.W., Langen, H. (2000) Mass spectrometry: a tool for the identification of proteins separated by gels, *Electrophoresis* **21**, 2105–2114.

Leavell, M.D., Leary, J.A., Yamasaki, R. (2002) Mass spectrometric strategy for the characterization of lipooligosaccharides from neisseria gonorrhoeae 302 using FTICR, *J. Am. Soc. Mass Spectrom.* **13**, 571–576.

Lee, Y., Mrksich, M. (2002) Protein chips: from concept to practice, *A TRENDS Guide to Proteomics II.* **20**, s14–s18.

Li, J., Steen, H., Gygi, S.P. (2003) Protein profiling with cleavable isotope-coded affinity tag (cICAT) reagents: the yeast salinity stress response, *Mol. Cell. Proteomics* **2**, 1198–1204.

Link, A. (2002) Multidimensional peptide separations in proteomics, *A TRENDS Guide to Proteomics II.* **20**, s8–s13.

Link, A.J., Eng, J., Schieltz, D.M., Carmack, E., Mize, G.J., Morris, D.R., Garvik, B.M., Yates, J.R. III (1999) Direct analysis of protein complexes using mass spectrometry, *Nat. Biotechnol.* **17**, 676–682.

Lipton, M.S., Pasa-Tolic, L., Anderson, G.A., Anderson, D.J., Auberry, D.L., Battista, J.R., Daly, M.J., Fredrickson, J., Hixson, K.K., Kostandarithes, H., Masselon, C., Markillie, L.M., Moore, R.J., Romine, M.F., Shen, Y., Stritmatter, E., Tolic, N., Udseth, H.R., Venkateswaran, A., Wong, K.K., Zhao, R., Smith, R.D. (2002) Global analysis of the deinococcus radiodurans proteome by using accurate mass tags, *Proc. Natl. Acad. Sci. U.S.A.* **99**, 11049–11054.

Liska, A.J., Shevchenko, A. (2003) Expanding the organismal scope of proteomics: cross-species protein identification by mass spectrometry and its implications, *Proteomics* **3**, 19–28.

Liu, P., Regnier, F. (2002) An isotope coding strategy for proteomics involving both amine and carboxyl labeling, *J. Proteome Res.* **1**, 443–450.

Mamyrin, B.A., Karataev V.I., Shmikk D.V., Zagulin V.A. (1973) Reflectron for TOF-MS, *Sov. Phys. JETP* **37**, 45–48.

Mamyrin, B.A., Shmikk D.V. (1979) Linear reflection, *Sov. Phys. JETP* **49**, 762–764.

Mann, M., Wilm, M. (1994) Error-tolerant identification of peptides in sequence databases by peptide sequence tags, *Anal. Chem.* **66**, 4390–4399.

McLachlin, D.T., Chait, B.T. (2001) Analysis of phosphorylated proteins and peptides by mass spectrometry, *Curr. Opin. Chem. Biol.* **5**, 591–602.

von Mering, C., Kruse, R., Snel, B., Cornell, M., Oliver, S.G., Fields, S., Bork, P. (2002) Comparative assessment of large scale data sets of rotein-protein interactions, *Nature* **417**, 399–403.

Mirgorodskaya, O.A., Kozmin, Y.P., Titov, M.I., Korner, R., Sonksen, C.P., Roepstorff, P. (2000) Quantitation of peptides and proteins by matrix-assisted laser desorption/ionization mass spectrometry using 18O-labeled internal standards, *Rapid. Commun. Mass Spectrom.* **14**, 1226–1232.

Molloy, M.P. (2000) Two-dimensional electrophoresis of membrane proteins using immobilized pH gradients, *Anal. Biochem.* **280**, 1–10.

Moskovets, E. (2000) Optimization of the mass reflector parameters for direct ion extraction, *Rapid. Commun. Mass Spectrom.* **14**, 150–155.

Moskovets, E., Karger, B.L. (2003) Mass calibration of a matrix-assisted laser desorption/ionization time-of-flight mass spectrometer including the rise time of the delayed extraction pulse, *Rapid. Commun. Mass Spectrom.* **17**, 229–237.

Mrksich, M., Dike, L.E., Tien, J., Ingber, D.E., Whitesides, G.M. (1997) Using microcontact printing to pattern the attachment of mammalian cells to self-assembled monolayers of alkanethiolates on transparent films of gold and silver, *Exp. Cell. Res.* **235**, 305–313.

Munchbach, M., Quadroni, M., Miotto, G., James, P. (2000) Quantitation and facilitated de novo sequencing of proteins by isotopic N-terminal labeling of peptides with a fragmentation-directing moiety, *Anal. Chem.* **72**, 4047–4057.

Muszynska, G., Andersson, L., Porath, J. (1986) Selective adsorption of phosphoproteins on gel-immobilized ferric chelate, *Biochemistry* **25**, 6850–6853.

Noordewier, M.O., Warren, P.V. (2001) Gene expression microarrays and the integration of biological knowledge, *Trends Biotechnol.* **19**, 412–415.

Oda, Y., Huang, K., Cross, F.R., Cowburn, D., Chait, B.T. (1999) Accurate quantitation of protein expression and site-specific phosphorylation, *Proc. Natl. Acad. Sci. U.S.A.* **96**, 6591–6596.

O'Farrell, P.H. (1975) High resolution two-dimensional electrophoresis of proteins, *J. Biol. Chem.* **250**, 4007–4021.

Oh-Ishi, M., Satoh, M., Maeda, T. (2000) Preparative two-dimensional gel electrophoresis with agarose gels in the first dimension for high molecular mass proteins, *Electrophoresis* **21**, 1653–1669.

Ong, S.E., Blagoev, B., Kratchmarova, I., Kristensen, D.B., Steen, H., Pandey, A., Mann, M. (2002) Stable isotope labeling by amino acids in cell culture, SILAC, as a simple and accurate approach to expression proteomics, *Mol. Cell. Proteomics* **1**, 376–386.

Pappin, D.J. (2003) Peptide mass fingerprinting using MALDI-TOF mass spectrometry, *Methods Mol. Biol.* **211**, 211–219.

Pasa-Tolic, L., Harkewicz, R., Anderson, G.A., Tolic, N., Shen, Y., Zhao, R., Thrall, B., Masselon, C., Smith, R.D. (2002) Increased proteome coverage for quantitative peptide abundance measurements based upon high performance separations and DREAMS FTICR mass spectrometry, *J. Am. Soc. Mass Spectrom.* **13**, 954–963.

Peters, E.C., Horn, D.M., Tully, D.C., Brock, A. (2001) A novel multifunctional labeling reagent for enhanced protein characterization with mass spectrometry, *Rapid. Commun. Mass Spectrom.* **15**, 2387–2392.

Petricoin, E.F., Ardekani, A.M., Hitt, B.A., Levine, P.J., Fusaro, V.A., Steinberg, S.M., Mills, G.B., Simone, C., Fishman, D.A., Kohn, E.C., Liotta, L.A. (2002) Use of proteomic patterns in serum to identify ovarian cancer, *Lancet* **359**, 572–577.

Puig, O., Caspary, F., Rigaut, G., Rutz, B., Bouveret, E., Bragado-Nilsson, E., Wilm, M., Seraphin, B. (2001) The tandem affinity purification (TAP) method: a general procedure of protein complex purification, *Methods* **24**, 218–229.

Rademacher, T.W., Parekh, R.B., Dwek, R.A. (1988) Glycobiology, *Annu. Rev. Biochem.* **57**, 785–838.

Reynolds, K.J., Yao, X., Fenselau, C. (2002) Proteolytic 18O labeling for comparative proteomics: evaluation of endoproteinase Glu-C as the catalytic agent, *J. Proteome Res.* **1**, 27–33.

Rosenblatt, K.P., Bryant-Greenwood, P., Killian, J.K., Mehta, A., Geho, D., Espina, V., Petricoin, E.F., Liotta, L.A. (2004) Serum proteomics in cancer diagnosis and management, *Annu. Rev. Med.* **55**, 97–112.

Rout, M.P., Aitchison, J.D., Suprapto, A., Hjertaas, K., Zhao, Y., Chait, B.T. (2000) The yeast nuclear pore complex: composition, architecture, and transport mechanism, *J. Cell. Biol.* **148**, 635–652.

Rudiger, H., Gabius, H.J. (2001) Plant lectins: occurrence, biochemistry, functions and applications, *Glycoconj. J.* **18**, 589–613.

Sanchez, J.C., Hochstrasser, D.F. (1999) High-resolution, IPG-based, mini two-dimensional gel electrophoresis, *Methods Mol. Biol.* **112**, 227–233.

Santoni, V., Molloy, M., Rabilloud, T. (2000) Membrane proteins and proteomics: un amour impossible? *Electrophoresis* **21**, 1054–1070.

Schwartz, J.C., Jardine, I. (1996) Quadrupole ion trap mass spectrometry, *Methods Enzymol.* **270**, 552–586.

Schwartz, J.C., Senko, M.W., Syka, J.E. (2002) A two-dimensional quadrupole ion trap mass spectrometer, *J. Am. Soc. Mass Spectrom.* **13**, 659–669.

Sheeley, D.M., Reinhold, V.N. (1998) Structural characterization of carbohydrate sequence, linkage, and branching in a quadrupole Ion trap mass spectrometer: neutral oligosaccharides and N-linked glycans, *Anal. Chem.* **70**, 3053–3059.

Siuzdak, G. (1994) The emergence of mass spectrometry in biochemical research, *Proc. Natl. Acad. Sci. U.S.A.* **91**, 11290–11297.

Smith, R.D., Anderson, G.A., Lipton, M.S., Pasa-Tolic, L., Shen, Y., Conrads, T.P., Veenstra, T.D., Udseth, H.R. (2002) An accurate mass tag strategy for quantitative and high-throughput proteome measurements, *Proteomics* **2**, 513–523.

Smith, R.D., Loo, J.A., Edmonds, C.G., Barinaga, C.J., Udseth, H.R. (1990) New developments in biochemical mass spectrometry: electrospray ionization, *Anal. Chem.* **62**, 882–899.

Snow, D.M., Hart, G.W. (1998) Nuclear and cytoplasmic glycosylation, *Int. Rev. Cytol.* **181**, 43–74.

Sunyaev, S., Liska, A.J., Golod, A., Shevchenko, A. (2003) MultiTag: multiple error-tolerant sequence tag search for the sequence-similarity identification of proteins by mass spectrometry, *Anal. Chem.* **75**, 1307–1315.

Uetz, P., Giot, L., Cagney, G., Mansfield, T.A., Judson, R.S., Knight, J.R., Lockshon, D., Narayan, V., Srinivasan, M., Pochart, P., Qureshi-Emili, A., Li, Y., Godwin, B., Conover, D., Kalbfleisch, T., Vijayadamodar, G., Yang, M., Johnston, M., Fields, S., Rothberg, J.M. (2000) A comprehensive analysis of protein-protein interactions in Saccharomyces cerevisiae, *Nature* **403**, 623–627.

Unlu, M., Morgan, M.E., Minden, J.S. (1997) Difference gel electrophoresis: a single gel method for detecting changes in protein extracts, *Electrophoresis* **18**, 2071–2077.

Varshavsky, A. (1996) The N-end rule: functions, mysteries, uses, *Proc. Natl. Acad. Sci. U.S.A.* **93**, 12142–12149.

Verma, R., Chen, S., Feldman, R., Schieltz, D., Yates, J., Dohmen, J., Deshaies, R.J. (2000) Proteasomal proteomics: identification of nucleotide-sensitive proteasome-interacting proteins by mass spectrometric analysis of affinity-purified proteasomes, *Mol. Biol. Cell.* **11**, 3425–3439.

Vosseller, K., Wells, L., Hart, G.W. (2001) Nucleocytoplasmic O-glycosylation: O-GlcNAc and functional proteomics, *Biochimie* **83**, 575–581.

Wall, D.B., Kachman, M.T., Gong, S., Hinderer, R., Parus, S., Misek, D.E., Hanash, S.M., Lubman, D.M. (2000) Isoelectric focusing nonporous RP HPLC: a two-dimensional liquid-phase separation method for mapping of cellular proteins with identification using MALDI-TOF mass spectrometry, *Anal. Chem.* **72**, 1099–1111.

Wall, D.B., Kachman, M.T., Gong, S.S., Parus, S.J., Long, M.W., Lubman, D.M. (2001) Isoelectric focusing nonporous silica reversed-phase high-performance liquid chromatography/electrospray ionization time-of-flight mass spectrometry: a three-dimensional liquid-phase protein separation method as applied to the human erythroleukemia cell-line, *Rapid. Commun. Mass Spectrom.* **15**, 1649–1661.

Washburn, M.P., Koller, A., Oshiro, G., Ulaszek, R.R., Plouffe, D., Deciu, C., Winzeler, E., Yates, J.R. III (2003) Protein pathway and complex clustering of correlated mRNA and protein expression analyses in Saccharomyces cerevisiae, *Proc. Natl. Acad. Sci. U.S.A.* **100**, 3107–3112.

Washburn, M.P., Ulaszek, R., Deciu, C., Schieltz, D.M., Yates, J.R. III (2002) Analysis of quantitative proteomic data generated via multidimensional protein identification technology, *Anal. Chem.* **74**, 1650–1657.

Washburn, M.P., Wolters, D., Yates, J.R. III (2001) Large-scale analysis of the yeast proteome by multidimensional protein identification technology, *Nat. Biotechnol.* **19**, 242–247.

Washburn, P.W., Yates, J.R. III (2000) New methods of proteome analysis: multidimensional chromatography and mass spectrometry, *Trends Biotechnol.* **18**, S27–S30.

Wasinger, V.C., Cordwell, S.J., Cerpa-Poljak, A., Yan, J.X., Gooley, A.A., Wilkins, M.R., Duncan, M.W., Harris, R., Williams, K.L., Humphery-Smith, I. (1995) Progress with gene-product mapping of the Mollicutes: Mycoplasma genitalium, *Electrophoresis* **16**, 1090–1094.

Watson, J.T. (1997) *Introduction to Mass Spectrometry*, Lippincott-Raven, Philadelphia, NY.

Wells, L., Whalen, S.A., Hart, G.W. (2003) O-GlcNAc: a regulatory post-translational modification, *Biochem. Biophys. Res. Commun.* **302**, 435–441.

Weng, S., Gu, K., Hammond, P.W., Lohse, P., Rise, C., Wagner, R.W., Wright, M.C., Kuimelis, R.G. (2002) Generating addressable protein microarrays with PROfusion covalent mRNA-protein fusion technology, *Proteomics* **2**, 48–57.

Wilkins, M.R., Sanchez, J.C., Gooley, A.A., Appel, R.D., Humphery-Smith, I., Hochstrasser, D.F., Williams, K.L. (1996) Progress with proteome projects: why all proteins expressed by a genome should be identified and how to do it, *Biotechnol. Genet. Eng. Rev.* **13**, 19–50.

Wilkins, M.R., Williams, K.L., Appel, R.D., Hochstrasser, D.F. (Eds.) (1997) *Proteome Research: New Frontiers in Functional Genomics (Principles and Practice)*, Springer-Verlag.

Wilson, N.L., Schulz, B.L., Karlsson, N.G., Packer, N.H. (2002) Sequential analysis of N- and O-linked glycosylation of 2D-PAGE separated glycoproteins, *J. Proteome Res.* **1**, 521–529.

Winslow, R.L., Boguski, M.S. (2003) Genome informatics: current status and future prospects, *Circ. Res.* **92**, 953–961.

Wolters, D.A., Washburn, M.P., Yates, J.R. III (2001) An automated multidimensional protein identification technology for shotgun proteomics, *Anal. Chem.* **73**, 5683–5690.

Yao, J., Burton, J.L., Saama, P., Sipkovsky, S., Coussens, P.M. (2001) Generation of EST and cDNA microarray resources for the study of bovine immunobiology, *Acta Vet. Scand.* **42**, 391–405.

Yates, J.R., Eng, J.K., McCormack, A.L., Schieltz, D. III (1995) Method to correlate tandem mass spectra of modified peptides to amino acid sequences in the protein database, *Anal. Chem.* **67**, 1426–1436.

Yost, R.A., Boyd, R.K. (1990) Tandem mass spectrometry: quadrupole and hybrid instruments, *Methods Enzymol.* **193**, 154–200.

Yu, J., Hu, S., Wang, J., Wong, G.K., Li, S., Liu, B., Deng, Y., Dai, L., Zhou, Y., Zhang, X., Cao, M., Liu, J., Sun, J., Tang, J., Chen, Y., Huang, X., Lin, W., Ye, C., Tong, W., Cong, L., Geng, J., Han, Y., Li, L., Li, W., Hu, G., Li, J., Liu, Z., Qi, Q., Li, T., Wang, X., Lu, H., Wu, T., Zhu, M., Ni, P., Han, H., Dong, W., Ren, X., Feng, X., Cui, P., Li, X., Wang, H., Xu, X., Zhai, W., Xu, Z., Zhang, J., He, S., Xu, J., Zhang, K., Zheng, X., Dong, J., Zeng, W., Tao, L., Ye, J., Tan, J., Chen, X., He, J., Liu, D., Tian, W., Tian, C., Xia, H., Bao, Q., Li, G., Gao, H., Cao, T., Zhao, W., Li, P., Chen, W., Zhang, Y., Hu, J., Liu, S., Yang, J., Zhang, G., Xiong, Y., Li, Z., Mao, L., Zhou, C., Zhu, Z., Chen, R., Hao, B., Zheng, W., Chen, S., Guo, W., Tao, M., Zhu, L., Yuan, L., Yang, H. (2002) A draft sequence of the rice genome (Oryza sativa L. ssp. indica), *Science* **296**, 79–92.

Zhu, H., Bilgin, M., Bangham, R., Hall, D., Casamayor, A., Bertone, P., Lan, N., Jansen, R., Bidlingmaier, S., Houfek, T., Mitchell, T., Miller, P., Dean, R.A., Gerstein, M., Snyder, M. (2001) Global analysis of protein activities using proteome chips, *Science* **293**, 2101–2105.

Matrix, Extracellular and Interstitial

Abebe Akalu and Peter C. Brooks
New York University School of Medicine, NY, USA

Encyclopedia of Molecular Cell Biology and Molecular Medicine, 2nd Edition. Volume 8
Edited by Robert A. Meyers.
Copyright © 2005 Wiley-VCH Verlag GmbH & Co. KGaA, Weinheim
ISBN: 3-527-30550-5

Keywords

Basement Membrane (BM)
A highly ordered thin sheetlike sub-compartment of the ECM composed of networks of glycoproteins and proteoglycans that separate epithelia and endothelia from the underlying stroma.

Cryptic ECM Epitope
A specific amino acid sequence of an ECM protein that is normally not accessible for cellular interactions under normal physiological conditions.

Extracellular Matrix (ECM)
A complex acellular network of adhesive collagenous and noncollagenous glycoproteins embedded in a hydrated gel composed of a mixture of proteoglycans.

Interstitial Matrix
A loosely organized sub-compartment of the ECM composed of glycoproteins, proteoglycans, and matricellular proteins that surround individual cells and tissues.

Matricellular Proteins
ECM-associated proteins that modulate cellular functions but which do not contribute significantly to structural architecture of the matrix.

■ The requirement of bidirectional communication between cells and their local acellular microenvironment for proper development and homeostasis in multicellular organisms has been appreciated for decades. Interestingly, intriguing new studies have suggested that some of this molecular information may be cryptic or hidden within the three-dimensional architecture of proteins and requires structural modification of the ECM for efficient transmission and utilization by cells and tissues. Thus, the dynamic interplay between cells and both intact and structurally altered ECM proteins help establish specificity, coordinate, and refine the complex interconnected biological responses critical for normal physiological processes. In this regard, unique insight will likely be gained into pathological events by a more complete understanding of the integrated network of molecular signals regulated by cellular interactions with the ECM.

1
Introduction

It has been known for decades that multicellular organisms depend on information within the local acellular microenvironment for proper structural organization, development, and homeostasis. The complex interconnected networks of molecules that compose the extracellular environment impacts nearly every aspect of normal physiology and disease, ranging from molecular control of gene expression and cell cycle progression to mechanical and architectural organization of tissues and organs. The major interests of many early investigators within the fields of molecular and developmental biology were primarily focused on the cellular compartment, including membrane, cytoplasmic, and nuclear components. However, as more sophisticated technology and tools were developed, it quickly became apparent that cells were not only physically linked to the insoluble scaffolding proteins outside

the cell, but that cell shape, function, and survival depended on unique interactions and the bidirectional flow of information from outside the cell to the cells interior. Thus, the stage was set to begin the fascinating journey into an in-depth understanding of the biological responses controlled by the extracellular matrix (ECM).

2
Structural Organization and Molecular Composition of the Extracellular Matrix

For the purpose of our discussion, the ECM can be thought of as being composed of two general compartments, including the basement membrane (BM) or basal lamina and the interstitial matrix (Fig. 1). While these two compartments can be studied as separate features, they do not exist in isolation, but rather are interconnected by a variety of anchoring and

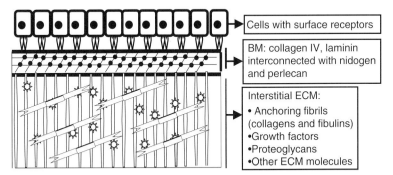

Fig. 1 Schematic representation of the structural organization and composition of ECM. The extracellular matrix can be organized into two compartments, including the basement membrane (BM) and the interstitial matrix. These two compartments are connected by a series of anchoring fibers, which consist of distinct forms of collagen, fibulins, as well as other ECM macromolecules. The major components of the BM include collagen type IV, laminin, nidogen, and perlecan. The interstitial matrix is composed of a gel-like composition of ECM proteins, including fibronectin, distinct forms of collagen, vitronectin, and various matricellular proteins such as thrombospondin. Growth factors, proteases, and a number of proteoglycans can also be found in the interstitial matrix.

interconnecting fibrils. Thus, a highly integrated continuum exists between the BM and interstitial matrix. In general terms, the BM is organized into a highly ordered meshlike network composed of structural glycoproteins, proteoglycans, and other macromolecules. BMs can be found underlying mostly all epithelial sheets and blood vessels, thus contributing to physical and functional compartmentalization. In contrast, the interstitial matrix can be characterized as a loosely organized gel-like composition of collagenous and noncollagenous glycoproteins, proteoglycans, and matricellular proteins that surround cells, tissues, and organs. The molecular composition of the BM and interstitial matrix can vary widely, depending on the repertoire of cells present within a given tissue. Moreover, the composition and structural integrity can also vary, depending on the particular physiological and/or pathological processes that may be occurring at any given time.

While heterogeneous in composition, examples of the major components of the BM include collagen type IV, laminin, nidogen, and perlecan. In addition to cross connections mediated by nidogen, collagen, and laminin, these molecules can interact with each other via a number of unique functional domains. While these macromolecules make up the majority of the BM, other minor, but no less important, components contribute to its structure, including genetically distinct forms of collagen (i.e. collagen type VIII, XV, and XVIII) and numerous proteoglycans. Examples of the major components of the interstitial matrix include an array of fiber forming collagens, including collagen type I, II, and III. Other major components that have been characterized include fibronectin, vitronectin,

Tab. 1 Partial list of the major components of the basement membrane and interstitial matrix.

---

1. Composition of basement membrane
   Collagen IV
   Laminin
   Nidogen
   Perlecan

2. Composition of interstitial matrix
   Collagen type I, II, III
   Fibronectin
   Thrombospondin
   Vitronectin
   SPARC
   Syndecan
   Proteoglycans
   Fibrinogen/Fibrin
   Elastin

---

fibrinogen, thrombospondin, and elastin (Table 1).

The complex sets of matrix macromolecules are not randomly organized, but are specifically ordered and arranged into supramolecular assembles. The physical interactions that facilitate the unique architectural organization are mediated, in large part, by distinct protein–protein binding domains, glucosaminoglycan (GAG) side chains, and a variety of carbohydrate moieties. For example, distinct functional domains found within nidogen bind collagen, and laminin, while unique domains within collagen can self-associate with other collagen molecules as well as other ECM proteins. These structural glycoproteins and proteoglycans also provide functional binding sites for a variety of growth factors, cytokines, proteolytic enzymes, and cell adhesion receptors. Thus, it is clear that besides participating in the structural assembly of mechanical scaffolds, ECM molecules facilitate localization and storage of a number of crucial regulatory factors. ECM localization of these regulatory molecules may serve to protect them from degradation and/or induce conformational changes necessary for optimal presentation to cell surface receptors. As can be appreciated from the vast complexity of protein–protein interactions, it would be well beyond the scope of our discussion to analyze in-depth the vast number of molecular association that contribute to the assembly of the ECM.

While it was once thought that the ECM was merely an insoluble scaffold that primarily functioned in providing mechanical support for cells and tissues, it is now known that the ECM can regulate a variety of processes. Some of the more well-characterized events (Fig. 2) regulated by the ECM include adhesion, migration, invasion, cell cycle progression, proliferation, cell survival, differentiation, gene expression, tissue compartmentalization, and filtration. The ECM is not simply a static structural scaffold, but is rather dynamic, with changes occurring in molecular composition and structure. This dynamic feature is critical in helping to integrate the complex sets of regulatory signals transferred to cells from the ECM. In fact, fine-tuning these matrix-derived signals can be accomplished by structurally altering matrix components, which may modify existing binding sites for growth factors, cytokines, proteases, and cell surface receptors. Conformational changes in matrix molecules may expose new cryptic binding sites for secreted molecules along with creating novel binding sites for cell surface receptors such as integrins. Notably, the compositional and structural changes occurring within the ECM can be facilitated by changes in the secretion, structural organization, glycation, and cross-linking of proteins. The wealth of biochemical and molecular information stored within the ECM can be

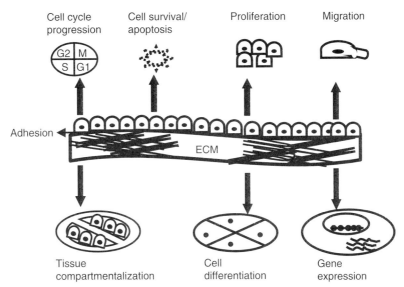

Fig. 2 Cellular processes regulated by the ECM. Cellular interactions with the extracellular matrix can regulate a wide array of important cellular processes. A partial list of the well-characterized events include cell cycle progression, cell survival, and apoptosis, proliferation, migration, adhesion, tissue compartmentalization, differentiation, and gene expression.

transmitted to the cells by binding to a number of cell surface receptors, including integrins, cell surface proteoglycans, and an emerging class of receptor tyrosine kinases. While examples of all these classes of receptors have been shown to mediate cellular interactions with the ECM, it could be argued that integrins are the most well-studied group of molecules that mediate cell–ECM interactions.

2.1
Molecular Composition and Structural Organization of the Basement Membrane

As mentioned above, the BM or basal lamina is a highly ordered meshlike network of glycoproteins and proteoglycans that separate epithelium and endothelium from the underlying stroma. Given the diversity of cell types that secrete ECM molecules, it is not surprising that the molecular composition of BMs can vary from tissue to tissue. While molecular diversity is evident, a number of major components are common to most BMs and include collagen type IV, laminin, nidogen, and perlecan. These molecular components are not meant to represent a complete list of the BM molecules, but rather, examples of well-characterized components to help illustrate its compositional and structural organization.

2.2
Collagen Type IV

The major collagenous component of the BM is collagen type IV. A variety of other minor collagen types are also present, including collagen types VIII, XV, and XVIII. Collagen type IV molecules are composed of a combination of 3 α chains organized in a triple helical manner. Six different collagen type IV chains ($\alpha 1$–$\alpha 6$) have been identified, each of which represents a distinct gene product. One of the most widely distributed isoforms of collagen type IV present in nearly all BMs is the triple helical molecule composed of 2 $\alpha 1$(IV) chains and 1 $\alpha 2$(IV) chain. Interestingly, other isotypes of type IV collagen have been detected in the BMs from distinct tissues, suggesting tissue-specific distribution patterns. Collagen type IV is organized into functional domains, including a large central triple helical region composed of repeating units with glycine at every third position. While this repeated pattern and triple helical structure exists throughout the central domain, interruptions on this unique organization also exist. The triple helical region has been shown to be critical for mediating cell adhesion, as a number of integrin-binding sites have been localized to this area. Moreover, several binding sites for other ECM proteins are also located within the triple helical region, suggesting an important role for this region in the collagen network assembly. The amino terminus of collagen type IV contains a globular region termed the *7S domain*. The 7S domain facilitates association of collagen molecules into tetramers, which is important in the formation of higher-order collagen networks. A third important structural and functional feature is the C-terminal globular noncollagenous (NC1) domain. These NC1 domains are thought to facilitate selection of individual α chains during network assembly. Interestingly, recent studies have indicated that specific integrin-binding sites are not only confined to the triple helical domain but are also located in NC1 domains. For example, studies have indicated that the NC1 domain from $\alpha 1$ chain can interact with integrin $\alpha 1\beta 1$, while the NC1 domain from the $\alpha 2$ chain supports interactions with integrins $\alpha 3\beta 1$, $\alpha v\beta 3$, and $\alpha v\beta 5$.

Moreover, NC1 domains from the $\alpha3$ and $\alpha6$ chains of collagen type IV specifically bind to integrin $\alpha v\beta3$, while NC1 domains form $\alpha4$ and $\alpha5$ chains fail to interact with integrin receptors. Given the facts that ligation of distinct integrin receptors can initiate unique signaling cascades, integrin interactions with noncollagenous (NC) domains may regulate a diversity of cellular processes. To this end, it is becoming increasingly clear that NC1 domains from distinct forms of collagens, such as types IV, VIII, XV, and XVIII, may possess unique regulatory properties since potent antiangiogenic and antitumorgenic modules have been identified from these regions.

Several families of molecules have been shown to mediate cellular interactions with collagen (18–20, 29, 30). However, integrin receptors likely represent the predominate class of cell surface molecules, facilitating interactions with collagen type IV. A number of distinct integrin heterodimers interact with triple helical collagen IV, including integrins $\alpha1\beta1$, $\alpha2\beta1$, and $\alpha3\beta1$. During invasive cellular events, such as tumor invasion, metastasis, and angiogenesis, the triple helical structure of collagen can be altered, leading to disruption of $\beta1$ integrin binding, resulting in exposure of cryptic $\alpha v\beta3$ bindings sites. This shift in integrin-mediated recognition could result in the initiation of unique signaling cascades. Thus, minor changes in the three-dimensional structure of collagen could profoundly impact signaling cascades, thereby modifying the biological response to the ECM.

2.3
Laminin

A second large glycoprotein that forms an interconnected sheetlike network with collagen type IV is laminin. Laminin represents a family of adhesive glycoproteins comprised of high molecular weight (about 800 kDa) disulfide-linked heterotrimers with chains designated α, β, and γ. Since the original discovery of laminin many years ago, five different α-chains ($\alpha1$ to $\alpha5$), three different β chains ($\beta1$ to $\beta3$), and three different γ-chains ($\gamma1$ to $\gamma3$) have been identified. Although the theoretical combinations of the three individual chains could allow up to 45 different trimeric isoforms, to date 12 isoforms have been described in detail. Of the 12 isoforms known, it could be argued that laminin-1 ($\alpha1\beta1\gamma1$) and laminin-5 ($\alpha3\beta3\gamma2$) are perhaps the most well characterized. The individual chains of laminin share a significant degree of structural similarity, and can be organized into a number of functional domains, including globular domains (IV and VI), epidermal growth factor (EGF)-like laminin-repeats (LE-module) (III and V), and an alpha-helical coiled-coil domain.

In general, laminin is organized into a crosslike structure containing three short arms and one long arm. The short arms are formed by the N-terminal region of one of the three chains, whereas the long arm is a rod-like coiled-coil structure comprised of the carboxyl termini of all three subunits. The long arm ends in a globular structure called the *G-domain*. The G-domain can be further subdivided into five globular subdomains (G1–G5). The G-domains contain major binding-sites for cellular receptors, such as integrins, syndecans, and α-dystroglycan. Moreover, laminin also possesses binding sites for minor collagen species such as collagen type XVIII. In addition to a number of proteoglycans, the BM protein nidogen directly interacts with laminin. Many of these unique protein–protein interactions are thought to be critical for structural assembly of the BM.

While distinct forms of laminin are found mainly in the BM, studies have also identified isoforms present within the matrix of embryonic mesenchyme, loose connective tissue, and developing central nervous system. Interestingly, throughout embryonic development and in the adult organisms, studies have identified different patterns of cell and tissue-specific expression. In fact, distinct isoforms of laminin have their own temporally and spatially regulated expression profiles. For example, while laminin-10 exhibits wide tissue distribution, the expression of other isoforms is tightly regulated. In fact, laminin-2 is expressed in the BM of skeletal muscle and plays a role in neurite outgrowth, while laminin-5 is a constituent of the BM of many epithelial tissues where it regulates stable adhesion and promotes migration. Thus, it is clear that interactions with laminin can regulate a variety of cellular and physiological processes.

Laminin interacts with cells through a variety of mechanisms, including proteoglycans and integrin receptors. The major integrins known to interact with laminin include $\alpha1\beta1$, $\alpha2\beta1$, $\alpha3\beta1$, $\alpha6\beta1$, $\alpha6\beta4$, $\alpha7\beta1$, whereas examples of nonintegrin receptors include the 67-kDa laminin receptors and dystroglycan. Recent studies have suggested that laminin the $\alpha4$ chain is a specific high affinity ligand for the $\alpha v \beta3$ and $\alpha3\beta1$. Notably, distinct integrin binding sites have been identified within laminin, including the classical RGD amino acid recognition sequence as well as a number of non RGD integrin binding sites. As was seen with collagen type IV, the molecular and structural integrity of laminin can impact integrin recognition and modify cellular responses. For example, studies have identified a cryptic site exposed following proteolytic cleavage of laminin-5 that promotes breast carcinoma cell migration.

2.4
Nidogen

As discussed above, the two major structural glycoproteins collagen type IV and laminin are organized into two sheetlike networks. These complex meshlike assemblies are interconnected by the protein Nidogen. Nidogen, also known as *entactin*, is a ubiquitous component of the BM and is conserved in all metazoans. Interestingly, invertebrates, including *Caenorhabditis elegans*, possess only a single gene, whereas in mammals, two nidogen isoforms have been identified termed *nidogen-1* (150 kDa) and *nidogen-2* (200 kDa). Nidogen-2 is predominately located in endothelial BM, whereas nidogen-1 can be ubiquitously distributed within the basal lamina from a variety of tissues. These tissue distribution patterns suggest that different forms of nidogen may play distinct roles in tissue organization, BM assembly, and in the regulation of physiological processes such as blood vessel development. Studies have indicated that nidogen-2 binds to endostatin, an antiangiogenic module of the NC1 domain of collagen type XVIII (41). The two nidogen isoforms share substantial domain and sequence identity with an overlapping binding repertoire for laminins, collagen IV, perlecan, and fibulin. Nidogen consists of three globular domains (G1 to G3). Domain G2 of nidogen-1 contains the binding site for collagen IV, perlecan, and fibulins. In contrast, domain G3 contains high affinity binding sites for the laminin γ chain, collagen IV, and weaker binding sites for fibulins.

Studies have suggested that nidogen may play an important role in BM organization by facilitating connections between laminin and collagen IV networks. Interestingly, *C. elegans*, harboring mutant forms of nidogen-1, exhibited relatively normal assemble of type IV collagen, suggesting that nidogen-1 may not be essential for type IV collagen assembly under these conditions. These findings are somewhat surprising, given the previous studies indicating the importance of nidogen in the structural organization of collagen IV and laminin networks. The nidogen-binding site within laminin has been localized to the laminin epidermal growth factor–like (LE) module γ1III4. Antibodies against γ1III3-5 modules were shown to disrupt early kidney, lung, and salivary gland development. In addition to nidogen's function in stabilizing the BM, it has also been implicated in other protein integration functions such as facilitating interactions with ECM components such as perlecan and fibulin. As with many ECM proteins, nidogen plays roles in signal transduction through integrins $\alpha v\beta 3$ and $\alpha 3\beta 1$, thereby influencing the biological response of the cells to the ECM.

2.5
Perlecan

Besides the major glycoprotein components of the basal lamina, a second family of molecules that contribute to the formation of the BM includes the proteoglycans. Proteoglycans are not only important components of the ECM but are also found on the cell surface. Proteoglycans can be characterized as consisting of a core protein with negatively charged polysaccharide glucosaminoglycan (GAG) side chain. On the basis of their GAG side chains, proteoglycans can be classified into four main groups, including chondrotin sulfate, dermatan sulfate, keratin sulfate, and heparan sulfate. The cellular and tissue distribution profiles along with their biological functions have been suggested to depend on both the GAG chain and the core protein content. Interestingly, the GAG side chains attached to the core proteins have been shown to vary depending on the tissue. For example, phosphacan, a chondroitin sulfate proteoglycans expressed abundantly in neural tissue contains keratin sulfate in rat embryonal cerebellum, but lack these GAGs in rat retinal tissues. Thus, variations in the structural features of these molecules may modify their functions, and thus alter cellular behavior in a tissue-specific manner.

In addition to the structural diversity associated with the GAG chains, the core proteins also consist of structurally distinct domains. In fact, proteoglycans can interact with various biologically active molecules through specific domains of the core protein, thus expanding their diverse capacity to impact biological responses. Recent studies have identified more than 30 structurally distinct proteoglycans from vertebrate tissues. Many proteoglycans bind several different ECM components and growth factors, and play critical roles in processes such as cell adhesion, migration, and differentiation. Notably, heparan sulfate proteoglycans can be subclassified as those present on the cell surface, such as syndecan, those anchored to the cell surface by glycosylphosphatidylinositol such as glypican, and those present in the BM, including perlecan and agrin.

Perlecan is a multidomain heparan sulfate proteoglycan with a wide distribution in the ECM. Perlecan has been shown to be an important element for assembly and function of BMs. Mammalian per-lecan consists of a 480-kDa core protein containing three to four heparin sulfate or chondroitin sulfate side chains. The molecular weight of the fully gly-cosylated protein is between 700 and 800 kDa. Structural studies have provided evidence that perlecan consists of five distinct domains. Domain I, has a molec-ular weight of 20 kDa and consists of 172 amino acids containing three consec-utive Ser-Gly-Asp triplets that represent the attachment sites for GAGs. Domain II of mammalian perlecan is homolo-gous to the low-density lipoprotein (LDL) receptor ligand-binding domain and con-tains the pentapeptide DGSDE. While domain III is homologous to the in-ternal segments of laminin A, domain IV is homologous to neural cell adhe-sion molecule (N-CAM). Domain V is located at the C-terminus and consists of tandem arrays of three laminin G-type modules and four epidermal growth factor–like modules. This domain can bind to the laminin–nidogen complex, fibulin-2, sulfatides, $\beta 1$ integrins, and α-dystroglycan, thereby contributing to the BM organization and structure. Domain V also contains two SGxG tetrapeptides, which may serve as attachment sites for GAG chains other than heparan sulfate. In addition, this domain contains two Leu-Arg-Glu tripeptides that may facili-tate binding of neurons to S-laminin in the synaptic BM. Finally, perlecan inter-acts with collagen type IV and fibronectin, further implicating it in the structural or-ganization of the ECM.

Perlecan is produced during early and late stages of embryonic development and is localized in BMs, vessel walls, carti-lage matrix, and other extracellular spaces. The importance of perlecan to mammalian development has been demonstrated in mouse knockout experiments. Although the BM can develop in the absence of perlecan, nearly half of all perlecan null mice died at embryonic day 10.5 with widespread cephalic, skeletal abnormali-ties, and aberrant BM formation in the heart. Studies have indicated that perlecan can regulate a variety of biological func-tions. Some of these well-characterized functions include its role in mainte-nance of the integrity of BM, and the storage of growth factors and cytokines within the ECM. Interestingly, proteolytic cleavage of perlecan may result in the release of proangiogenic factors such as fibroblast growth factors and/or antian-giogenic factors such as endorepellin, a newly identified antiangiogenic fragment of perlecan.

3
Molecular Composition and Structural Organization of the Interstitial Matrix

The ECM can be viewed as being com-posed of two compartments, including the BM and the interstitial matrix. While the general architecture of the BM is identifi-able as a thin highly organized sheetlike structure. The interstitial matrix can be de-scribed as a loose organization of collage-nous and noncollagenous glycoproteins, matricellular proteins, and proteoglycans that surrounds individual cells, tissues, and organs. It is important to note that while these compartments can be viewed separately, they are indeed intimately con-nected by anchoring fibers of collagen and fibulins. The components of interstitial matrix from a given microenvironment can vary, depending on the cellular reper-toire of the tissue. Given the diversity of cell types that secrete ECM molecules, it would be well beyond the scope of our discussion to attempt to describe in-depth all the components that contribute to the interstitial matrix. Thus, we have chosen to focus on a few examples of the major components, including collagen type I, the glycoproteins fibronectin, vitronectin, and the matricellular protein thrombospondin. These molecular components are not meant to represent a complete list of interstitial matrix molecules, but rather are examples of well-characterized components to help illustrate the composition, structural organization, and function of the interstitial matrix.

3.1
Collagen

The great importance of collagen in physiology, development, and tissue homeostasis has been known for decades. Its singular importance in all metazoan organisms is evident, given its high degree of conservation throughout species as diverse as sponges and man. Over the past several decades, a wealth of biochemical and genetic studies have demonstrated critical roles for collagen in structural organization and function of the ECM. In fact, a wide array of human diseases have been linked to aberrant synthesis, mutations, and biochemical modification of collagen, including Osteogenesis Imperfecta, Ehlers-Danlos Syndrome type V11, and Alport Syndrome, to name just a few. Collagen represents the most abundant ECM protein in the body. It has been estimated that at least 30 distinct genes code for specific α chains of collagen. In fact, at least 21 genetically distinct forms of collagen have been identified. The numerous forms of collagen can be grouped according to their general molecular structure, organization, and function, which include fibrillar collagens, nonfibrillar network

forming collagen, FACIT (fibril-associated collagens with interrupted triple helices) collagen, and multiplexes. Examples of the multiplexin subgroup include collagen type XV and XVIII. Interestingly, these types of collagen have received much attention recently due to the discovery of potent antiangiogenic activity within isolated fragments (Restin and Endostatin) of their respective NC domains. As described above, collagen type IV is an important example of the nonfibrillar networking forming collagen, while collagen type IX and XII are examples of FACIT collagens. Some of the earliest forms of collagen to be described belong to the fibrillar collagen subgroup such as collagen type I, II, and III.

It has been estimated that collagen type I may represent up to 25% of the total protein in the body and the majority of the protein within the interstitial matrix. Therefore, we have chosen to focus on collagen type I as an example of a major collagenous protein within the interstitial matrix. Collagen type I is composed of three α chains, including two $\alpha1$ chains and one $\alpha2$ chain organized into a triple helical structure. These collagen type I chains exhibit the characteristic amino acid triplet repeat G-X-Y with Glycine (G) at every third position. The amino acids proline and hydroxyproline are also often represented at the X and/or Y positions within the triplet. The relatively small size of the glycine residue at the third position has been suggested to be critical for proper folding of the triple helical structure. Secretion of procollagen is followed by cleavage of both the N- and C-terminal propeptides. N- and C-terminal propeptide processing is crucial in facilitating collagen fibril formation and higher-order supra molecular assemblies. Interestingly, soluble propeptide fragments have been detected in serum, and these correlate with enhanced collagen synthesis and bone metastasis. Moreover, propeptide collagen fragments have also been shown to exhibit a number of biological activities, including promoting chemotactic activity as well as inhibiting collagen synthesis under specific conditions. Thus, the maturation and synthesis of collagen type I appears to regulate a number of cellular processes.

Similar to what was discussed with other ECM molecules, collagen type I can interact with a variety of proteins, including other ECM molecules such as fibronectin, thrombospondin, secreted protein acidic rich in cysteine (SPARC) and a number of proteoglycans. These distinct protein–protein interactions can be altered by structural and biochemical changes, including altered hydroxylation, cross-linking, glycation, and the degree of triple helical formation. For example, structural denaturation of the triple helical region can expose unique cryptic binding sites for additional molecules. Moreover, collagen interactions with distinct proteoglycans such as decorin have been shown to specifically alter collagen structure, fibril, and matrix formation. In addition, collagen type I can interact with many other molecules, including proteolytic enzymes and cell surface receptors such as integrins, which facilitate communication between cells and the matrix. The major integrin receptors mediating cellular interactions with collagen include members of the $\beta 1$ integrin subfamily such as $\alpha 1\beta 1$ and $\alpha 2\beta 1$. Shifts in integrin interactions with collagen may result in the initiation of unique signaling pathways, leading to alterations in biological responses, including cell adhesion, migration, proliferation, survival, and gene expression.

3.2
Fibronectin

The large glycoprotein fibronectin is a multidomain molecule found in the ECM, on the surface of cells, and in plasma and other body fluids. Fibronectin can exist in multiple soluble and insoluble forms. Soluble forms of fibronectin represents a major component in the plasma and other bodily fluids, while insoluble forms are either associated with the cell surface or are incorporated into the interstitial matrix. The fibrillar form of fibronectin is of considerable interest to the present discussion because of its capacity to mediate an array of interactions with other ECM proteins. In particular, this multidomain glycoprotein exists as a dimer composed of a two 250-kDa monomers linked covalently near the C-termini by a pair of disulfide bonds. Each monomer is comprised of multiple regions termed *fibronectin type repeats*. Fibronectin is organized into multiple domains containing 12 type I, 11 type II and 15 to 17 type III repeats.

The complex structural organization of fibronectin helps facilitate interactions with a variety of molecules. For example, fibronectin contain binding sites for fibrin, native and denatured collagens, heparin and a number of growth factors. Interestingly, recently studies have indicated that fibronectin type II modules can interact with collagen types II, III, IV, V, and X. Moreover, fibronectin also contains heparin-binding domains that interact with heparin and chondroitin sulfate proteoglycans. Fibronectin type III repeats can bind to the matricellular protein tenacin. Taken together, these observations imply that the biological activities of fibronectin involve a number of critical interactions with a broad range of molecules.

Besides providing critical binding sites for many components of the interstitial matrix, fibronectin is well known for its capacity to facilitate cell adhesion, migration, proliferation, and cell survival. These important cellular processes are regulated in large part by integrins. Examples of the integrins known to bind to fibronectin include $\alpha5\beta1$, $\alpha4\beta1$, $\alpha9\beta1$, $\alpha4\beta7$, $\alpha v\beta3$, and $\alpha v\beta6$. One of the most common integrin receptors for fibronectin is $\alpha5\beta1$, which binds to fibronectin repeats I_{1-9}, $II_{1,2}$, III_9, and III_{10}. Importantly, the ability of fibronectin to facilitate cellular interactions mediated by multiple integrin binding events provides an opportunity to initiate distinct signaling cascades.

3.3
Vitronectin

Another important glycoprotein associated with the interstitial matrix is vitronectin. Vitronectin is a 75-kDa adhesive glycoprotein that is present in both the plasma and the ECM. It circulates in blood in a monomeric (native) form, but is converted into disulfide-linked multimeric (denatured) form when incorporated into the ECM. The native vitronectin molecules contain 459 amino acid residues organized into functionally distinct domains. The N-terminal domain (residues 1–44) of vitronectin is rich in cystine and is termed the *somatomedin B* (SMB) domain. Importantly, the well-known integrin recognition sequence Arg-Gly-Asp (R-G-D) is located in the C-terminus of the SMB domain residues. The most C-terminal portion (residues 348–370) of vitronectin contains sequences homologous to hemopexin and has a positively charged heparin-binding segment.

Although vitronectin is predominately produced in the liver and found in the circulation, it is also found in a variety of tissues, particularly during wound healing, inflammation, angiogenesis, and during tumor growth. Interestingly, some of the unique protein-binding properties of vitronectin depend on its structural conformation and degree of multimerization. For example, multimeric forms of vitronectin exhibit enhanced capacity to bind to a number of molecules, including other ECM proteins and specific proteoglycans. Moreover, vitronectin has been shown to interact with growth factors and growth factor–binding proteins present within the ECM such as transforming growth factor-β (TGFβ) and insulin-like growth factor–binding protein-5 (IGFBP-5). Vitronectin binding to IGFBP-5 may act to modulate its bioavailability, thus impacting the function of insulin-like growth factor-1 (IGF-1), which in turn could alter cellular migration and proliferation. Adding to the diversity of molecules known to interact with vitronectin include factors such as the C5b-8 complement complex, the thrombin-antithrombin III complex, plasminogen activator inhibitor 1 (PAI-1), urokinase plasminogen activator receptor (uPAR), heparin, plasminogen, and β-endorphin. Vitronectin has also been shown to regulate cell adhesion, migration, proliferation, and cell survival. The vast majority of these cellular functions are regulated by integrins. The major integrin receptors known to interact with vitronectin include the classical vitronectin receptor $\alpha v\beta3$ as well as a variety of other αv-containing integrins such as $\alpha v\beta1$, $\alpha v\beta5$, $\alpha v\beta6$, and $\alpha v\beta8$. Again, the variety of signaling pathways initiated by distinct integrin ligation events can serve to coordinate, transmit, and regulate cellular responses associated with vitronectin.

3.4
Thrombospondin

A number of matricellular proteins are also important components of the interstitial matrix. Matricellular proteins have been defined as those ECM-associated proteins that modulate cell–ECM interactions and/or cellular functions, but which do not contribute significantly to the structural architecture of the matrix. Some examples of well-characterized matricellular proteins include tenacin, osteopontin, thrombospondin, and SPARC. We have chosen to focus primarily on thrombospondin as one example of the important matricellular protein family.

Members of the thrombospondin family can be subdivided into at least two groups on the basis of their general structure and domain organization. For example, thrombospondin-1 (TSP1) and thrombospondin-2 (TSP2) are trimeric molecules, which contain a procollagen-like domain and type 1, and type III repeat (TRS) domains. In contrast, TSP3, TSP4, and TSP5 are pentameric molecules, which lack the procollagen-like domain. In general, members of the thrombospondin family can be characterized as multimeric, calcium-binding glycoproteins that regulate a variety of cellular functions. As an example, thrombospondin-1 is a 420-kDa homotrimeric glycoprotein, which contains N- and C-terminal globular domains. The N- and C-terminal globular domains interact with heparin, and heparin sulfate proteoglycans along with facilitating cell attachment, spreading, migration, and platelet aggregation. The procollagen-like domain plays a role in subunit assembly along with exhibiting the capacity to inhibit angiogenesis, while the type III repeat facilitates both calcium and cell binding.

While this multifunctional protein can interact with other ECM molecules such as heparin sulfate proteoglycans, it is not considered to play a major role in structural organization of the interstitial matrix. Thrombospondin is secreted by a variety of cells, including macrophages, fibroblasts, and endothelial cells, and is distributed within a variety of tissues. Interestingly, in the adult organism, thrombospondin has been observed predominately at sites of injury and tissue remodeling.

As has been observed with most ECM proteins, thrombospondin can interact with a variety of distinct proteins, thereby playing roles in the regulation of many cellular processes. The functions of TSP1 are thought to be mediated by direct binding to cell surface receptors, including integrins. In fact, TSP1 binds to a number of integrins such as, $\alpha v \beta 3$, $\alpha IIb\beta 3$, $\alpha 3\beta 1$, $\alpha 4\beta 1$, and $\alpha 5\beta 1$. Moreover, a number of nonintegrin-binding proteins or receptors have also identified, including the low-density lipoprotein receptor-related protein (LRP), proteoglycans, CD36, and CD47. Beyond the ability of thrombospondin to interact with cell surface receptors, studies have revealed a complex array of other binding proteins ranging from other ECM molecules such as fibrinogen, fibronectin, and heparin sulfate proteoglycans to a variety of proteases, including thrombin, elastase, and certain MMPs. Given the capacity of thrombospondin to interact with such a diversity of distinct protein, it is not surprising that it has been shown to regulate many important cellular events, including attachment, spreading, motility, cytoskeletal organization, proliferation, cell–cell aggregation, apoptosis, transcriptional regulation, differentiation, and angiogenesis. Thus, as can be clearly appreciated from the relatively short discussion of thrombospondin, this ECM protein plays major roles in regulating a diversity of biological responses to the interstitial ECM.

4
Cellular Receptors for ECM Components

The ability of cells to respond to their extracellular environment depends on bidirectional communication with stromal elements. In fact, this bidirectional flow of information is crucial to the development and homeostasis of multicellular organisms. Thus, to detect, integrate, and transmit information from outside the cell to the inside and, in turn, to respond to specific microenvironmental changes, cells need a network of unique sensor and responder-type receptor molecules with the capacity to assimilate, organize, and

use the information within the ECM. To this end, at least five families of cell adhesion receptors have been identified with the ability to mediate either cell–cell or cell–ECM interactions (Fig. 3). The major groups of cell adhesion molecules include selectin, cadherins, members of the immunoglobulin super gene family, cell surface proteoglycans, and integrins. Interestingly, recent studies have also identified an emerging group of receptor tyrosine kinases that may facilitate cellular interactions with the ECM. These groups of cell adhesion receptors have the capacity to transmit biochemical and molecular information from the stromal environment. With some exceptions, the members of the selectin, cadherin, immunoglobulin, and proteoglycans families mediate cell–cell interactions, while integrins are

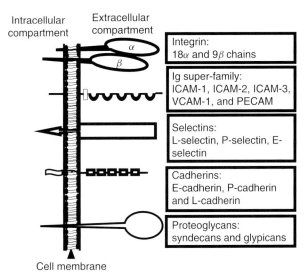

Fig. 3 Major families of cell adhesion molecules mediating cell–cell and/or cell–ECM interactions. In order for cells to sense and, in turn, respond to the acellular microenvironment, they require cell surface receptors with the capacity to facilitate the bidirectional flow of information between cells and their local ECM. Examples of the five major classes of cell adhesion receptors include integrins, Ig superfamily, selectins, cadherins, and cell surface proteoglycans.

the major receptors facilitating cell–ECM interactions. Some interesting examples of nonintegrin receptors mediating cellular communication with the ECM include the cell surface proteoglycans (Glypicans and Syndecans) as well as specific receptor tyrosine kinases.

Syndecans are transmembrane heparin sulfate proteoglycan. At least four distinct syndecans have been identified and studies have suggested that syndecans 1, 2, and 4 can bind to a number of ECM molecules. The syndecans are multidomain transmembrane molecules with an extracellular domain containing heparin sulfate GAG chains. It is thought that syndecan interactions with ECM components such as laminin and fibronectin are mediated primarily by interactions with the GAG chains rather than with core protein sequences. In fact, ECM binding specificity can be modulated by the degree of GAG sulfation as well as the length and composition of the disaccharide moieties. Syndecans have a conserved transmembrane domain and a short cytoplasmic domain containing three subdomains termed C1, V, and C2. The cytoplasmic domain has been shown to interact with cytoplasmic proteins such as src, cortactin, PIP2, and PKCα. The capacity of the syndecans to bind to these cytoplasmic proteins is thought to facilitate signal transduction. In this regard, syndecans contribute to the regulation of cellular proliferation and survival. Moreover, syndecans interact with a number of growth factors, cytokines, and protease inhibitors. These unique interactions can facilitate the regulation of protease and cytokine activity in close proximity to the cell surface.

As mentioned above, integrins are the major family of cell adhesion receptors that mediate cellular interactions with the ECM (Table 2). However, several $\beta1$ and $\beta2$ subunit containing integrins mediate cell–cell interactions by interacting with counter receptors and other binding ligands expressed on adjacent cells. Integrins represent a large family of heterodimeric transmembrane proteins composed of noncovalently associated α and β chains. The α and β chains represent distinct gene products with alternatively spliced forms identified. To date, at least 18 α and 9 β subunits have been described. Distinct combinations of the α and β chains lead to the formation of at least 24 distinct integrins with different ligand and tissues specificities. The functional integrin heterodimer is composed of a ligand-binding globular extracellular domain containing repeated regions with cation binding capacity. In addition, integrins have a transmembrane domain and a short cytoplasmic tail that has the capacity to bind to a wide array of cytosolic proteins known to be crucial in facilitating signal transduction. Studies have suggested that ligand-binding specificity

Tab. 2 Examples of selected integrins and their ECM ligands.

ECM ligand	Integrin
Collagens	$\alpha1\beta1, \alpha2\beta1, \alpha3\beta1$
Laminins	$\alpha1\beta1, \alpha2\beta1, \alpha6\beta1, \alpha7\beta1, \alpha6\beta4, \alpha v\beta3, \alpha3\beta1$
Nidogen	$\alpha3\beta1, \alpha v\beta3$
Fibronectin	$\alpha5\beta1, \alpha4\beta1, \alpha4\beta7, \alpha3\beta1, \alpha v\beta3$
Vitronectin	$\alpha v\beta1, \alpha v\beta3, \alpha v\beta5, \alpha IIb\beta3, \alpha v\beta6, \alpha v\beta8$
Thrombospondin	$\alpha3\beta1, \alpha v\beta3, \alpha IIb\beta3, \alpha4\beta1, \alpha v\beta1$

and relative affinity is dictated in part by structural conformation, cation binding, and specific contact residues within both the α and β chains. Thus, the affinity and ligand-binding specificity can be regulated at multiple levels, allowing fine-tuning of interactions.

Historically, integrins were thought to function as molecular glue, facilitating physical interactions between cells and the ECM. However, over the last decade, experimental evidence has expanded the repertoire of the molecular ligands and integrin functions (Fig. 4). For example, some of the more well-characterized processes regulated by integrins include cell adhesion, migration, proliferation, cell survival, differentiation, signal transduction, gene expression, and extracellular matrix assembly. Importantly, the multiplicity of functions regulated by integrins is attributable in large part to the vast number of distinct integrin-binding partners. Most of the classical integrin ligands initially identified were ECM proteins, including collagen, laminin, fibronectin, vitronectin, and thrombospondin. Early work on integrin ligands lead to the identification of the tripeptide RGD as an important sequence mediating cell adhesion. However, it has now been confirmed that a wide array of distinct non-RGD-containing proteins can interact with integrin receptors. In some respects, integrins can be thought of as molecular organizational centers on the cell surface since integrins can bind and regulate the activity of a number of different families of molecules (Fig. 4). For

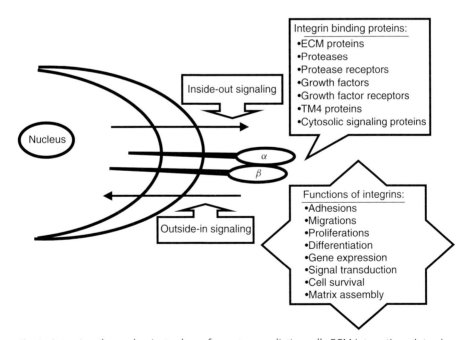

Fig. 4 Integrins: the predominate class of receptors mediating cell–ECM interactions. Integrins represent a family of multifunctional heterodimeric transmembrane receptors, which can bind a wide array of regulatory proteins. Integrin receptors can mediate "inside-out" signaling and "outside-in" signaling. These distinct signaling cascades can facility bidirectional communication between cells and the local microenvironment.

example, besides ECM proteins, recent studies have indicated that integrins can bind to and associate with such diverse sets of regulatory molecules as proteolytic enzymes, protease receptors, growth factors, growth factor receptors, and TM4 proteins. Thus, it is not surprising that integrins play major roles in regulating the structural assembly, integrity, and remodeling of the ECM. In turn, integrins serve as sensors and molecular responder proteins that facilitate integration and transmission of complex arrays of biochemical and mechanical information from the ECM to the cells. Thus, the capacity of integrin to act as bidirectional signaling receptors, facilitators, sensors, and responders makes them uniquely suited for the job of regulating biological responses to the ECM.

5
Mechanisms of ECM Modification and Remodeling

It is well known that the structural integrity of the BM and interstitial matrix can have a significant impact on cellular behavior. Thus, compositional, biochemical, and structural modification of the ECM can provide mechanisms to alter diverse sets of biological responses. In order to have a significant appreciation of how the ECM influences biological responses, one must first have some insight into the mechanisms by which the ECM is modified and how the unique modifications are recognized and/or interpreted by cells. There is a great diversity of potential mechanisms by which the various components of the ECM can be altered, and thus would be well beyond the scope of our discussion to analyze all of them in detail. Therefore, we have chosen to focus on a few interesting mechanisms as examples of how the ECM can be modulated (Fig. 5).

The molecular composition of distinct ECM compartments can vary substantially, depending on the tissue environment and physiological or pathological events that may be occurring at a particular time. The complex interplay between growth factors, integrins, proteases, and matrix components function cooperatively

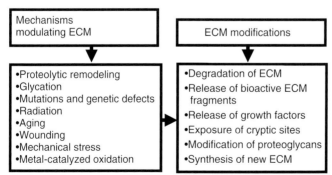

Fig. 5 Mechanisms by which ECM modification can altered cellular behavior. The extracellular microenvironment is not static, but is rather dynamic with numerous changes in structure, integrity, and composition, depending on the cellular repertoire and physiological processes that may be occurring. These important structural and biochemical modifications may lead to altered cellular interactions with the matrix, and thus modify biological responses to the ECM.

to regulate the structure and molecular composition of the ECM. During physiological processes such as embryo development, organogenesis, and wound healing responses, a variety of growth factors and cytokines, including members of the fibroblast growth factor (FGF) and TGFβ families can be released. These growth factors as well as many others can bind cognate receptors on the cell surface, initiating signaling cascades that regulate synthesis of specific ECM components, including collagen, laminin, and fibronectin. Recent studies using mutant embryoid bodies harboring mutations in FGF receptor-2 resulted in altered synthesis of collagen type IV and laminin. Interestingly, this FGF-regulated synthesis of collagen type IV and laminin may be controlled in part by the PI3 kinase and Akt/PKB signaling pathway. Altered ECM synthesis can lead to disruption of BM formation and epithelial differentiation. FGF signaling is also known to play important roles in regulating wound healing and angiogenesis. During these normal physiological responses, elevated levels of FGF have been observed, which, in turn, could lead to significant modification of the preexisting ECM.

A second important example of growth factor regulation of ECM protein synthesis involves TGFβ. It is well known that members of the TFGβ family play important roles in regulation of the expression of collagen and laminin. In fact, differentiation, morphogenesis, and dorsoventral patterning during embryo development are regulated by members of the TGFβ family. Ligation of TGFβ receptors initiate signaling through the SMAD protein family, which can ultimately translocate to the nucleus where they participate in transcriptional regulation by direct DNA-binding events or facilitating formation of transcriptionally active complexes. Thus, developmentally regulated expression patterns and concentration gradient of these growth factors can have a profound impact on the molecular composition of the ECM during physiological development. In turn, the compositional changes occurring during development could provide different binding sites for integrin receptors, other growth factors, matricellular proteins, proteoglycans, and matrix-degrading enzymes. These changes could function to alter signaling pathways, leading to differential regulation of a number of biological responses. It is important to point out that these interconnected events leading to ECM compositional changes can occur not only during normal physiological processes but also play significant roles during pathological events. For example, significant changes in ECM composition occur during inflammatory reaction, pathological angiogenesis, and tumor development. In fact, a reduction in fibronectin expression during tumor development has been observed, while an increase in fibrinogen/fibrin has been documented in association with some tumors. Moreover, studies have shown an increase in vitronectin synthesis in association with glioblastoma, in which the microenvironment is typically characterized with relatively low levels of structural ECM proteins. Finally, another well-documented example of ECM compositional changes involves the matricellular proteins thrombospondin and SPARC. While these proteins are not generally considered structural components, they are upregulated during wounding responses. These matricellular proteins bind components of the ECM and modulate integrin responses such as adhesion, migration, and proliferation.

As has been appreciated for decades, growth factor signaling can influence a wide array of molecules that modify the ECM, including integrin receptors and proteolytic enzymes. Two major classes of proteolytic enzymes known to modify the structural integrity of the ECM include the serine and matrix metalloproteinase (MMP) families. These classes of enzymes can be regulated at multiple levels, including regulation of transcription, translation, and post translation events, including conformational activation, secretion, glycation, dimerization, and inactivation by inhibitors. Members of the serine and metalloproteinase families are secreted in a latent form and are subsequently activated by a number of mechanisms. Members of the serine protease family such as uPA, tPA, and plasmin can cleave fibronectin, fibrinogen, fibrin, and a number of proteoglycans. The metalloproteinase family, of which at least 24 members have been defined so far, can modify nearly all the components of the ECM, including triple helical collagen, denatured collagen (gelatins), laminin, fibronectin, and proteoglycans. As with most complex biological systems, there exist interconnected regulatory activities between the fibrinolytic cascade and the metalloproteinase system. For example, members of the serine protease family such as plasmin may function to activate latent procollagenase, converting it into an active enzyme. Moreover, certain MMPs such as MMP-9 may degrade serpin protease inhibitors. Interestingly, active proteases from both the fibrinolytic and metalloproteinase systems not only cleave structural components of the ECM but also activate certain growth factors and growth factor receptors, release matrix sequestered molecules, modify proteoglycans, and bind integrins.

Notably, proteolytic enzymes impact the structure and integrity of the ECM in several ways. First, and perhaps the most obvious way, is the degradation of the ECM. Both serine proteases and MMPs degrade components of both the BM and the interstitial matrix. Degradation of major ECM proteins can lead to shifts in molecular composition, alterations in integrin, proteoglycan, and other interaction sites, leading to architectural disorganization. In addition, alterations in ECM proteins can lead to modified integrin signaling, thereby altering gene expression. Second, proteolytic enzymes may not only remove restrictive matrix barriers but, in addition, may also selectively facilitate small conformational changes exposing cryptic regions of ECM proteins that modulate cellular functions. In fact, recent work has lead to the development of Mabs that specifically react with denatured or proteolyzed collagen. Importantly, small alterations in the three-dimensional structure of proteins may impact their susceptibility to biochemical modifications such as glycation, cross-linking or sensitivity to proteolysis. Proteolytic modification of the ECM may also lead to the release of matrix-bound growth factors, protease inhibitors, or other regulatory molecules that are sequestered physically within the matrix. Finally, proteolytic activity could release biologically active fragments of larger ECM proteins that, in turn, might impact cellular responses by binding to the cell surface.

Besides protease-mediated modification of ECM proteins, posttranslational changes of matrix components can also occur. For example, conditions associated with hyperglycemia can lead to the formation of advanced glycation end products or AGEs. The formation of AGEs is associated with nonenzymatic glycation of proteins. Specifically, high glucose levels lead to glucose interacting with free amino groups, forming an unstable shift base. These functional groups undergo molecular re-arrangements, resulting in the formation of Amedori products that ultimately result in the formation of AGEs. The glycation processes have been shown to modify a wide variety of ECM proteins, including distinct forms of collagen, laminin, vitronectin, and fibronectin. Studies have suggested the existence of specific cell surface receptors for these AGEs, which may regulate distinct processes such as chemotaxis, proliferation apoptosis, and angiogenesis. Thus, it is likely that glycation of ECM proteins modulate their functions. In fact, glycation alters their ability to polymerize and associate with other ECM proteins as well as disrupt their capacity to facilitate cell adhesion. In this regard, studies have demonstrated that glycation of the integrin recognition sequence RGD in collagen and fibronectin altered its capacity to promote cell adhesion and migration. In still further studies, glycation of ECM proteins was shown to disrupt processes such matrix assembly, MMP activation, integrin recognition, and signal transduction. Thus, it is clear that glycation of ECM proteins could modify both the structural and functional properties of matrix components.

Interestingly, recent experimental evidence has been provided that both ionizing and nonionizing radiation can cause structural alterations of ECM proteins. For example, it is known that UV irradiation can cause structural cross-linking of proteins such as collagen. Aberrant cross-linking could alter proteolytic and thermal stability, induce changes in protein–protein interactions, and modify integrin recognition. All these structural changes could lead to significant modification of biological responses. In addition, studies have demonstrated that ionizing radiation can cause changes in ECM protein production,

thus altering the composition and integrity of the ECM. It was recently shown that a dose of ionizing radiation as small as 20cGy could result in significant loss of the exposure of the HUIV26 cryptic epitope within collagen type IV, both *in vitro* and *in vivo.* While functional expression of the HUIV26 cryptic epitope was reduced by approximately 50%, the exposure of a second cryptic epitope defined by the monoclonal antibody HU177 exhibited little change. These interesting results suggest that small doses of ionizing radiation may have a differential and significant impact on the structure and function of specific ECM proteins. These studies provide a few interesting examples of the potential mechanism by which the ECM can be altered.

6
Modulation of Cellular Behavior by Intact and Structurally Altered ECM Components

The ECM is known to regulate an array of molecular, cellular, and biochemical processes. Crucial to this regulation is the ability of cells to sense and, in turn, respond to their surrounding ECM. Integrin receptors are the major group of molecules that mediate bidirectional communication with the ECM. At least 24 different integrin heterodimers have been characterized with distinct and overlapping binding specificities. By virtue of their ability to interact with an array of cytosolic proteins, integrins are known to transmit signals from outside the cell to the inside. Biochemical changes inside the cell can also be transmitted to the outside by integrins, facilitating bidirectional communication.

Given the diversity of potential integrin signaling events, it would be well beyond the scope of our discussion to analyze them all in detail. Thus, we will limit our description to a few general concepts crucial for a basic appreciation of integrin signaling and its importance in regulating biological responses to the ECM. Integrin recognition of ECM ligands depends on a variety of molecular parameters, including α and β subunit composition, cation binding, substrate composition, and amino acid sequences. In general, integrin interactions can lead to subunit conformational changes and multimerization. In turn, these conformational changes can lead to the association of a number of cytosolic proteins such as α-actinin and talin with integrin cytoplasmic tails. These initial protein–protein interactions lead to the recruitment of sets of adaptor molecules and kinases, leading to formation of signaling complexes. In addition to direct integrin signaling cascades, integrins can cooperate and cross talk with growth factors and growth factor receptors, thus enhancing the level of complexity and potential mechanisms controlled by integrins.

Some important examples of signaling cascade regulated by integrin–ECM ligation include the Ras-Raf-Mek pathways, the PI3K/AKT pathway, and the JNK pathway. Interestingly, integrin ligation of fibronectin has been shown to regulate apoptosis by initiating specific signaling cascades that modified the expression and activation states of effector molecules such as Bcl-2, Bax, and a number of caspases. Moreover, studies have also shown that integrin–ECM interactions regulate expression of molecules involved in cell cycle control such as cyclins and cyclin-dependent kinases. In addition, studies have documented that integrin signaling controls actin cytoskeleton reorganization and formation of focal adhesions that are involved in the control of cellular migration and invasion. Finally, integrins

regulate the expression and activation of matrix altering proteases such as MMPs. Thus, it is clear from the variety of interconnected signaling cascades impacted by integrin binding that the ECM can control a diversity of biological responses.

The diversity of responses to the ECM from a given cell can be altered, depending on the structural integrity of the matrix. For example, it is known that proteolytic degradation of epithelial basement membranes can lead to induction of apoptosis in epithelial cells. MMP-mediated cleavage of interstitial collagen type I promotes melanoma cell survival and suppresses apoptosis by regulating Bcl-2 and Bax expression. Structural changes in ECM proteins can also alter other cellular processes such as migration. For example, studies have shown the MMP-mediated cleavage of laminin-5 resulted in the exposure of a cryptic epitope that promoted breast tumor cell migration. Moreover, MMP-9-dependent cleavage of collagen type IV can expose a cryptic epitope that regulates the migration of retinal endothelial cells. Finally, studies have suggested that MMP-2-mediated cleavage of fibronectin can lead to the exposure of a unique cryptic site that possesses antiadhesive properties.

A wide array of proteolytic fragments of collagen has been shown to modulate distinct biological responses. For example, proteolytic release of the endostatin molecule from the NC domain of collagen type XVIII was shown to inhibit endothelial cell proliferation and migration *in vivo*. Fragments of other NC1 domains from collagen type IV (α1NC1, α2NC1, α3NC1, and α6NC1) have all been shown to modulate cellular responses, including adhesion, migration, invasion, and proliferation. The ability of these collagen fragments to alter cellular events may be due, in part, to their capacity to bind integrin receptors.

Besides specific modules of collagen, proteolytic fragments of other ECM proteins have also exhibited modulatory effects on cells. Structurally altered fibronectin has been shown to modulate matrix assembly by exposing cryptic assembly sites. Fragments of laminin, elastin, SPARC, thrombospondin, and fibrinogen have all been shown to regulate processes such as adhesion, proliferation, motility, and protease production. Thus, cellular interactions with both native as well as structurally altered ECM components play major roles in regulating cellular behavior.

7
Control of Physiological and Pathological Processes by Interactions with the ECM

The rapid growth of gene-targeting techniques has lead to a greater understanding of the functional importance of the ECM in regulating complex sets of biological responses. In fact, greater than 40 mutant and/or ECM knockout mice and 26 integrin null mice have been developed. Studies utilizing these mice have provided unique insight into the interconnected roles of integrins and ECM proteins in the bidirectional flow of information between the ECM and cells. Owing to the number and diversity of studies using these genetic approaches, we have chosen to discuss only a few examples to illustrate the importance of the ECM in regulating biological responses.

A variety of transgenic and mutant mice have been developed that harbor alterations within collagen genes. Interestingly, studies involving mutations or gene knock outs of the $\alpha 1$ chains of collagen types I, II, and III exhibited abnormal blood vessel development, vessel rupture, skin defects, and chondrodysplasia. Mutations

have also been introduced into genes encoding collagen types IV, V, VI, VII, IX, X, and XI. The mutant phenotypes of these mice vary from renal failure, skin, skeletal, and cartilage defects to hearing loss. Interestingly, mutations in the $\alpha 1$, $\alpha 2$, and $\alpha 3$ chains of the collagen-binding integrins have exhibited some of the same defects, including skin blistering, renal defects, as well as reduced tumor associated angiogenesis. Thus, alterations in expression of collagen or specific integrins clearly impact crucial biological events.

In addition to collagen, the functional roles of other ECM glycoproteins have been studied using gene-targeting techniques. Mice harboring mutations in the gene coding for fibronectin exhibited abnormal development and defects such as improper somite formation. Similar defects were also observed in mice defective in the fibronectin receptor $\alpha 5 \beta 1$, including abnormal vascular and somite development. Mice carrying mutations in specific chains of laminin exhibited a variety of defects, including abnormal BM assembly as well as renal, neuromuscular, central nervous system, and vascular abnormalities. Interestingly, mice defective in laminin-binding integrins ($\alpha 3$, $\alpha 6$, and $\beta 4$ chains) showed similar sets of abnormalities, including renal defects and skin blistering and epithelial malformations. In further studies, mice harboring a mutation in perlecan exhibited defects in heart and brain development as well as chondrodysplasia. Finally, studies with thrombospondin-2 mutant mice suggested an important role for TSP-2 in control of vascular development and collagen matrix assembly.

During early embryonic development, many of the major ECM components are being produced. In fact, studies have shown that by the blastocyst stage, ECM components such as laminin, distinct isoforms of collagen, fibronectin, and per-lecan are already present. Mice lacking the laminin 1γ chain failed to develop proper BM structures and subsequently died at E5.5. This lack of a functional BM was thought to contribute to the failure of en-dodermal cells to differentiate, ultimately leading to extensive apoptosis and embryo death. During gastrulation, a variety of reg-ulatory molecules and morphogens help establish specific gradients that are critical for pattern formation. Mice deficient in fibronectin exhibit defects in neural tube closure and cardiovascular abnormalities. A critical role for cellular communica-tion with fibronectin is evident since mice lacking $\alpha5\beta1$ show similar cardiovascu-lar defects.

Members of the TGF-β family such as BMP-4 are thought to play important roles in early pattern formation through signal transduction cascades mediated by activation of the SMAD family members. Inhibitors of BMP signaling, such as Chordin, contain CR-1 domains. CR-1 domains are thought to facilitate binding of these inhibitors to BMP, thereby blocking receptor binding. Interestingly, these CR-1 domains have been found to be present within a number of proforms of collagen, including procollagen type I II, III, and IV. During the maturation of collagen, the prodomains are enzymatically released. Thus, the CR-1-containing domains may bind to BMP, thereby serving as a regulator of embryonic patterning.

Recent studies have also provided con-vincing data that invasive and patholog-ical events, such as angiogenesis, tumor growth, and metastasis, are tightly con-trolled by cellular interaction with the ECM. As with most complex physiological events, angiogenesis or the development of blood vessels from preexisting vessels require the precisely coordinated activi-ties of numerous families of regulatory molecules. Some well-characterized exam-ples of these regulatory proteins include growth factors and their receptors, cell ad-hesion receptors, proteolytic enzymes, and ECM components. It is interesting to point out that interactions with the ECM can regulate the production, activation, and biodistribution of all these molecular regu-lator of angiogenesis. In addition, studies have demonstrated that endothelial cells secrete and, in turn, interact with specific components of the ECM such as collagen laminin and fibronectin. In fact, endothe-lial cell interactions with collagen and laminin were required to initiate morpho-logical reorganization of the endothelial cells into tubelike structures. More recent work has demonstrated that mice harbor-ing mutations in genes coding for specific isoforms of collagen exhibit abnormal vasculature, resulting in embryo death. Moreover, several studies from transgenic mice harboring defects in collagen binding integrins also exhibited defective vascular development. Interestingly, studies have shown that administration of function-blocking antibodies directed to collagen binding integrins such as $\alpha2\beta1$ and $\alpha1\beta1$ potently blocked angiogenesis *in vivo*. One of the most well-characterized adhesion receptors known to regulate angiogenesis is the $\alpha v\beta3$ integrin. Importantly, $\alpha v\beta3$ is highly upregulated on the surface of acti-vated angiogenic blood vessels, while little is expressed on normal quiescent vessels. Antagonists of $\alpha v\beta3$, including functional blocking antibodies and peptides, have been shown to inhibit angiogenesis in multiple models. In fact, a humanized monoclonal antibody termed *vitaxin* is now being evaluated in human clinical trials for cancer therapy.

The molecular mechanisms by which blocking integrins inhibit angiogenesis are numerous. For example, blocking integrin interactions with ECM components can disrupt cellular adhesion and migration that are known to be required for specific steps during angiogenesis. Besides blocking physical interactions with ECM, disruption of integrin signaling events are also likely to contribute to the inhibitory activity. Integrin ligation of distinct domains within fibronectin differentially regulated the expression of matrix-degrading proteases such as MMPs. In fact, mice lacking either MMP-2 or MMP-9 have been shown to exhibit reduced angiogenesis *in vivo*. Further studies have shown that specific integrins such as $\alpha v \beta 3$ can contribute to the localization of proteolytically active forms of MMP-2 to the surface of invasive cells, thus modulating ECM remodeling. Modulation of protease-mediated ECM remodeling can impact angiogenesis in several ways. In particular, proteolytic release of NC domains of collagen type XVIII (endostatin), collagen type XV (restin), collagen type VIII (vastatin), and various NC domains from collagen type IV (arrestin, tumstatin, and canstatin) have all been shown to inhibit angiogenesis. A wide array of other bioactive ECM fragments have also been shown to either inhibit or induce angiogenesis, including fragments from laminin, fibronectin, fibrinogen, and thrombospondin, to name just a few. Finally, recent studies suggest that proteolytic remodeling of the ECM can expose specific matrix-immobilized cryptic sites. In fact, cellular interactions with the HUIV26 cryptic epitope are mediated by the integrin $\alpha v \beta 3$. A function-blocking monoclonal antibody directed to the HUIV26 cryptic epitope was shown to potently inhibit angiogenesis in several models. Finally, integrin ligations of ECM

components are also known to regulate the production of angiogenesis, inducing growth factors such as VEGF and bFGF. Thus, cellular interactions with the ECM can have profound impact on new blood vessel development.

Recent studies have provided evidence that many of the same cellular and molecular processes that regulate angiogenesis also play important roles in tumor growth and metastasis. The concept that modulating ECM interactions with tumor cells can alter their malignant behavior has been extensively evaluated. In fact, blocking tumor cell interactions with the ECM can inhibit tumor cell proliferation, invasion, and metastasis. Thus, while contact independent cell proliferation can be a characteristic of malignant tumor cells, ECM contact is still of great importance.

Interestingly, studies suggest that loss on fibronectin expression correlates with enhanced tumor growth. These studies along with many others provided some of the early evidence that modulation of ECM contacts could influence malignant transformation and the spread of tumor cells to distant sites. For example, a number of $\beta 1$-containing integrins have been shown to play important roles in tumor growth as well as metastasis in multiple murine tumor models. In this regard, transfection of $\alpha 2$ integrin subunit allowing functional expression of $\alpha 2 \beta 1$ resulted in enhanced rhabdosarcoma tumor metastasis *in vivo*. Besides $\beta 1$ integrins, a wealth of evidence suggests that many other integrins contribute to tumor growth and progression, including $\alpha 6 \beta 4$ and $\alpha v \beta 5$. In studies with human melanoma biopsies, it was shown that nonmalignant skin lesions and radial growth phase melanoma had little, if any, $\alpha v \beta 3$ expression, while vertical growth phase melanoma and metastatic lesions expressed elevated levels of $\alpha v \beta 3$. Importantly, studies have utilized a number of $\alpha v \beta 3$ antagonists to block tumor growth, including function-blocking Mabs, small peptides, nonpeptidic mimetics, and antisense approaches. The mechanisms by which blocking $\alpha v \beta 3$ function inhibits tumor growth and progression are likely to be many, and perhaps similar to those previously discussed in the context of angiogenesis. These mechanisms might include disruption of physical contact with the ECM, thereby altering traction necessary for tumor cell movement and disruption of signaling cascades. Other potential mechanisms might include altering protease production and growth factor secretion. In fact, it is likely that a combination of many of these mechanisms contribute to the inhibitory activity observed with antagonists of $\alpha v \beta 3$ integrin. Thus, it is clear from the wealth of experimental evidence that tumor cell interactions with the ECM play an important role in regulating many aspects of tumor growth and progression.

8
Conclusions

As can be appreciated from our discussion above, the development of new molecular tools, imaging techniques, as well as gene-targeting procedures have lead to unique insight into the complex structural features of the ECM and how changes within this microenvironment can dramatically alter a plethora of biological events. In fact, over the last decade, investigators have become keenly aware of the great importance of the molecular information contained within the ECM and the critical requirement of transferring this information to cells in order to regulate both physiological as well as pathological processes. Our expanding knowledge of integrin biology has provided important mechanistic understanding of the bidirectional flow of information between cells and their local microenvironment. This insight, in conjunction with the emerging concept that a substantial amount of information is functionally hidden within ECM and only exposed following structural remodeling, has opened up an interesting "Cryptic" microenvironment in which to study. This cryptic ECM environment may provide unique molecular insight and provide potential therapeutic targets for the control of genetic defects, angiogenesis, tumor growth, and metastasis. Thus, future studies into the dynamic and fascinating nature of the acellular compartment of multicellular organisms will likely fuel our quest for a more complete understanding of the impact that the ECM has on biological responses.

Acknowledgments

This work was supported in part from NIH grant ROI-CA91645 and a grant from BioStratum Inc., to PCB.

Bibliography

Books and Reviews

Adams, J.C. (2001) Thrombospondin: multifunctional regulators of cell interactions, *Annu. Rev. Cell Dev. Biol.* **17**, 25–51.
Collen, D. (2001) Role of the plasminogen system in fibrin-homeostasis and tissue remodeling, *Hematology (Am. Soc. Hematol. Educ. Program).* **1**, 1–9.
Iozzo, R.V. (1998) Matrix proteoglycans: from molecular design to cellular function, *Annu. Rev. Biochem.* **67**, 609–652.
Roberts, A.B., Caestecker, M.P., Lechleider, R.J. (2000) Signaling from TGF-β-receptors, in:

The Molecular Basis of Cancer and Other Diseases, Humana Press, Totowa, NJ, 39–51.

Simpson-Haidaris, P.J., Rybarczyik, B. (2001) Tumors and fibrinogen. The role of fibrinogen as an extracellular matrix protein, *Ann. N. Y. Acad. Sci.* **93**, 406–425.

Xu, J., Brooks, P.C. (2002) Extracellular matrix in the regulation of angiogenesis, in: *Assembly of the Vasculature and its Regulation*, Birkhauser, Boston.

Primary Literature

Adachi, E., Hopkinson, I., Hayashi, T. (1997) Basement-membrane stromal relationships: interactions between collagen fibrils and the lamina densa, *Int. Rev. Cytol.* **173**, 73–155.

Adams, J.C., Tucker, R.P. (2000) The thrombospondin type-1 repeat (TSR) superfamily: diverse proteins with related roles in neuronal development, *Dev. Dyn.* **218**, 280–299.

Adams, J.C., Watt, F.M. (1993) Regulation of development and differentiation by the extracellular matrix, *Development* **117**, 1183–1198.

Agah, A., Kyriakides, T.R., Lawler, J., Bornstein, P. (2002) The lack of thrombospondin-1 (TSP1) dictates the course of wound healing in double-TSP1/TSP2-null mice, *Am. J. Pathol.* **161**, 831–839.

Akiyama, S.K., Olden, K., Yamada, K.M. (1995) Fibronectin and integrins in invasion and metastasis, *Cancer Met. Rev.* **14**, 173–189.

Arikawa-Hirasawa, E., Watanabe, H., Takami, H., Hassell, J.R., Yamada, Y. (1999) Perlecan is essential for cartilage and cephalic development, *Nat. Genet.* **23**, 354–358.

Badylak, S.F. (2002) The extracellular matrix as a scaffold for tissue reconstruction, *Semin. Cell Dev. Biol.* **13**, 377–383.

Barnes, D.W., Reing, J.E., Amos, B. (1985) Heparin-binding properties of human serum spreading factor, *J. Biol. Chem.* **260**, 9117–91122.

Belkin, A.M., Stepp, M.A. (2000) Integrins as receptors for laminins, *Micr. Res. Tech.* **51**, 280–301.

Berditchevski, F., Zuttter, M.M., Hemler, M.E. (1996) Characterization of novel complexes on the cell surface between integrins and proteins with 4 transmembrane domains TM4 proteins), *Mol. Biol. Cell.* **7**, 193–207.

Bernfield, M., Gotte, M., Park, P.W., Reizes, O., Fitzgerald, M.L., Lincecum, J., Zako, M. (1999) Functions of cell surface heparan sulfate proteoglycans, *Annu. Rev. Biochem.* **68**, 729–777.

Bevilacqua, M.P., Nelson, R.M. (1993) Selectins, *J. Clin. Invest.* **91**, 379–387.

Birkedal-Hansen, H. (1995) Proteolytic remodeling of extracellular matrix, *Curr. Opin. Cell Biol.* **7**, 728–735.

Bokel, C., Brown, N.H. (2002) Integrins in development: moving on, responding to, and sticking to the extracellular matrix, *Dev. Cell.* **3**, 311–321.

Bootle-Wilbraham, C.A., Tazzyman, S., Marshall, J.M., Lewis, C.E. (2000) Fibrinogen E-fragment inhibits the migration and tubule formation of human dermal microvascular endothelial cells in vitro, *Cancer Res.* **60**, 4719–4724.

Bornstein, P., Sage, E.H. (1994) Thrombospondins, *Methods Enzymol.* **245**, 62–85.

Bornstein, P., Sage, H.E. (2002) Matricellular proteins: extracellular modulators of cell function, *Curr. Opin. Cell Biol.* **14**, 608–616.

Brassart, B., Fuchs, P., Huet, E., Alix, A.J., Wallach, J., Tamburro, A.M., Delacoux, F., Haye, B., Emonard, H., Hornebeck, W., Debelle, L. (2001) Conformational dependence of collagenase (matrix metalloproteinase-1) up-regulation by elastin peptides in cultured fibroblasts, *J. Biol. Chem.* **276**, 5222–5227.

Brightman, A.O., Rajwa, B.P., Sturgis, J.E., McCallister, M.E., Robinson, J.P., Voytik-Harbin, S.L. (2000) Time-lapse confocal reflection microscopy of collagen fibrillogenesis and extracellular matrix assembly in vitro, *Biopolymers* **54**, 222–234.

Brooks, P.C. (1996) Role of integrins in angiogenesis, *Eur. J. Cancer* **32A**, 2423–2429.

Brooks, P.C. (1997) Integrin avb3: A therapeutic target, *DN&P* **10**, 456–461.

Brooks, P.C., Clark, R.A., Cheresh, D.A. (1994) Requirement of vascular integrin alpha $\alpha v\beta 3$ for angiogenesis, *Science* **264**, 569–571.

Brooks, P.C., Klemke, R.L., Schon, S., Lewis, J.M., Schwartz, M.A., Cheresh, D.A. (1997) Insulin-like growth factor receptor cooperates with integrin $\alpha v\beta 5$ to promote tumor cell dissemination in vivo, *J. Clin. Invest.* **99**, 1390–1398.

Brooks, P.C., Roth, J.M., Lymberis, S.C., De-Wyngaert, K., Broek, D., Formenti, S.C. (2002) Ionizing radiation modulates the exposure of the HUIV26 cryptic epitope within collagen type-IV during angiogenesis, *Int. J. Rad. Oncol. Biol. Phys.* **54**, 1194–1201.

Brooks, P.C., Stromblad, S., Klemke, R., Visscher, D., Sarkar, F.H., Cheresh, D.A. (1995) Anti-integrin $\alpha v \beta 3$ blocks human breast cancer growth and angiogenesis in human skin, *J. Clin. Invest.* **96**, 1815–1822.

Brooks, P.C., Montgomery, A.M., Rosenfeld, M., Reisfeld, R.A., Hu, T., Klier, G., Cheresh, D.A. (1994) Integrin $\alpha v \beta 3$ antagonists promote tumor regression by inducing apoptosis of angiogenic blood vessels, *Cell* **79**, 1157–1164.

Brooks, P.C., Stromblad, S., Sanders, L.C., von Schalscha, T.L., Aimes, R.T., Stetler-Stevenson, W.G., Quigley, J.P., Cheresh, D.A. (1996) Localization of matrix metalloproteinase MMP-2 to the surface of invasive cells by interaction with integrin $\alpha v \beta 3$, *Cell* **85**, 683–693.

Brown, J.C., Sasaki, T., Gohring, W., Yamada, Y., Timpl, R. (1997) The C-terminal domain V of perlecan promotes beta1 integrin-mediated cell adhesion, binds heparin, nidogen and fibulin-2 and can be modified by glycosaminoglycans, *Eur. J. Biochem.* **250**, 39–46.

Casaroli Marano, R.P., Preissner, K.T., Vilaro, S. (1995) Fibronectin, laminin, vitronectin and their receptors at newly-formed capillaries in proliferative diabetic retinopathy, *Exp. Eye Res.* **60**, 5–17.

Chan, B.M., Matsuura, N., Takada, Y., Zetter, B.R., Hemler, M.E. (1991) In vitro and in vivo consequences of VLA-2 expression on rhabdomyosarcoma cells, *Science* **251**, 1600–1602.

Chang, C., Werb, Z. (2001) The many faces of metalloproteinases: cell growth, invasion, angiogenesis and metastasis, *Trends Cell Biol.* **11**, S37–S43.

Charonis, A.S., Tsilbary, E.C. (1992) Structural and functional changes of laminin and type-IV collagen after nonenzymatic glycation, *Diabetes* **2**, 49–51.

Chen, H., Herndon, M.E., Lawler, J. (2000) The cell biology of thrombospondin-1, *Matrix Biol.* **19**, 597–614.

Chung, A.E., Durkin, M.E. (1990) Entactin: structure and function, *Am. J. Respir. Cell Mol. Biol.* **3**, 275–282.

Colognata, H., Yurcheco, P.D. (2000) Form and function: the laminin family of heterotrimers, *Dev. Dyn.* **218**, 213–234.

Colorado, P.C., Torre, A., Kamphaus, G., Maeshima, Y., Hopfer, H., Takahashi, K., Volk, R., Zamborsky, E.D., Herman, S., Sarkar, P.K., Ericksen, M.B., Dhanabal, M., Simons, M., Post, M., Kufe, D.W., Weichselbaum, R.R., Sukhatme, V.P., Kalluri, R. (2000) Anti-angiogenic cues from vascular basement membrane collagen, *Cancer Res.* **60**, 2520–2526.

Cooper, A.R., MacQueen, H.A. (1983) Subunits of laminin are differentially synthesized in mouse eggs and early embryos, *Dev. Biol.* **96**, 467–447.

Cornelissen, M., Thierens, H., Ridder, L. (1996) Effects of ionizing radiation on cartilage: emphasis on effects on the extracellular matrix, *Scanning Microsc.* **10**, 833–840.

Costell, M., Gustafsson, E., Aszodi, A., Morgelin, M., Bloch, W., Hunziker, E., Addicks, K., Timpl, R., Fassler, R. (1999) Perlecan maintains the integrity of cartilage and some basement membranes, *J. Cell Biol.* **147**, 1109–1122.

Dallas, S.L., Rosser, J.L., Mundy, G.R., Bonewald, L.F. (2002) Proteolysis of latent transforming growth factor-beta (TGF-beta) binding protein-1 by osteoclasts. A cellular mechanisms for release of TGF-beta from bone matrix, *J. Biol. Chem.* **14**, 21352–21360.

Damsky, C.H., Ilic, D. (2002) Integrin signaling: its where the action is, *Curr. Opin. Cell Biol.* **114**, 594–602.

Damsky, C.H., Werb, Z. (1992) Signal transduction by integrin receptors for extracellular matrix: cooperative processing of extracellular information, *Curr. Opin. Cell Biol.* **4**, 772–781.

Davis, G.E. (1992) Affinity of integrins for damaged extracellular matrix: alpha v beta 3 binds to denatured collagen type-I through RGD sites, *Biochem. Biophys. Res. Commun.* **182**, 1025–1031.

Davis, G.E., Bayless, K.J., Davis, M.J., Meininger, G.A. (2000) Regulation of tissue injury responses by the exposure of matricryptic sites within extracellular matrix molecules, *Am. J. Pathol.* **156**, 1489–1498.

De Arcangelis, A., Mark, M., Kreidberg, J., Sorokin, L., Georges-Labouesse, E. (1999) Synergistic activities of alpha 3 and alpha 6 integrins are required during apical ectodermal ridge formation and organogenesis in the mouse, *Development* **126**, 3957–3968.

Di Lullo, G.A., Sweeney, S.M., Korkko, J., Ala-Kokko, L., San Antonio, J.D. (2002) Mapping the ligand-binding sites and disease-associated mutations on the most abundant protein in the human, type-I collagen, *J. Biol. Chem.* **277**, 4223–4231.

Dogic, D., Eckes, B., Aumailley, M. (1999) Extracellular matrix, integrins and focal adhesions, *Curr. Top. Pathol.* **93**, 75–85.

Dumin, J., Dickeson, K.S., Stricker, T.P., Bhatta-charyya-Pakrasi, M., Roy. Santoro, S.A., Parks, W.C. (2001) Pro-collagenase-1 (Matrix Metalloproteinase-1) binds the $\alpha 2\beta 1$ integrin upon release from keratinocytes migrating on type-I collagen, *J. Biol. Chem.* **276**, 29368–29374.

Duraisamy, Y., Slevin, M., Smith, N., Bailey, J., Zweit, J., Smith, C., Ahmed, N., Gaffney, J. (2001) Effects of glycation on basic fibroblast growth factor induced angiogenesis and activation of associated signal transduction pathways in vascular endothelial cells: possible relevance to wound healing in diabetes, *Angiogenesis* **4**, 277–288.

Durkin, M.E., Wewer, U.M., Chung, A.E. (1995) Exon organization of the mouse entactin gene corresponds to the structural domains of the polypeptide and has regional homology to the low-density lipoprotein receptor gene, *Genomics* **26**, 219–228.

Eliceiri, B.P. (2001) Integrin and growth factor receptor crosstalk, *Circ. Res.* **89**, 1104–1110.

Emsley, J., Knight, C.G., Frandale, R.W., Barnes, M.J., Liddington, R.C. (2000) Structural basis of collagen recognition by integrin $\alpha 2\beta 1$, *Cell* **100**, 47–56.

Erynck, R., Zhang, Y., Feng, X.-H. (1998) Smads: transcriptional activators of TGF-β responses, *Cell* **95**, 737–740.

Exposito, J.-Y., Cluzel, C., Garrone, R., Lethias, C. (2002) Evolution of collagens, *Anat. Rec.* **268**, 302–316.

Fox, J.W., Mayer, U., Nischt, R., Aumailley, M., Reinhardt, D., Wiedemann, H., Mann, K., Timpl, R., Krieg, T., Engel, J. (1991) Recombinant nidogen consists of three globular domains and mediates binding of laminin to collagen type IV, *EMBO J.* **10**, 3137–3146.

French, M.M., Smith, S.E., Akanbi, K., Sanford, T., Hecht, J., Farach-Carson, M.C., Carson, D.D. (1999) Expression of the heparan sulfate proteoglycan, perlecan, during mouse embryogenesis and perlecan chondrogenic activity in vitro, *J. Cell Biol.* **145**, 1103–1115.

Friedlander, M., Brooks, P.C., Shaffer, R.W., Kincaid, C.M., Varner, J.A., Cheresh, D.A. (1995) Definition of two angiogenic pathways by distinct alpha v integrins, *Science* **270**, 1500–1502.

Furuyama, A., Iwata, M., Hayashi, T., Mochitate, K. (1999) Transforming growth factor-beta-1 regulates basement membrane formation by alveolar epithelial cells in vitro, *Eur. J. Cell Biol.* **78**, 867–875.

Garcia Abreu, J., Coffinier, C., Larrain, J., Oelgeschlager, M., De Robertis, E.M. (2002) Chordin-like CR domains and the regulation of evolutionarily conserved extracellular signaling systems, *Gene* **287**, 39–47.

George, E.L., Georges-Labouesse, E.N., Patel-King, R.S., Rayburn, H., Hynes, R.O. (1993) Defects in mesoderm, neural tube and vascular development in mouse embryos lacking fibronectin, *Development* **119**, 1079–1091.

Ghosh, S., Stack, S.M. (2000) Proteolytic modification of laminins: functional consequences, *Micro. Res. Tech.* **51**, 238–246.

Giannelli, G., Falk-Marzillier, J., Schiraldi, O., Stetler-Stevenson, W.G., Quaranta, V. (1997) Induction of cell migration by matrix metalloproteinase-2 cleavage of laminin-5, *Science* **277**, 225–228.

Gladson, C.L., Cheresh, D.A. (1991) Glioblastoma expression of vitronectin and the alpha V beta 3 integrin. Adhesion mechanism for transformed glial cells, *J. Clin. Invest.* **88**, 1924–1932.

Goldsmith, E.C., Borg, T.K. (2002) The dynamic interaction of the extracellular matrix in cardiac remodeling, *J. Card. Fail.* **8**, S314–S318.

Gorden, M.K., Olsen, B.R. (1990) The contribution of collagenous proteins to tissue-specific matrix assemblies, *Curr. Opin. Cell Biol.* **2**, 833–838.

Grumet, M., Milev, P., Sakurai, T., Karthikeyan, L., Bourdon, M., Margolis, R.K., Margolis, R.U. (1994) Interactions with tenascin and differential effects on cell adhesion of neurocan and phosphacan, two major chondroitin sulfate proteoglycans of nervous tissue, *J. Biol. Chem.* **269**, 12142–12146.

Gustafsson, E., Fassler, R. (2000) Insights into extracellular matrix functions from mutant mouse models, *Exp. Cell Res.* **261**, 52–68.

Gutheil, J.C., Campbell, T.N., Pierce, P.R., Watkins, J.D., Huse, W.D., Bodkin, D.J., Cheresh, D.A. (2000) Targeted antiangiogenic therapy for cancer using Vitaxin: a humanized monoclonal antibody to the integrin $\alpha v \beta 3$, *Clin. Cancer Res.* **6**, 3056–3061.

Hangai, M., Kitaya, N., Xu, J., Chan, C.K., Kim, J.J., Werb, Z., Ryan, S.J., Brooks, P.C. (2002) Matrix metalloproteinase-9-dependent exposure of a cryptic migratory control site in collagen is required before retinal angiogenesis, *Am. J. Pathol.* **161**, 1429–1437.

Hayashi, M., Akama, T., Kono, I., Kashiwagi, H. (1985) Activation of vitronectin (serum spreading factor) binding of heparin by denaturing agents, *J. Biochem.* **98**, 1135–1138.

Hedstrom, L. (2002) Serine protease mechanism and specificity, *Chem. Rev.* **102**, 4501–4524.

Heissig, B., Hattori, K., Friedrich, M., Rafii, S., Werb, Z. (2003) Angiogenesis: vascular remodeling of the extracellular matrix involves metalloproteinases, *Curr. Opin. Hematol.* **10**, 136–141.

Hieken, T.J., Ronan, S.G., Farolan, M., Shilkaitis, A.L., Das Gupta, T.K. (1996) Beta 1 integrin expression: a marker of lymphatic metastases in cutaneous malignant melanoma, *Anticancer Res.* **16**, 2321–2324.

Hohenester, E., Engel, J. (2002) Domain structure and organization in extracellular matrix proteins, *Matrix Biol.* **212**, 115–128.

Holtkotter, O., Nieswandt, B., Smyth, N., Muller, W., Hafneer, M., Schulte, V., Krieg, T., Eckes, B. (2002) Integrin alpha 2-deficient mice develop normally, are fertile, but display partial defective platelet interaction with collagen, *J. Biol. Chem.* **277**, 10789–10794.

Hood, J.D., Bednarski, M., Frausto, R., Guccione, S., Reisfeld, R.A., Cheresh, D.A. (2002) Tumor regression by targeted gene delivery to the neovasculature, *Science* **296**, 2404–2407.

Hunter, D.D., Cashman, N., Morris-Valero, R., Bulock, J.W., Adams, S.P., Sanes, J.R. (1991) An LRE (leucine-arginine-glutamate)-dependent mechanism for adhesion of neurons to S-laminin, *J. Neurosci.* **11**, 3960–3971.

Hutter, H., Vogel, B.E., Plenefisch, J.D., Norris, C.R., Proenca, R.B., Spieth, J., Guo, C., Mastwal, S., Zhu, X., Scheel, J., Hedgecock, E.M. (2000) Conservation and novelty in the evolution of cell adhesion and extracellular matrix genes, *Science* **287**, 989–994.

Hynes, R.O. (2002) Integrins: bidirectional, allosteric signaling machines, *Cell* **110**, 673–687.

Iivanainen, E., Kahari, V.M., Heino, J., Elenius, K. (2003) Endothelial cell-matrix interactions, *Microsc. Res. Tech.* **60**, 13–22.

Inatani, M., Tanihara, H. (2002) Proteoglycans in retina, *Prog. Retin. Eye Res.* **21**, 429–447.

Inatani, M., Tanihara, H., Oohira, A., Honjo, M., Kido, N., Honda, Y. (2000) Spatiotemporal expression patterns of $\alpha 6 \beta 4$ proteoglycan/phosphacan in the developing rat retina, *Invest. Ophthalmol. Vis. Sci.* **41**, 1990–1997.

Ingber, D.E. (2002) Mechanical signaling and the cellular response to extracellular matrix in angiogenesis and cardiovascular physiology, *Circ. Res.* **91**, 877–887.

Iozzo, R.V. (2001) Heparan sulfate proteoglycans: intricate molecules with intriguing functions, *J Clin. Invest.* **108**, 165–167.

Iozzo, R.V., Cohen, I.R., Grassel, S., Murdoch, A.D. (1994) The biology of perlecan: the multifaceted heparan sulphate proteoglycan of basement membranes and pericellular matrices, *Biochem. J.* **302**, 625–639.

Itoh, T., Tanioka, M., Yoshida, H., Yoshioka, T., Nishimoto, H., Itohara, S. (1998) Reduced angiogenesis and tumor progression in gelatinase A-deficient mice, *Cancer Res.* **58**, 1048–1051.

Jones, J.C.R., Dehart, G.W., Gonzales, M., Goldfinger, L.E. (2000) Laminins: an overview, *Micro. Res. Tech.* **51**, 211–213.

Jung, Y.D., Ahmad, S.A., Liu, W., Reinmuth, N., Parikh, A., Stoeltzing, O., Fan, F., Ellis, L.M. (2002) The role of the microenvironment and intercellular cross-talk in tumor angiogenesis, *Semin. Cancer Biol.* **12**, 105–112.

Kadler, K.E., Holmes, D.F., Trotter, J.A., Chapman, J.A. (1996) Collagen Fibril formation, *Biochem. J.* **316**, 1–11.

Kadoya, Y., Salmivirta, K., Talts, J.F., Kadoya, K., Mayer, U., Timpl, R., Ekblom, P. (1997) Importance of nidogen binding to laminin gamma-1 for branching epithelial morphogenesis of the submandibular gland, *Development* **124**, 683–691.

Kamphaus, G.D., Colorado, P.C., Panka, D.J., Hopfer, H., Ramchandran, R., Torre, A., Maeshima, Y., Mier, J.W., Sukhatme, V.P., Kalluri, R. (2000) Canstatin, a novel matrix-derived inhibitor of angiogenesis and tumor growth, *J. Biol. Chem.* **275**, 1209–1215.

Karumanchi, S.A., Jha, V., Ramchandran, R., Karihaloo, A., Tsiokas, L., Chan, B., Dhanabal, M., Hanai, J.-I., Venkataraman, G., Shriver, Z., Keiser, N., Kalluri, R., Zeng, H.,

Mukhopadhyay, D., Chen, R.L., Lander, A.D., Hagihara, K., Yamaguchi, Y., Sasisekharan, R., Cantley, L., Sukhatme, V.P. (2001) Cell surface glypicans are low affinity endostatin receptors, *Mol. Cell.* **7**, 811–822.

Kauppila, S., Stenback, F., Kacinski, B.M., Carcangiu, M.-L., Risteli, J., Risteli, L. (1999) Characterization of type-I collagen synthesis and maturation in uterine carcinosarcomas, *Cancer* **86**, 1299–1306.

Kern, A., Marcantonio, E.E. (1998) Role of the I-domain in collagen binding specificity and activation of the integrins $\alpha 1 \beta 1$ and $\alpha 2 \beta 1$, *J. Cell Physiol.* **176**, 634–641.

Kohfeldt, E., Sasaki, T., Gohring, W., Timpl, R. (1998) Nidogen-2: a new basement membrane protein with diverse binding properties, *J. Mol. Biol.* **282**, 99–109.

Kresse, H., Schonherr, E. (2001) Proteoglycans of the extracellular matrix and growth control, *J. Cell Physiol.* **189**, 2666–2674.

Kumar, C.C., Malkowski, M., Yin, Z., Tanghetti, E., Yaremko, B., Nechuta, T., Varner, J., Liu, M., Smith, E.M., Neustadt, B., Presta, M., Armstrong, L. (2001) Inhibition of angiogenesis and tumor growth by SCH221153, a dual alpha (v) beta (3) and alpha (v) beta (5) integrin receptor antagonist, *Cancer Res.* **61**, 2232–2238.

Kuzuya, M., Asai, T., Kanda, S., Maeda, K., Cheng, X.W., Iguchi, A. (2001) Glycation cross-links inhibits matrix metalloproteinase-2 activation in vascular smooth muscle cells cultured on collagen lattice, *Diabetologia* **44**, 433–436.

Kyriakides, T.R., Zhu, Y.H., Yang, Z., Huynh, G., Bornstein, P. (2001) Altered extracellular matrix remodeling and angiogenesis in sponge granulomas of thrombospondin 2-null mice, *Am. J. Pathol.* **159**, 1255–1262.

Labat-Robert, J. (2002) Fibronectin in malignancy effects of aging, *Semin. Cancer Biol.* **12**, 187–195.

Lafrenie, R.M., Yamada, K.M. (1996) Integrin-dependent signal transduction, *J. Cell. Biochem.* **61**, 545–553.

Lane, T.F., Iruela-Arispe, M.L., Johnson, R.S., Sage, E.H. (1994) SPARC is a source of copper-binding peptides that stimulate angiogenesis, *J. Cell Biol.* **125**, 929–943.

Lee, J.E., Park, J.C., Hwang, Y.S., Kim, J.K., Kim, J.G., Sub, H. (2001) Characterization of UV-irradiated dense/porous collagen membranes: morphology, enzymatic degradation, and mechanical properties, *Yonsei Med. J.* **42**, 172–179.

Lester, B.R., McCarthy, J.B. (1992) Tumor cell adhesion to the extracellular matrix and signal transduction mechanisms implicated in tumor cell motility, invasion and metastasis, *Cancer Metastasis Rev.* **11**, 31–44.

Li, G., Fridman, R., Kim, H.R. (1999) Tissue inhibitor of metalloproteinases-1 inhibits apoptosis of human breast epithelial cells, *Cancer Res.* **59**, 6267–6275.

Li, J., Zhang, Y.P., Kirsner, R.S. (2003) Angiogenesis in wound repair: angiogenic growth factors and the extracellular matrix, *Microsc. Res. Tech.* **60**, 107–114.

Li, X., Chen, Y., Scheele, S., Arman, E., Haffner-Krausz, R., Lonai, P. (2001) Fibroblast growth factor signaling and basement membrane assembly are connected during epithelial morphogenesis of the embryoid body, *J. Cell Biol.* **153**, 811–822.

Li, X., Talts, U., Talts, J.F., Arman, E., Ekblom, P., Lonai, P. (2001) Akt/PKB regulates laminin and collagen IV isotypes of the basement membrane, *Proc. Natl. Acad. Sci. U. S. A.* **98**, 14416–14421.

Lichtner, R.B., Howlett, A.R., Lerch, M., Xuan, J.A., Brink, J., Langton-Webster, B., Schneider, M.R. (1998) Negative cooperativity between alpha (3) beta (1) and alpha (2) beta (1) integrins in human mammary carcinoma MDA MB 231 cells, *Exp. Cell. Res.* **240**, 368–376.

Lin, C.Q., Bissell, M.J. (1993) Multi-faceted regulation of cell differentiation by extracellular matrix, *FASEB J.* **7**, 737–743.

Liu, X., Wu, H., Byrne, M., Jeffrey, J., Krane, S., Jaenisch, R. (1995) A targeted mutation at the known collagenase cleavage site in mouse type I collagen impairs tissue remodeling, *J. Cell Biol.* **130**, 227–237.

Liu, Z., Zhou, S.D., Sharipo, J.M., Shipley, S.S., Twining, L.A., Diaz, R.M., Scnior, R.M., Werb, Z. (2000) The serpin a1-proteases inhibitor is a critical substrate for gelatinase B/MMP-9 in vivo, *Cell* **102**, 647–655.

Lode, H.N., Moehler, T., Xiang, R., Jonczyk, A., Gillies, S.D., Cheresh, D.A., Reisfeld, R.A. (1999) Synergy between an antiangiogenic integrin alpha v antagonist and an antibody-cytokine fusion protein eradicates spontaneous tumor metastases, *Proc. Natl. Acad. Sci. U. S. A.* **96**, 1591–1596.

Lohler, J., Timpl, R., Jaenisch, R. (1984) Embryonic lethal mutation in mouse collagen I gene causes rupture of blood vessels and is associated with erythropoietic and mesenchymal cell death, *Cell* **38**, 597–607.

Luo, Z.J., King, R.H.M., Lewin, J., Thomas, P.K. (2002) Effects of nonenzymatic glycosylation of extracellular matrix components on cell survival and sensory neurite extension in cell culture, *J. Neurol.* **249**, 424–431.

Maeshima, Y., Manfredi, M., Reimer, C., Holthaus, K.A., Hopfer, H., Chandamuri, B.R., Kharbanda, S., Kalluri, R. (2001) Identification of the anti-angiogenic site within vascular basement membrane-derived tumstatin, *J. Biol. Chem.* **276**, 15240–15248.

Marneros, A.G., Olsen, B.R. (2001) The role of collagen-derived proteolytic fragments in angiogenesis, *Matrix Biol.* **20**, 337–345.

Mayer, U., Zimmermann, K., Mann, K., Reinhardt, D., Timpl, R., Nischt, R. (1995) Binding properties and protease stability of recombinant human nidogen, *Eur. J. Biochem.* **227**, 681–686.

Mayer, U., Nischt, R., Poschl, E., Mann, K., Fukuda, K., Gerl, M., Yamada, K., Timpl, R. (1993) A single EGF-like motif of laminin is responsible for high affinity nidogen binding, *EMBO J.* **12**, 1879–1885.

McCarthy, A.D., Etcheverry, S.B., Bruzzone, L., Lettier, G., Barrio, D.A., Cortizo, A.M. (2001) Non-enzymatic glycosylation of type-I collagen matrix: effects on osteoblastic development and oxidative stress, *BMC Cell Biol.* **2**, 16–26.

Mercurio, A.M., Rabinovitz, I. (2001) Towards a mechanistic understanding of tumor invasion: lesions from the $\alpha6\beta4$ integrin, *Semin. Cancer Biol.* **11**, 129–141.

Mitjans, F., Sander, D., Adan, J., Sutter, A., Martinez, J.M., Jaggle, C.S., Moyano, J.M., Kreysch, H.G., Piulats, J., Goodman, S.L. (1995) An anti-alpha v-integrin antibody that blocks integrin function inhibits the development of a human melanoma in nude mice, *J. Cell Sci.* **108**, 2825–2838.

Mongiat, M., Sweeney, S.M., San Antonio, J.D., Fu, J., Iozzo, R.V. (2003) Endorepellin, a novel inhibitor of angiogenesis derived from the C-terminus of perlecan, *J. Biol. Chem.* **278**, 4238–4249.

Mosher, D.F., Sottile, J., Wu, C., McDonald, J.A. (1992) Assembly of the extracellular matrix, *Curr. Opin. Cell Biol.* **4**, 810–818.

Nakamura, T., Sato, K., Hamada, H. (2002) Effective gene transfer to human melanomas via integrin-targeted adenoviral vectors, *Hum. Gene Ther.* **13**, 613–626.

Nakashimqa, J., Sumitomo, M., Miyajima, A., Jitsukawa, S., Saito, S., Tachibana, M., Murai, M. (1997) The value of serum carboxyterminal propeptide of type-I procollagen in predicting bone metastasis in prostate cancer, *J. Urol.* **157**, 1736–1739.

Nam, T., Moralez, A., Clemmons, D. (2002) Vitronectin binding to IGF binding protein-5 (IGFBP-5) alters IGFBP-5 modulation of IGF-I actions, *Endocrinology* **143**, 30–36.

Noonan, D.M., Fulle, A., Valente, P., Cai, S., Horigan, E., Sasaki, M., Yamada, Y., Hassell, J.R. (1991) The complete sequence of perlecan, a basement membrane heparan sulfate proteoglycan, reveals extensive similarity with laminin A chain, low density lipoprotein receptor, and the neural cell adhesion molecule, *J. Biol. Chem.* **266**, 22939–22947.

O'Reilly, M., Boehm, T., Shing, Y., Fukai, N., Vasies, G., Lane, W.S., Flynn, E., Birkhead, J.R., Olsen, B.R., Folkman, J. (1997) Endostatin: an endogenous inhibitor of angiogenesis and tumor growth, *Cell* **88**, 277–285.

Olsen, B.J. (1995) New insight into the function of collagens from genetic analysis, *Curr. Opin. Cell Biol.* **7**, 720–727.

Ornitz, D.M. (2000) FgFs, heparan sulfate and FGFRs: complex interactions essential for development, *Bioessays* **22**, 108–112.

Ortega, N., Werb, Z. (2002) New functional roles for non-collagenous domains of basement membrane collagen, *J. Cell Sci.* **115**, 4201–4214.

Ottani, V., Raspanti, M., Ruggeri, A. (2001) Collagen structure and functional implications, *Micron* **32**, 251–260.

Palmieri, D., Camardellqa, L., Ulivi, V., Guasco, G., Manduca, P. (2000) Trimer carboxyl propeptide of collagen type-I produced by mature osteoblasts is chemotactic for endothelial cells, *J. Biol. Chem.* **275**, 32658–32663.

Pankov, R., Yamada, K.M. (2002) Fibronectin at a glance, *J. Cell Sci.* **115**, 3861–3863.

Parise, L.V., Lee, J.W., Juliano, R.L. (2000) New aspects of integrin signaling in cancer, *Semin. Cancer Biol.* **10**, 407–414.

Pepper, M.S. (2001) Role of the matrix metalloproteinase and plasminogen activator-plasmin system in angiogenesis, *Arterioscler. Thromb. Vasc. Biol.* **21**, 1104–1117.

Perez-Moreno, M., Jamora, C., Fuchs, E. (2003) Stick business: orchestrating cellular signals at adherence junctions, *Cell* **112**, 535–548.

Petitclerc, E., Stromblad, S., von Schalscha, T.L., Mitjans, F., Piulats, J., Mongomery, A.M.P., Cheresh, D.A., Brooks, P.C. (1999) Integrin αvβ3 promotes M21 melanoma growth in human skin by regulating tumor cell survival, *Cancer Res.* **59**, 2724–2730.

Petitclerc, E., Boutaud, A., Prestakyo, A., Xu, J., Sado, Y., Ninomiya, Y., Sarras, M.P., Hudson, B.G., Brooks, P.C. (2000) New Functions for non-collagenous domains of human collagen type-IV: novel integrin ligands inhibiting angiogenesis and tumor growth in vivo, *J. Biol. Chem.* **275**, 8051–8061.

Ponce, M.L., Nomizu, M., Kleinman, H.K. (2001) An angiogenic laminin site and its antagonist bind through the alpha(v)beta(3) and alpha(5) beta(1) integrins, *FASEB J.* **15**, 1389–1397.

Prockop, D.J., Kivirikko, K.I. (1995) Collagen: molecular biology, disease, and potentials for therapy, *Annu. Rev. Biochem.* **64**, 403–434.

Pupa, S.M., Menard, S., Forti, S., Tagliabue, E. (2002) New insights into the role of extracellular matrix during tumor onset and progression, *J. Cell Physiol.* **192**, 259–267.

Qian, F., Vaux, D.L., Weissman, I.L. (1994) Expression of the integrin alpha (4) beta (1) on melanoma cells can inhibit the invasive stage of metastasis formation, *Cell* **77**, 335–347.

Ramchandra, R., Dhanabal, M., Volk, R., Waterman, M.J.F., Segal, M., Lu, H., Knebelmann, B., Sukhatme, V.P. (1999) Antiangiogenic activity of restin, NC10 domain of human collagen XV: comparison to endostatin, *Biochem. Biophys. Res. Commun.* **255**, 735–739.

Ries, A., Gohring, W., Fox, J.W., Timpl, R., Sasaki, T. (2001) Recombinant domains of mouse nidogen-1 and their binding to basement membrane proteins and monoclonal antibodies, *Eur. J. Biochem.* **268**, 5119–5128.

Ross, J.J., Shimmi, O., Vilmos, P., Petryk, A., Kim, H., Gaudenz, K., Hermanson, S., Ekker, S.C., O'Conner, M.B., Marsh, J.L. (2001) Twisted gastrulation is a conserved extracellular BMP antagonist, *Nature* **410**, 479–483.

Salmivirta, K., Talts, J.F., Olson, M., Sasaki, T., Timpl, R., Ekblom, P. (2002) Binding of mouse nidogen-2 to basement membrane components and cells and its expression in embryonic and adult tissues suggest complementary functions of the two nidogens, *Exp. Cell Res.* **279**, 188–201.

Sasaki, M., Kleinman, H.K., Huber, H., Deutzmann, R., Yamada, Y. (1988) Laminin, a multidomain protein. The A chain has a unique globular domain and homology with the basement membrane proteoglycan and the laminin B chains, *J. Biol. Chem.* **263**, 16536–16544.

Sasaki, T., Gohring, W., Miosge, N., Abrams, W.R., Rosenbloom, J., Timpl, R. (1999) Tropoelastin binding to fibulins, nidogen-2 and other extracellular matrix proteins, *FEBS Lett.* **460**, 280–284.

Schittny, J.C., Yurchenco, P.D. (1989) Basement membranes: molecular organization and function in development and disease, *Curr. Opin. Cell Biol.* **1**, 983–988.

Schlessinger, J. (1997) Direct binding and activation of receptor tyrosine kinase by collagen, *Cell* **91**, 809–720.

Schneller, M., Vuori, K., Ruoslahti, E. (1997) αvβ3 integrin associates with activated insulin and PDGF beta receptors and potentiates the biological activity of PDGF, *EMBO J.* **16**, 5600–5607.

Schreiner, C.L., Fisher, M., Bauer, J., Juliano, R.L. (1993) Defective vasculature in fibronectin-receptor-deficient CHO cell tumors in nude mice, *Int. J. Cancer* **55**, 436–441.

Schwartz, M.A., Ginsberg, M.H. (2002) Networks and crosstalk: integrin signaling in cell spreading, *Nat. Cell Biol.* **4**, E65–E68.

Sechler, J.L., Rao, H., Cumiskey, A.M., Vega-Colon, I., Smith, M.S., Murata, T., Schwarzbauer, J.E. (2001) A novel fibronectin binding site required for fibronectin fibril growth during matrix assembly, *J. Cell Biol.* **154**, 1081–1088.

Senger, D.R., Claffey, K.P., Benes, J.E., Perruzzi, C.A., Sergiou, A.P., Detmar, M. (1994) Angiogenesis promoted by vascular endothelial growth factor: regulation through alpha(1)beta(1) and alpha(2)beta(1)integrins, *Proc. Natl. Acad. Sci. U. S. A.* **94**, 13612–13617.

Senger, D.R., Perruzzi, C.A., Streit, M., Koteliansky, V.E., de Fougerolles, A.R., Detmar, M. (2002) The alpha(1)beta(1) and alpha(2)beta(1) integrins provide critical support for vascular endothelial growth factor signaling, endothelial cell migration, and tumor angiogenesis, *Am. J. Pathol.* **160**, 195–204.

Siebold, B., Qian, R.-Q., Glanville, R.W., Hofmann, H., Deutzmann, R., Kuhn, K. (1987) Construction of a model for the aggregation and cross-linking region (7S domain) of type-IV collagen based upon an evaluation of the primary structure of the α1 and α2 chains in this region, *Eur. J. Biochem.* **168**, 569–575.

Simons, M., Horowitz, A. (2001) Syndecan-4-mediated signaling, *Cell Signal.* **13**, 855–862.

Sipes, J.M., Guo, N., Negre, E., Vogel, T., Krutzsch, H.C., Roberts, D.D. (1993) Inhibition of fibronectin binding and fibronectin-mediated cell adhesion to collagen by a peptide from the second type I repeat of thrombospondin, *J. Cell Biol.* **121**, 469–477.

Smyth, N., Vatansever, H.S., Murray, P., Meyer, M., Frie, C., Paulsson, M., Edgar, D. (1999) Absence of basement membranes after targeting the LAMC1 gene results in embryonic lethality due to failure of endoderm differentiation, *J. Cell Biol.* **144**, 151–160.

Somg, Q.H., Klepeis, V.E., Nugent, M.A., Trinkaus-Randall, V. (2002) TGF-beta-1 regulates TGF-beta-1 and FGF mRNA expression during fibroblast wound healing, *Mol. Pathol.* **55**, 164–176.

Sottile, J., Hocking, D.C. (2002) Fibronectin polymerization regulates the composition and stability of extracellular matrix fibrils and cell-matrix adhesions, *Mol. Cell Biol.* **13**, 3546–3559.

Stad, R.K., Buurman, W.A. (1994) Current views on structure and function of endothelial adhesion molecules, *Cell Adhes. Commun.* **2**, 261–268.

Stenback, F., Wasenius, V.M. (1985) Basement membranes in ultraviolet light-induced skin lesions and tumors, *Photodermatol.* **2**, 347–358.

Stromblad, S., Becker, J.C., Yebra, M., Brooks, P.C., Cheresh, D.A. (1996) Suppression of p53 activity and p21WF1/CIP1 expression by vascular cell integrin $\alpha v \beta 3$ during angiogenesis, *J. Clin. Invest.* **98**, 426–433.

Stupack, D.G., and Cheresh, D.A. (2002) ECM remodeling regulates angiogenesis: endothelial integrins look for new ligands, *Sci.STKE* **119**, PE7.

Sundaramoorthy, M., Meiyappan, M., Todd, P., Hudson, B.G. (2002) Crystal structure of NC1 domains: structural basis for type-IV collagen assembly in basement membranes, *J. Biol. Chem.* **277**, 31142–31253.

Talts, J.F., Andac, Z., Gohring, W., Brancaccio, A., Timpl, R. (1999) Binding of the G domains of laminin alpha1 and alpha2 chains and perlecan to heparin, sulfatides, alpha-dystroglycan and several extracellular matrix proteins, *EMBO J.* **18**, 863–870.

Tapanadechopone, P., Hassell, J.R., Rigatti, B., Couchman, J.R. (1999) Localization of glucosaminoglycan substitution sites on domain V of mouse perlecan, *Biochem. Biophys. Res. Commun.* **265**, 680–690.

Timpl, R. (1989) Structure and biological activity of basement membrane proteins, *Eur. J. Biochem.* **180**, 487–502.

Timpl, R. (1993) Proteoglycans of basement membranes, *Experientia* **49**, 417–4282.

Tomasini, B.R., Mosher, D.F. (1991) Vitronectin, *Prog. Hemost. Thromb.* **10**, 269–305.

Tomono, Y., Naito, I., Ando, K., Yonezawa, T., Sado, Y., Irakawa, S., Arata, J., Okigaki, T., Ninomiya, Y. (2002) Epitope-defined monoclonal antibodies against multiplexin collagen demonstrates that type-XV and XVIII collagens are expressed in specialized basement membranes, *Cell Struct. Funct.* **27**, 9–20.

Tremble, P., Chiquet-Ehrismann, R., Werb, Z. (1994) The extracellular matrix ligands fibronectin and tenascin collaborate in regulating collagenase gene expression in fibroblasts, *Mol. Biol. Cell.* **5**, 439–453.

Trikha, M., De Clerck, Y.A., Markland, F.S. (1994) Contortrostatin, a snake venom disintegrin, inhibits beta 1 integrin-mediated human metastatic cell adhesion and blocks experimental metastasis, *Cancer Res.* **54**, 4993–4998.

Tuckwell, D.S., Ayad, S., Grant, M.E., Takigawa, M., Humphries, M.J. (1994) Conformation dependence of integrin-type II collagen binding: inability of collagen peptides to support $\alpha 2 \beta 1$ binding, and mediation of adhesion to denatured collagen by a novel $\alpha 5 \beta 1$-fibronectin bridge, *J. Cell Sci.* **107**, 993–1005.

Tunggal, P., Smyth, N., Paulsson, M., Ott, M.-C. (2000) Laminins: structure and genetic regulation, *Microsc. Res. Tech.* **51**, 214–227.

Van Belle, P.A., Elenitsas, R., Satyamoorthy, K., Wolfe, J.T., Guerry, D., Schuchter, L., Van Belle, T.J., Albelda, S., Tahin, P., Herlyn, M.,

Elder, D.E. (1999) Progression-related expression of beta 3 integrin in melanomas as nevi, *Hum. Pathol.* **30**, 562–567.

Vernon, R.B., Angello, J.C., Iruela-Arispe, M.L., Lane, T.F., Sage, E.H. (1992) Reorganization of basement membrane matrices by cellular traction promotes the formation of cellular networks in vitro, *Lab. Invest.* **66**, 536–547.

Vu, T.H., Werb, Z. (2000) Matrix metalloproteinases: effects on development and normal physiology, *Gene. Dev.* **14**, 2123–2133.

Vu, T.H., Shipley, J.M., Bergers, G., Berger, J.E., Helms, J.A., Hanahan, D., Shapiro, S.D., Senior, R.M., Werb, Z. (1998) MMP-9/gelatinase B is a key regulator of growth plate angiogenesis and apoptosis of hypertrophic chondrocytes, *Cell* **93**, 411–422.

Wagner, D., Ivatt, R., Destree, A., Hynes, R. (1981) Similarities and differences between fibronectins of normal and transformed hamster cells, *J. Biol. Chem.* **25**, 11708–11715.

Watanabe, K., Takahashi, H., Habu, Y., Kamiya-Kubushiro, N., Kamiya, S., Nakamura, H., Yajima, H., Ishii, T., Katayama, T., Miyazaki, K., Fukai, F. (2000) Interaction with heparin and matrix metalloproteinase 2 cleavage expose a cryptic anti-adhesive site of fibronectin, *Biochemistry* **39**, 7138–7144.

Watt, F.M. (2002) Role of integrins in regulating epidermal adhesion, growth and differentiation, *EMBO J.* **21**, 3919–3926.

Werb, Z., Tremble, P.M., Behrendtsen, O., Crowley, E., Damsky, C.H. (1998) Signal transduction through the fibronectin receptor induces collagenase and stromelysin gene expression, *J. Cell Biol.* **109**, 877–889.

Wever, U.M., Shaw, L.M., Albrechtsen, R., Mercurio, A.M. (1997) The integrin alpha (5) beta (1) promotes survival of metastatic human breast carcinoma cells in mice, *Am. J. Pathol.* **151**, 1191–1198.

Woods, A. (2001) Syndecans: transmembrane modulators of adhesion and matrix assembly, *J. Clin. Invest.* **107**, 935–941.

Xu, J., Rodriguez, D., Kim, J.J., Brooks, P.C. (2000) Generation of monoclonal antibodies to cryptic collagen sites by using subtractive immunization, *Hybridoma* **19**, 375–385.

Xu, J., Rodriguez, D., Petitclerc, E., Kim, J.J., Mangai, M., Moon, S.Y., Davis, G.E., Brooks, P.C. (2001) Proteolytic exposure of a cryptic site within collagen type-IV is required for angiogenesis and tumor growth in vivo, *J. Cell Biol.* **154**, 1069–1079.

Yurchenco, P.D., Ruben, G.C. (1987) Basement membrane structure in situ: evidence for lateral associations in the type-IV collagen network, *J. Cell Biol.* **105**, 2559–2568.

Ziober, B.L., Chen, Y.Q., Ramos, D.M., Waleh, N., Kramer, R.H. (1999) Expression of the alpha (7) beta (1) laminin receptor suppresses melanoma growth and metastatic potential, *Cell Growth Differ.* **10**, 479–490.

Medicinal Chemistry

David J. Triggle
State University of New York, Buffalo, NY, USA

Encyclopedia of Molecular Cell Biology and Molecular Medicine, 2nd Edition. Volume 8
Edited by Robert A. Meyers.
Copyright © 2005 Wiley-VCH Verlag GmbH & Co. KGaA, Weinheim
ISBN: 3-527-30550-5

Keywords

Combinatorial Chemistry
The discipline of high output synthesis of chemical structures in solution, in the solid phase, or by the use of biological expression technologies.

High-throughput Screening
Technologies complementary to combinatorial chemistry that permit the rapid and efficient screening in assays of the biological activities of compounds in an automated manner.

Ligand
A molecule that binds to a receptor to initiate (agonist) or block (antagonist) response. Ligands may be small natural or synthetic molecules (neurotransmitters and their analogs) or large peptides, proteins, or nucleic acids.

Peptidomimetic
A synthetic molecule that mimics the biological properties of peptides and proteins but does not include typical peptide structures.

Receptor
A component of a cell that interacts specifically with (receives) other molecules and, in appropriate combination, initiates (or blocks) biological response. Receptors may be protein, nucleic acid, lipid, or carbohydrate in composition.

Stereoselectivity
The ability of molecules to interact with receptors and other macromolecules in an asymmetric manner such that one structure exhibits preferential interaction or activity. Stereoselectivity is extremely common in biological systems because the building blocks of nature – amino acids, sugars etc. – are chiral.

Structure–Activity Relationships
The definition of the relationship between chemical structure and biological activity. This may be achieved in qualititative terms or, and increasingly, in quantitative terms that involve the determination of the 3D structure of the ligand–receptor complex.

■ Medicinal chemistry defines one component of a sequence of events in the process of drug discovery and development. The critical steps of lead structure identification and refinement have been, and continue to be, the major contributions of medicinal chemistry to the drug discovery process: medicinal chemistry may be thought of as comprising three stages – a discovery step, an optimization step, and a production step (Fig. 1). Historically, drug development has been associated with biologically active products from natural sources, principally plants, and the development of natural product chemistry. Indeed, some 50% of currently available drugs have their origins, directly or indirectly, in natural products. With the ascendancy of synthetic organic chemistry, emphasis was increasingly placed on the screening of synthesized molecules in a variety of biological test systems that moved progressively from *in vivo* whole-animal systems to single expressed protein systems and to targets derived from reading of the genome. This approach has acquired increasingly mechanistic underpinnings with the development and quantitation of the receptor concept and the availability of receptor-based assays. The processes of lead structure identification and refinement have become progressively more sophisticated in partnership with developments in molecular and structural biology that have led to the characterization of new potentially drugable targets, the availability of human proteins as research tools with which to study drug–receptor interactions and as biopharmaceuticals, and the availability of three-dimensional structures of human receptors and enzymes with which to perform *in silico* medicinal chemistry. Finally, the new technologies embodied in high-throughput screening and combinatorial chemistry have led to an extraordinary level of biological activity determination: the translation of this productivity into therapeutically available medicines remains, however, a significant problem.

The genomics era, initiated in 1953 with the publication of the Watson and Crick paper on the structure of DNA, has had a major impact not only on biology but also on other scientific disciplines, including chemistry. The basic themes that govern biology – diversity, replication, evolution, and self-organization – are increasingly major components of medicinal chemistry, in the development of combinatorial chemistry, self-replicating molecules, *in vitro* evolution of molecules to biological fitness, and the synthesis of molecules around a template.

The elucidation of the human genome, as well as the genomes of other species including many bacteria, was the second major event of the genomics era. From the perspective of drug discovery, the human genome was anticipated to yield an enormous increase in the number of potential drug targets and thus greatly expedite the availability of new drugs. Together with the technologies of combinatorial chemistry and high-throughput screening, the production line for new drugs was anticipated to be faster and more capacious by several orders of magnitude. However, the number of human genes, approximately 30 000 (but still an uncertain number), is relatively small by comparison with the number of genes in other species, suggesting that our complexity arises not from a simple increase in the number of proteins, but rather from an expansion in the complexity of the signaling and regulatory networks. Thus, the number of potential drug targets is likely to be significantly smaller than was originally anticipated: nonetheless, the anticipated number of 1000 to 2000 potential targets is still significantly larger than the number of current targets, estimated at approximately 120 in all. The issue remains, however, as to how many of these new targets will actually be drugable.

Knowledge of the human genome will also lead to the introduction of *personalized* medicine whereby the drug will be matched more precisely to the patient, thus generating better response, facilitating clinical trial development, and reducing drug withdrawal and related misadventures that are typically due to the reactions with an extremely small percentage of patients, but nonetheless prominent because of the large patient base with many drugs.

1
Background

Active principles from natural sources were the original drugs available to man and still remain a very significant source. References to such products date back thousands of years to the Chinese, Greek and Roman civilizations, and "natural medicines" have been used therapeutically since prerecorded times, as part of religious ceremonies, and doubtless also to relieve a life that was "nasty, short and brutal". Thus, the acetycholinesterase inhibitor physotigmine, isolated from the Calabar bean was used in the West African tribal ritual of *trial by ordeal*, and various psychoactive substances found in plants have been used for thousands of years. The nature of these principles, which were invariably crude mixtures, was not known and their activity doubtless varied considerably. However, as their chemical constitutions came to be elucidated, they became available as pure materials and also served as leads for synthetic analogs with improved properties. Among the natural products that have served as leads in this manner are the antimalarial

quinine, the analgesic morphine, the antihypertensive drug reserpine and the stimulant cocaine (Fig. 2). Introductions in the twentieth century from plant natural sources include many antibiotics such as the penicillins, cephalosporins, and tetracyclines and, more recently, the cholesterol-lowering agent compactin (a lead to the statin family), the immunomodulator cyclosporin, the antimalarial agent artemisinin, and anticancer agents such as paclitaxel, camptothecin, and the bryostatins (Fig. 3). Finally, the discovery of the multiple regulatory and signaling properties of nitric oxide (NO) in neuronal, cardiovascular, immune, and inflammatory responses emphasizes the route to drug discovery through the elucidation of endogenous pathways and structures (Fig. 4).

Despite the early success of plant-based medicinal chemistry, the advent of synthetic organic chemistry resulted in a significant displacement of botanicals as drugs. There is now, however, a significant resurgence of interest in natural products as pharmaceuticals, and due to recent successes, this has led to an increasing

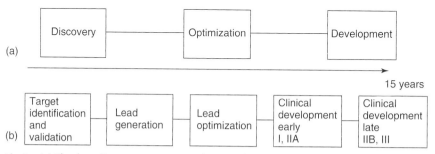

Fig. 1 (a) The drug discovery process may be viewed as composed of three principal stages – *discovery, optimization, development*. (b) The discovery phase may be viewed as composed of five stages: following target identification and validation that may arise from phenotypic or genotypic observations of the two stages of medicinal chemistry – *lead generation and lead optimization* – are initiated. During these two stages, compounds having appropriate pharmacodynamic and pharmacokinetic properties are identified and are subsequently optimized for these properties.

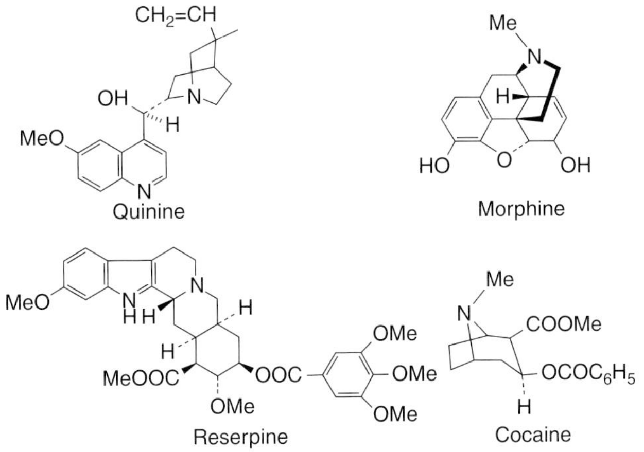

Fig. 2 Examples of potent naturally occurring pharmacologic entities.

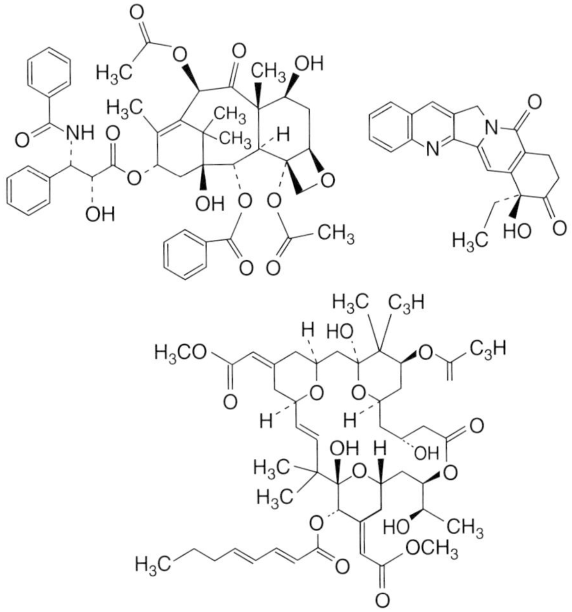

Fig. 3 Molecules derived from plant natural sources in the twentieth century including paclitaxel, camptothecin, and bryostatin 1.

understanding of indigenous peoples' healing practices, and to the increasing realization that habitat destruction is causing the loss of potentially valuable drugs and drug leads. For example, it is likely that fertility control has long been practiced by humans through the use of plant-based extracts, including

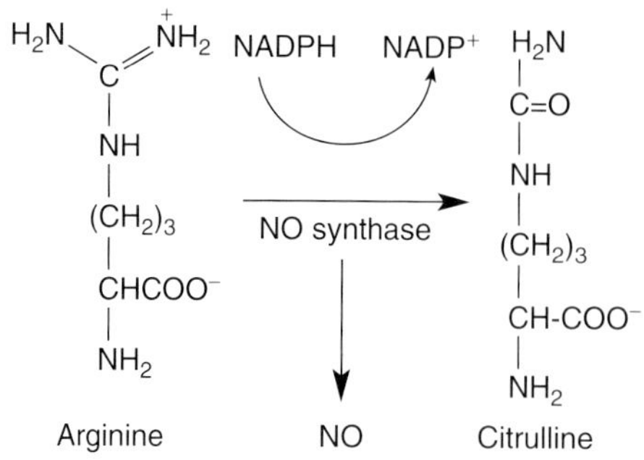

$$H_2N \overset{+}{\diagdown} \diagup NH_2$$
$$| \quad C$$
$$NH$$
$$|$$
$$(CH_2)_3$$
$$|$$
$$CHCOO^-$$
$$|$$
$$NH_2$$

NADPH   NADP⁺

H₂N
|
C=O
|
NH
|
(CH₂)₃
|
CH-COO⁻
|
NH₂

NO synthase

→

NO

Arginine   NO   Citrulline

Fig. 4 The nitric oxide (NO synthase) pathway.

one from *Silphion* during the Greek and Roman civilizations:this plant species was apparently harvested to extinction.

2
The Receptor Concept

The concept that drugs act at specific sites, though long recognized, is critical to current processes of drug discovery and to the elucidation of the mechanisms of drug action. Over the past century, the contributions of Paul Ehrlich to immunology and the chemotherapy of protozoan infections, and those of John Newton Langley to drug action in the nervous system are of particular note. Ehrlich viewed receptors as the protoplasmic side chains of cells, and Langley defined a receptor as "a specially excitable constituent [which] receives the stimulus and, by transmitting it, causes contraction" [response]. These considerations, together with Emil Fischer's concept of a *lock-and-key* process of small molecule–macromolecule interactions, categorized drug action as specific rather than nonspecific, with defined structure–activity relationships including stereoselectivity, and serving to stimulate or block receptor activity, agonism and antagonism, respectively. The mutual complementarity of the drug molecule

and the receptor-binding site provides the underlying basis of structure–activity relationships, the definition and analysis of dose–response relationships and the elucidation, through structural biology techniques, of new ligand structures.

Receptors are classified into several major families, the individual members of which show substantial structural and functional homology. The major classes include the following:

1. The G-protein-coupled receptors: for neurotransmitters (dopamine etc), many polypeptide hormones (substance P, etc.), odorants and light.
2. Ligand-gated ion channels: ion channels with integral receptors for neurotransmitters (acetylcholine, glutamic acid, etc.).
3. Voltage-gated ion channels: ion channels that are activated by changes in cellular electrical potential (Na⁺, K⁺ and Ca²⁺).
4. Enzyme-associated receptors: tyrosine kinase, and guanylyl cyclase.
5. Steroid hormone receptors: nuclear transcriptional regulators.

The G-protein-coupled receptor (GPCR) family is one of the largest gene families constituting more than 1% of the human genome, and more than half of all modern drugs have been estimated

Fig. 5 The receptor, depicted as three discrete entities, as an information generating and processing system.

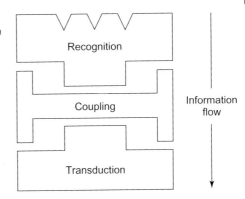

to be directed at these receptors. This family encompasses receptors for biogenic amines, polypeptide hormones, taste and olfactory stimuli, adhesion ligands and other signaling systems.

Receptors may be viewed as informational machines mediating a vectorial flow of information to the cellular machinery from the ligand, the receptor, or both (Fig. 5). Consistent with this scheme, the majority of receptors are multimeric protein assemblies and their activity may in principle be controlled by drug action at a variety of sites in the complex. Thus, in the G-protein-coupled receptor system depicted in Fig. 6, activity may be regulated by ligand interaction at the receptor-recognition site, at allosteric sites associated with the receptor, through drugs regulating the G-proteins themselves, or through drugs acting on the effectors. The several components of the total receptor complex, and the interfaces between the components, offer multiple opportunities for drug action in addition to the classically defined receptor itself.

Several properties including drug selectivity, affinity, biological response, coupling components, gene sequence, and protein structure determine the overall classification of pharmacological receptors. Thus, in the major family of G-protein-coupled receptors, characterized by a homologous structure with seven transmembrane sequences, there are five principal subclasses of muscarinic receptors (m1–m5) characterized by selective drugs and coupled

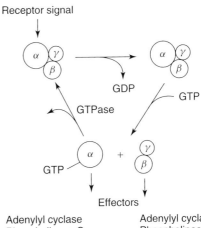

Fig. 6 The organization of heteromeric guanine nucleotide (G) binding proteins as a GTP/GDP exchanger and their coupling to a variety of biological effectors.

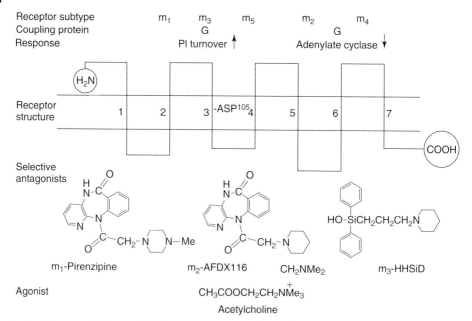

Receptor subtype m_1 m_3 m_5 m_2 m_4
Coupling protein G G
Response PI turnover ↑ Adenylate cyclase ↓

Receptor structure 1 2 3 -ASP^{105} 4 5 6 7

Selective antagonists

m_1-Pirenzipine

m_2-AFDX116 CH_2NMe_2

m_3-HHSiD

Agonist $CH_3COOCH_2CH_2\overset{+}{N}Me_3$
Acetylcholine

Fig. 7 The muscarinic acetylcholine receptor as a member of the G-protein-coupled family existing as five subtypes – m_1 to m_5 – recognized by selective antagonists and differentially coupled to stimulation of phosphatidylinositol (PI) hydrolysis or inhibition of adenylate cyclase.

differentially to pharmacological effectors (Fig. 7).

The traditional route to the classification, characterization, and eventual isolation of pharmacological receptors has stemmed from the recognition that a specific hormone, neurotransmitter, or drug, produce a defined biological response, that the action of the agent can be mimicked by structurally related compounds with

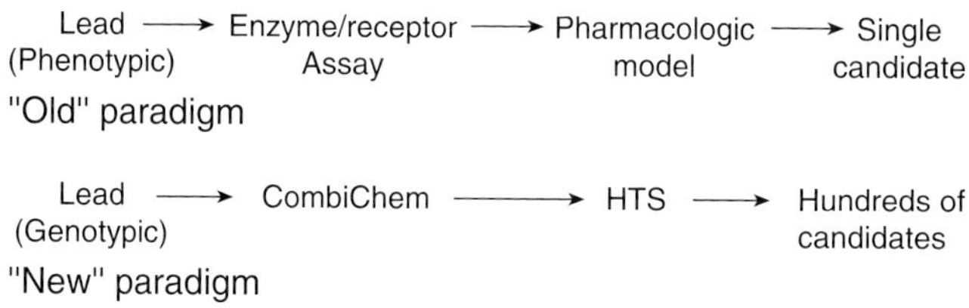

Lead ⟶ Enzyme/receptor ⟶ Pharmacologic ⟶ Single
(Phenotypic)   Assay   model   candidate
"Old" paradigm

Lead ⟶ CombiChem ⟶ HTS ⟶ Hundreds of
(Genotypic)   candidates
"New" paradigm

Fig. 8 The "old" and "new" paradigms of drug discovery. In the old paradigm, a lead target observation, typically a biological (phenotypic) observation such as excess gastric acid secretion or vasoconstriction was used as the basis for an assay into which were fed potential drug candidates typically synthesized "one molecule at a time" to generate a single clinical candidate. In the new paradigm, a lead target is derived from a genome database, and high-throughput chemistry and high-throughput screening are applied to generate, in principle, hundreds of clinical candidates.

a definable structure–activity relationship including stereoselectivity and that both agonists and antagonists exist. The actions of a natural product such as morphine led early to the conclusion that endogenous opiods exist for which morphine is a mere surrogate and that there must exist one or more opiod receptors through which the actions of endogenous, naturally occurring, and synthetic opiods occur. From the combined knowledge of receptor structure and the human and other genomes, it is now possible to identify receptors through homology searches in the absence of any knowledge of the endogenous ligand or of the biological responses controlled. The identification of such *orphan* receptors represents a reversal of the classical receptor characterization scheme and provides a further source of potential drug targets (Fig. 8).

3
Chemistry

Traditionally, the medicinal chemistry component of drug discovery has been initiated from a lead structure, either a natural product of plant or animal origin, an endogenous hormone or neurotransmitter, or a synthetic molecule exhibiting some degree of the appropriate biological activity. The lead structure is then optimized to produce the desired potency and selectivity of action and the required pharmacokinetic characteristics. Acquisition of the latter is of particular importance since many agents fail in development because of inadequate solubility, absorption or metabolic properties. This process, essentially a "one-molecule-at-a-time" scheme, has achieved many major successes including the development of the β-adrenergic blockers from isoproterenol, the development of angiotensin-converting enzyme (ACE) inhibitors from a peptide teprotide and the development of the angiotensin II receptor antagonists, including losartan, from a weakly active imidazole derivative (Fig. 9). In this process, the biological target, be it blood pressure, acid secretion, tumor growth, anxiety, or depression, has been derived from a biological or phenotypic response.

In the new postgenomic era, it is, however, possible to define targets through analysis of gene structure from the approximately 30 000 genes currently believed to comprise the human genome as well as from the genomes of bacteria, viruses or parasitic species. Although the genomic approach offers, in principle, many new targets, the issue of target validation looms far larger than in the classical process. It is vitally necessary to associate a particular gene with a particular phenotypic expression *and* to demonstrate the relevance of this to the disease process. The techniques of target validation are several and include the analysis of gene and protein expression in normal and diseased tissues. Knockout and conditional knockout animals, the creation of mutant (ethyl-nitrosourea-induced) mice, the use of model organisms including *Drosophila melanogaster* and *Caenorhabditis elegans*, the use of small interfering RNA, and small molecule–based chemical–biological techniques.

With increasing knowledge of the target structure, there is greater emphasis on structure-based drug design whereby X ray, NMR, and computational approaches are employed to provide a physical interpretation of the drug–receptor complex and, in principle, to provide a more rapid and efficient route to drug design. There are still significant limitations

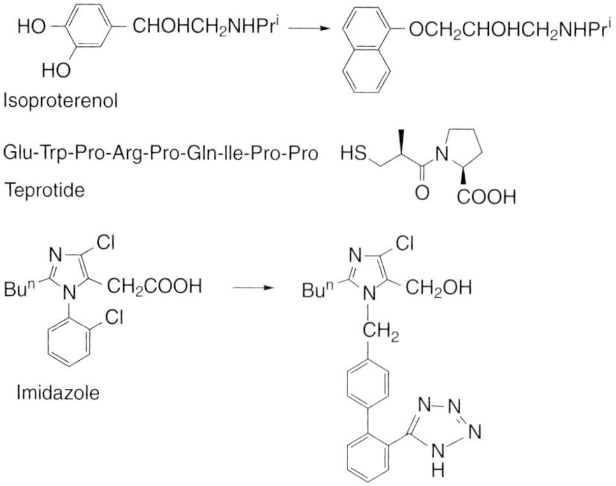

Fig. 9 Examples of lead structures (left) that gave rise to the synthetic antagonists propranolol, captopril, and losartan, active at β-adrenergic receptors, angiotensin-converting enzyme, and angiotensin II receptors, respectively.

to this approach, although it will doubtless continue to assume greater importance. First, the structures of many targets are still unknown and the structures of ligands in solution or in the solid state, although frequently easily accessible, may have no relationship to the structure of the receptor-bound ligand.

At the other extreme of drug discovery has been the development of combinatorial chemistry and high-throughput screening, whereby the ability to synthesize vast chemical libraries and to use automated techniques to screen them against a large number of (typically) genomically derived targets was intended to greatly automate and render more productive the drug discovery process (Fig. 8). This promise remains to be completely fulfilled. However, the synergies obtained by the combined application of combinatorial chemistry and structure-based designs are fueling the new approaches to drug discovery, designed to combine the intelligence of rationality with the efficiency of technology.

3.1
Druglike Molecules

Despite the ability to make chemical libraries of virtually unlimited size, the issue remains that the number of "druglike" molecules is relatively small and that quite limiting chemical and physical properties are necessary to ensure that molecules are both biologically active *and* orally available. A number of approaches have been employed in efforts to quantitate these properties. The best known of these approaches is probably the empirical "rule-of-five" derived from a consideration of over 2000 therapeutically active drugs that indicates a likely problem when a molecule obeys any two of the following conditions:

- Molecular weight >500
- Number of H-bond acceptors >10
- Number of H-bond donors >5
- Calculated logP >5.0.

A number of refinements of this system have been implemented. For example, a

Fig. 10 Examples of *pharmacophoric* structures or building blocks that are commonly found (with appropriate molecular decoration) in pharmacologically active agents.

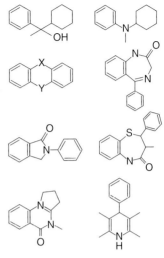

survey of over 1100 drug candidates at GlaxoSmithKline for which oral bioavailability measurements are available suggests that molecules with reduced flexibility (10 or fewer rotatable bonds) and a polar surface area of less than 140Å^2 (or 12 or fewer H-bond donors or acceptors) are important predictors of good oral bioavailability. These and related approaches have been reasonably successful in predicting "druglike" molecules present in large chemical databases such as the Available Chemicals Directory and the Comprehensive Medicinal Chemistry Database.

A second and related issue is that of the chemical structures of active molecules: consistent with the above considerations, these also tend to be relatively limited in terms of potential chemical space and component structures – *pharmacophores* – tend to be conservatively employed. A number of basic pharmacophoric structures or *scaffolds* exist (Fig. 10) onto which are grafted the functional groups that define specific drug families. It is proposed that these structures are those that resist "hydrophobic collapse" (the tendency to

self-associate to inactive species) through their conformational nonflexibility. Prominent examples of such scaffolds include the 1,4-dihydropyridine and benzodiazepine nuclei, both of which give rise to potent and biologically diverse classes of drugs (Figs. 11 and 12). An interesting example of combinatorial chemistry practiced by Nature, the supreme combinatorial chemist, centers around the venomous snails of the *Conus* genus. These snails produce disulfide-bridged peptide toxins that exhibit both high affinity and selectivity for a variety of neurotransmitter receptors and ion channels. The peptides are synthesized as precursors with a hypervariable C-terminus region, the latter permitting amino acid changes in discrete regions to tailor pharmacological specificity.

3.2
Combinatorial Chemistry and Compound Libraries

Chemical libraries have long been an important proprietary component of the drug discovery process in the search for "lead" structures. Such libraries

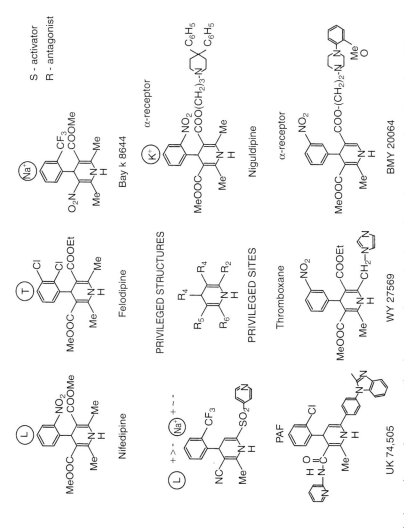

Fig. 11 The 1,4-dihydropyridine nucleus as a *privileged* structure that can be directed at a diverse set of ion channels and pharmacological receptors.

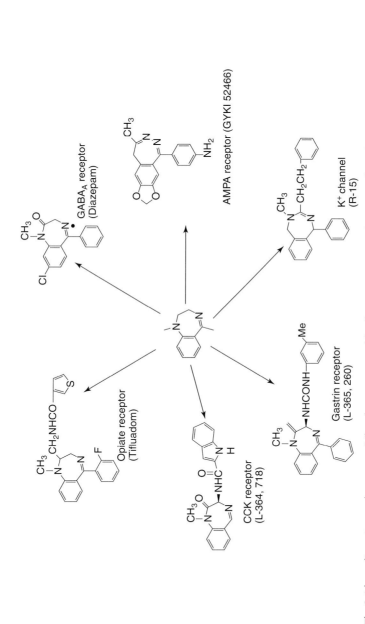

Fig. 12 The 1,4-benodiazepine nucleus as a *privileged* structure capable of interacting at a wide variety of ion channels and pharmacologic receptors.

were typically derived from the traditional "one-molecule-at-a-time" approach of drug discovery, and were typically between 1000 and 100 000 molecules in size: these libraries were based to a very significant extent on the previous chemical experience of the company and could vary widely in both molecular diversity and "drug-likeness". The technologies of parallel and combinatorial synthesis have dramatically increased library scale. A typical organic chemist can perhaps synthesize 50 to 100 compounds per year. The technique of parallel synthesis in which, with the aid of automation, multiple products can be synthesized in discrete reaction vessels can increase this chemical productivity by a 100-fold factor or more. The combinatorial approach, whether in solution or, more typically, in the solid phase, expands this scale further by performing multiple reactions in a single reaction vessel, creating all possible products from a precursor set of building blocks. A simple calculation illustrates the power of the combinatorial approach. From a basic set of 1000 building blocks, it is possible to construct 1000^5 pentapeptides, or one quadrillion molecules! In general, if the number of available building blocks is a, the number of synthetic steps is x and the same number of building blocks is used per step, then N, the number of individual compounds is given by

$$N = a^x \qquad (1)$$

And if the number of building blocks varies in each step (e, f, g in a three-step reaction), then

$$N = e.f.g \qquad (2)$$

In principle, one can make more compounds by combinatorial chemistry than there is mass in the universe! Although the original thrust of combinatorial chemistry

(coupled with the simultaneous development of high-throughput screening) was designed to produce as many molecules as possible and was a retreat from rational discovery processes, the paucity of success in such large random libraries led to a refocusing of combinatorial thinking to factor in both diversity and druglike properties in the construction of smaller libraries.

The selection and choice of building blocks *and* the linking chemistry are the critical steps in the combinatorial process. For example, the development of peptide and oligonucleotide libraries is quite straightforward since the building blocks and the coupling chemistry are both well known and easily implemented. However, because of the limitations to the use of these structures as drugs, increasing attention is being directed to the use of building blocks that create small-molecule libraries. A second consideration is whether the library is to be used for random screening or for lead development. Increasingly, combinatorial chemistry is used in the generation of "focused" libraries that are designed either to cover wide areas of chemical space in as small a collection as possible or to exploit chemical space around a particular structure. Furthermore, the need for such libraries to contain "druglike" molecules is increasingly recognized. The choice of natural products as libraries has much to commend it since these molecules, complex as they appear, have evolved and have been "forged in the crucible of evolution" to have specific biological functions and to bind to proteins or nucleic acids. A number of such natural product libraries are shown in Fig. 13. In any event, library design and development increasingly includes consideration of those molecular features that provide appropriate pharmacokinetic characteristics.

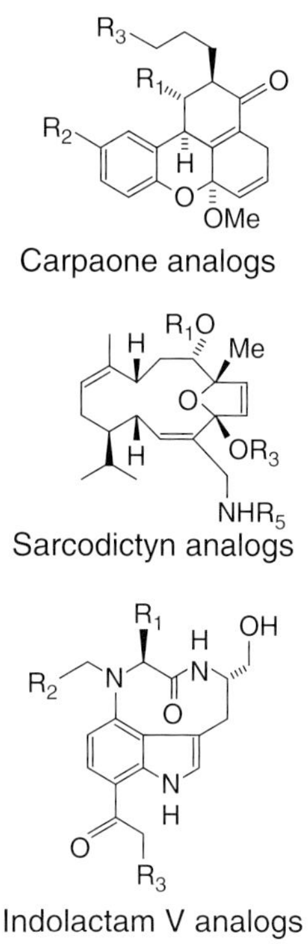

Fig. 13 Combinatorial libraries based on natural product lead structures.

The solid-phase approach of combinatorial chemistry is generally performed on resin particles or on two-dimensional arrays. An illustration of one such process, the split synthesis technique, is provided in Fig. 14. At each synthetic state, the resin support is pooled and then split: this technique overcomes the different reactivities of individual building blocks and ensures the equimolar distribution of the product, which is critical to the quantitation of the subsequent screening process. At the completion of the synthetic sequence, the compounds are cleaved from the resin support. A variety of techniques can be used to deconvolute the library and to identify the active structure(s). One approach is the encoded library in which a code or identifier tag is added at each stage of the synthetic process (Fig. 15). The library can then be "read" from the code sequence that is unique to each member of the library. Varying codes may be employed including sequencable, (nucleotides and peptides), nonsequencable, (fluorophores) or nonmolecular (radiofrequency) codes.

The process of *dynamic combinatorial chemistry* relies on reversible reactions between the building blocks of the system: thus, the composition of the library is under thermodynamic rather than kinetic control (Table 1). This is a critical differentiation from conventional

Fig. 14 The split synthesis method for the generation of combinatorial libraries in which each *monomer* is reacted to completion with the solid support and then pooled and split repeatedly to generate the resin-bound library. (From S. Hobbs DeWitt, *Pharmaceutical News* 1 : 11, 1994.)

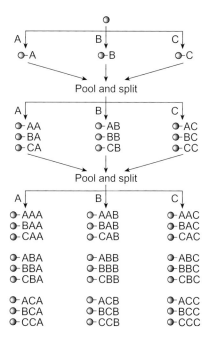

Carpaone analogs

Sarcodictyn analogs

Indolactam V analogs

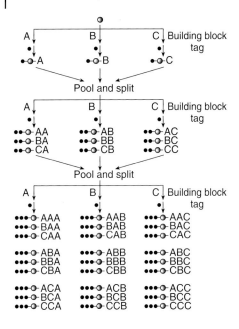

Fig. 15 The encoded combinatorial library in which the addition of each *monomer* is accompanied by a *tag*. The tag sequence that may be chemical or electronic specifically encodes the oligomeric sequence of interest. (From S. Hobbs DeWitt, Pharmaceutical News 1 : 11, 1994.)

Tab. 1 Combinatorial chemistry and dynamic combinatorial chemistry.

Combinatorial libraries	*Dynamic libraries*
Unaffected by recognition	Amplified by recognition
Selected compounds to be made independently	Selected compounds isolable from library
Complex topologies difficult to access	Complex topologies easier to access
Greater chemical access	More limited chemical access
Solubility not important	All members of library must be soluble
Under kinetic control	Under thermodynamic control

combinatorial chemistry because it indicates that the presence of a recognition process specific for one member of the equilibrium will stabilize that member by an equilibrium shift between the reactants and the products (Fig. 16). Accordingly, an *amplification* of yield is obtained. In principle, dynamic combinatorial chemistry offers the heady vision of potent and selective molecules self-assembling in the presence of the biological target and thus automating the entire process of drug discovery. This approach has been applied to a number of systems of biological interest, including the development of inhibitors against neuraminidase using a scaffold reacting with a series of aldehydes to yield a dynamic library of Schiff bases and heminanimals (Fig. 17). In the presence of the target neuraminidase, significant amplification of only the selected components occurred. In a related example, the enzyme acetylcholinesterase was used as a target: a series of acylhydrazides containing a charged amine function (to interact with the enzyme anionic site) and aldehydes yielded a dynamic library from which a compound was isolated with nanomolar inhibitory potency (Fig. 18). In practice, although proof of principle has been obtained in a number of systems, there are limitations, first in the number of suitable reversible reactions available and in both the stability of the target macromolecule under the reaction conditions, and in the presence of undesirable cross reactions between the protein and the chemical reactants.

The process of *target-accelerated synthesis* of active molecules is related to dynamic combinatorial chemistry whereby the coupling of building blocks is accelerated

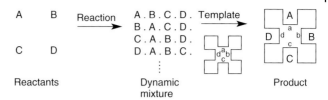

Fig. 16 In the process of dynamic combinatorial chemistry the reactant fragments, A–D, establish an equilibrium of products, one of which is selected by the presence of the template thus driving the overall reaction to select for that specific product.

Fig. 17 The use of dynamic combinatorial chemistry to develop a series of inhibitors against neuraminidase. The scaffold entity is reacted with a variety of ketones in the presence of the enzyme neuraminidase to produce by selection the active products.

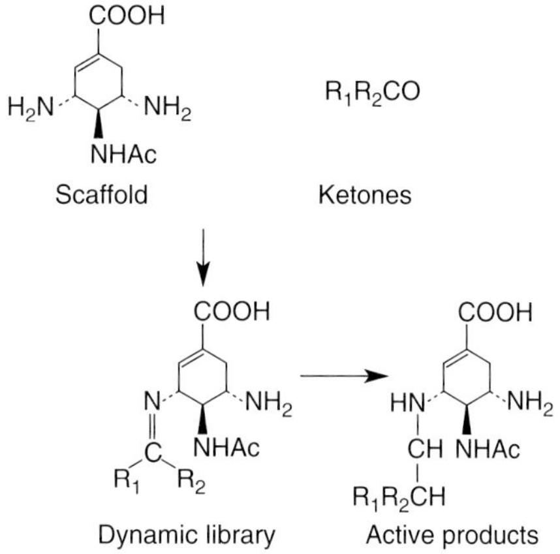

by assembling them on a target template: since the coupling is covalent, the system is under kinetic rather than under thermodynamic control. An example of target-accelerated synthesis is provided in Fig. 19, which depicts the assembly of an extremely potent (femtomolar KI) acetylcholinesterase inhibitor by 1,3-dipolar cycloaddition chemistry from the depicted starting materials.

Coming full circle back to Nature, combinatorial chemistry can also be achieved in laboratory-based *in vivo* systems through genetic manipulations of bacterial biosynthetic pathways to create new antibiotics and through phage display to generate peptide libraries. Both the polyketide and non-ribosomal classes of antibiotics are synthesized from their carboxylic acid or amino acid precursors, respectively, by large multifunctional polyketide synthases and peptide synthases. The modular construction of these enzymes makes possible their selective modification and thus the biosynthesis of novel products, including active antibiotics. Phage display technologies permit the display of very large peptide libraries that can be used *via* affinity-selection processes to define and target specific recognition sites and can be used to define new potential drug targets in cells and organs via phage targeting.

3.3
Structure-based Approaches to Drug Design

Increasingly, the intuition-based (and highly successful) approach to drug

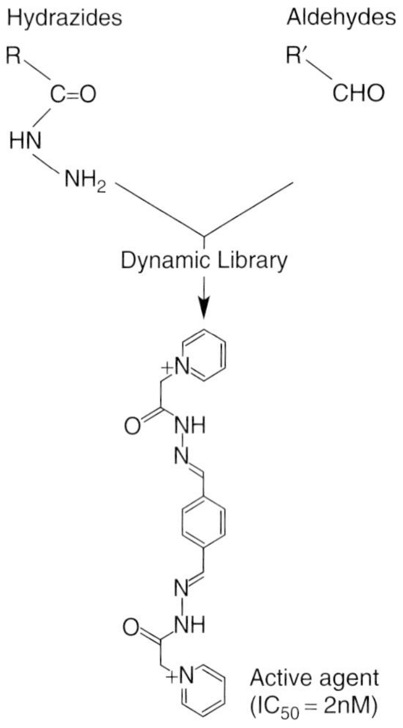

Hydrazides

R
　　C=O
HN
　　　NH₂

Aldehydes

R′
　　CHO

Dynamic Library

Active agent
(IC₅₀ = 2nM)

Fig. 18 The use of dynamic combinatorial chemistry to generate a dynamic library that in the presence of the target enzyme, acetylcholinesterase, yielded an active inhibitor with nanomolar affinity.

discovery is assuming a more quantitative character with the addition of structure-based components (Fig. 20). In particular, the understanding of the three-dimensional structure of the target proteins and of their ligand complexes with drugs is proving to be critical for the generation of docking processes to fit new molecules to active sites, *in silico* screening of potential drug structures, the elucidation of mechanisms of drug action, and the development of structure–activity relationships. This approach is frequently referred to as *structure-based drug design*, and with the increasing availability of macromolecular structures via X ray and NMR techniques or homology modeling, it is a rare contemporary drug discovery project that does not have inputs from these design principles. Additionally, the limitations caused by the lack of availability of protein three-dimensional structures are rapidly being resolved by detailed resolution of such membrane proteins as ion channels. With high-throughput protein crystallization processes now in place, it is likely that the next decade will see the majority of protein structures solved, either directly or indirectly, through the use of increasingly sophisticated computational methods.

The antagonists of HIV-1 protease, an enzyme critical to the process of viral assembly and maturation, provides one example of how structure-based design has led to the development of clinically effective and orally active drugs. Homology-based reasoning originally suggested that HIV-1 protease was related mechanistically to the aspartate protease family. This family is characterized by a two-domain active site, with each domain contributing an aspartate residue to the catalytic process. Direct X ray analysis confirmed this, showing the enzyme to be a homodimer with C2 symmetry (Fig. 21). Some symmetry-based antagonists designed from this structure are depicted in Fig. 22.

Most recently, the crystal structure of the angiotensin-converting enzyme complexed with the ACE inhibitor lisinopril has been obtained. ACE inhibitors are important anti-hypertensive agents that play an important role in the control of congestive heart failure. These inhibitors were developed without the knowledge of the structure of the human ACE enzyme but rather were designed on the assumption that the enzyme was mechanistically similar to carboxypeptidase A. There is, however, little three-dimensional similarity, thus indicating a comparatively common

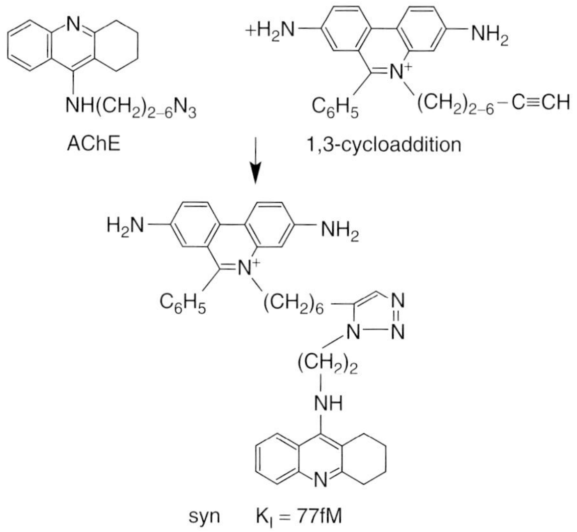

Fig. 19 An example of target-accelerated synthesis of an active molecule. In this example, acetylcholinesterase is the target macromolecule and small-molecule fragments that interact at discrete subsites of the enzyme become linked through 1,3-cycloaddition to yield the active inhibitor with femtomolar potency.

facet of scientific discovery, namely, that significant discoveries can be made on the basis of incorrect assumptions! However, the availability of the ACE structure should facilitate the development of new ACE inhibitors.

The structures of membrane-bound receptors and ion channels have, until recently, not been available at a resolution helpful to the generation of structure-based drug design. Such knowledge is, however, becoming available. In its absence, sequence data, homology modeling, and site-directed mutagenesis have permitted important conclusions to be drawn concerning the residues and topographies that are critical to drug–receptor interactions. The β-adrenergic receptor is a member of the G-protein coupled protein family characterized by the presence of 7 transmembrane helices. Site-directed mutagenesis has revealed the presence of residues critical to both agonist binding and receptor activation. A model for the interaction of the agonist isoproterenol is provided in Fig. 23. The interactions between transmembrane domains V and VI are consistent with observations that the cytoplasmic loop between these domains is critical to the coupling of the receptor and G-protein. Thus, a simple structural model can provide a mechanistic basis for receptor activation and also indicate that molecules that do not make these specific interactions are likely not to function as activators, but rather (as in the case of propranolol) as antagonists. Similar studies with receptors for the polypeptide hormone substance P and its nonpeptide antagonist CP 96 345 reveal that they occupy binding sites that are partially distinct (Fig. 24). Residues in transmembrane

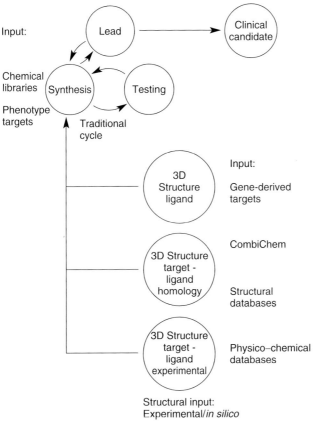

Fig. 20 The incorporation of structure-based drug design into the process of drug discovery. The traditional cycle of discovery (top left) is aided by the components of structure-based design that include inputs from both experimental and *in silico* structural design and from a structural and physicochemical databases.

domains 5 and 6 and the extracellular epitope between these domains are clearly critical for the nonpeptide antagonist, but not for peptide–agonist binding.

Ion channels are an important class of pharmacological receptors and are the loci of action of many drugs and toxins including, tetrodotoxin, pyrethroids, calcium channel antagonists, snake venoms and type II antidiabetic drugs (Fig. 25). The understanding of both channel function and drug action is being greatly facilitated by the availability of three-dimensional structures of several potassium channels. These studies, generally confirmatory of some fifty years of electrophysiological work, have elucidated the nature of the channel pore through which K^+ ions selectively flow, the subunit organization of the channel, the receptor site for quaternary ammonium inhibitors (NEt_4^+), and the nature of the voltage-sensing mechanism (Fig. 26).

These structure-based approaches are increasingly complemented by "expert" computer-based algorithms that both

Fig. 21 The ribbon backbone of HIV-1 protease, depicting the vertical twofold axis of symmetry and the active site aspartate residues in the middle. (Reproduced with permission from the American Chemical Society from Greer, J., Erickson, J.W., Baldwin, J.J. and Varney, M.D. (1994) *J. Med. Chem.* **37**, 1035.)

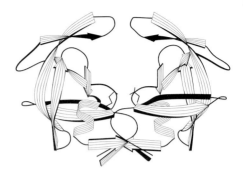

generate *and* screen, on a virtual basis, structures that fit protein-receptor sites. Complemented with algorithms that predict pharmacokinetic characteristics (ADME) true *in silico* drug design is likely to be an important component of actual drug discovery in the coming decade.

3.4
Peptidomimetics

The physiologically important ligands for many receptors and enzymes are peptides,

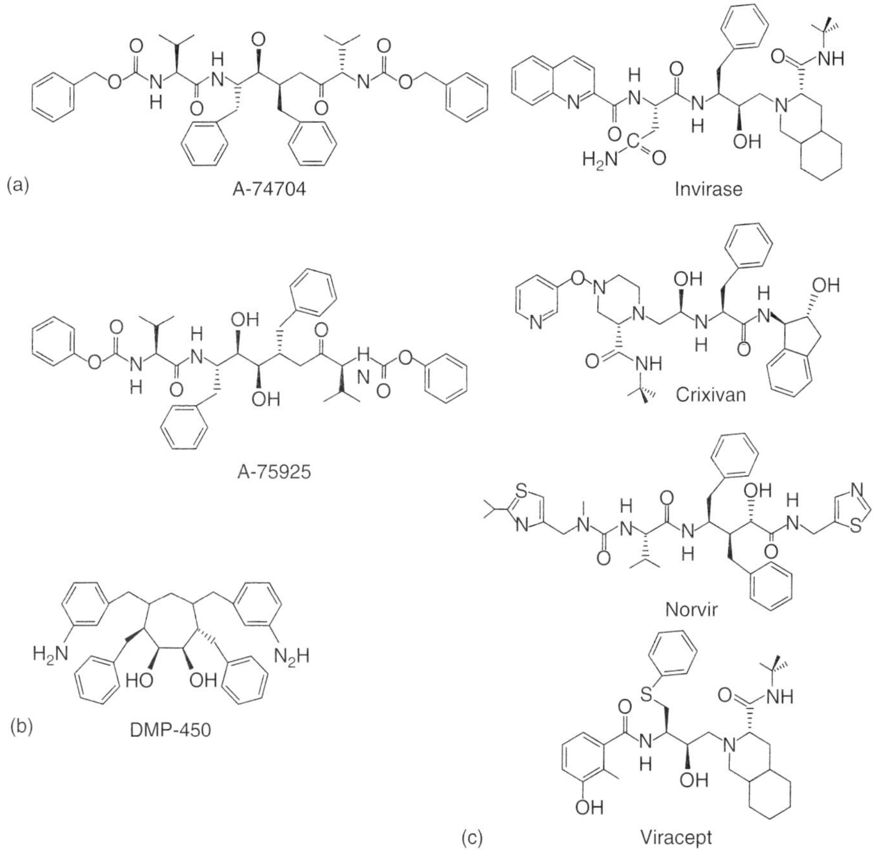

(a)   A-74704

Invirase

A-75925

Crixivan

(b)   DMP-450

Norvir

(c)   Viracept

Fig. 22 HIV-1 protease inhibitors.

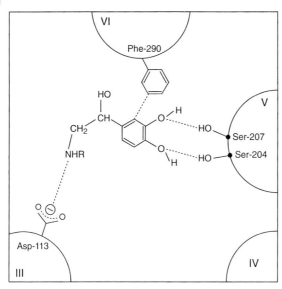

Fig. 23 Schematic representation of the interaction of a catecholamine with the β-adrenergic receptor and depicting specific functional interactions with residues on transmembrane helices III, V, and VII.

	Wild type NK-1	CR NK1 (NK3-TM7)	CR NK1 (NK3-TM6-7)	CR NK1 (NK3-TM5-7)	Wild type NK-3
Substance P	0.14 ± 0.03	0.49 ± 0.04	1.1 ± 0.2	0.9 ± 0.1	300 ± 80
Eledoisin	16 ± 5	4.6 ± 1.3	12 ± 3.5	4.5 ± 0.6	4.7 ± 1.1
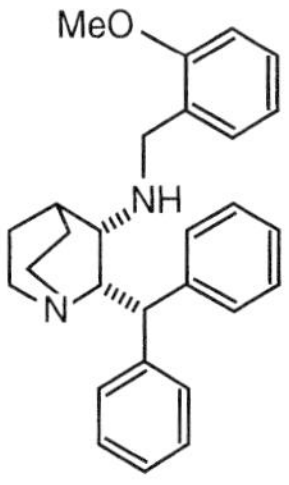 (CP 96345)	14 ± 6	5.4 ± 0.9	330 ± 60	>>10 000	>>10 000

(CP 96345)

Fig. 24 Substance P and the nonpeptide antagonist CP 96 345 interactions at chimeric NK-1/NK-3 receptors. (Data from Getler, Nature 362 : 345, 1994.)

and the impact of biotechnology on drug discovery has given further emphasis to peptides, both as drugs and as lead structures. However, the disadvantages of short biological half-life, poor bioavailability, and inability to penetrate the central nervous system generally outweigh the potential advantages of potency and selectivity. The search for peptidomimetics – small synthetic molecules that mimic peptide activity – is an increasingly important component of medicinal chemistry. The opiate alkaloids, including morphine, are examples of such peptidomimetics: they

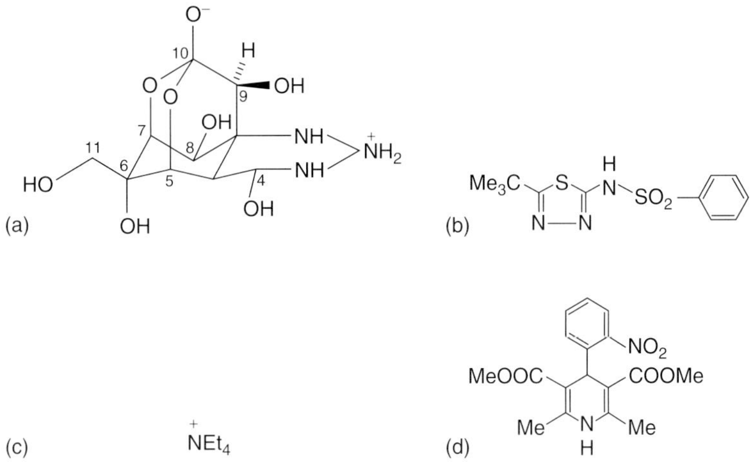

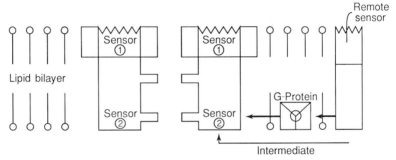

Fig. 25 Drugs that interact with ion channels (a) tetrodotoxin at Na^+ channels, (b) glybuzole at K_{ATP}^+ channels, (c) tetraethylammonium at K^+ channels and (d) nifedipine at Ca^{2+} channels.

Fig. 26 Schematic representation of an ion channel depicting the pore with "gates" that open and close in response to chemical or physical stimuli, intra- and extracellular sensors associated with the subunits of the ion channel (which is invariably an oligomeric construct) and a remote sensor that is coupled to the channel via cytosolic or membrane signals.

illustrate the principles of peptidomimetic design and demonstrate that peptides and peptidomimetics are frequently structurally quite distinct (Fig. 27). Indeed, the activities of morphine and related compounds were well described long before the endogenous opiod peptides had been discovered!

Peptidomimetics may be considered in three principal categories. *Reaction pathway mimetics* are chemical species that mimic an intermediate, typically a transition state, in an enzyme-catalyzed pathway. *Peptide topographic intermediates* are chemical entities that mimic the topography of non-adjacent peptide regions, and *peptide backbone mimetics* mimic local peptide topography (amide bonds, β-turns etc).

The naturally occurring pepsin inhibitor, pepstatin, contains the amino acid statine.

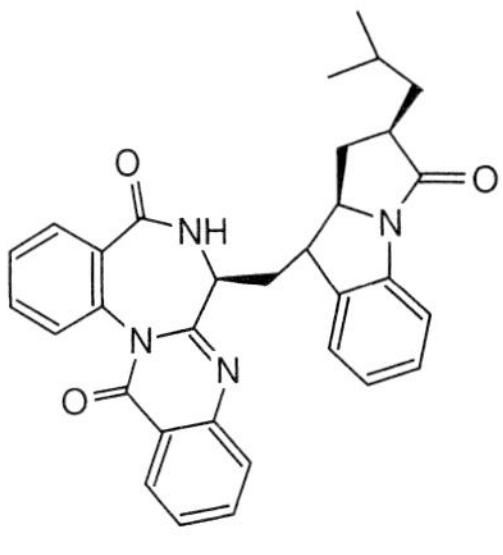

Fig. 27 Examples of nonpeptides that interact at peptide receptors: asperlicin at the cholecystokinin receptor; an oxytocin antagonist interacting at the oxytocin receptor; morphine acting at the opiate receptor; losartan interacting at the angiotensin II receptor.

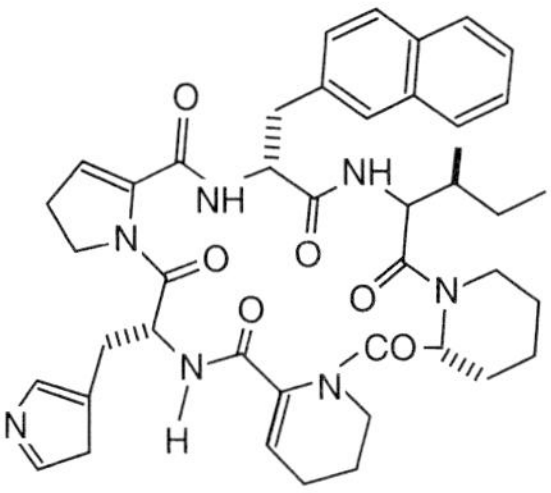

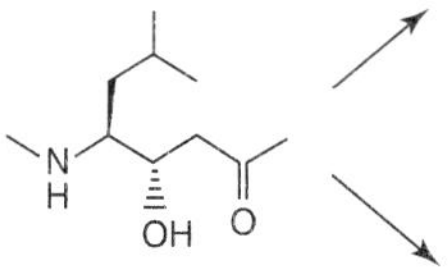

Asperlicin

Oxytocin antagonist

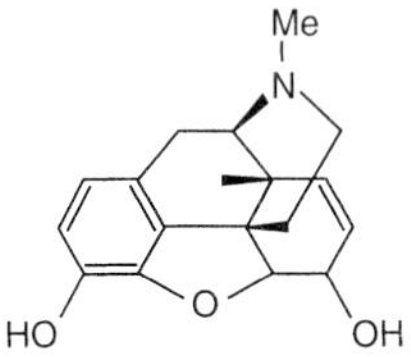

Morphine

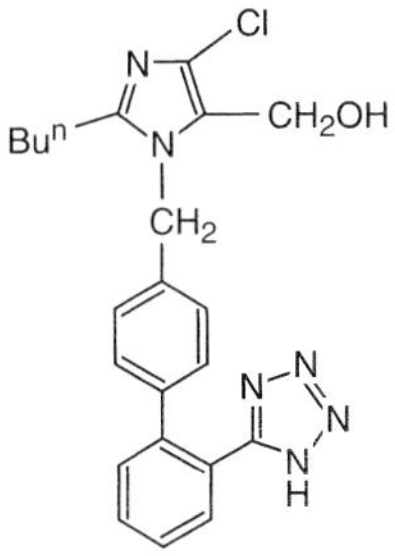

Losartan

Statine

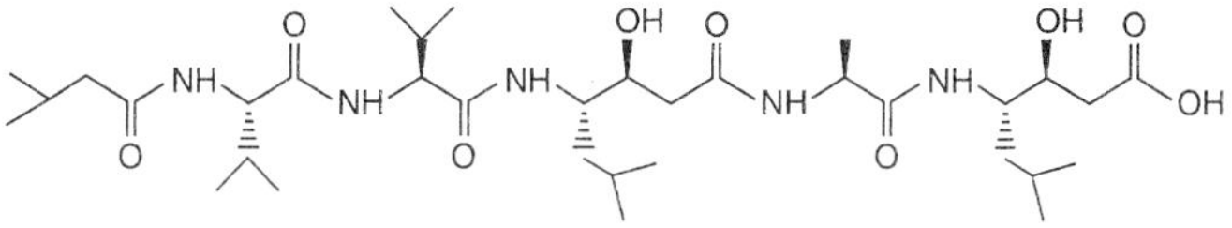

Pepstatin (Pepsin inhibitor)

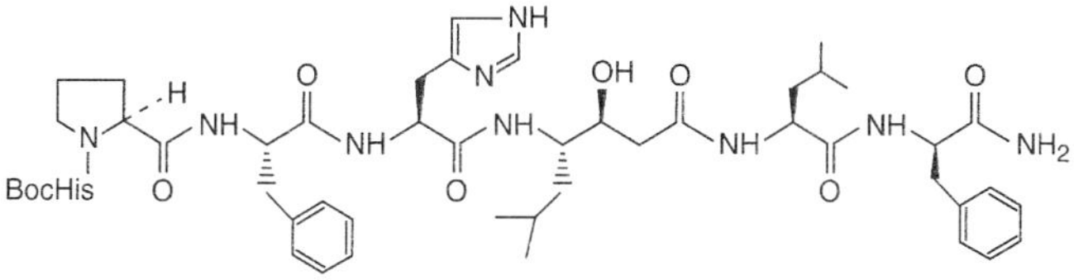

BocHis

L-363,564 (Renin inhibitor)

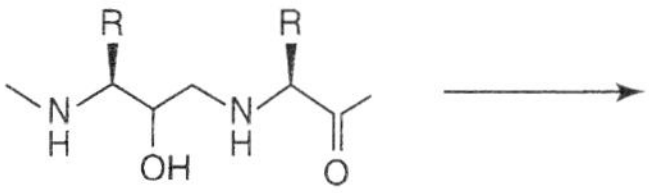

Hydroxyethylamine

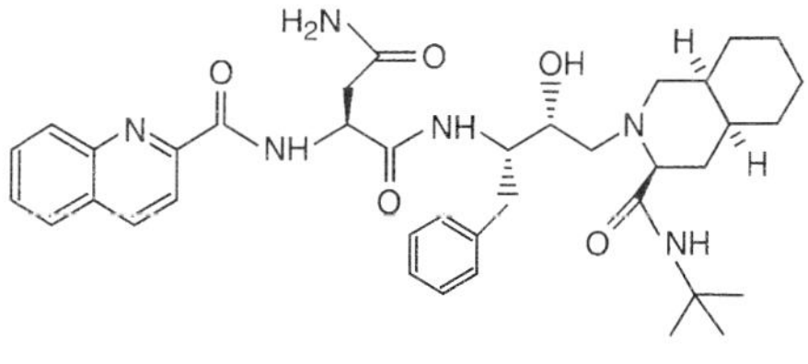

Ro-31-8959 (HIV-1 Protease inhibitor)

Fig. 28 The statine and hydroxyethylamine structures as peptide mimetics incorporated into the enzyme inhibitors pepstatin, L-363,564, and Ro-31-8959.

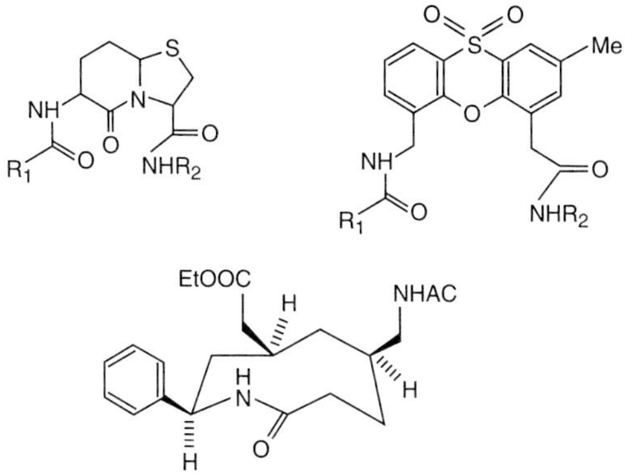

Fig. 29 Examples of peptidomimetic scaffolds.

Statine mimics a dipeptidyl tetrahedral hydrolytic transition state: this mimicking of a transition state likely underlies the ability of pepstatin to inhibit a variety of aspartic proteases, including renin and human immunodeficiency virus (HIV-1). The incorporation of the statin residue into the renin sequence was an important first step in the generation of potential renin inhibitors (Fig. 28). Other replacements serving a similar role in peptidomimetic design include the hydroxyethylene and dihydroxyethylene functionalities, which have generated potent renin inhibitors, and the hydroxyethylamine dipeptide mimetics that are used for generating potent HIV-1 protease inhibitors.

Peptide topographic intermediates are represented by an increasingly large group of compounds, frequently of natural product origin that compete with or mimic peptide actions. Prominent examples include asperlicin, which interacts at cholecystokinin receptors, imidazoles that interact at angiotensin II receptors, and, of course, morphine and its analogs that interact at opiate receptors (Fig. 27).

As part of the process of drug discovery, considerable effort has been devoted to the discovery of compounds that may mimic local protein conformations including β-turns and α-helices. The availability of such structures facilitates the construction of peptidomimetic scaffolds that can be decorated with the appropriate functional groups to generate receptor and enzyme selectivity. Some examples are shown in Fig. 29. The presence of the benzodiazepine nucleus in a variety of receptor-selective ligands is a likely example of such a peptide backbone–mimetic process (Fig. 12).

3.5
Protein–Protein Interactions

The traditional targets of drug design and discovery have been receptors and enzymes, typically viewed as acting independently and communicating through soluble biochemical intermediates, and in which small molecules are the effectors of the function. Increasingly, it is being realized that protein–protein interactions serve as critical mediators of cellular responses. Many receptors, enzymes, and ion channels function as modules composed of communicating subunits, and cells possess anchoring structures that serve to localize and spatially coordinate these protein–protein signaling complexes. Additionally, many receptors such

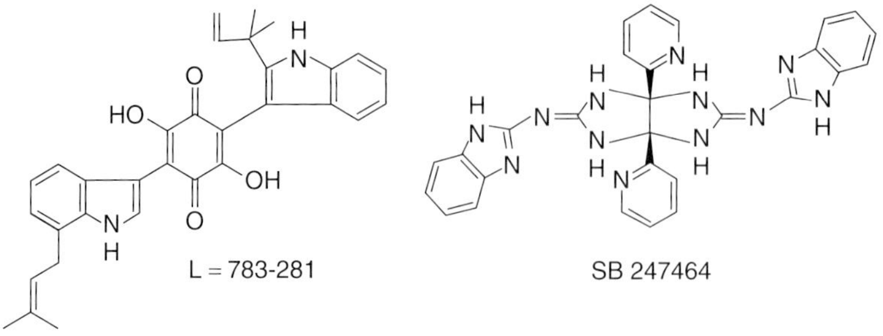

L = 783-281

SB 247464

YXCXXGPXTWXCXP

Fig. 30 Small-molecule mimetics of: insulin (L-783-381); erythropoietin (SB 24 7464) and granulocyte-stimulating hormone (cyclic peptide).

as tyrosine kinases that exist as monomers form oligomers under the influence of the activating hormone. The analysis of protein–protein interactions has been greatly facilitated by such technologies as the yeast two-hybrid system. And an increasing number of protein–protein interactions will be identified through the genome project. Hence, protein–protein interactions have become important drug targets. However, the relatively large size of interacting domains in such interactions has raised questions about whether small molecules can effectively interfere with such broad interactions, and whether such molecules would also possess sufficient specificity of action. However, an increasing number of such small molecules are known. For example, Aggrestat and Integrilin mimic the adhesive RGD epitope of fibrinogen interaction with GPIIb/IIIa, and are employed to reduce restenosis in coronary angioplasty. It can be anticipated that advances in our understanding of protein folding and construction will facilitate the design of small-molecule modulators of such interactions.

Of particular interest are studies that have demonstrated that protein hormones can be effectively reduced to small-molecule surrogates that effectively

mimic the activating properties of the parent hormones. These include small-molecule mimics of insulin, erythropoietin, and granulocyte-stimulating hormone (Fig. 30).

An increasing number of diseases are now known to be associated with protein misfolding, including such degenerative disorders as Alzheimer's and the spongiform encephalopathies. Similarly, in cancer, the p53 gene is frequently mutated giving rise to mutant p53 protein. Small molecules that can restore the wild-type protein conformation may therefore serve as useful therapeutic agents.

3.6
Quantitative Structure–Activity Relationships

The application of quantitative structure–activity relationships rests on the accepted thesis that a mutual structural complementarity exists between the ligand and its binding site and that the biological properties of the ligand are a function of its discrete physicochemical properties, including pKa, electronic configuration, lipophilicity, polar surface area, molar volume, and stereochemistry. An early application of this principle is the

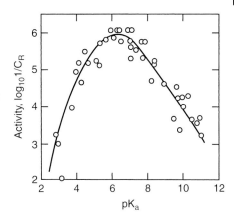

Fig. 31 The relationship between antibacterial activity of sulfonamides (log1/CR) and the pKa of the sulfonamide NH group. (Reproduced with permission from the American Chemical Society from Bell, P.H. and Roblin, R.O., (1942) *J. Amer. Chem. Soc.* **64**, 2905.)

inverse linear relationship observed by Overton and Meyer between water solubility and narcosis. A similarly simple relationship between structure and function was observed in the antibacterial sulfonamides, in which the bell-shaped curve of Fig. 31 indicates that the sulfonamide is active as the anionic species, but permeates the bacterial cell as the neutral molecule.

The definition of QSARs owes much to the pioneering work of Corwin Hansch who, building on the linear free energy relationship expressed in the Hammett equation of physical–organic chemistry, was able to correlate biological activity with hydrophobic, steric, and electronic substituent constants, through multiple linear regression analyses. Thus, for the inhibition of the enzyme acetylcholinesterase by a series of diethyl (substituted phenyl)phosphates:

$$\log 1/C_{50} = 5.77\sigma + 2.71 \quad (3)$$

indicating that the activity in this homologous series of compounds is determined solely by the electronic properties of the substituent in the phenyl ring. More commonly, multiparameter equations are used, giving, for example,

the following relationship for the local anesthetic activity of a series of 2-dialkylaminoethylbenzoates:

$$\log 1/C_{50} = 0.58\pi - 1.26\sigma + 0.96 \quad (4)$$

A more complex relationship is shown in Fig. 32 for a series of substituted 1,4-dihydropyridine calcium channel blockers, selected members of which are widely employed as cardiovascular agents. Activity is dependent upon the hydrophobic, electronic, and steric properties of the aryl substituents.

The QSAR approach in several variations has become a routine component of the practice of medicinal chemistry. There are, however, weaknesses associated with the method: substituent constants must be available and a large and varied experimental data set is necessary to have a statistically powerful analysis. Furthermore, the method defines only the properties of the ligand and usually within a restricted structural set. Thus, the method tends to be retrospective and interpolative rather than predictive and extrapolatable. Nonetheless, its usefulness should not be underestimated, particularly in conjunction with other components of rational drug design.

Port

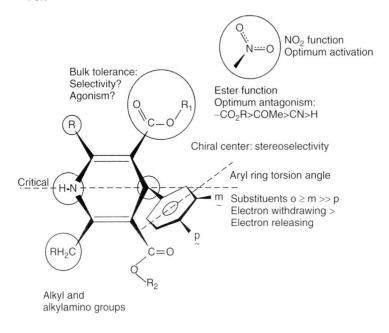

NO_2 function
Optimum activation

Bulk tolerance:
Selectivity?
Agonism?

Ester function
Optimum antagonism:
$-CO_2R > COMe > CN > H$

Chiral center: stereoselectivity

Aryl ring torsion angle

Critical

m Substituents $o \geq m \gg p$
Electron withdrawing >
Electron releasing

p

Alkyl and
alkylamino groups

Starboard

$$\log 1/C = 0.62\pi + 1.96\sigma_{meta} - 0.44L_{meta} - 3.26B_{1para} - 1.51L_{meta} + 14$$
$$n = 46, r = 0.90$$

Fig. 32 The structure–activity relationship for aryl-1,4-dihydropyridines active at the L-type voltage-gated Ca^{2+} channel. The structural formula depicts the general structural features that confer antagonist or activator properties. The equation describes the antagonist activity of a series of 1,4-dihydropyridines with the general formula -2,6-disubstituted-3,5-dicarboalkoxy-4-[substituted phenyl]-1,4-dihydropyridine.

3.7
Stereochemistry

That drug–receptor interactions exhibit stereoselectivity has been well recognized since the time of Pasteur. Indeed, so common is such stereoselectivity that it is generally considered to be an important index of the specificity of such interactions. This stereoselectivity of interaction represents a challenge to our definition of drug actions long recognized chemically, but now increasingly encountered at the clinical and regulatory levels. Clinical significance derives from considerations of the efficacy of a single enantiomer versus a racemate, from considerations of stereoselective metabolism and disposition, and from the impact of the route of administration and interpatient variability of response. Regulatory issues derive from considerations that racemic drugs can represent separate agents in fixed combinations, and development issues derive from considerations of costs, including those for chemical synthesis, and of pursuing a single enantiomer versus a racemate. (We consider here only optically active isomerism: the same principles apply, however, to geometrical and other isomers. Also, for the sake of brevity, we consider only the case of molecules with a single chiral (stereogenic) center).

Enantiomers can differ both quantitatively and qualitatively in their biological activities. At one extreme, one enantiomer may be totally devoid of activity and at the other extreme both enantiomers may have potent, but distinct, activities. Thus, the following categories of activity are possible:

1. Both the isomers are equally active and there is no stereoselectivity of action. This situation is not common.
2. The isomers differ quantitatively in their activity. Most drugs that occur as stereoisomers fall into this category (in the limiting situation one enantiomer possesses all of the activity).
3. The isomers differ qualitatively in their activity and have discrete biological activity at the same or distinct receptors. Such behavior may translate to agonist and antagonist behavior at the same receptor, subtype selectivity, or activity at distinct receptor classes.
4. The behavior of the isomers in the racemate is different from that of the enantiomers alone.

Examples of racemates in which the enantiomers differ quantitatively in their activity are quite common. They include the positive inotrope dobutamine, the anti-inflammatory ibuprofen, the anticoagulant warfarin, and the local anesthetic bupivacaine. The differences in activity may be large enough to regard one enantiomer as effectively inactive: examples include the β-blocker propranolol in which the S-enantiomer is active and the cholinergic agonist bethanchol in

which the S-enantiomer is also the active entity. Enantiomers that differ qualitatively in their activities include: nicotine where the R- and the S-enantiomers are substrates and inhibitors respectively of the enzyme N-methyltransferase; the 1-methyl-5-phenyl-5-propylbarbituric acid enantiomers in which the S- and R-forms are convulsant and anticonvulsant respectively; propoxyphene in which the enantiomers are analgesic and antitussive; sotalol in which the enantiomers are a β-adrenergic blocker and a K^+ channel blocker; and the enantiomers of Bay K 8644 which are activator and antagonist. Some examples are shown in Fig. 33.

In principle, stereoselectivity of drug action is derived from *both* pharmacodynamic and pharmacokinetic processes and can occur at any or all of the stages depicted in Fig. 34: transport to and from the site of action, storage in depots, interaction with binding proteins, interaction with receptors, metabolic enzymes, and transport in excretory pathways. The Ca^{2+} channel antagonist verapamil is available only as the racemate and is employed as an antianginal, anti-hypertensive and as a selective Class IV antiarrhythmic. The more active S-enantiomer is cleared more rapidly through first-pass metabolism. Accordingly, verapamil offers a different clinical profile according to whether it is administered orally or intravenously, being more active in the latter case. Even more subtle distinctions can be made in a comparison of the enantiomers of verapamil on cardiac and vascular smooth muscle: stereoselectivity is very low in the latter and significantly higher in the former (Table 2).

Important differences can exist between the toxicologic or side-effect profiles of enantiomers. For L-DOPA, the levo (available) form is less toxic, vomiting

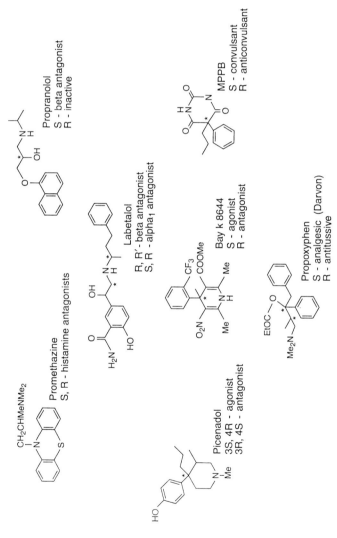

Fig. 33 The stereoselectivity of drug action depicting the quantitative and qualitative differences in biological activity that can exist between enantiomers.

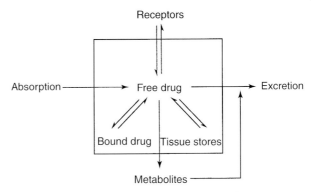

Fig. 34 The observed pharmacological activity of a drug depends upon both pharmacodynamic and pharmacokinetic properties as depicted. Each of the processes depicted has its own structural and stereochemical requirements. These requirements are not necessarily identical at each site and, in fact, may well conflict.

Tab. 2 Stereoselectivity of Verapamil in the cardiovascular system.

	Vasodilation	*Cardiodepression*
	−log EC50 (M)	−log EC50 (M)
Enantiomer		
R(+)−Verapamil	7.31	5.7
S(−)−Verapamil	7.70	7.2

is associated with the D-isomer of levamisole, and the teratogenic effects of thalidomide may be associated with the *S*-enantiomer (however, this has no clinical significance since the enantiomers epimerize very rapidly in solution). Stereoselectivity of general anesthesia has been demonstrated for the volatile agents isoflurane and halothane; although modest, this difference may be of clinical significance and expand the safety margin of these agents.

These considerations are part of an increasing scientific, clinical, and regulatory pressure to evaluate the chirality of drug action at all stages of the drug development process. The use of single enantiomer drugs continues to increase as do conversions of racemates to single enantiomers. Even those in which a racemate may be introduced clinically, this will be done only after the properties of the enantiomers have been thoroughly investigated, and in which it can be demonstrated clinically that there is no disadvantage in the use of a racemate rather than a single enantiomer.

4
The Influence of Molecular Biology

The principles and tools of molecular biology exert an increasing influence on the theory, practice, and application of medicinal chemistry. Molecular biology provides not only the structures of the receptors and targets for drug action, but validates these targets, defines disease states, provides the logic of chemical design, generates the rapid screening protocols, and directly provides drugs and drug candidates. These principles have been elaborated upon previously, but some brief additional comments follow. More comprehensive discussion will be found elsewhere in this Encyclopedia.

4.1
The Expanded Genetic Code

Our existing genetic code contains 4 nucleotides that specify 64 unique nucleotide triplets that in turn encode the 20 amino acids that are the building blocks for all the proteins in at least this corner of the universe, apart from the rare organism that incorporates selenocysteine and pyrrolysine. The ability to incorporate novel amino acids into proteins has the potential to answer important questions concerning the evolution of life, produce new proteins with novel biological functions and activities, *and* to produce organisms that are biologically unique. Amino acid analogs have been incorporated into proteins by a variety of techniques including posttranslational modification and the use of chemically aminoacylated t-RNA that can be recognized by existing stop codons. However, the generation of autonomous bacteria with an expanded genetic code that can incorporate nonnatural amino acids with efficiency and fidelity represents a major step in harnessing the genetic code for the *in vivo* design and production of novel potential protein therapeutics.

4.2
Genes as Drugs

In principle, gene insertion to replace a defective gene or genes or to enhance expression of a particular gene product provides a permanent curative treatment. The critical issues here for effective gene delivery are the nature of the disease itself, the delivery system and its specificity, the extent and duration of the resultant gene

expression, and the safety of the process. Additionally, there are significant ethical issues surrounding gene therapy that are absent with conventional drugs. A number of the potential targets for gene therapy are given in Table 3: clearly, the most plausible targets are monogenic diseases where the disease is due to the presence or misexpression of a single gene.

Gene therapy may be carried out *in vivo* or *in vitro* and the insertion methods are of several kinds. They include adenoviruses and retroviruses that are used as insertional carriers, liposomes to transfer plasmid DNA, and other physical methods including conjugation of the DNA with a protein that targets a specific receptor through which it is internalized.

Although there are in excess of some six hundred protocols, the majority of them being in the field of cancer, clinical success has been limited thus far. In particular,

Tab. 3 Targets and potential targets for gene therapy.

Monogenic diseases
X-Linked severe combined immunodeficiency
ADA deficiency
Cystic fibrosis
Mucopoylaccharidosis
Factor IX deficiency (Hemophilia B)
Familial hypercholesterolemia
Chronic granulomatous disease

Gene types transferred in clinical trials
Cytokines 26%
Antigen 14%
Tumor suppressor 12%
Suicide 8%
Deficiency 8%
Drug resistance 6%
Receptor 4%
Replication inhibitor 3%

Source: Data from Journal of Gene Medicine Database, 2004:
www.wiley.co.uk/genmed/clinical/

the issue of the specificity of the gene insertion process has assumed particular importance, with observations that individuals being treated for X-linked severe combined immunodeficiency syndrome ("bubble boy" disease) have developed leukemia as a consequence of random insertion of the therapeutic gene near the LM02 gene, a gene linked to leukemia.

4.3
Antisense and Small Interfering RNA

Oligonucleotides bind with great specificity to a variety of nucleic acid and protein targets. Additionally, nucleic acids are particularly well suited for combinatorial chemistry *and* selection processes: oligonucleotides selected from such a process and referred to as *aptamers* can serve both as targeting species (research reagents) to identify small-molecule ligands, and as actual drug species. For example, by using an iterative cycle of RNA library generation, target incubation, removal of non-binding RNA, recovery of bound RNA, and amplification of binding molecules by RT-PCR (Fig. 35), these techniques have led to the identification of small-molecule inhibitors of HIV-1 replication and platelet-derived growth factor binding.

In principle, oligonucleotides with specificity conferred by Watson–Crick pairing will constitute highly specific drugs for defined nucleic acid RNA targets. Thus, an oligonucleotide of some 15–19 bases will have a unique sequence relative to the entire human genome and could, theoretically, offer absolute specificity of interaction. Such oligonucleotides will base-pair to a specific segment of RNA to block translation, block splicing, or facilitate degradation through RNAseH. This is not necessarily realized in practice since the oligonucleotide does not have equal access to all sites on the target RNA. Additionally, oligonucleotides also interact with proteins in a non-antisense manner. Many derivatives of phosphodiester oligonucleotides have been examined, particularly the backbone-modified methylphosphonate, phosphorothioate, and phosphorodithioate analogs (Fig. 36). These modifications serve to enhance chemical stability, reduce non-specific interactions, and enhance cell permeation. However, only one antisense therapeutic, Vitavene$^{(TM)}$, is currently clinically available.

RNA interference is a gene-silencing mechanism originally discovered in lower organisms, but now shown to be widely applicable and to have considerable potential

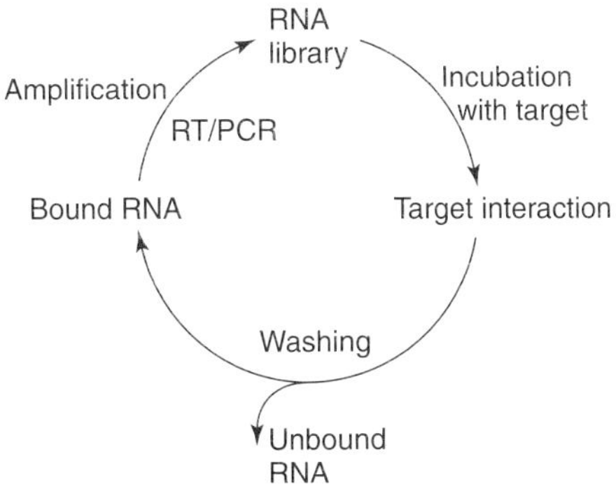

Fig. 35 An iterative cycle whereby RNA molecules that bind to a selected target are identified, amplified and subject to a further round of binding and selection to amplify the desired property.

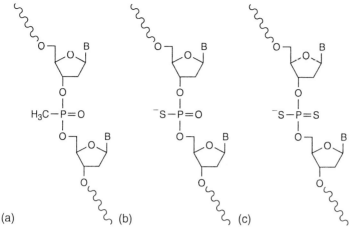

Fig. 36 Backbone-modified oligonucleotides (a) methyphosphonates, (b) phosphorothioates, and (c) phosphorodithioates.

for both target validation and for the generation of actual drugs. In lower organisms, the introduction of double-stranded RNA is followed by enzymatic processing to yield short, 21–22 double-stranded (ds) sRNAs, that guide sequence-specific degradation of the homologous mRNAs. The process is designed to recognize the introduction of RNA from foreign genes during, for example, virus infection, or when transposons and transgenes are randomly inserted. RNA interference (RNAi) can be produced in vertebrates by the introduction of short dsRNA, and serves as a mechanism for specifically silencing gene function. RNAi offers major promise for target validation, being faster than the production of animal knockouts, and the ultimate promise for actual drug delivery. It offers, in principle, significant advantages over antisense technology, since much lower concentrations of the dsRNA are needed, the effects are longer lasting and, of particular significance, siRNA can be produced intracellularly from RNA polymerase III promoters.

Despite the attractiveness of these nucleic acid–based technologies, it cannot be assumed that they are automatically superior to the existing small-molecule techniques. In particular, in protein targets that are part of an oligomeric complex, elimination of one protein could lead to the destruction of that complex and a cascade of effects that would not be produced by the action of a single small-molecule alone acting on the complex.

4.4
Gene-based Drug Targeting and Delivery

Gene-based mechanisms can also be used to confer or to enhance drug sensitivity in an unresponsive cell type. The transfer to cells, for example, of the herpes simplex thymidine kinase(HSV-tk) confers sensitivity to such nucleoside analogs as the antiherpes drug, ganciclovir. Such transfer can be mediated by a retrovirus that incorporates only into actively dividing cells and thus offers particular selectivity for malignant cells. HSV-tk converts the nontoxic ganciclovir into the phosphorylated derivative, which is a chain terminator of DNA synthesis.

5
Summary

Medicinal chemistry, a discipline firmly rooted in synthetic organic chemistry, has increasingly close relationships to structural and computational chemistry, particularly to molecular biology at the discovery interface, to toxicology and pharmacology at the development interface, and to medicine at the clinical interface. The linkage to molecular biology has assumed major significance in the past decade, and chemistry has increasingly been adopting characteristics that are dictated by the basic themes that drive biology – *diversity, replication, evolution,* and *self-organization*. It is entirely conceivable that, as this biological paradigm increasingly permeates science and, in particular, medicinal chemistry, we shall ultimately have Paul Ehrlich's "magic bullet" – self-synthesizing, self-replicating, self-evolving, and self-targeting nanomachines that circulate in our body, constantly repairing damage and destroying biological intruders.

See also Antitumor Agents: Taxol and Taxanes – Production by Yew Cell Culture; Gene Targeting; Oncology, Molecular; Pharmacogenomics and Drug Design.

Bibliography

Books and Reviews

Anderson, A. (2003) The process of structure-based drug design, *Chem. Biol.* **10**, 787–797.

Drug Discovery Today. (2002) *20th Anniversary of Combinatorial Chemistry*. Parts (1, 2). 7: Issues 1 and 2.

Gohlke, H., Klebe, G. (2002) Approaches to the description and prediction of the binding affinity of small-molecule ligands to macromolecular receptors, *Angew. Chem., Ed. Int.* **41**, 2644–2676.

Gordon, E.M., Kerwin, J.F. (Eds.) (1998) *Combinatorial Chemistry and Molecular Diversity in Drug Discovery*, Wiley-Liss, New York.

Houk, K.N., Leach, A.G., Kim, S.P., Zhang, Y. (2003) Binding affinities of host-guest, protein-ligand, and protein-transition-state complexes, *Angew. Chem., Int. Ed.* **42**, 4872–4897.

Lough, W.J., Wainer, I.W. (Eds.) (2003) *Chirality in Natural and Applied Science*, Blackwell Publishing, Oxford, UK.

Rothstein, M.A. (Ed.) (2003) *Pharmacogenomics. Social, Ethical and Clinical Dimensions*, Wiley-Liss, New York.

Venter, J.C., Adams, M.D., Myers, E.W., Li, P.W., Mural, R.J., Sutton, G.G.,. Smith, H.O., Yandell, M., Evans, C.A., Holt, R.A., et al. (2001) The sequence of the human genome, *Science* **291**, 1304–1351.

Wess, G., Urmann, M., Sickenburger, B. (2001) Medicinal chemistry: challenges and opportunities, *Angew. Chem., Int. Ed.* **40**, 3341–3350.

Wiesner, J., Ortmann, R., Jomaa, H., Schlitzer, M. (2003) New antimalarial drugs, *Angew. Chem., Int. Ed.* **42**, 5274–5293.

Primary Literature

Breinbauer, R., Vetter, I.R., Waldmann, H. (2002) From protein domains to drug candidates – natural products as guiding principles in the design and synthesis of compound libraries, *Angew. Chem., Int. Ed.* **41**, 2878–2890.

Carroll, F.I. (2003) Monoamine transporters and opiod receptors. Targets for addiction therapy, *J. Med. Chem.* **46**, 1775–1794.

Chanda, S.K., Caldwell, J.S. (2003) Fulfilling the promise: drug discovery in the post-genomic era, *Drug Discovery Today* **8**, 168–174.

Crooke, S.T. (2000) Potential roles of antisense technology in cancer chemotherapy, *Oncogene* **19**, 6651–6659.

Croston, G.E. (2002) Functional cell-based uHTS in chemical genomic drug discovery, *Trends Biotechnol.* **20**, 110–115.

Daly, J.W. (2003) Ernest Guenther award in chemistry of natural products. Amphibian skin: a remarkable source of biologically active arthropod alkaloids, *J. Med. Chem.* **46**, 45–452.

Famulok, M., Mayer, G., Blind, M.. (2000) Nucleic acid aptamers – from selection *in vitro* to applications *in vivo*, *Acc. Chem. Res.* **33**, 591–599.

Flower, D.R. (1999) Modeling G protein-coupled receptors and signal transduction for drug design, *Biochim. Biophys. Acta* **1422**, 207–234.

Fredrickson, R., Lagerstrom, M.C., Lundin, L.-G., Schioth, H.B. (2003) The G-protein-coupled receptors in the human genome form five main families. Phylogenetic analysis, paralogon groups, and fingerprints, *Mol. Pharmacol.* **63**, 1256–1272.

Freidinger R.M. (2003) Design and synthesis of novel bioactive peptides and peptidomimetics, *J. Med. Chem.* **46**, 5553–5566.

Gadek, T.R., Nicholas, J.B. (2003) Small molecule antagonists of proteins, *Biochem. Pharmacol.* **65**, 1–8.

Giannis, A., Rubsma, F. (1997) Peptidomimetics in drug design, *Adv. Drug Res.* **29**, 1–78.

Gohlke, H., Klebe, G. (2002) Approaches to the description and prediction of the binding affinity of small-molecule ligands to macromolecular receptors. *Angew. Chem., Int. Ed.* **41**, 2644–2676.

Jiang, Y., Lee, A., Chen, J., Ruta, V., Cadene, M., Chait, B.T., MacKinnon, R. (2003) X-ray structure of a voltage-dependent K^+ channel, *Nature* **423**, 33–41.

John, V., Beck, J.P., Bienkowski, M.J., Sinha, S., Heinrikson, R.L. (2003) Human β-secretase (BACE) and BACE inhibitors, *J. Med. Chem.* **46**, 4625–4630.

Kay, B.K., Kurakin, A.V., Hyde-DeRuyscher, R. (1998) From peptides to drugs via phage display, *Drug Discovery Today* **3**, 370–378.

Kolb, H.C., Finn, M.G., Sharpless, K.B. (2001) Click chemistry: diverse chemical function from a few good reactions, *Angew. Chem., Int. Ed.* **40**, 2004–2021.

Lane, D.P., Hupp, T.R. (2003) Drug discovery and p53, *Drug Discovery Today* **8**, 347–355.

Lipinski, C.A., Lombardo, F., Dominy, B.W., Feeney, P.J. (2001) Experimental and computational approaches to estimate solubility and permeability in drug discovery and development, *Adv. Drug Delivery Rev.* **46**, 3–26.

Mehl, R., Anderson, C., Santoro, S.W., Wang, L., Martin, A.B., King, D.S., Horn, D.M., Schultz, P.G. (2003) Generation of a bacterium with a 21 amino acid genetic code, *J. Am. Chem. Soc.* **125**, 935–939.

Natesh, R., Schwager, L.U., Sturrock, E.D., Acharya, K.R. (2003) Crystal structure of the human angiotensin-converting enzyme-lisinopril complex, *Nature* **421**, 551–554.

Olivera, B.M., Hillyard, D.R., marsh, M., Yoshikami, D. (1995) Combinatorial peptide libraries in drug design: lessons from venomous cone snails, *Trends Biotechnol.* **13**, 422–426.

Patchett A.A. (2002) Natural products and design: interrelated approaches in drug discovery, *J. Med. Chem.* **45**, 5609–5616.

Patchett, A.A., Nargund, R.P. (2000) Privileged structures – an update, *Annu. Rep. Med. Chem.* **35**, 269–298.

Pearlstein, R., Vaz, R., Rampe, D. (2003) Understanding the structure-activity relationship of the human ether-a-go-go-related gene cardiac K^+ channel. A model for bad behavior, *J. Med. Chem.* **46**, 2017–2023.

Ruoslahti, E. (2000) Special delivery of drugs by targeting to tissue-specific receptors in the vasculature, *Pharm. News* **7**(4) 35–40.

Tan, D.S., Foley, M.A., Stockwell, B.R., Shair, M.D., Schreiber, S.L. (1999) Synthesis and preliminary evaluation of a library of polycyclic small molecules for use in chemical genetic assays, *J. Am. Chem. Soc.* **121**, 9073–9087.

Thompson, J.D. (2002) Applications of antisense and siRNAs during preclinical drug development, *Drug Discovery Today* **7**, 912–917.

Triggle, D.J. (2003) Medicines in the 21st century. Pills, potions, politics and profits. Where is public policy? *Drug Dev. Res.* **59**, 269–291.

Triggle, D.J. (2002) Ion channels: structure, function and pharmacology, in: Krogsgaard-Larsen, P., Liljefors, T., Madsen, U. (Eds.) *Textbook of Drug Design and Discovery*, Francis & Taylor, London, pp 173–204.

Tuschl, T. (2001) RNA interference and small interfering RNAs, *ChemBioChem* **2**, 239–245.

Walsh, C.T. (2002) Combinatorial biosynthesis of antibiotics: challenges and opportunities, *ChemBioChem* **3**, 124–134.

Weng, Z., DeLisi, C. (2002) Protein therapeutics: promises and challenges for the 21st century, *Drug Discovery Today* **20**, 29–35.

Wiley, R.A., Rich, D.H. (1993) Peptidomimetics derived from natural products, *Med. Res. Rev.* **13**, 327–384.

Zaman, G.J.R., Michiels, P.J.A., van Boeckel, C.A.A. (2003) Targeting RNA: new opportunities to address drugless targets, *Drug Discovery Today* **8**, 297–306.

Membrane Traffic: Vesicle Budding and Fusion

Michael J. Clague and Sylvie Urbé
University of Liverpool, Liverpool, UK

Encyclopedia of Molecular Cell Biology and Molecular Medicine, 2nd Edition. Volume 8
Edited by Robert A. Meyers.
Copyright © 2005 Wiley-VCH Verlag GmbH & Co. KGaA, Weinheim
ISBN: 3-527-30550-5

Keywords

Clathrin
A large protein complex with a triskelion structure built from three heavy and three light chains, capable of self-associating to form lattices, which coat various vesicles.

Coated Vesicle
60- to 100-nm diameter membrane vesicle that buds from a subcellular compartment, encased in a regular protein coat, which selects vesicle cargo and drives budding. Examples include clathrin-coated and COP-coated vesicles; COP refers to coat protein complex.

Endocytosis
Process by which material is taken up into the cell from the external environment, through incorporation into vesicles, which bud from the plasma membrane.

Membrane Fusion
Merger of lipid bilayer membranes, which is accompanied by the establishment of continuity between the aqueous interiors that are bounded by the originally distinct membranes.

Rab Protein
A member of a large family of small GTPases, which regulate vesicle trafficking, notably through the recruitment of tethering molecules.

Secretory Pathway
A series of compartments that are traversed by material, destined to be, secreted from the cell and which are connected by vesicular transport. Principally comprising the endoplasmic reticulum (ER), ER-Golgi intermediate compartment (ERGIC) Golgi apparatus, and trans-Golgi network (TGN).

SNARE Protein
Member of a conserved family of proteins believed to impart specificity to and catalyze intracellular vesicle fusion.

Tight control of the flux of membrane components through vesicle budding and fusion is essential to cellular organization, giving rise to subcellular compartments with distinct functions. Cargo is generally selected for inclusion into vesicles by specific sorting signals, which are recognized by adaptor proteins, that link to protein components of vesicle coats. The membrane deformation required for vesicle formation is promoted by the coat complex in conjunction with accessory proteins, which may partially insert into the bilayer membrane. Specific targeting of vesicles is initially mediated through "tethering" molecules and then cemented through interactions between SNARE proteins, which also catalyze fusion. The essential mechanisms are conserved throughout the cell and from yeast to man.

1
Overview of Cellular Trafficking Pathways

1.1
Maintaining Cellular Compartmentalization

The advent of electron microscopy revealed the complexity of subcellular organization, allowing a first glimpse of a bewildering array of membrane-bounded compartments (Fig. 1). This raised deep questions that are still pertinent to modern day cell biology. How are distinct organelles generated and maintained in the face of a constant flux of material? To what degree is this dependent on spatial organization versus biochemical specification? How should compartments be defined? by morphology, function, or biochemical composition? Much is accomplished through vesicle budding and fusion. In order for this to be so, there must be very specific mechanisms of cargo selection and targeting. Delivery of lysosomal degradative enzymes to the cell nucleus, for example, would court disaster.

1.2
Secretory Pathway

As its name implies, this pathway is taken by molecules destined for cellular secretion. Following translocation of nascent proteins across the ER membrane, proteins must traverse a series of compartments (Fig. 1), including the Golgi stack, where successive posttranslational processing events, such as glycosylation, may take place. The striking stacked cisternae of the Golgi represent a polarized structure, such that the cis-most and trans-most cisternae are highly enriched in distinct enzymes. The number of Golgi stacks may vary dramatically between cell types, but its function is nevertheless conserved.

The trans-Golgi network (TGN) represents the principal sorting station of the secretory pathway. From this compartment, material may be constitutively secreted (e.g. extracellular matrix proteins) or packaged into regulated secretory vesicles (e.g. insulin), which normally undergo

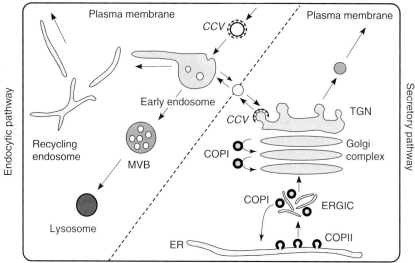

Fig. 1 Schematic diagram of cellular vesicular trafficking pathways in the eukaryotic cell. Newly synthesized proteins translocate into the endoplasmic reticulum (ER) and then progress along the secretory pathway, first by inclusion into COPII coated vesicles, which deliver material to the ER-Golgi intermediate compartment (ERGIC). Tubular elements of the ERGIC may coalesce to form a cis-Golgi cisterna. Anterograde transport toward the plasma membrane, is thought to proceed by maturation of Golgi cisternae, followed by budding of secretory vesicles from the trans-Golgi network (TGN). ER and Golgi resident enzymes are continually retrieved by inclusion into COPI vesicles, which move in a retrograde direction. Material internalized from the plasma membrane may be included in clathrin-coated vesicles (CCVs), which use the AP2 adaptor. The CCVs fuse with the early endosome from where internalized cargo can either be recycled back to the plasma membrane through the tubular recycling endosome or progress toward lysosomes by inclusion in multivesicular bodies (MVBs). Crossover between these two principal highways is mediated by vesicular transport between the TGN and endosomes. Note that other subcellular organelles, such as the nucleus and mitochondria, which do not form part of this endomembrane system are not depicted.

fusion with the plasma membrane (exocytosis) in response to elevated Ca^{2+} levels. Some proteins are diverted at this stage from the secretory pathway toward the endocytic pathway (e.g. lysosomal enzymes).

Progression along the secretory pathway is mediated by vesicular transport. This involves the selection and packaging of cargo into vesicles, which bud off from the limiting membrane and then fuse with specific target membranes. This mechanism can be used to designate forward (anterograde) transport of cargo or retrieval (retrograde transport) of specific components that function earlier in the pathway. Consider transport through the polarized Golgi stacks: according to one view, this is comprised of cisternae, each representing individual stable entities maintained by a delicate balance of retrograde and anterograde vesicular transport. However, it is currently considered more likely that individual cisternae progress through the Golgi complex, gradually undergoing maturation through the selective retrieval of "earlier" Golgi markers by incorporation into COP-coated vesicles. This "cisternal maturation" model of Golgi transport has the advantage that it can account for the secretion of large molecules, such as algal scales or collagen, which are too big to be packaged into individual COP-coated vesicles.

1.3
Endocytic Pathway

The best-understood pathway for internalization of material at the plasma membrane is the clathrin-coated vesicle (CCV) pathway, first described by Roth and Porter in mosquito oocytes. However, it has now become clear that several other non-clathrin-mediated pathways operate in parallel. Once internalized, most material is transported to a tubulovesicular compartment, referred to as the *early or sorting endosome*. From this point, endocytosed proteins can enter a pathway for recycling (e.g. transferrin receptor) or be sorted toward the lysosome (e.g. EGF receptor), the major degradative compartment of the cell (Fig. 1). Lysosomal sorting involves the inward budding of cargo containing vesicles from the limiting membrane of sorting endosomes to create multivesicular bodies (MVBs), which then fuse either with late endosomes or directly with lysosomes.

The recycling pathway from the early endosome probably involves tubular structures, which break off and direct material through a perinuclear recycling compartment back to the plasma membrane.

1.4
Principle Steps of Vesicular Transport

This article is limited to consideration of membrane trafficking mediated by small vesicles, for which common principles can be extracted. Other trafficking events, for example, generation of tubules for endosomal recycling or inward invagination of vesicles into the endosome lumen will not be considered here. The general mechanism of vesicle transport normally includes the following steps, which are depicted in Fig. 2. A coat assembles on an intracellular membrane through interactions with cargo proteins. Membrane deformation results in vesicular profiles, which must undergo scission to create coated vesicles. Upon uncoating, initial contact with target membranes is made through "tethering" molecules. Finally, the vesicle is committed to fusion through further molecular associations, which may provide the energy to overcome

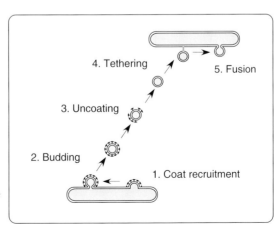

Fig. 2 Basic steps of the vesicle transport cycle (see text for details).

the barriers to membrane fusion, resulting in continuity of both lipid bilayers and lumenal contents.

2
Methodologies

2.1
Subcellular Fractionation

De Duve and colleagues pioneered the use of analytical centrifugation to separate subcellular compartments derived from homogenized tissue. By examining the concomitant separation of specific enzymatic activities, they were able to propose that biochemical pathways were more compartmentalized than previously imagined. Centrifugation techniques have been vital in the purification of various vesicle fractions, including clathrin-coated vesicles, COP-coated vesicles, and synaptic vesicles, thus allowing biochemical analysis of their composition. The introduction of mass spectroscopic techniques for proteomic analysis now allows for in depth profiling of vesicle-associated proteins. A recent study has identified 209 distinct proteins associated with highly purified brain-derived clathrin-coated vesicles.

2.2
Microscopy

The biochemical work of De Duve was brilliantly complemented by the morphological studies of Palade, Claude, Porter, and coworkers, who harnessed the power of electron microscopy to visualize the interior of the cell with a particular focus on the secretory pathway of the pancreatic acinar cell. Jamieson and Palade introduced an *in vitro* "pulse-chase" protocol to follow the secretory pathway of nascent proteins. A short incubation of cells with [^{3}H]-leucine selectively labels newly synthesized proteins, which can be visualized by autoradiography and subsequent electron microscopic examination. By varying the time of incubation in the absence of label (chase period) following the labelling "pulse," the progression of proteins through successive secretory compartments could be visualized. Palade frequently observed vesicular profiles in his micrographs and proposed that these may be intermediate vessels for transport between compartments. The introduction of immunocytochemistry and, in particular, the introduction of cryo-electron microscopy by Tokuyasu and Singer paved the way for a quantitative analysis of specific cargo proteins and structural components of vesicles. More recently, the ability to tag proteins with multiple colored fluorescent proteins has made it possible to directly observe protein dynamics through fluorescence microscopy of live cells. Thus, the flux of material in response to a given perturbation can now be followed in real time at the subcellular level.

2.3
Yeast Genetics

A genetic screen in yeast for mutants defective in secretion of specific enzymes identified 23 complementation groups that provided the first set of candidate genes (*sec* genes) for regulation of membrane traffic. The first *sec* gene to be cloned and characterized, *Sec4*, is a ras-like small GTPase belonging to the rab family that acts to target mature secretory vesicles to the plasma membrane. Further screens for mutants defective in vacuolar protein sorting (*vps* genes) or endocytosis (*end* genes) have produced equally informative inventories of genes responsible for specific regulation of membrane traffic.

2.4
Cell-free Assays of Membrane Transport

James Rothman pioneered the development of *in vitro* assays for the reconstitution of membrane trafficking steps. Rothman and coworkers first established a biochemical assay that reconstitutes a specific intracellular fusion event, namely, transport between cis and medial cisternae of the Golgi stack (Fig. 3). They combined wild-type Golgi fractions isolated from CHO cells with Golgi fractions isolated from mutant cells lacking a sugar-modifying (GlcNAc transferase) activity, but containing a model cargo protein, vesicular stomatitis virus glycoprotein (VSVG protein) by virtue of prior infection.

This protein is a substrate for GlcNAc, but can only acquire [^{3}H]-radiolabeled GlcNAc by transfer between heterologous Golgi stacks. Rothman and coworkers were able to show that this transfer requires both cytosol and ATP, and thus embarked on a dissection of the cytosolic factors required to support the assay. A major insight derived from this work was that specificity of fusion is principally determined biochemically, rather than by relying on the architecture of the cell interior.

One drawback of this original assay was that it required generation of a vesicular intermediate, and essential factors could therefore operate at the level of vesicle budding, vesicle docking, or vesicle fusion. Subsequent refinements led to the

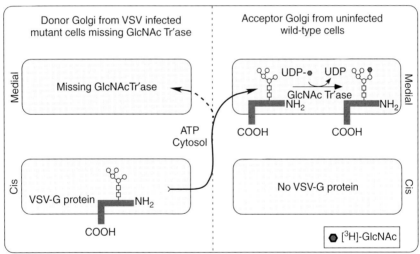

Fig. 3 Schematic diagram of the *in vitro* reconstitution of Golgi transport. The first *in vitro* assay of membrane transport opened the way for a biochemical dissection. Golgi fractions are prepared from two sets of homogenized tissue culture cells, one wild type and the other a mutant cell line lacking the medial-Golgi enzyme GlcNAc transferase, which adds a GlcNAc modification to the carbohydrate chain on secretory proteins. The mutant cells are infected with vesicular stomatitis virus (VSV), for which the spike glycoprotein VSVG provides a model secretory protein. VSVG is accumulated in the cis-Golgi cisternae prior to homogenization. The only way VSVG can acquire [^{3}H] labelled GlcNAc is through contents mixing between the medial-Golgi of wild-type cells and cis-Golgi of mutant cells. Under most conditions, this mixing requires COPI vesicle budding and fusion. This assay was used to fractionate cytosol for factors necessary to obtain efficient radiolabeling of VSVG and led to the identification of NSF, α-SNAP, and characterization of COPI coat proteins.

dissection of individual steps. Assays that reconstitute other transport steps have since been established and developed to the point where both vesicle budding and fusion can be reconstituted with simple model membranes and a minimal set of components. One powerful strategy has been to combine a biochemical approach with yeast genetics, notably in an assay of yeast vacuole fusion. This has allowed the manipulation of membrane proteins in the assay by deletion of specific genes in the parent yeast strains, which then enables the testing of the roles of individual proteins and the topological constraints for their participation in fusion.

3
Evolutionary Conservation of Fundamental Mechanisms

One of the earliest proofs of evolutionary conservation was the isolation of the NEM-sensitive factor (NSF), identified as a critical protein component in cytosol necessary to support mammalian Golgi transport *in vitro*. Rothman and colleagues showed that this protein is functionally equivalent to the *Sec18* gene product of the yeast *Saccharomyces cerevisiae*, which was known to be essential for vesicle-mediated transport from the endoplasmic reticulum to the Golgi apparatus. At one stroke, this not only suggested that NSF might operate at the heart of a fusion mechanism conserved from yeast to mammals but also that it may be required in multiple membrane transport events. Confirmation of this general role of NSF was obtained using *in vitro* assays of endosome fusion and of ER to Golgi transport. This led to the attractive idea that NSF may act as a component of a universal fusion machinery required for every intracellular

fusion event. This notion, which is still held to be broadly true (only less direct than initially envisaged), raises the question of how specificity of fusion events is attained.

NSF is recruited to membranes by α-SNAP (Sec17), which in turn binds to a family of proteins called *SNAREs*, that are also conserved from yeast to mammals. SNAREs are believed to be key players in determining specificity of intracellular fusion events and are also the best candidates for catalyzing the membrane fusion event itself. Their intracellular distribution is such that each membrane-bounded compartment is associated with a specific complement of SNAREs that form the basis of a combinatorial code for fusion (see Sect. 7). Recent genome-wide analysis suggests that there are 24 yeast SNAREs with similar numbers in flies and worms. Thirty-five SNAREs have been identified in the human genome, wherein the additional SNAREs predominantly represent tissue-specific isoforms. The current view of NSF function is that it is an ATPase, which configures SNARE proteins into a fusogenic conformation, but is not actively engaged in the fusion step.

This evolutionary conservation has meant that biochemical studies carried out with mammalian cell fractions have dovetailed with yeast genetic studies. The large-scale yeast screens can identify proteins that unerringly have mammalian counterparts, which are identifiable through homology.

4
Signal-mediated versus Bulk Flow Transport

Selective inclusion into transport vesicles is thought to be a major mechanism for establishing discrete compartments. An alternative "bulk flow" model posits that resident proteins are selectively retained, whilst vesicle contents are simply samples of the residual components of the parent organelle. To begin to understand this process, one must be able to compare the contents of the vesicle membrane with the parent membrane. Two approaches have been applied to this problem, quantitative immunoelectron microscopy and biochemical analysis of vesicle fractions. Thus, many studies have established that cargo is actively concentrated in clathrin-coated vesicles budding from the plasma membrane. This may be constitutive, as in the case of the transferrin receptor or regulated, as is the case for epidermal growth factor receptor (EGFR), which requires activation through ligand binding.

This issue has been more contentious when considering anterograde transport through the secretory pathway. Measurements using a glycosylated acyltripeptide as a "bulk flow" marker showed a high rate of secretion, suggesting that secretory proteins were nonselectively transported between the ER and the cell surface. However, subsequent analysis indicated that this may reflect a direct efflux pathway, following transport of the glycosylated peptide across the ER membrane to the cytosol.

Immunoelectron microscopical analysis of a SNARE protein (rBet1) has shown that it is enriched approximately sixfold in coat protein complex II (COPII) vesicles that bud from the ER. In contrast, in the same study, two soluble secretory proteins of pancreatic acinar cells, amylase, and chymotrypsinogen were not enriched in these vesicles and were instead concentrated further along the pathway in the tubular structures of the ER-Golgi intermediate compartment (ERGIC), where they are probably enriched by exclusion from COPI vesicles that mediate retrograde transport to the ER. In a separate study utilizing biochemical analysis, Malkus et al. have compared the content of COPII vesicles generated *in vitro* from yeast microsomal fractions with that of the starting material. They clearly show that a soluble protein, glycosylated pro-α-factor, is 20-fold enriched in these vesicles relative to bulk flow markers. A plasma membrane protein, the general amino acid permease (Gap1p), is also enriched in these vesicles by virtue of a di-acidic sorting signal in its COOH-terminal cytosolic domain. Hence, some, but not all, soluble secretory proteins may be selected for entry into COPII vesicles. Others may leave the ER by bulk flow, only to be concentrated in later compartments by selective exclusion from the retrograde COPI vesicle pathway.

Vesicles also contain lipids, which are more than just a permeability barrier to hydrophilic molecules. Brügger et al. have used nanoelectrospray ionization tandem mass spectroscopy to conduct a quantitative analysis of COPI-coated vesicle lipids compared to the parent Golgi composition. Vesicles were found to be de-enriched in both cholesterol and sphingomyelin, indicating that lipids too are subject to selection for inclusion into vesicles.

5
Coated Vesicle Formation

5.1
Coats

Morphological examination of vesicles frequently reveals protein dense coatings on the cytosolic surface. Coat assembly may drive vesicle formation. The first coat to be identified is comprised of hexagonal arrays of clathrin triskelia on the plasma membrane, which must rearrange to a mixture of hexagons and pentagons as the spherical vesicle forms and buds. Isolated clathrin triskelia can self-assemble into these "basket" structures *in vitro*. The principal sites of clathrin-coated vesicle formation are the plasma membrane and the TGN, from which both are delivered to the early endosome.

The other major class of coated vesicle so far characterized are the COP (coat protein complex) coated vesicles, which mediate retrograde trafficking between the Golgi and the ER, and between Golgi stacks (COPI), and anterograde trafficking from the ER to Golgi (COPII). The COPI coat is formed by Arf-dependent recruitment of "coatamer" (a preassembled heteroheptameric complex) from the cytosol. Arf is a small GTPase that recruits coatamer when it is in a GTP-bound active configuration. Similarly, the Arf family GTPase, Sar1p, choreographs the sequential recruitment of Sec23/24 and Sec13/31 complexes that comprise the COPII coat. Arf family proteins associate with membranes in their GTP-bound state by insertion of a hydrophobic anchor. This is withdrawn upon GTP hydrolysis and Arf dissociation can promote coat disassembly. The COPII component Sec23, is a GTPase-activating protein (GAP) for Sar1, and, so, upon recruitment of the Sec23/24 complex, the vesicles have only a brief opportunity to recruit cargo before coat disassembly begins.

5.2
Adaptors

Adaptor molecules link cargo-selected membrane proteins to the vesicle coat proteins. The first adaptors to be identified were the AP -1 and AP-2 adaptors (adaptor protein), which are highly enriched in TGN- and plasma membrane-derived clathrin-coated vesicles respectively. AP-1 and AP-2 are both heterotetramers of related proteins (AP-1: $\gamma\beta\mu$ and σ, AP-2: $\alpha\beta\mu$ and σ which promote clathrin self-assembly *in vitro*. Phosphoinositide lipids play a role in recruitment of these adaptor complexes; thus AP-1 binds to the TGN-enriched lipid PtdIns4P, whereas AP-2 binds to the plasma membrane enriched lipid PtdIns(4,5)P_2. The crystal structure of the AP-2 complex has been solved, revealing potential binding sites for phosphoinositides. Two further adaptor complexes, AP-3 and AP-4 were found by searching sequence databases for homologs of the AP-1 and AP-2 subunits. In distinction to AP-1 and AP-2, these are not enriched in clathrin-coated vesicles and may act independently or in conjunction with an unidentified scaffold to mediate membrane trafficking between the TGN and endosomes. Another class of adapter protein has recently been identified as a family of monomeric proteins called *GGAs* (<u>G</u>olgi-localized γ ear-containing, <u>A</u>RF-binding proteins), which contain a domain homologous to γ-adaptin. It is clear that they can bind to cargo receptors and clathrin, but, mysteriously, they are not enriched in clathrin-coated vesicles.

5.3
Principles of Cargo Selection

Specific signals mediate sorting of material into clathrin-coated vesicles. These signals may be embedded in the amino acid sequence or otherwise result from either constitutive or regulated posttranslational modification. The first sorting signal to be defined was the NPXY motif in LDL receptor, which resulted from Brown and Goldstein's pioneering studies of familial hypercholesterolemia. Mutation of the tyrosine residue within this motif leads to failure of LDL receptor endocytosis. Other well-defined signals have since been determined including YXXø where ø is a bulky hydrophobic residue that determines transferrin receptor internalization amongst others. Constitutive posttranslational modification serving as a lumenal sorting signal is exemplified by lysosomal enzymes that are phosphorylated on sugar residues in the cis-Golgi, leading to capture in the TGN by the cation-independent mannose 6-phosphate receptor (ci-M6PR). ci-M6PR receptors are then packaged into AP1 clathrin-coated vesicles destined for the endocytic pathway, through cooperative interactions between GGA proteins and AP-1. A good example of regulated sorting is the ligand-dependent phosphorylation of tyrosine kinase receptors, such as EGFR, which leads to recruitment of accessory factors such as the Cin85/Cbl complex that downregulates receptor from the plasma membrane through incorporation into clathrin-coated vesicles. The number of molecules capable of fulfilling an adaptor function has steadily extended beyond the AP complexes, leading to the notion that there is both redundancy between and specialization of adaptor proteins. Examples include β-arrestin, which couples to seven-transmembrane G-protein-coupled receptors and ARH, which is mutated in patients with autosomal recessive hypercholesterolemia (ARH). ARH patients have a clinical phenotype almost indistinguishable from familial hypercholesterolemia but have normal *LDLR* alleles.

Binding of cargo to Sec24 appears to be the principal means of selection into COPII vesicles budding from the ER. However, the diverse array of cargo molecules utilize more than one binding site, two of which have now been mapped through elegant structural and biochemical studies. Vesicles must also incorporate specific targeting molecules; an interesting study of SNARE incorporation into COPII vesicles has suggested that only fusogenic forms of the SNAREs are selected for transport.

5.4
Induction of Curvature

Producing vesicles requires bending the membrane, which in turn requires energy. A standard bilayer bending modulus is estimated to be around 20 $k_B T$, where k_B is Boltzmann's constant and T the absolute temperature in Kelvin. One view is that the polymerization of the coat on a nascent vesicle induces membrane curvature. Several accessory proteins such as endophilin, amphiphysin, and epsin, implicated in vesicle formation, will tubulate or vesiculate liposomes when incubated *in vitro*. Recently, it has been proposed that BAR (Bin/Amphiphysin/Rvs) domain containing proteins can induce curvature through tight electrostatic association with acidic membranes in the absence of substantial hydrophobic insertion. The structure of the dimeric BAR domain from *Drosophila amphiphysin* is rigid, elongated, and will present a concave face to the membrane, which could accommodate a curved membrane of 11 nm outer radius. This, of course, does not fit well with the radius of coated vesicles (50 – 100 nm). It may be that rather than inducing curvature under physiological conditions, these proteins sense curvature of membranes, as their affinity for preexisting curved membranes will be higher. Bigay et al. have proposed that the BAR domain protein ArfGAP may act as a curvature sensor in COPI vesicle coat disassembly.

An alternative scenario, by which curvature may be induced, is through insertion of hydrophobic groups into one leaflet of the bilayer membrane, which increase the lateral area of this monolayer relative to the other and, consequently, promote curvature. An interesting case is the endocytic regulator epsin, which upon binding the lipid PtdIns(4,5)P_2 undergoes a conformational change that induces an amphipathic α-helix, which can insert into one leaflet of the bilayer. Clathrin-coated vesicle formation at the plasma membrane is promoted by generation of this phosphoinositide, which may also play a role in recruitment of AP-1.

5.5
Vesicle Scission

Budding vesicles can frequently be seen using electron microscopy as omega-shaped profiles protruding from parent membranes. The final step of vesicle scission will require membrane fusion to produce a continuous bilayer. The *shibire* mutation in Drosophila was identified as a temperature-sensitive mutation that caused paralysis through inhibition of synaptic vesicle recycling. It was shown that this defect leads to an accumulation of vesicles at the synaptic membrane that are unable to complete the final scission step. Later, the *shibire* gene was shown to encode a GTPase protein called *dynamin*. Dynamin is able to undergo large-scale oligomerization, wrapping itself around lipid tubules *in vitro*, or around the necks of budding vesicles in *vivo*, to form a helical collar. The pitch of this helix increases as dynamin oligomers undergo highly cooperative GTP hydrolysis. This has led to the "poppase" model of vesicle formation, in which this cooperative extension of a dynamin collar is used to provide the force to pop off a budding vesicle.

6
Vesicle Targeting

Once formed, a vesicle must identify its target membrane. The ultimate arbiters of the commitment to vesicle fusion are the SNARE proteins (Sect. 7.3), but prior to their involvement, looser associations are made that allow exploration of the propriety of any union.

6.1
Rab Proteins

One identifier of membrane compartments are the small GTPases of the rab family, of which there are more than 40 members in the human genome. Attention was first focused on this family when the budding yeast protein Sec4 was shown to be a member. Shortly thereafter, work on early endosome fusion in mammalian cells implicated rab5 as a key regulator. It has now been realized that the function of rab proteins is extremely complex. The GTP-bound form of rab5, for example, interacts with upwards of 22 proteins. However, for our purposes, their preeminent role is in their conformation-dependent recruitment of a family of extended coiled-coil proteins or multisubunit complexes, which are believed to act as tethers between interacting membranes (see next section). By recruiting membrane tethers through the GTP-bound form of rab proteins, which possess an intrinsic decay rate due to GTP hydrolysis, it is ensured that the tethering interaction is finite. Thus, if the appropriate downstream interactions, such as SNARE complex formation, are not available, the relationship can be severed before any damage can be done. This provides a form of kinetic proof reading as a potential means of ensuring fidelity.

6.2
Tethering Molecules

Tethering factors are heterogeneous in sequence and structure, but they do fall into distinct classes. A number of extended coiled-coil proteins, which exist as flexible rod-shaped structures of parallel homodimers, have been linked with tethering at the Golgi complex (Golgins) and on endosomes (EEA1, rabaptin 5) and may confine the search for trans-SNARE partners to the tethered membranes. Only a couple of examples for which there is reasonable experimental evidence for this role will be discussed here. EEA1 is proposed to play a role in tethering of endosomes prior to fusion. It is recruited to endosomes through a dual interaction with PtdIns3*P* and GTP-rab5 at its C-terminus, and also has an additional rab5-binding site at its N-terminus. Depletion of EEA1 from the cytosol markedly reduces the efficiency of endosome fusion *in vitro*. The C-terminus of EEA1 also interacts with the SNARE proteins syntaxin 6 and syntaxin 13, both of which have been implicated in endosome fusion.

Attachment of ER-derived vesicles to yeast Golgi *in vitro* was found to require the cytosolic factor Uso1 (a homolog of p115 in mammalian cells) and the small GTPase Ypt1, a member of the Rab family. Removal of Ypt1 from membranes led to a corresponding loss of Uso1, although a direct interaction between the two proteins has so far not been demonstrated. Ypt1 may either bind directly to Uso1 or regulate availability of an alternative receptor-binding site. Interestingly, both factors are also implicated in sorting of cargo at the ER, suggesting a coupling between cargo selection and vesicle tethering.

Another class of tethering molecules exists as multisubunit complexes. Seven large conserved complexes have been proposed to have roles in vesicle tethering at distinct transport steps. These can be divided into two families. One, termed the *quatrefoil complexes*, shares a common N-terminal domain in the majority of complex subunits, which all assemble in multiples of four. They include the Sec6/8 complex (or exocyst), the conserved oligomeric Golgi complex (COG) and the Golgi associated retrograde protein (GARP) complex. The exocyst binds to plasma membrane in budding yeast through recruitment by the Sec4 GTPase, although it is also an effector of other small GTPases. The COG complex localizes to the Golgi complex, although its function is not precisely clear. The GARP complex localizes to the TGN, binds to the SNARE protein Tlg1p, and is implicated in retrograde traffic from endosomes to the Golgi. Other multisubunit complexes, which have rather hazily been ascribed tethering functions, include the TRAPP (transport protein particle) complexes I and II, the class C Vps complex and Dsl1p. Several of these particles contain guanine nucleotide exchange factors (GEFs) that will activate rab GTPases. Thus, subsequent to tethering COPII vesicles to the Golgi, TRAPPI is proposed to activate Ypt1, which, in turn, recruits a second, downstream tether, the coiled-coil protein Uso1p, which has been discussed above.

7
Vesicle Fusion

7.1
Understanding Energy Barriers to Fusion

A condition of membrane fusion is that the membranes must become very closely apposed. Measurement of changes in bilayer spacing in multilamellar vesicles, as a function of externally applied osmotic or hydrostatic pressure, revealed a surprisingly strong exponentially growing (characteristic decay length of 0.1–0.2 nm) repulsive force between phospholipid bilayers. Under normal circumstances then, bilayers do not spontaneously fuse. Insensitivity of this force to salt concentration and hence membrane electrostatics led to the initial conclusion that the repulsive force is due to hydration of the polar lipid head-groups by water. However, it is now generally believed that other forces also contribute to this repulsion, including steric interactions along with elastic pressures. In addition, an attractive force between closely apposed model membranes has been observed by measurements of adhesion of bilayers supported on mica surfaces in aqueous solutions. This force is believed to arise from exposure of hydrophobic moieties of phospholipids as a result of the stress exerted on bilayer surfaces that approach within about 1 nm.

7.2
Membrane Destabilization

A theoretical analysis of interactions between apposed membranes proposed that out-of-plane thermal fluctuations of the bilayers lead to the formation of close (less than 0.5 nm) contact between the membranes within a small area (approximately 10 nm^2). According to this theory, increasing hydration repulsion between apposed polar heads of lipid molecules in this area causes the rupture of interacting monolayers. This rupture results in monolayer fusion of the membranes, that is, in the formation of a bridge connecting the monolayers, which is usually named the "stalk" or hemifusion intermediate that can further evolve to form a hemifusion

diaphragm, which then ruptures leading to full fusion.

7.3
SNARE Proteins

The "SNARE hypothesis" was proposed in 1993 by Rothman and co-workers to explain the Specificity of intracellular fusion events. It proposes that the core of the fusion machinery is comprised of SNARE proteins that are localized to specific subcellular compartments. Only cognate SNAREs on partner membranes can form a complex that promotes fusion. Following vesicle tethering, partner membranes have the opportunity to form trans-SNARE complexes. The first SNARE complex to be characterized was the one that specifies fusion of synaptic vesicles with the plasma membrane. This interaction, which is believed to be characteristic of all SNARE complexes, involves the formation of a parallel four-helix bundle, termed a "*SNAREpin*," in which one helix is contributed by the vesicle-associated SNARE (v-SNARE) synaptobrevin, whilst the other three are contributed by SNAREs on the target membrane (t-SNAREs). In this case, two of the t-SNARE helices are provided by SNAP-25 and one by syntaxin 1. Sequence analysis has shown that all SNAREs probably derive from a single ancestral gene that has been duplicated in the case of SNAP-25.

SNARE complexes do not simply lock vesicles into a closely associated state; they also drive membrane fusion. Substitution of the transmembrane domains of syntaxin and synaptobrevin with lipid anchors allows the docking interaction of vesicles to proceed, but fusion is no longer observed. The resolution of the SNARE complex structure immediately suggested an attractive mechanism by which fusion is promoted that has been termed *the zipper model.* In this model, the assembly of the parallel four-helix bundle occurs initially at the N-termini of the helices and zips up toward the membrane anchors, consequently pulling the partner membranes into close apposition. The SNAREs themselves must tilt toward the membrane during this process, and it is attractive to speculate that this is coupled to deformation of the membrane through the transmembrane anchor (Fig. 4). Insertion of a flexible linker between the transmembrane domain and the coiled-coil domain reduces SNARE-dependent fusion efficiency systematically with increasing length of the linker. Note that after fusion, a SNARE complex will persist, but now in a cis configuration until it is disassembled by the action of NSF, which utilizes ATP hydrolysis.

Evidence that SNAREs participate in a late stage of physiological fusion reactions has been provided by an assay of regulated secretion of dense core vesicles in PC12 cells. In this system, the assembly of the SNARE complex occurred after the rise in Ca^{2+}, and could not be experimentally dissociated from the fusion process. In adrenal chromaffin cells, an antibody that inhibits SNARE assembly also reduces the initial fast component of exocytosis, indicating that even the vesicles poised for the quickest release require the SNARE assembly step.

A finding that the three synaptic SNAREs, when reconstituted into model liposomes and distributed according to cellular physiology, could promote liposome fusion led to a highly influential proposal that the SNAREpin complex represented the minimal fusion machinery. McNew et al. further developed this form of assay by taking advantage of the functional identification of yeast SNARE complexes associated with three specific transport steps, namely fusion of ER-derived transport vesicles with the Golgi, homotypic vacuole fusion, and fusion

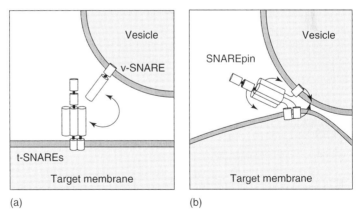

(a)   (b)

Fig. 4 Model for membrane deformation and fusion mediated by SNARE proteins. (a) The four helical sections of a SNARE complex in a prefusion state are arranged in a 3 : 1 distribution between partner membranes. (b) Progressive zipping of the four helical sections of SNARE proteins leads to the assembly of a tight parallel four-helix bundle (SNAREpin). This process forces membranes into close membrane apposition and will generate stresses at the membrane that are coupled to imposed bending of the participating proteins (indicated by arrows).

of secretory vesicles with the plasma membrane. They tested the fusion activity of liposomes bearing the set of t-SNAREs associated with each of these three steps with liposomes bearing any one of 11 actual or potential v-SNAREs in the yeast genome. The *in vitro* system recapitulated the cellular specificity to a remarkable degree, testifying to the high level of discrimination encoded within the SNARE complex.

These experiments built a three-component t-SNARE on one vesicle population to interact with a unitary v-SNARE on another. It is known that the four-helix bundle can also be formed in solution from soluble cytoplasmic domains of SNARE proteins. However, there are topological constraints on SNARE interactions when confined to membranes, which are not imposed in solution. On varying the distribution of each of four cognate SNAREs between donor and acceptor vesicles, Parlati et al. found that the original combination of Bet1 (v-SNARE) liposomes with Sed5–Bos1–Sec22 (t-SNARE complex) is uniquely fusogenic. Thus, the distribution of cognate SNAREs between partner membranes determines fusion specificity. This is important because the cellular distribution of SNARE proteins is not static.

7.4
Regulation of SNARE Complex Assembly

What is the effect of including a fifth SNARE in the liposome fusion system? In most cases, this has no effect, but instances have been found in which the additional SNARE can compete with and substitute for a fusogenic subunit, thereby inhibiting fusion. Components of the cis-Golgi fusion machinery are inhibited by components of the trans-Golgi machinery

and vice versa. This has led to the concept of the inhibitory SNARE (i-SNARE). If SNAREs can regulate fusion in both a positive and negative manner, then this will provide tighter control of specificity at biological membranes containing overlapping SNARE complements. The ratio of fusogenic SNAREs to i-SNARES will determine propensity for specific fusion.

The N-terminal domain of the syntaxin-like SNARE proteins, such as neuronal syntaxin1 and yeast Sso1p, fold into three-helix bundles and inhibit SNARE pairing when folded back against their SNARE domains. In this configuration, syntaxin can form a tight complex with the neuronal SM protein (Sec1/Munc18 homolog) Munc18-1, but it must change to the "open" conformation to engage in SNARE complex assembly. This interaction with Munc18-1 has thus been proposed to inhibit syntaxin SNARE function, but this remains controversial. The proteins Munc13 and RIM have been proposed to promote synaptic vesicle fusion through displacement of Munc18-1 from syntaxin. Likewise, some tethering factors such as EEA1, known to interact with syntaxin family members, may also act to displace SM proteins and promote an open syntaxin configuration.

8
Vesicle Trafficking and Disease

The origins of many human pathologies lie in defective vesicle trafficking. In many cases, this is due to a specific failure of cargo selection. For example, hypercholesterolemia leading to atherosclerosis can result from mutations in the cytoplasmic domain of the LDL receptor required for endocytosis or in the LDL receptor–specific adaptor protein, ARH. The hypertensive Liddle's syndrome is due to failure to endocytose epithelial Na+ channels. It has also been argued that various proteins are oncogenic because of crucial roles that they may play in downregulating activated growth factor receptors through endocytic trafficking to lysosomal degradative compartments. For example, the oncogene c-Cbl is an E3-ubiquitin ligase, which mediates monoubiquitination of activated tyrosine kinase receptors, such as EGFR. Ubiquitinated receptors are specifically selected for downregulation by inclusion into lumenal vesicles of endosomal multivesicular bodies that routes them to lysosomes away from the recycling pathway. Failure to direct receptors along this route will lead to retention of activated receptors and increased mitogenic signaling.

One would expect defects in the core components of vesicle budding and fusion to be lethal, but there are examples where heterozygosity or point mutations lead to specific pathologies. Knockout mice heterozygous for syntaxin 4 develop muscle insulin resistance, consistent with a role for this SNARE molecule in GLUT4 (glucose transporter) exocytosis. A point mutation in α-SNAP leads to a developmental defect in mice, known as *hydrocephaly with hop gait*, which is characterized by a reduced size in the cerebral cortex, owing to premature production of cerebral cortical neurons and depletion of the progenitor pool. This is accompanied by abnormal localization of many apical proteins implicated in regulation of cell fate in neuroepithelial cells.

A number of bacterial pathogens subvert the cellular trafficking machinery to fashion intracellular compartments conducive to their requirements. For example, *Legionella pneumophila* is a bacterial pathogen that infects eukaryotic host cells and replicates inside a specialized organelle that is morphologically similar to the endoplasmic reticulum (ER). The Legionella-containing vacuole acquires rab1 and the v-SNARE protein Sec.22b, which facilitate fusion of the vacuole with ER derived COPII vesicles, allowing formation of a specialized compartment that can support bacterial replication. The clostridial neurotoxins, botulinum B and tetanus toxin are zinc endopetidases, which cause paralysis through specific cleavage of the neuronal v-SNARE, synaptobrevin.

9
Conclusions

The basic components of the intracellular budding and fusion machineries have been identified through a combination of biochemistry and yeast genetics. Current frontiers are concerned with understanding the regulation of these components. For example, what special mechanisms allow a synaptic vesicle to be released within milliseconds of an elevated Ca^{2+} concentration and what then limits the release of vesicles at the synapse? There is an increasing awareness of the role of membrane traffic in disease due to inherited or acquired mutations and due to utilization of established routes or subversion of trafficking machineries by various pathogens such as viruses and bacteria.

See also Cell Junctions, Structure, Function, and Regulation; Cell Nucleus Biogenesis, Structure and Function; Intracellular Fatty Acid Binding Proteins and Fatty Acid Transport; Membrane Transport; Phagocytosis.

Bibliography

Books and Reviews

Antonny, B., Schekman, R. (2001) ER export: public transportation by the COPII coach, *Curr. Opin. Cell. Biol.* **13**, 438–443.

Blumenthal, R., Clague, M.J., Durell, S.R., Epand, R.M. (2003) Membrane fusion, *Chem. Rev.* **103**, 53–69.

Boal, D. (2002) *Mechanics of the Cell*, Cambridge University Press, ISBN: 0521796814, http://titles.cambridge.org/catalogue.asp?isbn=0521796814, http://www.sfu.ca/~boal/moc.html.

Bonifacino, J.S., Traub, L.M. (2003) Signals for sorting of transmembrane proteins to endosomes and lysosomes, *Annu. Rev. Biochem.* **72**, 395–447.

Chernomordik, L.V., Zimmerberg, J. (1995) Bending membranes to the task: structural intermediates in bilayer fusion, *Curr. Opin. Struct. Biol.* **5**, 541–547.

Meresse, S., Steele-Mortimer, O., Moreno, E., Desjardins, M., Finlay, B., Gorvel, J.P. (1999) Controlling the maturation of pathogen-containing vacuoles: a matter of life and death, *Nat. Cell Biol.* **1**, E183–E188.

Pfeffer, S.R. (2001) Rab GTPases: specifying and deciphering organelle identity and function, *Trends Cell Biol.* **11**, 487–491.

Rizo, J., Sudhof, T.C. (2002) Snares and Munc18 in synaptic vesicle fusion, *Nat. Rev. Neurosci.* **3**, 641–653.

Robinson, M.S. (2004) Adaptable adaptors for coated vesicles, *Trends Cell Biol.* **14**, 167–174.

Whyte, J.R.C., Munro, S. (2002) Vesicle tethering complexes in membrane traffic, *J. Cell Sci.* **115**, 2627–2637.

Primary Literature

Antonny, B., Madden, D., Hamamoto, S., Orci, L., Schekman, R. (2001) Dynamics of the COPII coat with GTP and stable analogues, *Nat. Cell Biol.* **3**, 531–537.

Balch, W.E., Dunphy, W.G., Braell, W.A., Rothman, J.E. (1984) Reconstitution of the transport of protein between successive compartments of the Golgi measured by the coupled incorporation of N-acetylglucosamine, *Cell* **39**, 405–416.

Bankaitis, V.A., Johnson, L.M., Emr, S.D. (1986) Isolation of yeast mutants defective in protein targeting to the vacuole, *Proc. Natl. Acad. Sci. U.S.A.* **83**, 9075–9079.

Beckers, C.J., Block, M.R., Glick, B.S., Rothman, J.E., Balch, W.E. (1989) Vesicular transport between the endoplasmic reticulum and the Golgi stack requires the NEM-sensitive fusion protein, *Nature* **339**, 397–398.

Bickford, L.C., Mossessova, E., Goldberg, J. (2004) A structural view of the COPII vesicle coat, *Curr. Opin. Struct. Biol.* **14**, 147–153.

Bigay, J., Gounon, P., Robineau, S., Antonny, B. (2003) Lipid packing sensed by ArfGAP1 couples COPI coat disassembly to membrane bilayer curvature, *Nature* **426**, 563–566.

Blondeau, F., Ritter, B., Allaire, P.D., Wasiak, S., Girard, M., Hussain, N.K., Angers, A., Legendre-Guillemin, V., Roy, L., Boismenu, D., Kearney, R.E., Bell, A.W., Bergeron, J.J., McPherson, P.S. (2004) Tandem MS analysis of brain clathrin-coated vesicles reveals their critical involvement in synaptic vesicle recycling, *Proc. Natl. Acad. Sci. U. S. A.* **101**, 3833–3838.

Bock, J.B., Matern, H.T., Peden, A.A., Scheller, R.H. (2001) A genomic perspective on membrane compartment organization, *Nature* **409**, 839–841.

Bonifacino, J.S. (2004) The GGA proteins: adaptors on the move, *Nat. Rev. Mol. Cell. Biol.* **5**, 23–32.

Brown, M.S., Goldstein, J.L. (1986) A receptor-mediated pathway for cholesterol homeostasis, *Science* **232**, 34–47.

Brügger, B., Erben, G., Sandhoff, R., Wieland, F.T., Lehmann, W.D. (1997) Quantitative analysis of biological membrane lipids at the low picomole level by nanoelectrospray ionization tandem mass spectrometry, *Proc. Natl. Acad. Sci. U.S.A.* **94**, 2339–2344.

Burri, L., Lithgow, T. (2004) A complete set of SNAREs in yeast, *Traffic* **5**, 45–52.

Cao, X., Ballew, N., Barlowe, C. (1998) Initial docking of ER-derived vesicles requires Uso1p and Ypt1p but is independent of SNARE proteins, *EMBO J.* **17**, 2156–2165.

Chac, T.H., Kim, S., Marz, K.E., Hanson, P.I., Walsh, C.A. (2004) The hyh mutation uncovers roles for alpha Snap in apical protein localization and control of neural cell fate, *Nat. Genet.* **36**, 264–270.

Chen, W.J., Goldstein, J.L., Brown, M.S. (1990) NPXY, a sequence often found in cytoplasmic tails, is required for coated pit-mediated internalization of the low density lipoprotein receptor, *J. Biol. Chem.* **265**, 3116–3123.

Chen, Y.A., Scales, S.J., Patel, S.M., Doung, Y.-C., Scheller, R.H. (1999) SNARE complex formation is triggered by Ca2+ and drives membrane fusion, *Cell* **97**, 165–174.

Christoforidis, S., McBride, H.M., Burgoyne, R.D., Zerial, M. (1999) The rab5 effector EEA1 is a core component of endosome docking, *Nature* **397**, 621–625.

Chvatchko, Y., Howald, I., Riezman, H. (1986) Two yeast mutants defective in endocytosis are defective in pheromone response, *Cell* **46**, 355–364.

Clague, M.J. (1998) Molecular aspects of the endocytic pathway, *Biochem. J.* **336**, 271–282.

Clague, M.J. (1999) Membrane transport: take your fusion partners, *Curr. Biol.* **9**, R258–R260.

Collins, B.M., McCoy, A.J., Kent, H.M., Evans, P.R., Owen, D.J. (2002) Molecular architecture and functional model of the endocytic AP2 complex, *Cell* **109**, 523–535.

Conibear, E., Cleck, J.N., Stevens, T.H. (2003) Vps51p mediates the association of the GARP (Vps52/53/54) complex with the late Golgi t-SNARE Tlg1p, *Mol. Biol. Cell.* **14**, 1610–1623.

Conner, S.D., Schmid, S.L. (2003) Regulated portals of entry into the cell, *Nature* **422**, 37–44.

De Duve, C. (1965) The separation and characterization of subcellular particles, *Harvey Lect.* **59**, 49–87.

Di Fiore, P.P., Gill, G.N. (1999) Endocytosis and mitogenic signalling, *Curr. Opin. Cell Biol.* **11**, 483–488.

Diaz, R., Mayorga, L., Weidman, P.J., Rothman, J.E., Stahl, P.D. (1989) Vesicle fusion following receptor-mediated endocytosis requires a protein active in Golgi transport, *Nature* **339**, 398–400.

Doray, B., Ghosh, P., Griffith, J., Geuze, H.J., Kornfeld, S. (2002) Cooperation of GGAs and AP-1 in packaging MPRs at the trans-Golgi network, *Science* **297**, 1700–1703.

Dumas, J.J., Merithew, E., Sudharshan, E., Rajamani, D., Hayes, S., Lawe, D., Corvera, S., Lambright, D.G. (2001) Multivalent endosome targeting by homodimeric EEA1, *Mol. Cell* **8**, 947–958.

Farsad, K., De Camilli, P. (2003) Mechanisms of membrane deformation, *Curr. Opin. Cell Biol.* **15**, 372–381.

Fasshauer, D. (2003) Structural insights into the SNARE mechanism, *Biochim. Biophys. Acta* **1641**, 87–97.

Ford, M.G., Mills, I.G., Peter, B.J., Vallis, Y., Praefcke, G.J., Evans, P.R., McMahon, H.T. (2002) Curvature of clathrin-coated pits driven by epsin, *Nature* **419**, 361–366.

Fries, E., Rothman, J.E. (1980) Transport of vesicular stomatitis virus glycoprotein in a cell-free extract, *Proc. Natl. Acad. Sci. U. S. A.* **77**, 3870–3874.

Gillingham, A.K., Munro, S. (2003) Long coiled-coil proteins and membrane traffic, *Biochim. Biophys. Acta* **1641**, 71–85.

Gorvel, J.P., Chavrier, P., Zerial, M., Gruenberg, J. (1991) Rab 5 controls early endosome fusion in vitro, *Cell* **64**, 915–925.

Guo, W., Roth, D., Walch-Solimena, C., Novick, P. (1999) The exocyst is an effector for Sec4p, targeting secretory vesicles to sites of exocytosis, *EMBO J.* **18**, 1071–1080.

Haas, A., Wickner, W. (1996) Homotypic vacuole fusion requires sec17p (yeast a-SNAP) and sec18p (yeast NSF), *EMBO J.* **15**, 3296–3305.

Hanson, P.I., Roth, R., Morisaki, H., Jahn, R., Heuser, J. (1997) Structure and conformational changes in NSF and its membrane receptor complexes visualized by quick-freeze/deep-etch electron microscopy, *Cell* **90**, 523–535.

He, G., Gupta, S., Yi, M., Michaely, P., Hobbs, H.H., Cohen, J.C. (2002) ARH is a modular adaptor protein that interacts with the LDL receptor, clathrin, and AP-2, *J. Biol. Chem.* **277**, 44044–44049.

Helm, C.A., Israelachvili, J.N., McGuiggan, P.M. (1989) Molecular mechanisms and forces involved in the adhesion and fusion of amphiphilic bilayers, *Science* **246**, 919–922.

Hirst, J., Lui, W.W., Bright, N.A., Totty, N., Seaman, M.N., Robinson, M.S. (2000) A family of proteins with gamma-adaptin and VHS domains that facilitate trafficking between the trans-Golgi network and the vacuole/lysosome, *J. Cell Biol.* **149**, 67–80.

Hong, H.K., Chakravarti, A., Takahashi, J.S. (2004) The gene for soluble N-ethylmaleimide sensitive factor attachment protein alpha is mutated in hydrocephaly with hop gait (hyh) mice, *Proc. Natl. Acad. Sci. U. S. A.* **101**, 1748–1753.

Huttner, W.B., Schiebler, W., Greengard, P., De Camilli, P. (1983) Synapsin 1 (protein 1), a nerve terminal-specific phosphoprotein. (III)

Its association with synaptic vesicles studied in a highly purified synaptic vesicle preparation, *J. Cell Biol.* **96**, 1374–1388.

Jamieson, J.D., Palade, G.E. (1967) Intracellular transport of secretory proteins in the pancreatic exocrine cell. I. Role of the peripheral elements of the Golgi complex, *J. Cell Biol.* **34**, 577–596.

Kagan, J.C., Stein, M.P., Pypaert, M., Roy, C.R. (2004) Legionella subvert the functions of rab1 and sec22b to create a replicative organelle, *J. Exp. Med.* **199**, 1201–1211.

Katzmann, D.J., Odorizzi, G., Emr, S.D. (2002) Receptor downregulation and multivesicular-body sorting, *Nat. Rev. Mol. Cell. Biol.* **3**, 893–905.

Leikin, S.L., Kozlov, M.M., Chernomordik, L.V., Markin, V.S., Chizmadzhev, Y.A. (1987) Membrane fusion: overcoming of the hydration barrier and local restructuring, *J. Theor. Biol.* **129**, 411–425.

Lippincott-Schwartz, J., Patterson, G.H. (2003) Development and use of fluorescent protein markers in living cells, *Science* **300**, 87–91.

Lipschutz, J.H., Mostov, K.E. (2002) Exocytosis: the many masters of the exocyst, *Curr. Biol.* **12**, R212–R214.

Malkus, P., Jiang, F., Schekman, R. (2002) Concentrative sorting of secretory cargo proteins into COPII-coated vesicles, *J. Cell Biol.* **159**, 915–921.

Martinez-Menarguez, J.A., Geuze, H.J., Slot, J.W., Klumperman, J. (1999) Vesicular tubular clusters between the ER and Golgi mediate concentration of soluble secretory proteins by exclusion from COPI-coated vesicles, *Cell* **98**, 81–90.

McNew, J.A., Weber, T., Engelman, D.M., Sollner, T.H., Rothman, J.E. (1999) The length of the flexible SNAREpin juxtamembrane region is a critical determinant of SNARE-dependent fusion, *Mol. Cell* **4**, 415–421.

McNew, J.A., Weber, T., Parlati, F., Johnston, R., Melia, T.J., Sollner, T.H., Rothman, J.E. (2000b) Close is not enough: SNARE-dependent membrane fusion requires an active mechanism that transduces force to membrane anchors, *J. Cell Biol.* **150**, 105–117.

McNew, J.A., Parlati, F., Fukuda, R., Lohnston, R.J., Paz, K., Paumet, F., Sollner, T.H., Rothman, J.E. (2000a) Compartmental specificity of cellular membrane fusion encoded in SNARE proteins, *Nature* **407**, 153–159.

Mills, I.G., Jones, A.T., Clague, M.J. (1998) Involvement of the endosomal autoantigen EEA1 in homotypic fusion of early endosomes, *Curr. Biol.* **8**, 881–884.

Mills, I.G., Urbe, S., Clague, M.J. (2001) Relationships between EEA1 binding partners and their role in endosome fusion, *J. Cell Sci.* **114**, 1959–1965.

Mironov, A.A., Beznoussenko, G.V., Nicoziani, P., Martella, O., Trucco, A., Kweon, H.S., Di Giandomenico, D., Polishchuk, R.S., Fusella, A., Lupetti, P., Berger, E.G., Geerts, W.J., Koster, A.J., Burger, K.N., Luini, A. (2001) Small cargo proteins and large aggregates can traverse the Golgi by a common mechanism without leaving the lumen of cisternae, *J. Cell Biol.* **155**, 1225–1238.

Mossessova, E., Bickford, L.C., Goldberg, J. (2003) SNARE selectivity of the COPII coat, *Cell* **114**, 483–495.

Nicholls, B.J., Ungermann, C., Pelham, H.R.B., Wickner, W., Haas, A. (1997) Homotypic vacuolar fusion mediated by v- and t-SNAREs, *Nature* **387**, 199–202.

Novick, P., Field, C., Schekman, R. (1980) Identification of 23 complementation groups required for post-translational events in the yeast secretory pathway, *Cell* **21**, 205–215.

Orci, L., Palmer, D.J., Amherdt, M., Rothman, J.E. (1993) Coated vesicle assembly in the Golgi requires only coatomer and ARF proteins from the cytosol, *Nature* **364**, 732–734.

Padron, D., Wang, Y.J., Yamamoto, M., Yin, H., Roth, M.G. (2003) Phosphatidylinositol phosphate 5-kinase Ibeta recruits AP-2 to the plasma membrane and regulates rates of constitutive endocytosis, *J. Cell Biol.* **162**, 693–701.

Painter, R.G., Tokuyasu, K.T., Singer, S.J. (1973) Immunoferritin localization of intracellular antigens: the use of ultracryotomy to obtain ultrathin sections suitable for direct immunoferritin staining, *Proc. Natl. Acad. Sci. U. S. A.* **70**, 1649–1653.

Palade, G. (1975) Intracellular aspects of the process of protein secretion, *Science* **189**, 347–358.

Parlati, F., McNew, J.A., Fukuda, R., Miller, R., Sollner, T.H., Rothman, J.E. (2000) Topological restriction of SNARE-dependent membrane fusion, *Nature* **407**, 194–198.

Parsegian, V.A., Fuller, N., Rand, R.P. (1979) Measured work of deformation and repulsion of lecithin bilayers, *Proc. Natl. Acad. Sci. U. S. A.* **76**, 2750–2754.

Pearse, B., Robinson, M.S. (1990) Clathrin, adaptors and sorting, *Ann. Rev. Cell Biol.* **6**, 151–172.

Pearse, B.M.F. (1975) Coated vesicles from pig brain: purification and biochemical characterization, *J. Mol. Biol.* **97**, 93–98.

Pelham, H.R.B. (1999) SNAREs and the secretory pathway-lessons from yeast, *Exp. Cell Res.* **247**, 1–8.

Piper, R.C., Luzio, J.P. (2001) Late endosomes: sorting and partitioning in multivesicular bodies, *Traffic* **2**, 612–621.

Poodry, C.A., Edgar, L. (1979) Reversible alteration in the neuromuscular junctions of Drosophila melanogaster bearing a temperature-sensitive mutation, shibire, *J. Cell Biol.* **81**, 520–527.

Robinson, M.S., Bonifacino, J.S. (2001) Adaptor-related proteins, *Curr. Opin. Cell. Biol.* **13**, 444–453.

Romisch, K., Ali, B.R. (1997) Similar processes mediate glycopeptide export from the endoplasmic reticulum in mammalian cells and Saccharomyces cerevisiae, *Proc. Natl. Acad. Sci. U. S. A.* **94**, 6730–6734.

Roth, M.G. (2004) Phosphoinositides in constitutive membrane traffic, *Physiol. Rev.* **84**, 699–730.

Roth, T.F., Porter, K.R. (1964) Yolk protein uptake in the oocyte of the mosquito Aedes Aegypti L, *J. Cell Biol.* **20**, 313–332.

Rothman, J.E. (1994) Mechanisms of intracellular protein transport, *Nature* **372**, 55–63.

Sacher, M., Barrowman, J., Wang, W., Horecka, J., Zhang, Y., Pypaert, M., Ferro-Novick, S. (2001) TRAPP I implicated in the specificity of tethering in ER-to-Golgi transport, *Mol. Cell.* **7**, 433–442.

Salminen, A., Novick, P.J. (1987) A ras-like protein is required for a post-Golgi event in yeast secretion, *Cell* **49**, 527–538.

Schiavo, G., Benfenati, F., Poulain, B., Rossetto, O., Polverino de Laureto, P., Das-Gupta, B.R., Montecucco, C. (1992) Tetanus and botulinum-B neurotoxins block neurotransmitter release by proteolytic cleavage of synaptobrevin, *Nature* **359**, 832–835.

Serafini, T., Stenbeck, G., Brecht, A., Lottspeich, F., Orci, L., Rothman, J.E., Wieland, F. (1991) A coat subunit of Golgi-derived non-clathrin-coated vesicles with homology to the clathrin-coated vesicle coat protein B-adaptin, *Nature* **349**, 215–220.

Sheetz, M.P., Singer, S.J. (1974) Biological membranes as bilayer couples. A molecular mechanism of drug-erythrocyte interactions, *Proc. Natl. Acad. Sci. U.S.A.* **71**, 4457–4461.

Snyder, P.M. (2002) The epithelial Na+ channel: cell surface insertion and retrieval in Na+ homeostasis and hypertension, *Endocr. Rev.* **23**, 258–275.

Sollner, T., Whiteheart, S.W., Brunner, M., Erdjument-Bromage, H., Geromanos, S., Tempst, P., Rothman, J.E. (1993) SNAP receptors implicated in vesicle targeting and fusion, *Nature* **362**, 318–324.

Soubeyran, P., Kowanetz, K., Szymkiewicz, I., Langdon, W.Y., Dikic, I. (2002) Cbl-CIN85-endophilin complex mediates ligand-induced downregulation of EGF receptors, *Nature* **416**, 183–187.

Stowell, M.H., Marks, B., Wigge, P., McMahon, H.T. (1999) Nucleotide-dependent conformational changes in dynamin: evidence for a mechanochemical molecular spring, *Nat. Cell Biol.* **1**, 27–32.

Sutton, R.B., Fasshauer, D., Jahn, R., Brunger, A.T. (1998) Crystal structure of a SNARE complex involved in synaptic exocytosis at 2.4 Åresolution, *Nature* **395**, 347–353.

Thien, C.B.F., Langdon, W.Y. (2001) Cbl:many adaptions to regulate protein tyrosine kinases, *Nature Rev.* **2**, 294–305.

Toonen, R.F., Verhage, M. (2003) Vesicle trafficking: pleasure and pain from SM genes, *Trends Cell Biol.* **13**, 177–186.

Traub, L.M. (2003) Sorting it out: AP-2 and alternate clathrin adaptors in endocytic cargo selection, *J. Cell Biol.* **163**, 203–208.

van der Bliek, A.M., Meyerowitz, E.M. (1991) Dynamin-like protein encoded by the drosophilia shibire gene associated with vesicular traffic, *Nature* **351**, 411–414.

Varlamov, O., Volchuk, A., Rahimian, V., Doege, C.A., Paumet, F., Eng, W.S., Arango, N., Parlati, F., Ravazzola, M., Orci, L., Sollner, T.H., Rothman, J.E. (2004) i-SNAREs: inhibitory SNAREs that fine-tune the specificity of membrane fusion, *J. Cell Biol.* **164**, 79–88.

Weber, T., Zemelman, B.V., McNew, J.A., Westermann, B., Gmachl, M., Parlati, F., Sollner, T.H., Rothman, J.E. (1998) SNAREpins: minimal machinery for membrane fusion, *Cell* **92**, 759–772.

Weimbs, T., Mostov, K., Low, S.H., Hofmann, K. (1998) A model for structural similarity between different SNARE complexes based on sequence relationships, *Trends Cell Biol.* **8**, 260–262.

Whiteheart, S.W., Schraw, T., Matveeva, E.A. (2001) N-ethylmaleimide sensitive factor (NSF) structure and function, *Int. Rev. Cytol.* **207**, 71–112.

Wieland, F.T., Gleason, M.L., Serafini, T.A., Rothman, J.E. (1987) The rate of bulk flow from the endoplasmic reticulum to the cell surface, *Cell* **50**, 289–300.

Wilson, D.W., Wilcox, C.A., Flynn, G.C., Chen, E., Kuang, W.-J., Hazel, W.J., Block, M.R., Ullrich, A., Rothman, J.E. (1989) A fusion protein required for vesicle mediated transport in both mammalian cells and yeast, *Nature* **339**, 355–359.

Xu, T., Rammner, B., Margittai, M., Artalejo, A.R., Neher, E., Jahn, R. (1999) Inhibition of SNARE complex assembly differentially affects kinetic components of exocytosis, *Cell* **99**, 713–722.

Yang, B., Gonzalez, L., Prekeris, R., Steegmaier, M., Advani, R.J., Scheller, R.H. (1999) SNARE interactions are not selective, *J. Biol. Chem.* **274**, 5649–5653.

Yang, C., Coker, K.J., Kim, J.K., Mora, S., Thurmond, D.C., Davis, A.C., Yang, B., Williamson, R.A., Shulman, G.I., Pessin, J.E. (2001) Syntaxin 4 heterozygous knockout mice develop muscle insulin resistance, *J. Clin. Invest.* **107**, 1311–1318.

Zimmerberg, J., McLaughlin, S. (2004) Membrane curvature: How BAR domains bend bilayers, *Curr. Biol.* **14**, R250–R252.

Membrane Transport

Caroline Engvall and Per Lundahl
Uppsala University, Uppsala, Sweden

Encyclopedia of Molecular Cell Biology and Molecular Medicine, 2nd Edition. Volume 8
Edited by Robert A. Meyers.
Copyright © 2005 Wiley-VCH Verlag GmbH & Co. KGaA, Weinheim
ISBN: 3-527-30550-5

Keywords

Amphiphilic Molecule
A molecule with both polar and nonpolar elements.

Cell Membrane or Plasma Membraneff
The boundary layer of a cell, composed of a mosaic of proteins in phospholipid bilayers, which is supplemented with cholesterol in animal cells.

Hydrophilic Molecule or Atom
A highly water-soluble polar molecule or atom.

Hydrophobic Molecule or Atom
A nonpolar molecule or atom that is highly soluble in organic solvents.

Liposome
A body consisting of an artificial lipid bilayer (or bilayers) enclosing an aqueous compartment (or aqueous layers).

Membrane Protein
An integral membrane protein is inserted into a lipid bilayer, and is called a transmembrane protein if exposed on both membrane faces, whereas a peripheral membrane protein is bound to one of the membrane surfaces.

Permeability (cm^3 s^{-1}) of a Membrane for a Solute
The ratio between the flow across the membrane (mol s^{-1}) and the concentration difference (mol cm^{-3}).

Phospholipid Bilayer
Thin lamellae composed of two layers of phospholipid molecules oriented to contain a hydrophobic, oil-like interior and expose hydrophilic structures toward the aqueous phases.

Phospholipid Molecule
An amphiphilic molecule with two hydrophobic chains and a polar headgroup containing a phosphate group and choline, ethanolamine, serine, inositol, or a similar moiety.

Proteoliposome
Liposome containing one or more membrane proteins inserted into the bilayers.

■ Small hydrophobic or amphiphilic solutes, such as oxygen or drugs, partition into the lipid bilayer of a cell membrane, diffuse within it, and partition out from it. More polar molecules, for example, glucose, require the help of a membrane protein to traverse the membrane. Porins contain pores of passage, whereas facilitative transporters change conformation to let the substrate through the membrane. Ion channels allow regulated passage of anions or cations. These modes of passive transport are thermally driven toward lower substrate concentration. Active transport proteins concentrate a solute by the use of adenosine triphosphate (ATP) hydrolysis or by means of an ion gradient whereby ions passing through the transporter drive the uphill transport of the substrate. Xenobiotics are expelled from cells by active transporters, for example, P-glycoprotein. Endo- or exocytosis transports compounds across the cell membrane in vesicles budding off from or fusing with the membranes, respectively. Partitioning and transport are analyzed with cells or cell models.

1
Cell Membranes

1.1
Structure

A membrane (schematically shown in Fig. 1) separates the cell from its surroundings, yet allows contact across the barrier, and membranes form organelles within cells. The thin, flexible membranes are composed of membrane proteins imbedded in lipid bilayers essentially impermeable to ions and large molecules. Protons, small molecules such as water (H_2O) and oxygen (O_2), and larger amphiphilic molecules, such as anesthetics and drugs can pass the bilayers slowly, whereas transmembrane proteins control the selective uptake of nutrients and the export of metabolic products, regulate the balance of ions and solutes between the exterior and interior of the cell or transmit signals.

Certain membranes collect or transform energy.

Phosphatidylcholine, -ethanolamine, and -serine are major lipid components of mammalian cell membranes (Fig. 2A), and contain two fatty acyl chains of 12 to 24 carbons, one saturated and one unsaturated, esterified to carbons 1 and 2, respectively, of a glycerol backbone, whereas carbon 3 is connected to a phosphate group, which in turn carries an alcohol (X in Fig. 2A) to form a polar headgroup. Other lipid components are the phosphatidylcholine analogue sphingomyelin, which contains amide bonds instead of ester bonds, and cholesterol (Fig. 2B), with four fused sterol rings bearing a single hydroxyl group on ring A and a branched hydrophobic tail on ring D. The phospholipid/cholesterol bilayer of an animal cell membrane is approximately 4.5-nm thick and exposes the polar phospholipid headgroups on each face of the

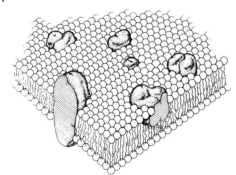

Fig. 1 The fluid mosaic model of the lipid-protein structure of a biological membrane. Reprinted, with permission, from Singer, S.J., Nicolson, G.L. (1972). The fluid mosaic model of the structure of cell membranes. Cell membranes are viewed as two-dimensional solutions of oriented globular proteins and lipids, *Science* **175**, 720–731. Copyright (1972) American Association for the Advancement of Science.

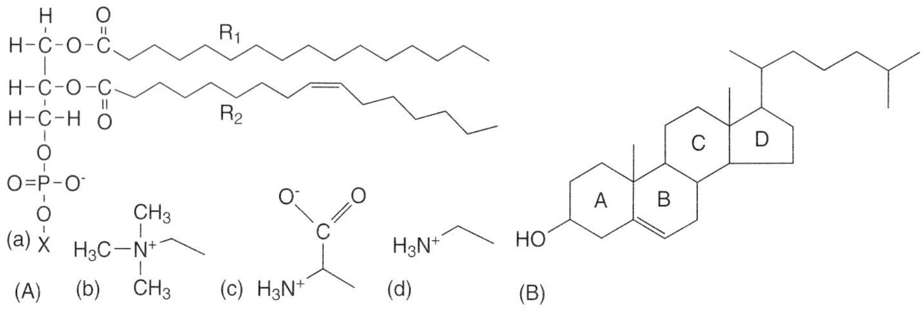

Fig. 2 (A) (a) Chemical structure of a glycerophospholipid with two 16-carbon chains where R_1 is saturated and R_2 is unsaturated with a double bond. X can be, for example, (b) choline, (c) serine, or (d) ethanolamine. (B) Cholesterol. The four fused rings are denoted A to D.

membrane. The overall structure of a lipid bilayer resembles the structure of a double layer of crystalline phospholipid, such as that determined by X-ray diffraction in Fig. 3, except that the lipids rotate, move their chains, and diffuse laterally in the liquid-crystalline phase, which solidifies to a gel-like consistence when the temperature is decreased below a transition temperature that depends mainly on the degree of unsaturation. Refinement of X-ray and neutron scattering data show the mobility of the various parts of the phospholipid molecules in the different regions of the membrane. They move in and out from their average position within certain limits. The fluidity of the membrane is modulated by cholesterol molecules, each

one with its hydroxyl group in one of the two outer polar membrane regions and its rigid sterol skeleton in one half of the hydrophobic oil-like liquid in between the polar surfaces.

Integral or transmembrane proteins (which may be monomeric, dimeric, or oligomeric) are inserted into the phospholipid bilayer. Most mammalian integral membrane proteins are designed as a hydrophobic bunch of α-helices, each one containing about 20 amino acid residues and connected by hydrophilic loops. An individual three-dimensional structure allows a specific function for each protein, such as signal transduction, nutrient transport, or transport of xenobiotics across the membrane. Other membrane

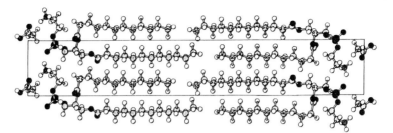

Fig. 3 The structure of crystalline dilauroylphosphatidylethanolamine acetic acid. The acetic acid molecules are seen outside the phospholipid molecules. The crystallographic unit cell is shown. Reprinted with permission from M. Karplus and the copyright owner from Biological membranes: a molecular perspective from computation and experiment, Eds. K.M. Merz, B. Roux, An empirical potential energy function for phospholipids: Criteria for parameter optimization and applications, by M. Schlenkrich, J. Brickmann, A.D. MacKerell Jr., M. Karplus, p. 31–81, Figure 14C, Birkhäuser, Boston. Copyright (1996) Springer.

proteins of entirely different structures (peripheral membrane proteins) are coupled to the membrane surfaces one by one or in chains forming cytoskeletal fibrous networks.

1.1.1 Measurements Indicating the Membrane Structure

The osmotic properties of plant cells were studied during the last half of the nineteenth century. Overton suggested (in 1899–1900) that the osmotic barrier of the cell consisted of a membrane "impregnated" with lipids of properties similar to cholesterol esters and phospholipids. The lipid–protein structure of cell membranes was debated during the first seven decades of the twentieth century. In 1925, Gorter and Grendel used a Langmuir trough to determine the surface area of a monolayer of the lipids extracted from one red blood cell. Calculations involving two errors that fortuitously cancelled each other indicated that the lipids formed a bilayer in the cell membrane. In 1972, Singer and Nicolson proposed, on the basis of thermodynamic considerations, that the hydrophilic and hydrophobic properties of

membrane components and experimental data on cell membranes were consistent with a "fluid mosaic" membrane model with proteins inserted into and floating around in a phospholipid bilayer (Fig. 1). One of the observations supporting this model is the random spreading of the $Rh_0(D)$ integral membrane protein molecules of the human red blood cell over the membrane as revealed by electron microscopy following ferritin staining of the $Rh_0(D)$ proteins by using antibodies. This view has essentially persisted since then, although peripheral membrane proteins are associated with the membrane, some integral membrane proteins are coupled to cytoskeletal networks of proteins at the membrane surface or in the cytoplasm, and many integral membrane proteins are dimers or oligomers of one or more subunits. In 1975, Henderson and Unwin described the three-dimensional structure of the small light-driven proton pump, bacteriorhodopsin, according to electron diffraction data collected from crystalline patches, called purple membranes, of this transmembrane protein in

Halobacterium halobium. The first detailed X-ray crystallographic three-dimensional structure of a transmembrane protein, the big photosynthetic reaction center, which is involved in photosynthesis in *Rhodopseudomonas viridis*, was presented in 1985 by Deisenhofer, Huber, and Michel. The development of genetic techniques and the high stability and availability of bacterial membrane proteins has transferred the emphasis of the crystallographic membrane research to such proteins. These show close resemblance to human transmembrane proteins in the few cases that, so far, have been more closely studied. One example is the similarity between the *Escherichia coli* lactose permease, a membrane protein that mediates lactose transport across the inner bacterial membrane, and the human facilitated glucose transporter GLUT1.

2
Transport Across Cell Membranes

Solutes enter or leave a cell (Fig. 4) by partitioning into and diffusing across the bilayer, by protein-mediated transport or by endo- or exocytosis. Small amphiphilic molecules, for example, most orally administered drugs, may use the nonspecific partition-diffusion transport (Fig. 4a,b), which slows down with decreasing temperature but cannot be stopped. Small hydrophilic solutes diffuse through passages between the cells in cell layers; this is called the paracellular route (not illustrated). Ions and polar molecules partition only weakly into the hydrophobic region of the membrane and are helped by transmembrane proteins to pass the membranes at a high rate. Such transport proteins (Fig. 4c) either provide nonselective pores or offer specific binding, passage, and release of the substrate. Depending on the type of organism or organ, the transport is either passive, toward a lower solute concentration, or active, that is, concentrating the solute by use of an energy source, for example, adenosine triphosphate (ATP). Alternatively, a gradient of protons or other ions in one direction can be utilized to drive a cotransport of the solute toward higher concentration and the ion toward lower concentration.

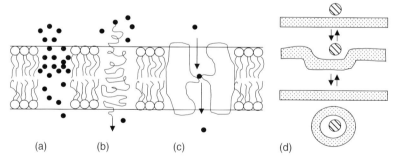

(a)   (b)   (c)   (d)

Fig. 4 Schematic illustrations of the permeation of a solute across a cell membrane by (a,b) partition–diffusion through the lipid bilayer (for clarity the lipid molecules are not drawn where the solute resides), where (a) exemplifies the concentration gradient formed, and (b) depicts the Brownian motion of a single solute molecule, (c) diffusion through a pore or channel of a membrane protein or facilitated diffusion mediated by a substrate-binding membrane protein, and (d) endo- or exocytosis.

Active multisubstrate transporters, such as the P-glycoprotein, expel xenobiotics from cells and cell organelles. Endo- and exocytosis are involved in the import and export, respectively, of certain molecules, for example, lipoproteins (Fig. 4d). The research on the many essential functions of membrane lipids and proteins has been hampered by methodological problems due to the insolubility of most membrane components in water.

2.1
Partition-diffusion

The molecules or atoms in gases or liquids move and collide randomly because of their thermal energy. Molecules that are nonrandomly distributed will move from a higher concentration toward a lower concentration because a collision is more probable where the molecules are closer to each other. This is diffusion, which also occurs in the semiliquid interior of a biological membrane. Solute diffusion in and across lipid bilayers is sometimes called simple diffusion, partition-diffusion, or nonfacilitated diffusion, to distinguish the process from facilitated diffusion; a thermally driven solute transport by a membrane protein toward a lower concentration.

2.1.1   Theory and Examples
The diffusional flux, J, (mol s^{-1}), of molecules from a region of high concentration toward regions of lower concentration can be expressed by Fick's first law of diffusion,

$$J = -D\frac{dc}{dx} \qquad (1)$$

with D, the diffusion coefficient (cm^2 s^{-1}); and dc/dx, the concentration gradient, that is, the derivative of the solute

concentration with regard to the distance across the membrane. In the case of diffusion, across a membrane forming a hydrophobic barrier, Equation 1 can be written

$$J = \frac{A \times D_m \times \Delta C_m}{d} \qquad (2)$$

with A, the area (cm^2) of the membrane; d, the thickness (about 3×10^{-7} cm) of the hydrophobic region; D_m, the diffusion coefficient of the solute in the hydrophobic region; and ΔC_m, the concentration difference of the solute across the hydrophobic region.

A solute partitions between the hydrophobic region of a membrane and the water (Fig. 4a). At equilibrium, the solute concentrations within and outside the membrane balance the energy change as the solute crosses the boundary. The partition coefficient, K, is defined as the ratio between the solute concentrations in the nonpolar and aqueous phases. Therefore,

$$\Delta C_m = K \times \Delta C_{aq} \qquad (3)$$

with ΔC_{aq}, the concentration difference between the water on each side of the membrane.

The permeation of a small, hydrophilic molecule is slow because of weak partitioning into the membrane, creating a shallow concentration gradient within the membrane. More hydrophobic compounds partition more strongly into the membrane, but may suffer from a low aqueous solubility. Amphiphilic compounds thus require a certain balance between the amount and distribution of hydrophilic and hydrophobic structural features in order to permeate a membrane quickly.

A correlation between the partition and permeability coefficients for many solutes has been observed for a long time. In the partition–diffusion model, K and D_m determine the rate of solute migration across a membrane. The permeability coefficient, P_s (cm s^{-1}), for the solute S is given by

$$P_s = \frac{D_m \times K}{d} \qquad (4)$$

and the permeability, $P_s \times A$, equals the flux J (mol s^{-1}) of S divided by the concentration difference, ΔC_{aq} (mol cm^{-3}). For example, the permeability coefficient for glucose is 2×10^{-10} cm s^{-1} and that for O_2 is 2×10^{-3} cm s^{-1}. D_m is approximately 10^{-8} cm^2 s^{-1} for glucose and 0.5×10^{-5} cm^2 s^{-1} for O_2. Equation 4 thus requires that the membrane–water partition coefficient of oxygen is 10^4-fold that of glucose.

A solute that enters or leaves the membrane interacts with the electric charges and dipole moments of the polar headgroup region of the lipid bilayers, of oriented water molecules, of the carbonyl oxygens of the fatty acyl chains and of membrane proteins. Upon steady state transport, the solute concentration, therefore, varies nonlinearly across the membrane and also laterally, which complicates the situation described by Equations 2–4 and affects different transported compounds in different ways. Very small molecules may diffuse at an enhanced speed in the membrane by jumping through empty spaces between the lipid molecules.

The partition coefficient, K, can be expressed in the free energy change, $-\Delta G$:

$$K = e^{\frac{-\Delta G}{RT}} \qquad (5)$$

with R, the general gas constant and T, the temperature. ΔG can be described by a hydrophobic-effect term and four electrostatic terms: the Born energy of transfer of an ion or dipole into a hydrophobic solvent; the image energy

owing to the thinness of the membrane; the dipole-potential energy caused by the polar headgroup region; and the energy change upon the formation or breakage of hydrogen bonds and the binding or release of hydrating water. The partition–diffusion model works for hydrated ions, although cations may pass thin bilayers through transient pores and H$^+$ may be transferred along water strands in transient defects of the bilayer.

The equilibrium ratio between the concentrations of an ionic compound on each side of a membrane is directly affected by the transmembrane potential. A 10-fold concentration ratio of a univalent ion balances a 59-mV potential difference and a 100-fold ratio corresponds to twice that potential difference.

2.1.2 Measurements

The partitioning of solutes into lipid bilayers or cell membranes can be determined by chromatography on liposomes, proteoliposomes, membrane vesicles or cells adsorbed to or entrapped in gel beads packed into columns. This is called immobilized-liposome chromatography or immobilized-biomembrane chromatography (ILC, IBC). The chromatographic retardations of the solutes divided by the lipid amount in the column provide partition data, which correlate with octanol-water (P_{oct}) partition data, as do solute partition data obtained by pH titration of the solute in a liposome suspension, or by optical surface-plasmon resonance analysis of solute binding to liposomes coupled to a planar surface. These techniques reveal details of solute interactions with membranes. The partitioning of drugs into lipid bilayers decreases in the order neutral, positively charged, and negatively charged drugs compared to partitioning into octanol. Cholesterol in the bilayers

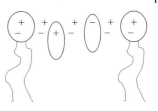

Fig. 5 Proposed distribution of positively and negatively charged amphiphilic solute molecules at the interface between the polar and nonpolar regions of the membrane. The lower half of the solute molecule is thought to be predominantly hydrophobic. Electrostatic effects cause different distributions of the two charged solutes, and thereby stronger or weaker hydrophobic effect.

decreases the partitioning in proportion to its amount. Membrane proteins enhance or decrease the partitioning by electrostatic interactions in combination with hydrophobic partitioning; and positively charged drugs partition relatively strongly into the hydrophobic region of the bilayer, whereas negatively charged drugs tend to dissociate more rapidly from liposome surfaces (Fig. 5). The simpler model of drug partitioning into a layer of (oxyethylene)$_{23}$ dodecyl ether molecules adsorbed to a stationary phase to mimic the membrane structure upon micellar liquid chromatography shows strikingly high correlations with ILC/IBC partitioning data obtained for a set of drugs on liposomes, red blood cell membrane vesicles, red blood cells, and ghosts.

A number of methods for permeability measurements are available. The permeabilities of cell membranes for solutes determined by observations of osmotic swelling or shrinking of cells by microscopy or weighing correlate rectilinearly with the oil–water partitioning or with P_{oct} of the osmotically active solutes. The permeability of liposomes for protons has been determined by measuring the pH-dependent fluorescence of pyranine, a cation-sensitive electrode has been used for potassium ions, and light-scattering analysis of osmotic swelling by use of a stop-flow apparatus has been used for water, glycerol, and urea. The liposomal

and proteoliposomal permeabilities for radioactively labeled glucose, tyrosine, and tryptophan have been estimated by incubation followed by size-exclusion chromatographic separation monitored by flow-scintillation detection. The permeability coefficients for α-substituted p-toluic acids and glucose have been determined by incorporation into liposomes, removal of the free compound by size-exclusion chromatography, collection of the compound released during incubation by ultrafiltration and determination of the amount of the transported compound by high performance liquid chromatography (HPLC).

In the pharmaceutical industry, the use of sealed monolayers of cultured epithelial tumor cells, for example, Caco-2 cells, has achieved a widespread practical application in the determination of permeabilities. The drawbacks are inconsistency among permeability values determined in different laboratories and the time-consuming culturing of the cell layers.

2.2
Diffusion through Porins

Pore-forming proteins, called porins, provide openings that allow diffusional passage of polar molecules of moderate size. There are 10 families of aquaporins that let water pass rapidly through membranes. Such a protein, in red blood cells is a tetramer of M_r 28 000-subunits, each containing six transmembrane α-helices forming a barrel that leads to a water-selective pore, which does not allow ion passage. The pore contains a hydrophobic section with three widely separated hydrophilic sites. Aquaporins in kidney cell membranes are involved in the reabsorption of 150 l of water per day.

In the outer membrane of gram-negative bacteria, for example, *E. coli*, porins provide openings that allow diffusional passage of polar molecules of M_r up to about 600. These porins are trimers of hollow cylinders, β-barrels, with a hydrophobic belt on the outside in contact with the fatty acyl chains of the surrounding phospholipids. The pores are hydrophilic and water-filled with protruding amino acid side chains that limit the size of the solutes that can pass by diffusion toward lower concentration. Similar size-selective porins are also found in the outer membrane of mitochondria.

The efflux of a variety of antibacterial components, including protein toxins, from *E. coli* and other gram-negative bacteria takes place by interaction between inner-membrane proteins and an outer-membrane TolC protein that spans both the membrane and the periplasmic space with a 14-nm tunnel, offering a transport mechanism similar to that of the porins, which span the outer membrane only. However, the tunnel can be sealed and unsealed to regulate the transport.

2.3
Diffusion through Ion Channels

2.3.1 Chloride Channels
Chloride channels allow the passage of different anions, of which chloride is physiologically the most common. The channels are divided into three classes, the dimeric

chloride channels (ClC), the cystic fibrosis transmembrane conductance regulator (CFTR) and the ligand-gated GABA and glycine receptors.

The chloride/hydrogen carbonate (anion) exchange protein of human red cells has been studied for decades and the sequence was deduced in 1985. A bacterial ClC has been crystallized and shows a complex pattern of 18 tilted α-helices. The ClC in the electric fish *Torpedo* contains two gated barrels that open and close independently to give three conductance levels, zero (both closed), medium (one open), and high (both open). The same behavior is shared by the muscle chloride channel expressed in oocytes from the African tree frog, *Xenopus laevis*.

The special CFTR type of chloride channel that regulates salt and water transport across epithelial tissue is an ATP binding cassette (ABC) transport protein activated by phosphorylation of a cytoplasmic domain by protein kinase A. Cystic fibrosis is caused by decreased CFTR activity in the lung and secretory diarrhea is caused by increased CFTR activity in the gut.

2.3.2 Na⁺, Ca²⁺, and K⁺ Channels
Voltage-dependent channel proteins that allow passage of Na^+, Ca^{2+}, or K^+ are tetramers of identical or similar subunits. The three-dimensional structures of potassium channel domains have recently been solved by X-ray crystallography and NMR spectroscopy. In the known K^+ channels, the "pore-forming region" of each subunit joins with its neighbors in the tetramer to constitute a channel. In *Streptomyces lividans*, the short K^+-ion selective part of the channel is lined with carbonyl oxygen atoms and accommodates two non-hydrated K^+ ions at a time, whereas Na^+ ions are too small to fit. A wider cavity of

the K^+ pathway is surrounded by helix-end dipoles.

2.4
Facilitated Diffusion

Many solutes are transported passively and selectively by so-called facilitated transporters. Such transmembrane proteins seem to consist of multi-α-helical hydrophobic regions with connecting hydrophilic loops, and the various transporters form large families of related membrane proteins from different organisms and organs with various substrate specificities. As for passage through porins and ion channels, the driving process is thermal diffusion toward lower concentration. However, the facilitated transport involves recognition and binding of the substrate, with a strictly limited range of allowed substrates for each protein. Facilitated diffusion reaches a saturation rate at high substrate concentration and can be competitively inhibited. The basic kinetic equation resembles the Michaelis–Menten equation for the rate of enzymatic reactions. The flux, J, depends hyperbolically on the substrate concentration, C:

$$J = \frac{J_{max} \times C}{K_m + C} \qquad (6)$$

with J_{max}, the maximal flux and K_m, the substrate concentration at half the maximal flux, equivalent to the Michaelis–Menten constant.

2.4.1 Examples
Detailed studies of the kinetics of transport of certain solutes have been carried through in red blood cells and model systems, for example, with GLUT1, which transports D-glucose and other monohexoses, particularly in red blood cells,

epithelial cells, and the blood-brain barrier. Low expression of GLUT1 or truncation of the protein, in two infants, has been shown to lead to a low glucose concentration in the cerebrospinal fluid and delayed development and seizures; owing to insufficient transport of glucose across the blood-brain barrier. The single subunit of this oligomeric protein probably contains 12 transmembrane α-helices. GLUT1's basic functional unit is the dimer, in which both the subunits switch 100 to 1000 times per second between two alternative conformations in a synchronized way, such that one subunit can bind a substrate molecule while the other can release another substrate molecule (Fig. 6). The thermal energy seems to be too low to drive the oscillations between the two states in a single, separate monomer, but the coupled mechanism of the conformational switches in the dimer may provide a much lower activation energy. Another mechanism involving evaporation and condensation of water in the glucose-binding spaces has been proposed. The three-dimensional structure of the protein is unknown, but a model structure has been designed by combining biochemical and mutagenesis data for GLUT1 with established similar data for a related active transporter, the lactose permease from *E. coli*. The packing of the 12 helices was deduced mainly from site-specific mutagenesis experiments. The GLUT1 model indicated the existence of a hydrophilic pore with two deeply embedded glucose-binding sites.

The blood glucose level is partly regulated by insulin. In muscle and fat cells, the GLUT4 glucose transporter, which is closely related to GLUT1, can be recruited from internal membrane systems to the cell membrane to enhance the rate of glucose transport. Insulin binds to a receptor in the cell membrane and

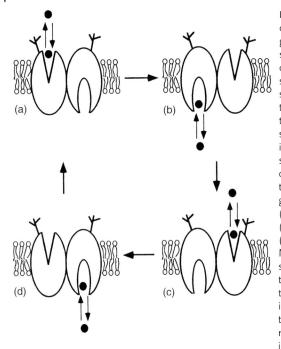

Fig. 6 The dimeric alternating-conformation model of facilitated glucose transport by GLUT1. Each subunit is shown with its oligosaccharide and, in each moment, a single substrate-binding site. The subunits are coupled to each other in the way that the change from outward- to inward-facing binding site of one subunit occurs together with a change in the opposite direction for the other subunit. (a) Glucose binds to the outward-facing site of one subunit (left), the conformation switches, and (b) the glucose molecule is released. (c) Glucose binds to the other subunit (right), the conformation switches, and (d) the glucose molecule is released. Next, the binding in (a) is repeated, and so on. Also, in the absence of glucose, the conformation alternates between the two conformations. The competitive inhibitor cytochalasin B binds only to the inward-facing opening and the number of cytochalasin B-binding sites is one per dimer in the native protein.

triggers the migration of vesicular internal membranes containing GLUT4 to the cell membrane, with which the vesicles then fuse. Too low insulin concentrations, or failure of the insulin receptor to convey the insulin signal to the vesicle migration system may cause diabetes of type 1 in children, or type 2 in adults or elderly persons, respectively. The response to insulin is also modulated in other ways, for example, by the lack of a membrane protein called FAT/CD36, whereby fatty acid uptake in various tissues decreases strongly.

The transport of cholesterol and phospholipids from the high-density lipoprotein (HDL) into the brush border membranes of the small intestine is mediated by receptor proteins. HDL binds to proteins in the membranes for the transfer of the lipids. HDL also absorbs cholesterol diffusing from cell membranes or transported out of the cell by facilitated diffusion, thereby protecting against cardiovascular diseases. Low-density lipoprotein (LDL) transfers cholesterol into cells by receptor-mediated endocytosis (Sect. 2.6).

Certain "equilibrative" nucleoside transporters are very similar to the GLUTs. Two cDNAs from placenta encoding two such transporters have been expressed. These proteins are predicted to contain 11 hydrophobic transmembrane α-helices. One of the transporters was inhibited by nitrobenzylthioinosine and facilitated the transport of adenosine and chemotherapeutic nucleosides.

Nucleotide sugars are transported across the membranes of the endoplasmic reticulum and the Golgi compartments in eukaryotic cells in exchange for nucleoside monophosphates to serve in the glycosylation of proteins and lipids.

Among amino acid transporters, the system L for large, neutral amino acids is an unusual, facilitated system, which appears as a heterodimer containing a cell-surface glycoprotein. Transport of the essential amino acid tryptophan and of tyrosine, both of which are precursors of signal substances in the brain, is mediated by the recently discovered TAT protein.

2.4.2 Measurements

Facilitated diffusion can be analyzed by the use of cells, cell membrane vesicles, or proteoliposomes, which are incubated with radioactively labeled substrates in most types of analyses. This closely resembles permeability measurements discussed briefly in Sect. 2.1.2, although the facilitated diffusion usually is much more rapid and allows the use of inhibitors. Both these circumstances make the determinations relatively easy. Human red blood cells or *X. laevis* oocytes have often been used. The oocytes allow expression of membrane proteins that they normally do not contain. The transport is stopped by the use of an inhibitor, if available, and the cells and so on that contain the transported substrate are separated from free substrate by centrifugation, filtration, or size-exclusion chromatography. The radioactive substrate may be added outside the cells and so on, for determination of inward transport, or loaded inside before washing to determine outward transport. For equilibrium exchange measurements, the substrate is present on both sides of the membrane at the same concentration, whereas the radioactive tracer is added to one side only.

2.5
Active Transport

In contrast to the passive facilitated diffusion, which uses thermal energy only, active transport of certain substances, for example, sodium and calcium ions, amino acids, and sugars, against their concentration gradients requires chemical energy provided by hydrolysis of ATP; light energy, as in the purple membrane (primary active transport), or energy supplied indirectly by the electromotive force of an ion gradient (secondary active transport). In the latter case, the concentrated substrate is cotransported with an ion. The two transported entities can either migrate in the same direction through a transporter called symport, or in opposite directions through an antiport.

2.5.1 Examples

A primary active transporter is the P-type (Na^+, K^+)-ATPase, which transports sodium ions in one direction, out from the cell, and potassium ions in the opposite direction, both against the concentration differences, by use of energy obtained by hydrolysis of ATP. Similarly, as for facilitated diffusion, a Michaelis–Menten constant, K_m, can be assigned to each of the ions and to each of the components, ATP, adenosine diphosphate (ADP), and phosphate ions. The protein is composed of α/β-heterodimers of an α-subunit containing 10 transmembrane α-helices with the longest connecting loop between helices 4 and 5, and a β-subunit with a single transmembrane helix and a relatively long periplasmic polypeptide chain with three Cys–Cys cross-links. The α-subunit seems to alternate between two distinctly different conformations.

The (Na^+, K^+)-ATPases form one group of cation transporters together with the closely related Ca^{2+}-ATPases (Fig. 7). One of these autophosphorylated, P-type, ATPases of M_r 110 000 pumps Ca^{2+} released in skeletal muscles back into the sarcoplasmic reticulum. This protein has

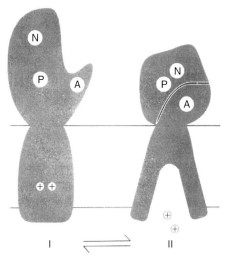

Fig. 7 Mechanism of a P-type ATPase. A schematic drawing of the conformational changes in the α-subunit of the sarcoplasmic reticulum Ca^{2+}-ATPase upon ion transport. This transporter has been crystallized and the structure was solved. It is thought to be closely similar to Na^+, K^+-ATPases. P, phosphorylation domain, A, actuator domain, and N, nucleotide binding domain. To the left (I) the protein is open to the cytoplasm but closed to the A domain. In (II), the protein is open to the membrane but closed to the cytoplasm. The change from I to II involves motion of the N and P domains toward each other and a rotation of the A domain. Reprinted with permission from J.H. Kaplan and the copyright owner. From Kaplan, J.H. (2002) Biochemistry of Na,K-ATPase, *Annual Reviews of Biochemistry.* **71**, 511–535. Copyright 2002 by Annual Reviews, www.annualreviews.org.

been determined by X-ray diffraction analysis of protein crystals to consist of a single subunit with 10 transmembrane α-helices. The phosphorylation and adenosine-binding sites reside on separate cytoplasmic domains.

Another type of primary active transporters, the ABC transporters, belong to a family of proteins that translocate various substrates across cell membranes. They contain two transmembrane domains forming a pathway for the substrate and two ABCs, which cause a conformational change of the pathway, effecting translocation of the substrate, by use of the energy dissipated upon hydrolysis of ATP to form ADP.

There are several ABC multidrug-resistance proteins, for example, the P-glycoprotein and MRP1 (multidrug resistance protein 1). The P-glycoprotein, a membrane protein with M_r of about 170 000, works as a dimer of two homologous subunits with six transmembrane helices and one ATP binding domain per subunit. The protein transports many substances, such as xenobiotics, which have in common that they are quite hydrophobic and, therefore, accumulate in the membrane, have an M_r of 300 to 2000, and are neutral or positively charged at pH 7. At least four transport and regulatory binding sites exist, which are able to switch between high and low affinities. Only one of the ABCs hydrolyzes ATP at a time and ADP release may be rate limiting. These transporters cause problems in cancer therapy. The cancer cells become resistant to certain drugs upon overexpression of the P-glycoprotein, or other such transporters, which extrude the drugs from the cells. These proteins are also widely distributed in normal tissues, for example, in liver, intestine, and kidney.

The crystal structure has been determined for another ABC transporter, the BtuCD protein in *E. coli*, which mediates vitamin B_{12} uptake through a transmembrane dimer containing, in all, 20 helices. ATP binding to each of the cassettes and ATP hydrolysis cause an opening of the vitamin B_{12} pathway toward the cytoplasm.

The lactose permease from *E. coli* belongs to the secondary active transporters and is a functionally monomeric H^+/lactose symport. One neutral and five charged amino acid side chains in transmembrane helices cannot be replaced in order to preserve the transport as was shown by genetic-engineering replacement of all amino acid residues, one by one, by a cysteine residue. The expression of the protein is regulated by the presence of glucose and other sugars via a mechanism involving cyclic adenosine monophosphate (cAMP). Two-dimensional crystals have been formed of both lactose permease with cytochrome b_{562} introduced in a cytoplasmic loop and of a transmembrane subunit of the *E. coli* glucose transporter. Trimers and dimers, respectively, were observed by electron microscopy techniques.

Two-dimensional subunit crystals have also been prepared of another type of sugar transporter, the phosphoenolpyruvate:carbohydrate phosphotransferase system (PTS), which transports hexoses and releases them as phosphorylated sugars that cannot leave the cell. The PTS transporter is preferentially used for sugar transport into the cell, and the amount of other sugar transporters is regulated accordingly.

Glucose can be actively taken up in the small intestine by a secondary Na^+/glucose symport. A Na^+ gradient created by a Na^+, K^+-ATPase drives the accumulation of glucose by the Na^+/glucose transporter. Another member of this Na^+/solute symport family

is the high affinity choline transporter, which pumps choline into the presynaptic terminals of cholinergic neurons and thereby regulates the rate of acetylcholine synthesis. Changes in the function of cholinergic neurons are thought to be involved in Alzheimer's disease, Down's syndrome, Parkinson's disease, and schizophrenia.

2.5.2 Measurements

The primary active transport of nutrients and metabolites is determined similarly as passive transport, commonly by use of radioactive tracers. The ATP hydrolysis can be monitored by use of coupled reactions, for example, with pyruvate kinase and lactate dehydrogenase, or by use of separation followed by phosphorus analysis. Ion transport may be determined by the use of an ion-sensitive electrode, and the transmembrane potential and current can be monitored in the case of large cells or planar membranes. Fluorescent and light-absorbing probes sensitive to various parameters, such as pH, are useful tools.

2.6 Endo-, Exo-, and Transcytosis

Endocytosis imports molecules into the cell from the surrounding medium and exocytosis exports molecules from the cytoplasm out to the surroundings by budding off vesicles containing the molecules from the cell membrane or fusing such vesicles with the membrane (Fig. 4d). Receptor-mediated endocytosis is encountered in many cases, for example, upon internalization of LDL lipoproteins bound to LDL receptors in pits coated with the protein clathrin. The binding is followed by an invagination of the cell membrane followed by the budding off of a membrane vesicle within the cell. In another example, receptors activated by the epidermal growth factor are internalized through clathrin-coated pits. Endothelial cells allow transcytosis of proteins by so-called cargo-vesicles and in transendothelial channels formed by such vesicles. These processes are mediated by protein machines. Exocytosis in eukaryotic cells involves an "exocyst," a complex of eight proteins that enables the secretory vesicle to dock on and fuse with the cell membrane.

3
Transport Across Epithelia

3.1
Intestinal Epithelia

The gut absorbs water, nutrients, and certain ions while excluding bacteria and unwanted molecules. Active and passive transporters allow the uptake of sugars, amino acids, nucleosides, and small peptides. For example, glucose is actively absorbed into the epithelial cell by the Na^+/glucose symport and passes on into the blood by facilitated diffusion mediated by GLUT5. The gaps between the apical parts of the epithelial enterocytes are sealed by tight junctions, allowing the passage of small molecules only. The elimination of xenobiotics, such as drugs, is handled by P-glycoproteins that use ATP to drive expulsion of the foreign compounds that have generally entered the cells by partition–diffusion. Drugs are also metabolized by enzymes, for example, cytochrome P450 proteins. Cell lines, such as Caco-2, are cultured to form monolayers that resemble the intestinal epithelia,

and P-glycoprotein expulsion of drugs can be observed as a lower permeation from the apical to the basolateral side (corresponding to transport from the gut to the blood), than in the opposite direction.

3.2
The Blood-brain Barrier

The blood-brain barrier (BBB) consists of the endothelial cells lining the capillary blood vessels in the brain and separates the blood from the extracellular matrix around the astrocytes. Tight junctions of high electrical resistance between the cells limit the paracellular flow, and the amount of endocytotic vesicles is low in the cells, which restricts the transcellular flow. Certain molecules are allowed into the brain by facilitated diffusion via membrane proteins, notably the brain fuel D-glucose, which enters through GLUT1, L-ascorbic acid (vitamin C), and certain amino acids and similar molecules (L-3,4-dihydroxyphenylalanine, L-DOPA). Vitamin C is transported into the brain in the choroid plexus by Na^+-dependent active transporters. Dehydroascorbic acid can be transported by GLUT1 to be reduced in the brain, but this pathway is physiologically unimportant. Even certain macromolecules are believed to enter the brain via a receptor-mediated mechanism, and *Streptococcus pneumoniae* and six more bacterial pathogens can interact with the endothelial cells in an unknown way that allows the bacteria to cross the barrier. The P-glycoprotein is present in substantial amounts in the endothelial cells and tries to eject back into the blood unwanted lipophilic molecules that have diffused into the brain or just into endothelial cells. The embryonic and postnatal development of the BBB is still unclear, and the use of cultured endothelial cells to mimic the physiological properties of the BBB is haunted by difficulties.

4
Solute Affinities for Membrane Proteins

Biospecific interaction between a solute, for example, an inhibitor, and a membrane protein in cells, membrane vesicles, proteoliposomes or other lipid environments can be quantified by immobilized-biomembrane affinity chromatography. Entrapment in gel bead cavities and covalent binding of phospholipid analogues are examples of the immobilization methods used. The retardation of the front of an eluting large-volume sample of the solute reflects the solute affinity for the protein and the solute concentration. In simple cases, a series of runs at different solute concentrations allow calculation of the equilibrium constant and the number of binding sites, provided that the number of binding sites matches the affinity. The glucose transporter from human red blood cells and a recombinant P-glycoprotein showed constant interactant affinities for months in the columns. The solute retardation is caused both by partitioning into the membrane and by binding to active sites of the membrane proteins. Both effects can be measured. A competitor for the solute–protein interaction in the system will decrease the retention volume of the solute to a degree determined by its affinity for the protein. Ligand binding to membrane proteins can also be analyzed by surface-plasmon resonance, and, in some cases, by analysis of changes in the intrinsic fluorescence of tryptophan side chains or other spectral properties upon ligand binding to the protein. A number of related techniques exist, but the fact that most membrane proteins require a lipid environment for activity limits the choice.

5
Prediction and Modeling of Drug Absorption Across Membranes

Some basic requirements for oral administration of new drug candidates are sufficient solubility in water, reasonable partitioning into membranes, and a fair diffusion coefficient. The partitioning is commonly modeled in the pharmaceutical industry by octanol/water partitioning, expressed as the logarithm of the partition coefficient, log P_{oct}. The advantages of log P_{oct} are the vast amount of available data and the extensive experience of the applicability of these data. Lipinski introduced the "rule of five," which can be used as a filter in early drug development. This rule predicts that poor absorption or permeability properties become likely when the drug molecule has more than five H-bond donors, more than 10 H-bond acceptors, when the M_r exceeds 500 and when the log P_{oct} is larger than 5.

Methods for experimental assessment of partitioning into membranes and permeability across membranes are required, although computer calculations are becoming increasingly popular for theoretical analysis. Computational methods for determination of absorption or permeability properties by statistical analysis often use molecular descriptors, such as the polar surface area of the drug and the number of H-bond donors/acceptors.

Molecular dynamics simulations are still at a primitive stage. For example, computer modeling indicates that the diffusion coefficients of O_2 and H_2O are highest in the center of the membrane, 0.9×10^{-5} and 1.4×10^{-5} cm^2 s^{-1}, respectively. An analog (M_r 315) of the drug nifedipine, obtained by replacing NO_2 by CH_3, had a diffusion coefficient of 1.3×10^{-6} cm^2 s^{-1} all across the hydrophobic region of the membrane. Furthermore, pharmacokinetic problems tend to override the shortcomings of the computer predictions of drug absorption. Absorption is a complex process dependent on factors such as the permeability and solubility of the drug, and physiological factors such as pH, intestinal enzymology and motility, so the absorption is therefore not so easily predicted from the structure alone or from computer simulations of drug-membrane interaction.

6
Conclusions and Visions

The permeability of lipid bilayers or biological membranes for a given solute is generally difficult to determine experimentally and can only be estimated approximately from the chemical structure of the solute or by the use of partition data with either octanol–water or membrane–water. These approaches do not truly mimic the interactions between solutes and membranes. Partition analyses with water–membrane systems do not reveal the detailed solute distribution within the membrane that determines the gradient in solute concentration and thereby the solute flux across the membrane. The solute permeability of membranes may, in the future, be determined by use of sensitive detection in miniaturized devices. Calculations by use of the detailed

structures of the solute and the membrane may become more accurate after advances in quantum chemical or physical–chemical methods. In the future, the transport kinetics and transport mechanisms of transmembrane proteins may be evaluated by small-scale permeability assays and advances in crystallization technology, or single-protein molecule diffraction analyses may allow determination of the three-dimensional structures of membrane proteins.

See also Intracellular Fatty Acid Binding Proteins and Fatty Acid Transport; Membrane Traffic: Vesicle Budding and Fusion.

Bibliography

Books and Reviews

Barrett, M.P., Walmsley, A.R., Gould, G.W. (1999) Structure and function of facilitative sugar transporters, *Curr. Opin. Cell Biol.* **11**, 496–502.

Burton, P.S., Goodwin, J.T., Vidmar, T.J., Amore, B.M. (2002) Predicting drug absorption: how nature made it a difficult problem, *J. Pharmacol. Exp. Ther.* **303**, 889–895.

Deamer, D.W., Kleinzeller, A., Fambrough, D.M. (Eds.) (1999) *Current Topics in Membranes*, Vol. 48, *Membrane Permeability – 100 Years since Ernest Overton*, Academic Press, San Diego, CA.

Gerardy-Schahn, R., Oelmann, S., Bakker, H. (2001) Nucleotide sugar transporters: biological and functional aspects, *Biochimie* **83**, 775–782.

Jentsch, T.J., Stein, V., Weinreich, F., Zdebik, A.A. (2002) Molecular structure and physiological function of chloride channels, *Physiol. Rev.* **82**, 503–568.

Kaback, H.R., Sahin-Tóth, M., Weinglass, A.B. (2001) The kamikaze approach to membrane transport, *Nat. Rev. Mol. Cell Biol.* **2**, 610–620.

Kaplan, J.H. (2002) Biochemistry of Na,K-ATPase, *Annu. Rev. Biochem.* **71**, 511–535.

Lin, G. (2002) Insights of high-density lipoprotein apolipoprotein-mediated lipid efflux from cells, *Biochem. Biophys. Res. Commun.* **291**, 727–731.

Mannhold, R., Kubinyi, H., Folkers, G. (Eds.) (2002) *Methods and Principles in Medicinal Chemistry*, Vol. 15, *Drug-Membrane Interactions – Analysis, Drug Distribution, Modeling*, Wiley-VCH, Weinheim.

Merz, K.M. Jr., Roux, B. (Eds.) (1996) *Biological Membranes: A Molecular Perspective from Computation and Experiment*, Birkhäuser, Boston, MA.

Rubin, L.L., Staddon, J.M. (1999) The cell biology of the blood-brain barrier, *Annu. Rev. Neurosci.* **22**, 11–28.

Sauna, Z.E., Smith, M.M., Müller, M., Kerr, K.M., Ambudkar, S.V. (2001) The mechanism of action of multidrug-resistance-linked P-glycoprotein, *J. Bioenerg. Biomembr.* **33**, 481–491.

Simionescu, M., Gafencu, A., Antohe, F. (2002) Transcytosis of plasma macromolecules in endothelial cells: a cell biological survey, *Microsc. Res. Tech.* **57**, 269–288.

Sorkin, A., von Zastrow, M. (2002) Signal transduction and endocytosis: close encounters of many kinds, *Nat. Rev. Mol. Cell Biol.* **3**, 600–614.

Stein, W.D. (1990) *Channels, Carriers, and Pumps – An Introduction to Membrane Transport*, Academic Press, San Diego, CA.

Primary Literature

Appleman, J.R., Lienhard, G.E. (1985) Rapid kinetics of the glucose transporter from human erythrocytes. Detection and measurement of a half-turnover of the purified transporter, *J. Biol. Chem.* **260**, 4575–4578.

Artursson, P. (1990) Epithelial transport of drugs in cell culture. I: A model for studying the passive diffusion of drugs over intestinal absorptive (Caco-2) cells, *J. Pharm. Sci.* **79**, 476–482.

Artursson, P., Karlsson, J. (1991) Correlation between oral drug absorption in humans and apparent drug permeability coefficients in human intestinal epithelial (Caco-2) cells, *Biochem. Biophys. Res. Commun.* **175**, 880–885.

Avdeef, A., Box, K.J., Comer, J.E.A., Hibbert, C., Tam, K.Y. (1998) pH-Metric logP 10. Determination of liposomal membrane-water partition coefficients of ionizable drugs, *Pharm. Res.* **15**, 209–215.

Baird, C.L., Courtenay, E.S., Myszka, D.G. (2002) Surface plasmon resonance characterization of drug/liposome interactions, *Anal. Biochem.* **310**, 93–99.

Beigi, F., Yang, Q., Lundahl, P. (1995) Immobilized-liposome chromatographic analysis of drug partitioning into lipid bilayers, *J. Chromatogr. A* **704**, 315–321.

Cheng, A., van Hoek, A.N., Yeager, M., Verkman, A.S., Mitra, A.K. (1997) Three-dimensional organization of a human water channel, *Nature* **387**, 627–630.

Danelian, E., Karlén, A., Karlsson, R., Winiwarter, S., Hansson, A., Löfås, S., Lennernäs, H., Hämäläinen, M.D. (2000) SPR biosensor studies of the direct interaction between 27 drugs and a liposome surface: correlation with fraction absorbed in humans, *J. Med. Chem.* **43**, 2083–2086.

Doyle, D.A., Morais Cabral, J., Pfuetzner, R.A., Kuo, A., Gulbis, J.M., Cohen, S.L., Chait, B.T., MacKinnon, R. (1998) The structure of the potassium channel: molecular basis of K^+ conduction and selectivity, *Science* **280**, 69–77.

Gottschalk, I., Gustavsson, P.-E., Ersson, B., Lundahl, P. (2003) Improved lectin-mediated immobilization of human red blood cells in superporous agarose beads, *J. Chromatogr. B* **784**, 203–208.

Gottschalk, I., Lundqvist, A., Zeng, C.-M., Lagerquist Hägglund, C., Zuo, S.-S., Brekkan, E., Eaker, D., Lundahl, P. (2000) Conversion between two cytochalasin B-binding states of the human GLUT1 glucose transporter, *Eur. J. Biochem.* **267**, 6875–6882.

Griffiths, M., Yao, S.Y.M., Abidi, F., Phillips, S.E.V., Cass, C.E., Young, J.D., Baldwin, S.A. (1997) Molecular cloning and characterization of a nitrobenzylthioinosine-insensitive (*ei*) equilibrative nucleoside transporter from human placenta, *Biochem. J.* **328**, 739–743.

Griffiths, M., Beaumont, N., Yao, S.Y.M., Sundaram, M., Boumah, C.E., Davies, A., Kwong, F.Y.P., Coe, I., Cass, C.E., Young, J.D., Baldwin, S.A. (1997) Cloning of a human nucleoside transporter implicated in the cellular uptake of adenosine and chemotherapeutic drugs, *Nat. Med.* **3**, 89–93.

Hajri, T., Han, X.X., Bonen, A., Abumrad, N.A. (2002) Defective fatty acid uptake modulates insulin responsiveness and metabolic responses to diet in CD36-null mice, *J. Clin. Invest.* **109**, 1381–1389.

Hamill, S., Cloherty, E.K., Carruthers, A. (1999) The human erythrocyte sugar transporter presents two sugar import sites, *Biochemistry* **38**, 16974–16983.

Harik, S.I., Hall, A.K., Richey, P., Andersson, L., Lundahl, P., Perry, G. (1993) Ontogeny of the erythroid/HepG2-type glucose transporter (GLUT-1) in the rat nervous system, *Dev. Brain. Res.* **72**, 41–49.

Hitchcock, P.B., Mason, R., Thomas, K.M., Shipley, G.G. (1974) Structural chemistry of 1,2 dilauroyl-DL-phosphatidylethanolamine: molecular conformation and intermolecular packing of phospholipids, *Proc. Natl. Acad. Sci. U.S.A.* **71**, 3036–3040.

Koronakis, V., Sharff, A., Koronakis, E., Luisi, B., Hughes, C. (2000) Crystal structure of the bacterial membrane protein TolC central to multidrug efflux and protein export, *Nature* **405**, 914–919.

Kreusch, A., Pfaffinger, P.J., Stevens, C.F., Choe, S. (1998) Crystal structure of the tetramerization domain of the *Shaker* potassium channel, *Nature* **392**, 945–948.

Lagerquist, C., Beigi, F., Karlén, A., Lennernäs, H., Lundahl, P. (2001) Effects of cholesterol and model transmembrane proteins on drug partitioning into lipid bilayers as analysed by immobilized-liposome chromatography, *J. Pharm. Pharmacol.* **53**, 1477–1487.

Lipinski, C.A. (2000) Drug-like properties and the causes of poor solubility and poor permeability, *J. Pharmacol. Toxicol. Methods* **44**, 235–249.

Locher, K.P., Lee, A.T., Rees, D.C. (2002) The *E. coli* BtuCD structure: a framework for ABC transporter architecture and mechanism, *Science* **296**, 1091–1098.

Lux, S.E., John, K.M., Kopito, R.R., Lodish, H.F. (1989) Cloning and characterization of band 3, the human erythrocyte anion-exchange protein (AE1), *Proc. Natl. Acad. Sci. U.S.A.* **86**, 9089–9093.

Martin, C., Berridge, G., Higgins, C.F., Mistry, P., Charlton, P., Callaghan, R. (2000) Communication between multiple drug binding sites on P-glycoprotein, *Mol. Pharmacol.* **58**, 624–632.

Mastroberardino, L., Spindler, B., Pfeiffer, R., Skelly, P.J., Loffing, J., Shoemaker, C.B.,

Verrey, F. (1998) Amino-acid transport by heterodimers of 4F2hc/CD98 and members of a permease family, *Nature* **395**, 288–291.

Molero-Monfort, M., Martín-Biosca, Y., Sagrado, S., Villanueva-Camañas, R.M., Medina-Hernández, M.J. (2000) Micellar liquid chromatography for prediction of drug transport, *J. Chromatogr. A* **870**, 1–11.

Mueckler, M., Caruso, C., Baldwin, S.A., Panico, M., Blench, I., Morris, H.R., Allard, W.J., Lienhard, G.E., Lodish, H.F. (1985) Sequence and structure of a human glucose transporter, *Science* **229**, 941–945.

Naren, A.P., Cormet-Boyaka, E., Fu, J., Villain, M., Blalock, J.E., Quick, M.W., Kirk, K.L. (1999) CFTR chloride channel regulation by an interdomain interaction, *Science* **286**, 544–548.

Noda, M., Shimizu, S., Tanabe, T., Takai, T., Kayano, T., Ikeda, T., Takahashi, H., Nakayama, H., Kanaoka, Y., Minamino, N., Kangawa, K., Matsuo, H., Raftery, M.A., Hirose, T., Inayama, S., Hayashida, H., Miyata, T., Numa, S. (1984) Primary structure of *Electrophorus electricus* sodium channel deduced from cDNA sequence, *Nature* **312**, 121–127.

Okuda, T., Okamura, M., Kaitsuka, C., Haga, T., Gurwitz, D. (2002) Single nucleotide polymorphism of the human high affinity choline transporter alters transport rate, *J. Biol. Chem.* **277**, 45315–45322.

Papazian, D.M., Schwarz, T.L., Tempel, B.L., Jan, Y.N., Jan, Y.L. (1987) Cloning of genomic and complementary DNA from *Shaker*, a putative potassium channel gene from *Drosophila*, *Science* **237**, 749–753.

Paula, S., Volkov, A.G., Van Hoek, A.N., Haines, T.H., Deamer, D.W. (1996) Permeation of protons, potassium ions, and small polar molecules through phospholipid bilayers as a function of membrane thickness, *Biophys. J.* **70**, 339–348.

Pessino, A., Hebert, D.N., Woon, C.W., Harrison, S.A., Clancy, B.M., Buxton, J.M., Carruthers, A., Czech, M.P. (1991) Evidence that functional erythrocyte-type glucose transporters are oligomers, *J. Biol. Chem.* **266**, 20213–20217.

Ramsden, J.J. (1993) Partition coefficients of drugs in bilayer lipid membranes, *Experientia* **49**, 688–692.

Saviane, C., Conti, F., Pusch, M. (1999) The muscle chloride channel ClC-1 has a double-barreled appearance that is differentially affected in dominant and recessive myotonia, *J. Gen. Physiol.* **113**, 457–467.

Schoenmakers, R.G., Stehouwer, M.C., Tukker, J.J. (1999) Structure-transport relationship for the intestinal small-peptide carrier: is the carbonyl group of the peptide bond relevant for transport? *Pharm. Res.* **16**, 62–68.

Seidner, G., Garcia Alvarez, M., Yeh, J.-I., O'Driscoll, K.R., Klepper, J., Stump, T.S., Wang, D., Spinner, N.B., Birnbaum, M.J., De Vivo, D.C. (1998) GLUT-1 deficiency syndrome caused by haploinsufficiency of the blood-brain barrier hexose carrier, *Nat. Genet.* **18**, 188–191.

Singer, S.J., Nicolson, G.L. (1972) The fluid mosaic model of the structure of cell membranes. Cell membranes are viewed as two-dimensional solutions of oriented globular proteins and lipids, *Science* **175**, 720–731.

Sui, H., Han, B.-G., Lee, J.K., Walian, P., Jap, B.K. (2001) Structural basis of water-specific transport through the AQP1 water channel, *Nature* **414**, 872–878.

Sultzman, L.A., Carruthers, A. (1999) Stop-flow analysis of cooperative interactions between GLUT1 sugar import and export sites, *Biochemistry* **38**, 6640–6650.

Suzuki, K., Kono, T. (1980) Evidence that insulin causes translocation of glucose transport activity to the plasma membrane from an intracellular storage site, *Proc. Natl. Acad. Sci. U.S.A.* **77**, 2542–2545.

Tanabe, T., Takeshima, H., Mikami, A., Flockerzi, V., Takahashi, H., Kangawa, K., Kojima, M., Matsuo, H., Hirose, T., Numa, S. (1987) Primary structure of the receptor for calcium channel blockers from skeletal muscle, *Nature* **328**, 313–318.

Toyoshima, C., Nakasako, M., Nomura, H., Ogawa, H. (2000) Crystal structure of the calcium pump of sarcoplasmic reticulum at 2.6 Å resolution, *Nature* **405**, 647–655.

Tsukaguchi, H., Tokui, T., Mackenzie, B., Berger, U.V., Chen, X.-Z., Wang, Y., Brubaker, R.F., Hediger, M.A. (1999) A family of mammalian Na$^+$-dependent L-ascorbic acid transporters, *Nature* **399**, 70–75.

Walz, T., Hirai, T., Murata, K., Heymann, J.B., Mitsuoka, K., Fujiyoshi, Y., Smith, B.L.,

Agre, P., Engel, A. (1997) The three-dimensional structure of aquaporin-1, *Nature* **387**, 624–627.

Wang, Y., Mackenzie, B., Tsukaguchi, H., Weremowicz, S., Morton, C.C., Hediger, M.A. (2000) Human vitamin C (L-ascorbic acid) transporter SVCT1, *Biochem. Biophys. Res. Commun.* **267**, 488–494.

Werder, M., Han, C.-H., Wehrli, E., Bimmler, D., Schulthess, G., Hauser, H. (2001) Role of scavenger receptors SR-BI and CD36 in selective sterol uptake in the small intestine, *Biochemistry* **40**, 11643–11650.

Widdas, W.F. (1998) The glucose transporter of human erythrocytes – working hypothesis for its functional mechanism, *Exp. Physiol.* **83**, 187–194.

Winiwarter, S., Bonham, N.M., Ax, F., Hallberg, A., Lennernäs, H., Karlén, A. (1998) Correlation of human jejunal permeability (in vivo) of drugs with experimentally and theoretically derived parameters. A multivariate data analysis approach, *J. Med. Chem.* **41**, 4939–4949.

Xiang, T.-X., Xu, Y.-H., Anderson, B.D. (1998) The barrier domain for solute permeation varies with lipid bilayer phase structure, *J. Membr. Biol.* **165**, 77–90.

Zhang, J.-Z., Ismail-Beigi, F. (1998) Activation of Glut1 glucose transporter in human erythrocytes, *Arch. Biochem. Biophys.* **356**, 86–92.

Zhang, Y., Leonessa, F., Clarke, R., Wainer, I.W. (2000) Development of an immobilized P-glycoprotein stationary phase for on-line liquid chromatographic determination of drug-binding affinities, *J. Chromatogr. B* **739**, 33–37.

Zhuang, J., Gutknecht, R., Flükiger, K., Hasler, L., Erni, B., Engel, A. (1999) Purification and electron microscopic characterization of the membrane subunit (IICB^{Glc}) of the *Escherichia coli* glucose transporter, *Arch. Biochem. Biophys.* **372**, 89–96.

Zhuang, J., Privé, G.G., Werner, G.E., Ringler, P., Kaback, H.R., Engel, A. (1999) Two-dimensional crystallization of *Escherichia coli* lactose permease, *J. Struct. Biol.* **125**, 63–75.

Zuniga, F.A., Shi, G., Haller, J.F., Rubashkin, A., Flynn, D.R., Iserovich, P., Fischbarg, J. (2001) A three-dimensional model of the human facilitative glucose transporter Glut1, *J. Biol. Chem.* **276**, 44970–44975.

Österberg, T., Svensson, M., Lundahl, P. (2001) Chromatographic retention of drug molecules on immobilised liposomes prepared from egg phospholipids and from chemically pure phospholipids, *Eur. J. Pharm. Sci.* **12**, 427–439.

Memory: *see* Immunologic Memory

Metabolic Basis of Cellular Energy

Meredith F. Ross and Michael P. Murphy
MRC Dunn Human Nutrition Unit, Cambridge, UK

Encyclopedia of Molecular Cell Biology and Molecular Medicine, 2nd Edition. Volume 8
Edited by Robert A. Meyers.
Copyright © 2005 Wiley-VCH Verlag GmbH & Co. KGaA, Weinheim
ISBN: 3-527-30550-5

Keywords

Adenosine Triphosphate (ATP)

The "energy currency" of the cell. The energy derived from metabolism can be stored by synthesizing ATP from ADP and P_i, displacing the ATP hydrolysis reaction from equilibrium by many orders of magnitude. The energy stored in this displacement from equilibrium can then be used to do work as ATP is hydrolyzed back to ADP and P_i. Thus, the synthesis and hydrolysis of ATP link catabolic reactions with anabolic and work reactions in metabolism, and ATP acts as a short-term energy store.

ATP Synthase

The enzyme complex that catalyzes the synthesis of ATP from ADP and P_i. The ATP synthase makes use of the energy stored in the protonmotive force across the mitochondrial inner membrane. Protons move through the complex from the intermembrane space back into the matrix, catalyzing rotation of the central stalk (γ subunit) with respect to the catalytic α and β subunits, which synthesize ATP from ADP and P_i.

Glycolysis

The sequence of enzyme-catalyzed reactions in the cytosol that is the first stage in breaking down glucose to release its energy. The end product of glycolysis is two molecules of pyruvate per glucose, along with a small amount of ATP and reduced nicotinamide adenine dinucleotide (NADH). Pyruvate and electrons from NADH may be transported into mitochondria to be used as substrates for the TCA cycle and the respiratory chain, respectively.

Mitochondria

The central organelles of energy metabolism. Mitochondria are responsible for producing most of the cell's adenosine triphosphate (ATP). The tricarboxylic acid (TCA) cycle and β oxidation of fatty acids take place in mitochondria, as does electron transport by the respiratory chain, the process that drives ATP synthesis by the establishment of a protonmotive force across the mitochondrial inner membrane.

β Oxidation

The process by which fatty acids are broken down in mitochondria. Fatty acids are converted to fatty acyl CoA before entering the β oxidation cycle, which involves the sequential shortening of the fatty acyl chain by two carbons at a time, with the release of acetyl CoA and the concomitant reduction of the electron carriers NAD^+ and FAD to NADH and $FADH_2$, respectively. The acetyl CoA molecules enter the TCA cycle, while the NADH and $FADH_2$ are oxidized by the respiratory chain.

Oxidative Phosphorylation

The process by which the oxidation of carbohydrates and fatty acids by oxygen is coupled to the phosphorylation of adenosine diphosphate (ADP) to ATP. Oxidative phosphorylation occurs in mitochondria when the NADH and $FADH_2$ produced by the

TCA cycle and β oxidation are oxidized by the respiratory chain. The redox energy is conserved as a protonmotive force across the mitochondrial inner membrane, which is used to drive ATP synthesis by the ATP synthase in the mitochondrial inner membrane.

Protonmotive Force
A measure (in mV) of the proton electrochemical potential gradient across the mitochondrial inner membrane. The protonmotive force is generated by the pumping of protons by the mitochondrial respiratory chain, which in turn is driven by the release of redox energy from electrons as they pass through the chain. The protonmotive force is composed of a membrane potential ($\Delta\psi$) and a pH gradient (ΔpH) component. It is used by the ATP synthase to synthesize ATP and by the adenine nucleotide transporter (ANT) and phosphate (P_i) carrier to transport ATP, ADP, and P_i across the mitochondrial inner membrane.

Respiratory Chain
A series of protein complexes and mobile electron carriers in the mitochondrial inner membrane that translocates electrons down a gradient of increasing redox potential. The respiratory chain oxidizes the carriers NADH and $FADH_2$ and passes the electrons down the chain toward oxygen, using the redox energy differences to pump protons across the mitochondrial inner membrane. This enables the redox energy to be stored as a protonmotive force, which is then used to synthesize ATP from ADP and P_i.

Tricarboxylic Acid (TCA) Cycle
Also known as the *Krebs cycle* or the *citric acid cycle*. The TCA cycle is a sequence of enzyme-catalyzed reactions that occur in the mitochondrial matrix. Acetyl CoA (produced by the oxidative decarboxylation of pyruvate) acts as the entry point for the TCA cycle by fusing its two-carbon acetyl group with the four-carbon oxaloacetate to form the six-carbon citrate. Citrate is then sequentially oxidized by the various enzymes of the TCA cycle, losing two carbons as CO_2 to regenerate oxaloacetate while producing NADH and reduced flavin adenine dinucleotide ($FADH_2$), and also generating GTP by substrate level phosphorylation. Many other small molecules connect with the TCA cycle at various points, making the TCA cycle central for most of metabolism. The NADH and $FADH_2$ produced are oxidized by the respiratory chain to generate ATP by oxidative phosphorylation.

■ All cell processes require energy to carry out work, whether it be synthesis, replication, secretion, movement, or signaling. Consequently, a continuous supply of energy to the cell is essential for life. Ultimately, the energy for mammalian cells is provided by the breakdown of food. In order to make this chemical energy available to the cell in a form that can do many different kinds of work, a common intermediate is required. This energy intermediate is the displacement from equilibrium of the

ATP $\rightleftharpoons$ ADP + P_i reaction. Within our cells, this reaction is displaced from equilibrium by increasing the ATP concentration relative to that of ADP by several orders of magnitude, making approximately 50 kJ.mol^{-1} of energy available from the hydrolysis of ATP to do work. Maintenance of this displacement from equilibrium requires the continued expenditure of energy supplied by food. Carbohydrate derivatives and fatty acids are broken down and oxidized by a series of reactions within mitochondria. Released electrons are transferred to NAD$^+$ and FAD, reducing them to NADH and FADH$_2$. In turn, these electron carriers are oxidized by the respiratory chain in the mitochondrial inner membrane, using the oxygen we breathe as the final electron acceptor. The action of the respiratory chain pumps protons across the mitochondrial inner membrane so that the redox potential energy is stored as an electrochemical potential gradient, or protonmotive force, across the mitochondrial inner membrane. The protonmotive force is then used to drive the synthesis of ATP from ADP and P_i and to export the ATP from the mitochondria into the cytoplasm. This process, called *oxidative phosphorylation*, maintains the displacement from equilibrium of the ATP $\rightleftharpoons$ ADP + P_i reaction and is central to cellular energetics.

1
Cellular Energetics: Displacement from Equilibrium, Coupling, and the Role of ATP

Living things are distinguished by the displacement from equilibrium of a myriad biochemical reactions. Indeed, one way of describing death is as a return to equilibrium. Why is displacement from equilibrium a necessity for life? Consider how potential energy is stored in a hydroelectric dam or in a tightly coiled spring. In a similar way, displacement from equilibrium provides a means of energy storage within the cell; indeed, it is only by being displaced from equilibrium that biochemical reactions can store energy that can be used to carry out cellular work such as movement, replication, growth, and repair.

Living things require large amounts of energy to maintain the displacement from equilibrium of chemical reactions within the cell. In mammals, the energy source is food, and the chemical energy stored in food is released through the reactions of metabolism. Metabolism comprises a series of catabolic reactions that break down food, releasing precursor molecules and energy for maintaining the displacement from equilibrium that drives anabolic reactions and work. The energy released by the catabolic reactions of metabolism must be available in a form that can do the work of the cell. A central concept underlying this process, and all of energy metabolism, is "coupling," by which the energy released from food drives otherwise unfavorable reactions. How this energy coupling occurs is the focus of this chapter.

1.1
Coupling Drives Energetically Unfavorable Reactions

By coupling together favorable and unfavorable chemical reactions, it is possible to use the energy stored in food to drive the cellular processes that comprise life. Thermodynamics indicates whether or not a reaction will occur spontaneously and gives the energy change associated with the reaction. In biology, the Gibbs free energy change (ΔG; kJ.mol^{-1}) is used. For spontaneous reactions, there is a release of free energy, so $\Delta G < 0$. In contrast, reactions that do not occur spontaneously have a ΔG greater than zero and require energy input to take place. In biological systems, there are many such processes, including synthesis, movement, replication, or signaling, that are thermodynamically unfavorable and cannot occur without an energy input. Fortunately, such unfavorable reactions can be driven by coupling them to a reaction with a favorable (negative) ΔG. This is illustrated below where the unfavorable reaction 1 is driven by reaction 2.

Reaction 1	A → B	$\Delta G_1 > 0$
Reaction 2	C → D	$\Delta G_2 < 0$
Overall	A + C → B + D	$\Delta G_1 + \Delta G_2 < 0$

This process, by which an unfavorable reaction is made possible by coupling it to a favorable reaction, is central to the bioenergetics of living organisms. The energy flowing through our cells and bodies (derived from favorable catabolic reactions) is coupled to drive unfavorable

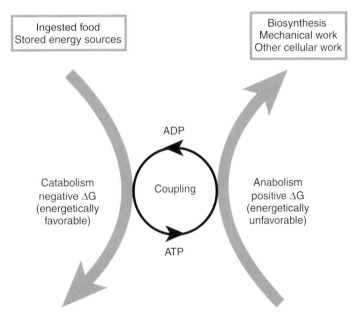

Fig. 1 The concept of coupling in mammalian energy metabolism. Energy is released by the breakdown of carbohydrate, fat, and protein, either from ingested food or from stored energy sources. Released energy is "coupled" to the otherwise unfavorable formation of adenosine triphosphate (ATP) by the addition of inorganic phosphate (P_i) to adenosine diphosphate (ADP). In turn, ATP hydrolysis to ADP and P_i (a highly favorable reaction) is coupled to unfavorable anabolic reactions, enabling them to take place. The Gibbs free energy change (ΔG) is used here to indicate whether a reaction is favorable (spontaneous) or unfavorable (requiring energy input through coupling).

reactions (anabolic reactions and work) (Fig. 1). In principle, many different reactions could be coupled together in this manner. However, biology has evolved to exploit a few general reactions that are used to couple most catabolic and anabolic reactions, and the most important of these is the synthesis and hydrolysis of adenosine triphosphate (ATP) (Fig. 2):

$$ATP \rightleftharpoons ADP + P_i$$

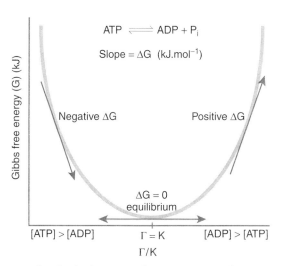

Fig. 2 Energy is stored in the displacement of ATP $\rightleftharpoons$ ADP $+$ P$_i$ from equilibrium. Energy from catabolic reactions is used to drive the synthesis of ATP from ADP and P$_i$, causing the ATP/ADP ratio to be many orders of magnitude greater than its equilibrium value (put another way, the mass action ratio (Γ) for the hydrolysis reaction is many times lower than the equilibrium constant (K)). Under these conditions, ATP hydrolysis becomes a highly favorable reaction and has a negative ΔG (indicated in the graph as the negative slope of the curve under conditions of high ATP concentration). When the ATP/ADP ratio is lower than its equilibrium value, ATP hydrolysis is not favored (ΔG > 0); rather, the reverse reaction is energetically favorable. At equilibrium, the ATP/ADP ratio is equal to its equilibrium value (the mass action ratio equals the equilibrium constant), the ΔG $= 0$, and neither the forward nor the reverse reactions are favored.

1.2
ATP Synthesis and Hydrolysis as a Coupling Reaction

The energy made available from the catabolism of food is funneled into a few forms that can be used in a wide range of anabolic and work reactions. The main reaction "chosen" by evolution for this purpose is the $ADP + P_i \rightleftharpoons ATP$ reaction. The energetically favorable reactions of catabolism are coupled to the synthesis of ATP from adenosine diphosphate (ADP) and P_i, and, in turn, the energy released by hydrolysis of ATP to ADP is coupled to work and anabolic reactions (Figs. 1 and 2). There are a number of reasons why this reaction plays a central role in biology. The equilibrium constant of the reaction enables it to straddle both the highly exergonic catabolic reactions and the highly endergonic anabolic reactions. Furthermore, coupling requires that the reaction be easily combined with many others within the active sites of enzymes, and phosphate transfer or ATP hydrolysis can be coupled to a wide range of reactions. Finally, while the hydrolysis of ATP at the biological concentrations of ATP, ADP, and P_i must be thermodynamically favorable in order to act as an energy store, it must not occur without the action of enzymes capable of harnessing the released energy for useful functions. The phosphodiester bonds in ATP fulfill this role as they are kinetically stable *in vivo*, but the activation energy for hydrolysis of the phosphodiester bond is easily decreased by appropriate arrangements within the active sites of enzymes to stabilize the pentagonal bipyramidal transition states. For these and other reasons, the $ATP \rightleftharpoons ADP + P_i$ reaction is central to bioenergetics.

1.3
Energy is Stored in the Displacement from Equilibrium of the ATP Hydrolysis Reaction

The ATP hydrolysis reaction ($ATP \rightleftharpoons ADP + P_i$) couples catabolic and anabolic reactions by storing energy in its displacement from equilibrium. As is the case for all chemical reactions, the amount of energy stored is determined by the mass action ratio (Γ; a measure of the relative concentrations of reactants and products) and its relationship to the equilibrium constant (K; a measure of the relative concentrations of reactants and products *at equilibrium*).

For ATP hydrolysis the mass action ratio is

$$\Gamma = [ADP][P_i]/[ATP]$$

The equilibrium constant is the special case of this reaction in which the reactants and products are at equilibrium ($K \sim 10^5$ M at pH 7). To see how energy is stored in this reaction, ΔG can be calculated from the following equation:

$$\Delta G = RT \ln(\Gamma/K)$$

This equation shows clearly the relationship between the displacement from equilibrium and the amount of energy stored, with $\Delta G = 0$ when the reaction is at equilibrium. This equation can be easily rearranged to give the more usual textbook presentation:

$$\Delta G = RT \ln \Gamma - RT \ln K$$

or, as $-RT \ln K$ is often written as ΔG^o,

$$\Delta G = \Delta G^o + RT \ln \Gamma$$

This equation shows that ΔG^o is merely another way of stating the equilibrium constant, and illustrates how misleading are discussions of cellular thermodynamics based on $\Delta G°$. It also emphasizes that the relevant concentrations of the reactants and products must be used in calculating ΔG.

In vivo, catabolic reactions hold the cytoplasmic concentrations of ATP $\sim$ 5 mM and ADP $\sim$ 100 to 500 µM, corresponding to a displacement from equilibrium of 8 to 10 orders of magnitude. The role of this displacement from equilibrium in energy storage is further illustrated in Fig. 2, which shows ΔG as a function of Γ/K. The energy stored *in vivo* in this reaction is about $-50\ \text{kJ.mol}^{-1}$. Note that this does not refer to the *y*-axis scale, or the energy change that would occur if the system were able to move from the *in vivo* position to that of $K = \Gamma$. Instead, the relevant ΔG is the slope of the curve, which is about $-50\ \text{kJ.mol}^{-1}$. This is the amount of work made possible when 1 mol of ATP is hydrolyzed without affecting the overall concentrations of ATP, ADP, and P_i. Equally, the ΔG for synthesis of 1 mol of ATP under these conditions is $+50\ \text{kJ.mol}^{-1}$; the energy change is the same as for hydrolysis, but the sign is opposite. *In vivo*, other reactions are constantly maintaining a ΔG of about $-50\ \text{kJ.mol}^{-1}$ for ATP hydrolysis by synthesizing ATP through reactions that require an energy input of about $50\ \text{kJ.mol}^{-1}$.

A corollary of this continual synthesis and breakdown of ATP is that ATP molecules within the cell (5 mM; approximately 100 g in the average male human) are turned over very rapidly. Indeed, it is estimated that a resting human will hydrolyze the equivalent of his or her own body weight of ATP every day. Furthermore, the turnover of ATP is increased greatly during periods of intense activity: a marathon runner may hydrolyze up to 10 g of ATP per second. Were we to rely on our ATP stores for energy, they would run out within about 2 min. So, it is clear that ATP acts as a go-between rather than as an energy store, ferrying energy from a primary source to drive a wide range of otherwise unfavorable reactions.

2
Energy Sources that Displace the ATP Hydrolysis Reaction from Equilibrium

2.1
Carbohydrate, Fat, and Protein as Energy Sources

The energy stored in the displacement from equilibrium of the ATP $\rightleftharpoons$ ADP + P_i reaction can act only as a transient energy intermediary. Other energy sources must be used to maintain the displacement from equilibrium of this reaction as ATP is used to do work. For mammals, this energy comes from food, specifically carbohydrate, fat, and protein. In the short term, carbohydrate and fat from our last meal circulates in the bloodstream for a few hours and can be used to provide energy; however, as we do not eat continually we have to store energy within our bodies. Fat in the form of triacylglycerides in adipose tissue is the most energy-dense store ($\sim$39 kJ.g^{-1}) and is consequently used for long-term energy storage. Carbohydrate ($\sim$17 kJ.g^{-1}) in the form of liver glycogen can be broken down to maintain glucose in the blood for 12 to 24 h; in the longer term, blood glucose levels are maintained by gluconeogenesis. Dietary protein is broken down to amino acids and those not required for protein synthesis are metabolized by pathways similar to those used for carbohydrates. During extended starvation, amino acids can be released from the body's protein to provide energy. All of these energy stores are broken down by sequential catabolic reactions that use the stored chemical potential energy to displace the ATP hydrolysis reaction from equilibrium. How these occur is considered next (Fig. 3).

2.2
Glycolysis

When glucose enters the cell, it is converted to two molecules of pyruvate by a sequence of linked enzymatic reactions termed *glycolysis*. Two reactions in this sequence, catalyzed by phosphoglycerate kinase and pyruvate kinase, are coupled to the synthesis of ATP from ADP and P_i and so contribute to the displacement from equilibrium of the ATP hydrolysis reaction (although only a fraction of the chemical potential energy stored in glucose is released at this stage). During glycolysis, electrons are removed from the carbohydrate intermediates and are transferred to the electron carrier NAD^+, reducing it to NADH (Fig. 4). However, to maintain a flux through the glycolytic pathway, it is necessary to reoxidize the NADH continually to NAD^+. In mammals, this can occur in the absence of oxygen through the reduction of pyruvate by NADH; this reaction produces lactic acid and is catalyzed by lactate dehydrogenase. In the presence of oxygen, it is possible to extract far more ATP from glucose, but, for this, both the pyruvate and the electrons from NADH must be transported into mitochondria.

2.3
Mitochondria, the TCA Cycle and β Oxidation

The mitochondrion is the organelle at the heart of energy metabolism and ATP synthesis in eukaryotic cells. Most

eukaryotic cells contain hundreds of mitochondria, which are filamentous organelles comprising a matrix surrounded by an invaginated inner membrane and a semipermeable outer membrane. Mitochondria arose from free-living bacteria that set up home within the cytoplasm of the protoeukaryotic cell, a process called *endosymbiosis*. These organelles break down fats and carbohydrate-derived metabolites using oxygen to maximize the amount of energy released and then use this energy to synthesize ATP.

To extract energy from carbohydrates, the pyruvate produced by glycolysis is taken up into the mitochondrial matrix. There, the three-carbon pyruvate is converted to a two-carbon acetyl group, which is carried by Coenzyme A as acetyl CoA. Acetyl CoA then reacts with the four-carbon oxaloacetate; together they generate the six-carbon citrate. In this way, carbohydrates enter the tricarboxylic acid (TCA) cycle (Fig. 3). In the course of the TCA cycle, oxaloacetate is reformed from citrate through a series of reactions that split off CO_2 and oxidize cycle intermediates, with the concomitant reduction of the electron carriers NAD^+ and FAD to NADH and $FADH_2$, respectively (Fig. 4). A single GTP molecule is also produced by each "rotation" of the TCA cycle; GTP is readily converted to ATP through the action of nucleoside diphosphate kinase. The oxaloacetate is then available to accept a further acetyl group from acetyl CoA and thus keeps the TCA cycle turning. When amino acids derived from the breakdown of proteins are used as energy sources, they are broken down to derivatives that are also fed into the TCA cycle.

Mitochondria are also the site of oxidation of the fatty acids derived from triacylglyceride, the storage form of fat. Within the cytoplasm, fatty acids are first

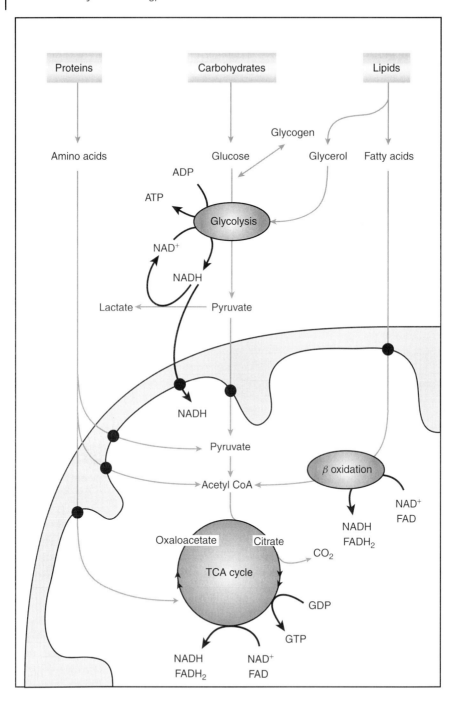

linked to CoA to form acyl CoA and then converted to acyl carnitine. This is taken up into mitochondria through the inner membrane in exchange for carnitine. The acyl carnitine then reacts with CoA to form an acyl CoA within mitochondria. Typically, fatty acids are 16 to 18 carbons in length and within mitochondria they are broken down and oxidized by a process called β *oxidation*. During β oxidation, the acyl CoA goes through a continuous cycle of degradation and oxidation. In each cycle, two carbon units are removed from the acyl CoA to form acetyl CoA, and electrons are removed from the shortening fatty acid to reduce NAD^+ and FAD to NADH and $FADH_2$. This cycling persists until the fatty acid has been completely converted to acetyl CoA, which is then fed into the TCA cycle.

The endpoint of the metabolism of carbohydrates, fats, and protein within mitochondria is the formation of CO_2, NADH, and $FADH_2$. The CO_2 will diffuse out of the cells and be expired through the lungs; NADH and $FADH_2$ will be reoxidized. While the major function of the breakdown of fats and carbohydrates by mitochondria is to provide chemical energy to displace the ATP hydrolysis reaction from equilibrium, at this point only a fraction of the potential energy available has been used to make ATP (through glycolysis and by the formation of GTP). It is the reoxidation of NADH and $FADH_2$ that supplies the majority of the chemical potential energy for ATP synthesis. The means of coupling these two processes is discussed in the following sections.

3
Oxidation of NADH and $FADH_2$ by Mitochondria

Through the action of the TCA cycle and β oxidation within the mitochondrial matrix, there is an accumulation of electrons in the reduced carriers, NADH and $FADH_2$. The NADH pool within the matrix is soluble, and, typically, the ratio of $NAD^+/NADH$ within mitochondria is about 5 to 10. The NADH built up in the cytoplasm by glycolysis can also be transported indirectly to the mitochondrial matrix through the malate–aspartate shuttle, or can donate electrons to the α-glycerophosphate

Fig. 3 Metabolic sources of the reducing equivalents NADH and $FADH_2$, and ATP. A small amount of NAD^+ is reduced to NADH with each round of glycolysis. This NADH can either be reoxidized to enable glycolysis to continue, or imported indirectly into mitochondria where it donates electrons to the mitochondrial respiratory chain. Pyruvate, the product of glycolysis, is transported to mitochondria where it is converted to acetyl CoA and enters the TCA cycle. The sequential oxidation of citrate enables the reduction of NAD^+ and FAD to NADH and $FADH_2$. Each "rotation" of the TCA cycle also produces a single GTP molecule (energetically equivalent to ATP). Triacylglyceride lipids are broken down into glycerol and fatty acids. Glycerol feeds into glycolysis, while fatty acids are transported to mitochondria and undergo multiple cycles of β oxidation, with each cycle yielding acetyl CoA (which feeds into the TCA cycle) and reducing NAD^+ and FAD to NADH and $FADH_2$. Proteins are metabolized through breakdown into amino acid monomers and the subsequent conversion of amino acids to pyruvate, acetyl CoA or TCA cycle intermediates. Black circles indicate either direct transport into mitochondria (e.g. the pyruvate transporter), or net transport through several molecular interconversions (e.g. the malate–aspartate shuttle for the "transport" of NADH). In summary, the catabolic reactions of metabolism act to reduce the electron carriers NAD^+ and FAD to NADH and $FADH_2$, and produce a small amount of ATP.

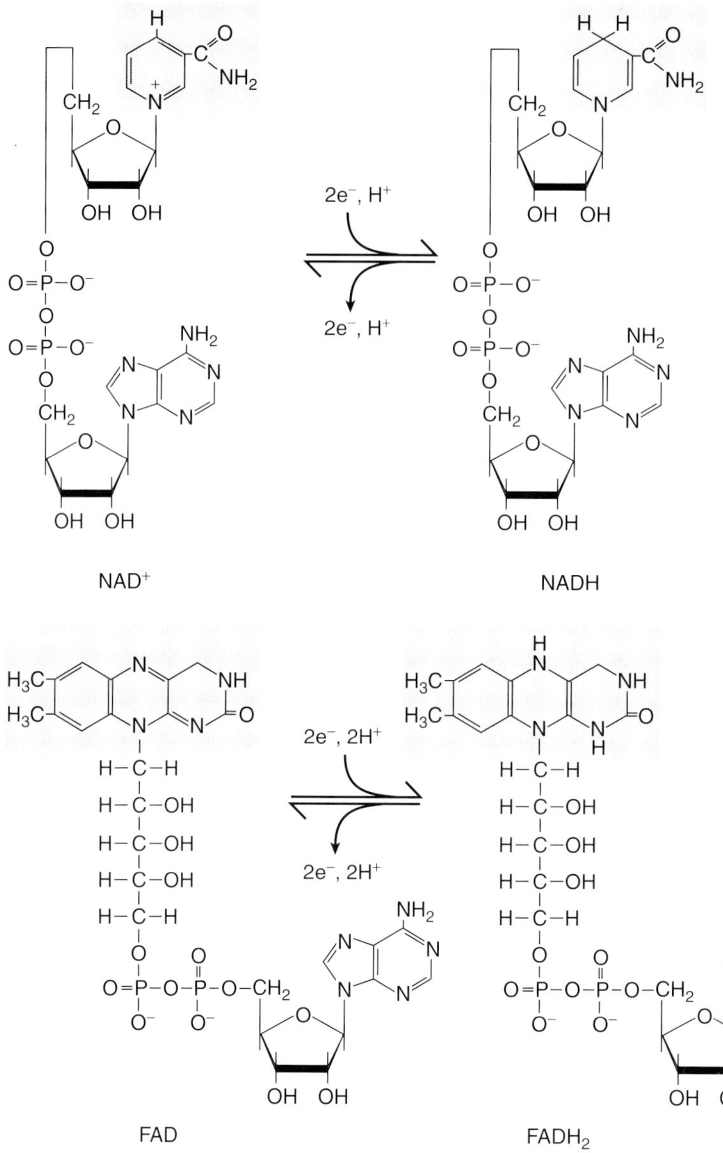

Fig. 4 Reduction of the electron carriers NAD^+ and FAD to their reduced forms NADH and $FADH_2$. Both carriers accept two electrons in conversion to the reduced form.

dehydrogenase on the outer surface of the mitochondrial inner membrane (Fig. 5). In contrast, the $FADH_2$ groups are not free in solution but are bound within the active sites of a number of different proteins, such as succinate dehydrogenase (for electrons derived from the TCA cycle), or on the electron transfer flavoprotein (ETF) (for electrons derived from β oxidation).

The reduction potential (E_h) of the $NAD^+/NADH$ couple can be derived from the midpoint potential of the couple $(E_m, \sim -320\ mV)$ as follows:

$$E_h = E_m + \frac{RT}{2F} \ln \left(\frac{[NAD^+]}{[NADH]} \right)$$

$$E_h \sim -290\ mV$$

As the energy stored in this redox couple will be released by the reduction of oxygen, it should be compared with the E_h value for the O_2/water couple, which is about $+790\ mV$ for air-saturated water. This gives a ΔE_h of 1080 mV for the oxidation of NADH by O_2, corresponding to a ΔG of $-208\ kJ.mol^{-1}$. In comparison, $FADH_2$ passes electrons directly to the ubiquinone/ubiquinol couple, which has a somewhat higher E_h $(\sim +10\ mV)$ than NADH, and thus a lower ΔE_h $(\sim 780\ mV)$ and less negative ΔG $(\sim -150\ kJ.mol^{-1})$ for electron transfer to O_2.

The redox energy stored by NADH and $FADH_2$ is ultimately released on reaction with oxygen. However, were this to be done by direct reaction with O_2, the energy would be dissipated uncontrollably as heat and could not be harnessed for ATP synthesis. Instead, the release of the redox potential energy stored in NADH and $FADH_2$ is controlled by the action of the mitochondrial respiratory chain.

3.1
The Mitochondrial Respiratory Chain

The respiratory chain is a sequence of electron carriers embedded within the mitochondrial inner membrane that catalyzes electron movement from NADH and $FADH_2$ to oxygen (Fig. 5). The first component of the respiratory chain is complex I, a massive membrane protein complex of about 980 kDa. This protein takes electrons from NADH and passes them to a flavin mononucleotide within complex I. From there, the electrons pass through a series of iron–sulfur centers and are then used to reduce ubiquinone to ubiquinol; these are the two redox forms of Coenzyme Q (CoQ), a small, mobile electron carrier within the mitochondrial inner membrane, which is not a protein but a hydrophobic organic molecule.

A number of other respiratory complexes also pass electrons to CoQ. Complex II, or succinate dehydrogenase, catalyzes the oxidation of succinate to fumarate in the TCA cycle. The electrons initially reduce a bound FAD within complex II; the resultant $FADH_2$ passes the electrons through a series of iron–sulfur centers to the CoQ binding site within the membrane where they reduce ubiquinone to ubiquinol. α-Glycerophosphate dehydrogenase (GPDH) is an enzyme on the outer surface of the mitochondrial inner membrane that is expressed in some tissues. This enzyme oxidizes α-glycerophosphate to dihydroxyacetone phosphate and uses the electrons to reduce a bound FAD within the enzyme, which then reduces ubiquinone to ubiquinol. The oxidation of fatty acids by β oxidation leads to the reduction of an FAD in a small mobile protein within the mitochondrial matrix called *electron transfer flavoprotein* (ETF). This electron carrier is reoxidized by the respiratory chain through an inner membrane protein called *ETF–ubiquinone oxidoreductase* (ETF-OR). This enzyme reduces an internal FAD and subsequently reduces ubiquinone to ubiquinol, possibly *via* an iron–sulfur center. Therefore, these, and other, enzymes catalyze electron flow into the CoQ pool. *In vivo*, the CoQ pool is generally about 70 to 80% in the reduced (ubiquinol) form.

The fate of ubiquinol is oxidation by complex III (also known as the cytochrome bc₁ complex), the next component of the respiratory chain. Complex III takes two electrons from the ubiquinol of the CoQ pool and passes them through a branching pathway where one electron goes through the two b-type cytochromes before being passed back to reduce ubiquinone. The other electron reduces an iron−sulfur

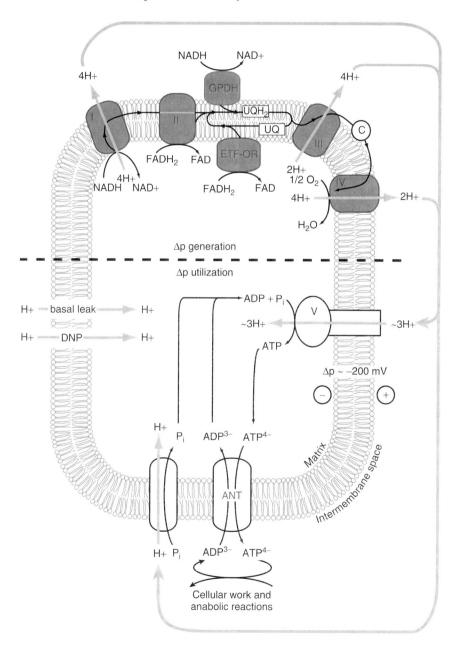

center, called the *Rieske iron-sulfur center*, and from there reduces a cytochrome c_1. This cytochrome is the last electron carrier in the complex and passes its electron to cytochrome c, a small protein that acts as a mobile electron carrier on the cytoplasmic surface of the mitochondrial inner membrane. The enzymes sulfite oxidase and NADH-cytochrome b_5 reductase (via cytochrome b_5) can also pass electrons to cytochrome c, but the flux through these reactions *in vivo* is thought to be small.

Cytochrome c carries electrons to complex IV (or cytochrome oxidase), the final stage of their journey toward oxygen. It is at cytochrome oxidase that about 95% of the oxygen we breathe is reduced to water. The electron from cytochrome c is initially passed to the heme *a* site within cytochrome oxidase and then to the Cu_a site. From there, the electrons pass to the heme a_3 site, which also contains the Cu_b center. Here the terminal electron acceptor of the respiratory chain, O_2, binds and is reduced to water.

So far, we have outlined the pathway taken by electrons as they move through the respiratory chain. In the next section,

the mechanism of coupling this movement to ATP synthesis is explained.

3.2
Energy Storage by the Protonmotive Force (Δp)

The respiratory chain provides a complex pathway for the gradual movement of electrons from the low potential of NADH (~ -290 mV) to the high potential of oxygen ($\sim +790$ mV). This pathway enables the gradual release of potential energy and is best illustrated by Fig. 6, which shows the gradual increase in reduction potential (E_h) as electrons pass through the respiratory chain. Three significant changes in E_h are apparent within the respiratory chain, at complexes I, III, and IV; the large increases in E_h – which signify large drops in the potential energy of the electrons – suggest that at these three points the redox energy may be conserved as an intermediate that can be used to make ATP.

The mechanism of coupling of electron transport and ATP synthesis was one of the major challenges to biochemistry

Fig. 5 NADH and $FADH_2$ are oxidized by the mitochondrial respiratory chain, giving rise to the protonmotive force and ATP synthesis. NADH and $FADH_2$ donate electrons to the respiratory chain through a number of inner membrane proteins and multiprotein complexes: complex I (mitochondrial NADH), α-glycerophosphate dehydrogenase (GPDH; cytosolic NADH), complex II (TCA cycle-derived $FADH_2$) and electron transfer flavoprotein–ubiquinone oxidoreductase (ETF-OR; β oxidation-derived $FADH_2$). Each of these proximal electron acceptors feeds electrons to CoQ, reducing ubiquinone (UQ) to ubiquinol (UQH_2). Electrons are subsequently passed through complex III, cytochrome c, and complex IV, finally reducing O_2 to H_2O. Complexes I, III, and IV all derive sufficient energy from the drop in redox potential of electrons to catalyze the pumping of protons out of the matrix and into the cytoplasm. This generates a protonmotive force (Δp) across the mitochondrial inner membrane. Driven by Δp, protons flow back into the matrix through the ATP synthase (complex V), catalyzing the formation of ATP from ADP and P_i. Some protons can also return to the matrix by mechanisms that do not involve passage through the ATP synthase, such as the basal leak through the inner membrane in all mitochondria, or via uncouplers such as 2,4-dinitrophenol (DNP) that render the inner membrane permeable to protons. Newly synthesized ATP^{4-} is transported out of the mitochondria and into the cytosol by exchange with ADP^{3-}, which in turn is converted to ATP. The P_i required for ATP synthesis is transported into the matrix in symport with a proton.

in the 1970s and early 1980s. The two processes are now known to be linked by "chemiosmotic coupling," a concept developed by Peter Mitchell in the early 1960s and established over the subsequent two decades. The energy available as the electrons pass through complexes I, III, and IV is used to pump protons out of the matrix across the mitochondrial inner membrane. For every pair of electrons that passes through the respiratory chain, 4 H^+ are translocated at complex I, 4 H^+

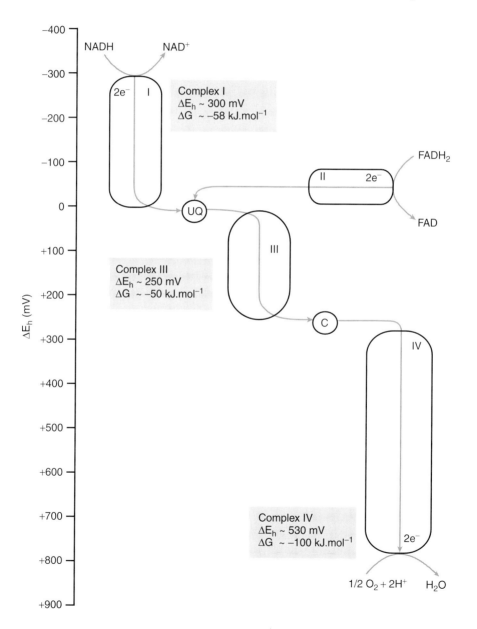

at complex III, and 2 H^+ at complex IV (although the movement of electrons with respect to the membrane at complexes III and IV means that the charges translocated by complexes I, III, and IV are 4, 2, and 4, respectively). The reduction of O_2 to water requires 4 e^-, so a pair of electrons corresponds to the reduction of a single O atom. As the two electrons go through the respiratory chain from NADH, 10 protons and charges are pumped across the inner membrane, while for electrons coming from $FADH_2$, six protons and charges are translocated.

Proton pumping builds up a charge gradient, or membrane potential, across the inner membrane ($\Delta\psi$ in mV, negative inside) (Fig. 5). In addition, proton translocation also leads to a proton concentration gradient across the inner membrane, which is usually expressed as a pH gradient (ΔpH, basic inside). Both the membrane potential and the pH gradient raise the potential energy of protons outside the mitochondrial inner membrane relative to those inside. The contribution of the two components to this potential energy difference is conveniently expressed as the protonmotive force. For this, the pH gradient is multiplied by $-2.303RT/F$ to express its value in mV; this can be added to the membrane potential to give the protonmotive force (Δp):

$$\Delta p = \Delta\psi - (2.303RT/F)\Delta pH$$

The advantage of this formulation is that the relative contributions of $\Delta\psi$ and ΔpH are clear. In isolated mitochondria, the maximum $\Delta\psi$ is about 170 to 190 mV and the pH gradient can be up to -0.5 to -1 pH unit (basic inside, hence the negative sign in the equation). At $37\,^\circ$C, $2.303RT/F = 61$ mV, so the maximum Δp is typically around 180 to 200 mV. In cells, the average values tend to be lower because of the utilization of Δp to synthesize ATP, as well as limiting electron and O_2 supply.

To maintain the Δp across the mitochondrial inner membrane, the membrane has to be topologically intact and the phospholipid bilayer must be relatively impermeant to protons. A further implication of the closed membrane is that mechanisms must exist for the transport of polar metabolites across the membrane, such as pyruvate, ATP, ADP, and P_i, among many others. As this movement has to occur without dissipating the Δp, the metabolite carriers in the mitochondrial inner membrane must be able to catalyze transport without rendering the membrane permeable to protons. In summary, the redox energy stored by NADH

Fig. 6 Sequential increases in redox potential (E_h) of respiratory chain components are coupled to proton pumping. NADH, a major electron donor to the mitochondrial respiratory chain, has a low E_h (~ -290 mV) under conditions found in the matrix. After oxidation of NADH by complex I, electrons are passed through a series of electron carriers in the complex (not shown) before reducing the mobile electron carrier ubiquinone ($E_h \sim +10$ mV). The large increase in E_h of respiratory components between NADH and ubiquinone – and hence the drop in redox potential energy of the transported electrons – provides sufficient energy for translocation of $4H^+$ (per $2e^-$) by complex I into the cytosol. Similarly, increases in E_h between ubiquinol and oxidized cytochrome c ($E_h \sim +260$ mV), and between reduced cytochrome c and O_2 ($E_h \sim +790$ mV), provide sufficient energy for proton translocation at complexes III ($4H^+$) and IV ($2H^+$), respectively. ΔE_h values shown for complexes I, III, and IV have been calculated from the increases in redox potential between NADH and ubiquinone, ubiquinol and oxidized cytochrome c, and reduced cytochrome c and O_2, respectively.

and $FADH_2$ is conserved as a protonmotive force that comprises a $\Delta\psi$ and a ΔpH across the mitochondrial inner membrane. This protonmotive force can now be used for the synthesis and transport of ATP.

3.3
The Protonmotive Force Drives ATP Synthesis and Transport

The energy stored in Δp is used to drive the synthesis of ATP through the action of the ATP synthase. This protein complex is also found in the mitochondrial inner membrane and catalyzes the movement of protons down their proton electrochemical gradient from the cytoplasm, back into the mitochondrial matrix. The drop in potential energy as the protons pass through the complex provides the energy for ATP synthesis by the synthase. This remarkable enzyme has a hydrophobic domain (the F_O domain) embedded in the lipid bilayer, which facilitates proton movement through the lipid bilayer, and a connected F_1 component in the mitochondrial matrix where the ATP is made (Fig. 7). The F_O subcomplex comprises a ring of 10 to 14 identical hydrophobic proteins called *subunit c*. This ring is attached to a stator arm that links together the F_O and F_1 subcomplexes. As a proton passes through the membrane, it forces the rotation of F_O relative to the stator arm, with the concomitant rotation of a second protein subunit, γ. The γ subunit is attached at its base to F_O; the remainder of the protein is inserted into the center of the F_1 complex, which comprises 3 α and 3 β subunits arranged alternately like the segments of an orange. This assembly of 3 α and 3 β subunits is held in place by the stator arm, so proton movement drives

the rotation of F_O and the γ subunit relative to F_1. The rotation of the γ subunit against the α and β subunits forces the three catalytic sites at the interfaces of the α and β subunits to change shape, with each site switching between three conformations that alternately bind ADP and P_i, synthesize ATP, and finally release ATP. The Δp is used to drive the rotation of the γ subunit relative to the α and β subunits at about 50 to 100 Hz *in vivo*; thus, Δp is converted to mechanical energy and then back to chemical energy in the displacement from equilibrium of the ATP hydrolysis reaction. Each complete rotation of the γ subunit within the ATP synthase enables the synthesis of three ATP molecules. This is associated with the complete rotation of F_O. It is likely that the number of protons translocated for each complete F_O rotation equals the number of c subunits; since there are thought to be at least 10 subunit c components in mammals, this suggests that 3.3 or more H^+ are transported per ATP synthesized. However, direct measurement of the number of H^+ translocated per ATP synthesized puts the ratio at about 3.

The ATP synthase makes ATP in the mitochondrial matrix, but, as most ATP is consumed in the cytoplasm, ATP must be exported from the mitochondrion. In addition, P_i and ADP must be imported into mitochondria from the cytoplasm. Much of the anion transport across the mitochondrial inner membrane is accomplished by a family of carriers of about 30 to 35 kDa. Among these are the adenine nucleotide translocator (ANT) and the phosphate carrier (Fig. 5). The ANT catalyzes exchange of ATP^{4-} in the mitochondrial matrix with ADP^{3-} in the cytoplasm, hence transport involves the net movement of a positive charge into the mitochondrion, and ATP export and ADP

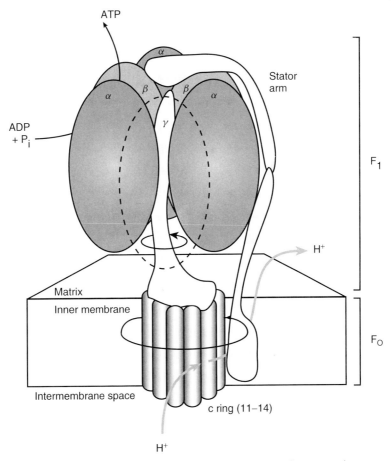

Fig. 7 The mitochondrial ATP synthase. To synthesize ATP, the ATP synthase harnesses the energy stored in the Δp across the mitochondrial inner membrane. Protons move down their electrochemical gradient from the intermembrane space via the ring of c subunits (F_O subcomplex) embedded in the membrane, catalyzing the rotation of both the F_O subcomplex and the neighboring γ subunit (part of the F_1 subcomplex). Rotation of the γ subunit relative to the static $\alpha_3\beta_3$ subunits (held in place by the stator arm) alters the shape of the three catalytic sites at interfaces between α and β subunits, promoting sequential ADP and P_i binding, ATP synthesis, and finally the release of ATP from the enzyme.

import is driven by the membrane potential component of Δp. A consequence of this is a higher ATP/ADP ratio in the cytoplasm than in the mitochondrial matrix. So, the displacement from equilibrium of the ATP hydrolysis reaction is driven by Δp acting through both the ATP synthase and the ANT. The phosphate carrier takes up the P_i required for ATP synthesis in electroneutral symport with a proton; hence, P_i uptake is driven by the ΔpH component of Δp. In sum, the import of ADP and P_i and the export of ATP is associated with the movement of one proton through the mitochondrial inner membrane. So, if it takes approximately three protons to synthesize an ATP molecule in the matrix, it takes about four protons to synthesize ATP in the cytoplasm.

A final consideration is the number of oxygen atoms that must be consumed to synthesize an ATP molecule and export it to the cytoplasm. As the oxidation of NADH or FADH$_2$ pumps 10 and 6 protons respectively, the theoretical P/O ratios are 10/4 for NADH and 6/4 for FADH$_2$. The true values will be less because of inefficiencies such as leakage of protons through the mitochondrial inner membrane.

3.4
Respiratory Control and Uncoupling

As mitochondria make ATP, they consume Δp, enabling them to sense and respond to the rate of ATP consumption. ATP hydrolysis increases the ADP concentration, thereby elevating the rate of ATP synthesis. Greater ATP synthase activity involves the consumption of Δp, which is then matched by a rise in respiration rate, as the respiratory chain responds to the decreased Δp by increasing electron flow and proton pumping to maintain Δp. This phenomenon is known as *respiratory control* and enables mitochondria to match ATP synthesis to ATP consumption, maintaining the displacement from equilibrium of the ATP hydrolysis reaction under conditions of variable ATP demands.

The link between respiration rate and Δp is also indicated by the phenomenon of uncoupling. To maintain Δp, the mitochondrial inner membrane is relatively impermeable to protons. Uncouplers counteract this by making the inner membrane permeable to protons. Uncouplers are typically lipophilic weak acids in which the conjugate base is also lipophilic. An example is 2,4-dinitrophenol (DNP; Fig. 5) where the charge on the phenolate anion is delocalized and, consequently, both the anion and its conjugate acid are lipid soluble. This enables DNP to partition into the phospholipid bilayer of the mitochondrial inner membrane, where it can shuttle across the membrane, picking up a proton on the intermembrane space side and depositing it in the matrix. Thus, uncouplers catalyze proton movement through the lipid bilayer; in the presence of an uncoupler, the mitochondrial inner membrane is effectively permeant to protons and consequently cannot maintain a Δp. In response, the mitochondria respire rapidly to pump out protons in a vain attempt to maintain a Δp, which is instead rapidly dissipated as heat (Fig. 8).

3.5
Alternative Uses of the Protonmotive Force

While the principal function of Δp is the synthesis and export of ATP from the mitochondrion, there are other processes that exploit Δp to carry out work. These include metabolite transporters that catalyze movement across the mitochondrial inner membrane by coupling to charge or proton translocation (analogous to ADP/ATP antiport and P$_i$ transport). The movement of calcium across the mitochondrial inner membrane is also driven by Δp. There is a calcium uniporter in the mitochondrial inner membrane that enables calcium uptake into mitochondria. As only the calcium dication is translocated, uptake is driven solely by the membrane potential. In addition, there is a separate

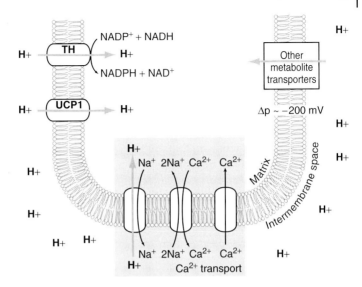

Fig. 8 Alternative fates of Δp. While the major function of the Δp is to generate ATP, proton movement down the electrochemical potential gradient across the mitochondrial inner membrane can be coupled to other activities. Such Δp-dependent processes include the action of the transhydrogenase (TH), which exploits Δp to drive reduction of $NADP^+$ to NADPH, at the expense of NADH; thermogenesis in brown adipose tissue by uncoupling protein 1 (UCP1), which dissipates the Δp as heat by increasing the proton permeance of the inner membrane; the activity of an Na^+-H^+ antiporter and Ca^{2+} uniporter, which (in concert with a $2Na^+/Ca^{2+}$ antiporter) exploit the ΔpH and $\Delta\psi$ components of the Δp, respectively, to catalyze mitochondrial uptake of calcium ions; and the transport of a number of other metabolites.

exchange activity involving electroneutral exchange of Ca^{2+} for $2H^+$ or $2Na^+$. In conjunction with the Ca^{2+} uniporter, this exchange can lead to the continual cycling of calcium across the mitochondrial inner membrane as cytosolic calcium increases. The overall effect is that increased cytoplasmic calcium concentration leads to a proportionate increase in calcium concentration in the mitochondrial matrix. As changes in cytosolic calcium are important signals that activate various aspects of metabolism, this mechanism enables such signals to be relayed to the mitochondrial matrix. The changes in matrix calcium activate dehydrogenases, thereby

increasing electron supply to the respiratory chain and elevating ATP synthesis. The reason for this is thought to be that calcium signals are often associated with increased ATP turnover; Δp-dependent Ca^{2+} uptake enables mitochondria to up-regulate ATP synthesis to match this increased demand.

A further use of Δp is in thermogenesis by brown adipose tissue. Under certain conditions, mitochondria can become uncoupled; that is, the energy stored in Δp can be dissipated as heat by providing a pathway for protons to flow back through the inner membrane to the matrix without carrying out work. Certain molecules, such as DNP, can act as uncouplers and dissipate Δp as heat (Fig. 5). Indeed, historically, DNP was used as a slimming agent as it led to the rapid breakdown of fat stores. Physiologically regulated uncoupling is used by mitochondria in brown adipose tissue mitochondria to generate heat. Mitochondria in brown adipose tissue contain uncoupling protein 1 (UCP1), which catalyzes proton movement through the mitochondrial inner membrane. When this protein is activated, it partially uncouples mitochondria, increasing the rate of respiration such that the fat stored in brown adipose tissue is broken down and oxidized by the mitochondria. However, instead of the energy in Δp being used to make ATP, it is instead dissipated as heat. This activity enables brown adipose tissue to be a site of nonshivering thermogenesis in neonates and in small mammals exposed to the cold.

A final example of the use of Δp is the transhydrogenase (TH) enzyme in the mitochondrial inner membrane. Within the mitochondrial matrix, the $NADH/NAD^+$ ratio is about 0.1 to 0.2, while that of the $NADPH/NADP^+$ couple is approximately 99% reduced. This is thought to be because NADPH is required to maintain the mitochondrial glutathione pool and thioredoxin in a reduced state, through NADPH-dependent glutathione reductase and thioredoxin reductase. A reduced glutathione pool and reduced thioredoxin are important in protecting mitochondria from oxidative damage. One of the ways the high $NADPH/NADP^+$ ratio is maintained is through the transhydrogenase, which couples proton movement down the Δp gradient to drive electrons from NADH to $NADP^+$ and thus maintains the disparity in redox states.

4
Overview of the Metabolic Basis of Cellular Energetics

A central feature of cellular bioenergetics is the displacement from equilibrium of the ATP hydrolysis reaction. This displacement is at the heart of metabolism where it acts as a coupling reaction between catabolism and anabolic or work reactions. While there is continual turnover of ATP, the energy stored in this displacement from equilibrium is usually held at about $-50\ kJ.mol^{-1}$ through the expenditure of energy from fat, carbohydrate, or protein. As these molecules are broken down, the chemical energy is converted to redox potential energy in the reduced electron carriers NADH and $FADH_2$. This redox potential energy is gradually released as electrons pass through the mitochondrial respiratory chain and is conserved at three regions in the respiratory chain by pumping protons across the mitochondrial inner membrane. This builds up a Δp that is then used to drive the synthesis of ATP in the mitochondrion and its subsequent transport to the cytosol. Thus, there is a linked series of energy conversions, from chemical energy to redox energy and protonmotive force, followed by a return to chemical energy, which is stored in the displacement from equilibrium of the ATP hydrolysis reaction. These energy interconversions allow ATP to drive the many otherwise unfavorable reactions that are necessary for life.

See also Adipocytes; Intracellular Fatty Acid Binding Proteins in Metabolic Regulation; Oncology, Molecular.

Bibliography

Books and Reviews

Brown, G.C. (2000) *The Energy of Life*, Flamingo, London.

Harold, F.M. (1986) *The Vital Force: A Study Of Bioenergetics*, W. H. Freeman, New York.

Nicholls, D.G., Ferguson, S.J. (2002) *Bioenergetics 3*, Academic Press, London.

Rich, P. (2003) Chemiosmotic coupling: the cost of living, *Nature* **421**, 583.

Scheffler, I.E. (1999) *Mitochondria*, Wiley-Liss, New York.

Primary Literature

Saraste, M. (1999) Oxidative phosphorylation at the *fin de siecle*, *Science* **283**, 1488–1493.

Wallace, D.C. (1999) Mitochondrial diseases in man and mouse, *Science* **283**, 1482–1488.

Metabonomics and Metabolomics

David J. Grainger[1] *and Jeremy K. Nicholson*[2]
[1] *Department of Medicine, University of Cambridge, Cambridge, UK*
[2] *Department of Biological Chemistry, Imperial College, London, UK*

Encyclopedia of Molecular Cell Biology and Molecular Medicine, 2nd Edition. Volume 8
Edited by Robert A. Meyers.
Copyright © 2005 Wiley-VCH Verlag GmbH & Co. KGaA, Weinheim
ISBN: 3-527-30550-5

Keywords

Chromatography
A range of methods designed to separate the molecular components of a complex mixture on the basis of their physical properties.

Mass Spectrometry
An analytical chemistry technique that can be used to identify molecules by fragmenting them into ions, which are then separated on the basis of mass.

Nuclear Magnetic Resonance
The context-dependent resonance of various atomic nuclei (such as the hydrogen nucleus) when placed in a magnetic field, which can be used to obtain structural information about the molecular components of a solution.

Regression Modeling
Mathematical methods designed to predict one (or more) variables from the values of many other measured variables.

Metabonomics is the study of systemic biochemical profiles and regulation of function in whole organisms by analyzing biofluids and tissues. Like genomics (the study of the complete repertoire of genes in an organism) and proteomics (the study of the protein complement of a tissue or cell), metabonomics can provide a holistic overview of the current physiological status of an organism, and its response to external stressors. Here we review the technological approaches to generating metabolic profiles, highlighting the advantages and disadvantages of each methodology, as well as the various strategies for extracting useful conclusions from the very large datasets that can be generated by such profiling. Metabonomics can be applied to a wide range of biological applications, including predictive toxicology, probing the physiology of disease both in animal models and in man, and to make clinically useful diagnoses of disease. With examples of each of these applications, we illustrate the potential of metabonomics to contribute to our understanding of complex biological systems in a post-genomic era where we understand many of the components of living systems, but few of their dynamic interactions.

1
What is Metabonomics?

The suffix "omics" is now routinely applied in many fields to the holistic study of an entire system, as opposed to a reductionist description of each of its parts independently. Thus, metabonomics is the name given to the holistic study of metabolic systems in living organisms. In principle, a metabolic profile is therefore a simple list of all the low molecular weight metabolites (such as sugars, amino acids, and lipids) present in a biological system, together with the concentration of each metabolite present (the generation of such profiles is one of the definitions in use for metabolomics). Clearly, such a profile is analogous to a genomic profile (a list of all the genes composing an organism, perhaps with their levels of expression also) or a proteomic profile (a list of the proteins in an organism).

Like genomics and proteomics, however, metabonomics is also much more than a simple list. The metabolic profile of a particular biological sample is just a snapshot of a complex, dynamic network that reflects the physiological activity of the organism. Enzymes are rapidly interconverting metabolites; new compounds are being absorbed from the environment; waste products are being excreted. The science of metabonomics, therefore, is not only about capturing metabolic profiles (described in Sect. 2 below) but also about extracting an understanding of the underlying biological system from the resulting dataset (described in Sect. 3). In particular, metabonomics is a global metabolic regulation approach based on understanding complex system behavior, designed to reveal the response of an organism to an external stressor or stimulus.

The relationship between metabonomics and other systems biology disciplines is illustrated in Fig. 1. Genetic information (coupled with the pre-existing levels of various other proteins) is the major determinant of the mRNA (or transcriptomic) profile, which in turn is a key determinant of the proteomic profile. Proteins (in the form of enzymes) are an important determinant of the metabolic profile, which, in turn, feeds back to modulate gene expression patterns. This "homeostatic loop" is then modulated by external inputs from the environment. The major input of low molecular weight compounds from the diet can have a significant effect on the metabonomic profile, which subsequently affects the gene expression and protein profiles of the organism. Other environmental factors can also affect the metabolic profile (for example, the amount of light determines the rate of vitamin D formation), as well as directly altering gene expression (the amount of exercise modulates skeletal muscle gene expression and ultimately protein content). This position at the interface between genetic and environmental determinants makes metabolic profiling a uniquely powerful tool for probing the dynamic physiological status of an organism.

Compared with genomic and proteomic profiles, the metabonomic profile is also more dynamic, reflecting the current physiological status of the organism as well as its future behavior. Polymorphisms, in particular genes (a component of the genomic profile), may allow the risk of developing a disease to be estimated, but they cannot determine whether the organism is suffering from that disease at a given point in time.

To properly understand the metabolic network of a multicellular organism would require continuous measurement of the

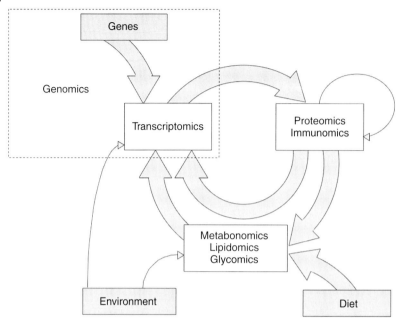

Fig. 1 Relationship between the "omics." Much of the variation in protein levels (measured in proteomics) is due to variations in the expression of the mRNA encoding that protein (measured in transcriptomics, a subset of genomics). In turn, the variation in the levels of metabolites (measured in metabonomics) is determined by the levels of various enzymes (proteins). Metabolites can then feedback and regulate gene expression patterns, closing a "homeostatic loop." The environment interacts with this homeostatic loop primarily through the influence of diet on the metabolic profile, although other environmental factors can also have a direct effect on both metabolism and gene expression (for example, exposure to UV light directly affects vitamin D3 levels, and exercise can affect gene expression patterns).

levels of all the metabolites present in all of the cells and tissues that compose the organism. Such an "ideal" metabolic profile is unlikely to be practicable attainable (at least in the near future). As a result, practical metabonomics involves selection of both a sample of the whole organism (for example, a blood specimen) and a time point (or series of time points) at which to make the observations. Clearly, the extent to which one can hope to understand the metabolism of the organism as a whole from a (potentially poorly representative) sample is unclear and care should be taken in drawing broad conclusions

from limited measurements. Similarly, available analytical chemistry techniques (such as spectroscopy or chromatography) do not allow accurate measurements of the levels of every low molecular weight compound present in a given biological specimen. These practical limitations are illustrated in Fig. 2.

Even with the current practical limitations, metabonomics is a powerful new tool for studying complex biological systems. It has already been used successfully to monitor the physiological response to xenobiotics (such as new pharmaceuticals under development), and its use in

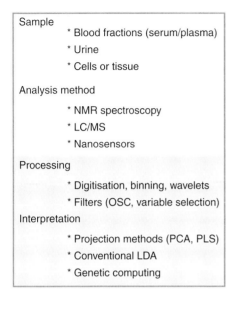

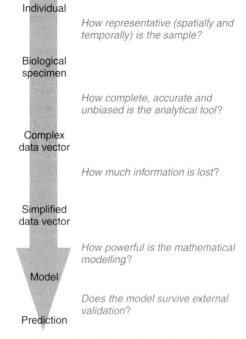

Fig. 2 A generic "omics" experiment. The aim of "omics" biology is to make measurements on individuals and then deduce predictive rules about the organism. In order to go from individual observations to a prediction, various steps must be followed: a sample must be taken on which many measurements are made to generate a complex data vector. Various mathematical modeling tools are then used to build a regression model which can be used to make a prediction that can be validated. Examples of each step in a typical metabonomics experiment are shown in the left panel. Various assumptions made at each step are listed in the right panel.

toxicology is increasing. Metabonomics can also be used to diagnose the presence of diseases, both for those where in-born errors of metabolism are responsible for the symptoms, and also in diseases where metabolic disregulation is less obviously involved in the pathogenesis of the disease. Using a metabolic profile to diagnose disease is not just useful in the clinic: it can also provide important new information about the physiological processes that are misregulated in the disease, which might ultimately assist in the search for new treatments.

The "homeostatic loop" illustrated in Fig. 1 also emphasizes the importance of integrating the genetic, protein, and metabolic profiles if we are to maximize our understanding of the organism. The boundaries between the "omics" represent technical limits in our methodologies for making measurements rather than being any useful dividing line between the applications of the information that has been generated. Fortunately, the profiles generated can be merged (at least in principle, although the bioinformatics challenges in doing so are significant) yielding a composite (or "multi-omics") profile of the organism. Any question partially addressed by the separate genomic, proteomic, and metabonomic profiles will likely be more fully answered through the careful construction and analysis of such "multi-omics" profiles.

2
Methods for Generating a Metabonomic Profile

2.1
Criteria for Judging Metabonomic Profiling Methods

A range of different analytical chemistry approaches have been used to generate metabonomic profiles. Unfortunately, none of the profiles generated even approximate to the "ideal" profile, and it is necessary to make an informed choice of analytical tool depending on various trade offs. There are three important criteria that contribute to the "ideal" profile:

1. *Completeness.* If we assume that the "ideal" profile consists of a list of all the different low molecular weight compounds present in a biological sample, then any profiling method can be judged on the fraction of all the metabolites present that contribute to the profile. In absolute terms, this can be difficult to assess, since without a "gold standard" complete profile it is impossible to know what components have been missed. In practice, however, applying multiple analytical approaches to the same sample soon throws up examples of components missed by the other techniques.

Two factors contribute to the completeness of a profile. Firstly, the general sensitivity limit of the technique sets the threshold below which no components are detected. While an "ideal" profile would include even components present as just a single molecule, practicality suggests that components present at such low levels are unlikely to have a biologically significant effect, and that a profile with sensitivity threshold in the pM or nM range would be adequate for all but the most demanding applications. Unfortunately, several techniques are more insensitive still, and will miss components that are biologically relevant. It is worth noting, however, that sensitivity can be a double-edged sword – many of the low abundance metabolic components are derived from symbiotic or xenobiotic organisms (such as gut microflora) not directly related to mammalian metabolism. It is unclear whether gathering such enlarged metabolic datasets using highly sensitive analytical approaches is useful – indeed, it may simply make the task of extracting a meaningful picture from the resulting dataset more difficult (see Sect. 3).

The second factor is the invisibility of particular compounds, or more likely classes of compounds, to a particular analytical technique. An obvious example would be the inability of non-volatile components to contribute to a gas chromatograph. Less obvious might be the inability of a technique to separate or distinguish components with closely related structures, such that a single "entry" in the profile is in fact a composite measure of two or more compounds.

2. *Bias.* The "ideal" profile is not merely a list of the components present in the sample but also an indicator of the relative amounts of each component. Thus, if certain components are more readily detected than others (on a molar basis) the analytical technique will display a bias, suggesting some components are present at relatively higher levels than they actually are.

3. *Cost.* The resource implications of any data-gathering exercise must be properly considered as part of the scientific experimental design. This is particularly true of any nonselective, high data density "omics" experiment, where the amount of understanding about the system which is ultimately gained will likely depend on how many different (that is, uncorrelated) components of the system are analyzed, rather than whether any particular class of components (be it metabolites, proteins or genes) has been exhaustively investigated. Consequently, the resource implications of selecting any given analytical methodology rightly forms a part of the experimental design: does the additional information gained by adding a particular technique to the portfolio of analyses to be performed on a given sample add sufficient uncorrelated variables to justify the resources employed, or could the same resources generate more uncorrelated information through application of a different technique?

No technique currently available is complete, unbiased, and inexpensive, but each has certain advantages for particular applications. Application of multiple techniques to the same sample may improve completeness and reduce bias, but only at increased resource implication.

2.2
Nuclear Magnetic Resonance (NMR) Spectroscopy

NMR (nuclear magnetic resonance) spectroscopy has been widely used to generate metabonomic profiles, particularly of serum and urine samples. The primary advantages of NMR spectroscopy are the intrinsic reproducibility of the generated spectrum and the complete lack of bias. Across the information-dense region of the spectrum the coefficient of variation between replicate measures made on different days is below 1%. Reproducibility of this nature allows even small differences between profiles to be interpreted as significant, increasing the power of the experiment, particularly when relatively small numbers of profiles are being compared.

The NMR spectrum depends on the context-dependent resonance of hydrogen nuclei within the various molecules that compose the biological sample. As a result, any molecular structure containing at least one hydrogen nucleus is in principle represented within the spectrum. Since all biological molecules fall into this category, essentially every metabolite can contribute to an NMR-derived metabonomic profile. Furthermore, the intensity of the signal due to each hydrogen nucleus is of the same strength irrespective of its molecular context. Consequently, there is absolutely no bias in the estimated relative amounts of each of the metabolites detected.

Despite this lack of bias, NMR spectroscopy cannot be considered to generate a complete metabonomic profile because of the inherent insensitivity of the approach. Although the high reproducibility allows even very small peaks to be distinguished from baseline noise, these peaks still represent relatively abundant molecular components of the sample. Although the absolute sensitivity cutoff varies depending on the particular implementation of the technique (for example, the use of cryoprobes significantly improves the sensitivity with nanogram quantities of compounds detected), nevertheless, very low abundance molecules are difficult to detect. Since many biological molecules of importance (such as vitamins and signaling molecules like prostaglandins or cyclicAMP) rarely achieve concentrations above the nM range, they are effectively absent from NMR-derived metabonomic datasets.

Although NMR spectroscopy is the tool of choice for most chemical structure determination problems, it also struggles to distinguish certain classes of closely related molecular structures. For example, it is difficult to study complex mixtures of fatty acids of different chain lengths by NMR spectroscopy, because the signals from the different molecular structures are overlaid in the resulting spectrum. Although ever more complex NMR-based approaches have been devised to aid separation and unique identification of given molecular structures (such as 2D-TOCSY or various heteronuclear NMR approaches), these approaches are resource intensive and may still be less informative than mass spectrometry for certain molecular classes.

2.3
Chromatography and Mass Spectroscopy (LC-MS and GC-MS)

Chromatography followed by mass spectrometry is the other major analytical tool that has been used to generate metabonomic profiles. Both liquid chromatography (LC) and gas chromatography (GC) have been used. GC may offer superior resolving power, at least for certain classes of molecules, but it is limited by the lack of volatility of many metabolites. Although this can be overcome to some extent by covalent modification of the sample prior to chromatography, a range of important metabolites still fail to enter the gas phase and hence do not contribute to the resulting metabonomic profile. As a result, the effective size limit for GC-MS is about 700 Da, whereas much larger molecules can contribute to the NMR-, and to some extent, to the LC-MS-derived profiles.

The major advantage of GC-MS and LC-MS is the sensitivity of the technique, which can detect component compounds down to the likely limit of biological relevance (and, indeed, may be too sensitive in some cases, populating the metabolic dataset with large numbers of minor contaminants of the sample, which are not the products of mammalian metabolism). As a result, components such as vitamins and signaling intermediates, which were invisible to the NMR spectrometer, now contribute to the GC-MS-derived metabonomic profile.

The combination of chromatography and mass spectroscopy (MS) can also allow unambiguous assignment of structure to the components of the biological sample in a way that is difficult, though not impossible, with NMR spectroscopy. Comparison of fragmentation patterns with databases, allows deconvolution of peaks with overlapping retention times, and as many as 1000 different molecular components to be unambiguously identified from a single complex biological fluid. This is particularly true when comparing different members of the same homologous series (such as fatty acids of different chain lengths). However, LC-MS and GC-MS usually poorly resolve structural isomers that are readily distinguished by NMR spectroscopy. It is also important to note that, unlike NMR spectroscopy, mass spectrometry cannot be used to obtain the structure of unknown components contributing the profile – if the fragmentation pattern is not among the database of "known" metabolites, no assignment is possible.

Furthermore, MS-based detection is inherently biased: some molecules do not ionize, or only ionize poorly, under any given set of conditions and as a result are either completely invisible or detected only weakly. Consequently, an MS-derived metabonomic profile may be highly biased, and little information can be gained from the relative signals due to different components in the biological sample. However, it is still possible to compare the levels of the same component across different samples in a reliable and reproducible way. Again, variations in the implementation of the technique (such as using both positive and negative ionization modes and varying the cone voltage) can alleviate, though not eliminate, this problem at the cost of increased resource implication.

2.4
Nanosensors

More recently, it has become clear that various designs of the nanosensor can also be applied to generating metabonomic profiles. Nanosensors coated with various hydrophobic coatings adsorb a wide range of metabolites differentially, allowing a profile to be generated that is related in a complex fashion to the molecular composition of the fluid under study. Although to date, no metabonomic profile

that has been generated using nanosensor technology has been reported in the scientific literature, it seems likely that such datasets will appear imminently.

Nanosensor-derived metabonomic profiles will presumably have the advantage of low cost (being rapid and high throughput, and not requiring capital intensive reading equipment). However, the complex nature of the relationship between the resulting data vector and the molecular composition of the sample will likely preclude any straightforward listing of the component molecules or estimation of the relative amounts. Such a profile will be neither substantially complete nor unbiased.

Yet, at least for clinical diagnostic applications, such profiles are not without utility. Although it may be difficult (or indeed impossible) to understand precisely which molecular components contribute to the systematic difference in profiles between two groups of interest (such as diseased individuals versus healthy individuals), nevertheless, the very presence of a systematic difference may be diagnostically useful.

2.5
Other Approaches

Many other analytical chemistry techniques can be applied to complex biological fluids to generate a metabonomic profile, although (like nanosensors) such profiles cannot usually be translated into a list of molecular components with associated relative concentrations. For example, infrared spectroscopy provides a low-cost approach to generating a metabonomic profile, which may be useful in some circumstances. It may be possible, depending on the nature of the biological sample and the particular molecular components of interest, to generate a limited list of molecular components from an infrared spectrum.

Ultimately, however, the utility of a metabonomic profile obtained with any given technique depends on the application. Many studies (particularly aimed at diagnostic applications) may not require the immense structural detail that can be generated using the resource-intensive NMR and MS techniques. Equally, attempts to identify biomarkers associated with a particular phenotype will require a metabonomic profile that more nearly corresponds to the theoretical "ideal" profile. There remains, therefore, a considerable amount of trial and error in the selection of the analytical toolkit best suited to answering a particular question.

3
Methods for Interpreting Metabolic Profiles

3.1
The Problems of Interrogating Very Large Datasets

Much of the power of the metabonomic approach stems from the generation of very large datasets, which are essentially unselected in terms of the contributing components (that is, there was no pre-existing hypothesis governing the selection of variables to be measured). As a consequence, special techniques are required to handle the resulting datasets, since extracting meaningful conclusions from datasets with millions of datapoints is a daunting exercise.

The basic aim in interpreting a metabonomic dataset is no different from any conventional multivariate analysis. The value of a dependent (or Y-) variable is estimated from a collection of measured X-variables (Fig. 3). For example, one might wish to estimate blood pressure from a range of physiological (or metabonomic) measures. While the Y-variable to be estimated (or "modeled") may be continuous (like blood pressure), it may equally be a discrete classification variable (which divides each observation into two or more groups, such as the presence and absence of a particular disease).

The problems associated with analyzing very large datasets are basically twofold. Firstly, a metabonomic profile might typically be composed of thousands of datapoints for each individual, yet such a profile may only have been generated from a relatively small number of individuals (tens or at most hundreds per group). Such datasets are typically described as "short and fat," having many more variables than observations (illustrated in Fig. 3). Analyzing such short and fat datasets using conventional multivariate statistics is dangerous, because it becomes increasingly likely that you can construct a model that correctly describes the phenotypic classification of the individuals by chance alone as the number of variables exceeds the number of observations.

Secondly, many of the variables that compose the metabolic profile may be highly correlated with each other. This is a particular problem with spectroscopic data, where neighboring variables are integrals of a continuous spectrum, forcing a relationship between nearby spectral regions. Conventional statistical approaches to multivariate analysis assume that all the predictor variables are independent, and the widespread colinearity of metabonomic profile variables renders the conventional multivariate models error prone.

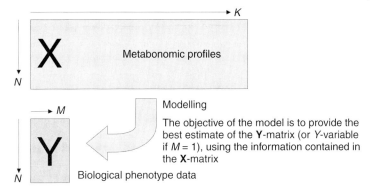

Fig. 3 The principle of regression modeling. The principle of regression modeling is to predict a range of features about a complex system (such as a biological organism) from a collection of unrelated measurements. In metabonomics, this consists of predicting phenotype or behavior (a **Y**-matrix of *M* features of *N* individuals) from a metabonomic dataset consisting of *K* measured metabolic variables from the same *N* individuals. Unlike conventional regression modeling, metabonomics datasets are typically "short and fat" with many more measurements per observation than individuals (*K* ≫ *N*). The **Y**-matrix to be predicted might be a single continuous variable (*M* = 1), or even a categorical description (diseased or healthy, for example), whereupon the modeling is usually termed *a discriminant analysis*. Figure reproduced with permission from www.graingerlab.org.

3.2
Conventional Statistical Approaches (LDA)

As noted above, a conventional multivariate model (in which a dependent variable (*Y*) is predicted from a matrix of independent variables) is based on the assumption that the number of *X*-variables is less than the number of observations (*n*), and that all the *X*-variables are uncorrelated. In most experiments, a series of raw metabonomic profiles violates both assumptions and as a general rule conventional multivariate statistics should not be applied to metabonomic datasets.

However, various preprocessing steps can convert the raw metabonomic dataset into a form amenable to conventional statistical analysis. For example, there exists a range of prefilter algorithms (see Sect. 3.6 below) that allow the number

of *X*-variables to be substantially reduced, for example, by retaining only those *X*-variables that are most significantly correlated with the dependent variable *Y*. Such variable-selection algorithms may also incorporate rules to eliminate intercorrelated *X*-variables. After variable selection, the resulting dataset can be analyzed by conventional multivariate statistics, such as linear discriminant analysis (LDA).

Although such variable-selection approaches circumvent the limitations of conventional multivariate statistics, they can limit the power of the analysis to identify associations between the **X**-matrix and the dependent *Y*-variable, since much of the information in the **X**-matrix has been discarded. In general, therefore, the other approaches below (which have been developed specifically for the analysis of very large, or megavariate, datasets) tend to be more powerful.

3.3
Projection Methods (PCA and PLS)

Instead of selecting a subset of variables from the **X**-matrix, it is possible instead to combine the *X*-variables in linear combinations to generate a smaller number of composite variables. This approach (termed *projection*) deals effectively with both of the limitations to megavariate analysis: variable number is reduced, and intercorrelation is minimized.

The principle of projection is illustrated in Fig. 4. Here, a complex 3-dimensional object is represented by a simpler 2-dimensional shadow. In the left panel, the axis of projection is chosen such that the 2-D shadow poorly retains the information of the original object, whereas in the right panel the optimum projection is chosen, which retains most of the information encoded in the original 3-D object. The mathematical algorithms that underlie projection methods such as principal component analysis (PCA) or projection to latent structures using partial least squares (PLS) work by selecting the best projections. A major difference between PCA and PLS is that PLS is a supervised method, which means that

the dependent *Y*-variable is used in the process of locating the best projection of the data. Application of these algorithms to high-dimensional datasets (such as metabonomic profiles) can yield just a handful of composite variables (principal components), which are simple linear combinations of the original matrix of *X*-variables. Unlike simpler variable-selection techniques, projection retains as much as possible of the information in the original **X**-matrix, while reducing the dimensionality of the dataset to manageable proportions.

One potential problem with projection methods is overfitting the model. With a sufficiently large number of *X*-variables, it will always be possible to generate combinations of the *X*-variables that predict the dependent variable very well. As a result, it is essential to include a robust external model validation step in the analysis protocol. One example of a useful validation step is scrambling the dependent variable and demonstrating that the **X**-matrix predicts the real dependent variable better than the scrambled variable. An alternative approach is to use the generated model to make predictions about an external dataset not used during model generation. Exhaustive validation is essential for models built using supervised techniques, such as PLS, where the dependent variable was

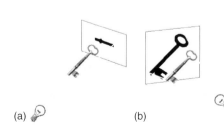

(a)   (b)

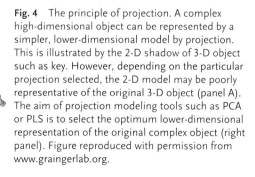

Fig. 4 The principle of projection. A complex high-dimensional object can be represented by a simpler, lower-dimensional model by projection. This is illustrated by the 2-D shadow of 3-D object such as key. However, depending on the particular projection selected, the 2-D model may be poorly representative of the original 3-D object (panel A). The aim of projection modeling tools such as PCA or PLS is to select the optimum lower-dimensional representation of the original complex object (right panel). Figure reproduced with permission from www.graingerlab.org.

used during the construction of the principal components.

3.4
Genetic Computing

Another approach to building models that optimally describe a dependent *Y*-variable from a very large **X**-matrix is to generate a large pool of random models (that are linear or nonlinear combinations of the *X*-variables), a few of which will be acceptable but most of which will be poor, and then apply an evolutionary algorithm to recombine the models and select for improved description of the dependent variable. The evolution is continued through a number of generations, and models can emerge that explain a good proportion of the variation in the *Y*-variable.

This approach can offer a number of advantages over projection-based methods. In particular, the ability to easily include rules that combine *X*-variables in nonlinear ways can be useful when modeling biological systems that have inherent nonlinearities. Another advantage of the evolved models is that they are generally easier to interpret than the optimized projection models. Combinations of simple rules can be more intuitively obvious than lists of principal components

One disadvantage of genetic computing approaches, however, can be "premature convergence" whereby the pool of models undergoing evolution rapidly converges on a local maximum of fitness and poorly explores the entire model space. Improvements are continually being made to the basic genetic computing algorithms, and recent advances such as multiobjective fitness functions can alleviate the premature convergence problem.

3.5
Other Approaches

The explosion of high data density analytical techniques in genomics and proteomics, as well as metabonomics, has stimulated the development and refinement of a wide range of other bioinformatics tools to assist in the interpretation of very large datasets. Hierarchical cluster analysis (HCA) is particularly popular for the analysis of gene expression datasets, but it is considerably less powerful than both projection methods and genetic computing algorithms, while offering few advantages. As a result, HCA has not been extensively used to interpret metabonomic datasets.

Another approach, which has been used in metabonomics, is neural network analysis. The neural net is set up with the *X*-variables as inputs and the dependent *Y*-variable as output, with a network of nodes in between. The mathematical function applied at each node is then iteratively varied during a learning phase to optimize the successful prediction of the *Y*-variable. Neural nets can be useful for performing discriminant analysis (that is, classifying observations into two or three groups on the basis of their **X**-matrix values) and hence used for clinical diagnostic purposes, but the models they generate are inherently difficult to interpret, because a large number of structurally different models can yield almost identical predictive power.

3.6
Data Preprocessing and Data Filters

Model building with very large datasets may be best performed in two or even more steps. Rather than directly applying one or more of the model building tools described above to the raw **X**-matrix, it may be more powerful to perform a preprocessing step first. In fact, as with the model building tools themselves, the object of most of the preprocessing steps is to reduce dimensionality. However, empirical observation suggests that more powerful models can be generated if the dimensionality is reduced in stepwise fashion, perhaps using more than one tool, rather than in a single leap.

Consequently, the most straightforward preprocessing step would be to apply one of the model building tools twice in succession. For example, with projection methods, it is possible (and sometimes useful) to perform an initial principal component analysis reducing the dimensionality of the raw **X**-matrix from, say 8096 variables to tens or hundreds of principal components, and then treat the resulting matrix of principal components as the **X**-matrix for a second round of PCA to reduce the dimensionality down to two or three components, which can be more readily interpreted. The extent to which such hierarchical PCA improves the model compared with application of a single round of PCA depends very much on the particular data structure.

When dealing with metabonomic profiles derived from continuous spectra (such as an NMR spectrum), the neighboring variables (derived from integration of contiguous regions of the spectrum) can be very highly correlated. The shorter the interval along the abscissa that composes each variable, the higher is the degree of local intercorrelation. This can be reduced by integrating over wider intervals (or by averaging neighboring integrals, a process termed *binning*). This effectively reduces the dimensionality of the spectral data, while retaining much of the important information encoded in the spectrum.

Binning relies on the fact that the most highly intercorrelated variables are related to each other by their position along the abscissa in linear fashion. This may not be true for all types of data; in some cases, the most intercorrelated variables may fall cyclically (for example, in the untransformed free-induction decay signal from an NMR spectrometer). For analyzing such datasets, wavelet transformation, rather than binning, will provide the most efficient dimensionality reduction prior to model construction.

Finally, noise filters can be applied to the dataset prior to model building, and for NMR-derived metabonomic datasets this has been shown to significantly improve the performance of projection-based modeling tools. Noise filters (such as orthogonal signal correction; OSC) aim to remove variation in the **X**-matrix, which is uncorrelated with the dependent *Y*-variable, simplifying the dataset for subsequent modeling. It is important to remember that noise filters like OSC are therefore supervised methods, and exhaustive validation of models built on the filtered dataset will be required irrespective of the nature of the modeling tool subsequently used.

4
Applications of Metabonomics

4.1
Probing Normal Human Metabolism

The most obvious application for metabonomics is to aid understanding of normal human metabolism. It is possible to build up a detailed picture of metabolic pathways by taking a time series of metabolic profiles and constructing quantitative models describing the metabolic flux through various pathways. Although the properties of key metabolic pathways (such as glycolysis or tricarboxylic acid cycle) have been extensively investigated for years, metabonomics can still reveal important aspects of metabolism, which had previously gone unnoticed. In particular, the interaction between endogenous metabolic pathways and the products of symbiotic bacterial metabolism have been extensively investigated using metabonomics.

4.2
Probing the Pathophysiology of Human Disease

Metabonomics is particularly powerful for analyzing the metabolic changes associated with the development of a particular disease state. By comparing metabonomic profiles taken from diseased individuals with profiles from healthy control individuals, it is possible to identify the systematic differences associated with the presence of the disease. What remains challenging, however, is determining which (if any) of these metabolic changes are causes of the disease pathology, as opposed to direct or even indirect consequences of the disease progression.

For example, metabonomics has been used to investigate the metabolic changes associated with the development of osteoporosis. Serum samples from women with low bone mineral density and from healthy control women were subjected to ^{1}H-NMR analysis, and the resultant profiles compared using the projection method PLS-DA following an OSC prefilter. One of the most important metabolites responsible for the separation of the diseased and healthy individuals was the amino acid proline. Women with pathological low bone mineral density had lower levels of proline in their serum; an observation that

was later confirmed used conventional biochemical assays. While this does not prove that low serum proline is responsible for the lower bone density, it is a plausible hypothesis for further investigation: proline is a key constituent of the collagen component of bone matrix, and inadequate proline supplies for collagen biosynthesis could represent an entirely novel pathophysiological mechanism in the development of osteoporosis.

4.3
Investigating and Validating Animal Models of Disease

Animal models of human diseases are important tools in scientific and pharmaceutical research and development. The development of genetic manipulation techniques has allowed good models of many monogenic disorders (such as muscular dystrophy) to be developed, confident in the knowledge that the underlying cause of the disease is similar in the animal as in man. In contrast, for complex polygenic disorders such as atherosclerosis or Alzheimer's disease, it is unclear to what extent any given animal model mimics the molecular mechanisms underlying disease susceptibility in man, even if the phenotype of the animals faithfully mirrors the human disease.

Metabonomics offers an opportunity to address this question. Metabolic profiles of animal models can be compared directly with the profiles from human sufferers, allowing a comparison of the physiological perturbations accompanying disease development in the two species. Perhaps more powerfully, it should be possible to track the metabolic trajectory of both animals and humans as the disease progresses, with similar trajectories increasing the confidence in the likely validity of the model. Studies of this nature are currently underway for a range of animal models of disease, but to date none have been published.

4.4
Clinical Diagnosis of Disease

In many senses, using metabonomics to perform clinical diagnoses is more straightforward than its application to biomarker identification and pathophysiological analysis. Providing a clinically useful diagnosis of a disease on the basis of a serum sample only requires the identification of a robust metabolic signature that always accompanies the disease and is rarely, if ever, present in healthy control individuals. It is not necessary to be able to identify any of the molecular components contributing to the disease-associated metabolic signature.

For example, projection analysis using PLS-DA of ^{1}H-NMR-derived metabonomic profiles of serum samples from individuals with coronary artery disease and healthy control individuals demonstrated clear separation of the two groups. After application of the OSC prefilter, it was possible to predict the disease status of individuals with at least 90% sensitivity and specificity. This metabonomics diagnostic test outperforms all existing noninvasive tests for coronary heart disease by a considerable margin. Although such a test is potentially useful in the clinic (since, at present, the gold-standard diagnostic test for heart disease is an invasive angiography procedure that is expensive and carries a small risk to the patient), it does not readily identify the particular molecular species responsible for separating the two groups. Much of the discriminatory power of the test falls in the region of the NMR spectrum due to

lipid components (unsurprisingly, given our knowledge of the mechanisms underlying heart disease) but NMR poorly resolves these closely related lipid structures and considerable further work will be required before the precise molecular basis for the success of the test is known.

4.5
Selection of Subjects for Clinical Trials

At present, a substantial impediment to the testing of new therapeutics for certain diseases is the ability to identify potential sufferers ahead of time. For example, to test a drug proposed to reduce the incidence of myocardial infarction requires the recruitment of subjects at high risk of suffering a myocardial infarction during the trial. Unfortunately, current methods of identifying such subjects are poor, and trials of this nature can require the study of thousands of individuals for three years or more to accumulate sufficient myocardial infarction events for the impact of the drug to be detectable. Although metabonomics-based diagnostics have yet to be used in such applications, it is likely that their widespread adoption will rapidly follow the first successful demonstration of such use.

4.6
Monitoring Efficacy of Therapeutic Interventions

One of the most exciting applications of megavariate diagnostics, whether based on metabonomic, genomic, or proteomic profiles, is the prospect of personalized therapeutic interventions. At present, many drugs are used on broad swathes of the population (for example, statins to lower circulating LDL cholesterol) without any clear indication as to whether they are equally effective in all individuals. Pilot studies already indicate that it is possible to predict the response of an individual to statin therapy from their metabonomic profile measured prior to beginning therapy. Such "pharmacometabonomics" could be extended to optimize the dose and delivery route of a wide range of drugs for each individual, with likely improvements in the efficacy of treatment.

4.7
Toxicology

Perhaps the most mature application of metabonomics is the application of metabolic profiling to toxicology. By studying the metabolic response to a range of model toxins (with organ-specific toxicity), it has been possible to identify metabolic signatures associated with damage to a particular organ. Extensive studies following the metabolic trajectory of animals treated with these model toxins have been published, although commercial considerations mean that few studies of clinically relevant pharmaceutical compositions have reached the public domain.

While metabonomics may help improve the predictivity of animal toxicology studies (which are notoriously difficult to interpret using conventional physiological and histological end points), perhaps the most exciting possibility is the use of metabonomics to perform early stage toxicology directly in man. Because of the sensitivity of metabonomics to detect minute perturbations in the metabolic signature, it may be possible to get an indication of the mode of toxicity of novel chemical entities given in man at doses well below those at which any irreversible damage might occur. The ability to perform meaningful toxicology in man should improve the safety of our medications, and at the same time reduce

the number of promising pharmaceutical compounds dropped at a relatively late stage in development because of adverse, and possibly species-specific, side effects observed in the animal models currently used for toxicology.

4.8
Predicting Future Disease Risk

If a metabonomic profile can be used to diagnose the presence of an existing disease (such as coronary heart disease or osteoporosis), there is no reason in principle why it cannot be used to predict future disease susceptibility in the same way that genomic profiles are currently being used. To provide a useful indicator of future disease risk, there must be a component of the dynamic metabolic profile that is temporarily stable on a timescale of years and that is variable between individuals. We have already shown that such a stable interperson variance component exists in NMR-derived metabonomic profiles, and it will be interesting to see to what extent this stable element of the metabolic profile predicts the risk of a range of important diseases.

5
Future Prospects and Challenges

The discipline of metabonomics is expanding rapidly. Although selection from among the many combinations of analytical chemistry and mathematical modeling approaches for any given applications remains empirical, nevertheless, the number of metabonomics studies is growing quickly. Successful examples of the use of metabonomics to answer a broad array of scientific and practical questions are now plentiful in the literature (see

Sect. 4), and application to an ever-broader array of problems seems inevitable. Indeed, it is hard to think of a problem that would not be better addressed by a combination of high data density "omics" approaches than by hypothesis-driven reductionist experimentation. It seems that only a deep-seated unease among much of the scientific community for such exploratory, holistic approaches relegates the typical metabonomics experiment to "second-class status" and hampers an even more rapid expansion.

One of the challenges facing metabonomics over the coming years, therefore, is to better understand which of the many experimental approaches is optimum for a given use. Doing so will likely require the careful analysis of a single sample set by multiple different analytical chemistry techniques, and then each resulting dataset be interpreted using a range of mathematical modeling tools. In this way, the comparative power of the different approaches will begin to emerge.

Another important challenge is the integration of metabonomics datasets with the large profiles generated by other high data density techniques such as genomics or proteomics. In a sense, the division of "omics" science along the lines of the analytical techniques needed to make the measurements is entirely arbitrary. Ultimately, it should prove powerful to combine the profiles obtained from multiple different measurement approaches (whether gene expression, protein levels, or metabolite profiles) into a single "multi-omics descriptor." While there remains considerable debate as to exactly how this amalgamation should be performed, it is widely acknowledged that such a system-wide profile is likely to prove more powerful than a metabonomic or genomic profile alone for many applications. Only such a system-wide profile can allow a complete understanding of such a complex system as a biological organism.

See also Adipocytes.

Bibliography

Books and Reviews

Breiman, L. (2001) Statistical modelling: the two cultures, *Stat. Sci.* **16**(3), 199–231.

Eriksson, L., Johansson, E., Kettaneh-Wold, N., Wold, S. *Multi-and Megavariate Data Analysis: Principles and Applications*, Umetrics Academy, Umea, Sweden, pp 1–525.

German, J.B., Roberts, M.A., Watkins, S.M. (2003) Personal metabolomics as a next generation nutritional assessment, *J. Nutr.* **133**(12), 4260–4266.

Holmes, E., Antti, H. (2002) Chemometric contributions to the evolution of metabonomics: mathematical solutions to characterising and interpreting complex biological NMR spectra, *The Analyst* **127**(12), 1549–1557.

Kell, D.B. (2004) Metabolomics and systems biology: making sense of the soup, *Curr. Opin. Microbiol.* **7**(3), 296–307.

Kell, D.B., Oliver, S.G. (2004) Here is the evidence, now what is the hypothesis? The complementary roles of inductive and hypothesis-driven science in the post-genomic era, *BioEssays* **26**(1), 99–105.

Lindon, J.C., Holmes, E., Bollard, M.E., Stanley, E.G., Nicholson, J.K. (2004) Metabonomics technologies and their applications in physiological monitoring, drug safety assessment and disease diagnosis, *Biomarkers* **9**(1), 1–31.

Moolenaar, S.H., Engelke, U.F., Wevers, R.A. (2003) Proton nuclear magnetic resonance spectroscopy of body fluids in the field of inborn errors of metabolism, *Ann. Clin. Biochem.* **40**(Pt 1), 16–24.

Nicholson, J.K., Wilson, I.D. (2003) Understanding 'global' systems biology: metabonomics and the continuum of metabolism, *Nat. Rev. Drug Discov.* **2**(8), 668–676.

Weckwerth, W. (2003) Metabolomics in systems biology, *Annu. Rev. Plant Biol.* **54**, 669–689.

Primary Literature

Beckwith-Hall, B.M., Brindle, J.T., Barton, R.H., Coen, M., Holmes, E., Nicholson, J.K., Antti, H. (2002) Application of orthogonal signal correction to minimise the effects of physical and biological variation in high resolution 1H NMR spectra of biofluids, *The Analyst* **127**(10), 1283–1288.

Blake, G.J., Otvos, J.D., Rifai, N., Ridker, P.M. (2002) Low-density lipoprotein particle concentration and size as determined by nuclear magnetic resonance spectroscopy as predictors of cardiovascular disease in women, *Circulation* **106**(15), 1930–1937.

Boersma, M.G., Solyanikova, I.P., Van Berkel, W.J., Vervoort, J., Golovleva, L.A., Rietjens, I.M. (2001) 19F NMR metabolomics for the elucidation of microbial degradation pathways of fluorophenols, *J. Ind. Microbiol. Biotechnol.* **26**(1–2), 22–34.

Brindle, J.T., Nicholson, J.K., Schofield, P.M., Grainger, D.J., Holmes, E. (2003) Application of chemometrics to 1H NMR spectroscopic data to investigate a relationship between human serum metabolic profiles and hypertension, *The Analyst* **128**(1), 32–36.

Brindle, J.T., Antti, H., Holmes, E., Tranter, G., Nicholson, J.K., Bethell, H.W., Clarke, S., Schofield, P.M., McKilligin, E., Mosedale, D.E., Grainger, D.J. (2002) Rapid and noninvasive diagnosis of the presence and severity of coronary heart disease using 1H-NMR-based metabonomics, *Nat. Med.* **8**(12), 1439–1444; Epub 2002 Nov 25. Erratum in: *Nat. Med.* 2003 **9**(4), 477.

Bundy, J.G., Ramlov, H., Holmstrup, M. (2003) Multivariate metabolic profiling using 1H nuclear magnetic resonance spectroscopy of freeze-tolerant and freeze-intolerant earthworms exposed to frost, *Cryo. Letters* **24**(6), 347–358.

Burns, S.P., Woolf, D.A., Leonard, J.V., Iles, R.A. (1992) Investigation of urea cycle enzyme disorders by 1H-NMR spectroscopy, *Clin. Chim. Acta.* **209**(1–2), 47–60.

Bundy, J.G., Spurgeon, D.J., Svendsen, C., Hankard, P.K., Osborn, D., Lindon, J.C., Nicholson, J.K. (2002) Earthworm species of the genus Eisenia can be phenotypically differentiated by metabolic profiling, *FEBS Lett.* **521**(1–3), 115–120.

Choi, Y.H., Kim, H.K., Hazekamp, A., Erkelens, C., Lefeber, A.W., Verpoorte, R. (2004) Metabolomic differentiation of cannabis sativa cultivars using 1H NMR spectroscopy and principal component analysis, *J. Nat. Prod.* **67**(6), 953–957.

Coen, M., Lenz, E.M., Nicholson, J.K., Wilson, I.D., Pognan, F., Lindon, J.C. (2003) An integrated metabonomic investigation of acetaminophen toxicity in the mouse using NMR spectroscopy, *Chem. Res. Toxicol.* **16**(3), 295–303.

Eads, C.D., Furnish, C.M., Noda, I., Juhlin, K.D., Cooper, D.A., Morrall, S.W. (2004) Molecular factor analysis applied to collections of NMR spectra, *Anal. Chem.* **76**(7), 1982–1990.

Engelke, U.F., Liebrand-van Sambeek, M.L., de Jong, J.G., Leroy, J.G., Morava, E., Smeitink, J.A., Wevers, R.A. (2004) N-acetylated metabolites in urine: proton nuclear magnetic resonance spectroscopic study on patients with inborn errors of metabolism, *Clin. Chem.* **50**(1), 58–66.

Freedman, D.S., Otvos, J.D., Jeyarajah, E.J., Shalaurova, I., Cupples, L.A., Parise, H., D'Agostino, R.B., Wilson, P.W., Schaefer, E.J. (2004) Sex and age differences in lipoprotein subclasses measured by nuclear magnetic resonance spectroscopy: the Framingham study, *Clin. Chem.* **50**(7), 1189–1200.

Gavaghan, C.L., Holmes, E., Lenz, E., Wilson, I.D., Nicholson, J.K. (2000) An NMR-based metabonomic approach to investigate the biochemical consequences of genetic strain differences: application to the C57BL10J and Alpk:ApfCD mouse, *FEBS Lett.* **484**(3), 169–174.

Gavaghan, C.L., Nicholson, J.K., Connor, S.C., Wilson, I.D., Wright, B., Holmes, E. (2001) Directly coupled high-performance liquid chromatography and nuclear magnetic resonance spectroscopic with chemometric studies on metabolic variation in Sprague-Dawley rats, *Anal. Biochem.* **291**(2), 245–252.

Griffin, J.L. (2004) Metabolic profiles to define the genome: can we hear the phenotypes? *Philos. Trans. R. Soc. Lond. B Biol. Sci.* **359**(1446), 857–871.

Griffin, J.L., Cemal, C.K., Pook, M.A. (2004) Defining a metabolic phenotype in the brain of a transgenic mouse model of spinocerebellar ataxia 3, *Physiol. Genomics* **16**(3), 334–340.

Griffin, J.L., Troke, J., Walker, L.A., Shore, R.F., Lindon, J.C., Nicholson, J.K. (2000) The biochemical profile of rat testicular tissue as measured by magic angle spinning 1H NMR spectroscopy, *FEBS Lett.* **486**(3), 225–229.

Griffin, J.L., Williams, H.J., Sang, E., Clarke, K., Rae, C., Nicholson, J.K. (2001) Metabolic profiling of genetic disorders: a multitissue (1)H nuclear magnetic resonance spectroscopic and pattern recognition study into dystrophic tissue, *Anal. Biochem.* **293**(1), 16–21.

Hammad, S.M., Powell-Braxton, L., Otvos, J.D., Eldridge, L., Won, W., Lyons, T.J. (2003) Lipoprotein subclass profiles of hyperlipidemic diabetic mice measured by nuclear magnetic resonance spectroscopy, *Metabolism* **52**(7), 916–921.

Harrigan, G.G., LaPlante, R.H., Cosma, G.N., Cockerell, G., Goodacre, R., Maddox, J.F., Luyendyk, J.P., Ganey, P.E., Roth, R.A. (2004) Application of high-throughput Fourier-transform infrared spectroscopy in toxicology studies: contribution to a study on the development of an animal model for idiosyncratic toxicity, *Toxicol. Lett.* **146**(3), 197–205.

Hirabayashi, Y., Matsumoto, Y., Matsumoto, M., Toida, T., Iida, N., Matsubara, T., Kanzaki, T., Yokota, M., Ishizuka, I. (1990) Isolation and characterization of major urinary amino acid O-glycosides and a dipeptide O-glycoside from a new lysosomal storage disorder (Kanzaki disease). Excessive excretion of serine-and threonine-linked glycan in the patient urine, *J. Biol. Chem.* **265**(3), 1693–1701.

Holmes, E., Foxall, P.J., Spraul, M., Farrant, R.D., Nicholson, J.K., Lindon, J.C. (1997) 750 MHz 1H NMR spectroscopy characterisation of the complex metabolic pattern of urine from patients with inborn errors of metabolism: 2-hydroxyglutaric aciduria and maple syrup urine disease, *J. Pharm. Biomed. Anal.* **15**(11), 1647–1659.

Jonsson, P., Gullberg, J., Nordstrom, A., Kusano, M., Kowalczyk, M., Sjostrom, M., Moritz, T. (2004) A strategy for identifying differences in large series of metabolomic samples analyzed by GC/MS, *Anal. Chem.* **76**(6), 1738–1745.

Joshi, L., Van Eck, J.M., Mayo, K., Di Silvestro, R., Blake Nieto, M.E., Ganapathi, T., Haridas, V., Gutterman, J.U., Arntzen, C.J. (2002) Metabolomics of plant saponins: bioprospecting triterpene glycoside diversity

with respect to mammalian cell targets, *OMICS* **6**(3), 235–246.

Kell, D.B. (2002) Metabolomics and machine learning: explanatory analysis of complex metabolome data using genetic programming to produce simple, robust rules, *Mol. Biol. Rep.* **29**(1–2), 237–241.

Ketchum, R.E., Rithner, C.D., Qiu, D., Kim, Y.S., Williams, R.M., Croteau, R.B. (2003) Taxus metabolomics: methyl jasmonate preferentially induces production of taxoids oxygenated at C-13 in Taxus x media cell cultures, *Phytochemistry* **62**(6), 901–909.

Keun, H.C., Beckonert, O., Griffin, J.L., Richter, C., Moskau, D., Lindon, J.C., Nicholson, J.K. (2002) Cryogenic probe 13 C NMR spectroscopy of urine for metabonomic studies, *Anal. Chem.* **74**(17), 4588–4593.

Keun, H.C., Ebbels, T.M., Bollard, M.E., Beckonert, O., Antti, H., Holmes, E., Lindon, J.C., Nicholson, J.K. (2004) Geometric trajectory analysis of metabolic responses to toxicity can define treatment specific profiles, *Chem. Res. Toxicol.* **17**(5), 579–587.

Khandelwal, P., Beyer, C.E., Lin, Q., Schechter, L.E., Bach, A.C., II. (2004) Studying rat brain neurochemistry using nanoprobe NMR spectroscopy: a metabonomics approach, *Anal. Chem.* **76**(14), 4123–4127.

Kikuchi, J., Shinozaki, K., Hirayama, T. (2004) Stable isotope labeling of Arabidopsis thaliana for an NMR-based Metabolomics approach, *Plant Cell. Physiol.* **45**(8), 1099–1104.

Kleno, T.G., Kiehr, B., Baunsgaard, D., Sidelmann, U.G. (2004) Combination of 'omics' data to investigate the mechanism(s) of hydrazine-induced hepatotoxicity in rats and to identify potential biomarkers, *Biomarkers* **9**(2), 116–138.

Kraus, W.E., Houmard, J.A., Duscha, B.D., Knetzger, K.J., Wharton, M.B., McCartney, J.S., Bales, C.W., Henes, S., Samsa, G.P., Otvos, J.D., Kulkarni, K.R., Slentz, C.A. (2002) Effects of the amount and intensity of exercise on plasma lipoproteins, *N. Engl. J. Med.* **347**(19), 1483–1492.

Lenz, E.M., Bright, J., Wilson, I.D., Morgan, S.R., Nash, A.F. (2003) 1H NMR-based metabonomic study of urine and plasma samples obtained from healthy human subjects, *J. Pharm. Biomed. Anal.* **33**(5), 1103–1115.

Li, Z., Lamon-Fava, S., Otvos, J., Lichtenstein, A.H., Velez-Carrasco, W., McNamara, J.R., Ordovas, J.M., Schaefer, E.J. (2004) Fish consumption shifts lipoprotein subfractions to a less atherogenic pattern in humans, *J. Nutr.* **134**(7), 1724–1728.

Mitchell, S., Holmes, E., Carmichael, P. (2002) Metabonics and medicine: the biochemical oracle, *Biologist (London)* **49**(5), 217–221.

Moolenaar, S.H., Engelke, U.F., Abeling, N.G., Mandel, H., Duran, M., Wevers, R.A. (2001) Prolidase deficiency diagnosed by 1H NMR spectroscopy of urine, *J. Inherit. Metab. Dis.* **24**(8), 843–850.

Moolenaar, S.H., Gohlich-Ratmann, G., Engelke, U.F., Spraul, M., Humpfer, E., Dvortsak, P., Voit, T., Hoffmann, G.F., Brautigam, C., van Kuilenburg, A.B., van Gennip, A., Vreken, P., Wevers, R.A. (2001) Beta-ureidopropionase deficiency: a novel inborn error of metabolism discovered using NMR spectroscopy on urine, *Magn. Reson. Med.* **46**(5), 1014–1017.

Mortishire-Smith, R.J., Skiles, G.L., Lawrence, J.W., Spence, S., Nicholls, A.W., Johnson, B.A., Nicholson, J.K. (2004) Use of metabonomics to identify impaired fatty acid metabolism as the mechanism of a drug-induced toxicity, *Chem. Res. Toxicol.* **17**(2), 165–173.

Nikiforova, V.J., Gakiere, B., Kempa, S., Adamik, M., Willmitzer, L., Hesse, H., Hoefgen, R. (2004) Towards dissecting nutrient metabolism in plants: a systems biology case study on sulphur metabolism, *J. Exp. Bot.* **55**(404), 1861–1870. Epub 2004 Jun 18.

Ohdoi, C., Nyhan, W.L., Kuhara, T. (2003) Chemical diagnosis of Lesch-Nyhan syndrome using gas chromatography-mass spectrometry detection, *J. Chromatogr. B Analyt. Technol. Biomed. Life. Sci.* **792**(1), 123–130.

Ohse, M., Matsuo, M., Ishida, A., Kuhara, T. (2002) Screening and diagnosis of beta-ureidopropionase deficiency by gas chromatographic/mass spectrometric analysis of urine, *J. Mass. Spectrom.* **37**(9), 954–962.

Ott, K.H., Aranibar, N., Singh, B., Stockton, G.W. (2003) Metabonomics classifies pathways affected by bioactive compounds, Artificial neural network classification of NMR spectra of plant extracts. *Phytochemistry* **62**(6), 971–985.

Otvos, J.D., Jeyarajah, E.J., Bennett, D.W., Krauss, R.M. (1992) Development of a proton nuclear magnetic resonance spectroscopic method for determining plasma lipoprotein concentrations and subspecies distributions from a single, rapid measurement, *Clin. Chem.* **38**(9), 1632–1638.

Pham-Tuan, H., Kaskavelis, L., Daykin, C.A., Janssen, H.G. (2003) Method development in high-performance liquid chromatography for high-throughput profiling and metabonomic studies of biofluid samples, *J. Chromatogr. B Analyt. Technol. Biomed. Life. Sci.* **789**(2), 283–301.

Plumb, R.S., Stumpf, C.L., Gorenstein, M.V., Castro-Perez, J.M., Dear, G.J., Anthony, M., Sweatman, B.C., Connor, S.C., Haselden, J.N. (2002) Metabonomics: the use of electrospray mass spectrometry coupled to reversed-phase liquid chromatography shows potential for the screening of rat urine in drug development, *Rapid Commun. Mass Spectrom.* **16**(20), 1991–1996.

Purohit, P.V., Rocke, D.M., Viant, M.R., Woodruff, D.L. (2004) Discrimination models using variance-stabilizing transformation of metabolomic NMR data, *OMICS* **Summer;** **8**(2), 118–130.

Raamsdonk, L.M., Teusink, B., Broadhurst, D., Zhang, N., Hayes, A., Walsh, M.C., Berden, J.A., Brindle, K.M., Kell, D.B., Rowland, J.J., Westerhoff, H.V., van Dam, K., Oliver, S.G. (2001) A functional genomics strategy that uses metabolome data to reveal the phenotype of silent mutations, *Nat. Biotechnol.* **19**(1), 45–50.

Robertson, D.G., Reily, M.D., Albassam, M., Dethloff, L.A. (2001) Metabonomic assessment of vasculitis in rats, *Cardiovasc. Toxicol.* **1**(1), 7–19.

Sato, S., Soga, T., Nishioka, T., Tomita, M. (2004) Simultaneous determination of the main metabolites in rice leaves using capillary electrophoresis mass spectrometry and capillary electrophoresis diode array detection, *Plant J.* **40**(1), 151–163.

Slim, R.M., Robertson, D.G., Albassam, M., Reily, M.D., Robosky, L., Dethloff, L.A. (2002) Effect of dexamethasone on the metabonomics profile associated with phosphodiesterase inhibitor-induced vascular lesions in rats, *Toxicol. Appl. Pharmacol.* **183**(2), 108–109.

Soedamah-Muthu, S.S., Chang, Y.F., Otvos, J., Evans, R.W., Orchard, T.J. (2003) Pittsburgh epidemiology of diabetes complications study. Lipoprotein subclass measurements by nuclear magnetic resonance spectroscopy improve the prediction of coronary artery disease in Type 1 diabetes. A prospective report from the Pittsburgh epidemiology of diabetes complications study, *Diabetologia* **46**(5), 674–682.

Tang, H., Wang, Y., Nicholson, J.K., Lindon, J.C. (2004) Use of relaxation-edited one-dimensional and two dimensional nuclear magnetic resonance spectroscopy to improve detection of small metabolites in blood plasma, *Anal. Biochem.* **325**(2), 260–272.

Tate, A.R., Foxall, P.J., Holmes, E., Moka, D., Spraul, M., Nicholson, J.K., Lindon, J.C. (2000) Distinction between normal and renal cell carcinoma kidney cortical biopsy samples using pattern recognition of (1)H magic angle spinning (MAS) NMR spectra, *NMR Biomed.* **13**(2), 64–71.

Van, Q.N., Chmurny, G.N., Veenstra, T.D. (2003) The depletion of protein signals in metabonomics analysis with the WET-CPMG pulse sequence, *Biochem. Biophys. Res. Commun.* **301**(4), 952–959.

Verhoeckx, K.C., Bijlsma, S., Jespersen, S., Ramaker, R., Verheij, E.R., Witkamp, R.F., Van Der Greef, J., Rodenburg, R.J. (2004) Characterization of anti-inflammatory compounds using transcriptomics, proteomics, and metabolomics in combination with multivariate data analysis, *Int. Immunopharmacol.* **4**(12), 1499–1514.

Viant, M.R. (2003) Improved methods for the acquisition and interpretation of NMR metabolomic data, *Biochem. Biophys. Res. Commun.* **310**(3), 943–948.

Wang, Y., Bollard, M.E., Keun, H., Antti, H., Beckonert, O., Ebbels, T.M., Lindon, J.C.,

Holmes, E., Tang, H., Nicholson, J.K. (2003) Spectral editing and pattern recognition methods applied to high-resolution magic-angle spinning 1H nuclear magnetic resonance spectroscopy of liver tissues, *Anal. Biochem.* **323**(1), 26–32.

Watkins, S.M., Reifsnyder, P.R., Pan, H.J., German, J.B., Leiter, E.H. (2002) Lipid metabolome-wide effects of the PPARgamma agonist rosiglitazone, *J. Lipid. Res.* **43**(11), 1809–1817.

Wevers, R.A., Engelke, U., Heerschap, A. (1994) High-resolution 1H-NMR spectroscopy of blood plasma for metabolic studies, *Clin. Chem.* **40**(7 Pt 1), 1245–1250.

Wevers, R.A., Engelke, U.F., Moolenaar, S.H., Brautigam, C., de Jong, J.G., Duran, R., de Abreu, R.A., van Gennip, A.H. (1999) 1H-NMR spectroscopy of body fluids: inborn errors of purine and pyrimidine metabolism, *Clin. Chem.* **45**(4), 539–548.

Xu, J., Chang, V., Joseph, S.B., Trujillo, C., Bassilian, S., Saad, M.F., Lee, W.N., Kurland, I.J. (2004) Peroxisomal proliferator-activated receptor alpha deficiency diminishes insulin-responsiveness of gluconeogenic/glycolytic/pentose gene expression and substrate cycle flux, *Endocrinology* **145**(3), 1087–1095.

Yu, H.H., Ginsburg, G.S., O'Toole, M.L., Otvos, J.D., Douglas, P.S., Rifai, N. (1999) Acute changes in serum lipids and lipoprotein subclasses in triathletes as assessed by proton nuclear magnetic resonance spectroscopy, *Arterioscler. Thromb. Vasc. Biol.* **19**(8), 1945–1949.

Metalloenzymes

Walther R. Ellis
Utah State University, Logan, UT, USA

Encyclopedia of Molecular Cell Biology and Molecular Medicine, 2nd Edition. Volume 8
Edited by Robert A. Meyers.
Copyright © 2005 Wiley-VCH Verlag GmbH & Co. KGaA, Weinheim
ISBN: 3-527-30550-5

Keywords

Cytochrome P-450
Refers to a family of heme monooxygenases, present in certain pseudomonads and most mammalian cell types, that catalyze the oxidation of a wide variety of structurally diverse compounds.

Electrophilic Catalysis
Refers to the electrostatic stabilization of a negative charge that develops in the transition states of certain reactions.

Metalloprotease
An enzyme, typically containing Zn, that catalyzes the hydrolysis of peptide bonds.

Oxidase
An enzyme that catalyzes an oxidation using O_2 as the electron acceptor; O atoms from dioxygen are not incorporated into the product of the oxidation.

Oxygenase
An enzyme that catalyzes the reaction of O_2 with an organic substrate in which oxygen atoms (one in the case of monooxygenases; two in the case of dioxygenases) from dioxygen are incorporated into the product.

Peroxidase
An enzyme that catalyzes the oxidation of an organic/inorganic substrate by hydrogen peroxide.

Superoxide Dismutase
A metalloenzyme that catalyzes the disproportionation of superoxide, forming hydrogen peroxide and O_2.

Urease
A nickel enzyme that hydrolyzes urea, forming ammonia and carbamate.

Metalloenzymes, which comprise approximately one-third of the known enzymes, require stoichiometric quantities of metal ions as cofactors, typically transition metal ions, for their catalytic activities. The roles of metal ions in enzyme active sites (aside from structure maintenance) include electron transfer, oxygen atom transfer, formation of coordinated hydroxide, electrophilic catalysis, as well as substrate binding. Metalloenzymes catalyze numerous reactions of physiological importance, including mitochondrial O_2 reduction, peptide bond cleavage, hydrocarbon hydroxylation, destruction of O_2^- and H_2O_2, and hydration of CO_2.

1
Occurrence

1.1
Discovery

The study of metalloenzymes has its roots in investigations that took place more than a century ago. In 1897, Gabriel Bertrand, working on laccase (a polyphenol oxidase), suggested for the first time that a metal ion was essential for the catalytic activity of an enzyme. At this time, the nature of enzymes was a matter of vociferous debate – Are enzymes protein-based catalysts, or is the catalysis traceable to low-level contaminants? In 1926, James Sumner presented a pivotal result: crystallization of jack bean urease, demonstration that urease is a protein, and that the dissolved crystals catalyze the hydrolysis of urea. Sumner, who failed to detect any metal ions in his urease preparations, expressed the view that metal ions are unlikely to play important roles in the enzymatic catalysis of biological reactions. It is ironic that 49 years after Sumner's pioneering crystallization of an enzyme, urease was subsequently found to contain nickel ions, which are essential for enzyme activity!

Prior to the development of modern methods for trace element analysis and spectroscopic instrumentation, many metalloenzymes (e.g. blue copper oxidases) were initially isolated simply because they were a colored component of a cell or tissue homogenate. During the last 40 years, numerous nonchromophoric metalloenzymes have been isolated as well. Our current understanding of metalloenzymes is largely based on enzymes containing iron, copper, or zinc. Numerous examples of enzymes containing other metals (Mg, Ca, Mn, Co, Ni, V, Mo, W) are also known; however, the structural and mechanistic information currently available for these is less detailed.

1.2
Biological Importance

Metalloenzymes play key roles in many processes central to human physiology, including the biosynthesis of DNA and certain amino acids, steroid metabolism, destruction of superoxide and hydrogen peroxide, biosynthesis of leukotrienes and prostaglandins, carbon dioxide hydration, neurotransmitter metabolism, digestion, collagen biosynthesis, and, of course, respiration. The latter could be viewed as a process of global bioenergetic importance, one that complements photosynthesis – the dioxygen that is evolved (via the Mn_4 oxygen-evolving complex of Photosystem II) by photosynthetic organisms is consumed by aerobic microbes and animals (Fe, Cu cytochrome oxidases catalyze the reduction of O_2 to H_2O).

Within the last 25 years, it has become widely appreciated that metal ions play important catalytic roles that frequently cannot be matched by protein side chains or organic prosthetic groups. The most significant observation of this type concerns the microbial fixation of molecular nitrogen, a very inert molecule:

$$N_2 + 6H^+ + 6e^- \longrightarrow 2NH_3 \qquad (1)$$

This reaction is catalyzed by nitrogenase, whose active site consists of a dissociable Mo- and Fe-containing cofactor. The catalysis of such multielectron redox reactions is frequently carried out by metalloenzymes containing multimetal centers (i.e. metal clusters).

1.3
Metal Ion Bioavailability

Only a small number of the metallic elements appear to be utilized in biology. Of these, Na^+, K^+, Ca^{2+}, and Mg^{2+} are considered macrominerals, or bulk elements; high concentrations of these ions are needed for osmotic homeostasis, neuromuscular transmission, and biomineralization (e.g. bone formation). As indicated in Table 1, other essential metals are present in trace quantities in humans and most other organisms. Even the most prominent biologically active transition metals (iron, zinc, and copper) are trace elements.

The selection of metal ions for incorporation into metalloenzymes is strongly influenced by bioavailability – a given element must be abundant in the environment and must be present in an extractable form. A striking exception to this generalization involves the nearly universal requirement for iron. Organisms have evolved selective uptake mechanisms for this element,

Tab. 1 Essential metal composition of a 70-kg human adult[a].

Element	Composition (wt%)
Ca	1.4
Na	0.63
K	0.26
Mg	4×10^{-2}
Fe	5×10^{-3}
Zn	3×10^{-3}
Cu	1×10^{-4}
Mn	2×10^{-5}
Mo	2×10^{-5}
Ni	4×10^{-6}
Cr	4×10^{-6}
V	3×10^{-6}
Co	2×10^{-6}

[a] Note: O,C,H,N,P,&S constitute 97.6 wt%.

the most abundant transition metal in the earth's crust, which forms insoluble ferric hydroxides in the presence of O_2. The incorporation of metal ions into metalloenzymes is also influenced by other, chemically oriented, parameters such as ionic radius, charge, preferred coordination geometry, ligand substitution and redox kinetics, aqueous solution chemistry, and thermodynamic stability.

1.4
Active Site Assembly

Incorporation of a metal ion, a posttranslational biosynthetic event, requires that the folding of the polypeptide chain permit several side chains to congregate to form an appropriate metal-binding site. The primary metal-binding amino acid side chains are imidazole (His), carboxylate (Asp and Glu), thiol (Cys), thioether (Met), and hydroxyl (Ser, Thr, and Tyr). Less frequently, indole (Trp), guanidinium (Arg), and amide (Asn and Gln) groups are used. Backbone carbonyl groups can also participate in metal binding. The side chain functional groups must usually be deprotonated in order for a donor atom (O, N, or S) to form a metal–ligand bond.

Some metal ions coordinate to their binding sites in apo-metalloenzymes as simple aqua ions. For example, Zn^{2+} binds to apo-carbonic anhydrase in a multidentate ligand reaction: the metal ion sheds coordinated water molecules as it binds to the active site of the apoenzyme, which could be viewed as an elaborate chelating agent. However, redox-active metal ions (e.g. Cu, Fe, Mn, Mo) pose special problems by virtue of their reactivities with O_2^- and H_2O_2. For example, it has become clear during the last decade that pools of soluble copper ions are not used in the physiological activation of eukaryotic

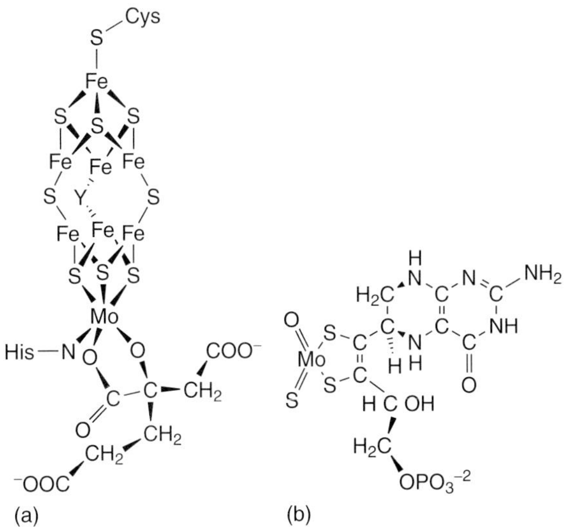

Fig. 1 Molybdenum cofactors:
(a) Structure of the FeMo cofactor of
Azotobacter vinelandii nitrogenase.
The cofactor is bound to the enzyme
via the indicated cysteine and
histidine residues. Homocitrate is
bound to the molybdenum center.
(b) Proposed structure of the oxidized
molybdopterin cofactor (Mo-co) of
mammalian molybdenum enzymes.

copper enzymes, such as Cu,Zn superoxide dismutase and cytochrome *c* oxidase. Copper ions are known Fenton reagents (OH· generators), and their sequestration is a straightforward way of circumventing unwanted redox reactions. Intracellular copper delivery is mediated by "chaperone" proteins that form complexes with their apoenzyme targets prior to metal-ion exchange. Emerging research results indicate that chaperones likely exist for other redox-active metals as well.

The biosyntheses of some metalloenzymes of physiological importance requires the prior synthesis of a specialized organometallic complex. Such species include vitamin B_{12} (a cobalt corrin), hemes, and molybdenum cofactors (Fig. 1). Molybdoenzymes contain unusual metal-containing cofactors that have been shown to be traceable to the participation of accessory biosynthetic genes. The FeMo cofactor (FeMoCo) of nitrogenase also illustrates a more complex phenomenon – the use of metal clusters as cofactors in enzymes. Such aggregates typically utilize sulfide or oxide (hydroxide)

ions as bridges between metal centers. Sometimes (e.g. the iron–sulfur cluster of aconitase) the cluster assembles in a stepwise fashion after apoenzyme biosynthesis. In other cases (e.g. the nitrogenase FeMo cofactor), accessory proteins, "molecular scaffolds", are needed for complete cluster assembly; the cluster is subsequently transferred to the apoenzyme to complete the biosynthesis of the mature metalloenzyme.

2
Active Site Characterization

2.1
Metal Content

Traditional methods of trace element analysis have relied on the use of techniques best suited for quantitative, rather than qualitative, analysis such as complexometric titrations, colorimetric procedures, or atomic absorption spectrometry. Determinations of the metal contents of putative metalloenzymes have frequently been in error, owing to impure enzyme preparations, cofactor loss during purification, or inadequate analytical techniques. The most suitable methods for qualitative metal analysis include X-ray fluorescence spectrometry, neutron activation analysis, and atomic emission plasma spectrometry. Many metalloenzymes contain more than one kind of metal ion; hence, both qualitative and quantitative metal analyses should be done whenever possible.

A key indicator of metal-activated enzyme turnover is the ability to abolish the activity by adding a metal chelator (usually EDTA) or a small molecule (e.g. CO, CN^-, N_3^-, carboxylic acids) that is known to bind to metal ions in proteins and coordination compounds. Further proof of the nature of the native metal ion often requires tedious trial-and-error studies to determine which metal gives the greatest enzyme activity.

2.2
Physical Methods

Going beyond the determination of metal content, to demonstrate that a particular metal ion is present at the active site of a metalloenzyme, requires spectroscopic and/or X-ray crystallographic experimentation. Detailed structures, derived from X-ray crystallographic or nuclear magnetic resonance analyses, are available for less than 5% of the known metalloenzymes. Other physical methods have therefore been used in obtaining information about physicochemical properties these enzymes, such as hydrodynamic radius, substrate-binding constants, or types of ligands bound to the metal(s).

Extended X-ray absorption fine structure (EXAFS) spectroscopy can provide information about a primary metal coordination environment that is comparable with that obtained from X-ray crystallography – numbers and types of ligands, metal–ligand bond lengths, and bond angles. Other spectroscopic methods (e.g. infrared, Raman, electron spin resonance, fluorescence, electronic absorption), while much less informative, can nonetheless yield valuable information about the ability of an active-site metal ion to bind exogenous ligands (e.g. inhibitors) and about the likely oxidation state(s) in cases involving redox-active metals. In cases involving redox-active metal ions, determining the metal-ion oxidation states is an important prelude to formulation of enzyme turnover mechanisms. In the past, controversies arose because of proposed metalloenzyme mechanisms that invoked unusual oxidation states (e.g. Cu(III)) that subsequently were shown to be incorrect. Sometimes, metal–ligand bonds (especially involving Cys, O_2, S^{2-}, O^{2-} ligands) have such a high degree of charge transfer that assignments of metal-ion oxidation states are difficult.

Redox metalloenzymes usually undergo pronounced color changes during enzyme turnover, making optical spectroscopic studies particularly valuable. This first became evident with the early twentieth-century publications by Michaelis and Menten of observations on the turnover of horseradish peroxidase, a heme enzyme. Changes in the visible region of the spectrum during turnover (due to formation of transient intermediates) led them to formulate the Michaelis–Menten model of enzyme action.

Determinations of redox potentials of resting forms of redox enzymes yields valuable information that can be used to set constraints on proposed enzyme mechanisms. In some cases, notably the heme enzymes horseradish peroxidase and yeast cytochrome *c* peroxidase, electrochemical methods have been used to generate unstable intermediates (ferryl hemes) for spectroscopic study and determination of their redox potentials.

2.3
Metal/Ligand Substitution

Metalloenzymes frequently contain metal cofactors that are spectroscopically useless. For example, zinc enzymes are diamagnetic and colorless – electron spin resonance or electronic absorption spectroscopy are therefore uninformative. However, information about the environment of the metal-binding site can be obtained by replacing the Zn^{2+} ion with a paramagnetic, chromophoric metal-ion probe. Co^{2+} readily replaces Zn^{2+} – they prefer similar coordination environments, have comparable ionic radii, and display similar reactivity patterns. The resultant cobalt-substituted enzyme can be characterized with regard to intermediates formed during enzymatic turnover. Furthermore, the use of metal derivatives can also indicate whether the metal binding site is rigid or flexible.

Obtaining the structure of a metalloenzyme is only the first step in developing an understanding of its physiological function. The production of recombinant derivatives affords opportunities to address additional issues, such as endogenous ligand exchange at the active-site metal(s) and participation of nearby amino acid residues in substrate binding/release, proton transfer, and/or electron transfer. Fast kinetic techniques (stopped-flow spectrophotometry, freeze-quench trapping of intermediates, flash photolysis, temperature-jump relaxation) are typically used to map out the individual steps that characterize the turnover

of a metalloenzyme. In many cases, individual steps are too fast to accurately measure. One commonly used application of such techniques is the study of absorption spectra of transient enzyme intermediates generated by photolyzing carbon monoxide adducts of reduced heme enzymes in the presence of O_2. What tends to be overlooked in interpretations of structure and reactivity data is the importance of protein fluctuations (particularly in substrate binding and product release) – metalloenzymes are much more than metal ions in disguise.

3
Representative Metalloenzymes of Medical Importance

3.1
Functional Roles of Metal Cofactors

There are two principal types of metal-dependent enzyme. Metalloenzymes contain at least one tightly bound metal ion, at the active site, that is required for activity. Metal-activated enzymes, on the other hand, generally lose catalytic activity during purification because their affinity for the required metal is rather low. Mg^{2+}-, K^+-, and (most) Ca^{2+}-dependent enzymes are metal-activated and will not be discussed further. Na^+-dependency has yet to be unequivocally demonstrated for an enzyme.

Examples of metalloenzymes are known for each class of enzyme designated by the International Union of Biochemistry and are as follows: oxidoreductases, transferases, hydrolases, lyases, isomerases, and ligases. The following properties of metal ions make them well suited as catalysts for these types of reactions:

1. Metal ions bind at least three (usually four or more) ligands, thereby promoting the organization of protein structure.
2. With the notable exception of zinc, all of the other known transition metal cofactors can potentially exist in two or more oxidation states; metalloenzyme catalysis of oxidation-reduction (redox) reactions is thus quite common.
3. Metal-ion cofactors are electrophilic and can serve as effective Lewis acids for binding and activating substrates.

Two types of metal-ion reactivity are of fundamental importance in considering the wide range of activities displayed by metalloenzymes: changes in oxidation state and changes in bound ligands. Numerous examples of electron-transfer metalloproteins (blue copper proteins, iron–sulfur proteins, cytochromes) are known; the metal coordination spheres of these electron carriers do not change during electron transfer. Metal ions possessing static coordination spheres could alternatively play purely structural roles, as do Ca^{2+} in calmodulin, horseradish peroxidase, or stromelysin.

Certain metals (e.g. Zn^{2+} in carboxypeptidase A) cannot undergo redox reactions within the constraints imposed by nature; instead, acid–base catalysis is the *raison d'être* for the metal in the enzyme. Lastly, both the metal-ion oxidation state and coordination sphere could change during enzymatic turnover. Regardless, it is clear that metal ions that directly participate in enzymatic catalysis must possess dynamic coordination environments.

3.2
Mammalian Iron Enzymes

With few exceptions (certain lactobacilli), all forms of life require iron as a biocatalyst. However, only small amounts are needed – a healthy 70-kg human adult possesses just 4 g of the metal. Most of the body's iron is found in hemoglobin and ferritin; taken together, the body's iron enzymes contain less than 300 mg of iron. The common oxidation states are 2+ and 3+; higher oxidation states are known to be operative during the turnover of some iron enzymes. The widespread occurrence of iron–porphyrin prosthetic groups, or hemes, in biology has prompted the subdivision of iron enzymes into nonheme iron and heme classes.

Nonheme iron enzymes, in turn, could be classified as mononuclear iron, binuclear iron, or iron–sulfur enzymes. Table 2 lists some of these enzymes that have been found in humans, together with the reactions catalyzed and metal contents. Iron can evidently be used to catalyze both redox and acid–base reactions. Unfortunately, the active site structures of the intermediates present during turnover are not yet well understood for many of these enzymes.

A representation, partly derived from an X-ray crystallographic analysis, of the active site of pig heart mitochondrial aconitase is displayed in Fig. 2. This iron–sulfur enzyme plays a key role in the Krebs cycle, where it catalyzes the isomerization, via a *cis*-aconitate intermediate, of citrate to isocitrate. Spectroscopic and X-ray diffraction experiments have convincingly shown that the substrate directly binds to the iron–sulfur cluster during enzyme turnover. Replacement of the Ser residue depicted in the figure with Ala does not alter binding of citrate, and X-ray structures indicate that the isomerization likely proceeds via a $180°$ inversion of *cis*-aconitate about the $C_\alpha - C_\beta$ bond.

Several nonheme iron enzymes play important roles in the metabolism of the

Tab. 2 Representative human nonheme iron enzymes.

Enzyme	Reaction Catalyzed
Aconitase (Fe_4S_4)	Citrate $\rightleftharpoons$ isocitrate
Homogentisate dioxygenase (Fe)	Homogentisate $\longrightarrow$ maleylacetoacetate
Lipoxygenase (Fe)	Unsaturated fatty acids $\longrightarrow$ conjugated fatty acid hydroperoxides
Phenylalanine hydroxylase[a] (Fe)	L-Phe $\longrightarrow$ L-Tyr
Purple acid phosphatases (Fe_2)	Phosphate monoester hydrolysis
Tryptophan hydroxylase[a] (Fe)	L-Trp $\longrightarrow$ 5-hydroxy-L-Trp

[a]Also uses a tetrahydrobiopterin as an active-site cofactor.

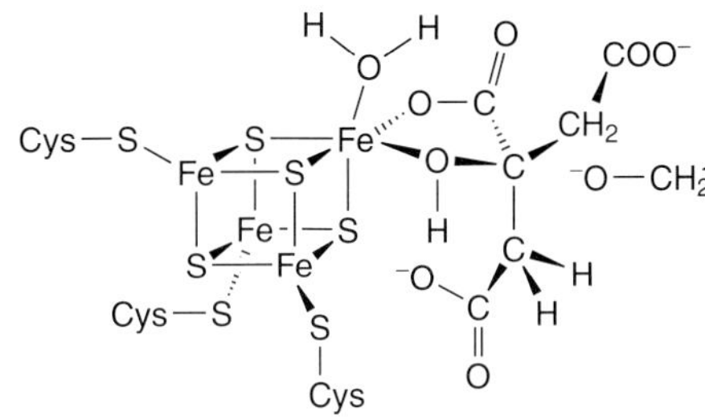

Fig. 2 Proposed structure of citrate bound to the active-site iron–sulfur cluster of pig heart mitochondrial aconitase.

aromatic amino acids phenylalanine, tyrosine, and tryptophan. These enzymes use nonheme iron and tetrahydrobiopterin (as redox cofactors) and O_2 in catalyzing regiospecific hydroxylations of these amino acids. Disfunctions in these enzymatic activities have been correlated with severe disorders, including phenylketonuria and psychiatric conditions attributed to altered serotonergic/dopaminergic neurotransmission. The mechanisms of these enzymes are believed to involve novel Fe^{2+}-O-O-pterin intermediates that heterolytically cleave to produce highly reactive $Fe^{4+}=O$ species that subsequently oxidize substrates bound nearby.

Heme proteins figure prominently in biochemistry – their functions include electron transfer, O_2 transport, O_2 activation, and H_2O_2 activation. The latter two functions are associated with several major subclasses of oxidoreductases: oxidases, oxygenases, peroxidases, and catalases. While such activities can also be displayed by enzymes containing nonheme iron, copper, manganese, or even vanadium, it has long been evident that the use of heme cofactors in dioxygen and peroxide metabolism overshadows other alternatives in eukaryotes. Representative heme enzymes of relevance to human physiology are set out in Table 3.

All heme enzymes share one notable feature: during turnover, an oxidized intermediate (oxoferryl iron) is formed from the reaction of peroxide bound to Fe^{3+}. Two-electron reduction of dioxygen yields peroxide; not surprisingly, oxoferryl species can also be generated using O_2 and additional reducing equivalents. Catalases are unique in that H_2O_2 functions as both donor and acceptor of electrons.

Cytochromes P-450 constitute a family of heme enzymes that comprise

Tab. 3 Representative human heme enzymes.

Enzyme	Reaction Catalyzed
Catalase	$2H_2O_2 \longrightarrow 2H_2O + O_2$
Cytochrome *c* oxidase[a]	$O_2 + 4H^+ + 4e^+ \longrightarrow 2H_2O$; "proton pumping"
Cytochrome *c* peroxidase	$H_2O_2 + 2H^+ + 2e^- \longrightarrow 2H_2O$
Cytochromes[b] P-450	$RH + 2e^- + O_2 + 2H^+ \longrightarrow ROH + H_2O$
Lactoperoxidase	$H_2O_2 + I^- \longrightarrow H_2O + IO^-$; others
Myeloperoxidase	$H_2O_2 + Cl^- \longrightarrow OCl^- + H_2O$
Nitric oxide synthase[c]	$\text{L-Arg} + 2O_2 + 2H^+ + 3e^- \longrightarrow \text{L-citrulline} + 2H_2O + \cdot NO$
Prostaglandin H synthase	Arachidonic acid $+ O_2 \longrightarrow PGH_2$
Thromboxane synthase	$PGH_2 \longrightarrow$ thromboxane A_2 + malondialdehyde + hydroxyheptadecatrienoic acid
Thyroid peroxidase	$\text{L-Tyr} + I^- + H_2O_2 \longrightarrow H_2O + OH^-$ + monoiodotyrosine; others
Tryptophan oxygenase	$\text{L-Trp} + O_2 \longrightarrow$ formylkynurinine

[a] Also contains active-site Cu_B, and noncatalytic metals (heme *a* and Cu_A redox centers, Na^+, Mg^{2+}, Zn^{2+}).
[b] A broad family of heme enzymes.
[c] Also uses a tetrahydrobiopterin as an active-site cofactor.

as many as 200 members in humans alone. Their roles in drug and steroid metabolism, as well as the activation of potential cancer-causing aromatic hydrocarbons, have led to their present status as the most intensely studied enzyme family. A particularly fascinating aspect of cytochrome P-450 biochemistry involves genetic (transcriptional) regulation – potential substrates can also serve as inducers *in vivo*.

Although the mammalian microsomal P-450's are of great clinical interest; they have proven difficult to purify and characterize in detail. Studies using a soluble cytochrome P-450 from *Pseudomonas putida* have led to the formulation of a detailed mechanism of P-450 enzyme action that remains the "benchmark" against which mechanistic features of other cytochromes P-450 are still compared. The availability of a crystal structure of the cytochrome P-450$_{cam}$–camphor complex affords an unusual opportunity to inspect the active site of a metalloenzyme–substrate complex (Fig. 3). What is evident from this structure is the clear origin of the regiospecific hydroxylation of the camphor substrate. The metal cofactor (heme) produces the oxidizing moiety, while the substrate-binding pocket restricts the orientation of the substrate vis-à-vis the heme iron and hence dictates *where* the substrate is hydroxylated. Recent studies on cytochrome P-450 mutants, having single mutations in the substrate-binding pocket, reinforce this view.

As indicated in Fig. 4, the ferric "resting" enzyme is inactive. Reduction of the heme to the ferrous state is required prior to O_2 binding; the necessary reducing equivalents taken up during enzyme turnover are supplied, via a series of electron carriers, by NADH. The key enzyme intermediate contains iron, in a formally 4+ oxidation state, and a porphyrin radical cation – this oxoferryl species oxidizes the bound substrate to form the

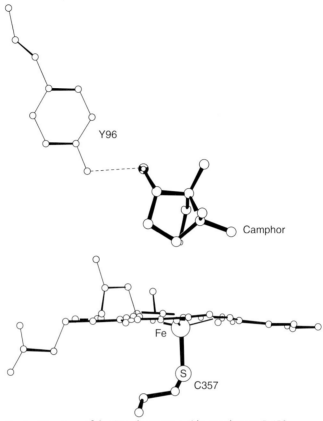

Fig. 3 Structure of the *Pseudomonas putida* cytochrome P-450$_{cam}$ active site. The substrate, camphor, is shown complexed to the enzyme, but does not coordinate to the iron.

corresponding alcohol product. The other oxygen atom from dioxygen dissociates as water; hence, all cytochromes P-450 are monooxygenases.

3.3
Mammalian Copper Enzymes

Copper ions have been found in dioxygen carriers, electron carriers, oxidases, superoxide dismutases, and oxygenases. Thus, copper and iron could be viewed as "cousins" insofar as their biological roles are concerned. Copper ions in biology evidently use just two oxidation states, 1+ and 2+; there is no compelling evidence for Cu^{3+} in any enzyme. Thus, catalysis of multielectron redox reactions requires more than one copper ion, or an additional redox unit – a tetrahydrobiopterin or 6-hydroxydopa quinone, for example.

Human copper enzymes (Table 4) are receiving increasing attention as a result of a growing awareness of their linkages to pathological conditions such as diabetes, cardiovascular disorders, and Alzheimer's disease. Serum amine oxidases catalyze the oxidative deamination of primary amines via a ping-pong mechanism composed of oxidative and reductive

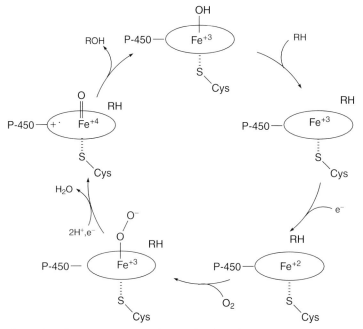

Fig. 4 Cytochrome P-450$_{cam}$ catalytic cycle. The substrate is designated as RH. Electrons are delivered to the heme in two discrete steps, the second of which is believed to limit the enzyme turnover rate.

Tab. 4 Representative human copper enzymes.

Enzyme	Reaction Catalyzed
Ceruloplasmin	$Fe(II) \longrightarrow Fe(III)$; other oxidations; oxidase activity
Dopamine β-hydroxylase	Dopamine $\longrightarrow$ norepinephrine
Hephaestin	$Fe(II) \longrightarrow Fe(III)$; oxidase activity
Peptidylglycine α-amidating monooxygenase	$P\text{-}NH\text{-}C\text{-}COO^- + O_2 + 2H^+ + 2e^- \longrightarrow P\text{-}NH_2 + H(CO)COO^- + H_2O$
Serum amine oxidase[b] B	$RCH_2NH_2^{[a]} + O_2 + H_2O \longrightarrow RCHO + H_2O_2 + NH_3$
Superoxide dismutase[c]	$2O_2^- + 2H^+ \longrightarrow O_2 + H_2O_2$

[a]Arylalkylamine substrates include serotonin and dopamine.
[b]Also uses topa quinone as a cofactor.
[c]The role of the active-site zinc is structural, rather than catalytic.

half-reactions (Eqs. 2a and 2b):

$$E_{ox} + RCH_2NH_2$$
$$\longrightarrow E_{red}\text{-}NH_2 + RCHO \quad (2a)$$

$$E_{red}\text{-}NH_2 + O_2 + H_2O$$
$$\longrightarrow E_{ox} + NH_3 + H_2O_2 \quad (2b)$$

These enzymes are dimeric, having one copper ion and one covalently bound cofactor, topa quinone (2,4,5-trihydroxyphenylalaninequinone) per monomer. Topa quinone is posttranslationally produced from a tyrosine residue during a multistep process involving the nearby copper ion and O_2. The Cu^{2+} active-site geometry in the resting enzyme is approximately square pyramidal, consisting of three histidine and two water ligands. In the reductive half-reaction (2a), a covalent bond (Schiff base complex) is formed between the substrate and topa quinone; chromophoric adducts of this type have been trapped and studied in detail. Hydrolysis of this adduct generates the aldehyde product and an E_{red}-NH_2, containing an aminoquinol form of topa quinone. Mechanistic details of the oxidative half-reaction (2b) are uncertain, and there is no unambiguous proof for redox cycling between Cu^{2+} and Cu^{1+} during enzyme turnover.

Also of intense interest is Cu,Zn superoxide dismutase (SOD), a catalyst for the disproportionation (Eq. 3) of the toxic superoxide anion:

$$2O_2^- + 2H^+ \longrightarrow O_2 + H_2O_2 \qquad (3)$$

X-ray crystal structures of the oxidized bovine erythrocyte enzyme reveal a binuclear active site: four histidine residues and a water molecule coordinate to the Cu^{2+}, while the Zn^{2+} is coordinated to three histidine and one aspartic acid. A bridging imidazolate, provided by His-61, holds the metal ions 6 Å apart. The role of the zinc appears to be structural; replacement of the Zn^{2+} by Co^{2+}, a redox-inactive spectroscopic probe, has been used to study active site changes that occur upon reduction of Cu^{2+} to Cu^{1+}. The copper bond to the imidazolate bridge is cleaved upon reduction, while the Co^{2+} bond to the bridge is retained. Subsequent X-ray crystal structures of the reduced native enzyme and

mutants have confirmed this view. These observations have been incorporated into a catalytic mechanism that is illustrated in Fig. 5.

The accepted Cu,Zn SOD mechanism involves two phases: reduction of Cu^{2+} to Cu^{1+} (accompanied by liberation of O_2) and then oxidation of Cu^{1+} to Cu^{2+} (releasing hydrogen peroxide). During turnover, His-61 breaks and then re-forms its bond to the active-site copper ion (there are no significant changes in Zn^{2+} geometry); these steps account for the ability of superoxide to act as both an oxidant and reductant during enzyme turnover. Situated on the rim of a deep channel, positively charged Arg-141 facilitates the catalysis by electrostatically steering and docking superoxide anions to the copper ion. Superoxide binding occurs at specific rates of ca. 2×10^9 $M^{-1}s^{-1}$, very near the limit of diffusion control.

3.4
Mammalian Zinc Enzymes

In 1939, Mann and Keilin discovered zinc in carbonic anhydrase, an enzyme whose function is to catalyze the hydration of carbon dioxide to the more soluble bicarbonate anion. Since that time, hundreds of zinc enzymes have been discovered; X-ray structures for dozens of these are now available. Table 5 lists some of the known human zinc enzymes, together with their activities. Divalent zinc plays a particularly conspicuous role in hydrolytic and group transfer biochemistry. Zinc is an excellent choice for such tasks because it is redox-inactive, easily exchanges ligands, and forms four-, five-, and six-coordinate complexes of comparable stability. Catalytic zinc sites always contain several endogenous (i.e. protein-derived) ligands and coordinated water.

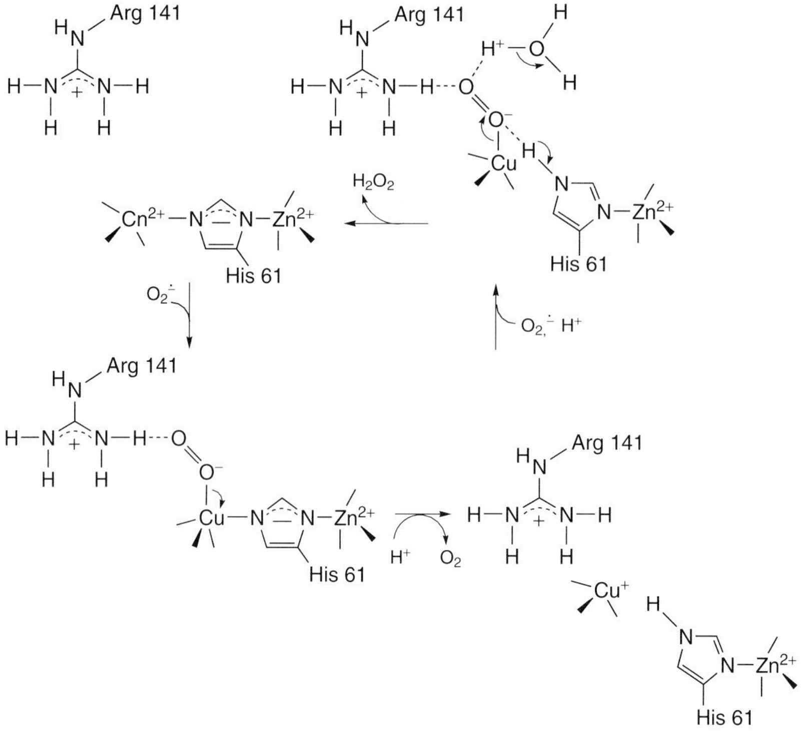

Fig. 5 Proposed mechanism of action of bovine erythrocyte Cu,Zn superoxide dismutase. The copper ion is involved in the catalysis, whereas the zinc plays a purely structural role.

Tab. 5 Representative human zinc enzymes.

Enzyme	Reaction Catalyzed
Alcohol dehydrogenase	$RR'CHOH + NAD^+ \longrightarrow RR'C{=}O + NADH + H^+$
Carbonic anhydrase	$CO_2 + H_2O \rightleftharpoons HCO_3^- + H^+$
Carboxypeptidases	C-terminal peptide bond hydrolysis
Glutaminyl cyclase	N-terminal Gln $\longrightarrow$ N-terminal pyroGlu + NH_3
Phospholipase C	Phospholipid hydrolysis
Ribonuclease A	RNA digestion (products contain 3′-phosphate termini)
RNA polymerase I	RNA biosynthesis (transcriptional elongation)
Stromelysin[a]	Peptide bond hydrolysis (ECM proteins)

[a] Also contains three noncatalytic Ca^{2+} ions.

The most important reactivity feature of divalent zinc is its Lewis acidity – it can polarize bound ligands and increase their susceptibility to attack by external nucleophiles or even increase the attacking power of a coordinated base. Water molecules coordinated to Zn^{2+} can deprotonate to form coordinated hydroxide at neutral pH. Alternatively, displacement of the coordinated water by a substrate could lead to electrophilic catalysis.

Human carbonic anhydrase II catalyzes the *reversible* hydration of CO_2 in several distinct phases, where A represents a proton-transfer relay in which a proton is believed to reach His-64 before one transfers to the solvent:

$$EZNOH^- + CO_2$$
$$\longleftrightarrow EZnHCO_3^- \qquad (4a)$$

$$EZnHCO_3^- + H_2O$$
$$\longleftrightarrow EznH_2O + HCO_3^- \quad (4b)$$

$$EznH_2O + A$$
$$\longleftrightarrow EznOH^- + AH^+ \qquad (4c)$$

The kinetics of the carbonic anhydrase reaction increase with increasing pH, and display a sigmoidal pH-activity profile (apparent pK_a of ca. 7.0). It is now agreed that this reflects the deprotonation of coordinated water (by Thr-199) to produce a zinc-bound hydroxide ion, as indicated in Fig. 6. Thr-199 may also stabilize the transition state of the reaction through a hydrogen bond, and proline substitution at this position reduces catalytic efficiency by a factor of 3000. Further support for the proposed cycle in the figure comes from X-ray crystal structures of bicarbonate complexes of the Zn^{2+} and Co^{2+} enzymes. Bicarbonate is a monodentate ligand in both structures. However, water is also bound to the metal in the Co^{2+} enzyme–substrate complex, indicating that a five-coordinate metal intermediate in the native enzyme in solution is possible.

Carboxypeptidase A has also been the subject of intense mechanistic scrutiny. Glu-270 appears to abstract a proton from

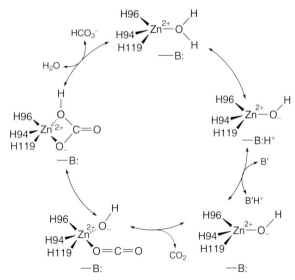

Fig. 6 Proposed mechanism for human carbonic anhydrase II. Basic side chains (B & B') assist the zinc ion in forming coordinated hydroxide, which attacks coordinated CO_2 to form an anhydride intermediate. Residue B, likely Thr-199, performs a "gatekeeper" function.

the Zn^{2+}-coordinated water to produce the nucleophile needed to attack a backbone carbonyl of the substrate. A nearby side chain, Arg-127, activates the carbonyl group of the scissile peptide bond, facilitating the attack of coordinated hydroxide to produce a tetrahedral intermediate. This intermediate then coordinates to the zinc to produce a five-coordinate complex, which subsequently collapses to produce the products (two peptide fragments).

Many other zinc-catalyzed hydrolytic reactions are believed to proceed via the use of coordinated hydroxide. Matrix metalloproteases (MMPs) constitute a large family of mononuclear zinc endoproteases that are involved in the degradation of collagen, fibronectin, laminin, and other components of the extracellular matrix of connective tissue. They are important medicinal targets because uncontrolled MMP activity can lead to pathological conditions that include osteoarthritis, rheumatoid arthritis, and tumor invasion. Most MMPs share the consensus sequence HEXXHXXGXXH, in which the Zn^{2+} is coordinated to the three histidine residues. Substrates (which may include peptide loops of other enzyme molecules), inhibitors, and H_2O/OH^- molecules, can provide the fourth (or possibly a fifth) ligand.

Other human zinc enzymes contain more complex active sites involving more than one metal ion in the catalysis. This is thought to occur in renal dipeptidase-catalyzed reactions, for example. For this enzyme, a step involving binding of the dipeptide substrate by one of the two active-site zinc ions has been suggested on the basis of an X-ray structure, while the second zinc ion generates the coordinated hydroxide that catalyzes the hydrolytic reaction. Alkaline phosphatases contain two Zn^{2+} and one Mg^{2+} arranged in an active-site cluster whose molecular mechanism of action remains uncertain.

3.5
Mammalian Cytochrome *c* Oxidase

The redox chemistry catalyzed by heme enzymes includes respiratory dioxygen reduction. This reaction is catalyzed by cytochrome *c* oxidase, a complex membrane enzyme, consisting of 13 different polypeptides, that contains heme and copper centers at the dioxygen binding site, for the purpose of vectorially translocating protons across the inner mitochondrial membrane. The resulting proton gradient is then used by ATP synthase to catalyze ATP synthesis from ADP and phosphate. The bovine heart enzyme is the best studied of all of the cytochrome *c* oxidases, having been the subject of intense study in numerous laboratories for more than 70 years. For decades, there was a great deal of controversy concerning even the most basic information – the enzyme's metal content. Purification is extremely difficult, because of the multisubunit nature of this enzyme, and this has contributed to conflicting experimental results. During the last decade, Shinya Yoshikawa and coworkers have successfully crystallized the bovine heart enzyme and produced medium/low-resolution (2.3–2.9 Å) X-ray structures of four forms of the enzyme, including azide- and carbon monoxide-bound derivatives, while Hartmut Michel and coworkers produced X-ray structures of a bacterial cytochrome *c* oxidase from *Paracoccus denitrificans*. These structures have clarified many controversial issues, and are guiding ongoing work on enzyme mutants and refinements of spectroscopic assignments.

Cytochrome *c* oxidase contains an unusual type of solvent-extractable heme cofactor, designated an *a*-type heme in contrast to the more familiar *b*-type heme found in hemoglobin and heme peroxidases (Fig. 7). Redox-active copper ions are also present, as well as sodium, zinc, and magnesium ions that appear to serve structural roles. Two of the redox cofactors, designated heme *a* and Cu$_A$, are believed to serve as electron-transfer agents, linking electron flow between the physiological reductant, cytochrome *c*, and the active site of the enzyme. Discrete cytochrome

c^{red}-binding events are needed to transfer electrons to the fully oxidized enzyme, enabling it to catalyze the four-electron reduction of one O_2 molecule to two water molecules. Figure 8 displays a simplified view of the active site, which contains heme *a*$_3$ and Cu$_B$ separated by ca. 5 Å. The X-ray structures revealed a surprising cross-link between Tyr-244 and His-240 that is of unknown origin, but is reminiscent of other cross-links found in copper enzymes like galactose oxidase. Another peculiar feature of the X-ray structures is the lack of an obvious channel for passage

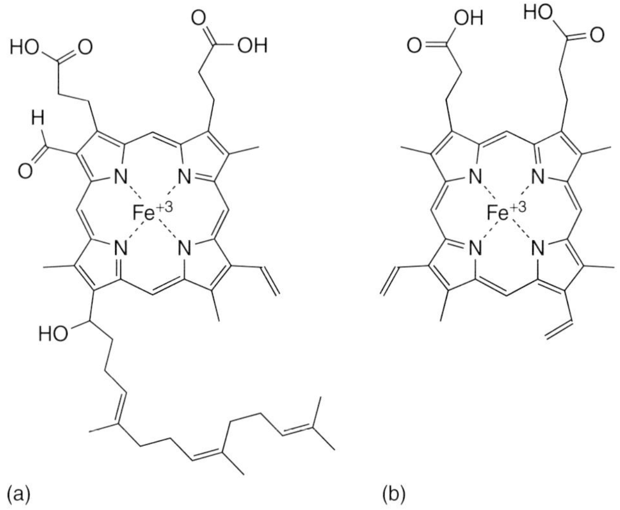

(a)   (b)

Fig. 7 Structures of active-site heme groups found in human metalloenzymes: (a) heme *a*$_3$ and (b) heme *b*.

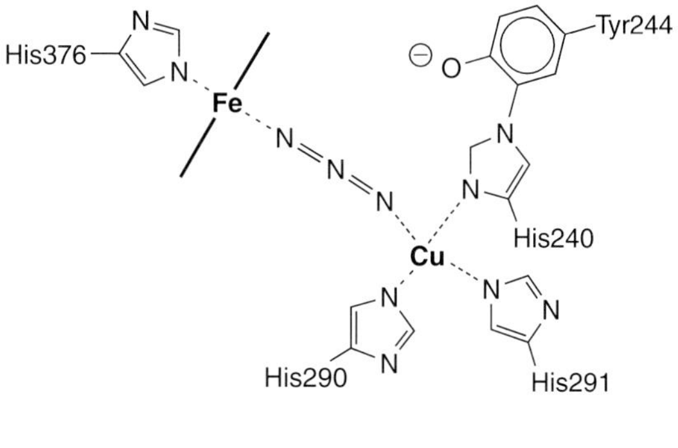

Fig. 8 Simplified cartoon of the binuclear active site of bovine cytochrome *c* oxidase, showing azide ion bound to both heme *a*$_3$ and Cu$_B$. Tyr-244 and His-240 are cross-linked, a result of a posttranslational self-processing event.

of O_2 to the binuclear active site during enzyme turnover, reinforcing the importance of protein fluctuations in the mechanism of action of cytochrome c oxidase.

While it is generally agreed that this heme a_3-Cu_B site is the active site for O_2 reduction, the details of this process remain unclear. Small molecules like azide and cyanide stabilize the active-site Fe(III)-Cu(II) oxidation states, locking the enzyme into a conformation that cannot react with O_2, and this is the molecular origin of the toxicities of these anions (the result is heart failure once the heart mitochondria begin to run out of ATP). CO, on the other hand, stabilizes the active-site Fe(II)-Cu(I) oxidation states, making another enzyme conformation amenable to study, and provides ready access to unstable intermediates when the CO adduct is photolyzed at low temperature in the presence of dissolved O_2. Resonance Raman spectroscopy has demonstrated that an early enzyme intermediate contains O_2 bound to reduced Fe_{a3}^{2+}. The X-ray structures also suggest that reduced Cu_B weakly interacts with bound O_2. Electron transfers from both metals to the bound O_2 would then produce a peroxo intermediate, long thought to be bridging Fe(III)-O-O-Cu(II). However, the X-ray structures raise the possibility of a hydrogen bond between protonated Tyr-244 and bound O_2, suggesting a peroxo intermediate that does not bind to Cu_B. Once a third electron is delivered to O_2, the O—O bond breaks, producing water (after protonation) and an oxoferryl heme a_3. Where does the third electron come from – the cross-linked Tyr-244, Cu_B^{1+}, or re-reduced heme a_3 (produced by fast electron transfer from heme a)? The fourth electron reduces the ferryl heme back to the ferric oxidation state, with the oxo group forming coordinated hydroxide that dissociates as water after protonation.

The mechanism of proton pumping by cytochrome c oxidase is even less clear because physical information, such as conserved amino acid residues or X-ray structures, cannot currently distinguish pumped protons from other dissociable protons. Hence, there has been much speculation regarding the proton pumping mechanism, particularly mechanistic proposals that link proton-translocation events to specific electron transfers.

3.6
Mammalian Enzymes Containing Less Common Metals

Much less is known about the active-site structural motifs and mechanisms of enzymes that contain metals other than iron, copper, or zinc. While most of these ''unusual'' enzymes (Table 6) are typically found in microbes and plants, there are known mammalian examples that play physiologically important roles.

Cobalt is found in vitamin B_{12}-dependent enzymes, and thus can play a role in radical reactions, as well as hydrolytic chemistry discussed earlier. There are two main roles for vitamin B_{12} (cyanocobalamin) in humans, both involving the formation of alkyl derivatives as enzyme intermediates. When cobalamin binds to its domain in nascent methionine synthase, it acquires an axial histidine ligand, which replaces benzimidazole, leaving a sixth coordination site available. The B_{12} domain of this complex enzyme must rapidly and efficiently interact with three other domains containing methyltetrahydrofolate, homocysteine, and S-adenosylmethionine, and methyl group transfer to form the methionine

Tab. 6 Representative human enzymes containing unusual metals.

Enzyme	Cofactor	Reaction Catalyzed
Arginase	2Mn	L-Arg + $H_2O \longrightarrow$ L-ornithine + urea
Methionine aminopeptidase	2Mn	Hydrolysis of N-terminal Met residue of nascent proteins
Methylmalonyl-CoA mutase	1AdoCbl[a]	Methylmalonyl-CoA $\longrightarrow$ succinyl-CoA
Secreted Phospholipases A_2	1Ca	Glycerophospholipid hydrolysis
Sulfite oxidase	1Mo-co, 1 heme/subunit	$SO_3^{2-} + H_2O \longrightarrow SO_4^{2-} + 2H^+ + 2e^-$
Superoxide dismutase	1Mn	$2O_2^- + 2H^+ \longrightarrow O_2 + H_2O_2$
Xanthine oxidase[b]	1Mo-co, 2Fe$_2$S$_2$, 1FAD/subunit	Xanthine $\longrightarrow$ uric acid

[a] 5-Adenosylcobalamin.
[b] Can catalyze other reactions: O_2^- generation from O_2, and NO and nitrite production from nitrate during anoxia.

and tetrahydrofolate products is catalyzed by an intermediary methylcobalamin complex (Co^{3+}). During catalysis, the cobalt cycles between Co^{3+} and Co^{1+}. A minor Co^{2+} form can appear as a result of oxidative processes, which must be reduced back to Co^{1+} to reactivate the enzyme.

The 5-deoxyadenosyl adduct of cobalamin is the most abundant derivative of vitamin B$_{12}$ in the liver, and is required for the catabolism of some branched-chain amino acids (using methylmalonyl-CoA mutase to generate succinyl-CoA from methylmalonyl-CoA). Deficiencies in this vitamin B$_{12}$, while rare, include neurologic disorders attributed to demyelination of nervous tissue. Succinyl-CoA formation occurs via an enzyme-catalyzed isomerization, rather than the methyl transfer process discussed above:

$$\text{CoA-SCOCH(CH}_3)\text{COOH}$$
$$\longleftrightarrow \text{CoA-SCOCH}_2\text{CH}_2\text{COOH} \quad (5)$$

The CoA-SCO- is transferred intramolecularly and the hydrogen that is transferred does not exchange with water during enzyme turnover.

Three oxidation states (2+, 3+, and 4+) are commonly observed for manganese cofactors in biology, making them well suited for the catalysis of many of the reactions discussed in the section on iron enzymes. Human manganese superoxide dismutase (SOD) contains a manganese coordinated to three histidine, one aspartic acid, and one water molecule in a trigonal bipyramidal geometry. Catalysis is thought to proceed via the following scheme:

$$\text{Mn(III)SOD} + O_2^- + H^+$$
$$\longrightarrow \text{Mn(II)SOD(H}^+) + O_2 \quad (6a)$$
$$\text{Mn(II)SOD(H}^+) + O_2^- + H^+$$
$$\longrightarrow \text{Mn(III)SOD} + H_2O_2 \quad (6b)$$
$$\text{Mn(II)SOD(H}^+) + O_2^-$$
$$\longrightarrow \text{Mn(III)(O}_2^{2-})\text{SOD(H}^+)$$
$$\quad (6c)$$
$$\text{Mn(III)(O}_2^{2-})\text{SOD(H}^+) + H^+$$
$$\longrightarrow \text{Mn(III)SOD} + H_2O_2 \quad (6d)$$

Steps (6c) and (6d) represent the formation and dissociation of a product-inhibited complex. Gln-143 is hydrogen-bonded to the coordinated water and this residue that is essential for the high activity of the enzyme (minimizing steps (6c) and (6d), which involve a presumed, intermediary peroxo complex).

Liver arginase I contains a binuclear manganese center (ca. 3.3 Å Mn–Mn distance), catalyzing the hydrolysis of L-arginine to form L-ornithine and urea. This reaction is the final cytosolic step of the urea cycle. It is generally agreed that both Mn^{2+} ions orient and polarize a bridging hydroxide, which attacks the guanidinium carbon of the arginine substrate (held in place by Asp-277 and His-141). An isozyme, arginase II, is found in nonhepatic tissues (e.g. gastrointestinal smooth muscle, penile corpus cavernosum smooth muscle), where it facilitates the regulation of NO biosynthesis by attenuating the levels of L-arginine available to nitric oxide synthase (a heme enzyme):

$$\text{Arginase I/II : L-arginine}$$
$$\longrightarrow \text{L-ornithine} + \text{urea} \quad (7)$$
$$\text{NO synthase : L-arginine}$$
$$\longrightarrow \text{L-citrulline} + \text{NO} \quad (8)$$

The latter process proceeds via an N^{ω}-hydroxy-L-arginine intermediate, which is itself a modest, competitive inhibitor of arginase.

Molybdenum is a moderately important element – it is found in nitrogenase, a variety of bacterial enzymes, and several mammalian enzymes (sulfite oxidase, aldehyde oxidase, and xanthine oxidase). Molybdoenzymes are large and typically contain other cofactors as well. Rather surprisingly, since oxides (e.g. MoO_4^{2-}) are found in aqueous solutions, the molybdenum centers are usually in sulfur-rich environments (see Fig. 1). Most molybdoenzymes hydroxylate their substrates, yet are not classified as oxygenases because they transfer oxo ligands that are *not* derived from O_2.

Mitochondrial sulfite oxidase is responsible for the oxidation of sulfite (SO_3^{2-}) to sulfate (SO_4^{2-}) and has been studied in detail. This enzyme catalyzes the final reaction in the catabolism of sulfur-containing amino acids. Each enzyme monomer has domains for the molybdenum-pterin cofactor (Mo-co), bound to the polypeptide via a cysteine ligand, and for a cytochrome b_5 (an electron-transfer center, represented as iron in the scheme below). The overall reaction can be viewed in terms of the following events:

$$Mo(VI),Fe(II) + \text{cyt } c^{ox}$$
$$\longrightarrow Mo(VI),Fe(III) + \text{cyt } c^{red} \quad (9a)$$
$$Mo(VI),Fe(III) + SO_3^{2-}$$
$$\longrightarrow Mo(IV),Fe(III) + SO_4^{2-} + 2H^+ \quad (9b)$$
$$Mo(IV),Fe(III) \longrightarrow Mo(V),Fe(II) \quad (9c)$$
$$Mo(V),Fe(II) + \text{cyt } c^{ox}$$
$$\longrightarrow Mo(V),Fe(III) + \text{cyt } c^{red} \quad (9d)$$
$$Mo(V),Fe(III) \longrightarrow Mo(VI),Fe(II) \quad (9e)$$

In step (9b), sulfite reacts with an oxo ligand coordinated to Mo(VI), forming sulfate coordinated to Mo(IV), and hydroxide (derived from water) coordinates to the molybdenum site after departure of sulfate. The oxo ligand (deprotonated hydroxide) forms in response to formation of Mo^{6+}.

3.7
Bacterial Metalloenzymes and Their Roles in Infection

Understanding molecular mechanisms underlying microbial virulence is critically important to the development of effective antimicrobial compounds and remedies for intoxication (e.g. botulism), particularly caused by food- and water-borne bacteria. Bacteria that must survive the acidic gastric environment (e.g. *Helicobacter pylori*, *Klebsiella aerogenes*) produce enzymes that provide a source of nitrogen for growth and help neutralize the pH immediately surrounding the cell. *Klebsiella aerogenes* urease has been the most intensively studied of these, and contains two Ni^{2+} ions in the enzyme active site, separated by ca. 3.6 Å. Assembly of the binuclear Ni^{2+} active site is complex, requiring a variety of accessory proteins (including at least one nickel chaperone, UreE) *in vivo*. These ions are bridged by a lysine residue that has been carbamylated as a result of a posttranslational self-processing reaction with CO_2. *In vitro* reconstitution of the apoenzyme in the absence of CO_2 does not generate any urease activity. The accepted mechanism comprises a step involving urea binding to Ni-1 (carbonyl oxygen coordination), and hydrogen bonding to His-320. Hydroxide coordinated to Ni-2 attacks the carbonyl carbon to form a tetrahedral intermediate. His-320 transfers a proton to an amide nitrogen of the urea intermediate, which then collapses into ammonia and carbamate (H_2NCOOH). The carbamate product is itself unstable, and spontaneously decomposes to form CO_2 and a second ammonia molecule. The net result of the urease-catalyzed and uncatalyzed reactions is the following:

$$CO(NH_2)_2 + H_2O \longrightarrow 2NH_3 + CO_2 \tag{10}$$

One host defense strategy involves the production of IgA molecules that bind to both bacterial surface antigens and mucin, trapping the invading bacterial cells in the mucin. Some strains of pathogenic bacteria (particularly streptococci) secrete a variety of zinc metalloproteases, including enzymes (IgA1 proteases) that cleave human IgA1 in the hinge region. These proteases all possess one Zn^{2+} ion coordinated to two histidine and one glutamic acid

Tab. 7 Representative bacterial metalloenzymes of medical importance.

Enzyme	Cofactor	Reaction Catalyzed
Botulinum toxin	1Zn	Peptide bond hydrolysis (long substrates[a], specific cleavages)
Lethal factor[b]	1Zn	Peptide bond hydrolysis (mitogen-activated protein kinases)
Phospholipase[c] C	3Zn	Hydrolysis of specific phospholipids
Pneumococcal IgA1 protease	1Zn	Peptide bond hydrolysis (hinge region of human IgA1)
Tetanus toxin	1Zn	Peptide bond hydrolysis (VAMP/synaptobrevin)
Urease	2Ni/subunit	$CO(NH_2)_2 + H_2O \longrightarrow NH_3 + H_2NCOOH$

[a] SNAP-25, VAMP (synaptobrevin), syntaxin.
[b] *Bacillus anthracis*.
[c] *Bacillus cereus*.

residue, with the remaining site occupied by hydroxide.

The causative agent of anthrax, *Bacillus anthracis*, produces a variety of secreted toxins, one of which (lethal toxin) is the primary cause of death in infected animals and is considered a key virulence factor. Lethal toxin consists of protective antigen and lethal factor, which is a zinc metalloprotease that cleaves members of themitogen-activated protein-kinase–kinase (MAPKK) family near their amino-termini. The X-ray structure of lethal factor shows that the Zn^{2+} is coordinated to His-686, His-690, Glu-735 and a water/hydroxide molecule in an approximately tetrahedral geometry. Glu-687 is positioned ca. 3.5 Å from the coordinated water to act as a proton acceptor and facilitate the formation of coordinated hydroxide. The active site of lethal factor is contiguous with a 40-Å long groove that serves as the substrate-binding pocket. The specificity for MAPKK substrates arises in primarily from this binding site, rather than the zinc site.

Botulism and tetanus are caused by Gram-positive, spore-forming bacteria (*Clostridium botulinum* and *Clostridium tetani* respectively) that require near-anaerobic conditions for growth. These organisms secrete dangerous toxins that are frequently fatal (death results from intoxication rather than infection *per se*). The toxins responsible for the symptoms are zinc metalloproteases that cleave proteins that are associated with neurotransmitter release. Strains of toxigenic *C. botulinum* produce a variety of antigenically distinct botulinum toxins, designated A, B, C1, D, E, F, and G. Botulinum toxins A, E, and C1 cleave synaptosomal-associated protein (SNAP-25) at different sites, the others (and tetanus toxin) cleave vessicle-associated membrane protein (VAMP, also known as synaptobrevin), and C1 additionally cleaves syntaxin. Types A, B, and E are the principal causes of botulism in humans. The catalytic domains contain the usual [HExxH + E] motif displayed by many other zinc endoproteases. Mature botulinum toxins are very large, composed of a heavy chain (ca. 100 kDa) and a light chain (ca. 50 kDa) that are held together by a disulfide bond. The heavy chain binds to the presynaptic membrane and helps to translocate the light chain across the membrane into the cytosol, where the light chain cleaves a VAMP/SNAP-25 target protein. Botulinum toxins, particularly type A toxins, are considered the most potent toxins known. Once a botulinum toxin A molecule enters the neuronal cytosol, it is estimated that a single enzyme molecule is sufficient to intoxicate a synapse.

The zinc active site is deeply buried in botulinum toxins, and is accessible to the solvent via a ca. 12 Å × 15 Å × 35 Å channel. The clostridial neurotoxins are atypical among zinc metalloproteases because of their stringent substrate specificities (substrates must be at least 16 amino acids in length, and cleavage is site-specific). The catalytic mechanism of the all of these toxins appears to involve formation of zinc-coordinated hydroxide, polarization of a carbonyl (amide) group of the substrate by a nearby amino acid residue, hydrogen bonding of an additional nearby residue to the amide nitrogen, and collapse to form the product peptides. Botulinum toxin A has been shown to form a dimer in aqueous solution at concentrations above 75 nM (the enzyme is a monomer in crystals studied to date). Endopeptidase activity is inhibited by 75% when the dimer forms, suggesting that substrate access to the channel is sterically restricted in the dimer.

4
Perspectives

In addition to the metals mentioned in this article, roles for other metals as enzyme cofactors will undoubtedly be elucidated in the near future. One fascinating, and perplexing, feature of biological systems concerns the rationale for the selection of a particular metal cofactor for a given biological task. Consider the superoxide dismutases. Four different kinds of enzyme active sites have been discovered in various prokaryotes and eukaryotes, containing either mononuclear iron, mononuclear manganese, a binuclear Cu,Zn dimer, or mononuclear nickel (in some bacteria). Why do all forms of life not use just one type of cofactor? Why are specialized kinds of human superoxide dismutases needed (an intracellular manganese enzyme, an intracellular Cu,Zn enzyme, and an extracellular Cu,Zn enzyme)?

As noted earlier in this article, many redox enzymes utilize O_2 or H_2O_2 as the source of oxidizing equivalents needed to hydroxylate a substrate. In particular, iron-peroxo or iron-oxo intermediates have been proposed for most iron enzymes discussed in this chapter. What keeps these kinds of enzymes from inactivating (self-destructing) during enzymatic turnover, a result of repeated oxidation of nearby amino acid residues or production of radicals that tunnel and carry out redox transformations elsewhere in the enzyme? Both hemoglobin and myoglobin function as O_2 carriers, but can also react with H_2O_2 and organic substrates in peroxidase-like fashion. Unlike peroxidases, the globins rapidly lose enzyme activity during turnover because their Compound I intermediates are unstable, ultimately destroying the heme chromophores, and forming protein oligomers. What are the structural constraints for a good heme peroxidase? Similar stability problems are posed by other broad classes of redox enzymes.

New, clinically important, roles for small-molecule interactions with metalloenzymes continue to emerge. For example, the gases NO and CO, both vertebrate hormones, are generated by metalloenzymes. Furthermore, some target cell responses to these gases are the result of binding to other metalloenzymes (e.g. NO binding to the heme enzyme guanylyl cyclase causes a large conformational change that results in a ca. 300-fold increase in activity). Many therapeutic drugs target metalloenzymes. Prominent examples include nonsteroidal anti-inflammatory drugs, such as aspirin or indomethacin. These drugs block prostaglandin production by inhibiting a heme enzyme, prostaglandin H synthase.

Many clinically important roles for enzyme oligomerization and folding pathways involving metalloproteins are only beginning to be understood. For example, mutations in the structural gene SOD1, encoding a Cu,Zn superoxide dismutase, figure prominently in current research on the molecular origins of amylotropic lateral sclerosis (ALS). Some mutant enzymes cannot tightly bind active-site metal ions, their apo-forms are partially unfolded, and are hence more likely to form insoluble aggregates.

Finally, it is important to note that copper ions have been shown to play key roles in posttranslational self-processing, leading to the production of new, covalently bound redox centers (thus far only known to be derived from endogenous tyrosine residues in the active site). Additional cases of this phenomenon are to be expected, with important mechanistic implications.

See also Bioinorganic Chemistry; Heme Enzymes; Hemoglobin.

Bibliography

Books and Reviews

Bertini, I., Gray, H.B., Lippard, S.J., Valentine, J.S., (Eds.) (1994) *Bioinorganic Chemistry*, University Science Press, Mill Valley, CA.

Crichton, R.R. (Ed.) (2001) *Inorganic Biochemistry of Iron Metabolism: from Molecular Mechanisms to Clinical Consequences*, 2nd edition, Wiley, New York.

da Silva, J.J.R.F., Williams, R.J.P. (2001) *The Biological Chemistry of the Elements*, 2nd edition, Oxford University Press, New York.

Dunford, H.B. (1999) *Heme Peroxidases*, Wiley-VCH, New York.

Eichhorn, G., Marzilli, L., Series (Eds.) (1979–1996) *Advances in Inorganic Biochemistry*, Vols. 1–11, Elsevier/Prentice Hall/VCH, New York.

Hooper, N.M. (Ed.) (1996) *Zinc Metalloproteases in Health and Disease*, Taylor & France, Bristol, PA.

Messerschmidt, A., Bode, M., Cygler, W. (Eds.) (2004) *Handbook of Metalloproteins*, Vol. 3, Wiley, New York.

Messerschmidt, A., Huber, R., Wieghardt, K., Poulos, T., (Eds.) (2001) *Handbook on Metalloproteins*, Vols. 1–2, Wiley, New York.

Ortiz de Montellano, P.R. (Ed.) (1995) *Cytochrome P-450: Structure, Mechanism, and Biochemistry*, 2nd edition, Plenum, New York.

Reedijk, J., Bouwman, E., (Eds.) (1999) *Bioinorganic Catalysis*, 2nd Edition, Dekker, New York.

Sigel, H., Sigel, A., Series (Eds.), (1974–2004) *Metal Ions Biological Systems*, Vols. 1–42, Dekker, New York.

Primary Literature

Andersen, O.A., Stokka, A.J., Flatmark, T., Hough, E. (2003) 2.0 Å resolution crystal structures of the ternary complexes of human phenylalanine hydroxylase catalytic domain with tetrahydrobiopterin and 3-(2-thienyl)-L-alanine or L-norleucine: substrate specificity and molecular motions related to substrate binding, *J. Mol. Biol.* **333**, 747–757.

Arnesano, F., Banci, L., Bertini, I., Ciofi-Baffoni, S., Molteni, E., Huffman, D.L., O'Halloran, T.V. (2002) Metallochaperones and metal-transporting ATPases: A comparative analysis of sequences and structures, *Genome Res.* **12**, 255–171.

Banerjee, R., Ragsdale, S.W. (2003) The many faces of vitamin B_{12}: catalysis by cobalamin-dependent enzymes, *Annu. Rev. Biochem.* **72**, 209–247.

Bartnikas, T.B., Gitlin, J.D. (2003) Mechanisms of biosynthesis of copper/zinc superoxide dismutase, *J. Biol. Chem.* **278**, 33602–33608.

Berry, C.E., Hare, J.M. (2004) Xanthine oxidoreductase and cardiovascular disease: molecular mechanisms and pathophysiological implications, *J. Physiol.* **555.3**, 589–606.

Borbulevych, O.Y., Jankun, J., Selman, S.H., Skrzypczak-Jankun, E. (2004) Lipoxygenase interactions with natural flavonoid, quercetin, reveal a complex with protocatechuic acid in its X-ray structure at 2.1 Å resolution, *Proteins* **54**, 13–19.

Bratton, M., Mills, D., Castleden, C.K., Hosler, J., Meunier, B. (2003) Disease-related mutations in cytochrome *c* oxidase studied in yeast and bacterial models, *Eur. J. Biochem.* **270**, 1222–1230.

Cai, S., Singh, B.R. (2001) A correlation between differential structural features and the degree of endopeptidase activity of type A botulinum neurotoxin in aqueous solution, *Biochemistry* **40**, 4693–4702.

Cama, E., Colleluori, D.M., Emig, F.A., Shin, H., Kim, S.W., Kim, N.N., Traish, A.M., Ash, D.E., Christianson, D.W. (2003) Human arginase II: crystal structure and physiological role in male and female sexual arousal, *Biochemistry* **42**, 8445–8451.

Cama, E., Pethe, S., Boucher, J.-L., Han, S., Emig, F.A., Ash, D.E., Viola, R.E., Mansuy, D.,

Christianson, D.W. (2004) Inhibitor coordination interactions in the binuclear manganese cluster of arginase, *Biochemistry* **43**, 8987–8999.

Choi, E.-Y., Stockert, A.L., Leimkühler, S., Hille, R. (2004) Studies on the mechanism of action of xanthine oxidase, *J. Inorg. Biochem.* **98**, 841–848.

Chopra, A.P., Boone, S.A., Liang, X., Duesbery, N.S. (2003) Anthrax lethal factor proteolysis and inactivation of MAPK kinase, *J. Biol. Chem.* **278**, 9402–9406.

Cioni, P., Pesce, A., Morozzo della Rocca, B., Castelli, S., Falconi, M., Parrilli, L., Bolognesi, M., Strambini, G., Desidero, A. (2003) Active-site copper and zinc ions modulate the quaternary structure of prokaryotic Cu,Zn superoxide dismutase, *J. Mol. Biol.* **326**, 1351–1360.

Collins, C.M., D'Orazio, S.E. (1993) Bacterial ureases: structure, regulation of expression and role in pathogenesis, *Mol. Microbiol.* **9**, 907–913.

Cooper, C.E. (1999) Nitric oxide and iron proteins, *Biochim. Biophys. Acta* **1411**, 290–309.

Costas, M., Mehn, M.P., Jensen, M.P., Que, Jr., L. (2004) Dioxygen activation at mononuclear nonheme iron active sites: enzymes, models, and intermediates, *Chem. Rev.* **104**, 939–986.

DeMatteis, G., Agostinelli, E., Mondovì, B., Morpurgo, L. (1999) The metal function in the reactions of bovine serum amine oxidase with substrates and hydrazine inhibitors, *J. Biol. Inorg. Chem.* **4**, 348–353.

Duda, D., Govindasamy, L., Agbandje-McKenna, M., Tu, C., Silverman, D.N., McKenna, R. (2003) The refined atomic structure of carbonic anhydrase II at 1.05 Å resolution: implications of chemical rescue of proton transfer, *Acta Cryst.* **D59**, 93–104.

Elam, J.S., Thomas, S.T., Holloway, S.P., Taylor, A.B., Hart, P.J. (2002) Copper chaperones, *Adv. Protein Chem.* **60**, 151–219.

Erlandsen, H., Patch, M.G., Gamez, A., Straub, M., Stevens, R.C. (2003) Structural studies on phenylalanine hydroxylase and implications toward understanding and treating phenylketonuria, *Pediatrics* **112**, 1557–1565.

Eswaramoorthy, S., Kumaran, D., Keller, J., Swaminathan, S. (2004) Role of metals in the biological activity of *Clostridium botulinum* neurotoxins, *Biochemistry* **43**, 2209–2216.

Fei, M.J., Yamashita, E., Inoue, N., Yao, M., Yamaguchi, H., Tsukihara, T., Shinazawa-Itoh, K., Nakashima, R., Yoshikawa, S. (2000) X-ray structure of azide-bound fully oxidized cytochrome *c* oxidase from bovine heart at 2.9 Å resolution, *Acta Cryst.* **D56**, 529–535.

Garrett, R.M., Johnson, J., Graf, T.N., Feigenbaum, A., Rajagopalan, K.V. (1998) Human sulfite oxidase R160Q: identification of the mutation in a sulfite oxidase-deficient patient and expression and characterization of the mutant enzyme, *Proc. Natl. Acad. Sci. USA* **95**, 6394–6398.

Hamza, I., Gitlin, J.D. (2002) Copper chaperones for cytochrome *c* oxidase and human disease, *J. Bioenerget. Biomembr.* **34**, 381–388.

Hanson, M.A., Stevens, R.C. (2000) Cocrystal structure of synaptobrevin–II bound to botulinum neurotoxin type B at 2.0 Å resolution, *Nat. Struct. Biol.* **7**, 687–692.

Jabri, E., Karplus, P.A. (1996) Structures of the *Klebsiella aerogenes* Urease Apoenzyme and two active-site mutants, *Biochemistry* **35**, 10616–10626.

Linzie, S.D., Thevis, M., Ngo, K., Whitelegge, J., Loo, J.A., Abu- Omar, M.M. (2003) Posttranslational hydroxylation of human phenylalanine hydroxylase is a novel example of enzyme self-repair within the second coordination sphere of catalytic iron, *J. Amer. Chem. Soc.* **125**, 4710–4711.

Kisker, C., Schindelin, H., Rees, D.C. (1997) Molybdenum-cofactor-containing enzymes: structure and mechanism, *Annu. Rev. Biochem.* **66**, 233–267.

Kisker, C., Schindelin, H., Pacheco, A., Wehbi, W.A., Garrett, R.M., Rajagopalan, K.V., Enemark, J.H., Rees, D.C. (1997) Molecular basis of sulfite oxidase deficiency from the structure of sulfite oxidase, *Cell* **91**, 973–983.

Kolberg, M., Strand, K.R., Graff, P., Andersson, K.K. (2004) Structure, function, and mechanism of ribonucleotide reductases, *Biochim. Biophys. Acta* **1699**, 1–34.

Kuchar, J., Hausinger, R.P. (2004) Biosynthesis of metal sites, *Chem. Rev.* **104**, 509–525.

Kuwabara, Y., Nishino, T., Okamoto, K., Matsumura, T., Eger, B.T., Pai, E.F., Nishino, T. (2003) Unique amino acids cluster for switching from the dehydrogenase to oxidase form of xanthine oxidoreductase, *Proc. Natl. Acad. Sci. USA* **100**, 8170–8175.

Lalli, G., Bohnert, S., Deinhardt, K., Verastegui, C., Schiavo, G. (2003) The journey of tetanus and botulinum neurotoxins in neurons, *Trends Microbiol.* **11**, 431–437.

Lauble, H., Kennedy, M.C., Emptage, M.H., Beinert, H., Stout, C.D. (1996) The reaction of fluorocitrate with aconitase and the crystal structure of the enzyme-inhibitor complex, *Proc. Natl. Acad. Sci. USA* **93**, 13699–13703.

Lecerof, D., Fodje, M., Hansson, A., Hansson, M., Al-Karadaghi, S. (2000) Structural and mechanistic basis of porphyrin metallation by ferrochelatase, *J. Mol. Biol.* **297**, 221–232.

Maaß, A., Scholz, J., Moser, A. (2003) Modeled ligand-protein complexes elucidate the origin of substrate specificity and provide insight into catalytic mechanisms of phenylalanine hydroxylase and tyrosine hydroxylase, *Eur. J. Biochem.* **270**, 1065–1075.

Maksos, K., Bode, W. (2003) Structural basis of matrix metalloproteinases and tissue inhibitors of metalloproteinases, *Mol. Biotechnol.* **25**, 241–266.

Martelin, E., Lapatto, R., Raivio, K.O. (2002) Regulation of xanthine oxidoreductase by intracellular iron, *Am. J. Physiol. Cell Physiol.* **283**, C1722–C1728.

Montecucco, C., Schiavo, G. (1995) Structure and function of tetanus and botulinum neurotoxins, *Q. Rev. Biophys.* **4**, 423–472.

Moriwaki, Y., Yamamoto, T., Higashino, K. (1997) Distribution and pathophysiologic role of molybdenum-containing enzymes, *Histol. Histopathol.* **12**, 513–524.

Nitanai, Y., Satow, Y., Adachi, H., Tsujimoto, M. (2002) Crystal structure of human renal dipeptidase involved in β-lactam hydrolysis, *J. Mol. Biol.* **321**, 177–184.

Noor, R., Mittal, S., Iqbal, J. (2002) Superoxide dismutase – applications and relevance to human diseases, *Med. Sci. Monit.* **8**, RA10–RA15.

Offioni, M.R., Memmi, G., Maggi, T., Chiavolini, D., Iannelli, F., Pozzi, G. (2003) Pneumococcal zinc metalloprotease zmpc cleaves human matrix metalloproteinase 9 and is a virulence factor in experimental pneumonia, *Mol. Microbiol.* **49**, 795–805.

Oliver, C.F., Modi, S., Sutcliffe, M.J., Primrose, W.U., Lian, L.-Y., Roberts, G.C.K. (1997) A single mutation in cytochrome P450 BM3 changes substrate orientation in a catalytic intermediate and the regiospecificity of hydroxylation, *Biochemistry* **36**, 1567–1572.

Pan, Y.H., Yu, B.-Z., Singer, A.G., Ghomashchi, F., Lambeau, G., Gelb, M.H., Jain, M.K., Bahnson, B.J. (2002) Crystal structure of human group X secreted phospholipase A$_2$, *J. Biol. Chem.* **277**, 29086–29093.

Pannifer, A.D., Wong, T.Y., Schwarzenbacher, R., Renatus, M., Petosa, C., Bienkowska, J., Lacy, D.B., Collier, R.J., Park, S., Leppla, R.S., Hanna, P., Liddington, R.C. (2001) Crystal structure of the anthrax lethal factor, *Nature* **414**, 229–233.

Pearson, M.A., Park, I.-S., Schaller, R.A., Michel, L.O., Karplus, P.A., Hausinger, R.P. (2000) Kinetic and structural characterization of urease active site variants, *Biochemistry* **39**, 8575–8584.

Petersen, S.V., Oury, T.D., Valnickova, S., Thøgersen, I.B., Crapo, J.D., Enghild, J.J. (2003) The dual nature of human extracellular superoxide dismutase: one sequence and two structures, *Proc. Natl. Acad. Sci. USA* **100**, 13875–13880.

Pinakoulaki, E., Pfitzner, U., Ludwig, B., Varotsis, C. (2002) The role of the cross-link His-Tyr in the functional properties of the binuclear center in cytochrome *c* oxidase, *J. Biol. Chem.* **277**, 13563–13568.

Pylypenko, O., Schlichting, I. (2004) Structural aspects of ligand binding to and electron transfer in bacterial and fungal P450s, *Annu. Rev. Biochem.* **73**, 991–1018.

Rosetto, O., Caccin, P., Rigoni, M., Tonello, F., Bortoletto, N., Stevens, R.C., Montecucco, C. (2001) Active-site mutagenesis of tetanus neurotoxin implicates TYR-375 and Glu-271 in metalloproteolytic activity, *Toxicon* **39**, 1151–1159.

Segelke, B., Knapp, M., Kadkhodayan, S., Balhorn, R., Rupp, B. (2004) Crystal structure of *Clostridium botulinum* neurotoxin protease in a product-bound state: evidence for noncanonical zinc protease Activity, *Proc. Natl. Acad. Sci. USA* **101**, 6888–6893.

Skrzypczak-Jankun, E., Bross, R.A., Carroll, R.T., Dunham, W.R., Funk, Jr, M.O. (2001) Three-dimensional structure of a purple lipoxygenase, *J. Amer. Chem. Soc.* **123**, 10814–10820.

Song, H.K., Mulrooney, S.B., Huber, R., Hausinger, R.P. (2001) Crystal structure of *Klebsiella aerogenes* UreE, a nickel-binding metallochaperone for urease activation, *J. Biol. Chem.* **276**, 49359–49364.

Su, Q., Klinman, J.P. (1998) Probing the mechanism of proton coupled electron transfer to dioxygen: the oxidative half-reaction of bovine serum amine oxidase, *Biochemistry* **37**, 12413–12525.

Swaminathan, S., Eswaramoorthy, S. (2000) Structural analysis of the catalytic and binding sites of *Clostridium botulinum* neurotoxin B, *Nat. Struct. Biol.* **7**, 693–700.

Thunnissen, M.M.G.M., Nordlund, P., Haeggström, J.Z. (2001) Crystal structure of human leukotriene A₄ hydrolase, a bifunctional enzyme in inflammation, *Nat. Struct. Biol.* **8**, 131–135.

Valentine, J.S., Hart, J.P. (2003) Misfolded CuZnSOD and amylotrophic lateral sclerosis, *Proc. Natl. Acad. Sci. USA* **100**, 3617–3622.

Vallee, B.L., Auld, D.S. (1993) New perspectives on zinc biochemistry: cocatalytic sites in multi-zinc enzymes, *Biochemistry* **32**, 6493–6500.

Wang, J., Sheppard, G.S., Lou, P., Kawai, M., Park, C., Egan, D.A., Schneider, A., Bouska, J., Lesniewski, R., Henkin, J. (2003) Physiologically relevant metal cofactor for methionine aminopeptidase-2 is manganese, *Biochemistry* **42**, 5035–5042.

Wang, L., Erlandsen, H., Haavik, J., Knappskog, P.M., Stevens, R.C. (2002) Three-dimensional structure of human tryptophan hydroxylase and its implications for the biosynthesis of the neurotransmitters serotonin and melatonin, *Biochemistry* **41**, 12569–12574.

Williams, P.A., Cosme, J., Vinković, D.M., Ward, A., Angove, H.C., Day, P.J., Vonrhein, C., Tickle, I.J., Jhoti, H. (2004) Crystal structures of human cytochrome P450 3A4 bound to metyrapone and progesterone, *Science* **305**, 683–686.

Wilmot, C.M. (2003) Oxygen activation in a copper-containing amine oxidase, *Biochem. Soc. Trans.* **31**, 493–496.

Wilmot, C.M., Hadju, J., McPherson, M.J., Knowles, P.F., Phillips, S.E.V. (1999) Visualization of dioxygen bound to copper during enzyme catalysis, *Science* **286**, 1724–1728.

Wilson, H.L., Rajagopalan, K.V. (2004) The role of tyrosine 343 in substrate binding and catalysis by human sulfite oxidase, *J. Biol. Chem.* **279**, 15105–15113.

Yoshikawa, S. (1999) X-ray structure and reaction mechanism of bovine heart cytochrome *c* oxidase, *Biochem. Soc. Trans.* **27**, 351–362.

Methanogens and the Archaebacteria, Molecular Biology of

Madeline E. Rasche[1] *and James G. Ferry*[2]
[1] *University of Florida, Gainesville, FL 32611, USA*
[2] *The Pennsylvania State University, University Park, PA 16801, USA*

Encyclopedia of Molecular Cell Biology and Molecular Medicine, 2nd Edition. Volume 8
Edited by Robert A. Meyers.
Copyright © 2005 Wiley-VCH Verlag GmbH & Co. KGaA, Weinheim
ISBN: 3-527-30550-5

Keywords

Domain
A phylogenetic taxon that is higher than the level of kingdom and is based on the classification of organisms according to similarities in 16S ribosomal RNA sequences. The current phylogenetic classification scheme divides organisms into three domains: the Archaea, Bacteria, and Eucarya.

Domain Archaea (Formerly Archaebacteria)
The domain composed of prokaryotic microorganisms with archaeal-type ribosomal RNA. Archaeal cellular membranes typically contain lipids with glycerol diether or diglycerol tetraether linkages. The domain Archaea consists of three phenotypic groups (methane producers, extreme halophiles, and sulfur-dependent extreme thermophiles), but it is formally divided into four kingdoms: the Euryarchaeota, Crenarchaeota, Korarchaeota, and Nanoarchaeota.

Domain Bacteria (Formerly Eubacteria)
The domain composed of prokaryotic microorganisms with bacterial-type ribosomal RNA. Bacterial cellular membranes contain lipids with glycerol diester linkages.

Domain Eucarya
The domain that includes all eukaryotic organisms, characterized by eucaryotic-type 16S ribosomal RNA. Eucaryal cells contain a defined nucleus and glycerol diester cell membrane lipids.

Prokaryote
A cellular organism that lacks a true nucleus.

Although members of the domain Archaea are prokaryotic in cell structure, 16S rRNA sequencing reveals that they form a phylogenetically coherent group, which is separate from both Bacteria (Eubacteria, domain Bacteria) and eucaryotes (Eucarya, domain Eucarya). The Archaea are further distinguished by the ether-lipid composition of cytoplasmic membranes and cell wall components not found in either the Bacteria or Eucarya domains. Like in Bacteria, transcription and translation in Archaea take place in the same cellular compartment. Unlike Bacteria, Archaea possess a complex DNA-dependent RNA polymerase and genes with promoter elements similar to the TATA box of eucaryotes. Archaea produce and utilize homologs to eukaryotic transcription-initiation factors, which are not found in the Bacteria domain. The methane-producing Archaea (methanoarchaea) are the most diverse and extensively studied group of the Archaea. The strictly anaerobic methanoarchaea are terminal organisms in the anaerobic phase of the global carbon cycle.

1
Introduction

For nearly half a century, life was classified at the highest level by two taxons, prokaryotes and eucaryotes, on the basis of phenotypic cellular characteristics that included the presence or absence of nuclear membranes, organelles, or cell walls. It was not until 1990 that a molecular approach, based on comparisons of 16S ribosomal RNA sequences, revealed that all life forms are divided into three primary divisions (domains) of life: the Eucarya, encompassing all eucaryotes, the Bacteria (or Eubacteria), including "typical" Eubacteria, and the Archaea (or Archaebacteria). The Archaea are comprised of four kingdoms: Euryarchaeota, Crenarchaeota, Korarchaeota, and the recently described Nanoarchaeota. In evolutionary terms, the two prokaryotic domains are more genetically distinct from each other than one of the groups, the Archaea, is from the Eucarya. The largest, most diverse, and well-studied representatives of the Archaea are the methane-producing anaerobes (methanoarchaea), which provided the first clues leading to the discovery of the three-domain concept. The methanoarchaea are members of the Euryarchaeota. Research on the methanoarchaea over the past two decades has made the greatest impact on our understanding of the ecology, physiology, and biochemistry of the Archaea. More recently, genomic sequencing, development of genetic systems, and investigations into the molecular biology of the methanoarchaea has significantly expanded our knowledge of the Archaea domain.

2
Biochemistry of Methanogenesis

The Italian physicist Alessandro Volta is most often credited with the discovery of biological methane (natural gas) production, methanogenesis, more than 125 years ago. In performing the "Volta experiment," he disturbed the sediment of lake Como in northern Italy with a pole releasing the trapped methane bubbles. Lighting the gas produced a flame that he called "combustible air." Figure 1 shows a modern-day Volta experiment in which gas from the sediment of a freshwater pond is collected in an inverted funnel submerged in the water column and ignited on release by tipping the funnel. The process of methanogenesis requires a consortium of at least three interacting metabolic groups of obligately anaerobic microbes (Fig. 2). The fermentative group decomposes cellulose and other complex molecules to volatile carboxylic

Fig. 1 A modern-day Volta experiment (see color plate p. xxxiv).

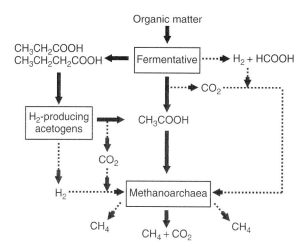

Fig. 2 A scheme showing carbon flow through the three major metabolic groups of anaerobes involved in the conversion of complex organic matter to methane in freshwater environments.

acids and hydrogen gas. Only acetate, carbon dioxide, hydrogen, and formate are the major substrates for the methanoarchaea; thus, the hydrogen-producing acetogenic group is necessary to further metabolize butyrate and propionate to substrates for the methanoarchaea. The methanoarchaea employ two separate pathways in which methane derives from either the methyl group of acetate (reaction 1) or the reduction of carbon dioxide with electrons from hydrogen or formate (reactions 2a and 2b). At least two-thirds of the total methane produced derives from the acetate and the remainder from the reduction of carbon dioxide.

$$CH_3COO^- + H^+ \longrightarrow CH_4 + CO_2 \tag{1}$$

$$CO_2 + 4H_2 \longrightarrow CH_4 + 2H_2O \tag{2a}$$

$$4HCO_2H \longrightarrow 3CO_2 + CH_4 \\ + 2H_2O \tag{2b}$$

2.1
Steps Common to the Two Major Pathways for Methanogenesis

The steps shown in Fig. 3 are common to the two major pathways for methane production in nature, carbon dioxide reduction and fermentation of acetate. The two pathways differ primarily in the manner by which a methyl group is generated and passed to the cofactor tetrahydromethanopterin (H_4MPT) to form CH_3-H_4MPT. Conversion of CH_3-H_4MPT to methane is essentially the same in both pathways. Step 1 in Fig. 3 is catalyzed by N^5-methyltetrahydromethanopterin:coenzyme M methyltransferase, which has

been recently reviewed. The methyltransferase is an integral membrane-bound complex that also functions to generate a sodium-ion gradient across the membrane coupled to the methyl transfer reaction. The enzyme as characterized from *Methanothermobacter marburgensis* and *Methanosarcina mazei* contains a corrinoid cofactor, cobalamine (5′-hydroxybenzimidazolyl cobamide), which accepts the methyl group from CH_3-H_4MPT in the first of the two partial reactions catalyzed by the enzyme. The second partial reaction involves transfer of the methyl group from CH_3-cobalamine to coenzyme M (HS-CoM) producing CH_3-S-CoM. The methyltransferase complex contains eight nonidentical subunits (MtrA-H). MtrA contains the cobalamine cofactor protruding into the cytoplasm that undergoes methylation and demethylation. It is proposed that methylation is catalyzed by MtrH and demethylation by MtrE, which also translocates sodium ions. The mechanism of translocation is thought to involve conformational changes induced in MtrA by methylation and demethylation that are transmitted to MtrE, which drives the

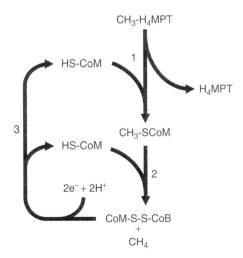

Fig. 3 Steps common to the two major pathways for methanogenesis.

translocation. Growth of the methanoar-chaea utilizing either pathway is dependent on sodium and is proposed to use sodium gradients for energetic functions.

The methyl-coenzyme M-reductase catalyzes step 2 in Fig. 3. The electron donor is coenzyme B (CoB) and the heterodisulfide CoM-S-S-CoB is produced in addition to methane. *Methanothermobacter marburgensis* and *Methanothermobacter thermoautotrophicus* contain two isozymes (MCRI and MCRII) with native molecular masses of approximately 300 kDa composed of three different subunits with molecular masses of 65 (α), 46 (β), and 30–35 (γ) kDa in an $\alpha_2\beta_2\gamma_2$ configuration. The crystal structure of the MCRI isozyme from *M. marburgensis* shows coenzyme F_{430} (F_{430}) positioned at the bottom of a narrow channel in two independent active sites separated by approximately 50 Å. The two active sites are formed by residues from the $\alpha\alpha'\beta\gamma$ or $\alpha'\alpha\beta'\gamma'$ subunits indicating that two trimers are necessary to form the active sites. Structures, complexed with either HS-CoM plus HS-CoB or CoM-S-S-CoB, have been the basis for a proposed reaction mechanism. The relative positions of CoM, CoB, and F_{430} in the crystal structure suggests a nucleophilic attack of $[F_{430}]$Ni(I) on CH_3-S-CoM which forms the $[F_{430}]$Ni(III)-CH_3 intermediate. In the next step, $[F_{430}]$Ni(III) oxidizes HS-CoM producing the thiyl radical ·S-CoM and $[F_{430}]$Ni(II)-CH_3. In the final step, protonolysis of $[F_{430}]$Ni(II)-CH_3 releases methane and the thiyl radical is coupled to $^-$S-CoB to form the heterodisulfide (CoB-S-S-CoM) accompanied by a one-electron reduction of Ni(II) to the active redox state. The binding of CoM induces specific conformational changes that ensures entry of CH_3-S-CoM adjacent to F_{430} and before entry of CoB in the narrow active-site channel.

Heterodisulfide reductase catalyzes the final step in Fig. 3 releasing the active sulfhydryl forms of the cofactors. The enzyme is comprised of two subunits (HdrDE). The smaller HdrE is a *b*-type cytochrome containing low-spin and high-spin hemes. HdrE accepts electrons from methanophenazine, a component of the membrane-bound electron transport chain. HdrE donates electrons to HdrD, which contains two Fe_4-S_4 centers, one of which is the proposed active site together with a redox-active disulfide of two cysteines. On the other hand, it has been proposed from kinetic experiments with the heterodisulfide reductase from *Methanosarcina thermophila* that the low-potential heme is involved in reduction of the heterodisulfide. In both pathways, heterodisulfide reduction is coupled to formation of an electrochemical proton gradient that drives ATP synthesis catalyzed by an A_1A_0-type ATP synthase. It is hypothesized that a proton gradient is generated across the membrane in a mechanism in which the protons are translocated upon electron transfer from methanophenazine to the heterodisulfide.

2.2
Reactions Involved in the Synthesis of CH_3-H_4 MPT

2.2.1 Carbon Dioxide Reduction Pathway
In most methanogenic environments, the six electrons needed for carbon dioxide reduction to CH_3-H_4MPT derive from the oxidation of either hydrogen gas or formate as shown in reactions (3–5).

$$H_2 + 2 \text{ ferredoxin}^{ox}$$

$$\longrightarrow 2 \text{ ferredoxin}^{red} + 2H^+ \quad (3)$$

$$2H_2 + 2F_{420} \longrightarrow 2F_{420}H_2 \quad (4)$$

$$2HCOOH + 2F_{420} \longrightarrow 2F_{420}H_2$$

$$+ 2CO_2 \qquad (5)$$

Reaction (3), catalyzed by the Ech hydrogenase, is endergonic given the low partial pressures of hydrogen gas in the native environment; however, it is proposed that the Ech hydrogenase utilizes the electrochemical proton gradient to drive the reaction. Coenzyme F_{420}-dependent hydrogenases (reaction 4) and a formate dehydrogenase (reaction 5) have been characterized. Coenzyme F_{420} is an obligate two-electron carrier involving hydride transfer. Six reactions are required for reduction of carbon dioxide to CH_3-H_4MPT (reactions 6–10).

$$CO_2 + MF + 2 \text{ ferredoxin}^{red} + 2H^+$$

$$\longrightarrow 2 \text{ ferredoxin}^{ox} + \text{formyl-MF}$$

$$+ H_2O \qquad (6)$$

$$\text{formyl-MF} + H_4MPT$$

$$\longrightarrow 5\text{-formyl-}H_4MPT + MF \qquad (7)$$

$$5\text{-formyl-}H_4MPT + H^+$$

$$\longrightarrow 5,10\text{-methenyl-}H_4MPT^+$$

$$+ H_2O \qquad (8)$$

$$5,10\text{-methenyl-}H_4MPT^+ + F_{420}H_2$$

$$\longrightarrow 5,10\text{-methylene-}H_4MPT$$

$$+ F_{420} + H^+ \qquad (9a)$$

$$5,10\text{-methenyl-}H_4MPT^+ + H_2$$

$$\longrightarrow 5,10\text{-methylene-}H_4MPT$$

$$+ H^+ \qquad (9b)$$

$$5,10\text{-methylene-}H_4MPT + F_{420}H_2$$

$$\longrightarrow 5\text{-methyl-}H_4MPT + F_{420} \qquad (10)$$

In the first reaction (reaction 6), carbon dioxide is attached to methanofuran (MF) as the carbamate and reduced to the formyl level by formylmethanofuran dehydrogenase with ferredoxin as the electron donor. Reaction (7) is catalyzed by formylmethanofuran:tetrahydromethanopterin formyltransferase. Reaction (8) is catalyzed by N^5,N^{10}-methenyltetrahydromethanopterin cyclohydrolase comprised of two identical subunits with no prosthetic groups. Two mechanistically distinct N^5,N^{10}-methylenetetrahydromethanopterin dehydrogenases catalyze reaction (9), one that oxidizes reduced coenzyme F_{420} (reaction 9a) and the other that oxidizes H_2 (reaction 9b). The F_{420}-dependent enzyme is composed of one type of subunit, either as a hexamer or octamer, with no detectable prosthetic group. The other dehydrogenase is essentially a hydrogenase that contains no metals in contrast to all known mechanisms for hydrogenases. The enzyme catalyses the stereospecific transfer of a hydride ion from hydrogen gas into the *pro-R* position of the methylene carbon involving a pentacoordinated carbonium ion ($CH_3^+=H_4MPT$) transition state intermediate in which the *pro-R* hydrogen bond is protonated and exchanges with water. N^5,N^{10}-Methylenetetrahydromethanopterin reductase catalyses reaction (10). The enzyme is F_{420}-dependent, contains one subunit with no discernible prosthetic groups, and exhibits a ternary complex kinetic mechanism suggesting direct hydride transfer.

2.2.2 Acetate Fermentation Pathway

Most of the methane produced in nature derives from the methyl group of acetate (reaction 1). In *M. thermophila*, CH_3-H_4MPT is synthesized by reactions (11–13).

$$CH_3COO^- + ATP$$

$$\longrightarrow CH_3CO_2PO_3^{-2} + ADP \qquad (11)$$

$$CH_3CO_2PO_3^{-2} + HS\text{-}CoA$$

$$\longrightarrow CH_3COSCoA + Pi \qquad (12)$$

$$CH_3COSCoA + H_4SPT$$

$$+ H_2O + ferredoxin^{ox}$$

$$\longrightarrow CH_3 - H_4SPT$$

$$+ ferredoxin^{red} + CO_2$$

$$+ HS\text{-}CoA \qquad (13)$$

Acetate kinase (reaction 11) and phosphotransacetylase (reaction 12) function together to convert acetate to acetyl-CoA. The crystal structure of acetate kinase from *M. thermophila* reveals an α_2 homodimer with overall features suggesting that the enzyme belongs to the Sugar Kinase/Hsc70/Actin superfamily of phosphotransferases and is possibly the original member of this superfamily. Kinetic and biochemical analyses of site-specific replacement variants indicates a direct in-line mechanism in which the transition state phosphate is stabilized by Arg91 and Arg241. The CO dehydrogenase/acetyl-CoA synthase (Cdh) five-subunit complex cleaves the C−C and C−S bonds in the acetyl moiety of acetyl-CoA, transfers the methyl group to H_4SPT, and oxidizes the carbonyl group to CO_2 (reaction 13). The complex is divisible into three enzyme components. The nickel/iron-sulfur (Ni/Fe-S) component contains the CdhA and CdhB subunits. The corrinoid/iron-sulfur (Co/Fe-S) component contains the CdhD and CdhE subunits. The CdhC subunit constitutes the third component. Three metal clusters (A, B, and C) contained in the complex have properties nearly identical to three metal clusters in the well-characterized clostridial synthase for which functions have been assigned. Thus, cluster A is the proposed site for cleavage of the C−C and/or C−S bonds

of acetyl-CoA, whereas the carbonyl group is thought to be oxidized on cluster C. Cluster B is a conventional Fe_4-S_4 center proposed to shuttle electrons from cluster C to the ferredoxin. Dissociation of CdhC from the complex results in loss of an EPR signal diagnostic of cluster A; a result identifying the location of cluster A in CdhC. Finally, the methyl group is transferred to the corrinoid cofactor III of the Co/Fe-S component in which H_4SPT is methylated.

3
Molecular Biology

Over the last decade, two significant advances in molecular biology have greatly facilitated our progress towards a more complete understanding of the biochemistry and physiology of methanoarchaea. First, the sequencing of complete methanoarchaeal genomes has revolutionized approaches to cloning methanoarchaeal genes and characterizing the functions of unknown genes. Second, the development of improved genetics tools has enhanced the ability of researchers to investigate methanogenic processes using genetics approaches. A third exciting discovery in methanoarchaeal genetics has been the expansion of the genetic code to include pyrrolysine, the twenty-second genetically encoded amino acid. This section of the review will briefly summarize the general features of methanoarchaeal genetics and then focus on recent advances in the molecular biology of the methanoarchaea.

3.1
General Features

Methanoarchaea and other Archaea share some genetic features in common with both Bacteria and Eucarya. Like Bacteria, Archaea lack a membrane-enclosed nucleus and have a relatively small genome (1.66 to 5.75 Mbp) with circular chromosomal DNA and extrachromosomal elements. Some methanoarchaea, particularly members of the family Methanosarcinaceae, possess operons as well as bacterial-like regulatory mechanisms for gene expression. Furthermore, messenger RNA molecules contain a ribosome-binding site that is similar to the Shine–Dalgarno sequence of Bacteria. However, although Archaeal and Bacterial ribosomes are similar in size (70 S), rooted trees comparing small subunit rRNA sequences indicate that Archaea and Eucarya may share a common lineage separate from that of Bacteria. As predicted from this relationship, some components of archaeal replication, transcription, and translation are more similar to those of Eucarya. For example, Archaea possess homologs of eucaryal histones as well as a single complex RNA polymerase that resembles eucaryotic RNA polymerase II in subunit composition. Archaeal transcripts tend to lack introns, but the promoters of methanoarchaea and other Archaea contain a Box A sequence that is similar to the TATA box of Eucarya. Homologs of the eucaryal transcription factors TFIIB, the TATA-binding protein, and TFIIS have been identified in Archaeal genomes. In both Archaea and Eucarya, translation is initiated using methionine rather than formyl-methionine, which is common in Bacteria.

Histone-like proteins purified from *Methanothermus fervidus* (Hmf) have been characterized, and these proteins share amino acid sequence similarity with a protein family that includes eucaryotic histone proteins (H2A, H2B, H3, and H4). X-ray crystallography of the recombinant forms of two methanoarchaeal histone homologs (HmfA and HmfB) shows that the three-dimensional structure of both proteins resembles the characteristic histone fold of eucaryal core histone proteins. HmfA and HmfB may form homodimers, heterodimers, and tetramers in solution, but have not been observed to associate into octamers, which are typical of eukaryotic nucleosomes. The crystal structure of the histone-like protein from *Methanopyrus kandleri* (Hmk) has also been determined. Hmk is twice as long as HmfA, HmfB, and eucaryal histones, and contains two histone folds per monomer. The precise function of methanoarchaeal histone-like proteins has not yet been determined, but tetramers of Hmf interact with DNA and may play roles in regulation of gene expression or protection of DNA against heat denaturation.

3.2
Transcription

Gene expression in Archaea is said to resemble Bacteria in form and Eucarya in function. Similarities between Archaea and Bacteria include circular chromosomal DNA, the occurrence of transcription and translation in the same cellular compartment, the organization of some genes into operons, the absence of introns in protein-encoding genes, minimal post-transcriptional modification of mRNA, similarity in ribosome size, and the presence of a Shine–Dalgarno-like sequence to direct translation initiation. Interestingly, there are also a number of similarities between eucaryal and archaeal transcription and translation including the complexity of DNA-dependent RNA polymerases, promoters that contain TATA-box-like elements, transcription factors, elongation factors II that can be ribosylated with ADP, introns in certain rRNA and tRNA genes, and translation beginning with methionine rather than formylated methionine.

3.2.1 RNA Polymerase

Bacteria contain a single DNA-dependent RNA polymerase composed of four different subunits with the stoichiometry of $\alpha_2\beta\beta'\sigma$. In contrast, Eucarya have three different nuclear RNA polymerases (I, II, and III), which function in rRNA synthesis (I), mRNA synthesis (II), and tRNA and small RNA synthesis (III). Each of the eucaryal RNA polymerases contains two large subunits and 8–12 smaller subunits. The large subunits A and B correspond to the β' and β subunits respectively of the bacterial RNA polymerase. Archaeal species possess a single DNA-dependent RNA polymerase, composed of about 12 subunits which may include B, A', A", D, E, F, G, H, I, K, L, and N. In some

Euryarchaeota, subunit B appears to be divided into two fragments, B' and B", which respectively correspond to the C-terminal and N-terminal halves of subunit B. The deduced amino acid sequences of Archaeal A' and A" components have similarity to eucaryal subunit A and bacterial subunit β', while B (or B' and B") are similar to eucaryal B and bacterial β. Archaeal subunit H bears homology to a eucaryal subunit (RPB5) that is present in all three eucaryal RNA polymerases but has no counterpart in the bacterial holoenzyme.

3.2.2 Promoters and Terminators

The general features of methanoarchaeal promoters are illustrated in Fig. 4. A consensus sequence for the Archaeal promoter sequence called Box A has been determined from analysis of numerous rRNA, tRNA, and protein-encoding genes. Box A consists of a stretch of

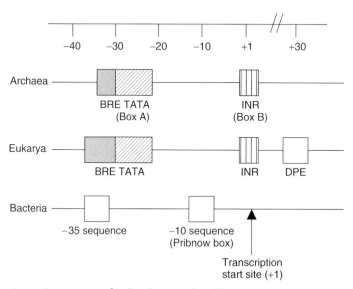

Fig. 4 Components of archaeal, eucaryal, and bacterial promoters. Archaeal and eucaryal promoters typically contain a TATA box sequence, a BRE, and an INR in the vicinity of the transcription start site. Eucarya may have a downstream promoter element (DPE). Bacterial promoters typically consist of a −35 and −10 consensus sequence.

Fig. 5 Proposed model for the archaeal preinitiation complex. The TBP and TFB bind to DNA and recruit the complex archaeal RNA polymerase. Transcription activator factor TFE α is present in all sequenced archaeal genomes. A TBP-interacting protein homolog (TIP49) is also found in the genomes of *M. kandleri* and several hyperthermophilic sulfur-dependent archaea. Adapted from.

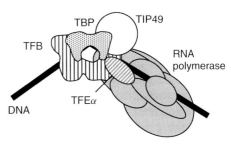

six nucleotides TTTA (T/A) A with similarity to the eukaryotic TATA box and is centered approximately 27 nucleotides upstream of the transcription start site. When the Archaeal consensus promoter sequence is replaced with the eukaryotic TATA box, transcription efficiency decreases to 13% of its original activity. Most methanoarchaea genes contain a second consensus sequence called box B or the initiator element (INR), which is located around the transcription start site. In the methanoarchaea, the INR consensus sequence (ATGC) transcription begins at the G position. In addition, a pair of adenines positioned approximately 33 and 34 nucleotides upstream of the transcription start site functions as a recognition element for transcription factor B (TFB) and is called the BRE (B recognition element) sequence. In Eucarya, the role of the TFB homolog (TFIIB) is to cooperate with transcription factor TFIID to assist in the binding between the TATA box and RNA polymerase II. The subunit of TFIID that specifically interacts with the eukaryotic TATA box is called the *TATA-binding protein* (TBP), and a homolog of TBP functions in Archaea. An archaeal homolog of transcription factor TFIIE has been shown to stimulate transcription from *M. thermoautotrophicus* promoters in an *in vitro* system. A proposed model for the archaeal preinitiation complex is shown in Fig. 5.

Two different potential termination signals have been found in Archaea. The first is an oligo-T sequence or a T-rich stretch of polypyrimidines. The second consists of short inverted repeats that may form stem-loop structures to direct termination. In thermophiles, transcription termination appears to be relatively inefficient, usually occurring at different sites within polypyrimidine stretches 4–30 nucleotides long.

3.2.3 Messenger RNA

Archaea contain polycistronic transcriptional units that resemble bacterial operons. Unlike most bacterial operons, however, Archaeal transcriptional units may contain internal promoters and terminators. In some cases, transcripts of different lengths are produced from the same operon, consistent with mRNA processing. No introns have been found in Archaeal or bacterial mRNAs that encode proteins; however, some Archaeal rRNA and tRNA transcripts contain sequences that are subsequently removed by posttranscriptional modification. At present, there is no evidence for posttranscriptional modification of mRNA in Archaea. RNAs with short poly(A) tails (about 12 nucleotides in length) have been found in Archaea, but they have not been experimentally identified as mRNA. The 5′ cap structures found in eucaryal mRNAs are not found in

Archaeal mRNAs. Ribosome-binding sites complementary to the 3′ terminus of the 16S Archaeal rRNA and similar to bacterial Shine–Dalgarno sequences are present in methanoarchaea mRNAs.

3.3
Translation

Archaea use the same universal genetic code as Bacteria and Eucarya. Codon usage generally depends on the GC content of the organisms, which varies from 26 to 71% in Archaea; thus GC-rich Archaea tend to prefer codons with G or C in the third position, although exceptions have been observed. The most common translation initiation codon is AUG, but GUG and UUG are also used. The ribosomes and other translational machinery used in the different domains exhibit distinctive qualities unique to each of the three lineages. Eucaryal translation differs from bacterial translation in several ways. Translation in Eucarya is posttranscriptional; mRNAs generally encode only a single gene; mRNAs are modified with a 5′ cap (7-methylguanosine) and a polyadenylated tail; the ribosome recognizes the 5′ cap structure of mRNA rather than a Shine–Dalgarno sequence; translation is initiated with methionine rather than formylated methionine; and initiation requires up to 10 additional protein factors. Translation in Archaea has some bacterial and some eucaryal features. Since no nuclear membrane is present, Archaeal transcription necessarily occurs in the same compartment as translation; mRNAs may be polycistronic, and posttranscriptional modification of mRNAs is not observed. Shine–Dalgarno sequences are utilized in some Archaea, but are unnecessary in others. As in Eucarya, translation is initiated with nonformylated methionine and uses

a diphtheria-sensitive elongation factor (EF-2).

3.3.1 Transfer RNA
The tRNAs of Archaea contain several modified nucleotides present in Bacteria and Eucarya (pseudouridine, 2′-*O*-methylcytidine, 1-methylguanosine, and $N^2 N^2$-dimethylguanosine), but lack ribothymidine and 7-methylguanosine, which are present in Bacteria and Eucarya. Archaeal tRNAs contain additional modified nucleotides unique to the Archaeal domain. Several aspects of Archaeal tRNA synthetases differ from the model systems described in Bacteria, including different mechanisms for charging asparagine, cysteine, glutamine, and lysine tRNAs.

3.3.2 Initiation and Elongation Factors
The number of protein factors required for archaeal initiation is unknown; however, it has been proposed that Archaea may possess initiation factors similar to those in Eucarya because hypusine, an unusual amino acid previously found only in eukaryotic initiation factor 5A, has been found in a *Sulfolobus* protein with sequence similarity to eukaryotic initiation factor 5A. The recombinant *M. jannaschii* initiation factor 5A homolog has been crystallized and contains a conserved lysine at position 40 that may be converted into hypusine *in vivo*. Archaeal elongation factors EF-1α and EF-2 contain regions that are highly conserved among elongation factors from all three domains. These conserved amino acid stretches are involved in GTP binding and hydrolysis. Interestingly, as in the eucaryal EF-2, one of the histidines at the C-terminus of the Archaeal EF-2 is posttranslationally modified to form

diphthamide, which makes it susceptible to ADP-ribosylation by the diphtheria
toxin. Phylogenetic trees based on comparisons of EF-2 homologs are consistent with
the clustering of Archaeal sequences on
a branch separate from branches leading
to eucarya or Bacteria; however, phylogenies based on EF-1α sequences are
less consistent.

3.3.3 Archaeal Ribosomes

Archaeal ribosomes bear superficial similarity to those of Bacteria. Both Archaeal
and bacterial ribosomes have a sedimentation coefficient of 70S and are composed
of two subunits with sedimentation coefficients of 50S and 30S respectively.
Archaeal rRNAs (23S, 16S, and 5S) are
similar in size to bacterial rRNAs, but
differ substantially in sequence. In terms
of overall structure, Archaeal ribosomes
appear intermediate in shape between
eucaryal and bacterial ribosomes. For example, the small ribosomal subunits of
Eucarya possess several bulges and gaps,
plus a "bill" that is not present in bacterial ribosomes (Fig. 6). All Archaeal
ribosomes possess a bill in their 30S ribosomes; however, only ribosomes from

sulfur-dependent hyperthermophiles possess the other eukaryotic features.

3.4 Complete Genome Sequences

Sequencing of the complete genomes
of five representative methanoarchaea
has led to genome-level comparisons
among organisms of different species
and domains and simplified strategies
for determining the function of unknown
methanoarchaea genes. The first Archaeal
genome to be sequenced was that of
the hyperthermophilic methanoarchaeon
Methanococcus jannaschii. The genome
consists of a 1.66-Mbp chromosome with
1682 predicted protein-encoding genes, a
large circular extrachromosomal element,
and a small circular extrachromosomal
element. Initially, putative cellular functions could be assigned to only 38% of
the predicted genes. Genome sequencing has confirmed the similarity of Archaeal and eucaryal proteins involved in
the processes of replication, transcription,
and translation. For example, the *M.
jannaschii* genome contains homologs
of eucaryal histones, RNA polymerase

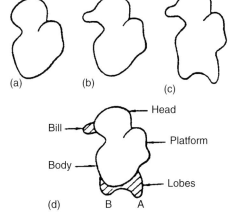

Fig. 6 Diagrammatic representation of
small ribosomal subunits from (a),
Bacteria, (b) Archaea, and (c) Eucarya.
The striped regions (d) show the
archaeal and eucaryal bill regions and
the lobes (A and B). [From J. A. Lake, E.
Henderson, M. W. Clark, and A. T.
Matheson, *Proc. Natl. Acad. Sci. U.S.A.*
79: 5948–5952 (1982).]

subunits, transcription-initiation factors, selected ribosomal subunits, and some aminoacyl–tRNA synthetases. However, homologs of the aminoacyl–tRNA synthetases for glutamine, asparagine, lysine, and cysteine are notably absent. The genome contains a tRNA for selenocysteine, which is encoded by the UGA stop codon, as well as four genes with internal UGA codons that are predicted to encode selenocysteine.

The genome sequence of *M. jannaschii* has provided an invaluable resource for cloning methanoarchaeal genes for expression in *Escherichia coli*. In many cases, the thermostable nature of recombinant *M. jannaschii* proteins provides a convenient means of protein purification by heating cell extracts to denature thermolabile *E. coli* proteins. Problems involving codon-usage bias have been overcome by expressing genes in *E. coli* strains that contain rare codon tRNAs. To a large extent, this strategy has obviated the need for growing large quantities of strictly anaerobic methanoarchaea for use in protein purification. The approach has been particularly successful for identifying the genes encoding low abundance proteins, such as enzymes involved in biosynthetic processes.

Sequencing the complete genome of *M. thermoautotrophicus* ΔH provided the first opportunity to make genomic comparisons of methanoarchaea with other Archaea, Bacteria, and Eucarya. Forty-six percent of the predicted proteins have been assigned putative functions based on homology to known enzymes, 28% have homology to genes of unknown function, and 27% lack significant homology to any known sequence. Substantial divergence of *M. thermoautotrophicus* from *M. jannaschii* has been postulated on the basis of a low degree of conservation in gene organization and the observation that less than

20% of the *M. thermoautotrophicus* proteins show greater than 50% identity with *M. jannaschii* proteins. Homologs of two-component regulatory systems and heat shock-70 response proteins are abundant in *M. thermoautotrophicus* but absent in *M. jannaschii*. As in *M. jannaschii*, tRNA synthetases for asparagine, glutamine, cysteine, and lysine are unrecognizable, but interestingly, the translation machinery for selenocysteine is lacking in *M. thermoautotrophicus*. The genome contains homologs of the eucaryal TATA-binding proteins, TFIIB and TFIIS.

Methanopyrus kandleri is a hyperthermophilic methanoarchaea that grows at temperatures up to 110 °C. Phylogenies based on rRNA sequencing have placed this methanoarchaeon in a deeply branching position separate from all other known methanoarchaea. Whole genome sequence comparisons have provided a different perspective on this evolutionary model. Although *M. kandleri* diverges from other methanoarchaea according to 16S rRNA phylogenies, it clusters with *M. jannaschii* and *M. thermoautotrophicus* when the genomes are compared on the basis of either conserved gene order and content or concatenated alignments of ribosomal proteins. In contrast to other methanoarchaea, the *M. kandleri* genome contains a high proportion of negatively charged amino acids that may assist in adapting proteins to the high intracellular salt concentrations of this organism. In addition, *M. kandleri* has relatively few putative regulatory and signaling proteins and less evidence for the lateral transfer of genes from Bacteria.

In contrast, the genomes of *Methanosarcina acetivorans* C2A and *Methanosarcina mazei* Go1 reflect the wide metabolic diversity of the *Methanosarcinaceae* family. The 5.75-Mbp genome of *M. acetivorans*

C2A is three times larger than the three previously sequenced methanoarchaea genomes and shows evidence of extensive lateral gene transfer. Similarly, in the 4.1-Mbp genome of *M. mazei* Go1 greater than 30% of the open-reading frames are more closely related to bacterial homologs than Archaeal homologs. In *M. acetivorans* C2A, predicted coding regions account for 74% of the genome, substantially lower than the gene densities of *M. jannaschii*, *M. thermoautotrophicus*, and *M. kandleri* that all exceed 90%. The genome contains multiple copies of numerous genes; including genes for methanogenesis related enzymes, putative corrinoid proteins, and methyltransferases, TATA-binding proteins, and homologs of bacterial regulator proteins such as sensory transduction histidine kinases. *M. acetivorans* contains both class I and class II lysyl–tRNA synthetases. One of these may be involved in translating a UAG amber codon into pyrrolysine.

3.5
Use of Methanoarchaeal Genomes to Identify Coenzyme Biosynthesis Genes

One field that has reaped tremendous benefit from the sequencing of methanoarchaeal genomes is functional genomics, including the identification of genes involved in the biosynthesis of specialized methanoarchaeal coenzymes. Progress has been made in the elucidation of the pathways of methanoarchaeal coenzyme biosynthesis. However, prior to genome sequencing progress in identifying the biosynthetic genes was limited, due in part to the low abundance of the enzymes and the lack of a genetics system.

Published genome sequences have opened up new strategies for identifying coenzyme biosynthesis genes. For example, the complete *Methanococcus jannaschii* genome sequence enabled use of a secondary structure analysis program to identify a proposed tetrahydromethanopterin (H_4MPT) biosynthesis gene encoding an enzyme that catalyzes a reaction similar in mechanism to that of dihydropteroate synthase. The secondary structure analysis program predicted that *M. jannaschii* gene *MJ0301* encodes a protein that would demonstrate secondary structure features similar those of dihydropteroate synthase. When *MJ0301* was expressed in *Escherichia coli*, the resulting protein catalyzed the predicted reaction of H_4MPT biosynthesis. *MJ0301* was the first H_4MPT biosynthesis gene to be identified.

A second effective strategy for identifying genes involved in coenzyme biosynthesis has been to propose a mechanism for the enzymatic reaction, and then search for genes that show homology to proteins that catalyze a similar reaction. For example, the formation of the α-keto acid (S)-sulfopyruvate during coenzyme M biosynthesis is proposed to occur by oxidation of (S)-sulfolactate. This mechanism resembles that of malate dehydrogenase, and the *M. jannaschii* genome contains only two genes annotated as malate dehydrogenase genes (*MJ1425* and *MJ0490*). Only the recombinant *MJ1425* gene product efficiently catalyzes the reaction of (S)-sulfolactate dehydrogenase, implicating a role for the gene *MJ1425* in coenzyme M biosynthesis. The *MJ1425* protein also catalyzes the reduction of α-ketoglutaric acid to (S)-hydroxyglutaric acid, a component of the coenzyme H_4MPT. Using mechanistic predictions, at least 16 genes involved in coenzyme, amino acid, or nucleotide biosynthesis have been assigned biochemical functions.

A third strategy to facilitate the assignment of coenzyme biosynthesis genes has been to take advantage of clustered genes in Bacteria that synthesize analogs of coenzymes. The methylotrophic bacterium *Methylobacterium extorquens* AM1 synthesizes an analog of H$_4$MPT and contains a cluster of genes that encode proteins involved in the H$_4$MPT-dependent oxidation of methanol. Some of these genes code for proteins with similarity to uncharacterized archaeal proteins and have been tested as candidates for H$_4$MPT biosynthesis genes. One of the key enzymes of H$_4$MPT biosynthesis is ribofuranosylaminobenzene 5′-phosphate synthase (RFAP), which catalyzes the first committed step of the pathway. The N-terminal sequence of purified RFAP synthase from the methanoarchaeon *M. thermophila* has been used to determine the corresponding gene in the genomes of *M. jannaschii* and *M. thermoautotrophicus*. Fortuitously, an RFAP synthase gene homolog also occurs among the clustered bacterial genes required for H$_4$MPT-dependent metabolism in *M. extorquens*. Expression in *E. coli* of the putative RFAP synthase genes from *M. thermoautotrophicus* (*MTH0830*) and *M. extorquens* (*orf4*) has produced enzymes with RFAP synthase activity.

The highly developed genetics system of *M. extorquens* has recently been used to identify seven additional genes of H$_4$MPT biosynthesis. Using an enzymatic assay to measure levels of H$_4$MPT coenzymes in *M. extorquens* cells, it has been shown that deletion mutants lacking the RFAP synthase gene are incapable of producing H$_4$MPT. Using this assay as a screen, seven additional genes implicated in H$_4$MPT biosynthesis have been found in *M. extorquens*. Six of these genes cluster together with *orf4* and encode Archaeal like proteins of unknown function. The seventh gene (*dmrA*) codes for an analog of bacterial dihydrofolate reductase that catalyzes the NADPH-dependent reduction of dihydromethanopterin to H$_4$MPT. These studies demonstrate how bacterial model systems can be combined with complete genome sequences to assist in determining the functions of uncharacterized genes.

3.6
Genetic Systems

Recent advances in the development of genetic systems have opened up the possibility of creating mutants to investigate the physiology, biochemistry, and regulation of gene expression in the methanoarchaea. Genetic tools for creating random and site-directed mutants, complementation strains, and gene reporters have been developed in combination with improved plating techniques, transformation procedures, and selectable markers. Model genetics systems have been developed for *Methanococcus maripaludis* and *Methanosarcina* species, and some of the genetics tools are effective in other methanoarchaea of the same genus.

3.6.1 Methanococcus

One of the challenges of developing a genetics system for methanoarchaea has been the lack of suitable antibiotics and antibiotics resistance markers. Because of fundamental differences between the ribosomal proteins of Bacteria and Archaea, typical bacterial antibiotics are often ineffective as inhibitors of methanoarchaeal protein synthesis. The first antibiotic to be used in a methanoarchaeal vector was puromycin, which inhibits the growth of *Methanococcus* and *Methanosarcina* species at concentrations less than 5 µg mL^{-1}.

Methanococcal integration vectors have been created using the puromycin acetyltransferase gene (*pac*) from *Streptomyces alboniger* under the control of the promoter and terminator for the methyl coenzyme M methylreductase gene. Vectors for transposon insertional mutagenesis are also available. Transformation procedures in *Methanococcus* species include polyethylene glycol treatment of *M. maripaludis* cells, which produces over 100 transformants per µg of plasmid, and transformation of *Methanococcus voltae* protoplasts by electroporation.

A shuttle vector (pDLT44) allowing plasmid replication in both *M. maripaludis* JJ and *E. coli* has been developed on the basis of the methanococcal plasmid pURB500. From pDLT44, an expression vector has been created by adding the strong methanococcal histone promoter and a ribosomal binding site. The sensitivity of *Methanococcus maripaludis* to neomycin at concentrations between 100 and 250 µg ml^{-1} and the use of neomycin resistance genes has expanded the range of methanococcal genetics tools and allowed for deletion of multiple genes in a single organism and complementation using expression vectors.

Studies on the regulation of gene expression in *M. maripaludis* have focused on processes involving nitrogen metabolism and formate utilization. Nitrogen fixation, the reduction of atmospheric nitrogen to ammonia, is highly regulated in methanoarchaea and has provided an excellent model system for identifying regulatory components in *M. maripaludis*. As in Bacteria, nitrogenase activity and the expression of nitrogen fixation (*nif*) genes in methanoarchaea decrease when fixed nitrogen is available as ammonia. Growth of *M. maripaludis* under nitrogen-fixing conditions requires molybdenum (but not vanadium)

and is inhibited by tungsten. These data are consistent with phylogenetic evidence that the *M. maripaludis* nitrogenase is related to the class of bacterial nitrogenases that contain iron-molybdenum cofactor. Nitrogenase is encoded by the structural genes *nifH*, *nifD*, and *nifK*, and accessory genes are required for assembly of the enzyme into an active form. The order of the *nif* genes in methanoarchaea (*nifHDKENX*) is the same as the order found in Bacteria, with the interesting exception that two homologs of the regulatory gene *glnB* are inserted between the *nifH* and *nifD* genes. The *glnB* homologs also occur in the *nif* gene clusters of *Methanosarcina barkeri*. In Bacteria, the *glnB* gene product is involved in regulation of nitrogen metabolism and can be uridylylated at a conserved tyrosine residue in the T-loop.

In contrast to the bacterial *nif* genes, the eight *nif* genes of *M. maripaludis* form a single operon that can be transcribed as one large transcript (or up to ten smaller transcripts depending on the growth conditions). Transposon insertion mutagenesis indicates that all of the genes except *nifX* are essential for nitrogen fixation activity. The function of *nifX* in both Bacteria and Archaea is unknown, but a role in either cofactor biosynthesis or regulation has been proposed.

Transcription of the *nif* genes of *M. maripaludis* is repressed by ammonia. Two sets of palindromic sequences have been found between the *nifH* promoter and the transcriptional start site. Directed mutagenesis of the *nifH* promoter region fused to a *lacZ* reporter gene indicates that the first palindrome, but not the second, is involved in the regulation of *nif* gene transcription by ammonia. A similar palindrome (GGAA-N6-TTCC) is also found in the promoter region of several other genes involved in nitrogen metabolism, including genes for glutamine synthetase (*glnA*), an ammonia-dependent NAD synthetase homolog, and an operon containing a putative ammonia transporter gene and a *glnB* homolog. Genes controlled by this operator may constitute a nitrogen metabolism regulon under the control of a single repressor protein. The Archaeal *nif* repressor protein is a negative regulatory element, in contrast to the bacterial regulatory proteins NifA and NtrC, which are positive regulators (activators) of *nif* expression. When alanine is the sole nitrogen source, alanine represses transcription of both the *nif* operon and the *glnA* gene to an intermediate level. Both the first and the second set of palindromic sequences in the *nifH* promoter region are needed for alanine repression of *nif* gene expression.

The availability of ammonia and alanine also regulates the transcription of *glnA*. The *glnA* promoter has only one operator for binding the nitrogen repressor protein, but there appear to be three TATA boxes and three transcriptional start sites. The first promoter provides a low, constitutive level of *glnA* expression, while the other two promoters are regulated through an operator that resembles the first palindromic sequence of the *nifH* promoter.

Nitrogenases from both bacterial and archaeal sources demonstrate the phenomenon of ammonia switch-off, which is a posttranslational inactivation of nitrogenase in the presence of ammonia. In *Rhodospirillum rubrum*, ammonia addition leads to the reversible ADP-ribosylation of the *nifH* gene product. However, in *M. maripaludis* ammonia switch-off occurs without ADP-ribosylation. Deletion mutagenesis and genetic complementation studies show that the process requires the two novel *glnB* homologs found in the methanoarchaeal *nif* operon. Although bacterial GlnB homologs can be modified by uridylylation of a conserved tyrosine in the T-loop, the Archaeal homologs might not become uridylylated because the methanoarchaeal homologs lack the conserved T-loop.

The ability of *M. maripaludis* to grow on either formate or H_2/CO_2 provides a second system for studying gene regulation in methanoarchaea. The genome of *M. maripaludis* has two clusters of formate dehydrogenase genes. The neomycin and puromycin resistance genes have been used to construct individual deletion strains and double mutants lacking both of the *fdhA* genes. Both genes are required for maximal growth rate on formate, and deletion of both *fdhA* genes eliminates the ability to grow on formate. Promoter–*lacZ* fusions show that H_2 represses expression of formate operons, but formate concentration has no effect on expression.

3.6.2 Methanosarcina

The metabolic diversity of *Methanosarcina* species and the higher abundance of regulated genes and operons make this genus an attractive target for genetic analysis. Early technical obstacles to developing a genetics system for *Methanosarcina* species included the tendency of the cells to grow in clumps called *sarcina*, the low transformation efficiency, a paucity of selectable markers, and a lack of suitable vectors for mutagenesis and expression. Progress in establishing a facile genetics system began with the observation that *Methanosarcina* cells grow as single cells in a moderate salt (marine) medium, which enables the plating of single colonies. A second breakthrough in *Methanosarcina* transformation has been the use of liposomes to deliver DNA into protoplasts. This procedure increases the transformation efficiency of

M. acetivorans by up to five orders of magnitude over polyethylene glycol-mediated transformation and electroporation. Nine out of eleven *Methanosarcina* strains tested could be transformed using the liposome-mediated technology.

Shuttle vectors for use in *E. coli* and *Methanosarcina* species have been developed by modifying plasmid pC2A from *M. acetivorans* to contain a puromycin resistance gene, a multiple cloning site, and *lacZα* for blue–white screening. Because neomycin apparently does not inhibit *Methanosarcina* strains, pseudomonic acid has been used to develop a second selectable marker system. Pseudomonic acid is a structural analog of isoleucine and inhibits the charging of isoleucyl tRNA synthetases. The resistance gene codes for a pseudomonic acid–resistant isoleucyl tRNA synthetase created by mutagenesis of *ileS* from *M. barkeri*.

A plasmid that can be used for random *in vivo* transposon mutagenesis of *Methanosarcina* is now available. The suicide vector contains the *Himar1* transposase under the control of the methyl coenzyme M methylreductase promoter and mini-*Himar1* elements. Using transposons, a mutation conferring resistance to the methanogenesis inhibitor bromoethanesulfonic acid (BES) has been localized to a putative membrane transport protein that may be needed for BES uptake. Random transposon mutagenesis has also shown that the resistance of *M. acetivorans* to fluoroacetic acid involves the acetate kinase–phosphotransacetylase operon needed for growth on acetate.

Directed mutagenesis of proline biosynthesis genes in *M. acetivorans* has been accomplished by replacing wild-type genes (*proA*, *proB*, or *proC*) with the genes disrupted by insertion of puromycin or pseudomonic acid resistance genes. Liposomes containing the mutated linearized DNA are used to transform *M. acetivorans*, and replacement of the wild-type gene by homologous recombination produces proline auxotrophs. These mutants can be complemented with plasmids containing the wild-type genes.

This suite of genetic tools has been used to investigate the physiological role of the Ech hydrogenase (Ech) of *M. barkeri* Fusaro. Ech catalyzes the reversible reduction of ferredoxin by H_2, and its association with the cell membrane has led to a proposed role in (i) energy conservation, (ii) the endergonic reduction of CO_2 to CHO-methanofuran during hydrogenotrophic growth, or (iii) steps in the synthesis of acetyl-CoA and pyruvate for anabolism. A *M. barkeri* Fusaro mutant has been created in which a puromycin resistance gene cassette replaces the entire *echABCDEF* operon. As anticipated, cell extracts from the *ech* deletion mutant lack ferredoxin-dependent hydrogenase activity. The mutant grows normally on methanol, but is unable to grow on acetate, H_2/CO_2, or H_2/CO_2/methanol. The mutant produces methane from methanol, but not from acetate or H_2/CO_2. Bypassing the first step of methanogenesis from H_2/CO_2 by providing H_2 and formaldehyde results in methane production, suggesting that the mutant is defective in the first step of hydrogenotrophic methanogenesis. The mutant is also unable to convert CO and H_2O to CO_2 and H_2, indicating that the oxidative half-reaction of methanogenesis from acetate is also nonfunctional. The mutant can be complemented for all phenotypes by inserting a functional copy of the *echABCDEF* operon in a neutral chromosomal location. These data are consistent with a role for Ech in the first step of methano-genesis from H_2/CO_2 and in methane production from acetate, which may both require a ferredoxin-dependent hydrogenase. An additional role for Ech in synthesizing pyruvate for anabolic pathways has been proposed because exogenous pyruvate restores the ability of the deletion mutant to grow on $H_2/CO_2/$methanol, consistent with a role for Ech in pyruvate biosynthesis. The experiments described demonstrate the usefulness of the improved genetic tools for functional genomics analysis and for genetics studies that address fundamental questions about methanoarchaeal physiology and biochemistry.

3.7
Pyrrolysine

An exciting recent discovery in methanoarchaeal genetics has been the expansion of the genetic code through characterization of pyrrolysine, the twenty-second genetically encoded amino acid. This molecule consists of a substituted pyrroline-5-carboxylate attached by an amide bond to the epsilon amino group of lysine

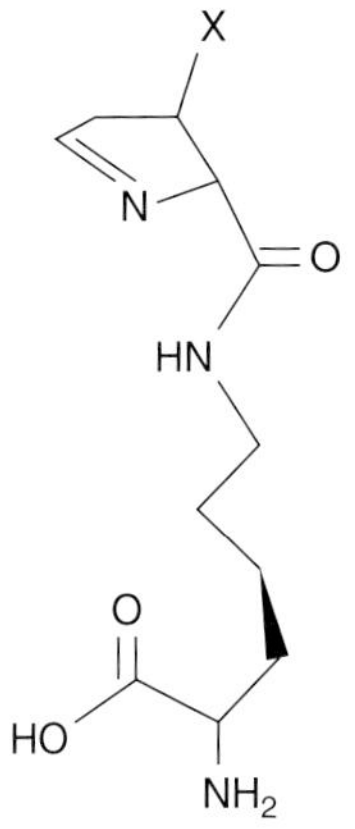

Fig. 7 Structure of pyrrolysine.

(See Fig. 7). Like selenocysteine, pyrrolysine is encoded by read-through of a stop codon (in this case, UAG rather than UGA). The N-terminal sequence of the purified monomethylamine methyltransferase from *M. barkeri* MS corresponds to the gene *mtmB1*. Although the molecular mass of the purified protein (MtmB) is 50 kDa, the *mtmB1* gene contains an apparent in-frame UAG stop codon that would truncate the protein of at 23 kDa. Internal sequencing of MtmB and mass spectrometric analysis of peptide fragments suggest that lysine occurs at the position corresponding to the UAG codon. The x-ray crystal structures of the monomethylamine methyltransferase show electron density consistent with a lysine linked to a 4-substituted-pyrroline-5-carboxylate moiety at the UAG position. The substitution at the 4-position may be an amine or a hydroxyl group. On the basis of its position in the protein active site, the novel amino acid has been proposed to play a specific role in transferring methyl groups during catalysis.

The mechanism of tRNA charging with pyrrolysine is currently being elucidated, and there is at present some controversy about the enzymes involved in the process. One investigation implicates the *pylTS-BCDL* gene cluster located near one of the methyltransferase gene clusters of *M. barkeri* Fusaro. The *pylT* gene codes for a tRNA with a CUA anticodon capable of base pairing with the UAG codon. The protein encoded by *pylS* shares some homology with the core catalytic domains of class II aminoacyl–tRNA synthetases. Recombinant *M. barkeri* MS *pylS* expressed with an N-terminal six-histidine tag (His$_6$) in *E. coli* has been tested for lysine aminoacyl–tRNA synthetase activity. When combined with [^{14}C]lysine and tRNA isolated from *M. barkeri*, His$_6$-PylS appears to catalyze the lysyl–tRNA synthetase reaction with an apparent K_m of 2.2 µM and an apparent k_{cat} of 1.6 min^{-1}. The K_m is comparable to the values found for other aminoacyl–tRNA synthetases. However, the k_{cat} value is between 12 and 5000 times lower than the rate obtained for other aminoacyl–tRNA synthetases. Given the low K_m for lysine, it is conceivable that the amber codon tRNA (PylT) is first charged with lysine and later modified to become pyrrolysyl-tRNA. A similar mechanism has been proposed for the biosynthesis of asparaginyl-tRNA, glutamyl-tRNA, and selenocysteinyl-tRNA. Because the proposed amber suppressor lysyl–tRNA synthetase (PylS) does not strongly resemble the typical class I or class II lysyl–tRNA synthetases, PylS is proposed to be a new class of lysyl–tRNA synthetase. In contrast, a second study presents contradictory evidence that His$_6$-PylS from *M. barkeri* Fusaro and *M. barkeri* MS cannot charge the amber suppressor tRNA (PylT) with lysine. Instead, it is proposed that PylT can be charged only if both the class I and class II lysyl–tRNA synthetases of *M. barkeri* are present in the reaction. Neither of the synthetases alone can catalyze the tRNA synthetase reaction. Further experiments will undoubtedly help resolve this inconsistency and provide further details on the genetic processes involved in incorporating pyrrolysine into proteins.

See also Bacterial Cell Culture Methods; Bacterial Growth and Division; Microbial Development; Organic Cofactors as Coenzymes; Plasmids.

Bibliography

Books and Reviews

Deppenmeier, U. (2002b) The unique biochemistry of methanogenesis, *Prog. Nucleic Acid Res. Mol. Biol.* **71**, 223–283.

Ferry, J.G. (1993) *Methanogenesis*, Chapman & Hall, New York, p. 536.

Ferry, J.G. (2003) One-carbon Metabolism in Methanogenic Anaerobes, in: Ljungdahl, L.G., Adams, M.W., Barton, L.L., Ferry, J.G., Johnson, M.K. (Eds) *Biochemistry and Physiology of Anaerobic Bacteria*, Springer-Verlag, New York, pp. 143–156.

Gottschalk, G., Thauer, R.K. (2001) The Na$^+$ translocating methyltransferase complex from methanogenic archaea, *Biochim. Biophys. Acta* **1505**, 28–36.

Hickey, A.J., Conway de Macario, E., Macario, A.J. (2002) Transcription in the archaea: basal factors, regulation, and stress-gene expression, *Crit. Rev. Biochem. Mol. Biol.* **37**, 537–599.

Lange, M., Ahring, B.K. (2001) A comprehensive study into the molecular methodology and molecular biology of methanogenic Archaea, *FEMS Microbiol. Rev.* **25**, 553–571.

Reeve, J.N., Nolling, J., Morgan, R.M., Smith, D.R. (1997) Methanogenesis: genes, genomes, and who's on first? *J. Bacteriol.* **179**, 5975–5986.

Soppa, J. (1999b) Transcription initiation in Archaea: facts, factors and future aspects, *Mol. Microbiol.* **31**, 1295–1305.

Thomm, M. (1996) Archaeal transcription factors and their role in transcription initiation, *FEMS Microbiol. Rev.* **18**, 159–171.

Primary Literature

Bartig, D., Lemkemeier, K., Frank, J., Lottspeich, F., Klink, F. (1992) The archaebacterial hypusine-containing protein. Structural features suggest common ancestry with eukaryotic translation initiation factor 5A, *Eur. J. Biochem.* **204**, 751–758.

Bartoschek, S., Vorholt, J.A., Thauer, R.K., Geierstanger, B.H., Griesinger, C. (2000) *N*-carboxymethanofuran (carbamate) formation from methanofuran and CO$_2$ in methanogenic archaea. Thermodynamics and kinetics of the spontaneous reaction, *Eur. J. Biochem.* **267**, 3130–3138.

Bechard, M.E., Chhatwal, S., Garcia, R.E., Rasche, M.E. (2003) Application of a colorimetric assay to identify putative ribofuranosylaminobenzene 5′-phosphate synthase genes expressed with activity in *Escherichia coli*, *Biol. Proced Online* **5**, 69–77.

Beifuss, U., Tietze, M., Baumer, S., Deppenmeier, U. (2000) Methanophenazine: structure, total synthesis, and function of a new cofactor from methanogenic Archaea, *Angew. Chem., Int. Ed.* **39**, 2470–2473.

Berghofer, Y., Klein, A. (1995) Insertional mutations in the hydrogenase *Vhc* and *Frc* operons encoding selenium-free hydrogenases in *Methanococcus voltae*, *Appl. Environ. Microbiol.* **61**, 1770–1775.

Bult, C.J., White, O., Olsen, G.J., Zhou, L., Fleischmann, R.D., Sutton, G.G., Blake, J.A., FitzGerald, L.M., Clayton, R.A., Gocayne, J.D. et al. (1996) Complete genome sequence of the methanogenic archaeon, Methanococcus jannaschii, *Science* **273**, 1058–1073.

Buss, K.A., Cooper, D.R., Ingram-Smith, C., Ferry, J.G., Sanders, D.A., Hasson, M.S. (2001) Urkinase: structure of acetate kinase, a member of the ASKHA superfamily of phosphotransferases, *J. Bacteriol.* **183**, 680–686.

Buurman, G., Shima, S., Thauer, R.K. (2000) The metal-free hydrogenase from methanogenic Archaea: evidence for a bound cofactor, *FEBS Lett.* **485**, 200–204.

Caccamo, M.A., Malone, C.M., Rasche, M.E. (2004) Biochemical characterization of dihydromethanopterin reductase, a tetrahydromethanopterin biosynthesis enzyme in *Methylobacterium extorquens* AM1, *J. Bacteriol.* (in Press).

Chistoserdova, L., Vorholt, J.A., Thauer, R.K., Lidstrom, M.E. (1998) C_1 transfer enzymes and coenzymes linking methylotrophic bacteria and methanogenic Archaea, *Science* **281**, 99–102.

Cohen-Kupiec, R., Blank, C., Leigh, J.A. (1997) Transcriptional regulation in Archaea: in vivo demonstration of a repressor binding site in a methanogen, *Proc. Natl. Acad. Sci. U.S.A.* **94**, 1316–1320.

Cohen-Kupiec, R., Marx, C.J., Leigh, J.A. (1999) Function and regulation of glnA in the methanogenic archaeon *Methanococcus maripaludis*, *J. Bacteriol.* **181**, 256–261.

Creti, R., Ceccarelli, E., Bocchetta, M., Sanangelantoni, A.M., Tiboni, O., Palm, P., Cammarano, P. (1994) Evolution of translational elongation factor (EF) sequences: reliability of global phylogenies inferred from EF-1 alpha(Tu) and EF-2(G) proteins, *Proc. Natl. Acad. Sci. U.S.A.* **91**, 3255–3259.

Creti, R., Sterpetti, P., Bocchetta, M., Ceccarelli, E., Cammarano, P. (1995) Chromosomal organization and nucleotide sequence of the fus-gene encoding elongation factor 2 (EF-2) of the hyperthermophilic archaeum *Pyrococcus woesei*, *FEMS Microbiol. Lett.* **126**, 85–90.

Darcy, T.J., Hausner, W., Awery, D.E., Edwards, A.M., Thomm, M., Reeve, J.N. (1999) *Methanobacterium thermoautotrophicum* RNA polymerase and transcription in vitro, *J. Bacteriol.* **181**, 4424–4429.

Decanniere, K., Babu, A.M., Sandman, K., Reeve, J.N., Heinemann, U. (2000) Crystal structures of recombinant histones HMfA and HMfB from the hyperthermophilic archaeon methanothermus fervidus, *J. Mol. Biol.* **303**, 35–47.

Deppenmeier, U., Johann, A., Hartsch, T., Merkl, R., Schmitz, R.A., Martinez-Arias, R., Henne, A., Wiezer, A., Baumer, S., Jacobi, C. et al. (2002) The genome of *Methanosarcina mazei*: evidence for lateral gene transfer between bacteria and Archaea, *J. Mol. Microbiol. Biotechnol.* **4**, 453–461.

Deppenmeier, U., Muller, V., Gottschalk, G. (1996) Pathways of energy conservation in methanogenic Archaea, *Arch. Microbiol.* **165**, 149–163.

Duin, E.C., Madadi-Kahkesh, S., Hedderich, R., Clay, M.D., Johnson, M.K. (2002) Heterodisulfide reductase from methanothermobacter marburgensis contains an active-site [4Fe-4S] cluster that is directly involved in mediating heterodisulfide reduction, *FEBS Lett.* **512**, 263–268.

Ermler, U., Grabarse, W., Shima, S., Goubeaud, M., Thauer, R.K. (1997) Crystal structure of methyl-coenzyme M reductase: the key enzyme of biological methane formation, *Science* **278**, 1457–1462.

Fahrner, R.L., Ferry Cascio, D., Lake, J.A., Slesarev, A. (2001) An ancestral nuclear protein assembly: crystal structure of the Methanopyrus kandleri histone, *Protein Sci.* **10**, 2002–2007.

Ferry, J.G. (1997) Methane: small molecule, big impact [comment], *Science* **278**, 1413–1414.

Galagan, J.E., Nusbaum, C., Roy, A., Endrizzi, M.G., Macdonald, P., FitzHugh, W., Calvo, S., Engels, R., Smirnov, S., Atnoor, D. et al. (2002) The genome of *M. acetivorans* reveals extensive metabolic and physiological diversity, *Genome Res.* **12**, 532–542.

Gernhardt, P., Possot, O., Foglino, M., Sibold, L., Klein, A. (1990) Construction of an integration vector for use in the archaebacterium *Methanococcus voltae* and expression of a eubacterial resistance gene, *Mol. Gen. Genet.* **221**, 273–279.

Grabarse, W., Mahlert, F., Duin, E.C., Goubeaud, M., Shima, S., Thauer, R.K., Lamzin, V., Ermler, U. (2001) On the mechanism of biological methane formation: structural evidence for conformational changes in methyl-coenzyme M reductase upon substrate binding, *J. Mol. Biol.* **309**, 315–330.

Graupner, M., Xu, H., White, R.H. (2000) Identification of an archaeal 2-hydroxy acid dehydrogenase catalyzing reactions involved in coenzyme biosynthesis in methanoarchaea, *J. Bacteriol.* **182**, 3688–3692.

Hanzelka, B.L., Darcy, T.J., Reeve, J.N. (2001) TFE, an archaeal transcription factor in methanobacterium thermoautotrophicum related to eucaryal transcription factor TFIIEalpha, *J. Bacteriol.* **183**, 1813–1818.

Hao, B., Gong, W., Ferguson, T.K., James, C.M., Krzycki, J.A., Chan, M.K. (2002) A new UAG-encoded residue in the structure of a methanogen methyltransferase, *Science* **296**, 1462–1466.

Huber, H., Hohn, M.J., Rachel, R., Fuchs, T., Wimmer, V.C., Stetter, K.O. (2002) A new phylum of Archaea represented by a nanosized hyperthermophilic symbiont, *Nature* **417**, 63–67.

Ide, T., Baumer, S., Deppenmeier, U. (1999) Energy conservation by the H_2:heterodisulfide oxidoreductase from *Methanosarcina mazei* Go1: identification of two proton-translocating segments, *J. Bacteriol.* **181**, 4076–4080.

Jarrell, K.F., Bayley, D.P., Florian, V., Klein, A. (1996) Isolation and characterization of insertional mutations in flagellin genes in the archaeon *Methanococcus voltae*, *Mol. Microbiol.* **20**, 657–666.

Kessler, P.S., Blank, C., Leigh, J.A. (1998) The nif gene operon of the methanogenic archaeon *Methanococcus maripaludis*, *J. Bacteriol.* **180**, 1504–1511.

Kessler, P.S., Daniel, C., Leigh, J.A. (2001) Ammonia switch-off of nitrogen fixation in the methanogenic archaeon *Methanococcus maripaludis*: mechanistic features and requirement for the novel GlnB homologues, NifI(1) and NifI(2), *J. Bacteriol.* **183**, 882–889.

Kessler, P.S., Leigh, J.A. (1999) Genetics of nitrogen regulation in *Methanococcus maripaludis*, *Genetics* **152**, 1343–1351.

Klein, A.R., Fernandez, V.M., Thauer, R.K. (1995a) H_2-forming N^5,N^{10}-methylenetetrahydromethanopterin dehydrogenase: mechanism of H_2 formation analyzed using hydrogen isotopes, *FEBS Lett.* **368**, 203–206.

Klein, A.R., Hartmann, G.C., Thauer, R.K. (1995b) Hydrogen isotope effects in the reactions catalyzed by H_2-forming N^5,N^{10}-methylenetetrahydromethanopterin dehydrogenase from methanogenic archaea, *Eur. J. Biochem.* **233**, 372–376.

Lake, J.A., Henderson, E., Clark, M.W., Matheson, A.T. (1982) Mapping evolution with ribosome structure: intralineage constancy and interlineage variation, *Proc. Natl. Acad. Sci. U.S.A.* **79**, 5948–5952.

Langer, D.H.J., Thuriaux, P., Zillig, W. (1995) Transcription in Archaea: similarity to that in eucarya, *Proc. Natl. Acad. Sci. U.S.A.* **92**, 5768–5772.

Leigh, J.A. (2000) Nitrogen fixation in methanogens: the archaeal perspective, *Curr. Issues Mol. Biol.* **2**, 125–131.

Lie, T.J., Leigh, J.A. (2002) Regulatory response of *Methanococcus maripaludis* to alanine, an intermediate nitrogen source, *J. Bacteriol.* **184**, 5301–5306.

Macario, A.J., Lange, M., Ahring, B.K., De Macario, E.C. (1999) Stress genes and proteins in the archaea, *Microbiol. Mol. Biol. Rev.* **63**, 923–967.

Madadi-Kahkesh, S., Duin, E.C., Heim, S., Albracht, S.P.J., Johnson, M.K., Hedderich, R. (2001) A paramagnetic species with unique EPR characteristics in the active site of heterodisulfide reductase from methanogenic archaea, *Eur. J. Biochem.* **268**, 2566–2577.

Marc, F., Sandman, K., Lurz, R., Reeve, J.N. (2002) Archaeal histone tetramerization determines DNA affinity and the direction of DNA supercoiling, *J. Biologic. Chem.* **277**, 30879–30886.

Metcalf, W.W., Zhang, J.K., Apolinario, E., Sowers, K.R., Wolfe, R.S. (1997) A genetic system for Archaea of the genus *Methanosarcina*:

liposome-mediated transformation and construction of shuttle vectors, *Proc. Natl. Acad. Sci. U.S.A.* **94**, 2626–2631.

Meuer, J., Kuettner, H.C., Zhang, J.K., Hedderich, R., Metcalf, W.W. (2002) Genetic analysis of the archaeon *Methanosarcina barkeri* fusaro reveals a central role for Ech hydrogenase and ferredoxin in methanogenesis and carbon fixation, *Proc. Natl. Acad. Sci. U.S.A.* **99**, 5632–5637.

Muller, V., Ruppert, C., Lemker, T. (1999) Structure and function of the A_1A_0-ATPases from methanogenic Archaea, *J. Bioenerg. Biomembr.* **31**, 15–27.

Murakami, E., Deppenmeier, U., Ragsdale, S.W. (2000) Characterization of the intramolecular electron transfer pathway from 2-hydroxyphenazine to the heterodisulfide reductase from *Methanosarcina thermophila*, *J. Biol. Chem.* **276**, 2432–2439.

Murakami, E., Ragsdale, S.W. (2000) Evidence for intersubunit communication during acetyl-CoA cleavage by the multienzyme CO dehydrogenase/acetyl-CoA synthase complex from *Methanosarcina thermophila*. Evidence that the beta subunit catalyzes C−C and C−S bond cleavage, *J. Biol. Chem.* **275**, 4699–4707.

Ouhammouch, M., Dewhurst, R.E., Hausner, W., Thomm, M., Geiduschek, E.P. (2003) Activation of archaeal transcription by recruitment of the TATA-binding protein, *Proc. Natl. Acad. Sci. U.S.A.* **100**, 5097–5102.

Polycarpo, C., Ambrogelly, A., Ruan, B., Tumbula-Hansen, D., Ataide, S.F., Ishitani, R., Yokoyama, S., Nureki, O., Ibba, M., Soll, D. (2003) Activation of the pyrrolysine suppressor tRNA requires formation of a ternary complex with class I and class II lysyl-tRNA synthetases, *Mol. Cell* **12**, 287–294.

Praetorius-Ibba, M., Ibba, M. (2003) Aminoacyl-tRNA synthesis in archaea: different but not unique, *Mol. Microbiol.* **48**, 631–637.

Qureshi, S.A.B.P., Rowlands, T., Khoo, B., Jackson, S.P. (1995) Cloning and functional analysis of the TATA binding protein from sulfolobus shibatae, *Nucl. Acids Res.* **23**, 1775–1781.

Rasche, M.E., Wyles, S.A., Rosenzvaig, M. (2004) Characterization of two methanopterin biosynthesis mutants of *Methylobacterium extorquens* AM1 using a tetrahydromethanopterin bioassay, *J. Bacteriol.* **186**, (in Press).

Reeve, J.N. (1993) Structure and Organization of Genes, in: Ferry, J.G. (Eds) *Methanogenesis: Ecology, Physiology, Biochemistry & Genetics*, Chapman & Hall, New York, pp. 493–527.

Rospert, S., Linder, D., Ellermann, J., Thauer, R.K. (1990) Two genetically distinct methyl-coenzyme M reductases in *Methanobacterium thermoautotrophicum* strain Marburg and delta H, *Eur. J. Biochem.* **194**, 871–877.

Scott, J.W., Rasche, M.E. (2002) Purification, overproduction, and partial characterization of beta-RFAP synthase, a key enzyme in the methanopterin biosynthesis pathway, *J. Bacteriol.* **184**, 4442–4448.

Shima, S., Warkentin, E., Grabarse, W., Sordel, M., Wicke, M., Thauer, R.K., Ermler, U. (2000) Structure of coenzyme F420 dependent methylenetetrahydromethanopterin reductase from two methanogenic archaea, *J. Mol. Biol.* **300**, 935–950.

Shima, S., Warkentin, E., Thauer, R.K., Ermler, U. (2002) Structure and function of enzymes involved in the methanogenic pathway utilizing carbon dioxide and molecular hydrogen, *J. Biosci. Bioeng.* **93**, 519–530.

Slesarev, A.I., Mezhevaya, K.V., Makarova, K.S., Polushin, N.N., Shcherbinina, O.V., Shakhova, V.V., Belova, G.I., Aravind, L., Natale, D.A., Rogozin, I.B. et al. (2002) The complete genome of hyperthermophile *Methanopyrus kandleri* AV19 and monophyly of archaeal methanogens, *Proc. Natl. Acad. Sci. U.S.A.* **99**, 4644–4649.

Smith, D.R., Doucette-Stamm, L.A., Deloughery, C., Lee, H., Dubois, J., Aldredge, T., Bashirzadeh, R., Blakely, D., Cook, R., Gilbert, K. et al. (1997) Complete genome sequence of methanobacterium thermoautotrophicum deltaH: functional analysis and comparative genomics, *J. Bacteriol.* **179**, 7135–7155.

Soppa, J. (1999a) Normalized nucleotide frequencies allow the definition of archaeal promoter elements for different archaeal groups and reveal base-specific TFB contacts upstream of the TATA box, *Mol. Microbiol.* **31**, 1589–1592.

Srinivasan, G., James, C.M., Krzycki, J.A. (2002) Pyrrolysine encoded by UAG in Archaea: charging of a UAG-decoding specialized tRNA, *Science* **296**, 1459–1462.

Tang, T.H., Rozhdestvensky, T.S., d'Orval, B.C., Bortolin, M.L., Huber, H., Charpentier, B., Branlant, C., Bachellerie, J.P., Brosius, J., Huttenhofer, A. (2002) RNomics in Archaea reveals a further link between splicing of archaeal introns and rRNA processing, *Nucl. Acids. Res.* **30**, 921–930.

Thiru, A., Hodach, M., Eloranta, J.J., Kostourou, V., Weinzierl, R.O., Matthews, S. (1999) RNA polymerase subunit H features a beta-ribbon motif within a novel fold that is present in archaea and eukaryotes, *J. Mol. Biol.* **287**, 753–760.

Tumbula, D.L., Whitman, W.B. (1999) Genetics of *Methanococcus*: possibilities for functional genomics in Archaea, *Mol. Microbiol.* **33**, 1–7.

Watanabe, H., Gojobori, T., Miura, K., Watanabea, H. (1997) Bacterial features in the genome of *Methanococcus jannaschii* in terms of gene composition and biased base composition in ORFs and their surrounding regions, *Gene* **205**, 7–18.

Wich, G., Hummel, H., Jarsch, M., Bar, U., Bock, A. (1986a) Transcription signals for stable RNA genes in methanococcus, *Nucl. Acids. Res.* **14**, 2459–2479.

Wich, G., Sibold, L., Bock, A. (1986b) Genes for tRNA and their putative expression signals in methanococcus, *Systemat. Appl. Microbiol.* **7**, 18–25.

Woese, C.R., Kandler, O., Wheelis, M.L. (1990) Towards a natural system of organisms. Proposal for the domains archaea, bacteria,

and eucarya, *Proc. Natl. Acad. Sci. U.S.A.* **87**, 4576–4579.

Wood, G.E., Haydock, A.K., Leigh, J.A. (2003) Function and regulation of the formate dehydrogenase genes of the methanogenic archaeon *Methanococcus maripaludis*, *J. Bacteriol.* **185**, 2548–2554.

Xu, H., Aurora, R., Rose, G.D., White, R.H. (1999) Identifying two ancient enzymes in Archaea using predicted secondary structure alignment, *Nat. Struct. Biol.* **6**, 750–754.

Zhang, J.K., Pritchett, M.A., Lampe, D.J., Robertson, H.M., Metcalf, W.W. (2000) In vivo transposon mutagenesis of the methanogenic archaeon *Methanosarcina acetivorans* C2A using a modified version of the insect mariner-family transposable element Himar1, *Proc. Natl. Acad. Sci. U.S.A.* **97**, 9665–9670.

Zhang, J.K., White, A.K., Kuettner, H.C., Boccazzi, P., Metcalf, W.W. (2002) Directed mutagenesis and plasmid-based complementation in the methanogenic archaeon *Methanosarcina acetivorans* C2A demonstrated by genetic analysis of proline biosynthesis, *J. Bacteriol.* **184**, 1449–1454.

Zillig, W., Palm, P., Reiter, W.-D., Gropp, F., Puhler, G., Klenk, H.-P. (1988) Comparative evaluation of gene expression in archaebacteria, *Eur. J. Biochem.* **173**, 473–482.

Methods and Model Organisms in Embryogenomics: *see* Principles and Applications of Embryogenomics

Microarray-Based Technology: Basic Principles, Advantages and Limitations

Rumiana Bakalova[1], Ashraf Ewis[1,2], and Yoshinobu Baba[1,3,4]
[1] *National Institute of Advanced Industrial Science and Technology, Takamatsu, Japan*
[2] *El-Minia University, El-Minia, Egypt*
[3] *University of Tokushima, Shomachi, Tokushima, Japan*
[4] *Nagoya University, Nagoya, Japan*

Encyclopedia of Molecular Cell Biology and Molecular Medicine, 2nd Edition. Volume 8
Edited by Robert A. Meyers.
Copyright © 2005 Wiley-VCH Verlag GmbH & Co. KGaA, Weinheim
ISBN: 3-527-30550-5

Keywords

Genomics
Analysis of gene expression of cell, tissue, or organ under given conditions.

Proteomics
Analysis of proteins of cell, tissue, or organ under given conditions.

Pharmacogenetics
Analysis of variability in drug responses attributed to hereditary factors in different populations.

Pharmacogenomics
Analysis of genome and its products (RNA and proteins) related to drug response.

Messenger RNA (mRNA)
The class of cellular RNA that undergoes extensive editing and that contains the protein-coding sequences of genes and functions as informational intermediate between DNA and protein.

Complementary DNA (cDNA)
The DNA version of the messenger RNA that is produced after a PCR reaction using reverse transcriptase (RT) enzyme (RT-PCR).

Gene Expression
The cellular process by which the genetic information flows from gene to messenger RNA, and finally to the protein.

DNA Microarrays
DNA (from bacterial clones or synthesized oligonucleotides) attached to a surface in an ordered, predetermined fashion at extremely high density; allow the monitoring of gene expression for thousands of genes simultaneously in a single hybridization experiment based on two principles of nucleic acid hybridization: (1) DNA and RNA (isolated from tissue specimens or cell lines) specifically interact with their complementary sequence attached on the array surface, and (2) this interaction happens in proportion to the amount of mRNA in the mixture.

Spotted DNA Arrays
Composed of cDNA clones arranged robotically on a nonporous surface (usually glass microscope slide or nylon membranes).

Oligonucleotide-based DNA Arrays

Composed of short (∼25 bp) or long (∼60–70 bp) oligodeoxynucleotides synthesized *in situ* directly onto a glass wafer, using a modification of semiconductor photolithography technology.

■ In April 2003, the 50th anniversary year of the discovery of the double-helical structure of DNA, a high-quality and comprehensive sequencing of the human genome was completely accomplished, and a revolution in molecular cell biology and molecular medicine had begun. Many complex questions arise (e.g. how genes contribute to normal human development, individual variability, and common diseases) and they need complex answers. It was necessary to develop techniques that permit complex answers. Microarray technology is a revolution technology that allows the simultaneous assessment of the transcription of tens of thousands of genes rapidly, as well as of their relative expression between normal and injured cells. There is widespread hope that microarrays will significantly impact on our ability to explore the genetic changes associated with etiology and development of many diseases, and to discover new biomarkers for disease diagnosis and prognosis prediction, and new therapeutic tools.

   The present article provides an overview of microarray technology with accents on its recent advantages and limitations. The first chapter is focused on the basic principles of microarrays, array platforms and fabrication, and advantages and restrictions of cDNA-based (spotted) and oligonucleotide-based arrays (GeneChips and Codelink). The second chapter describes the crucial points in microarray study design (e.g. choice of reference source, sample preparation and preservation, labeling, and hybridization). Finally, a brief description of microarray data normalization, mining, and validation is given at the end of the review, transferring to several basic web sites and software with detailed explanation of this most underappreciated challenge facing researchers working on microarray projects.

1
Introduction in Functional Genomics and Proteomics

All mammalian cells have an identical copy of one and the same genome. However, the human organism consists of different types of cells, realizing different functions. The cell variety is a result of the expression (transcription) of different genes from the identical genome, and subsequent translation of their genetic information into different types and quantity of proteins. Therefore, the status of contemporary genetics might be succinctly summarized as follows: many genes, few functions.

   Only 1 to 2% of the human genome codes expressed genes via messenger RNA (mRNA), which is translated into protein. The rest of the genome consists of repeated sequences, regulatory regions, or unique noncoding sequences with unknown function.

The Human Genome Project (HGP) was begun in 1990, and its goal is the complete mapping and understanding of all the genes of human beings. Two groups, The International Human Genome Sequencing Consortium and Celera Genomics, simultaneously reported the draft sequencing of human genome in February 2001 (www.genome.gov). In April 2003, the 50th anniversary year of the discovery of the double-helical structure of DNA, a high-quality and comprehensive sequencing of the human genome was completely accomplished. HGP covers about 99% of the human genome's gene-containing regions with a sequence accuracy of 99.99%. The number of genes in the human genome was revealed to be approximately 30 000, which is considerably less than the initial estimation. At present, all of the initial objectives of the HGP have been achieved at least two years ahead of expectation, and a revolution in biological research has begun.

Each gene can be present in different variants, called *polymorphisms* (mostly single-nucleotide polymorphisms). To date, 3×10^6 gene polymorphisms are described – a number that increases daily and is estimated to be greater than 11×10^6 for the whole human genome. Each individual is composed of a pair of inherited combinations of variants for each of the 30 000 genes, explaining the enormous genetic diversity. Only a fraction of these gene polymorphisms is important to human health and disease, and to find the important ones and to predict the diseases is a major challenge for the next decades. The discovery of microarray technique (allowing relatively inexpensive study of all genes expressed in

the genome/transcriptome in parallel) is a great promise in this field.

Before proceeding to the description of the principles and design of microarray studies, let us introduce the major definitions in molecular cell biology and molecular medicine as functional genomics, proteomics, pharmacogenetics, and pharmacogenomics (Table 1).

Not all of the genes in human genome are used simultaneously. Depending on the developmental stage, age of the individual, cell type, organ, and environmental factors, a different set of genes is transcribed into mRNA. The mRNA expression analysis under defined conditions is called *functional genomics*. It allows a comparison of different sets of genes used in different conditions – in norm and pathology. Most studies that apply microarrays are functional genomic studies.

Not all of the transcribed genes result in proteins. Moreover, practically all proteins are modified after the first assembly of amino acids. It has been established that a protein derived from the same gene can be

Tab. 1 Major definitions in molecular cell biology and molecular medicine.

Term	Definition
Genomics	Analysis of gene expression of cell, tissue, or organ under given conditions
Proteomics	Analysis of proteins of cell, tissue, or organ under given conditions
Pharmacogenetics	Analysis of variability in drug responses attributed to hereditary factors in different populations
Pharmacogenomics	Analysis of genome and its products (RNA and proteins) related to drug response

altered in 10 to 20 different splits and 3D forms. Some proteins interact directly with DNA, leading to gene silencing. Almost all of the proteins interact with other proteins within biochemical pathways and form protein networks. The analysis of the proteins of a cell, tissue, or organ under given conditions is called *proteomics*. Recently, techniques similar to microarrays have been introduced in proteomic studies.

There is a long gap from the genotype to phenotype. If the genotype would automatically lead to a specific condition, so-called phenotype, all identical twins would have exactly the same homeostasis and diseases. Although they considerably resemble each other, they also differ in many ways. Thus, a 100% match is not there between genotype and phenotype because of environment–gene interactions. On the other hand, many conditions are a result of pathogenic mechanisms, and despite the different genetic makeup of the individuals, the same phenotype arises, resulting in development of one and the same disease (for example, hypertension, asthma, diabetes, etc.). In these so-called complex diseases, a constellation of different susceptibility and disease-modifying genes need to be present. Research has therefore failed to identify specific disease genes, for example, an "asthma gene," a "hypertension gene," a "diabetes genes," and so on. To understand the functional aspects of the disease better and to bridge the long gap between genotype and phenotype, it is necessary to combine functional genomic and proteomic analyses.

For complex questions (e.g. how genes contribute to normal human development, individual variability, and common diseases), there might be complex answers. Thus, it is necessary to apply techniques that permit complex answers. Therefore,

novel techniques are necessary to screen thousands of genes more rapidly and to generate new hypotheses. This is the role of high-throughput technology, like microarrays. Although this technology is new, it has already resulted in significant research advances and its use is spreading rapidly. As a hypothesis-generating approach, high-throughput methods can lead to the identification of a set of potentially interesting genes associated with a certain condition, so-called candidate genes. Microarray techniques, however, will not replace the classical hypothesis-driven research. Identified candidate genes have to be validated for their function and relevance by classical approaches. It is necessary to have in mind that each method of investigation has its limits and must be interpreted properly. For example, although several genes in a particular metabolic pathway may show changes at the mRNA level, the actual function of that pathway may not be affected if none of these genes is a rate-limiting factor. Therefore, these high-throughput genome-level technologies must be interpreted within the appropriate biological context.

This review summarizes the current state of microarray technology and presents examples of its application for translation of genome from laboratory to clinic. All about microarrays can be found in www.arrayit.com.

2
Microarray Technique – Historical Background and Basic Principles

DNA microarrays had first been described by Patrick Brown's Lab in 1995. Since that original report, the number of microarray publications has increased exponentially, from less than 150 publications in the

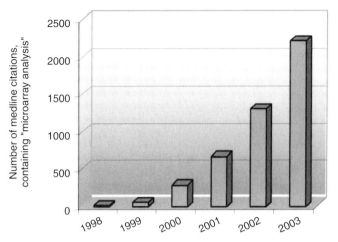

Fig. 1 Pubmed citations, containing "microarray analysis."

first 4 years (1995–1999) to about 2000 publications in 2003 (Fig. 1).

DNA microarrays – also called *gene chips* – consist of DNA attached to a surface in an ordered, predetermined fashion at extremely high density. More than 20 000 genes (almost a half of expressed human genes) can be arrayed on a surface with the size of a glass microscope slide. Microarrays allow the monitoring of gene expression for thousands of genes simultaneously in a single hybridization experiment because of two basic principles of nucleic acid hybridization: (1) DNA and RNA specifically interact with their complementary sequence attached on the array surface, and (2) this binding happens in proportion to the amount of mRNA (termed as *mRNA abundance*) in a mixture.

Although microarray technology has been originally designed for depositing DNA onto solid support for gene expression profiling (expression arrays), it has been extended to DNA copy number analysis (comparative genomic hybridization arrays), and, recently, a similar technology has been pioneered for protein and antibody spotting (Panorama Antibody Microarray Cell Signaling, www.sigma-aldrich.com).

Figure 2 provides an overview of a typical DNA microarray experiment from the selection of bacterial clones for manufacturing of the microarrays and nucleic acid extraction from tissue specimens or cell lines to *in silico* analysis, mining, and validation of the microarray data obtained. Optional steps, such as laser capture microdissection (LCM) and RNA amplification are also indicated.

The aim of the typical microarray experiment is to measure the amount of a given mRNA species (transcribed genes) of a tissue or cell type. For this purpose, the mRNA (isolated from the analyzed sample, e.g. drug-treated patients or cells – case sample, or untreated patients or cells – control sample) is usually transformed by reverse transcription into complementary DNA (cDNA), which is more stable. cDNA (or mRNA itself) from the sample is applied to the array surface and is further hybridized for several hours to the microarray. Each specific mRNA sequence (corresponding to the expressed gene) is able to bind only to a single complementary

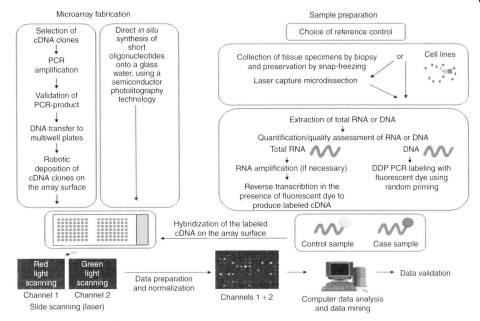

Fig. 2 Microarray study design.

oligonucleotide sequence attached on the array surface. If the mRNA sequence matches with the oligonucleotide sequence on a specific spot of the microarray, hybridization occurs. Direct information of how much of each gene is expressed in the cells is obtained by measurement of the amount of cDNA (corresponding to mRNA or expressed gene) bound to each spot on the array. The more abundant a given cDNA (or mRNA) is in a sample, the more it binds to its immobilized complement will occur on the array surface. The control and case samples are labeled with different fluorophores for detection. On each of the up to 450 000 spots, different binding intensities occur, depending on the concentration of the different cDNAs (equal to mRNAs or expressed genes) in the case and/or control samples tested. Thus, by comparison of the hybridization of cDNA from a control sample with that from a case sample, the relative amounts

of all the genes present on the array and expressed in both samples can be measured.

In contrast to the conventional Northern blot, which analyses 1, 2, or up to 20 mRNAs, a microarray technique allows the simultaneous analysis of the expression levels of hundreds, thousands, or even tens of thousands of genes in a single experiment. The latest chips carry up to 450 000 spots for the analysis of more than 20 000 genes and gene sequences on a small glass slide (1.2 × 1.2 cm).

3
Array Platforms and Fabrication

Production of arrays begins with the selection of the probes to be printed on the array surface. These are often chosen directly from gene databases (e.g. GenBank, dbESt, and UniGene – all available on http://www.ncbi.nlm.nih.gov). Two main methods are used to generate microarrays, and, currently, there are two main types of microarrays: cDNA arrays (also called *spotted arrays*) and oligonucleotide arrays (also called *GeneChips*).

3.1
cDNA-based (Spotted) Arrays – Advantages and Drawbacks

Spotted arrays had been originally developed in the Patrick Brown's Laboratory at Stanford University. They are composed of cDNA clones (selected from public databases and PCR-amplified) robotically arranged on a nonporous surface. A typical cDNA array is printed onto 30×15 mm glass microscope slide by a computer-controlled robotic cantilever arm; each spot is about 50 to 150 μm in diameter. cDNA clone sets of 15 000 to 20 000 mouse and 40 000 human genes are available for in-house arrays in addition to ready-made commercial chip sets. About 80% of all human genes are identified (the number increases daily) and can be queried on about two slides printed at high density. Other flexible and porous surfaces such as nylon membranes can be used as alternatives to glass. This approach allows the construction of arrays spotted with up to 80 000 spots.

The spotted arrays possess several advantages:

1. Affordability – they can be manufactured in-house.
2. Versatility – project-specific arrays (e.g. cancer arrays, immunoarrays, chromosome-specific arrays) can be custom designed.
3. Applicability to the simultaneous analysis of two different biological samples (e.g. treated vs untreated patient samples or cell lines, normal tissue vs malignant tissue, tumors of different

stages vs a common internal reference), using two or more different fluorophores.
4. Finally, they provide the opportunity for discovery of novel genes since expressed sequence tags of unknown function can be spotted on the DNA arrays in addition to well-characterized genes.

The major restriction of spotted arrays is their inability to measure absolute levels of gene expression in biological samples, a measurement that GeneChips can provide. Several drawbacks are related to clone collections selected for spotting:

1. Clone collections must be sequence verified, a process that can be costly and time consuming.
2. Clone annotation needs to be updated regularly as new information becomes available in public databases.
3. Clones must be periodically reamplified by polymerase chain reaction (PCR), as the spotting process consumes DNA.

3.2
Oligonucleotide-based Arrays (GeneChips and Codelink) – Advantages and Drawbacks

Affymetrix Inc. (Santa Clara, CA, USA) owns a registered trademark – GeneChip®, which refers to its high-density oligonucleotide-based DNA arrays. However, in some articles appearing in professional journals, popular magazines, and on the WorldWideWeb, the term "gene chip(s)" has been applied, generally referring to microarray technology. Affymetrix GeneChips® are manufactured by synthesizing of *short oligonucleotides* (about 25 bp) *in situ* directly onto a glass surface, using a modification of semiconductor photolithography technology.

These arrays provide the advantage of an integrated platform, in which analytical tools are provided to the user in addition to the arrays themselves. The choice of microarray platform remains largely a user's personal preference, although it is often directed by budget considerations and local access and availability. These privileges are the reason for Affymetrix ChenChips® to be widely used in cancer research. The main restrictions of Affymetrix GeneChips® are in the impossibility to compare two biological samples directly on the same array and in their high cost that can be prohibitive to small research laboratories.

Agilent Technology (Palo Alto, CA, USA) offers arrays based on *longer oligonucleotides*. They are manufactured *in situ* by use of ink-jet printing technology. Many investigators use longer (60–70 bp) *oligonucleotide arrays* (e.g. Codelink arrays, Amersham Bioscience, Piscataway, NJ, USA) that are made in a way similar to cDNA arrays. However, these longer oligonucleotide arrays have an advantage that they do not require PCR-amplification and clone-insert purification before arraying. Several technical issues are recurrent challenges, as the optimization of slide surface and oligonucleotide attachment, and the design of the most specific oligonucleotide probe(s) for each gene. Cross-platform comparisons often yield significantly nonoverlapping results, and the source of these discrepancies is unknown.

Other techniques for measurement of gene expression that involve large-scale sequencing (such as serial analysis of gene expression – SAGE, and massively parallel signature sequencing – MPSS) can be more sensitive. These methods involve highly efficient identification of the expressed genes by sequencing of short fragments and counting of cDNA clones that correspond to mRNA expressed in a particular cell or tissue. However, neither technique is as rapid as a microarray, and, therefore, they are less likely to be useful in clinical practice.

This review will focus mainly on spotted (cDNA) microarrays. Excellent reviews of Affymetrix technology are available in Bibliography.

3.3
Fabrication of Spotted Arrays

The spotted microarrays are most commonly fabricated on poly-L-lysine-coated microscope slides (or surfaces of similar dimension). Other surface coatings that are also available include various silane derivatives, acrylamide gels, nitrocellulose, and nylon. Some of the microarray surfaces require the use of modified oligonucleotides or PCR-products for efficient attachment, while others rely solely on molecular charge to maintain binding of nucleic acid to the surface. The DNA fragments are cross-linked by UV to the matrix. After fixation, residual amines on the slide surface react with succinic anhydride to reduce the positive charge at the surface.

Several commercial robotic machines are currently used for fabrication of spotted arrays, for example, MicroGrid (BioRobotics, Cambridge, UK), GMS-417 (Genetic Microsystems, Woburn, MA, USA), OmniGrid (GeneMachines, San Carlos, CA, USA), and PixSys PA series (Cartesian Technologies, Irvine, CA, USA).

Several techniques are applied to replace DNA probe on the slide surface: Pin-to-Pin, Pin-and-Ring, Quill or Slit Pin, and Jetting (Piezo Electric, Inc Jet, Bubble Jet).

The *pin-to-pin technology* (Fig. 3) has the longest history of use as a spotting tool. Briefly, a series of pins is aligned on a

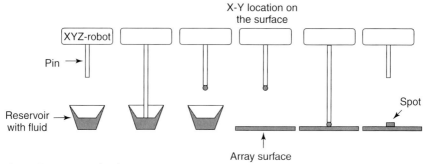

Fig. 3 Pin-to-pin technology.

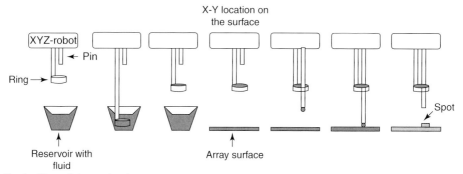

Fig. 4 Pin-and-ring technology.

head such that all the pins can be dipped into fluid reservoirs simultaneously (e.g. 96 pins, set in an array head, each pin at 9-mm center-to-center, for use with standard 96-well microplates). The pin-head module is attached to a 3D (x–y–z) robotic arm to move "into" and "out" of the wells, and to make contact with the spotting (array) surface. After dipping into the wells, the pins are lifted, and a small aliquot of fluid hangs on the tip of each pin by surface tension. These aliquots are transferred onto a slide surface through ensuring a pin–surface contact.

For fabrication of spotted microarrays on the microscopic slide surface, it is practically impossible to use more than 12 to 16 pins simultaneously, and the rate of the spotting is relatively slow. A crucial pint is the evaporation of the fluid aliquots during the time of transportation of the sample from the wells to the slide surface. This can provoke differences in the amount of fluid spotted and can introduce undesired variations into the arrays. Pin-to-pin alignment and consistency of pin dimensions are also both crucial points because of the necessity to place very small spots very close to each other, and to make at least a semiquantitative analysis of the hybridization to each spot.

The *pin-and-ring technology* (Fig. 4) has been specifically developed for fabrication of spotted DNA microarrays containing highly consistent elements. The key mechanical component consists of an open ring, which is oriented parallel to the wells (fluid reservoirs) and slide surface. The ring is fixed to a vertical rod, running perpendicular to the wells and slide surface.

A vertical pin is centered on the ring. Both the ring rod and the pin are attached to control devices so that each part can be moved separately in the z-axis. Both parts are kept in constant relation to each other in the $x-y$ plane. The spotting procedure includes several steps: the ring is dipped into the well (containing DNA solution) – the ring is small enough to fit easily into the well of either 96- or 384-well microplates; when the ring is lifted, it takes an aliquot of sample, which is held in the center of the ring by surface tension; the pin-ring device moves to any desired location in the $x-y$ plane and selects the place for spotting on the underlying surface; the pin is moved down through the ring; when the pin passes through the ring, a portion of the solution is transferred from the interior ring meniscus to the bottom of the pin – a pendant drop is formed on the pin tip; the pin continues to move down until the solution on the pin tip makes contact with the slide surface; the pin is then lifted; the combined forces of gravity and surface tension deposit the solution on the slide surface as a spot.

The pin-driving process can be repeated many times so that a very large number of similar spots can be created from a single moving ring. This consistency of spotting is the most important feature of the system. The spot size is dependent on pin dimensions and varies from 50 to 500 µm in diameter. It can be controlled by available hardware.

Special mechanisms have been designed to achieve a balance between the need to move the pin in z-axis rapidly, and to minimize the extent of contact between the pin and the slide surface. Thus, this technique is applicable for a wide variety of surface substrates – from solid glass to soft membranes and gels. The system appears to function well with a wide variety of fluids, ranging from volatile organic solvents to viscous aqueous solutions.

The *quills or slit pins* (Fig. 5) have been one of the most widely used devices for fabrication of spotted arrays. A quill is a solid pin with a slit at the end. The standard quill design looks very much like two tines of a fork, with a capillary between them. There are several variations of quills, but they all have the same principle of orientation. The quill is dipped into a fluid reservoir and the fluid is drawn up into the slit by capillary tension. The quill is moved to a desired position (in the $x-y$ plane) on the slide surface, then toward the slide surface (z-axis), and upon impact, the drop of fluid is placed on the surface. The quill

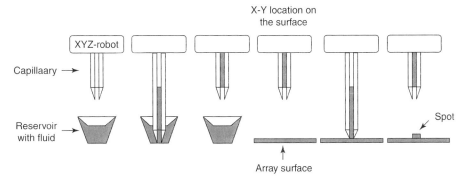

Fig. 5 Quill or slip pin technology.

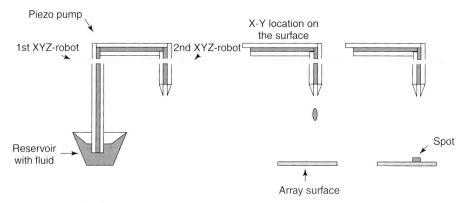

Fig. 6 Jetting technology.

is then withdrawn, moved to a subsequent location, and the process is repeated.

One of the advantages of the quill approach is that many quills can be attached to a single spotting head. One can spot 16 or more probes onto a surface simultaneously and without having to go back to the fluid reservoirs after each spotting. Therefore, this method enables rapid spotting. The quill has a relatively small void volume, and a little material is wasted. However, once the volume of the fluid in the quill becomes low, the amount of the spotted fluid begins to diminish with each subsequent spotting; this can compromise the quality of the arrays. The quill is also not highly effective for spotting of viscous solutions as a result of their difficult movement "into" and "out of" the capillary. The restriction of this technology is also due to the fact that it should be employed in a sterile environment because dust particles in the air or on the spotting surface can cause plugging of the quill. This characteristic of the quill limits its use to surfaces that are nonfibrous (as gels and membranes).

The *jetting technologies* (Fig. 6) had been originally developed for ink printing, and, nowadays, they are applied in microarray

fabrication. They are based on syringes or piezo pumps.

The syringes offer a possibility of extremely fast jetting rates and exact repeatable volume deposition. However, the minimum volume they can dispense is about two orders of magnitude greater than other spotting techniques described above.

The piezo pumps are unidirectional pumps attached to flexible capillary tubing. The end of the tubing needs to be transferred from fluid to fluid by a 3D (x–y–z) robot. A second 3D (x–y–z) robot moves the other end of the tubing to a location above the surface onto which a drop of fluid has to be deposit. The pumps work with supply and wash solutions.

The piezo pumps offer a high speed and an ability to operate in any environment. However, there are several limitations: they cannot work with fluids containing long molecules, as well as with viscous fluids, because the glass heads easily become clogged; the spotting head is made of fragile glass that needs frequent repairs; this method of deposition often gives artifacts known as "satellite" spots that can compromise the quantitative analysis; a large volume of fluid is necessary to operate, and the void volume of fluid is also high (100–500 µL). All these limitations make this technique less usable for microarray spotting.

4
Design of Microarray Experiment – Crucial Points

The most important factor for effective and successful microarray-based research projects is the careful experimental design, requiring a careful choice of the reference (control) source and unification of the experimental conditions. Even small variations in the conditions can induce significant changes in gene expression. For microarray analyses of human tissues, these variations include tissue-preservation methods, postmortem interval, dissection methods, and RNA quality. Frozen tissues usually provide the best material, although ethanol fixation can be also used. It is necessary to keep in mind that the differences in gene expression between samples may be due to differences in the genetic background (ethnicity).

Other general sources of variability include cDNA amplification methods, probe labeling, hybridization conditions, and washing procedures. However, when these issues are taken into account in the experimental design and proper technique is applied, microarray experiments provide reliable data on gene expression.

4.1
Choice of Reference (Control) Source

One of the most critical decisions involves the choice of reference (control) sample when using dual-channel DNA arrays.

Several limitations have to be borne in mind during the design of a microarray-based project, especially in the clinic.

For example, for studies in cancer etiology and progression, the better choice is to compare a tumor (case) sample with a healthy (control) sample directly. It is ideal to obtain the healthy tissue from the same patient (case–control study). However, this possibility is often problematic and restricts the feasibility of this type of design. In this case, it is better to choose

Tab. 2 Labeling protocols for microarray experiments (according to Macgregor, P.F. (2003)).

Amount of total RNA	Labeling mode	Protocols and kits
>10 µg total RNA	Direct labeling (no amplification required)	Bittner, M. (2003) www.microarrays.ca
4–10 µg total RNA	Indirect labeling (no amplification required)	Randolph, J. et al. (1997) www.microarrays.ca
<10 µg total RNA	Amplification required before labeling:	
	-PCR-amplification	Vernon, S. et al. (2000) Livesey, F. et al. (2000) BD Atlas SMART Fluorescent Probe Amplification
	-Linear amplification	Eberwine, J. et al. (1992) Wang, E. et al. (2000) Arcturus RiboAmp Ambion MessageAmp aRNA

a comparison between tissue specimens obtained from the same patient before and after chemotherapy. This design is also limited by the impossibility to obtain sufficient amount of high-quality necrosis-free tumor tissue after chemotherapy. Because of these reasons, there is now a global trend to use a common internal reference in large microarray projects as the investigation of gene expression profiles of tumor specimens obtained from many different patients. This design increases the challenge to obtain global expression profiling across a large sample cohort, and it is already successfully applied by several groups. Commercially available reference RNA is often a good choice because of wide-gene-expression representation (e.g. Stratagene – La Jolla, CA, USA, and BD Bioscience Clontech – Palo Alto, CA, USA) and virtually unlimited availability. Finally, this approach allows easier comparison of microarray data between different research groups, thus providing the possibility of more global studies.

4.2
Labeling and Hybridization

Once the choice of reference source is made and the number of replicates decided, the actual labeling protocol depends predominantly on the amount of sample RNA (Table 2).

In studies using cell lines, RNA availability is usually not a problem. However, great care should be taken to ensure maximum reproducibility and stability in cell culture conditions, especially if RNA has to be isolated from cells during a long time interval.

In studies using human tumor tissues, RNA availability can be problematic because of the impossibility to obtain a sufficient amount of sample tissues every time.

Despite the amount of sample RNA, its high quality (degradation-free RNA) is crucial for successful microarray experiments. Assessment of RNA quality has been greatly facilitated by the use of microcapillary-based devices (e.g. Agilent Bioanalyzer, Agilent Technologies, Palo Alto, CA, USA), which can evaluate RNA quality and quantity using a little amount (~200 ng) of total RNA. More traditional techniques (such as agarose gel electrophoresis and spectroscopy) can also be used when RNA quantity is not limiting.

There are basically two labeling techniques, depending on the amount of RNA prepared for microarray analysis: direct and indirect labeling.

If the amount of total RNA available is in the range of 10 to 20 µg per experiment, direct labeling is the simplest choice. Direct labeling involves direct incorporation of Cy3- or Cy5-labeled dCTP during reverse transcription of RNA to cDNA. Usually 10 µg of total RNA are sufficient to obtain high-quality microarray signals using direct labeling.

If the amount of total RNA is slightly lower (in the range of 4–10 µg per experiment), indirect labeling can be used. In this case, the labeling proceeds in two steps: (1) a modified nonfluorescent nucleotide (e.g. amino-allyl dUTP) is incorporated into cDNA during the reverse transcription step, and (2) subsequent coupling to fluorophores is performed in a separate reaction.

Detailed protocols for direct and indirect labeling can be found in www.microarrays.ca.

If the amount of the sample RNA is only submicrograms (e.g. from limiting clinical samples, such as fine needle biopsies or LCM specimens), RNA amplification is

Tab. 3 Microarray data analysis software (according to Macgregor, P.F. (2003)).

Software package	Company	Web site
Acuity[R]	Axon Instruments (commercial)	www.axon.com/GN_Acuity.html
Array-Pro Analyzer	Media Cybernetics (commercial)	www.mediacy.com/arraypro.htm
ArrayStat[TM]	Imaging Research (commercial)	www.imagingresearch.com/products/AST.asp
DNA-Chip Analyser	Wing Wong and Cheng Li of the Harvard School of Public Health (free)	www.dchip.org/
F-Scan	National Institute of Health, Bethesda, MD (free)	http://abs.cit.nih.gov/fscan/
Gene Cluster	Whitehead Institute for Genomic Research (free)	www.genome.wi.mit.edu/cancer/software/software.html
GeneLinker[TM] Gold	Molecular Mining Corporation (commercial)	http://microarray.genelinker.com/products.htm#GeneLinkerGold
GeneSight[TM]	Biodiscovery (commercial)	www.biodiscovery.com/genesight.asp
GeneSpring	Silicon Genetics (commercial)	www.silicongenetics.com/cgi/SiG.cgi/Products/GeneSpring/index.smf
Genesis	Institute for Biomedical Engineering, Graz University of Technology (IBMT-TUG) (free)	http://genome.tugraz.at/Software/GenesisCenter.html
GeneTraffic[TM]	Iobion Informatics, Stratagene (commercial)	www.iobion.com
J-express Pro	MolMine (commercial)	www.molmine.com/frameset/frm_jexpress.htm
MAExplorer	National Cancer Institute, Laboratory of Experimental and Computational Biology (free)	www.lecb.ncifcrf.gov/MAExplorer/
Partek Software Suites	Partek (commercial)	www.partek.com/html/products/products.html
TIGR Array Viewer	The Institute for Genomic Research (free)	www.tigr.org/software
Xpression NTI[TM]	Informax Inc. (commercial)	www.informaxinc.com/solutions/expression/index.html

required before to proceed to the labeling. This strategy now enables the molecular profiling of biological samples down to few cells. Two amplification approaches have been used and commercial kits are available: PCR-based amplification (e.g. from BD Bioscience Clontech) and linear amplification (e.g. from Arcturus, Mountain View, CA, USA). In the case of necessity of RNA amplification, microarray results need more extensive validation. Moreover, the use of RNA amplification significantly increases the overall time and cost of molecular profiling of the samples.

After fluorophore incorporation, the fluorescent cDNA sample (probe) is hybridized to the microarray (target) at a temperature determined by the hybridization buffer used, in the presence of nonspecific competitors (e.g. yeast tRNA and salmon sperm DNA). The microarray is then washed under stringent conditions to remove nonspecific binding and is air-dried. The dried array is then scanned using a confocal laser scanner and the images obtained are quantified using image acquisition software.

The labeling and hybridization procedures, described above, are typical for the expression arrays.

A protocol very similar to the one used for expression arrays can be used for comparative genomic hybridization (CGH) arrays. In CGH arrays, the genetic material hybridized to the array is not RNA but genomic DNA, and the purpose is to investigate changes in gene copy numbers in tumor samples. Genomic DNA is extracted from tumor samples or normal tissue, digested with a restriction enzyme, such as EcoR1, and fluorescently labeled with Cy3 or Cy5 using random priming. Labeled DNAs are cohybridized to the microarray in the presence of blocking agent, usually Cot-1 DNA, yeast tRNA, and poly(dA-dT).

After hybridization and stringent washes, the arrays are scanned and quantified in a process essentially identical to that described for expression arrays. CGH array experiments can be carried out using the same spotted arrays applied for expression analysis, thus providing the opportunity of a direct comparison on the same genes between gene copy numbers and expression levels. Alternatively, BAC arrays can be used, either manufactured in-house or commercially available, for example, from Spectral Genomics (Houston, TX, USA).

5
Data Analysis, Validation, and Interpretation

Perhaps the most underappreciated challenge facing researchers working on microarray projects is the extremely large quantity of data generated by this technology and the choice of an appropriate and effective database and data analysis system. In the early days of microarray technology, the basic data analysis was carried out using standard spread-sheet software, such as Microsoft Excell, or programs developed in-house. At present, there are a number of companies offering software for microarray data analysis. This is either offered free of charge on the Internet by academic institutions or it is commercially available (Table 3).

5.1
Normalization of Microarray Data

The first step in microarray data processing and analysis is the normalization. Normalization is used to combine the data from multiple microarray experiments into a single data set for analysis. It is absolutely necessary to account for and to minimize the artifacts and experimental variations in the calculation of gene expression ratios. The normalization is a complex and often controversial issue in microarray data analysis, and a discussion of the different strategies is beyond the scope of this review. Briefly, there are two main approaches to normalization: (1) *intensity-dependent normalization*, in which a normalization factor is calculated on the basis of all spots on the array, and (2) *intensity-independent normalization*, which uses normalization function to equalize both channels (as Locally Weighted Scatter Plot Smoother, LOWESS – http://stat-www.berkeley.edu /users/terry/zarry/Html/). The results obtained after normalization are generally displayed as normalized expression ratios, where each row represents one spot on the array and each column represents one hybridization experiment. If some genes are represented by several spots on the array, it is possible to combine and average the expression ratios obtained for these different spots.

5.2
Statistical Data Analysis

In traditional data analysis, there are few variables and great number of replicates for each data-point. In microarray data analysis, there are many variables and a small number of replicates for each data-point. Microarray data analysis is typically carried out by one of two approaches – supervised or unsupervised (or a combination of both).

In the *unsupervised statistical analysis*, the data are analyzed without *a priori* assumption about the identity, characteristics, or clinical attributes of the specimens analyzed. Typical examples of unsupervised analysis in microarray data are 2D hierarchical clustering (software is now available – http://rana.lbl.gov), K-means clustering, self-organization maps (SOM), and binary tree–structured vector quantization (BTSVQ). Unsupervised clustering of microarray data is widely used in tumor classification, for instance, in leukemia, lung cancer, ovarian cancer, colon cancer, prostate cancer, or breast cancer.

The *supervised statistical analysis* of microarray data integrates some of the biological and clinical attributes available for the specimens analyzed. Typical examples of supervised analysis are significance analysis of microarrays (SAM) and prediction analysis of microarrays (PAM). Supervised data analysis is widely used for identification of genes capable of stratifying tumor samples based on one particular attribute, such as clinical outcome or response to treatment.

Detailed explanation of microarray data normalization and statistical analysis can be found in several books and reviews cited in Bibliography.

5.3
Microarray Data Validation

Microarray data analysis is a multistep process that uses sophisticated statistical analysis tools and is prone to errors and sometimes to overinterpretations. In addition, quality issue in the raw microarray data, for example, high background or low reproducibility, as well as nonlinearity of microarray signal will significantly impact on the final results. Finally, with spotted arrays, misidentification of probes can also lead to errors. Validation of gene expression data by conventional techniques is essential. Usually, the validation of microarray data is carried out first at the RNA level by Northern blot, semiquantitative PCR, real-time quantitative (RT)-PCR, or *in situ* hybridization (ISH) on tissue sections or high-density microarrays (TMA). The microarray data can be also validated on the protein level by immunoblot analysis. Owing to the nonquantitative nature of immunoblotting, validation by other methods is required. An excellent example of such an extensive validation is presented in the study of Beer and colleagues, where the authors validated their microarray results both at the RNA level by Northern blot and at the protein level by IHC on tissue microarrays.

The development of high-throughput screening and validation technologies is an important priority because the number of genes identified through microarray studies rises exponentially.

5.4
Data Mining and Interpretation

Once gene data sets are identified and validated, data mining and interpretation is the ultimate challenge in microarray studies. It is not a trivial task to determine which of the hundreds of potentially differentially expressed genes identified are biologically relevant to the problem being investigated. Several bioinformatics tools and public databases provide information on the biological function, cellular localization, and gene ontology of the selected gene of interest: GenBank (www.ncbi.nlm.nih.gov), UniGene (http://ncbi.nlm.nih.gov/UniGene), PubMed (www.pubmed.com), US National Center for Biotechnology Information (NCBI) Databases (www.ncbi.nlm.nih.gov), Stanford Online Universal Resource for Clones and ESTs

(SOURCE) (http://source.stanford.edu), SWISS SPOT (http://ca.expasy.org/sprot/). Another computer application of interest is GenMAPP, available online at www.genmapp.org. GenMAPP allows a visualization of gene expression data such as fold changes in expression ratios on maps representing biological pathways. Similarly, Onto-Express (www.openchannelfoundation.org/projects/Onto-Express/) correlates the identified genes with their functional and biological characteristics. Finally, literature searches can be greatly facilitated by using Internet text-mining tools, such as MedMiner, available at http://discover.nci.nih.gov/textmining/filters.html.

However, it is important to keep in mind that the complexity of the cellular machinery is extraordinary, and it is impossible to provide a biological explanation for every one of the genes identified by microarray analysis only with one simple click on the computer mouse. In addition, incorrect gene annotation in databases is a significant problem. Even when the genes identified in microarray analysis are correctly annotated, fully identified, and have a known function, many of these genes will belong to several biological pathways involved both in normal cellular processes and pathogenesis. In this context, downstream functional analysis using classical molecular biology approaches will remain paramount for the thorough interpretation of microarray findings.

6
Major Technical Restrictions

In order to perform genome-wide expression profiling, it is necessary to have *genome-wide microarrays*. In the case of mouse and human, such whole genome arrays are on the horizon, although not yet available. This situation is likely to be resolved in the next two to three years, given the maturity of the genome and EST-sequencing projects in the mouse and humans. Certainly, the number of genes that can be investigated using microarrays increases all the time. It is likely that improvements in deposition or synthesis of DNA probes will allow a suitable density to be achieved, avoiding the requirement of hybridizing multiple arrays in order to achieve full genome coverage. The advent of spotted oligonucleotide arrays and related technologies also opens up the possibility of exon-specific probes for the detection of tissue-specific (or cell-specific) splice variants.

Perhaps one of the basic restrictions in the use of microarrays is *the limiting amount of RNA* available from standard tissue or cell specimens. The embryonic dissections used in developmental biology are a good example. A whole 7.5-dpc mouse embryo yields ~0.5 µg of total RNA. Usually, the conventional microarray hybridization analysis requires 5 to 100 µg of total RNA. It is clear that specific embryonic tissues or organs, especially from early stages, will yield only a fraction of the required target RNA. Embryo-sorting techniques have been employed, but target amplification appears to be the most widely applicable solution. A number of amplification protocols have been described that allow the generation of sufficient target from small amounts of staring RNA. It is not clear which protocol offers the "best" solution of the problem of limiting RNA because different scientists have different criteria for the assessment, including simplicity, reliability, and cost.

The last category of potential technical limitations of microarray analysis is closely related to *the possibility for identification of those incredibly rare transcripts* that encode those critical regulators of cellular homeostasis and development. A combination of variations in transcript abundance and cellular heterogeneity makes it difficult to perform expression profiling in standard tissue explants in a useful fashion. Subdissection and cell sorting are usually necessary; this makes labeling and hybridization practically impossible.

7
Application of Microarrays in Molecular Cell Biology and Molecular Medicine

What can be measured by microarrays?

Microarray technology can be used for three main applications:

1. Gene expression profiling – mRNA extracted from a biological sample is applied to the microarray. The result reveals the level of expression of tens of thousands of genes in the sample. This result is known as a gene expression "profile" or "signature."
2. Genotyping – DNA, extracted from a biological sample, is amplified by a PCR and applied to the microarray. The genotype for hundreds or thousands of genetic markers across the genome can be determined in a single experiment. This approach has considerable potential in disease risk assessment, both in research and clinical practice.
3. DNA sequencing – DNA extracted from a biological sample is amplified and applied to specific "sequencing" microarrays. Thousands of base pairs

of DNA can be screened on a single microarray for polymorphisms in specific genes whose sequence is already known. This greatly increases the scope for precise molecular diagnosis in single gene and genetically complex diseases.

In the medical practice, the precise diagnosis and effective treatment of disease depend on the ability of specialists to recognize the recurring constellations of clinical signs and symptoms that permit meaningful classification of the diseases. Unfortunately, the clinical signs and symptoms or even blood and pathological investigations are poor predictors of the clinical outcome or response to therapy. In the postgenomic era, studying of gene expression profiling will ultimately determine the biological behavior of both normal and injured tissues and cells, which may provide insights into disease mechanisms and help to identify novel candidates for therapeutic intervention. Gene profiling may also identify markers useful in diagnostic and prognostic purposes. Since thousands of genes are simultaneously and quantitatively analyzed on microarrays, this technology provides a resolution and precision not previously possible. Therefore, molecular phenotyping of diseases through gene expression profiling and sequencing may offer diagnostic, prognostic, and mechanistic insights that improve management of human diseases. Microarray technology has also a dramatic impact in the fundamental biological studies, providing an effective way to assess globally the transcriptional effects of specific genetic and pharmacological interventions, thus rapidly identifying possible up- and downstream effectors or alternative signaling pathways.

Bibliography

Books and Reviews

Beheshti, B., Park, P.C., Braude, I., Squire, J.A. (2002) Microarray CGH, *Methods Mol. Biol.* **204**, 191–207.

Bilban, M., Buehler, L.K., Head, S., Desoye, G., Ouaranta, V. (2002) Normalizing DNA microarray data, *Curr. Issues Mol. Biol.* **4**, 57–64.

Bittner, M. (2003) Fluorescent labeling of first-strand cDNA using reverse transcriptase, in: Botwell, D., Sambrook, J. (Eds.) *DNA Microarrays: a Molecular Cloning Manual*, Cold Spring Harbor Laboratory Press, Cold Spring Harbor, New York, pp. 178–186.

Cook, S.A., Rosenzweig, A. (2002) DNA microarrays: Implications for cardiovascular medicine, *Circ. Res.* **91**, 559–564.

Eyster, K.M., Lindahl, R. (2001) Molecular medicine: a primer for clinicians. Part XII: DNA microarrays and their application to clinical medicine, *S D J Med.* **54**, 57–61.

Geschwind, D.H. (2000) Mice, microarrays, and the genetic diversity of the brain, *Proc. Natl. Acad. Sci. U.S.A.* **97**, 10676–10678.

Geschwind, D.H. (2003) DNA microarrays: translation of the genome from laboratory to clinic, *Lancet Neurol.* **2**, 275–282.

Geschwind, D.H., Gregg, J. (Eds.) (2002) *Microarrays for the Neurosciences*, MIT Press, Cambridge, MA.

Goldsmith, Z.G., Dhanasekaran, N. (2004) The microevolution: applications and impacts of microarray technology on molecular biology and medicine, *Int. J. Mol. Med.* **13**, 483–495.

Greenberg, S.A. (2002) Bioinformatics in microarray expression analysis technology, in: Warrington, J.A., Todd, R., Wong, D. (Eds.) *Microarray and Cancer Research*, BioTechniques Press, Westborough, MA, pp. 25–42.

Hardiman, G. (2004) Microarray platforms – comparisons and contrasts, *Pharmacogenomics* **5**, 487–502.

Harkin, D.P. (2000) Uncovering functionally relevant signaling pathways using microarray-based expression profiling, *Oncologist* **5**, 501–507.

Harrington, C.A., Rosenow, C., Retief, J. (2000) Monitoring gene expression using DNA microarrays, *Curr. Opin. Microbiol.* **3**, 285–291.

Hartigan, J.A. (1975) *Clustering Algorithms*, John Wiley and Sons, New York.

Jain, K.K. (2004) Application of biochips: from diagnostics to personalized medicine, *Curr. Opin. Drug. Discov. Devel.* **7**, 287–289.

Joos, L., Eryuksel, E., Brutsche, M.H. (2003) Functional genomics and gene microarrays – the use in research and clinical medicine, *Swiss Med. Wklv.* **133**, 31–38.

Kohane, I.S., Kho, A., Butte, A.J. (2002) *Microarrays for an Integrative Genomics*, MIT Press, Cambridge, MA.

Leung, Y.F., Cavalieri, D. (2003) Fundamentals of cDNA microarray data analysis, *Trends Genet.* **19**, 649–659.

Liefers, G.J., Tollenaar, R.A. (2002) Cancer genetics and their application to individualized medicine, *Eur. J. Cancer* **38**, 872–879.

Lipshutz, R.J., Fodor, S.P., Gingeras, T.R., Lockhart, D.J. (1999) High density synthetic oligonucleotide arrays, *Nat. Genet.* **21**(Suppl. 1), 20–24.

Macgregor, P.F. (2003) Gene expression in cancer: the application of microarrays, *Expert Rev. Mol. Diagn.* **3**, 185–200.

Marcotte, E.R., Srivastava, L.K., Ouirion, R. (2003) cDNA microarray and proteomic approaches in the study of brain diseases: focus on schizophrenia and Alzheimer's disease, *Pharmacol. Ther.* **100**, 63–74.

Mengel, M., Kreipe, H., von Wasielewski, R. (2003) Rapid and large-scale transition of new tumor biomarkers to clinical biopsy material by innovative tissue microarray systems, *Appl. Immunohistochem. Mol. Morphol.* **11**, 261–268.

Nadon, R., Shoemaker, J. (2002) Statistical issues with microarrays: processing and analysis, *Trends Genet.* **18**, 265–271.

Noordewier, M.O., Warren, P.V. (2001) Gene expression microarrays and the integration of biological knowledge, *Trends Biotechnol.* **19**, 412–415.

Panda, S., Sato, T.K., Hampton, G.M., Hogenesch, J.B. (2003) An array of insights: application of DNA chip technology in the stuffy of cell biology, *Trends Cell Biol.* **13**, 151–156.

Petrocoin, E.F., Hackett, J.L., Lesko, L.J., Puri, R.K., Gutman, S.I., Chumakov, K., Woodcock, J., Feigal, D.W., Zoon, K.C., Sistare, F. (2002) Medical applications of microarray technologies: a regulatory science perspective, *Nat. Genet.* **32**(Suppl. 2), 474–479.

Quackenbush, J. (2002) Microarray data normalization and transformation, *Nat. Genet.* **32**(Suppl. 2), 496–501.

Smith, L., Greenfield, A. (2003) DNA microarrays and development, *Hum. Mol. Genet.* **12**, R1–R8.

Tineke, C.M.T., van der Pouw Kraan, T.C., Kasperkovitz, P.V., Verbeet, N., Verweij, L. (2004) Genomics in the immune system, *Clin. Immunol.* **111**, 175–185.

Van Deerlin, V., Ginsberg, S.D., Lee, V.M., Trojanowski, J.G. (2002) The use of fixed human postmortem brain tissue to study mRNA expression in neurodegenerative diseases: application of microdissection and amplification, in: Geschwind, D.H., Gregg, J. (Eds.) *Microarrays for the Neuroscience: an Essential Guide*, MIT Press, Cambridge, MA, pp. 201–236.

Velculescu, V.E., Vogelstein, B., Kinzler, K.W. (2000) Analyzing uncharted transcriptomes with SAGE, *Trends Genet.* **16**, 423–425.

Warrington, J.A., Todd, R., Wong, D. (Eds.) (2002) *Microarrays in Cancer Research*, BioTechniques Press, Westborough, MA.

Weinstein, J.N., Scherf, U., Lee, J.K., Nishizuka, S., Gwadry, F., Bussey, A.K., Kim, S., Smith, L.H., Tanabe, L., Richman, S., Alexander, J., Kouros-Mehr, H., Maunakea, A., Reinhold, W.C. (2002) The bioinformatics of microarray gene expression profiling, *Cytometry* **47**, 46–49.

Primary Literature

Alizadeh, A.A., Eisen, M.B., Davis, R.E., Ma, C., Lossos, I.S., Rosenwald, A., Boldrick, J.C., Sabet, H., Tran, T., Yu, X., Powell, J.I., Yang, L., Marti, G.E., Moore, T., Hudson, J., Jr., Lu, L., Lewis, D.B., Tibshirani, R., Sherlok, G., Chan, W.C., Greiner, T.C., Weisenburger, D.D., Armitage, J.O., Warnke, R., Levy, R., Wilson, W., Grever, M.R., Byrd, J.C., Botstein, D., Brown, P.O., Stuardt, L.M. (2000) Distinct types of diffuse large B-cell lymphoma identified by gene expression profiling, *Nature* **403**, 503–511.

Alon, U., Barkai, N., Notterman, D.A., Gish, K., Ybarra, S., Mack, D., Levine, A.J. (1999) Broad patterns of gene expression revealed by clustering analysis of tumor and normal colon tissue probed by oligonucleotide arrays, *Proc. Natl. Acad. Sci. U.S.A.* **96**, 6745–6750.

Bayani, J., Brenton, J.D., Macgregor, P.F., Beheshti, B., Albert, M., Nallainathan, D., Karaskova, J., Rosen, B., Murphy, J., Laframboise, S., Zanke, B., Squire, J.A. (2002) Parallel analysis of sporadic primary ovarian carcinomas by spectral karyotyping, comparative genomic hybridization and expression microarrays, *Cancer Res.* **62**, 3466–3476.

Beer, D.G., Kardia, S.L., Huang, C.C., Giordano, T.J., Levin, A.M., Misek, D.E., Lin, L., Chen, G., Gharib, T.G., Thomas, D.G., Lizyness, M.L., Kuick, R., Hayasaka, S., Taylor, J.M., Iannattoni, M.D., Orringer, M.B., Hanash, S. (2002) Gene-expression profiles predict survival of patients with lung adenocarcinoma, *Nat. Med.* **8**, 816–824.

Bhattacharjee, A., Richards, W.G., Staunton, J., Li, C., Monti, S., Vasa, P., Ladd, C., Beheshti, J., Bueno, R., Gillette, M., Loda, M., Weber, G., Mark, E.J., Lander, E.S., Wong, W., Johnson, B.E., Golub, T.R., Sugarbaker, D.J., Meyerson, M. (2001) Classification of human lung carcinomas by mRNA expression profiling reveals distinct adenocarcinoma subclasses, *Proc. Natl. Acad. Sci. U.S.A.* **98**, 13790–13795.

Bittner, M., Meltzer, P., Chen, Y., Jiang, Y., Seftor, E., Hendrix, M., Radmacher, M., Simon, R., Yakhini, Z., BenDor, A., Sampas, N., Dougherty, E., Wang, E., Marincola, F., Gooden, C., Lueders, J., Glatfelter, A., Pollock, P., Carpten, J., Gillanders, E., Leja, D., Dietrich, K., Beaudry, C., Berens, M., Alberts, D., Sondak, V. (2000) Molecular classification of cutaneous malignant melanoma by gene expression profiling, *Nature* **406**, 536–540.

Brenner, S., Johnson, M., Bridgham, J., Golda, G., Lloyd, D.H., Johnson, D., Luo, S., McCurdy, S., Foy, M., Ewan, M., Roht, R., George, D., Eletr, S., Albrecht, G., Vermaas, E., Williams, S.R., Moon, K., Burcham, T., Pallas, M., DuBridge, R.B., Kirchner, J., Fearon, K., Mao, J., Corcoran, K. (2000) Gene expression analysis by massively parallel signature sequencing (MPSS) on microbead arrays, *Nat. Biotechnol.* **18**, 630–634.

Brutsche, M.N., Brutsche, I.C., Wood, P., Brass, A., Morrison, N., Rattay, M., Mogulkoc, N., Simler, N., Craven, M., Custovic, A., Egan, J.J., Woodcock, A. (2001) Apoptosis signals in atopy and asthma measured with cDNA arrays, *Clin. Exp. Immunol.* **124**, 181–187.

Chiang, L.W., Grenier, J.M., Ettwiller, L., Jenkins, L.P., Ficenec, D., Martin, J., Jin, F., DiStefano, P.S., Wood, A. (2001) An orchestrated gene expression component of neuronal programmed cell death revealed by cDNA array analysis, *Proc. Natl. Acad. Sci. U.S.A.* **98**, 2814–2819.

Dahlquist, K.D., Salomonis, N., Vranizan, K., Lawlor, S.C., Conklin, B.R. (2002) GenMAPP, a new tool for viewing and analyzing microarray data on biological pathways, *Nat. Genet.* **31**, 19–20.

Datson, N.A., Van der Perk, J., DeKloet, E.R., Vreugdenhil, E. (2001) Expression profile of 30,000 genes in rat hippocampus using SAGE, *Hippocampus* **11**, 430–444.

De Risi, J., Penland, L., Brown, P.O., Bittner, M.L., Meltzer, P.S., Ray, M., Chen, Y., Su, Y.A., Trent, J.M. (1996) Use of a cDNA microarray to analyze gene expression patterns in human cancer, *Nat. Genet.* **14**, 457–460.

De Vos, J., Thykjaer, T., Tarte, K., Ensslen, M., Raynaud, P., Requirand, G., Pellet, F., Pantesco, V., Reme, T., Jourdan, M., Rossi, J.F., Orntoft, T., Klein, B. (2002) Comparison of gene expression profiling between malignant and normal plasma cells with oligonucleotide arrays, *Oncogene* **21**, 6848–6857.

Dhanasekaran, S.M., Barrette, T.R., Ghosh, D., Shah, R., Varambally, S., Kurachi, K., Pienta, K.J., Rubin, M.A., Chinnaiyan, A.M. (2001) Delineation of prognostic biomarkers in prostate cancer, *Nature* **412**, 822–826.

Dixon, A.K., Richardson, P.J., Lee, K., Carter, N.P., Freeman, T.C. (1998) Expression profiling of single cells using 3′ and amplification (TPEA) PCR, *Nucleic Acids Res.* **26**, 4426–4431.

Eberwine, J., Yeh, H., Miyashiro, K., Cao, Y., Nair, S., Finnell, R., Zettel, M., Coleman, P. (1992) Analysis of gene expression in single live neurons, *Proc. Natl. Acad. Sci. U.S.A.* **89**, 3010–3014.

Euer, N., Schwirzke, M., Evtimova, V., Burtscher, H., Jarsch, M., Tarin, D., Weidle, U.H. (2002) Identification of genes associated with metastasis of mammary carcinoma in metastatic versus non-metastatic cell lines, *Anticancer Res.* **22**, 733–740.

Furlong, E.E., Andersen, E.C., Null, B., White, K.P., Scott, M.P. (2001) Patterns of gene expression during Drosophila mesoderm development, *Science* **293**, 1629–1633.

Garber, M.E., Troyanskaya, O.G., Schluens, K., Petersen, S., Thaesler, Z., Pacyna-Gengelbach, M., van de Rijn, M., Rosen, G.D., Peron, C.M., Whyte, R.I., Altman, R.B., Brown, P.O., Botstein, D., Petersen, I. (2001) Diversity of gene expression in adenocarcinoma of the lung, *Proc. Natl. Acad. Sci. U.S.A.* **98**, 13784–13789.

Golub, T.R., Slonim, D.K., Tamayo, P., Huard, C., Gaasenbeek, M., Mesiroy, J.P., Coller, H., Loh, M.L., Downing, J.R., Caligiuri, M.A., Bloomfield, C.D., Lander, E.S. (1999) Molecular classification of cancer: class discovery and class prediction by gene expression monitoring, *Science* **286**, 531–537.

Haab, B.B., Dunham, M.J., Brown, P.O. (2001) Protein microarrays for parallel detection and quantitation of specific proteins and antibodies in complex solutions, *Genome Biol.* **2**, Research2004.

Hedenfalk, I., Duggan, D., Chen, Y., Radmacher, M., Bittner, M., Simon, R., Meltzer, P., Gusterson, B., Esteller, M., Kallioniemi, O.P., Wilfond, B., Borg, A., Trent, J. (2001) Gene-expression profiles in hereditary breast cancer, *N. Engl. J. Med.* **344**, 539–548.

Hughes, T.R., Mao, M., Jones, A.R., Burchard, J., Marton, M.J., Shannon, K.W., Lefkowitz, S.M., Ziman, M., Schelter, J.M., Meyer, M.R., Kobayashi, S., Davis, C., Dai, H., He, Y.D., Stephaniants, S.B., Cavet, G., Walker, W.L., West, A., Coffey, E., Shoemaker, D.D., Stoughton, R., Blanchard, A.P., Friend, S.H., Linsley, P.S. (2001) Expression profiling using microarrays fabricated by an ink-jet oligonucleotide synthesizer, *Nat. Biotechnol.* **19**, 342–347.

Iscove, N.N., Barbara, M., Gu, M., Gibson, M., Modi, C., Winegarden, N. (2002) Representation is faithfully preserved in global cDNA amplified exponentially from sub-picogram quantities of mRNA, *Nat. Biotechnol.* **20**, 940–943.

Jazaeri, A.A., Yee, C.J., Sotiriou, C., Brantley, K.R., Boyd, J., Liu, E.T. (2002) Gene expression profiles of BRCA1-linked, BRCA2-linked and sporadic ovarian cancers, *J. Natl. Cancer Inst.* **94**, 990–1000.

Kacharmina, J.E., Crino, P.B., Eberwine, J. (1999) Preparation of cDNA from single cells and subcellular regions, *Methods Enzymol.* **303**, 3–18.

Karsten, S.L., Geschwind, D.H. (2002) Gene expression analysis using cDNA microarrays, *Curr. Protoc. Neurosci.* Unit 4.25.

Karsten, S.L., Van Deerlin, V.M., Sabatti, C., Gill, L.H., Geschwind, D.H. (2002) An evaluation of tyramide signal amplification and archived fixed and frozen tissue in microarray gene expression analysis, *Nucleic Acids Res.* **30**, E4.

Livesey, F.J., Furukawa, T., Steffen, M.A., Church, G.M., Cepko, C.M. (2000) Microarray analysis of the transcriptional network controlled by the photoreceptor homeobox gene Crx, *Curr. Biol.* **10**, 301–310.

Lukasiuk, K., Pitkannen, A. (2004) Large-scale analysis of gene expression in epilepsy research: is synthesis already possible? *Neurochem. Res.* **29**, 1169–1178.

Luo, H., Salunga, R.C., Guo, H., Bittner, A., Joy, K.C., Galindo, J.E., Xiao, H., Rogers, K.E., Jackson, M.R., Erlander, M.G. (1999) Gene expression profiles of laser-captured adjacent neuronal subtypes, *Nat. Med.* **5**, 117–122.

Notterman, D.A., Alon, U., Sierk, A.J., Levine, A.J. (2001) Transcriptional gene expression profiles of colorectal adenoma, adenocarcinoma and normal tissue examined by oligonucleotide arrays, *Cancer Res.* **61**, 3124–3130.

Novoradoyskaya, N., Whitfield, M.L., Basehore, L.S., Novoradovsky, A., Pesich, R., Usary, J., Karaca, M., Wong, W.K., Aprelikova, O., Fero, M., Perou, C.M., Botstein, D., Braman, J. (2004) Universal reference RNA as a standard for microarray experiments, *BMC Genomics* **5**, 20.

Perou, C.M., Jeffrey, S.S., Van de Rijn, M., Rees, C.A., Eisen, M.B., Ross, D.T., Pergamenschikov, A., Williams, C.F., Zhu, S.X., Lee, J.C., Lashkari, D., Shalon, D., Brown, P.O., Botstein, D. (1999) Distinctive gene expression patterns in human mammary epithelial cells and breast cancers, *Proc. Natl. Acad. Sci. U.S.A.* **96**, 9212–9217.

Perou, C.M., Sorlie, T., Eisen, M.B., van de Rijn, M., Jeffrey, S.S., Rees, C.A., Pollack, J.R., Ross, D.T., Johnsen, H., Akslen, L.A., Fluge, O., Pergamenschikov, A., Williams, C., Zhu, S.X., Lonning, P.E., Borresen-Dale, A.L., Brown, P.O., Botstein, D. (2000) Molecular portraits of human breast tumors, *Nature* **406**, 747–752.

Pinkel, D., Segraves, R., Sudar, D., Clark, S., Poole, I., Kowbel, D., Collins, C., Kuo, W.L., Chen, C., Zhai, Y., Daikee, S.H., Ljung, B.M.,

Gray, J.W., Albertson, D.G. (1998) High resolution analysis of DNA copy number variation using comparative genomic hybridization to microarrays, *Nat. Genet.* **20**, 207–211.

Puskas, L.G., Zvara, A., Hackler, L., Jr., Van Hummelen, P. (2002) RNA amplification results in reproducible microarray data with slight ratio bias, *Biotechniques* **32**, 1330–1334.

Randolph, J.B., Waggoner, A.S. (1997) Stability, specificity and fluorescence brightness of multiply-labeled fluorescent DNA probes, *Nucleic Acids Res.* **25**, 2923–2929.

Rose, S.D. (1998) Application of a novel microarraying system in genomics research and drug discovery, *J. Assoc. Lab. Automat.* **3**, 53–57.

Sandberg, R., Yasuda, R., Pankratz, D.G., Carter, T.A., Del Rio, J.A., Wodicka, L., Mayford, M., Lockhart, D.J., Barlow, C. (2000) Regional and strain-specific gene expression mapping in the adult mouse brain, *Proc. Natl. Acad. Sci. U.S.A.* **97**, 11038–11043.

Schena, M., Shalon, D., Davis, R.W., Brown, P.O. (1995) Quantitative monitoring of gene expression patterns with a complementary DNA microarray, *Science* **270**, 467–470.

Selaru, F.M., Zou, T., Xu, Y., Shustova, V., Yin, J., Mori, Y., Sato, F., Wang, S., Olaru, A., Shibata, D., Greenwald, B.D., Krasna, M.J., Abraham, J.M., Meltzer, S.J. (2002) Global gene expression profiling in Barrett's esophagus and esophageal cancer: a comparative analysis using cDNA microarrays, *Oncogene* **21**, 475–478.

Snijders, A.M., Nowak, N., Segraves, R., Blackwood, S., Brown, N., Conroy, J., Hamilton, G., Hindle, A.K., Huey, B., Kimura, K., Law, S., Myambo, K., Palmer, J., Ylstra, B., Yue, J.P., Gray, J.W., Jain, A.N., Pinkel, D., Albertson, D.G. (2001) Assembly of microarrays for genome-wide measurement of DNA copy number, *Nat. Genet.* **29**, 263–264.

Sugita, M., Geraci, M., Gao, B., Powell, R.L., Hirsch, F.R., Johnson, G., Lapadat, R., Gabrielson, E., Bremnes, R., Bunn, P.A., Franklin, W.A. (2002) Combined use of oligonucleotide and tissue microarrays identifies cancer/testis antigens as biomarkers in lung carcinoma, *Cancer Res.* **62**, 3971–3979.

Sultan, M., Wigle, D.A., Cumbaa, C.A., Maziarz, M., Glasgow, J., Tsao, M.S., Jurisica, I.

(2002) Binary tree-structured vector quantization approach to clustering and visualizing microarray data, *Bioinformatics* **18**(Suppl.), S111–S119.

Tabuchi, M., Ueda, M., Kaji, N., Yamasaki, Y., Nagasaki, Y., Yoshikawa, K., Kataoka, K., Baba, Y. (2004) Nanospheres for DNA separation chips, *Nat. Biotechnol.* **22**, 337–340.

Takahashi, Y., Ishii, Y., Nagata, T., Ikarashi, M., Ishikawa, K., Asai, S. (2003) Clinical application of oligonucleotide probe array for full-length gene sequencing of TP53 in colon cancer, *Oncology* **64**, 54–60.

Tamayo, P., Slonim, D., Mesirov, J., Zhu, O., Kitareewan, S., Dmitrovsky, E., Lander, E.S., Golub, T.R. (1999) Interpreting patterns of gene expression with self-organizing maps: methods and application to hematopoietic differentiation, *Proc. Natl. Acad. Sci. U.S.A.* **96**, 2907–2912.

Tanabe, L., Scherf, U., Smith, L.H., Lee, J.K., Hunter, L., Weinstein, J.N. (1999) MedMiner: an Internet text-mining tool for biomedical information, with application to gene expression profiling, *Biotechniques* **27**, 1210–1214.

Tanabe, L., Scherf, U., Smith, L.H., Lee, J.K., Hunter, L., Weinstein, J.N. (1999) MedMiner: an Internet text-mining tool for biomedical information, with application to gene expression profiling, *Biotechniques* **27**, 1216–1217.

Tavazoie, S., Hughes, J.D., Campbell, M.J., Cho, R.J., Church, G.M. (1999) Systematic determination of genetic network architecture, *Nat. Genet.* **22**, 281–285.

The International Human Genome Sequencing Consortium (2001) The sequencing and analysis of the human genome, *Nature* **409**, 860–921.

Unami, A., Shinohara, Y., Kajimoto, K., Baba, Y. (2004) Comparison of gene expression profiles between white and brown adipose tissues of rats by microarray analysis, *Biochem. Pharmacol.* **67**, 555–564.

Velculescu, V.E., Zhang, L., Vogelstein, B., Kinzler, K.W. (1995) Serial analysis of gene expression, *Science* **270**, 484–487.

Venter, J.C., Adams, M.D., Myers, E.W., Li, P.W., Mural, R.J., Sutton, G.G. et al. (2001) The sequence of the human genome, *Science* **291**, 1304–1351.

Vernon, S.D., Unger, E.R., Rajeevan, M., Dimulescu, I.M., Nisenbaum, R., Campbell, C.E.

(2000) Reproducibility of alternative probe synthesis approaches for gene expression profiling with arrays, *J. Mol. Diagn.* **2**, 124–127.

Wang, D.G., Fan, J.B., Siao, C.J., Berno, A., Young, P., Sapolsky, R. et al. (1998) Large-scale identification, mapping, and genotyping of single-nucleotide polymorphisms in the human genome, *Science* **280**, 1077–1082.

Wang, E., Miller, L.D., Ohnmachr, G.A., Liu, E.T., Marincola, F.M. (2000) High-fidelity mRNA amplification for gene profiling, *Nat. Biotechnol.* **18**, 457–459.

Whitney, L.W., Becker, K.G., Tresser, N.J., Caballero-Ramos, C.I., Munson, P.J., Prabhu, V.V., Trent, J.M., McFarland, H.F., Biddison, W.E. (1999) Analysis of gene expression in multiple sclerosis lesions using cDNA microarrays, *Ann. Neurol.* **46**, 425–428.

Wigle, D.A., Jurisica, I., Radulovich, N., Pintilie, M., Rossant, J., Liu, N., Lu, C., Woodgett, J., Seiden, I., Johnston, M., Keshavjee, S., Darling, G., Winton, T.,

Breitkreutz, B.J., Jorgenson, P., Tyers, M., Shepherd, F.A., Tsao, M.S. (2002) Molecular profiling of non-small cell lung cancer and correlation with disease-free survival, *Cancer Res.* **62**, 3005–3008.

Yang, Y.H., Dudoit, S., Luu, P., Lin, D.M., Peng, V., Ngai, J., Speed, T.P. (2002) Normalization for cDNA microarray data: a robust composite method addressing single and multiple slide systematic variation, *Nucleic Acids Res.* **30**, e15.

Yue, H., Eastman, P.S., Wang, B.B., Minor, J., Doctolero, M.H., Nuttall, R.L., Stack, R., Becker, J.W., Montgomery, J.R., Vainer, M., Johnston, R. (2001) An evaluation of the performance of cDNA microarrays for detecting changes in global mRNA expression, *Nucleic Acids Res.* **29**, E41–E41.

Yuen, T., Wurmbach, E., Pfeffer, R.L., Ebersole, B.J., Sealfon, S.C. (2002) Accuracy and calibration of commercial oligonucleotide and custom cDNA microarrays, *Nucleic Acids Res.* **30**, E48.

Microarray Technology: *see* Microarray-Based Technology: Basic Principles, Advantages and Limitations

Microbial Development

Paul Robert Fisher
La Trobe University, VIC, Australia

Encyclopedia of Molecular Cell Biology and Molecular Medicine, 2nd Edition. Volume 8
Edited by Robert A. Meyers.
Copyright © 2005 Wiley-VCH Verlag GmbH & Co. KGaA, Weinheim
ISBN: 3-527-30550-5

Keywords

Differentiation
A stable change in global patterns of gene expression usually producing a recognizably different type of cell.

Morphogenesis
The creation or development of an organism's or cell's morphology, structure, and form.

Development
A genetically programmed combination of differentiation and morphogenesis.

Cell Cycle
The orderly progression of events during cell proliferation through phases of growth, DNA replication, chromosome segregation, and cell division (cytokinesis).

Protein Kinase
An enzyme that phosphorylates itself and/or other proteins. Major categories of protein kinase are named according to the amino acid that is specifically phosphorylated, for example, tyrosine protein kinases phosphorylate target proteins on tyrosine residues.

Sigma Factor
A subunit of bacterial RNA polymerase that directs binding of the polymerase to specific DNA sequences (promoters), resulting in the transcription of downstream genes. Sigma factors form a family of polypeptides, each of which recognizes a different promoter sequence and so directs transcription of a different set of genes.

Transcription Factor
A DNA-binding protein that is not part of the RNA polymerase, but which binds to specific DNA sequences and regulates the expression of nearby genes by activating or inhibiting transcription.

Signal Transduction
The process whereby extracellular or intracellular, physical or chemical signals elicit a series of intracellular biochemical events in which each event either activates or inhibits the next.

Holdfast
A specialized attachment organ (of a multicellular organism) or organelle (of a unicellular organism).

Flagellum
A long, thin cellular appendage that is responsible for swimming motility – in the case of bacteria a rigid, helical, hollow filament that functions as a propellor attached to a rotary motor in the cell membrane through a semiflexible coupling.

Pilus
A long, fiberlike appendage on a bacterial cell that is thinner than a flagellum.

Pseudopod or Pseudopodium
A region of an amoeboid cell that does not contain membrane-bounded organelles and is extended outward during amoeboid movement.

Spore
A specialized, dormant, resistant cell formed as a means of dispersal from and survival of harsh conditions.

Endospore
A spore formed from a daughter cell within the membrane of a mother cell in certain bacterial species (notably of the genera *Bacillus* and *Clostridium*).

Sorus
A cluster or mass of spores.

▪ Development is not only restricted to large multicellular animals and plants but is also a feature of the life cycles of microorganisms. The environment of a microorganism is rapidly changeable so that both eukaryotic and prokaryotic microbes have evolved developmental strategies to enhance their nutrient-scavenging abilities or in more extreme circumstances to form dormant, resistant cells (spores). The spores can survive under harsh conditions that do not support growth and can be dispersed to other environments that do. The endospores of pathogenic *Clostridium* species are amongst the most resistant cells on earth, and their destruction (or removal) is the key criterion for successful microbiological sterilization. Developmental programs in microbial pathogens are initiated in response to infection of the host and play important roles in pathogenesis. However, the best-understood examples of microbial development are not pathogens, but free-living microbes. The general principles underlying microbial development are being elucidated by their study.

1
Introduction

Biologists do not usually think of microbial model organisms when asking questions about the fundamental mechanisms of development. Yet, even in the most complex of metazoans, development involves cells doing things in an ordered way through space and time – growing, dividing, adhering to one another and to the substratum, differentiating, moving, secreting, and, above all, communicating – and the fundamental mechanisms of what cells do evolved very early in the history of life. So it should not be surprising to learn that the molecular mechanisms at the heart of these cellular processes in development are frequently similar.

At the same time, every organism in the tree of life has faced its own unique problems in using these primeval molecular mechanisms to meet the needs of its "chosen" life style. Each has accordingly evolved its own developmental programs using its own unique selection of molecular techniques to carry them out. So we are not surprised either that organisms belonging to different phylogenetic lineages have often opted for different molecular mechanisms to perform similar developmental functions or that in some lineages, particular molecular mechanisms have been elaborated upon, used, and reused in many different biological roles, while others are not used at all.

An example of these principles is to be found in the universal use of protein kinase cascades in signal transduction pathways controlling development. Yet there are differences. The histidine protein kinases are used in developmental signaling pathways in all of the major bacterial groups and in many eukaryotic lineages including the fungi, the slime moulds, and the plants,

but not, it seems, in metazoan animals. No one knows why the use of histidine kinases was abandoned in the animal lineage, but retained in the next most closely related eukaryotic lineages. The bacteria, on the other hand, appear to make little, if any, use of the serine/threonine protein kinases so ubiquitously found in eukaryotic organisms. Differences like this tell us that there is nothing about one class of kinase compared to the other that makes its use essential for particular signaling purposes. They also identify classes of molecules that could be potentially used as targets for drug or antibiotic therapy. Histidine kinase signaling pathways participate in bacterial pathogenesis and are therefore potential antibiotic targets.

What then might be the general principles to be gleaned from our current understanding of development in eukaryotic and bacterial microbes? To attempt to answer this question I have opted in the following sections to discuss microbial development by describing what we know about developmental pathways in five well-studied systems – three bacterial and two eukaryotic. The emphasis is on the signaling cascades that control cellular activities in development. Having examined each of these five systems in some detail, I conclude by noting general features that seem to be shared by all.

2
Bacteria

2.1
Cell Cycle Control and Morphogenesis in *Caulobacter crescentus*

Bacteria are usually thought of as exhibiting one of only a limited number of cellular morphologies – rods, cocci, or spirals – that is monotonously present in every cell. *Caulobacter crescentus* is different. An aquatic bacterium, common in freshwater environments characterized by low and patchy nutrient levels, *Caulobacter's* unique cell cycle involves polarized differentiation and cytofission to produce two morphologically distinct cells – a non-dividing, motile, swarmer cell, and a sessile, stalked cell that is able to progress through cell division (Fig. 1). The swarmer cell is able to disperse by both random and chemotactic swimming motility to nutrient-rich microhabitats where it differentiates, losing its polar flagella and pili to replace them with a stalk and holdfast. Through these, it attaches to the substratum and begins progression through the cell cycle just as the mother cell did from which it was derived. It is this cell cycle, with its two dramatically different cell types, that has made *Caulobacter* a favorite organism for studying the mechanisms of cell cycle control, polarized differentiation and morphogenesis in bacteria.

2.1.1 CtrA – A Master Regulator of Cell Cycle Progression and Polar Morphogenesis

At the heart of the molecular machinery controlling *Caulobacter* morphogenesis is a protein called CtrA, a DNA-binding protein that functions as a transcriptional activator and repressor to regulate the expression of genes involved in DNA synthesis and cytofission, flagellar motility, and chemotaxis. CtrA is the final response regulator in a multicomponent phosphorelay system of the histidine protein kinase type that is universally found in bacteria and archaea, as well as in some eukaryotes (e.g. *Dictyostelium discoideum*). The phosphorylation cascade in such systems is initiated by an autophosphorylating histidine protein kinase, which is activated (or inhibited) by reception of a signal in the form of a ligand binding to its sensor domain. The phosphate is initially attached to a conserved histidine residue in the transmitter domain of the kinase, from where it is subsequently transferred (sometimes via multiple phosphotransfers) to an aspartate residue in the receiver domain of the same or a different protein – the response regulator that mediates the final response. In multicomponent systems, multiple phosphotransfers proceed from the primary histidine to the final aspartate via alternating histidine and aspartate residues in transmitter and receiver domains respectively of what can be distinct polypeptides.

The sequencing of the entire genome of *C. crescentus* made possible a comparative microarray analysis of gene expression in a wild type and temperature-sensitive *CtrA* loss-of-function mutant. Nearly 20% of the ca. 3700 genes in *Caulobacter* are cell cycle regulated and the microarray studies revealed that about one-third of these are controlled by CtrA, either directly or indirectly. Many of the direct targets were found by coimmunoprecipitation of DNA fragments that had been cross-linked to CtrA. This approach identified 55 promoters (regulating 95 genes) that bind and are presumably controlled by the active, phosphorylated form of CtrA (CtrA~P). The binding site is a conserved nonameric DNA sequence motif that is found not only in CtrA-regulated promoters but also at five sites within the chromosomal origin of replication, *ori*. Binding of CtrA~P to *ori* prevents the initiation of DNA synthesis and entry to S-phase. CtrA~P also represses expression of the early cell division gene *ftsZ*, the product of which forms a ring that circumscribes the cell at the division site and assembles the cellular division proteins there at the inner face of the cytoplasmic membrane.

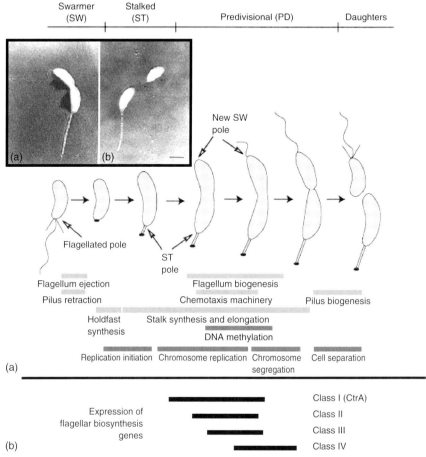

Fig. 1 The cell cycle and flagellar morphogenesis in *Caulobacter*. (a) The motile swarmer cell on the left loses its flagellum, retracts its polar pili and differentiates into a stalked cell, which enters the S-phase and progresses through the cell division cycle. The differentiation of two distinct daughter cells begins in the predivisional cell (inset, panel a) with distinct morphogenetic events occurring at the stalked (ST) and swarmer (SW) poles. After cytofission (inset, panel b), the stalked daughter cell is able to resume cell cycle progression, whereas the swarmer cell is arrested in growth phase (G1) until such time as it has differentiated into a stalked cell. The bars show the timing of various morphogenetic events during the cell cycle. (b) In the predivisional cell, a cascade of sequentially dependent inductions of flagellar gene expression and assembly events result in biogenesis of the flagellar apparatus at the swarmer pole. Horizontal bars show the timing of expression of the four sequentially induced classes of flagellar biosynthesis genes. Panel (a) is modified from Fig. 1 of Jacobs-Wagner, C. (2004) Regulatory proteins with a sense of direction: cell cycle signalling network in *Caulobacter, Mol. Microbiol.* **51**, 7–13. Panel (b) contains information presented in Fig. 1 of England & Gober (2001) Cell cycle control of morphogenesis in *Caulobacter Curr. Opin. Microbiol.* **4**, 674–680. Inset image courtesy of Jeanne S. Poindexter, Barnard College, Columbia University.

In keeping with its regulatory roles, the levels of CtrA in *Caulobacter* oscillate in synchrony with the cell cycle and differ in stalked and swarmer cells (Fig. 2). The levels of the phosphorylated form of CtrA change in concert with the levels of the protein itself. After accumulating in the late predivisional cell, CtrA~P remains at high levels in swarmer cells where it represses cell cycle progression at the point of entry into S-phase and also induces the gene expression cascades involved in swarmer cell differentiation (e.g. those involved in pilus and flagellar assembly). In the stalked cell at the onset of replication, CtrA~P levels are low, the protein having been both dephosphorylated and also degraded by ClpXP proteases. These two mechanisms of inactivating CtrA are redundant – mutant *ctrA* alleles that either are resistant to proteolysis or to inactivation by dephosphorylation are not lethal. The mutants progress through the cell cycle normally because in each case the alternative CtrA inactivation mechanism (dephosphorylation or degradation) is still present and is active at the appropriate cell cycle stage. As expected, a CtrA mutant that is both constitutively active and protease-resistant causes cell cycle arrest at the G1–S transition.

The disappearance and inactivation of CtrA at the end of G1 relieves its repression of replication, allowing the cell to enter S-phase and initiate DNA synthesis. Also relieved is the autoinhibitory repression of *ctrA* expression by high levels of CtrA~P bound to the low affinity *ctrA* promoter, P1. The *ctrA* expression that results is autocatalytic because of CtrA~P-mediated induction through a second, high affinity promoter, P2. Consequently, CtrA~P levels rise again to high levels in the late predivisional cell, where they prevent

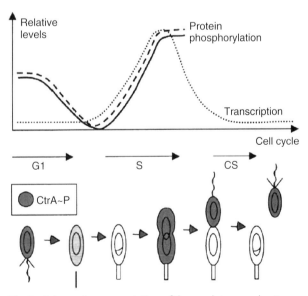

Fig. 2 Schematic representation of the regulatory mechanisms that modulate the level of CtrA~P (in grey) during the cell cycle. From Fig. 2 of Ausmees, N., Jacobs-Wagner, C. (2003) Spatial and temporal control of differentiation and cell cycle progression in *Caulobacter crescentus, Annu. Rev. Microbiol.* **57**, 225–247.

initiation of another round of replication and again repress *ctrA* expression through the P1 promoter site.

2.1.2 Morphogenesis in the Swarmer Cell – Flagellar Biogenesis

Having accumulated in the predivisional cell, CtrA~P persists in the swarmer cell, but is absent from the stalked cell. In the swarmer compartment of the predivisional cell, it initiates a gene expression cascade whose outcome is the differentiated swarmer cell and whose most obvious and best-understood feature is flagellar biogenesis. The ca. 50 gene products required for flagellar assembly in *Caulobacter* belong to four classes: Class I, the master regulator (CtrA); Class II, including a σ^{54} transcription factor (FlbD) and the first flagellar components to be assembled (the MS ring – FliF; the switch – FliG,M,N; the flagellum-specific secretion system – FlhH, FliI,J,O,P,Q,R); Class III, polypeptides forming outer parts of the basal body and hook; Class IV, the flagellar filament (Fig. 1).

Once initiated by CtrA~P, the assembly of the flagellar apparatus proceeds in three major stages defined by the onset of supply of the requisite polypeptides encoded by the Class II, III, and IV genes. Within each stage, the assembly of the parts progresses in a sequentially dependent manner – the incorporation of each polypeptide depending upon the assembly of those before it. The start of each major assembly stage is defined by a checkpoint at which the synthesis of polypeptides for that stage can begin only if the preceding stage has been completed. Thus, the Class II genes are induced in the predivisional cell by CtrA~P, the accumulation of which constitutes the Class I/II checkpoint.

The product of one of the Class II genes, *FlbD*, in its active phosphorylated form, in turn induces the Class III and IV genes in the swarmer pole of the predivisional cell. However, FlbD activity also depends upon two other Class II proteins – FlbE and FliX. FlbE was thought to be the kinase that phosphorylates FlbD, but recent evidence suggests that it activates FlbD by some other means. FliX functions as the checkpoint sensor that detects proper assembly of the inner parts of the flagellar basal body – FliX interacts directly with FlbD at its N terminus and only permits its activation when this Class II/III checkpoint has been reached. Active FlbD~P not only induces the Class III and IV genes but it also feeds back to repress early Class II genes.

Although both Class III and Class IV genes are induced by FlbD~P, the otherwise very stable Class IV mRNAs are targeted for degradation by the binding to them of the FlbT protein. This protein is present throughout the cell cycle, but its activity is inhibited by completion of the assembly of the flagellar motor, complete with hook. This constitutes the Class III/IV checkpoint. Only when this checkpoint has been reached are the Class IV mRNAs stabilized so that flagellin synthesis and completion of the entire flagellar apparatus can proceed.

2.1.3 Phosphorelays Regulating the Cell Cycle and Morphogenesis

CtrA is active only when phosphorylated at the conserved aspartate residue in its receiver domain (Asp51), and its activation by this means is controlled by at least three converging phosphorelay cascades (Fig. 3). The first of these involves CckA, a histidine protein kinase that is responsible for CtrA activation during the cell cycle. Consistent with its upstream role as a CtrA

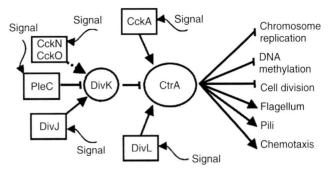

Fig. 3 Model of the signal transduction network controlling the activity of the master response regulator, CtrA. Response regulators and histidine kinases are shown as circles and boxes respectively. Barred ends indicate inhibition and arrowheads indicate activation. The dotted arrow indicates that the functions of the indicated upstream kinases are unknown. The molecular nature of the indicated signals that initiate the phosphorelays shown are unknown. Modified from Fig. 3 of Ausmees, N., Jacobs-Wagner, C. (2003) Spatial and temporal control of differentiation and cell cycle progression in *Caulobacter crescentus*, *Annu. Rev. Microbiol.* **57**, 225–247.

activator, CckA is essential for viability and temperature-sensitive mutants of this protein exhibit similar or identical phenotypes to temperature-sensitive CtrA mutants. However, CckA is rendered dispensable by expression of a constitutively active CtrA derivative (CtrAD51E) that does not require phosphorylation for activity. Because CckA contains its own response regulator receiver domain at its C terminus, it is anticipated that CckA passes phosphates to CtrA via an as yet unidentified intermediate histidine phosphotransferase in a multicomponent phosphorelay.

The second phosphorelay pathway regulating CtrA activity is mediated by DivL, a sensory kinase whose interactions with CtrA have been demonstrated genetically. Although it possesses all the conserved regions of the histidine kinase superfamily, DivL is unusual in possessing an autophosphorylatable tyrosine residue (Tyr550) instead of histidine. A cold sensitive *divL* allele is genetically suppressed

(i.e. the wild-type phenotype is restored) by a specific *ctrA* allele (called *sokA*). This phenomenon of suppression is a classical genetic indicator of interactions between the corresponding encoded proteins. It is supported in this case by *in vitro* biochemical evidence that DivL can phosphorylate CtrA.

DivK is a response regulator protein that directly or indirectly controls the levels of active, phosphorylated CtrA, probably by accelerating both its dephosphorylation and its degradation by ClpX proteases. The cold-sensitive *divK341* allele causes cell cycle arrest at the G1–S transition – initiation of replication is blocked, genes involved in replication fail to be induced, and the ClpX-mediated degradation of CtrA does not occur. The upstream histidine kinases that regulate DivK include the nonessential proteins PleC and DivJ. *In vitro*, they can catalyze both the phosphorylation and the dephosphorylation of the response regulator DivK. However, their *in vivo* actions appear to be opposed – DivJ phosphorylates and PleC dephosphorylates DivK so that its phosphorylation levels are reduced in a DivJ mutant, but increased in a PleC mutant. Two additional cytoplasmic histidine kinases of unknown function, CckN and CckO, have been identified in yeast two-hybrid screens as interacting partners of DivK.

The response regulator DivK and its upstream histidine kinases constitute the third phosphorelay cascade, leading to CtrA and regulating its activity. Just as the *sokA* allele of *ctrA* suppresses a conditional *divL* allele, it also suppresses mutant alleles of *divK, divJ,* and *pleC*. This provides a genetic demonstration of interactions between CtrA and the proteins encoded by these three genes. DivJ and PleC also interact with another response regulator protein, PleD, whose phosphorylated form

activates proteolytic degradation of FliF, leading to loss of the flagella, transition to the stalked form, and entry to S-phase. The phosphorylated form of PleD is an active di-guanylate cyclase that synthesizes cyclic di-guanosine monophosphate (c-di-GMP) from 2 GTP molecules. In some other bacteria, this small molecule acts as an allosteric regulator activating cellulose synthase, but its downstream targets in *Caulobacter* are not known.

2.1.4 Dynamic Subcellular Localization of Phosphorelay Proteins Controlling Cell Division and Morphogenesis

Fusing the gene encoding jellyfish green fluorescent protein (GFP) with that for specific *Caulobacter* signaling proteins made it possible to use fluorescence microscopy to study their dynamic localization in living cells during the cell cycle (Fig. 4). Each of

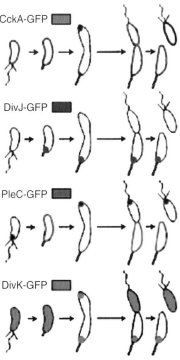

Fig. 4 Localization of some morphogenetic signaling proteins during the *Caulobacter* cell cycle. CckA and PleC are membrane-bound and cycle as shown between being dispersed throughout the membrane and concentrated at the indicated poles. DivJ, when present, is always concentrated at the stalked (ST) pole. DivK cycles between a diffuse cytoplasmic localization in swarmer cells and concentration at the indicated poles in predivisional and stalked cells. The length of the arrows indicates the time spent between each stage of differentiation. From Fig. 2 of Jacobs-Wagner, C. (2004) Regulatory proteins with a sense of direction: cell cycle signalling network in *Caulobacter, Mol. Microbiol.* **51**, 7–13. (See color plate. p. xxix)

these signaling proteins exhibits a characteristic pattern of localization that reflects its functions in morphogenesis and cell division. From being dispersed throughout the inner, cytoplasmic membrane, CckA translocates in the predivisional cell to the poles, particularly the swarmer pole where it can phosphorylate CtrA and initiate flagellar assembly. It returns to its original diffuse membrane distribution by the time cell division is complete. DivK translocates from the cytoplasm in the swarmer cell to both poles in the predivisional cell where it is regulated by either DivJ (at the stalked pole) or PleC (at the swarmer pole). By the time division is complete, DivK has been released from its polar location in the swarmer cell to return to its original distribution. However, it remains localized at the stalked pole in the stalked cell. At the stalked pole DivJ would activate DivK and so promote dephosphorylation and degradation of CtrA, favoring progression through the cell cycle. During early differentiation of the stalked cell, PleC is released from the newly forming stalked pole and moves to the opposite pole where it would keep DivK inactive and CtrA active in inducing flagellar biogenesis.

What controls these very striking localization patterns? For the most part, the answer to this question is not yet known, but some elements of the mechanisms involved have emerged. Firstly, correct polar localization of DivK depends upon phosphorylation of its conserved receiver aspartate and recruitment by DivJ at the stalked pole. Secondly, DivK release from the swarmer pole requires PleC, presumably because of the PleC-mediated dephosphorylation of DivK. Thirdly, the release of PleC itself from the future stalked pole requires PleC's own catalytic activity, while its localization to the swarmer pole

requires a protein with similar localization patterns called *PodJ*. Fourthly, a 61-residue sequence in the cytoplasmic linker region of the sensor domain of DivJ has been shown to be necessary for its proper localization. Finally, the localization of all of these proteins to the correct pole depends on MreB, a bacterial homolog of the eukaryotic cytoskeletal protein, actin. MreB forms a filament that oscillates during the cell cycle between a spiral form running the length of the cell and a medial ring. The ring could be an intermediate in a reversal of polarity of the spiral as the flagellated swarmer pole matures into a stalked pole.

Clearly, although much remains to be elucidated, the correct localization of signaling proteins in *Caulobacter* morphogenesis both controls and is itself dependent upon the phosphorelay cascades in which these proteins participate.

2.2
Sporulation and Differentiation in *Bacillus subtilis*

Caulobacter is not the only bacterium that differentiates into two cell types. Under conditions of high population density and starvation, many Gram-positive bacteria belonging most notably to the genera *Bacillus* (aerobic) and *Clostridium* (anaerobic) can differentiate to form an endospore enclosed within a mother cell, which subsequently perishes. The endospore is an extremely heat- and dessication-resistant quiescent cell that is of medical significance because some endospore-forming bacteria of both genera are human pathogens (e.g. *Bacillus anthracis*, causative agent of anthrax, and *Clostridium tetani*, causative agent of tetanus). It is the extreme resistance properties of bacterial endospores that determine the minimum requirements for successful sterilization of culture media, equipment, and surgical instruments. Phylogenetic studies have shown that the important signaling interactions and central mechanisms for sporulation are conserved amongst all endospore-forming bacteria. What seems to change the most in different lineages is the sensor domains of the receptor kinases that transmit relevant environmental information to the cells. This implies that different environmental signals relevant to endospore formation are monitored by the different species – as might be expected given the very varied habitats in which they live, including oil wells, salt lakes, soil, or infected mammalian hosts. Endospore formation has been most intensively studied in the saprophytic soil bacterium, *Bacillus subtilis*.

When *B. subtilis* cells reach a stationary phase, they enter a so-called transition state in which they must choose between several possible survival strategies – endospore formation, nutrient scavenging, and transformation. Nutrient scavenging is the strategy of first choice and involves expression of genes for chemotaxis toward nutrient sources, secretion of antibiotics that inhibit competitors, and the secretion of extracellular enzymes, such as proteases, that degrade environmental macromolecules into their constituents for uptake as nutrients. If the adverse conditions that elicit the nutrient scavenging strategy persist, but do not worsen significantly, 1–10% of the cells will opt for the transformation strategy. This involves the cell differentiating to become competent for the uptake of exogenous DNA (released into the environment by the lysis of other cells). This can lead to the formation of recombinant cells with new combinations of genetic information that could enhance survival. Because endospore formation is a complex energy-intensive process, the cell only commits itself to this path as a last resort in response to integrated information from multiple environmental and physiological signals. These signals converge on the phosphorylated form of a master regulator protein, SpoOA~P, whose levels determine whether the cell enters the sporulation pathway or not. At intermediate SpoOA~P levels, only the nutrient scavenging and (in some cells) transformation strategies are pursued. When

Fig. 5 The key stages of the sporulation cycle in *Bacillus subtilis* (from Fig. 1 of Errington, J. (2003) Regulation of endospore formation in *Bacillus subtilis*, *Nat. Rev. Microbiol.* **1**, 117–126). The inset (from "Todar's Online Textbook of Bacteriology", http://www.textbookofbacteriology.net/ Kenneth Todar, 2002) shows electron micrographs of sporulating cells at each of the major stages. The formation of endospores is a complex and highly regulated form of development in a relatively simple (procaryotic) cell. In all *Bacillus* species studied, the process of spore formation is similar, and can be divided into seven defined stages (0–VI). The vegetative cell (Stage 0) begins spore development when the DNA coils along the central axis of the cell as an "axial filament" (Stage I). The DNA then separates and one chromosome becomes enclosed in plasma membrane to form a protoplast (Stage II). The protoplast is then engulfed by the mother cell membrane to form an intermediate structure called a *forespore* (Stage III). Between the two membranes, the core (cell) wall, cortex, and spore coats are synthesized (Stage IV). As water is removed from the spore and as it matures, it becomes increasingly heat resistant and more refractile (Stage V). The mature spore is eventually liberated by lysis of the mother cell. The entire process takes place over a period of 6 to 7 h and requires the temporal regulation of more than 50 unique genes. Inset image and description of the stages were kindly provided by Kenneth Todar.

Spo0A~P reach high-enough levels, cells will abandon the nutrient scavenging strategy, forego the transformation strategy, and sporulate (see Fig. 5).

2.2.1 Spo0A and the Phosphorelay in Endospore Formation

In Fig. 6, which shows the basic outline of the sporulation phosphorelay mechanism, a sensor histidine-kinase autophosphorylates the conserved histidine in its transmitter domain and then passes this phosphate to an aspartate in the receiver domain of Spo0F. The phosphate is then passed via a histidine in Spo0B to an aspartate residue in the receiver domain of Spo0A, the final response regulator. Although only one sensor kinase is shown in Fig. 6, there are in fact five different sensor kinases (KinA to KinE) in *B. subtilis*

that send signals into this phosphorelay (Fig. 7). The two most important of these, in order of significance, are KinA and KinB. The other kinases on their own seem to be able only to generate intermediate levels of Spo0A~P, sufficient for the nutrient scavenging and competence pathways, but insufficient to induce sporulation. Although it is clear that sporulation is initiated primarily by starvation and high cell density, the molecular identities of the signals to which the sensor kinases respond are unknown.

Spo0A~P is an active DNA-binding protein that can function either as a positive or a negative regulator of transcription and has been shown in recent microarray studies to induce, directly or indirectly, the transcription of 349 genes (66 of them in combination with the sigma factor σ^F).

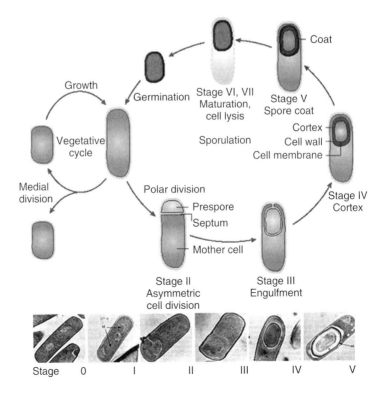

The *spoIIA, spoIIG,* and *spoIIE* promoters are amongst those that are directly induced by Spo0A~P, and the corresponding gene products play essential roles in Stage II of sporulation (see later). These genes are also induced indirectly by Spo0A~P via SinI. The synthesis of SinI is induced by Spo0A~P, and, once made, SinI interacts with SinR, preventing it from repressing its target genes.

SpoA~P also directly or indirectly represses expression of as many as 242 different genes, including *abrB*, which encodes a transcriptional repressor. AbrB is responsible for repressing stationary phase and sporulation functions during

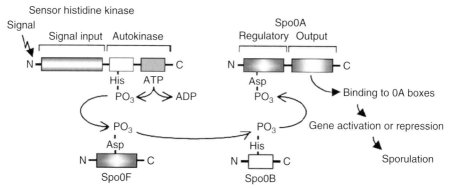

Fig. 6 The *Bacillus subtilis* sporulation phosphorelay (from Fig. 1 of Stephenson, K., Hoch, J.A. (2002) Evolution of signalling in the sporulation phosphorelay, *Mol. Microbiol.* **46**, 297–304).

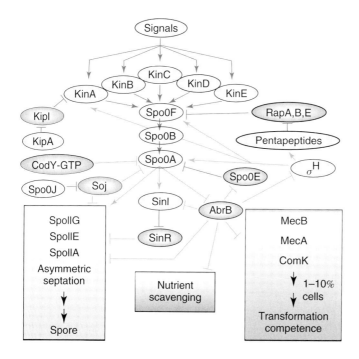

the growth phase. Its targets are upregulated by Spo0A~P indirectly by removal of AbrB-mediated repression. One of these target genes, *spo0H*, encodes the sigma factor σ^H, which targets RNA polymerase to at least 49 promoters controlling at least 87 early sporulation genes. These include the *kinA* promoter and secondary, highly active promoters for *spo0F* and *spo0A*. By this means, a positive feedback is established that results in a new, higher level of Spo0A~P, committing the cell to sporulation.

The phosphorelay leading to Spo0A~P accumulation in stationary phase cells is subject not only to the positive controls described but also to negative controls that prevent the cell from committing to sporulation inappropriately or prematurely. There are several such negative control mechanisms.

The first mechanism for limiting Spo0A~P accumulation is mediated by cellular GTP levels, which serve as the main indicator of the status of intermediary metabolism. During growth, GTP levels are high (1–3 mM), but they drop significantly upon entry to the stationary phase. Two proteins have been implicated

in sensing GTP levels – CodY (early stages) and Obg (later stages). CodY in its GTP-bound form is a repressor of stationary phase/sporulation genes including *spo0A*.

The second mechanism for restraining Spo0A~P accumulation is by the action of Spo0E, a phosphatase that specifically dephosphorylates Spo0A~P. Without Spo0E, Spo0A would become maximally phosphorylated even at low levels of activity of the phosphorelay. Spo0E expression is subject to AbrB-mediated repression.

The third mechanism for restricting Spo0A phosphorylation is mediated by pentapeptide pheromone signals. The precursors of the pentapeptides are small polypeptides of about 40 amino acids that are secreted and processed proteolytically outside the cell, then imported via specific transport systems (products of the *opp* and *app* operons) into the cell where they inhibit the Rap family phosphatases RapA, B, and E. The extracellular concentrations of the signaling pentapeptides depend upon the density of stationary phase cells, so that only at high cell density do they prevent the dephosphorylation of Spo0F and thereby allow unrestrained phosphorelay to produce maximum Spo0A~P levels.

Fig. 7 Regulation of the phosphorelay and the initiation of sporulation in *Bacillus subtilis*. During the growth phase, AbrB represses genes involved in the three stationary phase survival strategies (boxed) – sporulation, nutrient scavenging, and transformation. At the onset of stationary phase, multiple signals result in activity of the phosphorelay, producing intermediate levels of phosphorylated Spo0A. These levels of Spo0A~P are sufficient to support the nutrient scavenging and, eventually, in some cells the transformation strategy for survival. Under more extreme circumstances of nutritional stress and high density, several positive feedbacks can combine with the lifting of several negative restraints to elicit production of high levels of Spo0A~P. At these high levels, the expression of sporulation-specific genes is induced and the cell progresses down the pathway of endospore formation. Green ellipses indicate proteins whose activity stimulates the phosphorelay, while red ellipses indicate proteins whose activity inhibits it. Arrowheads indicate stimulatory interactions and barred ends indicate inhibitory interactions. Blue arrows stand for interactions at the transcriptional level (gene induction or repression) and red arrows stand for interactions at the protein level (activation or inhibition). MekR, MekB, and ComK are AbrB-repressible proteins involved in the development of transformation competence. Other proteins are described in the text. (See color plate. p. xxx)

The fourth mechanism for limiting the activity of the sporulation phosphorelay is inhibition of the KinA kinase by the KipI protein, which is itself antagonized by KipA. The expression of KipI and KipA is controlled by carbon and nitrogen sources, so that phosphorelay is inhibited by the availability of nutrients and the cell persists with the nutrient scavenging strategy rather than sporulate.

The fifth mechanism by which Spo0A∼P levels are limited is by Soj-mediated repression of expression of *spo0A*. Soj also represses the *spoIIA,E* and *G* operons, which are targets for Spo0A∼P induction and its repressive activity is antagonized by Spo0J. During growth, the interaction between Spo0J and Soj serves what appears to be a completely different function in partitioning chromosomes into daughter cells. However, these disparate roles may be intimately connected – during sporulation Spo0J is required to localize Soj at the poles, probably in association with the chromosomal origin of replication *oriC*. Key *spo* promoters elsewhere may be liberated from Soj repression by this means, which could thus constitute a final checkpoint, delaying progression of sporulation until the chromosome has been replicated and the *oriC* regions have segregated.

2.2.2 Mother–Daughter Communication in Endospore Formation – A Spatiotemporally Regulated Sigma Factor Cascade

Once the cell has embarked on endospore formation as a result of the accumulation of high enough levels of Spo0A∼P, a gene expression cascade is induced, which is coupled to morphogenetic events such as septation to form the forespore and its subsequent engulfment by the mother cell. This gene expression cascade is orchestrated by timed, serially dependent expression of specific sigma factors, each of which associates in turn with the RNA polymerase core enzyme, targeting it to a specific set of promoters (Fig. 8). The first of the sigma factors to be induced are σ^E and σ^F, whose genes belong to the *spoIIG* and *spoIIA* operons respectively (Fig. 7). Both proteins are synthesized prior to septum formation in an inactive form, and both operons also encode proteins that regulate the activities of their cognate sigma factors. Only after septum formation do these two sigma factors become active – σ^E in the mother cell and σ^F in the forespore.

Septum formation is thus a critical event in orchestrating other events in spore formation, and only when the septum has formed is the cell irrevocably committed to sporulation. The formation of the septum is a specialized cell division event that uses much of the same cellular machinery as normal cell fission, including the FtsZ ring, which forms, in this case, not at the middle of the cell, but near one end. FtsZ is one of the proteins whose synthesis is induced directly by Spo0A∼P and the resulting rise in its cellular concentration is necessary for ring formation during sporulation. During sporulation, FtsZ initially forms two rings located near the cell poles, one of which (usually the larger) is subsequently "selected" for septum formation. SpoIIE, a protein phosphatase also necessary for σ^F activation (see later), is required for correct placement of the ring. There, as in normal cell division, FtsZ acts as a scaffold for recruitment of other cell division proteins to the division site and possibly as a GTP-powered contractile structure.

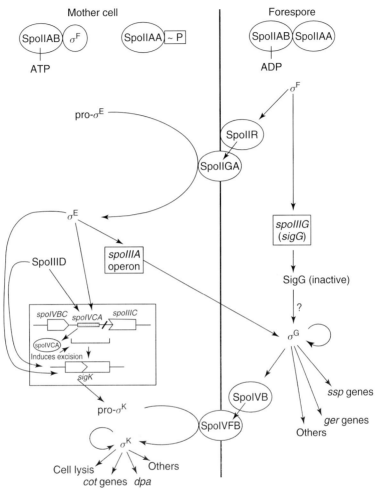

Fig. 8 The mechanisms involved in the regulation of sigma factors in the compartmentalized sporulating cell. The symbol σ and appropriate letter indicates the sigma factors, and precursor proteins of these sigma factors are indicated as pro-σ and the appropriate letter. The status of sigma F (σ^F) in each compartment post septation is indicated at the top of the figure. A vertical line represents the asymmetric septum, which divides the mother cell and the forespore. Arrows indicate activation. Circular arrows indicate autoregulation. Those proteins associated with the membrane are illustrated as embedded in the septum. Unknown regulatory mechanisms are indicated with a question mark. From Fig. 2 of Phillips, Z.E.V., Strauch, M.A. (2002) *Bacillus subtilis* sporulation and stationary phase gene expression, *Cell. Mol. Life Sci.* **59**, 392–402.

With the FtsZ ring in place, septation proceeds by a process of membrane invagination and initial synthesis of some cell wall material, which is later removed (this lends support to the idea that spore formation evolved as a modified form of normal cell division). Septation is complete before the entire forespore chromosome has been segregated to the spore compartment. During chromosome segregation, the chromosomal origin of replication, *oriC,* is translocated and anchored to the poles by a mechanism that is not understood, but involves several proteins – DivIVA, Spo0J, Soj, and RacA. However, only about 30% of the chromosome, centered around *oriC,* is initially located within the forespore. The rest is subsequently transported there from the mother cell over a period of 10 to 15 min by a DNA translocase, SpoIIIE. It is during this brief period when the forespore is bereft of about two-third of the genes, while the mother cell has two copies of them, that some of the critical asymmetries between mother and forespore are established. These asymmetries cause the two cells to embark on different programs of gene expression, as shown in Fig. 8.

One such asymmetry arises from the location of the SpoIIA operon on the opposite side of the chromosome from *oriC.* SpoIIAB is an antisigma factor (and protein kinase), which, in the mother cell, is bound to σ^F, keeping it inactive. However, SpoIIAB is labile so that its concentration drops rapidly in the forespore in the period before delivery of the SpoIIA operon to that compartment. This would result in activation of σ^F only in the forespore. The second, and perhaps the most important asymmetry, is that the SpoIIE protein phosphatase, which is associated with the division machinery, is delivered into the forespore after septation by a SpoIIIE-dependent mechanism. There it dephosphorylates the protein SpoIIAA in opposition to the kinase activity of SpoIIAB. Dephosphorylated SpoIIAA displaces σ^F from its complex with SpoIIAB, so that the sigma factor is free to associate with RNA polymerase. The combined effects of increased SpoIIE and decreased SpoIIAB concentrations in the forespore thus result in the activation of σ^F-mediated gene transcription there, but not in the mother cell.

Amongst the genes whose transcription is directed by σ^F is *spoIIR* whose gene product is secreted into the intermembrane space of the septum bounded by the forespore and mother cell membranes. There it activates the SpoIIGA protease, an integral membrane protein located in the septal membrane of the mother cell. On the mother cell side of the membrane, the activated SpoIIGA proteolytically cleaves pro-σ^E to release active σ^E and so initiates σ^E-dependent gene expression in the mother cell.

Another of the σ^F-dependent genes expressed in the forespore is *spoIIIG*, which encodes σ^G, the next sigma factor to be deployed in the forespore. However, σ^G is also synthesized in an inactive form. Its activation depends upon the proteins encoded in the *spoIIIA* operon transcribed only in the mother cell by σ^E-directed RNA polymerase. The SpoIIIA proteins are initially located in the mother cell septal membrane and, after engulfment, are found in the topologically equivalent outer forespore membrane. How, from this location, they activate σ^G is unknown, but having been thus activated, σ^G directs the transcription of a suite of genes in the forespore involved in spore maturation and germination. Amongst the σ^G-dependent genes is *spoIVB* whose protein product in the forespore somehow sends a signal to activate the SpoIVFB protease that resides in the mother cell septal membrane in an inhibitory complex with SpoIVFA and BofA.

The role of the activated SpoIVFB protease in the mother cell membrane is to activate the final mother cell–specific sigma factor, σ^K, encoded by *sigK*. Expression of *sigK* depends upon σ^E in two ways. Firstly, *sigK* is transcribed from a σ^E-driven promoter. Secondly, *sigK* in many

Bacillus strains is interrupted by a 48-kbp element called *skin* that is excised precisely by a site-specific recombinase (SpoIVCA) whose expression also depends on σ^E (as well as on SpoIIID). SpoIVCA is encoded within the *skin* element itself. Transcription in the mother cell from σ^K-driven promoters results in the assembly of the outer spore coat and production of dipicolinic acid, which is accumulated as its calcium salt by the maturing spore to very high concentrations and is responsible for many of its essential resistance properties. Some of the σ^K-dependent gene products are also responsible for the eventual lysis of the mother cell – a form of programmed cell death – that occurs after spore maturation is complete.

2.3
Differentiation and Morphogenesis in *Myxococcus xanthus*

Caulobacter differentiation occurs as part of the growth cycle and does not depend upon the presence of neighboring cells, whereas sporulation in *Bacillus* is initiated under starvation conditions at high cell densities when growth is no longer an option. The social bacterium *Myxococcus xanthus* takes this cooperativity one step further in its foraging behavior and fruiting body formation. *M. xanthus* is one species of the myxobacteria, gliding bacteria that live in soil environments and, under conditions of starvation and high density, aggregate together to form true multicellular structures that contain resistant, dormant spores (Fig. 9). *Myxococcus* cells exhibit two different forms of gliding motility that can be separated genetically and are mediated by completely different mechanisms. Adventurous or A motility is exhibited by isolated cells, while social or S motility requires close association between cells and is exhibited only in groups. The mechanism of A motility is still not understood. S motility is mediated by extension and retraction of Type IV pili and also requires the lipopolysaccharide

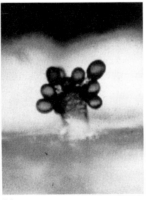
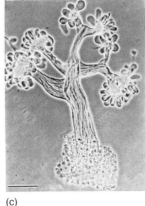

(a)   (b)   (c)

Fig. 9 Fruiting bodies of myxobacteria.
(A) *Myxococcus fulvus*. Phase contrast. Bar, 50 mm. (B) *Stigmatella aurantiaca*. Phase contrast. Fruiting body is about 150 mm tall. (C) *Chondromyces crocatus*. Slide mount, phase contrast. Bar, 100 mm. *Myxococcus xanthus* fruiting bodies are similar in morphology (but not pigmentation) to those of *M. fulvus*. From Fig. 1 of Dworkin, M. (1996) Recent advances in the social and developmental biology of the myxobacteria, *Microbiol. Rev.* **60**, 70–102.

O-antigen on the cell surface as well as polysaccharide/protein appendages called *fibrils*. During growth, S motility allows *M. xanthus* cells to migrate together in groups and prey on other bacteria by secreting enzymes that lyse and digest the macromolecules of their victims. This so-called "wolf pack" behavior supports faster growth because of the cooperative effect of many cells secreting the necessary lytic enzymes. Under starvation conditions, S motility allows myxobacteria to aggregate and form fruiting bodies, within which a proportion of the cells differentiate into round myxospores. Mutants deficient in S motility are therefore unable to aggregate and sporulate.

2.3.1 Initiation of Development by Starvation and Extracellular Signals

The primary signal that initiates development in *M. xanthus* is starvation. As in other bacteria, this elicits the so-called *stringent response*, whereby nutritional stress is sensed by the stalling of protein synthesis and accompanying activation of RelA, the ribosome-associated (p)ppGpp synthase. The resulting increase in (p)ppGpp concentrations regulates transcription, leading to decreased rRNA synthesis, cessation of growth, and increased expression of early developmental genes. Amongst these are genes encoding proteins whose activity leads to production of extracellular developmental signals (morphogens) that accumulate and thus convey cell density information to the developing myxobacteria.

The extracellular signals controlling early development were originally discovered genetically by testing the ability of pairs of nondeveloping mutants to restore one another's ability to develop normally when mixed together. This phenomenon was called *extracellular complementation* by analogy with genetic complementation in which two different mutant genomes are mixed within a cell (by appropriate genetic crosses) and tested for restoration of normal behavior. Extracellular complementation tests of pairwise combinations of mutants revealed five distinct groups, A to E. Whereas mixtures of cells from different mutants within the same group were unable to form fruiting bodies, mixtures of mutants from different groups could undergo normal development. These results indicated that mutants within each extracellular complementation group were deficient in the production of a specific extracellular factor or signal that could still be supplied by mutants from a different group.

The mutants within each group have been found to express developmentally regulated genes normally up to a characteristic time in development. Development was arrested at this time point, which therefore represented the time at which the extracellular signal missing in that group would normally have been deployed. Thus, the order of action in development of the five extracellular signals was found to be B, A, D, E, and C (Fig. 10).

The mutant genes associated with the extracellular complementation groups in *M. xanthus* have been identified and, in some cases, provide clues to the identity of the corresponding signal. The genes in the A group are *asgA*, *B*, *C*, and *D*. Two of these (*asgA* and *asgD*) encode histidine protein kinases with both kinase and response regulator domains. AsgB is a putative transcription factor with a helix-turn-helix DNA-binding motif near its C terminus, while AsgC is the RNA polymerase sigma factor σ^{70} (also known as SigA or RpoD). AsgA and/or D could directly or indirectly sense the accumulation of the alarm molecule (p)ppGpp and activate AsgB-

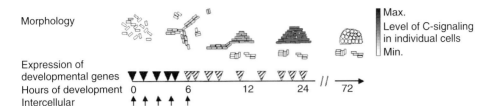

Morphology

Max.
Level of C-signaling
in individual cells
Min.

Expression of
developmental genes

Hours of development 0   6   12   24   //   72

Intercellular
signals   B A D E   C

Fig. 10 Multicellular morphogenesis in *Myxococcus xanthus*. On the developmental timeline, the black triangles indicate the expression time of developmental genes expressed in all cells before the initiation of C-signaling; the hatched triangles indicate the expression time of C-signal-dependent genes, all of which are only expressed in sporulating cells. The time of action of the A to E signals is indicated by the arrows below the timeline. Cellular arrangements during the different stages of fruiting body formation are shown above the timeline. The grayscale coloring of cells according to the intensity profile shown on the right indicates the level of C-signaling in individual cells during the different morphological stages. At 72 h, cells inside the fruiting bodies have matured to spores, and the cells remaining outside the fruiting bodies are the peripheral rods. From Fig. 1 of Søgaard-Anderson et al. (2003) Coupling gene expression and multicellular morphogenesis during fruiting body formation in *Myxococcus xanthus, Mol. Microbiol.* **48**, 1–8.

and AsgC-directed transcription of genes encoding the extracellular proteases that produce the A-signal. These extracellular proteases are secreted and hydrolyze cell surface proteins whose constituent amino acids accordingly increase locally in concentration to an average of about 25 μM. This is insufficient to support growth, but sufficient for a subset of six amino acids (tryptophan, proline, phenylalanine, tyrosine, leucine, and isoleucine) to jointly constitute the A-signal. Proteins involved in transducing the A-signal include SasS, a histidine kinase, and SasR, its cognate response regulator.

Apart from the A-signal, the best understood of the five extracellular *M. xanthus* developmental signals is the C-signal. The processed product of the only C group gene, *csgA*, the authentic C-signal appears to be the 17-kD C-terminal portion of the 25-kDa CsgA protein. CsgA is secreted by developing *M. xanthus* cells and processed by an extracellular protease to produce the active C-signal. The full length CsgA protein shares homology with short chain alcohol dehydrogenases (SCADs), its N-terminal region containing the NAD^+-binding motif and the C-terminal portion containing the catalytic domain. The absence of the NAD^+-binding site from the active C-signal suggests that the protein's putative enzymatic activity does not play a role in its C-signaling capabilities. Exogenous addition of both the active, N-terminally truncated form, and the full-length form of CsgA are able to rescue development in C-group mutants, the full-length form doing so by first being clipped by extracellular proteases, as in normal development.

The three remaining extracellular signals (B, D, and E) are less well characterized. The only gene in the B group is *bsgA*, which encodes a protein sharing homology with the *Escherichia coli* Lon protease. The nature of the B-signal is still not understood, but its generation requires the BsgA protein and so may depend, directly or indirectly, on a proteolytic event. As is the case with A-signal transduction, a histidine kinase (SpdS) and response regulator (SpdR) pair are involved in B-signal transduction. The only gene in the D group is *dsgA*, which encodes an initiation factor for translation, IF3. Point mutations in *dsgA* reduce the efficiency of sporulation, but do not completely prevent it, while disruption of the gene is lethal. Presumably, the generation of the D-signal depends more sensitively than other cellular functions on some aspect of IF3 activity in translation initiation. The generation of the E-signal depends upon two *esg* genes, which encode the E1α and E1β subunits of branched-chain keto acid dehydrogenase. This enzyme produces coenzyme A-linked forms of the fatty acids isovalerate, methyl butyrate, and isobutyrate, suggesting the possibility that these short-chain-branched fatty acids might constitute the E-signal. Consistent with this, the defect in E-group mutants can be rescued by exogenous supply of isovalerate.

2.3.2 C-signal Transduction and Sporulation After Aggregation

Four of the five *Myxococcus* morphogenetic signals are involved in early development – the first 6 h after the onset of starvation, during which time there is little overt morphological change in the lawn of starving bacteria. After six h of development, the cells begin to aggregate to form mounds that will develop into mature fruiting bodies. This change and the associated shifts in gene expression patterns are initiated by the C-signal. Although C-signal can rescue development in *csgA* mutants when supplied exogenously in solution, transmission of the C-signal between cells during normal development depends on direct end-to-end cell contacts because CsgA (full length and clipped) is membrane-bound and (presumably) localized at the poles. Such end-to-end contacts

between cells are more frequent in aggregating streams of cells and most frequent within mounds, so that the aggregation process results in an ordered increase in the level of C-signaling and provides a link between it and the stage of morphogenesis that has been reached.

C-signal transduction pathways lead from an unidentified receptor(s) to three different responses in the cells that are separated in space and time – C-signal amplification, aggregation, and sporulation in mounds (Fig. 11).

The first of the three C-signaling pathways leads to heightened induction of transcription of *csgA* and so to an increase in the level of C-signal. The proteins encoded by the *act* operon, ActA to D mediate this response. ActA is a histidine kinase and ActB is its cognate response regulator that activates *csgA* transcription, while ActC and D somehow control the correct developmental timing of *csgA* expression. A consequence of this positive feedback is that the level of C-signaling continues to increase once it has been initiated.

The second of the C-signal-mediated responses is aggregation and, as noted above, it also feeds back to increase the level of C-signaling by increasing the frequency of end-to-end contacts between cells. The earliest recognized component in the transduction pathway leading to aggregation is the activation (possibly by phosphorylation) of FruA, a DNA-binding response regulator whose expression was induced earlier in development by the A- and E-signals. FruA, in turn, activates the Frz chemotaxis-like signaling system and this leads to aggregation. The Frz proteins include homologs of the major chemotaxis-signaling proteins in *E. coli*. FrzCD is a methyl-accepting chemotaxis protein (MCP) homolog whose role is presumed to be analogous to the MCPs

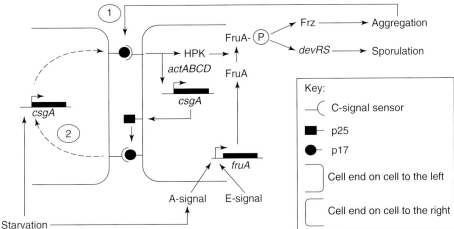

Fig. 11 Model of the C-signal transduction pathway. The schematic illustrates the C-signal transmission between two neighboring cells. For simplicity, the components in the pathway are only shown in the cell on the right. These components are also present in the cell on the left. In this cell, only *csgA* is shown to illustrate the signal amplification loop labelled "2." The second signal amplification loop is labelled "1." HPK indicates the hypothetical FruA histidine protein kinase. Modified from Fig. 2 of Søgaard-Anderson et al. (2003) Coupling gene expression and multicellular morphogenesis during fruiting body formation in *Myxococcus xanthus*, Mol. Microbiol. **48**, 1–8.

of *E. coli*, which function as receptors and transmembrane signal transducers. However, FrzCD differs from them in being a soluble, cytoplasmic protein, implying that the signals to which it directly responds must be intracellular, having either entered the cell or been produced in response to an extracellular signal. Cytoplasmic MCPs are known in other bacteria as well, *Rhodobacter sphaeroides*, for example. FrzE is a hybrid histidine kinase with catalytic and response regulator domains related respectively to the CheA and CheY chemotaxis proteins of *E. coli*. A second chemotaxis-like signaling system, the Dif system, also contributes to S motility and aggregation in response to lipidic (dioleoyl phosphatidyl ethanolamine) signals released by neighboring cells. It includes DifA, a transmembrane MCP whose extracellular domain seems too small to specifically bind an extracellular water-soluble ligand. The transmembrane

domains of DifA may interact directly with specific lipid ligands or indirectly via other macromolecules in the membrane such as the polysaccharide fibrils that adorn the surface of *M. xanthus* cells. Both the fibrils and the lipid seem to be required for aggregation based on S motility. Other proteins of the Dif signaling system controlling S motility are DifE (a CheA homolog) and DifD (a CheY homolog).

The third C-signal transduction pathway involves FruA-mediated induction of the genetic program, leading to sporulation. Phosphorylation of FruA by an unknown upstream histidine kinase leads to induction of FruA~P-dependent genes such as *devTRS* and repression of genes such as *dofA* whose expression depends upon the unphosphorylated FruA. It is this shift in gene expression at high C-signaling levels that leads to sporulation.

The three C-signaling pathways indicated are able to control the timing of morphogenesis because they differ in the threshold levels of C-signaling activity required for their activation. Thus, the early, low levels of *csgA* expression are sufficient to activate further expression via the Act proteins. By 6 h of development, the C-signaling levels have reached the threshold for eliciting aggregation. As the cells enter aggregation streams and then form mounds, the level of C-signaling through end-to-end contacts increases to the point where for some of the cells within the mounds sporulation is induced. Other cells within the mounds ultimately undergo autolysis in a form of programmed cell death, while the peripheral rods, cells that failed to enter the aggregates, embark on a completely different program of gene expression. Thus, in *M. xanthus*, it is contact-mediated C-signaling that provides the necessary checkpoints, allowing the extent of morphogenesis to regulate gene expression.

3
Eukarya

3.1
Cell Cycle Control and Differentiation in *Saccharomyces cerevisiae*

Eukaryotic microorganisms, like their prokaryotic neighbors, face fluctuating levels of nutrients and food supplies in the environment, and, like them, have evolved survival strategies to deal with these uncertainties. The baker's yeast *Saccharomyces cerevisiae* has a choice of different developmental pathways when faced with declining nutrient levels (Fig. 12).

At low nitrogen levels in the presence of an abundant, fermentable carbon source, the normally unicellular, diploid yeast switch to a filamentous form of growth during which the cells elongate, bud at the pole opposite the last site of cytokinesis, and fail to separate from daughter cells after budding. This results in the generation of multicellular pseudohyphae, so called because of their macroscopic similarity to the true hyphae of filamentous fungi or molds, which are composed of multinucleated (coenocytic) tubes of continuous cytoplasm. The pseudohyphal growth form of *Saccharomyces* can be considered a nutrient scavenging mode – the elongated form has a more favorable surface-to-volume ratio for nutrient uptake and the filamentous growth allows penetration of the substratum for potentially better access to nutrient sources.

When both nitrogen and fermentable carbon sources are lacking, diploid yeast cells undergo meiotic divisions to produce a packet (ascus) containing four haploid progeny called *ascospores* – two of each of the mating types **a** and **α**. The process of meiosis provides an opportunity for genetic recombination between the members of each pair of chromosomes in the diploid. Just as is the case for *B. subtilis* cells that become transformation competent, new combinations of genetic information could arise from this process that might better suit their bearers to altered environmental circumstances. The ascospores are resistant to dessication, heat, freezing, and some chemical agents (although not as resistant as bacterial endospores) and can be dispersed in that form to environments more suited to growth. They thus represent a potent survival strategy for yeast cells that have fallen on hard times.

Germination of the ascospores produces haploid yeast cells that are able to grow vegetatively just as the diploid parent did from which they were derived. However, if a nearby cell is of the alternative mating

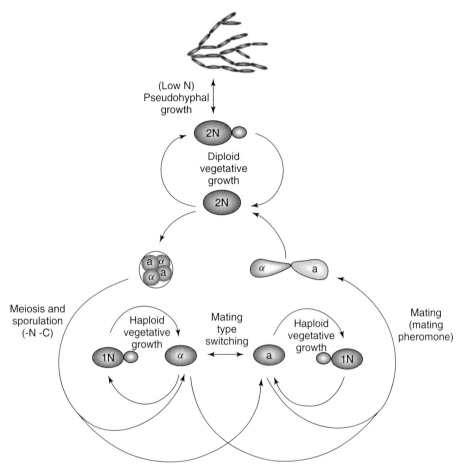

Fig. 12 Cell differentiation during the life cycle of the budding yeast *S. cerevisiae*. In the presence of abundant nutrients, yeast cells can proliferate as either haploid or diploid vegetative cells via a mitotic cell cycle. However, they can differentiate in several ways in response to both internal and environmental cues. Diploid cells, when starved for nutrients, can progress down either of the two pathways of differentiation. Cells starved for nitrogen in the presence of a fermentable carbon source can differentiate into a filamentous form known as *pseudohyphae*, which are capable of invading the substratum. Alternatively, cells starved for nitrogen on a poor carbon source can undergo meiosis and sporulation to generate haploid cells. Haploid cells of both mating types (a and α) can proliferate via a mitotic cell cycle during which they can convert to the other mating type via a programmed DNA rearrangement known as mating-type switching. In addition, when exposed to peptide mating pheromone secreted by cells of the other mating type, haploid cells can differentiate into gametes capable of mating to form a zygote competent to give rise to diploid vegetative cells. From Fig. 1 of Wittenburg, C., La Valle, R. (2003) Cell-cycle-regulatory elements and the control of cell differentiation in the budding yeast, *Bioessays* **25**, 856–867.

type, each of the two haploids responds to a pheromone released by its partner by differentiating into an elongated, pear-shaped gamete called a "shmoo". During shmooing, the cells elongate toward one another in a chemotaxis-like process, meet, and fuse to form a new diploid zygote. Additionally, during growth, yeast mother cells that have budded at least once undergo a process called *mating type switching*, whereby copies of the alternate mating type alleles are placed into the mating-type expression site on chromosome III. Thus, even a solitary ascospore that has by misfortune (or by the design of an experimenter!) found itself without sexually compatible neighbors will generate them during growth and go on to produce diploids homozygous at all except the expressed mating-type loci.

3.1.1 Pseudohyphal Growth

Of the two alternative differentiation pathways open to a starving diploid yeast cell, pseudohyphal growth is the less well studied. However, it is clear that one of the fundamental features of this form of growth is an alteration of the cell cycle – the G_1 phase of the cell cycle (following mitotic division (M phase) and preceding DNA replication (S-phase)) is shortened, while the G_2-phase (following S and preceding M) is lengthened. It is the prolonged period of polar cell growth in G_2 that generates the elongated morphology of the cell prior to unipolar budding. This shifting of the cell cycle results from modulation of the two major cell cycle checkpoints controlling entry to the S- and M-phases respectively. As in other eukaryotic cells, these checkpoints are controlled by cyclin-dependent protein kinases (CDKs), which consist of a catalytic (the kinase) and a regulatory subunit (cyclin). In *S. cerevisiae*,

a single catalytic subunit, Cdc28, is used in conjunction with several different cyclins, which are employed at different stages of the cell cycle. Three G_1 cyclins (Cln1–3) control passage through G_1 and entry to S, while four mitotic or B-type cyclins (Clb1–4) control entry to and passage through mitosis.

The pathways that induce pseudohyphal growth do so by activating the G_1 cyclins (firstly Cln3, which in turn induces synthesis of Cln1 and Cln2) and inhibiting the mitotic cyclins so that passage through G_1 and entry to S is accelerated, while entry to mitosis at the end of G_2 is delayed (Fig. 13). The two best-studied pathways involve a MAP kinase cascade or cAMP-activated protein kinase A. In the former case, the pathway contains most of the elements involved in mating pheromone responses (see later), but leads to phosphorylation of a different MAP kinase – Kss1. Both pathways lead to activation of transcription factors that would induce expression of G_1 cyclins. In addition to these two pathways, the transcription factor Xbp1 and the cyclin-inhibiting tyrosine kinase Swe1 play roles in filamentation, although precisely how they are integrated into the signaling pathways is not clear.

3.1.2 Meiosis and Sporulation

S. cerevisiae zygotes enter meiosis and sporulation when deprived of both a nitrogen source and a fermentable carbon source. The entry to meiosis depends both upon repression of the G_1 cyclins so that the mitotic cycle is arrested in G_1 and upon induction of the transcription factor Ime1, which elicits transcription of meiosis-specific genes (Fig. 14). Both are a response to starvation, but Ime1 is expressed only in diploid cells heterozygous at the mating type locus (i.e. *MAT*a/α). This is

Fig. 13 Proposed pathways regulating cyclin-dependent kinases during pseudohyphal differentiation in *Saccharomyces cerevisiae*. From the nutritional signals that elicit filamentation, the pathways proceed via activation of a MAP kinase phosphorylation cascade, cAMP-activated protein kinase (PKA), Swe1 tyrosine protein kinase (a Cln/CDK inhibitor), or the stress-induced transcription factor Xbp1. Ste11 is the yeast equivalent of mammalian MAPKKK (MAPKK kinase), Ste7 the equivalent of MAPKK (MAPK kinase) and Kss1 is a MAP kinase (MAPK). Flo8 and Tek1/Ste12 are transcription factors. Unphosphorylated Kss1 inhibits Ste12 and its phosphorylation relieves this inhibition. Flo8 and Tek1/Ste12 are transcription factors. The pathways activating Swe1 and Xbp1 are uncertain and both may be coupled to the MAP kinase pathway. A variety of other proteins whose activities influence

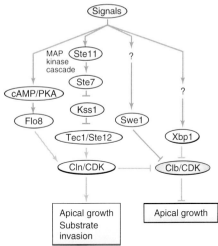

CDK activities also have corresponding effects on filamentation. Green ellipses indicate proteins whose activity stimulates apical growth, while red ellipses indicate proteins whose activity inhibits it. Arrowheads indicate stimulatory interactions and barred ends indicate inhibitory interactions. Blue arrows stand for interactions at the transcriptional level (gene induction or repression) and red arrows stand for interactions at the protein level (activation or inhibition). (See color plate. p. xxix)

because, in haploid cells, the transcriptional regulator Rme1 is active, both in repressing *IME*1 expression and in inducing G$_1$ cyclin expression. *IME*1 expression is also repressed by the G$_1$ CDKs containing cyclins Cln1 or Cln2 (Cln1–2/CDKs). In addition, Cln1–2/CDKs inhibit the nuclear localization of Ime1 with the result that it cannot interact with target DNA sequences. Accordingly, only starving, diploid cells embark upon meiosis.

During mitotic growth of both haploids and diploids, *IME*1 expression is repressed partly by Cln1–2/CDKs and partly by the transcriptional regulator SBF (which also induces the G$_1$ cyclins). Under these same conditions, the Ime1 polypeptide is also inactivated by virtue of being excluded from the nucleus and being inhibited from interacting with its transcriptional coactivator Ume6. Thus, mitotic growth and meiotic sporulation are mutually exclusive.

Amongst the genes whose expression is induced by Ime1/Ume6 upon entry into meiosis is *IME*2, which encodes a meiosis-specific protein kinase. *Ime2* phosphorylates and so promotes the proteolytic destruction of Sic1, an inhibitor of mitotic CDKs containing the cyclins Clb5 and Clb6 (Clb5–6/CDKs). At the same time, Ime2 activates Ndt80, the major transcriptional activator of *CLB*1–6 expression. As a result of these two activities of Ime1, the B-type cyclins are activated and elicit progression through the events of meiosis, including DNA replication, synapsis/recombination, and meiotic division. Along the way, Swe1 functions as a checkpoint kinase to inhibit Clb1–4/CDKs and meiotic progression until chromosome synapsis and recombination events are completed properly.

Swe1 also plays a checkpoint role in mitosis, restricting progression until

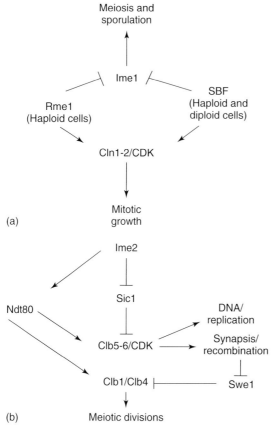

Fig. 14 Regulation of cyclin-dependent protein kinase (CDK) during meiosis and sporulation. Diploid cells undergo meiosis and sporulation when starved for nitrogen and a fermentable carbon source, but grow vegetatively when both nitrogen and carbon sources are abundant. A: Ime1, the primary inducer of meiosis-specific genes, and G1 cyclins, which promote entry into the mitotic cell cycle are subject to differential regulation by the Rme1, a haploid-specific repressor of meiosis, and SBF, the primary G1-specific transcriptional activator. B: The primary repressor of B type cyclin/CDK activity during meiosis, Sic1, and the major transcriptional activator of CLB genes, Ndt80, are differentially regulated by the meiosis-specific protein kinase Ime2. Clb-associated CDKs are also subject to inhibition by the Swe1 protein kinase until synapsis and recombination are complete. From Fig. 2 of Wittenburg, C., La Valle, R. (2003) Cell-cycle-regulatory elements and the control of cell differentiation in the budding yeast, *Bioessays* **25**, 856–867.

the actin cytoskeleton is properly assembled in preparation for budding and, as noted above, it monitors filamentation signals during pseudohyphal growth. How it monitors these disparate signals is not known and is one example of the gaps in our understanding of yeast cell differentiation pathways that remain to be filled.

3.1.3 One Locus, Three Cell Types

The result of meiotic sporulation is an ascus containing four haploid ascospores, two of each mating type (**a** and **α**), derived from one parental diploid cell. When it germinates, each haploid cell

can begin vegetative growth and mitotic proliferation. However, the presence of mating pheromones from cells of the opposite mating type will induce further differentiation – gametogenesis or shmooing – as a prelude to mating. For this to occur, the cells of each mating type must synthesize and release its corresponding pheromone and must also express a receptor for the pheromone of the opposite mating type. In other words, vegetatively growing yeast cells can be one of three distinct cell types – diploid, haploid **a**, and haploid **α**. The different patterns of gene expression in these three cell types result from the expression of

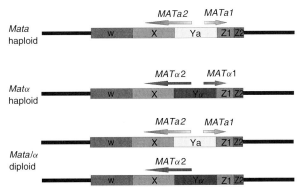

Fig. 15 The mating-type locus in *Saccharomyces cerevisiae*. The locus on Chromosome III can be divided into five regions on the basis of their presence in the two mating type alleles, *MAT*a and *MAT*α, and in the two silent copies on the left (*HML*) and right (*HMR*) arms of Chromosome III. The two *MAT* alleles differ only in the approximately 700 bp region Y, which thus has two alleles Ya and Yα. Each of the Y-regions contains divergent promoters for two transcripts, as indicated by the arrows. Most of the coding sequence of the encoded polypeptides is contained within the Y-region. The *MAT*α1 transcript is repressed in the *MAT*a/α diploid by the Mata1/Matα2 heterodimer. The X and Z1 regions are shared with *HMR*, while the W and Z2 regions are shared with *HML*. These provide regions of sequence identity that promote the strand alignment and annealing involved in the gene conversions that lead to mating type switching.

mating type–specific transcription factors encoded by the mating type alleles in residence at the mating-type (*MAT*) locus on chromosome III.

The genetic configuration of the mating-type locus in each of these three cell types is shown in Fig. 15. As described in the previous section, in haploid cells, the transcriptional repressor Rme prevents expression of Ime and diploid-specific genes. However, *RME* expression is itself repressed in the diploid by a heterodimer of the Mata1 corepressor and the Matα2 homeodomain DNA-binding protein. Cells of the two haploid mating types owe their differences to the activities of the *MAT*α gene products in combination with the Mcm1 polypeptide – Matα2/Mcm1 represses a-specific

genes, while Matα1/Mcm1 induces the expression of α-specific genes. In a haploid a cell, neither of the *MAT*α gene products is present so that the constitutive state prevails – a-specific genes are expressed and α-specific genes are not. Mata2 has no known function, while Mata1 exerts its function only in the diploid state where it can combine with Matα2.

3.1.4 Gametogenesis and Mating

The process of gametogenesis leading to mating is induced in haploid a and α cells by pheromones secreted by neighboring cells of the opposite mating type. The pheromones bind to their cognate receptors (Ste2 for the α-factor, Ste3 for the a-factor) of the ubiquitous eukaryotic GPCR

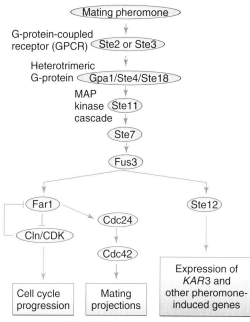

Fig. 16 Signal transduction pathways controlling gametogenesis in *Saccharomyces*. Binding of the pheromone to its cognate receptor (Ste2 = α factor receptor, Ste3 = a factor receptor) causes a conformational change and dissociation of the heterotrimeric G-protein (Gpa1 = α subunit, Ste4 = β subunit, Ste18 = γ subunit). The Ste4/Ste18 ($\beta\gamma$) heterodimer activates Ste20 a serine/threonine protein kinase, which initiates the MAP kinase phosphorylation cascade. Ste11 (MAPKKK), Ste7 (MAPKK) and Fus3 (MAPK) are each in their turn phosphorylated by the preceding kinase and thereby activated. At the end of the phosphorylation cascade, Fus3 phosphorylates and activates the Cln/CDK inhibitor Far1 as well as the transcription factor Ste12. As described in the text, the former causes cell cycle arrest in G1 and formation of mating projections, while the latter causes pheromone-inducible gene expression. Arrowheads indicate stimulatory interactions and barred ends indicate inhibitory interactions. Blue arrows stand for interactions at the transcriptional level (gene induction or repression) and red arrows stand for interactions at the protein level (activation or inhibition). (See color plate. p. xxxi)

(G-protein-coupled receptor) family and thereby elicit dissociation of the α and $\beta\gamma$ subunits of the associated G-protein. The $G_{\beta\gamma}$ heterodimer released in this way then interacts with and activates the Ste20 protein kinase, which phosphorylates Ste11, a MAPKK kinase (MAPKKK). The ensuing phosphorylation cascade culminates in activation of the MAP kinase Fus3, which, in turn, phosphorylates and activates two target proteins – the transcription factor Ste12 and the CDK inhibitor Far1 (Fig. 16).

Ste12~P initiates the characteristic pheromone-dependent changes in gene expression that accompany gametogenesis. For example, pheromone treatment induces a 20-fold increase in the expression of *KAR3*, which encodes a kinesin-like motor protein required for nuclear fusion. At the same time, Far1

that has been phosphorylated by Fus3 inhibits the G_1 CDKs – both Cln3/CDK-mediated induction of Cln1,2 expression and Cln1–2/CDK activities are suppressed, so that the cell cycle arrests in G1-phase. In addition, Far1~P binds to and activates Cdc24, a GTP exchange factor (GEF) for the small GTP-binding protein Cdc42. This promotes release of GDP and binding of GTP to Cdc42, activating its role in actin filament recruitment to and promotion of the mating projections that lend shmoos their characteristic elongated morphology. Cdc42 also plays an important role in actin recruitment during budding of vegetatively growing cells.

As is the case with so many critical developmental events, Far1 activation is coupled to the cell cycle and is autocatalytic. Thus, *FAR1* expression is limited to G1 and is induced severalfold by pheromone signaling. Furthermore, the Far1 protein is also phosphorylated by Cln/CDK, an event that promotes its ubiquitination and proteolytic degradation. By inhibiting Cln/CDK activity, Far1 thus facilitates its own stabilization in G1. After G1, when Cln/CDK levels are high, the protein becomes labile because of its CDK-mediated phosphorylation. This pattern of regulation means that Far1-mediated gametogenesis and mating are restricted to G1. Mutants that lack Far1 still exhibit normal pheromone-induced gene expression because of Ste12 activation, but gametogenesis and mating are impaired. Restricting mating to G1 coordinates it with nuclear events, ensures that each mating partner contributes a single spindle polar body (SPB) for participation in nuclear fusion, and guarantees correct stable ploidy after fusion of nuclei in the zygote. Furthermore, the formation of the mating projection in a shmoo may be effective only in G1 when the cell is not budding.

3.1.5 Silent Copies of the Mating-type Alleles and Mating-type Switching

In homothallic yeast strains, mating can occur in a clonal population of cells derived from a single haploid ascospore. This occurs because of the phenomenon of mating-type switching whereby experienced mother cells (i.e. those that have budded at least once) switch mating types before budding again. Mating-type switching involves copying the mating-type allele present at one of the silent loci *HML* or *HMR* into the mating-type locus *MAT* to replace the existing alleles at that site

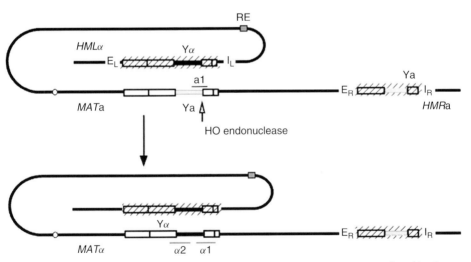

Fig. 17 Mating-type switching at the *MAT*a locus. The HO endonuclease creates a double-strand break at the Y/Z1 border of the *MAT* locus and initiates gene conversion by using the silent donor. The two donor loci (*HML* and *HMR*) are maintained in a transcriptionally inactive chromatin structure (indicated by diagonal stripes). Other shared regions of homology are indicated. The preference for the *HML* donor in *MAT*a cells is mediated by the recombination enhancer (RE) region located ~17-kb centromere-proximal to *HML*. The positions of *HML*, RE, the centromere (○), *MAT*, and *HMR* are indicated. Figure kindly provided by Prof. J.E. Haber, Brandeis University (http://www.bio.brandeis.edu/haberlab/jehsite/donorprf.html).

(Fig. 17). Unlike *MAT, HML* and *HMR* are each flanked by a pair of related, but distant silencer regions – *HML*-E, *HML*-I, *HMR*-E, *HMR*-I – sequences, which are binding sites for a complex of proteins that collaborate to produce a transcriptionally silent region of about 3 kb of highly ordered heterochromatin between the E and the I silencer sequences. These proteins include Sir1–4 (silent information regulator proteins), DNA-binding proteins Rap1 and Abf1, proteins of the origin recognition complex (ORC), which assemble at DNA replication origins, protein *trans*-acetylases and deacetylases, and chromatin assembly factors.

Each of the four silencer regions is able to act as an origin of replication when placed into an extrachromosomal plasmid, but does not function as such normally. Thus, although the ORC proteins in this case are needed for silencing, they do not normally initiate replication at these sites. The Sir proteins do not bind directly to the DNA, but bind indirectly via other DNA-binding proteins in the complexes. The Sir proteins also play other roles in the cells – in gene silencing at telomeres, in maintaining mitotic chromatin structure, in limiting rates of recombination, and in stabilizing rDNA. These functions also have repercussions for cell longevity – Sir proteins promote longevity in yeast cells and Sir2-like proteins (sirtuins – NAD$^+$-dependent deacetylases) do the same in animals. These and many other proteins are required for or affect silencing, enhancing, or weakening it, and their roles are not all understood.

Because of the transcriptional silencing of the *HML* and *HMR* loci, only the mating-type alleles at the *MAT* locus are expressed and thus control the gene expression patterns associated with the haploid mating type or with diploidy (see earlier). The switching of mating types that occurs in homothallic yeast strains is initiated by the activity of HO, a site-specific endonuclease that recognizes a 24-bp sequence at the border of the Y-region with the Z-region in *MAT* (Fig. 15). The silencing of *HML* and *HMR* also protects them from recognition and cleavage by HO, so that only the *MAT* locus is cut to leave a 4-bp, 3′-overhang on each side of the cleavage site. The 3′ ends are resistant *in vivo* to exonucleolytic digestion, but the 5′-ends are processed by 5′ to 3′ exonuclease activity, which includes a trio of interacting proteins (Rad50, Mre11, and Xrs2). This yields long, single stranded 3′ tails both sides of the cut site, one of which extends through the Y-region and the other of which extends through the Z-region. The latter invades the homologous region of *HML* or *HMR*, as the case may be, where it is extended across the HO cut site through the Y-region using the silent alleles as template, replicated and subsequently used to displace the remaining 3′ tail at the *MAT* locus. The nonreciprocal nature of this gene conversion means that a copy of the silent alleles has now replaced the original alleles at the *MAT* locus.

Because there are two silent loci containing mating-type alleles, the process of mating-type conversion involves a selection of either *HML* or *HMR* as the donor locus. This selection is not random – *MAT*a cells almost always use *HML* as the donor and *MAT*α cells use *HMR*. Since the great majority of strains carry α alleles at *HML* and a alleles at *HMR*, this results in high rates of mating-type switching. This donor selectivity results from activation of the left (*HML*) end of chromosome III for mating-type donation (and for recombination) in *MAT*a cells and its inactivation in *MAT*α cells (making *HMR* the only donor available). A 270-bp sequence called *RE* (recombination enhancer) that is responsible for donor selectivity has been identified on the left arm of chromosome III, 17 kb to the right of *HML*. This 270-bp sequence contains four regions (A to D) that are highly conserved amongst different *Saccharomyces* species.

Region C of *RE* includes a consensus binding site, an operator sequence, to which the Matα2/Mcm1 repressor binds. At other such sites in the genome, this represses the expression of nearby **a**-specific genes, but, in this case, in conjunction with the Tup1 and Ssn6 proteins, it inactivates *HML* selection, suppresses other recombinational events in the region, and elicits the assembly of a chromatin structure with tightly phased nucleosomes. In *MAT***a** cells, with Matα2 absent, Mcm1 binds to the same site and activates the region for recombination and *HML* selection. Thus, a 2-bp mutation in the operator eliminates the ability of Mcm1 to bind to it and abolishes the preference for *HML* in *MAT***a** cells. Furthermore, a mutant Mcm1 with reduced ability to bind to the operator also impairs *HML* preference in *MAT***a** cells.

Regions A and D, as well as another nearby sequence region (designated E), have been found to contain sites for *in vivo* binding of two transcriptional activators – the forkhead proteins Fkh1 and Fkh2 and the associated protein Ndd1. As is the case for the Mcm1 site in Region C, the Fkh1/Fkh2 binding sites in Regions A, D, and E do not function in transcription regulation, but in activation of *HML* donor preference.

The process of mating-type switching is one of the best-understood examples of a developmentally regulated DNA rearrangement that functions in the regulatory cascades involved in cellular differentiation. A prokaryotic example involving excision of an intervening sequence (*skin*) was described in the earlier section on *Bacillus* sporulation.

3.2
Differentiation and Morphogenesis in *Dictyostelium discoideum*

Yeast differentiation is fundamentally a unicellular process – even the pseudohyphae, although multicellular, do not contain specialized cells communicating with one another and serving different functions in the whole multicellular assemblage. In this respect, the social amoeba (or cellular slime mould) *Dictyostelium discoideum* is different. In the wild, *Dictyostelium* amoebae lead a predatory lifestyle, actively hunting bacteria chemotactically (bacterially secreted folate and other pterins serving as the attractants) and feeding on them by phagocytosis. Like the prokaryotic myxobacteria, *Dictyostelium* amoebae, when starved at high density, will differentiate and aggregate chemotactically (toward secreted cAMP signals) to eventually form a multicellular fruiting body within which the cells specialize to form different tissues with different functions (Fig. 18). The choices by cells as to which differentiation pathway to take have already been made at the mound stage of development when they first differentiate into recognizably different cell types – prespores and several subclasses of prestalk cells – as revealed by differential expression of cell-type-specific genes. The different cell types have been marked by expression of reporter genes such as *lacZ* under the control of cell-type-specific promoters and their developmental fates mapped (Fig. 18). Although cells in the mounds and later stages of development

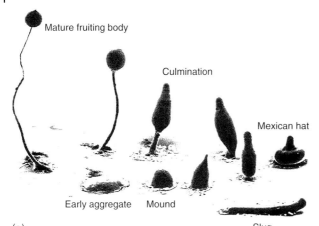

Mature fruiting body

Culmination

Mexican hat

Early aggregate   Mound

Slug

(a)

Fig. 18 Multicellular development in *Dictyostelium*. Panel A. A montage of scanning electron micrographs of stages in the *Dictyostelium* life cycle. Successive developmental stages are shown proceeding anticlockwise from the early aggregate formed by chemotactic aggregation of starving cells. The mature fruiting body is approximately 2 mm high. The original image was kindly provided by M.J. Grimson and R.L. Blanton, Biological Sciences Electron Microscopy Laboratory, Texas Tech University. Panel B. Diagrammatic representation of culmination where, for the sake of clarity, the band of pstB cells that will form the outer basal disc (see Panel C) is not shown. The *ecmA* promoter can be divided into two parts, a proximal part (the *ecmA* region) that directs expression predominantly in the cells within the tip (ie. in the pstA cells), and a distal part (the *ecmO* region) that directs expression in cells in the back of the prestalk region (the pstO cells) and in a subset (pstO:ALC cells) of the anterior-like cells (ALC). The whole *ecmA* promoter (the *ecmAO* promoter) directs expression in all these cell subtypes and has been termed the *pstAO population*. From Plate 6 of Maeda et al.(1997) *Dictyostelium discoideum – A Model Organism for Cell and Developmental Biology*. Universal Academy Press Inc., Tokyo, Japan. Image kindly provided by Prof. J.G. Williams, University of Dundee. Panel C. PstB cell behavior at culmination. The pstB cells are defined by selective staining with neutral red and because they express the *ecmB* gene at a high level relative to the *ecmA* gene. They have a complex movement pattern during slug migration. In this representation, for the sake of simplicity, separate pstA and pstO populations are not shown, but the behavior of the entire pstAO population is represented. From Plate 7 of Maeda et al. (1997). Image kindly provided by Prof. J.G. Williams, University of Dundee. (See color plate. p.xxxii)

are predestined to differentiate according to the developmental fate map shown in Fig. 18, they are not irrevocably committed to do so. In fact, right up until the final stages of differentiation with the deposition of the spore or stalk cell wall, they are able to transdifferentiate into the other cell types or even dedifferentiate and return to the vegetative state under appropriate conditions. This

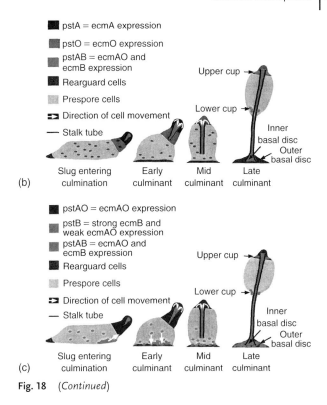

pstA = ecmA expression
pstO = ecmO expression
pstAB = ecmAO and
ecmB expression
Rearguard cells
Prespore cells
Direction of cell movement
Stalk tube

Upper cup →
Lower cup →
Inner
basal disc
Outer
basal disc

Slug entering
culmination
Early
culminant
Mid
culminant
Late
culminant

(b)

pstAO = ecmAO expression
pstB = strong ecmB and
weak ecmAO expression
pstAB = ecmAO and
ecmB expression
Rearguard cells
Prespore cells
Direction of cell movement
Stalk tube

Upper cup →
Lower cup →
Inner
basal disc
Outer
basal disc

Slug entering
culmination
Early
culminant
Mid
culminant
Late
culminant

(c)

Fig. 18 (*Continued*)

developmental plasticity is important in the changeable soil environment for a fragile organism that might easily suffer mechanical injury with tissue loss during development.

Unlike the myxobacteria, *Dictyostelium* interposes between the aggregation stage and the final formation of the fruiting body, a multicellular migratory stage – the "slug", so called because of its superficial resemblance to garden slugs. The slug is phototactic and thermotactic – behaviors which are controlled by a specialized region in its tip and designed to take the organism to the soil surface, the most advantageous place from which to disperse the spores of the fruiting body. The slug contains several different tissues and cell types, distinguishable on the basis of subtle morphological differences

and different patterns of gene expression (Fig. 18). Intercellular signals and associated intracellular signaling cascades control the initial transition from growth to development, the subsequent process of chemotactic aggregation, the formation of the slug and its motile behavior, the morphogenetic processes that transform the slug into a fruiting body as well as the spatial patterns, and proportions of its cell types. Not surprisingly, *Dictyostelium* devotes more of its ca. 12 000 genes to signal transduction than any other well-studied microorganism.

3.2.1 The Transition from Growth to Development

Like the prokaryotes *Bacillus* and *Myxococcus* and the eukaryote *Saccharomyces*, *Dictyostelium* differentiation is elicited in response to two major signals – shortage of nutrients and signals from neighbors. At the onset of starvation at high cell density, polysomes are disassembled (and later reassembled), the expression of growth-specific genes (such as those encoding UMP- and GMP-synthetases) is repressed, and the earliest developmentally induced genes are expressed. Although the precise molecular nature of the starvation signal is not understood, at least one signal has been characterized that informs the cells of their density in a growing culture. That signal is PSF (prestarvation factor), a 65-kD protease-sensitive and heat-labile glycoprotein that is secreted into the medium. Its concentration is proportional to the cell density and reaches the threshold for eliciting development about four generations prior to the end of the exponential phase of growth. PSF induces some of the first genes to be expressed during development when added to growing cells that would otherwise not express them. These include *carA* (encoding the major cAMP receptor used during chemotactic aggregation), *pdsA* (which encodes a secreted cAMP phosphodiesterase needed during aggregation), and members of the discoidin gene family (possibly used during aggregation for cell–substratum interactions). However, this response is inhibited by the presence of bacteria, so that the PSF response effectively measures the ratio of amoebal to bacterial cell density. When exhaustion of the bacterial food supply becomes imminent, the first events in the growth to development transition are induced.

The PSF receptor has not been identified, but several intracellular signaling proteins have been identified as playing roles in the PSF signal transduction pathway (Fig. 19). YakA is a protein kinase whose absence results in cells that fail

to aggregate and fail to repress growth-specific genes at the onset of starvation. YakA is a member of the family of Dyrk (Dual specificity Yak-related kinases) found in a variety of eukaryotes, including *Saccharomyces, Drosophila* and mammals, where their functions in the cell cycle and development, when known, appear to be analogous to those of YakA in *Dictyostelium*. The use of a temperature-sensitive YakA mutant allowed demonstration that it plays essential roles throughout development, downstream of G-protein-coupled receptors. PufA is a member of the pumilio/FBF family of proteins that bind to specific sequences at the 3′ ends of target RNA transcripts, preventing their translation. In this way, the *Drosophila* pumilio protein, together with the nanos protein, binds to nanos response elements (NREs) in the mRNA for the hunchback protein, inhibiting its expression. PufA in *Dictyostelium* binds to similar sequence elements found near the 3′ end of the coding sequence of the mRNA of PKA (protein kinase A, cAMP-dependent protein kinase). PufA was found by virtue of the fact that in a YakA-deficient mutant, disrupting the *pufA* gene restores starvation-induced PKA synthesis and development. PufA is expressed during growth, but disappears after a couple of hours of starvation. While present and active, it keeps PKA expression suppressed at the translational level even after transcription of the *pkaC* gene has been initiated in the earliest stages of development.

Intracellular cAMP signals, through PKA, act as master signals throughout *Dictyostelium* development and, like so many key developmental signals, are autocatalytic in nature. The genes whose transcription is induced by PKA-activated

Fig. 19 The Prestarvation Factor (PSF) signaling pathway for initiating *Dictyostelium* development. PSF binds to an unknown receptor presumably belonging to the GPCR (G-protein-coupled Receptor) superfamily and elicits a signaling cascade that activates the protein kinase YakA, which, in turn, phosphorylates and inhibits the RNA-binding protein PufA. In its active, nonphosphorylated form, PufA binds to the 3′ end of cAMP-dependent protein kinase (PKA) mRNA preventing its translation. Once made and activated, PKA induces downstream genes directly or indirectly by phosphorylation of target proteins, ultimately regulating specific transcription factors such as Myb2 or CRTF (cAMP responsive transcription factor). Growth phase-specific

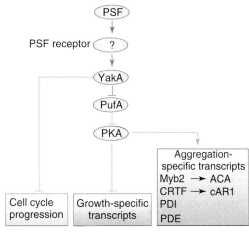

transcripts are repressed and aggregation-specific transcripts are induced. These include those required for synthesizing and secreting cAMP such as adenylyl cyclase A (ACA) and for sensing and responding to extracellular cAMP signals, such as the cAMP receptor (cAR1). This establishes an autoactivatory feedback loop for induction of early development. YakA also phosphorylates other targets to arrest cell cycle progression at the phase shift point PS where cells exit from the cell cycle and enter differentiation. Arrowheads indicate stimulatory interactions and barred ends indicate inhibitory interactions. Blue arrows stand for interactions affecting protein expression (at the transcriptional or posttranscriptional level), and red arrows stand for interactions affecting protein activity (activation or inhibition). (See color plate. p. xxxiii)

transcription factors include those encoding the major adenylyl cyclase involved in aggregation (ACA), the cAMP receptor (cAR1), the extracellular cAMP phosphodiesterase (PDE) whose activity keeps background extracellular cAMP levels low during aggregation, and the phosphodiesterase inhibitor (PDI) that prevents premature destruction of the extracellular cAMP signal in the earliest stages of aggregation. One of the responses to extracellular cAMP is activation of ACA to produce more cAMP, which is secreted as the extracellular attractant during aggregation, but which also functions as an intracellular signal activating PKA. PKA signaling also represses growth-specific transcripts. Two of the transcription factors that participate in these signaling pathways downstream of PKA are Myb2 and CRTF (cAMP response transcription factor).

Another action of PSF signaling through YakA is to cause cell cycle arrest. Conditional YakA overexpression causes growth arrest, while growth-specific genes including *PufA* continue to be expressed in YakA-deficient mutants. As is the case with some rapidly growing embryonic cells in metazoa, the *Dictyostelium* cell cycle includes only a very short or nonexistent G1 phase. Experiments with synchronously dividing cell populations have revealed that starving cells progress about two-thirds of the way through G2 before they exit the cell cycle and differentiate. The cells that are closest to this point at the onset of starvation differentiate first, are presumably the first to start extracellular cAMP signaling, and thereby become the cells at the centers of aggregation. During subsequent differentiation in the aggregates, they preferentially differentiate into prespore cells and ultimately into spores. The later arrivals, initially at the periphery of the aggregates, preferentially become prestalk cells, entering the tip and ultimately forming the stalk.

There are other proteins involved in the growth-to-development transition whose roles are not yet completely clear, including GdtA (a large, integral membrane protein) and AmiB, which appears to act upstream of Myb2 (as mutant phenotypes can be suppressed by ectopic expression of Myb2). Clearly, there is more to be learned about how development is initiated in *Dictyostelium*, but a solid foundation for future work has been laid.

3.2.2 Chemotactic Aggregation
PSF informs the cells of imminent depletion of the bacterial food supply and induces early gene expression in preparation for aggregation. Once the food supply has been exhausted and starvation has begun, PSF production declines and a second secreted protein called CMF (conditioned medium factor) assumes the role of a cell-density-dependent signal. CMF is an 80-kD glycoprotein whose presence is required above a threshold concentration for the onset of aggregation. Starved mutant cells lacking CMF are unable to respond to extracellular cAMP signals with the normal Ca^{2+} influx, cGMP synthesis, or cAMP synthesis, and they cannot aggregate unless CMF is supplied exogenously. However, they become responsive to cAMP within 10 s of exposure to CMF, suggesting that CMF's permissory action in aggregation is to regulate cAMP signal transduction. CMF also induces expression of a number of developmentally regulated genes including those encoding gp80 (a cell–cell adhesion molecule involved in aggregation) and SP70 (or CotB,

a protein expressed in prespore cells in aggregates). The details of the CMF signaling pathways remain unknown, but there appear to be at least two CMF receptors with affinities in the nanomolar range – a G-protein-coupled receptor regulating cAMP responsiveness and a 50-kD transmembrane protein with two or three predicted transmembrane domains (CMFR1) mediating CMF-induced gene expression.

The first hours of differentiation in *Dictyostelium* render the cell competent to begin synthesizing, secreting, and responding to the chemoattractant cAMP. The cAMP signals are detected by cAMP receptors belonging to the ubiquitous eukaryotic GPCR family with their seven transmembrane domains, extracellular ligand-binding regions and intracellular signaling domains that couple to heterotrimeric G-proteins. There are 4 cAMP receptors in *Dictyostelium*, cAR1-4 encoded by the genes *carA-D*, two with high affinities (cAR1,3 – in the nM range) and two with lower affinities (cAR2,4 – in the µM range). The major receptor involved in aggregation is cAR1. Binding of extracellular cAMP to cAR1 elicits three kinds of responses:

1. Synthesis and secretion of cAMP so that the chemoattractant signal is relayed from one cell to the next.
2. Chemotactic movement toward the source of cAMP so that the cells aggregate.
3. Expression of aggregation-specific genes including those required for cAMP responses so that once it has been initiated, aggregation is an autocatalytic process.

During aggregation, the cAMP signal is pulsatile with a period of about 6 min and it spreads through the field of aggregation-competent amoebae as waves that emanate from the aggregation center. This occurs because the cAMP signal elicits synthesis and secretion of further cAMP by neighboring cells, while at the same time extracellular phosphodiesterase degrades the cAMP left behind in the back of the passing wave. The signal transduction pathway involved in cAMP signal relay is shown in Fig. 20. Binding of cAMP to cAR1 elicits a conformational change in the receptor and dissociation of the heterotrimeric G-protein $G\alpha 2\beta\gamma$

from the receptor to release the $\alpha 2$ subunit and the $\beta\gamma$ heterodimer. The $\alpha 2$ subunit stimulates some of the downstream events involved in chemotaxis, such as cGMP synthesis, while the $\beta\gamma$ heterodimer stimulates phosphatidylinositol-3 kinases (PI3K1-3) to phosphorylate phosphatidyl inositol bisphosphate (PIP2), converting it to PIP3. PIP3 recruits the protein CRAC (cytosolic regulator of adenylyl cyclase) through its PH (pleckstrin homology) domain to the membrane from where

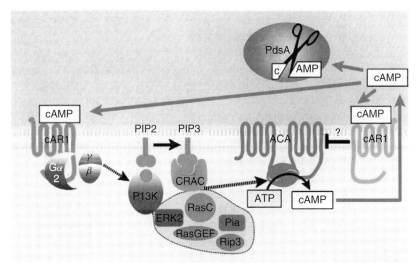

Fig. 20 The regulation of adenylyl cyclase A. Binding of cAMP to the serpentine receptor cAR1 induces dissociation of the heterotrimeric G-protein, G2, into its α and $\beta\gamma$ subunits. The $\beta\gamma$ subunits induce activation of phosphatidylinositol-3 kinase (PI3K) in a manner that is not yet understood. PI3K converts phosphatidyl inositol bisphosphate (PIP2) into phosphatidyl inositol trisphosphate (PIP3). PIP3 binds to the pleckstrin homology domain of CRAC, the cytosolic regulator of adenylyl cyclase and thereby recruits CRAC from the cytosol to the plasma membrane, where it can participate in adenylyl cyclase A (ACA) activation. A set of proteins, including the MAP kinase, ERK2, the small G-protein, RasC, the guanine nucleotide exchange factor, RasGEF, the Ras Interacting Protein (Rip3), and a novel protein, pianissimo (Pia) are also required for ACA activation. The interactions of these proteins with each other and with CRAC and ACA have not yet been clarified. cAMP produced by ACA is rapidly secreted to further activate ACA in a positive feedback loop. Binding of cAMP to cAR1 blocks ACA activation via a negative feedback loop, that is little understood. Extracellular cAMP is degraded by the phosphodiesterase PdsA, which terminates both loops and returns cells to the basal excitable state. From Fig. 2 of Saran, S., Meima, M., Alvarez-Curto, E., Weening, K.E., Rozen, D.E., Schaap, P. (2002) cAMP signaling in *Dictyostelium, J. Muscle Res. Cell Motil.* **23**, 793–802.

it activates ACA (adenylyl cyclase A). Other proteins that participate in this process include ERK2 (extracellular signal regulated kinase 2), RasC, Aimless (a guanine nucleotide exchange factor that activates Ras) and Rip3 (Ras interacting protein 3). The recruitment of CRAC to the plasma membrane occurs locally in the cell in the region of highest cAMP concentration, closest to the cAMP source. This would be the leading edge of an aggregating cell. However, the adenylyl cyclase is localized at the opposite end of the cell where synthesis and secretion of cAMP relays the attractant signal to cells further out from the aggregation center. Clearly, information must pass from the leading edge to the rear of the cell to activate ACA, but the nature of this information transfer is unknown.

Amoeboid motility involves different cytoskeletal events in different parts of the cell – pseudopod activation and extension at the leading edge, retraction in the rear. Chemotactic signals activate both of these processes in such a way as to translate a shallow extracellular attractant gradient into steep intracellular signaling and response gradients – a fast, local pseudopodium activation signal at the leading edge and a slower, global pseudopodium inhibition signal elsewhere. This spatially polarized response is not due to gradients of chemoreceptor or associated G-proteins, as fluorescently tagged forms of these are distributed uniformly or close to uniformly around the cell surface. Instead, it is the activation of PI3 kinases at the upgradient or leading edge, which exhibits the first dramatic asymmetry in chemotactic signaling events. Their action is opposed by that of PTEN, a phosphatase, which converts PIP3 (phosphatidyl inositol triphosphate) back to PIP2. After a chemotactic stimulus, the PI3 kinases phosphorylate PIP2 to form PIP3 in the membrane at the leading edge of the cell. At the same time PTEN, which is initially distributed in the cell membrane uniformly over the entire cell surface, is displaced from the front leaving it localized in the rear portions of the cell. This accentuates further the PIP3 gradient in the membranes created by the PI3 kinases. The phenotypes of null mutants show that PTEN and all three PI3 kinases in *Dictyostelium* contribute to chemotactic orientation – PI3K2 to the greatest extent, followed by PI3K1 and PI3K3. The activation of PI3 kinases involves their dynamic translocation from the cytoplasm to the membrane by an unknown recruitment mechanism that depends upon their N-terminal domains.

Once recruited to the membrane, the PI3 kinases are activated by a Ras protein through their Ras-binding domains to synthesize PIP3. This results in the recruitment of several PH domain proteins to the site including not just CRAC (see above) but also the docking protein PhDA and Akt/PKB protein kinase. PhDA acts at the leading edge to facilitate the assembly there of the necessary proteins for pseudopodial extension including F-actin. PhDA-null mutants are defective in actin assembly at the leading edge. Other proteins implicated in F-actin assembly and pseudopodium extension include the Arp2/3 complex SCAR and WASP (Wiskott-Aldrich Syndrome Protein). Amongst the target proteins for the Akt/PKB kinase is PAKa, itself a protein kinase homologous to the mammalian p21-activate kinases (PAKs). Surprisingly, in response to an attractant stimulus, PAKa translocates to the posterior of the cell where it associates with the actomyosin cytoskeleton. PAKa activity in the rear cortex of the cell inhibits myosin heavy chain kinase, causing dephosphorylation of the myosin heavy chain by uncharacterized phosphatases. This activates myosin assembly into bipolar thick filaments, its association with actin, and its participation in actomyosin contractions. Cortical contraction of the rear of the cell is inimical to pseudopodium extension. Thus, the relocation of PAKa to the rear after its PKB-mediated phosphorylation at the leading edge provides a means whereby localized pseudopodium activation (in the front) can be coupled to and accompanied by a pseudopodium inhibition signal elsewhere (in the rear).

PTEN and PAKa are not the only signaling molecules whose site of action in chemotaxis is in the rear cortex. Attractant binding also elicits an influx of extracellular Ca^{2+} that occurs primarily in the rear cortical regions of the cell via both a G-protein-dependent and G-protein-independent pathway. A number of the actin-binding proteins that regulate its assembly into a cross-linked network of filaments are Ca^{2+}-regulated. In particular, the actin cross-linking protein, α-actinin, is Ca^{2+}-inhibited, while the actin filament severing and capping protein, severin, is Ca^{2+}-activated. The combined effect of elevated Ca^{2+} on these proteins would be a shortening of the actin filaments and a reduction in the extent of their cross-linking. This could facilitate the myosin-mediated sliding of actin filaments required for contraction of the rear cortex. However, prevention of the Ca^{2+} influx either with Ca^{2+} channel blockers or by disruption of the *iplA* (IP3-receptor-like protein) gene has no obvious effect on chemotaxis, so that the possible role for Ca^{2+} fluxes in chemotaxis remains elusive.

Another small molecule whose concentrations increase in response to an attractant stimulus (a second messenger) is cGMP. In this case, there is strong genetic evidence in the form of mutant phenotypes that cGMP plays an important role in actomyosin-mediated contractile events that inhibit pseudopodium extension in the rear cortex. The intracellular concentrations of cGMP are controlled by two guanylyl cyclases (GCA and sGC) and three phosphodiesterases (PDE3, 5, and 6), one of which (PDE6) prefers cAMP over cGMP as a substrate. The guanylyl cyclases are activated either directly or indirectly by the $\alpha2$ subunit that has been liberated from its heterotrimeric G-protein-receptor complex by cAMP binding to the receptor. Their action is countered by the phosphodiesterases, one of which (PDE5) is itself activated by cGMP at an allosteric site separate from the catalytic site. cGMP is a small highly diffusible molecule highly suited to carrying a global pseudopodium inhibition signal to all parts of the cell upon attractant stimulation. It binds to two proteins (GbpD and GbpC) whose combined activities initiate a phosphorylation cascade terminating at the phosphorylation of the regulatory light chain (RLC) of myosin II (Fig. 21). RLC phosphorylation by MLCKA (myosin light chain kinase A) and other unidentified kinases activates both the association of myosin with the actin cytoskeleton in the rear cortex and its catalytic activity in contraction.

Chemoattractant stimulation of *Dictyostelium* amoebae has been reported to elicit a number of other molecular events whose roles in chemotaxis are unclear or controversial. These include K^+ efflux and H^+ efflux, phospholipase C-mediated release of inositol 1,4,5-triphosphate, and an associated release of Ca^{2+} from the endoplasmic reticulum into the cytoplasm. In addition, all of the described intracellular signaling processes that mediate both cAMP signal relay and chemotaxis are transient in the face of a constant extracellular

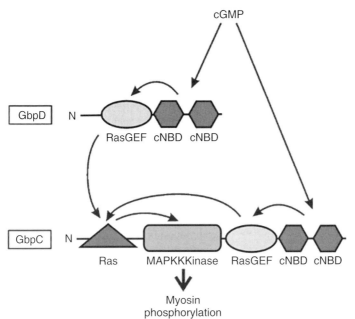

Fig. 21 Model of the function of GbpC and GbpD. The cyclic nucleotide–binding domains (cNBD) of the cGMP-binding proteins GbpC and GbpD bind cGMP, leading to the activation of the Ras guanine exchange factor domains (RasGEF), which activates the small G-protein domain (Ras) of GbpC. This results in the activation of the kinase domain of GbpC (MAPKKK), which stimulates a phosphorylation cascade eventually leading to the phosphorylation of myosin II and chemotaxis. Modified from Fig. 4 of Bosgraaf, L., van Haastert, P.J.M. (2002) A model for cGMP signal transduction in *Dictyostelium* in perspective of 25 years of cGMP research, *J. Muscle Res. Cell Motil.* **23**, 781–791.

cAMP stimulus. This is a result of adaptation processes that are poorly understood and include, but are not limited to, phosphorylation of the intracellular C-terminal domain of the cAMP receptor. The phosphorylated receptor has reduced affinity for cAMP. Adaptation not only renders the responses transient but also ensures that cells respond to relative changes in attractant concentration rather than to the absolute levels of the attractant.

The final response to pulsatile cAMP signaling during aggregation is induction of aggregation-specific gene expression. This begins with the early starvation-induced differentiation events as soon as intracellular cAMP and PKA levels are sufficient for PKA activation. However, with the arrival of each wave of extracellular cAMP, the ACA is activated and the intracellular levels of cAMP rise dramatically before most of it is secreted. The resulting pulsatile activation of PKA induces further expression of aggregation-related proteins including gp80 or contact sites A (csA), which mediates end-to-end contact of cells in aggregation streams. These contacts facilitate, but are not essential for, the natural aggregation process. Contact sites A was one of the earliest cell–cell adhesion molecules (CAMs) to be isolated from any organism and the methods used to identify it were subsequently applied successfully to mammalian systems. During the course of aggregation, the expression of another adenylyl cyclase (ACB) is induced and, although it is not activated by extracellular cAMP signals, it contributes to basal cAMP synthesis and secretion rates and thus to the activation of PKA. Both ACA and ACB contribute to the supply of intracellular cAMP to activate PKA beyond aggregation into the later stages of development.

3.2.3 Postaggregative Development and Cell Type Choice

The result of chemotactic aggregation is that the cells crowd together into aggregation streams and then into aggregates that with time become ever more tightly packed. One consequence of this is the accumulation of a suite of extracellular protein factors that sense cell density and act to break up aggregate streams, thereby preventing aggregates from becoming too large. One of these factors is CF (counting factor), which is a high molecular weight complex (~450 kD) of several polypeptides including countin (40 kD) and CF50 (50 kD). Another of the secreted proteins affecting aggregate size is a homolog of countin called countin2 that is not part of CF. The ability of CF to promote the disintegration of aggregation streams has been attributed to the combined effects of subtle changes in cAMP signal relay, cell–cell adhesion, and actomyosin-mediated chemotactic motility.

A second consequence of cells crowding together into aggregates is that the extracellular cAMP levels experienced by the cells continue to rise during aggregation to reach micromolar concentrations in the mound and slug stages. At these concentrations, binding to the high-affinity receptors cAR1 and cAR3 is saturated so that even in the face of pulsatile cAMP

signals, these receptors perceive a constant stimulus that induces expression of genes important for postaggregative development. These genes include GBF (G-box-binding factor) a transcription factor required for the expression of postaggregative genes, LagC (a cell surface signaling molecule), and RasD (a small GTPase of the Ras superfamily). GBF binds to a defined GC-rich sequence element in the promoters of these genes called the *G-box*, and is required for the transcription of both prespore-specific and prestalk-specific genes. GBF is activated by high cAMP concentrations through the cAR1 cAMP receptor in a G-protein-dependent manner. Extracellular cAMP thus acts as a cell density–sensing signal itself, coupling the progress of morphogenesis to expression of the appropriate genes.

The morphogenetic movements required for slug formation and migration as well as culmination are controlled by high concentration waves of cAMP that are relayed through the developing tissue in a manner analogous to the waves of cAMP that orchestrate the movements of aggregating cells earlier in development. These cAMP waves can be sensed by the low affinity cAMP receptors cAR2 and cAR4, which are expressed at these later stages of development.

Although high extracellular cAMP concentrations and active GBF are necessary for the expression of developmentally regulated genes during postaggregative development, they are not sufficient on their own to ensure proper morphogenesis and differentiation into the correct cell types in the correct spatiotemporal sequence. This requires additional information to inform cells of their appropriate developmental fate and position in the multicellular aggregate. This information takes two forms. The first is historical – cells that happened to be in the mid-to-late G2-phase of the cell cycle at the time of starvation begin to differentiate and enter the aggregate first, occupying its central regions where they preferentially differentiate into prespores and ultimately spores. The later arrivals initially occupy the outer layers of the mound and preferentially locate (sort) to the tip as prestalk cells. The second type of information informing cells as to their correct developmental choices takes the form of coupled extracellular signals. These induce cells to choose either the prestalk or prespore differentiation pathways with correspondingly different locations in the aggregate and different fates in the final fruiting body (Fig. 18). They are also responsible for regulation of cell-type proportions, which remain relatively constant both in the face of major variations in slug size and after mechanical trauma to the slug, resulting in removal of parts of the tissue (such as might occur accidentally in the soil environment or be deliberately performed by an experimenter!). Slugs that have been injured in this way reorganize their tissues to form normal slugs with the correct proportions and locations of the different cell types.

The best understood of these extracellular signals is differentiation-inducing factor (DIF), a dichlorinated hexaphenone that is produced in the mound and later stages of development and induces cells to differentiate into stalk cells. In the presence of high extracellular cAMP concentrations (required both for prespore and prestalk differentiation), DIF will cause most cells in a suspension culture to enter the stalk differentiation pathway, as assayed by expression under the control of the prestalk cell-specific *ecmA* and *ecmB* promoters. DIF does this both by repressing prespore-specific gene expression and inducing prestalk-specific gene

expression (although there are at least two genes expressed preferentially in prestalk cells that are not DIF-inducible, *tagB* and *carB*). The prestalk differentiation pathway leading to stalk cell formation is a form of programmed cell death that differs from, but shares some features with apoptotic and necrotic cell death pathways in mammalian cells. Interestingly enough, DIF has been found to induce cell death in mammalian cells as well, making it a potential antitumor agent.

Although the pathways involved in DIF biosynthesis in *Dictyostelium* have been elucidated, the signaling pathways involved in DIF-induced gene expression remain to be clarified – the only signaling molecule known to be involved is intracellular Ca^{2+}. DIF treatment elicits a slow prolonged elevation of intracellular free Ca^{2+} levels that temporally coincides with induction of *ecmB* expression, a marker for late prestalk differentiation. Artificially inducing a similar elevation of intracellular Ca^{2+} levels by pharmacological inhibition of ATP-driven Ca^{2+} pumps in the endoplasmic reticulum also causes *ecmB* expression. Conversely, inhibition of the DIF-induced Ca^{2+} elevation prevents *ecmB* expression. The induction of expression of an "early" prestalk gene, *ecmA*, by DIF precedes and is independent of the elevation of Ca^{2+} concentration. Thus, there are at least two different DIF signaling pathways – one involved in "early" prestalk gene expression that is independent of Ca^{2+}, and a Ca^{2+}-dependent pathway involved in "late" prestalk gene expression.

What is the relationship between the cell cycle–regulated and DIF-mediated choices of cell type? This was investigated by determining the DIF-sensitivity of cells starved at different stages of the cell cycle. Cells starved late in the cell cycle, which preferentially differentiate into pre-spores were less sensitive to DIF than "stalk-loving" cells starved earlier in the cell cycle. Thus, the cell cycle–regulated preferences for particular cell types may be a result of differential DIF sensitivity. On the other hand, a mutant that is DIF-deficient (but not totally DIF-less) still shows relatively normal development, suggesting that cell-type choice is not completely DIF-dependent. It is possible that in normal development, cell-type preferences are initially based on the stage of the cell cycle at which cells entered starvation and that the role of DIF is to reinforce this and to generate correct proportions of the two cell types.

How would DIF allow cells to sense cell number in such a way as to produce correct cell-type proportions in the multicellular tissue of the slug? DIF is produced by prespore cells which are, however, comparatively resistant to its stalk-inductive activity. Prestalk cells are much more sensitive to DIF and also produce a DIF dechlorinase, which degrades it so that DIF concentrations in the slug are lower in the tip region. These properties can explain how DIF regulates cell-type proportions as follows. An overabundance of prespore cells would result in overproduction of DIF and underproduction of the dechlorinase so that DIF levels would rise. This would cause some cells to transdifferentiate from prespore into prestalk cells and thereby lead to a decrease in DIF levels. At some point, the decreased DIF levels would no longer be sufficient to induce further transdifferentiation, but would be sufficient to prevent the more DIF-sensitive prestalk cells from converting back into prespore cells. Conversely, if prestalk cells were overabundant, DIF levels would drop too low to prevent some of them from transdifferentiating into prespores and producing

more DIF, until its levels rise sufficiently to stabilize the prestalk/prespore proportions. With DIF being responsible for regulating the proportions of prespore and prestalk cells in aggregates, their spatial separation in the slug results from the fact that prestalk (pstA) cells sort chemotactically to the tip. This differential sorting behavior of the different cell types couples the spatial arrangement of cells to their differentiation state.

The differentiation of the different cell types in *Dictyostelium* aggregates must require the activity of cell-type-specific transcription factors. Whereas GBF is a necessary transcription factor for both prestalk and prespore-specific genes, the transcription factors STATa and STATc regulate differentiation specifically in the stalk pathway. Both belong to a family of transcription factors that are activated by tyrosine phosphorylation. In mammalian cells, STAT phosphorylation is mediated by receptor tyrosine kinases (receptors with an intracellular tyrosine kinase catalytic domain) and Janus kinases (JAKs) coupled to ligand-induced receptor oligomerization. In *Dictyostelium*, no JAK homologs have been found, and the tyrosine kinases responsible for STAT phosphorylation remain unidentified. STATa induces some prestalk genes (e.g. *cudA*) and represses others (e.g. *ecmB*). The main defect in STATa deficient mutants is in precocious expression of late prestalk genes exemplified by the marker *ecmB*, so that they are expressed throughout the prestalk region. The phosphorylation of STATa is dependent on the cAR1 cAMP receptor (but independent of heterotrimeric G-proteins) and leads to its homodimerization, activation, and translocation to the nucleus. During normal development, STATa initially localizes to the nucleus in all cells in the mounds, but this is maintained subsequently only in tip cells. Within minutes of exposing cells to exogenous DIF, STATc is phosphorylated and translocated to the nucleus. It represses *ecmA* and is responsible for the differential expression of *ecmA* in different subpopulations of prestalk cells. Like STATa, STATc can also function as an activator of gene expression, for example, in induction of *gapA* and *rtoA* expression in response to hyperosmotic stress.

The major roles of both STATa and STATc in prestalk differentiation appear to be as repressors. There must also be a positive transcriptional regulator responsible for DIF-induced prestalk gene expression. DimA is a bZIP/bRLZ family transcription factor that is required for DIF induction of prestalk genes – a DimA-null mutant produces, but is unresponsive to both endogenous and exogenous DIF.

Our understanding of the signaling molecules and pathways required for regulating cell-type choice and proportions in *Dictyostelium* is still very incomplete. Other proteins for which mutant phenotypes indicate roles in these pathways include the putative transcriptional regulators rZIP (which contains a RING-type Zn^{2+}-binding domain, a leucine zipper, and an SH3 binding motif) and Wariai (a homeobox protein preferentially expressed at the mound stage). Disruption of the genes encoding rZIP or Wariai cause, respectively, increases or decreases in the proportion of prespore cells. The upstream signaling pathways controlling cell-type proportions include MEKKα (a MAPKK kinase), whose absence causes a decrease in the proportion of prespore cells.

3.2.4 Culmination and Terminal Differentiation

Dictyostelium slugs migrate for a variable period of time determined by genotype and by environmental conditions – low humidity, high ionic strength, overhead light, low cell density prior to aggregation, and small slug size are amongst the circumstances that favor an earlier decision to cease migration and culminate. During development, *Dictyostelium* cells use as a source of energy proteins from vegetative growth and earlier developmental stages that are no longer required. Ammonia is released as a waste product of this process and is used as a morphogenetic signal. Many of the conditions that affect the decision to culminate do so by influencing the concentrations of free ammonia in and around the tip of the migrating slug – low NH_3 levels induce culmination. Overhead light, for example, causes the slugs to spend more time with their tips lifted into the air as a result of phototactic responses. This allows gaseous NH_3 to escape more rapidly from the slug with a concomitant decrease in the concentrations in the tip.

Culmination proceeds by a series of morphogenetic movements in which pstA cells enter a stalk tube funnel in the central regions of the tip and differentiate into stalk cells as they move downward in the direction of the substratum relative to other cells in the aggregate. Terminal differentiation of stalk cells involves enlargement and vacuolization of the cell, laying down of a cellulosic cell wall and, ultimately, membrane breakdown and cell death. The beginning of this process is marked by expression of *ecmB* and other late prestalk genes as the cell enters the stalk funnel. The other cells ascend the nascent stalk as it forms, and the prespore cells amongst them undergo terminal differentiation to form a droplet of spores atop the stalk in the mature fruiting body (Fig. 18). The terminal differentiation of spores involves cell dehydration and shrinkage, the production of a peculiar cytoskeletal structure (the actin rod) responsible for the elongated, slightly curved shape of mature spores, and the laying down of the spore wall – an extracellular matrix of cellulose and proteins whose incorporation into the coat involves extrusion of the contents of specialized vacuoles found in prespore cells.

As with all of the other major stages of *Dictyostelium* development, intracellular cAMP signals activating PKA are essential for terminal differentiation. Thus, precocious maturation of both stalk and spore cells occurs in response to treatment with the active, membrane permeant cAMP analogue 8-bromo-cAMP and in mutants that overexpress the catalytic subunit of PKA (PKAC), or are deficient in the inhibitory, regulatory subunit of PKA (PKAR) or which lack RegA – a cAMP phosphodiesterase that limits the accumulation of intracellular cAMP. Conversely, maturation of both cell types is inhibited by overexpression of a mutant form of the regulatory subunit that permanently inhibits the catalytic subunit because it is no longer responsive to cAMP. These results show that a further elevation of intracellular cAMP concentrations and correspondingly higher levels of PKA activity act as the signal to permit or initiate terminal differentiation and culmination.

What causes cAMP levels to increase at the time when culmination is initiated? Current evidence indicates that extracellular ammonia signals detected by the cell surface receptor DHKC act to inhibit the initiation of culmination (Fig. 22). DHKC null mutants exhibit accelerated morphogenesis like mutants deficient in RegA or PKAR. Conversely, culmination is delayed by overexpression of a DHKC mutant form that lacks the DHKC sensor

domain and may be constitutively active in signaling. This arrest in development can be overcome either by addition of 8-bromo-cAMP or by a deficiency in the RegA cAMP phosphodiesterase. The precocious culmination of the DHKC null mutant cannot be rescued by ammonia (which normally delays culmination), presumably because the mutant is unable to sense the presence of ammonia. The decision to culminate is thus mediated in normal development by a drop in extracellular ammonia levels, which relieves the DHKC-mediated activation of RegA and thereby allows intracellular cAMP levels to rise.

DHKC is one of the five histidine protein kinases found in *Dictyostelium* – homologs of the histidine kinases ubiquitously found in bacterial signaling systems. The phosphotransfer from DHKC proceeds via the phosphoshuttle protein RdeA, which has been shown by genetic suppressor analysis to lie upstream of RegA. RegA is a homolog of the bacterial response regulators with an N-terminal response regulator domain and a C-terminal cAMP phosphodiesterase catalytic domain.

For culmination to proceed successfully, morphogenesis and gene expression in differentiating spores and stalk cells must be coordinated and correctly timed. This coordination is achieved by intercellular signals that mediate communication between cells and regulate the differentiation process (Fig. 22). Two such signals have been identified. They are secreted peptides called *spore differentiation factors 1 and 2* (SDF1, molecular mass 1100 Da and SDF2, molecular mass 1300 Da), which were identified on the basis of their ability to elicit spore encapsulation in an *in vitro* assay. The encapsulation response to SDF1 takes about 75 min and

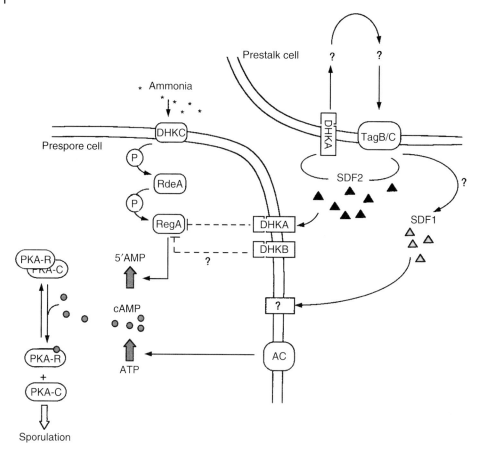

Fig. 22 Signaling pathways controlling sporulation. During culmination, protein kinase (PKA) is essential to trigger spore encapsulation. PKA activity is dependent on the level of intracellular cAMP (which controls the dissociation of the PKA-C and PKA-R subunits), which is the result of the balance between synthesis by the adenylyl cyclase (AC) and degradation by the hybrid protein phosphodiesterase/response regulator RegA. The pathway leading to the stimulation of RegA activity is proposed to be regulated by extracellular ammonia accumulation and involves a classical two component signaling phosphorelay, including the hybrid histidine kinase DHKC and the phosphodonor RdeA. Gene disruption of any of these components results in the interruption of the cascade initiated by DHKC and induces rapid sporulation. In addition to ammonia, two peptides, SDF1 and SDF2, released by the prestalk cells, have been proposed to activate intracellular pathways, the final output of which seems to be a modulation of PKA activity. SDF2 is probably processed by the protease/ABC transporter TagB/C and is probably a ligand for the histidine kinase DHKA. The downstream components of DHKA are so far unknown. DHKA shares partially redundant functions with a third histidine kinase, DHKB, a potential receptor for discadenine, an inhibitor of germination. From Fig. 5 of Aubry, L., Firtel, R. (1999) Integration of signaling networks that regulate *Dictyostelium* differentiation, *Annu. Rev. Cell Dev. Biol.* **15**, 469–517.

requires *de novo* protein synthesis, while prespore cells encapsulate in a matter of minutes in response to SDF2. SDF1 is produced at the onset of culmination when it induces the expression of culmination-specific genes, while SDF2 is made late in culmination when it elicits a wave of encapsulation that passes downward through the spore mass or sorus. SDF2 binds to the cell surface receptor DHKA, another of the *Dictyostelium* sensory histidine kinases. DHKA null mutants exhibit a specific defect in sporulation – they form fruiting bodies with long stalks and few spores. This phenotype can be rescued by 8-bromo-cAMP, by PKAC overexpression or by inactivation of either RegA or PKAR. The pathway downstream of DHKA connecting it to Reg A is unknown, but the upstream production of SDF2 has been shown genetically to involve the TagB and TagC membrane proteins expressed on prestalk cells. These proteins have a predicted protease domain, suggesting that their role in SDF2 production may be hydrolytic processing of a larger precursor polypeptide.

The final targets of signaling pathways in terminal differentiation are specific transcription factors, which are activated or inhibited as a result. In spore cell differentiation, these include Stalky and SrfA. Stalky null mutant prespore cells differentiate into stalk cells rather than spores. SrfA null mutants produce abnormal, non-viable spores. In stalk cell differentiation, one of the important transcription factors is STATa – STATa null mutants do not form stalks and fail to culminate.

3.2.5 Spore Maturation and Germination

Dictyostelium spore coats contain nine major glycoproteins (and a number of minor proteins) and the polysaccharide galuran, which are synthesized in prespore cells where they are stored in preformed complexes in specialized vesicles – prespore vesicles (PSVs). The PSV contents are the main markers for prespore cells. During encapsulation of the spores, the PSVs fuse with the plasma membrane extruding their contents, which then form the protein component of the spore coat. During subsequent spore maturation, cellulose is synthesized and extruded though the plasma membrane and the coat protein complexes bind to it via one member of the complex, SP96, which is a cellulose-binding protein. The final mature coat is 210 nm thick, and has a complex trilaminar structure with an outer protein layer, a double layer of cellulose fibrils, and an inner protein and galuran layer, with the galuran concentrated near the plasma membrane. Toward the end of spore maturation, the spore polysomes break down to single ribosomes and trehalose accumulates to high concentrations in the cytoplasm where it serves as a heat-shock protectant and as an energy source during later germination.

The spores are prevented from germinating prematurely in the sorus by extracellular accumulation of a specific germination inhibitor, discadenine (an adenine analogue), and of high concentrations of ammonium phosphate and other osmolytes. Germination is induced by removal of these (as would occur if the sorus were dispersed into the environment) in combination with a brief heat shock and in the presence of amino acids, indicating the existence of a food supply. Both discadenine and high osmolarity inhibit germination by maintaining high levels of cAMP and correspondingly high activities of PKA activity. The receptor for the discadenine signal that inhibits germination is DhkB, another of the *Dictyostelium* sensor histidine kinases that elicits high cAMP levels by inhibiting the RegA phospho-diesterase. The osmosensor in spores is the third adenylyl cyclase of *Dictyostelium*, ACG, which is directly activated by high osmolarity to synthesize cAMP. Another signaling protein involved in germination is phospholipase C, which cleaves PIP2 to form diacyl glycerol and inositol 1,4,5-triphosphate (IP3). Diacyl glycerol in the membrane and IP3 in the cytoplasm both act as small signaling molecules (second messengers) in eukaryotes, respectively activating protein kinase C and the opening of Ca^{2+} channels in the endoplasmic reticulum. Phospholipase C–deficient mutants are unable to abort germination once it has been initiated, even in appropriate conditions that are inimical to survival of the emerging amoeba. The upstream signal for phospholipase C activation in spores and the downstream events that control abortion of germination are unknown. Spore germination is itself a developmental program that is poorly understood, but includes expression of two cellulases, which hydrolyze cellulose in the spore coat to allow emergence of the amoeba within.

4
Conclusion

The following general principles of microbial development may be gleaned from what is known about well-studied bacterial and eukaryotic systems.

1. Microbial development is usually initiated in response to nutritional stress, frequently in the form of a shortage of nitrogen combined with high cell density, which is sensed *via* secreted signaling molecules. Developmental programs presumably evolved independently in multiple lineages as a means of escaping the deadly consequences of starvation. The exception, *Caulobacter*, in which development is a constitutive part of the cell division cycle, lives in nutrient-poor aquatic environments, may be under constant nutritional stress, and so may receive little advantage by having development as an optional alternative to growth.

2. The signaling pathways that initiate development involve protein kinase cascades. Bacteria primarily use phosphorelays involving histidine protein kinases and response regulators – the so-called two-component signaling systems. Eukaryotes use tyrosine and serine/threonine protein kinases as well as histidine kinases.

3. Signaling pathways for development tend to converge upon a master regulator whose activity controls entry to major developmental stages, for example, CtrA∼P in *Caulobacter*; Spo0∼P in *B. subtilis*; PKA in *Dictyostelium*.

4. Signaling pathways for initiating development usually terminate with changes in the phosphorylation state and activity of transcriptional regulators that either induce or repress developmentally regulated genes.

5. Differentiation is coupled to the cell cycle and cells embark upon differentiation only after cell cycle arrest at a specific stage of the cell cycle. The particular cell cycle stage at which this occurs can differ in different organisms, for example, entry into gametogenesis in late G1 by yeast; onset of differentiation in middle-to-late G2 by *Dictyostelium*.

6. Development involves serially dependent, coordinate changes in gene expression whereby genes are induced in major groups under the control of common transcription factors, and

each major shift in gene expression is induced by components of the previous one, for example, sigma cascades in *Bacillus*; cAMP-elicited PKA signaling in *Dictyostelium* after the onset of starvation.

7. Transitions between the major morphogenetic stages in a developmental program involves checkpoints that couple morphogenesis to differentiation. This ensures that morphogenetic events and gene expression are coordinated, for example, septation in *Bacillus* endospore formation and the sigma factor cascade in the mother cell and the forespore; initiation of culmination by ammonia loss from the *Dictyostelium* slug when the tip is raised from the substratum for an extended time.

8. Differentiation of two or more cell types is coordinated by intercellular signals, for example, cross talk between the *Bacillus* forespore and mother cell; cAMP-, DIF- and SDF2-mediated signaling interactions between differentiating spores and stalk cells in *Dictyostelium*.

Although the details differ, these fundamental principles of development seem applicable to organisms from widely divergent groups. The reason no doubt is a combination of convergent evolution and the original presence in ancestral cells of the central mechanisms for signal transduction and gene regulation.

Acknowledgments

I am grateful to Profs. U. Jenal and J.G. Williams for helpful comments on sections of this manuscript.

See also Bacterial Cell Culture Methods; Bacterial Growth and Division; Bacterial Pathogenesis, Molecular Basis of; Bacteriorhodopsin, Molecular Biology of; *E. Coli* Genome; Methanogens and the Archaebacteria, Molecular Biology of.

Bibliography

Books and Reviews

Aubry, L., Firtel, R. (1999) Integration of signaling networks that regulate *Dictyostelium* differentiation, *Annu. Rev. Cell Dev. Biol.* **15**, 469–517.

Ausmees, N., Jacobs-Wagner, C. (2003) Spatial and temporal control of differentiation and cell cycle progression in *Caulobacter crescentus*, *Annu. Rev. Microbiol.* **57**, 225–247.

England, J.C., Gober, J.W. (2001) Cell cycle control of morphogenesis in *Caulobacter*, *Curr. Opin. Microbiol.* **4**, 674–680.

Errington, J. (2003) Regulation of endospore formation in *Bacillus subtilis*, *Nat. Rev. Microbiol.* **1**, 117–126.

Haber, J.E. (1998) Mating-type switching in *Saccharomyces cerevisiae*, *Annu. Rev. Genet.* **32**, 561–599.

Kaplan, H.B. (2003) Multicellular development and gliding motility in *Myxococcus xanthus*, *Curr. Opin. Microbiol.* **6**, 572–577.

Kessin, R.H. (2001) Dictyostelium, *Evolution, Cell Biology, and the Development of Multicellularity*, Cambridge University Press, Cambridge, UK.

Maeda, Y., Inouge, K., Takeuchi, I. (eds.) (1997) Dictyostelium *A Model Organism for Cell and Developmental Biology*, Universal Academy Press Inc., Tokyo, Japan.

Phillips, Z.E.V., Strauch, M.A. (2002) *Bacillus subtilis* sporulation and stationary phase gene expression, *Cell. Mol. Life Sci.* **59**, 392–402.

Søgaard-Andersen, L., Overgaard, M., Lobedanz, S., Ellehauge, E., Jelsbak, L., Rasmussen, A.A. (2003) Coupling gene expression and multicellular morphogenesis during fruiting body formation in *Myxococcus xanthus*, *Mol. Microbiol.* **48**, 1–8.

Wittenburg, C., La Valle, R. (2003) Cell-cycle-regulatory elements and the control of cell differentiation in the budding yeast, *Bioessays* **25**, 856–867.

Primary Literature

Anderson, P.E., Gober, J.W. (2000) FlbT, the post-transcriptional regulator of flagellin synthesis in *Caulobacter crescentus*, interacts with the 5′ untranslated region of flagellin mRNA, *Mol. Microbiol.* **38**, 41–52.

Baker, D.A., Kelly, J.M. (2004) Structure, function and evolution of microbial adenylyl and guanylyl cyclases, *Mol. Microbiol.* **52**, 1229–1242.

Bath, J., Wu, L.J., Errington, J., Wang, J.C. (2000) Role of *Bacillus subtilis* SpoIIIE in DNA transport across the mother cell-prespore division septum, *Science* **290**, 995–997.

Bosgraaf, L., van Haastert, P.J.M. (2002) A model for cGMP signal transduction in *Dictyostelium* in perspective of 25 years of cGMP research, *J. Muscle Res. Cell Motil.* **23**, 781–791.

Britton, R.A., Eichenberger, P., Gonzalez-Pastor, J.E., Fawcett, P., Monson, R., Losick, R., Grossman, A.D. (2002) Genome-wide analysis of the stationary-phase sigma factor (sigma-H) regulon of *Bacillus subtilis*, *J. Bacteriol.* **184**, 4881–4890.

Chung, C., Firtel, R.A. (2002) Signaling pathways at the leading edge of chemotaxing cells, *J. Muscle Res. Cell Motil.* **23**, 773–779.

Colomina, N., Gari, E., Gallego, C., Herrero, E., Aldea, M. (1999) G$_1$ cyclins block the Ime1 pathway to make mitosis and meiosis incompatible in budding yeast, *EMBO J.* **18**, 320–329.

Dirick, L., Bohm, T., Nasmyth, K. (1995) Roles and regulation of Cln-Cdc28 kinases at the start of the cell cycle of *Saccharomyces cerevisiae*, *EMBO J.* **14**, 4803–4813.

Dohlman, H.G., Thorner, J.W. (2001) Regulation of G protein-initiated signal transduction in yeast: paradigms and principles, *Annu. Rev. Biochem.* **70**, 703–754.

Dworkin, M. (1996) Recent advances in the social and developmental biology of the myxobacteria, *Microbiol. Rev.* **60**, 70–102.

Eichenberger, P., Jenson, S.T., Conlon, E.M., van Ooij, C., Silvaggi, J., Gonzalez-Pastor, J.E., Fujita, M., Ben-Yehuda, S., Stragier, P., Liu, J.S., Losick, R. (2003) The σ^E regulon and the identification of additional sporulation genes in *Bacillus subtilis*, *J. Mol. Biol.* **327**, 945–972.

Fabret, C., Feher, V.A., Hoch, J.A. (1999) Two component signal transduction in *Bacillus subtilis*: how one organism sees its world, *J. Bacteriol.* **181**, 1975–1983.

Fawcett, P., Eichenberger, P., Losick, R., Youngman, P. (2000) The transcriptional profile of early to middle sporulation in *Bacillus subtilis*, *Proc. Natl. Acad. Sci. U. S. A.* **97**, 8063–8068.

Fukuzawa, M., Abe, T., Williams, J.G. (2003) The *Dictyostelium* prestalk cell inducer DIF regulates nuclear accumulation of a STAT protein by controlling its rate of export from the nucleus, *Development* **130**, 797–804.

Fukuzawa, M., Araki, T., Adrian, I., Williams, J.G. (2001) Tyrosine phosphorylation-independent nuclear translocation of a *Dictyostelium* STAT in response to DIF signaling, *Mol. Cell* **7**, 779–788.

Gitai, Z., Dye, N., Shapiro, L. (2004) An actin-like gene can determine cell polarity in bacteria, *Proc. Natl. Acad. Sci. U. S. A.* **101**, 8643–8648.

Gomer, R., Gao, T., Tang, Y., Knecht, D., Titus, M.A. (2002) Cell motility mediates tissue size regulation in *Dictyostelium*, *J. Muscle Res. Cell Motil.* **23**, 809–815.

Gronewold, T.M.A., Kaiser, D. (2001) The act operon controls the level and time of C-signal production for *Myxococcus xanthus* development, *Mol. Microbiol.* **40**, 744–756.

Grossman, A.D. (1995) Genetic networks controlling the initiation of sporulation and the development of genetic competence in *Bacillus subtilis*, *Annu. Rev. Genet.* **29**, 477–508.

Grunenfelder, B., Rummel, G., Vohradsky, J., Roder, D., Langen, H., Jenal, U. (2001) Proteomic analysis of the bacterial cell cycle, *Proc. Natl. Acad. Sci. U. S. A.* **98**, 4681–4686.

Jacobs, C., Hung, D., Shapiro, L. (2001) Dynamic localization of a cytoplasmic signal transduction response regulator controls morphogenesis during the *Caulobacter* cell cycle, *Proc. Natl. Acad. Sci. U. S. A.* **98**, 4095–4100.

Jacobs, C., Domian, I.J., Maddock, J.R., Shapiro, L. (1999) Cell cycle-dependent polar localization of an essential bacterial histidine kinase that controls DNA replication and cell division, *Cell* **97**, 111–120.

Jacobs-Wagner, C. (2004) Regulatory proteins with a sense of direction: cell cycle signalling network in *Caulobacter*, *Mol. Microbiol.* **51**, 7–13.

Jenal, U., Stephens, C. (2002) The *Caulobacter* cell cycle: timing, spatial organization and checkpoints, *Curr. Opin. Microbiol.* **5**, 558–563.

Jiang, M., Grau, R., Perego, M. (2000) Differential processing of propeptide inhibitors of Rap phosphatases in *Bacillus subtilis, J. Bacteriol.* **182**, 303–310.

Jiang, M., Shao, W., Perego, M., Hoch, J.A. (2000) Multiple histidine kinases regulate entry into stationary phase and sporulation in *Bacillus subtilis, Mol. Microbiol.* **38**, 535–542.

Julien, B., Kaiser, D., Garza, A. (2000) Spatial control of cell differentiation in *Myxococcus xanthus, Proc. Natl. Acad. Sci. U. S. A.* **97**, 9098–9103.

Kaiser, D. (2001) Building a multicellular organism, *Annu. Rev. Genet.* **35**, 103–123.

Kelly, A.J., Sackett, M.J., Din, N., Quardokus, E., Brun, Y.V. (1998) Cell cycle-dependent transcriptional and proteolytic regulation of FtsZ in *Caulobacter, Genes Dev.* **12**, 880–893.

Kim, S.K., Kaiser, D. (1990a) Cell alignment required in differentiation of *Myxococcus xanthus, Science* **249**, 926–928.

Kim, S.K., Kaiser, D. (1990b) C-factor: a cell-cell signalling protein required for fruiting body morphogenesis of *M. xanthus, Cell* **61**, 19–26.

Kirby, J.R., Zusman, D.R. (2003) Chemosensory regulation of developmental gene expression in *Myxococcus xanthus, Proc. Natl. Acad. Sci. U. S. A.* **100**, 2008–2013.

Laub, M.T., McAdams, H.H., Feldblyum, T., Fraser, C.M., Shapiro, L. (2000) Global analysis of the genetic network controlling a bacterial cell cycle, *Science* **290**, 2144–2148.

Lazazzera, B.A. (2000) Quorum sensing and starvation: signals for entry into the stationary phase, *Curr. Opin. Microbiol.* **3**, 177–182.

Lee, S., Parent, C.A., Insall, R., Firtel, R.A. (1999) A novel Ras-interacting protein required for chemotaxis and cyclic adenosine monophosphate relay in *Dictyostelium, Mol. Biol. Cell* **10**, 2829–2845.

Lobedanz, S., Søgaard-Andersen, L. (2003) Identification of the C-signal, a contact-dependent morphogen coordinating multiple developmental responses in *Myxococcus xanthus, Genes Dev.* **17**, 2151–2161.

Losick, R., Stragier, P. (1992) Crisscross regulation cell-type-specific gene expression during development in *B. subtilis, Nature* **355**, 601–604.

McBride, M.J., Weinberg, R.A., Zusman, D.R. (1989) 'Frizzy' aggregation genes of the gliding bacterium *Myxococcus xanthus* show sequence similarities to the chemotaxis genes of enteric bacteria, *Proc. Natl. Acad. Sci. U. S. A.* **86**, 424–428.

Montelone, B.A. (2002) Yeast Mating Type, *Nature Encyclopedia of Life Sciences*, Nature Publishing Group, London. http://www.els.net/ [doi:10.1038/npg.els.0000598].

Muir, R.E., Gober, J.W. (2001) Regulation of the late flagellar gene transcription and cell division by flagellum assembly in *Caulobacter crescentus, Mol. Microbiol.* **41**, 117–130.

Nebl, T., Kotsifas, M., Schaap, P., Fisher, P.R. (2002) Multiple signalling pathways connect chemoattractant receptors and calcium channels in *Dictyostelium, J. Muscle Res. Cell Motil.* **23**, 853–865.

Ogawa, M., Fujitani, S., Mao, X., Inouye, S., Komano, T. (1996) FruA, a putative transcription factor essential for the development of *Myxococcus xanthus, Mol. Microbiol.* **22**, 757–767.

Ohlson, K.L., Grimsley, J.K., Hoch, J.A. (1994) Deactivation of the sporulation transcription factor Spo0A by the Spo0E protein phosphatase, *Proc. Natl. Acad. Sci. U. S. A.* **91**, 1756–1760.

Parkinson, J.S., Kofold, E.C. (1992) Communication modules in bacterial signalling proteins, *Annu. Rev. Genet.* **26**, 71–112.

Paul, R., Weiser, S., Amiot, N.C., Chan, C., Schirmer, T., Giese, B., Jenal, U. (2004) Cell cycle-dependent dynamic localization of a bacterial response regulator with a novel di-guanylate cyclase output domain, *Genes Dev.* **18**, 715–727.

Peter, M., Herskowitz, I. (1994) Direct inhibition of the yeast cyclin-dependent kinase Cdc28-Cln by Far1, *Science* **265**, 1228–1231.

Peter, M., Gartner, A., Horecka, J., Ammerer, G., Herskowitz, I. (1993) FAR1 links the signal transduction pathway to the cell cycle machinery in yeast, *Cell* **73**, 747–760.

Quon, K.C., Marczynski, G.T., Shapiro, L. (1996) Cell cycle control by an essential bacterial two-component signal transduction protein, *Cell* **84**, 83–89.

Roberts, C.J., Nelson, B., Marton, M.J., Stoughton, R., Meyer, M.R., Bennett, H.A., He, Y.D. D., Dai, H.Y., Walker, W.L., Hughes, T.R., Tyers, M., Boone, C., Friend, S.H. (2000) Signalling and circuitry of multiple MAPK pathways revealed by a matrix of global gene expression profiles, *Science* **287**, 873–880.

Rua, D., Tobe, B.T., Kron, S.J. (2001) Cell cycle control of yeast filamentous growth, *Curr. Opin. Microbiol.* **4**, 720–727.

Rudner, D.R., Losick, R. (2001) Morphological coupling in development: lessons from prokaryotes, *Dev. Cell* **1**, 733–742.

Saran, S., Meima, M., Alvarez-Curto, E., Weening, K.E., Rozen, D.E., Schaap, P. (2002) cAMP signaling in *Dictyostelium*, *J. Muscle Res. Cell Motil.* **23**, 793–802.

Sciochetti, S.A., Lane, T., Ohta, N., Newton, A. (2002) Protein sequences and cellular factors required for polar localization of a histidine kinase in *Caulobacter crescentus*, *J. Bacteriol.* **184**, 6037–6049.

Shimkets, L.J. (1999) Intercellular signalling during fruiting-body development of *Myxococcus xanthus*, *Annu. Rev. Microbiol.* **53**, 525–549.

Søgaard-Andersen, L., Slack, F., Kimsey, H., Kaiser, D. (1996) Intercellular C-signalling in *Myxococcus xanthus* involves a branched signal transduction pathway, *Genes Dev.* **10**, 740–754.

Sonnenshein, A.L. (2000) Control of sporulation initiation in *Bacillus subtilis*, *Curr. Opin. Microbiol.* **3**, 561–566.

Spellman, P.T., Sherlock, G., Zhang, M.Q., Iyer, V.R., Anders, K., Eisen, M.B., Brown, P.O., Botstein, D., Futcher, B. (1998) Comprehensive identification of cell cycle-regulated genes of the yeast *Saccharomyces cerevisiae* by microarray hybridization, *Mol. Biol. Cell* **9**, 3273–3297.

Spormann, A. (1999) Gliding motility in bacteria: insights from studies of *Myxococcus xanthus*, *Microbiol. Mol. Biol. Rev.* **63**, 621–641.

Stephenson, K., Hoch, J.A. (2002) Evolution of signalling in the sporulation phosphorelay, *Mol. Microbiol.* **46**, 297–304.

Strauch, M.A. (1993) Regulation of *Bacillus subtilis* gene expression during the transition from exponential growth to stationary phase, *Prog. Nucleic Acids Res. Mol. Biol.* **46**, 121–153.

Strauch, M.A., Trach, K.A., Day, J., Hoch, J.A. (1992) Spo0A activates and represses its own synthesis by binding at its dual promoters, *Biochemie* **74**, 619–626.

Sun, K., Coïc, E., Zhou, Z., Durrens, P., Haber, J.E. (2002) *Saccharomyces* forkhead protein Fkh1 regulates donor preference during mating-type switching through the recombination enhancer, *Genes Dev.* **16**, 2085–2096.

Takamatsu, H., Watabe, K. (2002) Assembly and genetics of spore protective structures, *Cell. Mol. Life Sci.* **59**, 434–444.

Thomason, P., Kay, R. (2000) Eukaryotic signal transduction via histidine-aspartate phosphorelay, *J. Cell Sci.* **113**, 3141–3150.

Thompson, C.R.L., Kay, R.R. (2000) Cell-fate choice in *Dictyostelium*: intrinsic biases modulate sensitivity to DIF signaling, *Dev. Biol.* **227**, 56–64.

Thompson, C.R.L., Fu, Q., Buhay, C., Kay, R.R., Shaulsky, G. (2004) A bZIP/bRLZ transcription factor required for DIF signaling in *Dictyostelium*, *Development* **131**, 513–523.

Van Es, S., Weening, K.E., Devreotes, P.N. (2001) The protein kinase YakA regulates G-protein-linked signaling responses during growth and development of *Dictyostelium*, *J. Biol. Chem.* **276**, 30761–30765.

Wheeler, R.T., Shapiro, L. (1999) Differential localization of two histidine kinases controlling bacterial cell differentiation, *Mol. Cell* **4**, 683–694.

Williams, H.P., Harwood, A.J. (2003) Cell polarity and *Dictyostelium* development, *Curr. Opin. Microbiol.* **6**, 621–627.

Wood, J.G., Rogina, B., Lavu, S., Howitz, K., Helfand, S.L., Tatar, M., Sinclair, D. (2004) Sirtuin activators mimic caloric restriction and delay ageing in metazoans, *Nature* **430**, 686–689.

Micronutrients, Trace Elements

Robert J. Cousins
University of Florida, Gainesville, FL, USA

Keywords

Iron regulatory Protein (IRP)
Binds in response to metal occupancy to a stem loop of nucleotides called an IRE in the
5′ untranslated region of a specific mRNA to influence its translation.

Iron-responsive Element (IRE)
A specific mRNA stem-loop structure that binds an iron regulatory protein.

Encyclopedia of Molecular Cell Biology and Molecular Medicine, 2nd Edition. Volume 8
Edited by Robert A. Meyers.
Copyright © 2005 Wiley-VCH Verlag GmbH & Co. KGaA, Weinheim
ISBN: 3-527-30550-5

Metal-responsive Element (MRE)

Specific sequence of promoter region of gene that enhances transcription through a specific transcription factor (MRE-binding transcription factor) that requires metal occupancy for DNA binding.

Metal-responsive Transcription Factor (MTF)

In response to metal occupancy, translocates to the nucleus and binds to a specific MRE sequence of the DNA.

Metalloenzyme

An enzyme that requires one or more atoms of a metal for catalytic activity and/or for structure required for activity.

Trace Element Micronutrient

An essential nutrient metal required in the diet in small (µg or mg) amounts per day.

Zinc Finger

A zinc-binding domain of a protein that has an X-Cys-X_2-Cys-X_n-Cys-X_{2-5}-Cys-X (or His substituted for Cys) motif that, upon Zn(II) binding, forms a structure capable of binding DNA or RNA or interacting with another protein with a compatible zinc finger.

■ Trace element micronutrients is a term that refers to the metals (inorganic elements) in the diet that are essential for specific cellular processes and are required in only small amounts (mg or less) daily. In mammals, required dietary trace elements include Cu, Fe, Mn, Se, and Zn. They perform catalytic, regulatory, and/or structural roles related to gene expression. Zinc has a major role in maintaining conformations (i.e. zinc fingers) necessary for proteins to bind to specific DNA sequences (regulatory elements) or protein–protein interactions required for specific cellular functions. Zinc can also influence transcription through metal-binding proteins (metal-responsive transcription factors) that, during metal occupancy, initiate translocation to the nucleus for binding to specific DNA sequences (metal response elements) in the promoter/enhancer region of genes. Regulation of gene expression by iron can occur through metal-requiring, RNA-binding proteins. These metal-binding proteins influence translation of specific mRNAs via nucleotide sequences in the untranslated regions. Trace element micronutrients have potential applications in medicine and biotechnology either directly as signaling molecules for regulation of existing or chimeric genes or indirectly as therapeutic agents by targeting specific metal-binding sites in regulatory factors.

1
Scope of Trace Element Micronutrients in Molecular Biology

1.1
Chemical Considerations

Cellular constituents, particularly proteins, provide ligands (e.g. amine, thiolate, carboxyl, etc.) for metal binding, which follow approximately the Irving–Williams order of divalent ions: $Cu > Zn \geq Ni > Co > Fe > Mn > Mg > Ca$. Cellular transport systems regulate intracellular trace element concentrations through homeostatic control of uptake and efflux. These processes limit the influences of natural abundance and thermodynamic constraints in determining which elements participate in specific cellular functions. If ligand concentrations and metal concentrations were not controlled within cells, abnormal binding of Cu would predominate for most ligands. Only Cu, Fe, and Zn will be discussed here as these nutrients have received the bulk of attention with respect to mammalian molecular biology.

1.2
Nutritional Considerations

Micronutrients as a general term describes both essential organic molecules (vitamins) and metals (trace elements). In contrast to vitamins, metals are utilized by all living organisms to sustain cellular processes. Required trace elements include Cu, Fe, Mn, Se, and Zn. In humans, these are required in the diet in µg to mg amounts per day to maintain balance with endogenous losses. The composite cellular and extracellular need for each trace element to satisfy the needs of binding sites as balanced against turnover and endogenous losses comprises the dietary requirement for the micronutrient.

Food abundance in the developed world has minimized the incidence of deficiencies of these trace elements. Genetic variance in trace element utilization and function suggest individual requirements must be considered. This is particularly true where intakes are marginally adequate through selective diets and/or consumption of highly processed/unfortified foods. Nutritional inadequacy in the developing world is prevalent. Iron and zinc are the trace elements most likely to be in limited supply. How such nutritional deficits influence gene regulation and epigenetic effects that influence human health are active areas of investigation.

1.3
Classes of Involvement

There are three classes of involvement for trace element micronutrients (metals) in molecular biology. These are catalytic, structural, and regulatory. Metalloenzymes, where metals provide a catalytic function, comprise the first class. Another class is structural, with metals providing interaction among various binding groups of specific motifs to facilitate conformations necessary to achieve unique opportunities for specific interaction. The third class is regulatory, as exemplified by the metal binding to specific trans-acting proteins, which then provide signaling to initiate, enhance, or inhibit transcription of genes through interaction with specific DNA sequences, that is, metal-responsive elements (MRE). Metal-responsive RNA-binding proteins provide another example of regulatory involvement. These proteins, while performing a regulatory function, act through structural changes. Collectively, these classes of biological function of trace element micronutrients are such that they provide a vital essential link between the organism and its external nutrient supply.

Eukaryotic cells employ trace elements to differing extents. Overall, zinc has the greatest spectrum of gene-related involvement through catalytic, structural, and regulatory roles in all eukaryotes. Iron and copper have catalytic roles in lower eukaryotes and a complex array of regulatory roles through metal-responsive transcription factors or metal-binding elements of RNA binding proteins. These are directed at iron and copper uptake, trafficking, and homeostasis. Most such information has been obtained using yeast (a fungus). In higher eukaryotes, copper functions in catalytic roles, while iron has catalytic and structural roles, most related to oxygen utilization. Transcriptional regulation of genes in mammalian cells is limited to zinc. The teleological explanation most likely is that zinc is redox neutral with respect to the metal atom.

1.4
Cellular Distributions

Concentrations of trace elements in various cell types vary greatly but usually are no higher than the µg/g tissue range. Frequently, such concentrations are expressed on a protein, DNA, or dry weight basis. Cells differ with respect to which organelles trace elements are most abundantly distributed. However, metal occupancy at specific binding sites, rather than concentration, determines functional effects.

Lower eukaryotes vary widely in response to the trace element content of the environment. For example, metal-resistance genes with elaborate control mechanisms have evolved to deal with elevated levels of a particular metal present in the environment through regulated transport and compartmentalization. Mammalian cells within an integrative system are not subjected to such extremes, in part through powerful homeostatic mechanisms operative at both the cell and organ level. Therefore, the tissue metal concentration in cells of animals that consume diets deficient in that essential trace element or receive an excess of that element are frequently within normal limits. Functional deficiency may occur only after depletion of a critical pool or compartment of the trace element, which then affects specific binding sites. Unfortunately, our understanding of the cell biology of trace elements lags behind our understanding of roles of these micronutrients in molecular biology.

2
Structural/Catalytic/Regulatory Roles

2.1
Metalloenzymes and Metal-binding Domains

Characterization of individual metal-binding molecules has shown that stoichiometry is usually 1 to 4 metal atoms per binding molecule. There are exceptions, for example, ferritin (4500 atoms Fe/molecule) and metallothionein (7 atoms Zn or 12 atoms Cu/molecule). In addition, abundance of metal-binding molecules ranges from a few to thousands of molecules per cell. Trace elements are essential to the enzymology of molecular biology. For example, all of the RNA nucleotide transferases (RNA polymerases) are zinc-requiring metalloenzymes. Metalloenzyme systems are not uniformly susceptible to alteration in dietary trace element restriction or supplementation.

However, many manipulations in molecular biology use metal chelation (e.g. with EDTA) as a method of controlling activity of metal-requiring enzymes.

2.2
Metalloregulatory Proteins

Metalloproteins that act as DNA-binding transcription factors, trans-acting MRE-binding proteins, mRNA-binding proteins, and proteins exhibiting metal-dependent interactions with other proteins usually bind metals through thiolate groups from cysteine (Cys) and/or nitrogen from histidine (His). Very little is known about the ligand-exchange reactions that are necessary for metals to fulfill these regulatory roles. Progress has been made regarding how dietary abundance and transport mechanisms are necessary for trace elements to meet these needs.

3
Regulation of Gene Expression by Trace Element Micronutrients

Trace elements provide signals to systems that influence rates of either gene transcription or mRNA translation. They do not act directly. Rather, through recognition that requires specific coordination chemistry, metals bind to metalloregulatory proteins (sensors or receptors) and influence signaling pathways. This ultimately augments production of a specific protein. In the classical Newtonian sense, these micronutrients (metals) act as inducers (or repressors) of the system. Since the dietary supply provides the source of the inducer (or repressor), fluctuations in intake and/or physiologic factors influencing utilization or tissue distribution of the micronutrient are determinants of how

effective the trace element will be in regulating the metalloregulatory system.

3.1
Transcriptional Regulation

Examples of transcriptional regulation by trace elements are found in all eukaryotic cells, but detailed descriptions of the cellular apparatus involved have emerged more rapidly from systems in lower eukaryotes. For example, in yeast, over 100 genes that influence trace element homeostasis and cellular functions, such as defense, have been shown to be regulated by 13 different metal-responsive transcription factors. Metal transporters and chaperones comprise the largest group of these metal-regulated genes.

Metallothionein has been the most widely studied mammalian gene that is transcriptionally regulated by trace elements. Mammalian metallothionein is tissue-specific, with high expression in liver, intestine, bone marrow, pancreas, and kidney, but expression can be detected in most tissues. Depletion of the animals' dietary zinc supply has a negative effect on expression, while reduced copper intake has little effect. Similarly, elevation of dietary zinc intake level provides a positive stimulus for expression, but changes in the copper intake level have little effect. This difference between copper and zinc is an example of how mammalian systems are not analogous to metal-produced responses with unicellular organisms. In the latter, copper is an important inducer of *metallothionein*. Studies with intact animals in which trace metals are injected can show regulation by multiple inducer trace metals. These responses could be a direct action of the trace element on a sensor protein or an indirect action via transcriptional regulation signaled by metal-stimulated release of stress-related mediators or activation of oxidant-responsive factors. Therefore, an underlying question to answer in integrative systems is whether the trace metal is a direct or an indirect mediator of gene transcription. The body of evidence with many mammalian systems, including transfected cells, demonstrates that zinc and the toxic metal cadmium (not a nutrient) induce gene expression via transcription. The exact mechanism of how zinc enters mammalian cells, traverses the intracellular space, and enters the nucleus to initiate or repress transcription has been described in limited detail. Zinc homeostasis in mammals is controlled by genes regulating influx, efflux, and compartmentalization. Some appear to be nutritionally responsive. Some transporter proteins may have metal-sensing domains and provide a signal transduction function.

At least three components are considered necessary for transcriptional regulation by a trace element (Fig. 1). First, the responsive gene must have an MRE at one or more sites in regulatory regions. All mammalian DNA motifs responsible for metal regulation thus far identified have a consensus 13 to 15 nucleotide imperfect sequence. The consensus MRE sequence for the mammalian metallothionein promoter is CTC<u>TGCRCNCGGCCC</u>. The core of this consensus sequence is underlined. MREs are usually found in multiple copies of either orientation within the first few hundred bases of the promoter. They are most frequently upstream from the transcription initiation site. MREs are frequently located near other regulatory elements, perhaps for synergistic or differential regulatory purposes. A search of the genome reveals that a plethora of genes have MRE sequences within regulatory regions. The proportion of these that confer transcriptional activity is not known.

The second component needed for transcriptional regulation by trace element micronutrients is a trans-acting MRE-binding metalloregulatory protein(s) called a *metal-responsive transcription factor* (MTF). Such a metal-binding/sensing protein (depicted in Fig. 1) acts as a transcription factor to initiate or enhance transcription of an MRE-regulated gene. In mammalian cells, there are only two such factors, MTF-1 and MTF-2, currently identified. Only MTF-1 has been functionally characterized. Zinc binds to MTF-1 through unique (most likely tetrahedral) coordination chemistry. MTF-1 has six (C_2H_2-type) zinc-finger domains; three bind zinc ions more strongly than the other three. The more facile sites have the regulatory role and, upon occupancy, facilitate translocation to the nucleus and DNA binding to the MRE. This may constitute the mode by which zinc from the diet activates the MTF-1/MRE system. Evidence suggests MTF-1 DNA-binding activity is influenced by phosphorylation and the state of cellular redox. Shuttling of MTF-1 between cytoplasm and the nucleus could be a point of regulation. Similarly, the *MTF-1* gene is upregulated in zinc restriction, and the *MTF-2* gene is concomitantly downregulated. Multiple MTF proteins may yield complexes that produce negative as well as positive effects on transcription of a specific MRE-regulated gene. Furthermore, tissue-specificity of metal regulation requires that genes for the MTF protein are expressed to differing extents requisite to cell function and stage of development. Data are lacking on these points. Genome screening methods, including DNA microarray analysis, have shown that numerous genes are likely regulated by MTF-1. In mice, a null mutation of *MTF-1* causes death during fetal growth. In contrast, *MT* null mice exhibit normal growth. This difference suggests MTF-1 regulates a gene(s) other than *MT* that is critical for the developing mouse embryo. Numerous zinc transporter genes are believed to be regulated by MTF and may be among these regulated genes.

The third component for this metal-responsive unit is the trace element (metal), that is, the inducer. Intracellular concentrations of zinc, while held relatively constant, vary sufficiently with physiologic state and dietary intake to influence metal availability for binding to the MTF protein. The source (a low molecular weight chelate or a chaperone protein, possibly of lower binding affinity) and the cellular location (nuclear, cytoplasmic, or both) of metal destined for binding to MTF proteins remain to be defined (Fig. 1). It is likely that the intracellular pools for these roles may constitute only a few atoms per cell.

3.2
Translational Regulation

The initial concept of trace element control of translation was developed from the acute induction of ferritin synthesis by iron, a process that does not alter transcription of the *ferritin* gene. Iron circulates in the plasma, bound to transferrin (Tf). This protein binds to transferrin receptor (TfR) located on the cell surface, undergoes endocytosis, and contributes iron to a low molecular weight iron pool. When this cellular pool is large, ferritin synthesis is stimulated via increases in translatable ferritin mRNA. In contrast, when the cellular iron pool is low, for example, during dietary iron deficiency, translation of TfR mRNA increases (Fig. 2).

The translational regulation of TfR and ferritin mRNAs and those of other iron-regulated proteins is controlled by two cytoplasmic iron regulatory proteins (IRP-1 and IRP-2). A unique secondary structure of specific mRNAs are nucleotide loops that constitute iron-responsive elements (IRE). IREs for different mRNAs are not identical. Ferritin, TfR, the ferroportin transporter, and erythroid δ-aminolevulinate synthase are among the proteins of iron metabolism whose mRNAs have IREs. RNA-binding activity of IRP molecules is strongly regulated by dietary iron. IRP high-affinity binding to TfR mRNA causes stabilization at the A/U rich region of the 3' end of the mRNA by inhibiting degradation and stimulating its translation. This promotes cellular iron uptake when the iron supply is low. IRP-1 may bind and thus stabilize the mRNA for the iron transporter DMT-1. This would enhance synthesis of the transporter for increased apical iron transport by enterocytes when the dietary iron supply is low. Conversely, with an increased iron supply, IRP binding to ferritin and ferroportin mRNAs is reduced, and their translation increases. This facilitates iron storage. Of interest is that slight structural differences in the 3' end of ferroportin mRNA may cause low iron to enhance translation in the enterocytes but repress translation in liver. Iron binding is correlated with the c-aconitase activity of IRP-1. This mitochondrial TCA cycle enzyme may alter cellular citrate levels and, through iron-binding properties of citrate, influence the low molecular weight iron pool. Mitochondrial succinate dehydrogenase iron protein mRNA is also regulated by IRP-1. Iron metabolism makes further use of the IRE-IRP system for control of heme synthesis in erythroid cells via regulation of δ-aminolevulinate synthase mRNA translation.

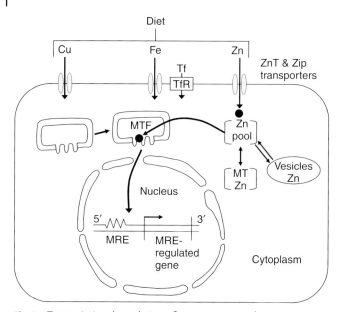

Fig. 1 Transcriptional regulation of gene expression by a trace element (metal) micronutrient. The diet provides copper, iron, and zinc for distribution to cells via the plasma. Of the three nutrients, only zinc provides a demonstrated stimulus for induction of transcription. Metallothionein gene expression by Zn provides an example of this mode of regulation. A requirement of this system is that the regulated gene(s) has a metal response element (MRE) sequence(s) in its promoter/enhancer region. The metal (●), following transport into the cell via ZnT and Zip transporters, interacts rapidly with a variety of ligands and/or is transported into vesicles, thus maintaining an extremely small pool. One of these ligands is an MRE-binding protein termed a *metal-responsive transcription factor* (MTF). Upon metal occupancy of specific zinc fingers, the MTF enters the nucleus for MRE binding. Zinc may influence the MTF translocation process and/or interaction with specific inhibitors. Phosphorylation and the cell redox state may influence these processes also. The rate of transcription is directly proportional to the dietary metal supply, the intracellular metal pool, and the extent of MTF/MRE binding. It is possible that some MRE-binding transcription factors act as negative rather than positive regulators of transcription.

4
Applications and Perspectives

The application of trace element micronutrients for control of biotechnological systems has been demonstrated. The dramatic growth of transgenic animals producing excess growth hormone through activation of chimeric metallothionein promoter-growth hormone constructs in response to zinc in the drinking water illustrated the potential of this approach two decades ago.

Chimeric constructs have been used to regulate plant genes by trace elements. The copper-responsive transcription factor

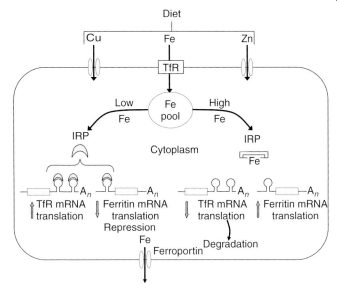

Fig. 2 Translational regulation of mRNA by a trace element (metal) micronutrient. The diet provides copper, iron, and zinc for distribution to cells via the plasma. Of the three nutrients, only iron provides a demonstrated stimulus for control of mRNA translation. Transferrin receptor mRNA (TfR mRNA) and ferritin mRNA are among the small number of mRNAs regulated by iron. The intracellular iron pool is reflective of dietary iron availability. Cellular iron is monitored by iron regulatory proteins (IRP). One IRP (IRP-1) has c-aconitase activity, which may help regulate the low molecular weight iron pool size via citrate availability. When cellular iron is low, less iron is bound to IRPs. This favors their binding to specific mRNAs via unique stem loops in the UTR of the mRNA (5′ and 3′ of the coding region for ferritin and TfR mRNAs, respectively). Binding protects TfR mRNA from degradation and increases its translation, thus transferrin receptor synthesis increases. Consequently, more iron enters the cell. Simultaneously, low iron also favors IRP binding to ferritin mRNA, which decreases translation of this mRNA. When cellular iron levels are high, iron binding to IRPs increases. This lowers the mRNA-binding affinity of IRPs, thus increasing ferritin mRNA translation and TfR mRNA degradation. As a result, ferritin synthesis increases, and TfR synthesis decreases. Erythroid δ-aminolevulinate synthase mRNA and succinate dehydrogenase mRNA translation decrease upon IRP binding with low iron in a fashion like mRNAs for H/L ferritin. Ferroportin mRNA translation in intestine is stimulated by low iron (like TfR mRNA) but, in liver, ferroportin mRNA translation is decreased. Since ferroportin is an iron export protein, this apparent difference stimulates intestinal absorption and hepatic retention.

from yeast (ACE1) was used to stimulate expression of a gene coding for a commercially important enzyme, via copper treatment of the plant. A DNA construct with the ACE1 binding site (MRE) was inserted into the promoter of the gene coding for the enzyme. Through copper treatment of the transgenic plant, production of the enzyme is increased.

A key to using trace elements for technological purposes is that those chosen must be nontoxic, for example, zinc, so as not to damage the organism but sufficiently bioavailable to promote robust expression of the transgene. The unique physiology, tissue-specific sequestration, and nutritional intake of each trace element would need to be considered in these strategies. A concern is that the metal-resistance genes of co-existing microbial systems could lead to overgrowth of endogenous microbial populations and induction of metal-inducible proteins, which may provide gratuitous changes in metabolic fates of xenobiotics. Similarly, efforts to improve micronutrient (iron and zinc) nutrition in the developing world through plants engineered to sequester more of these micronutrients could lead to overconsumption in humans and dysregulation of genes regulated by these nutrients.

The development of pharmaceuticals with actions related to metal chelation or metal donation is of major current interest. Many of these novel compounds will use trace element micronutrients as signaling systems for altering gene expression. Alternatively, a naturally occurring cellular component, for example, a zinc-finger transcription factor, could be the target of a specific trace element–containing therapeutic or sequestering agent. Similarly, engineering zinc-finger transcription factor delivered by viral vectors (gene therapy) to atypically regulate specific target genes by trace elements is possible.

See also Adipocytes.

Bibliography

Books and Reviews

Davis, S.R., Cousins, R.J. (2000) Metallothionein expression in animals: a physiological perspective on function, *J. Nutr.* **130**, 1085–1088.

Eisenstein, R.S. (2000) Iron regulatory proteins and the molecular control of mammalian iron metabolism, *Annu. Rev. Nutr.* **20**, 627–662.

Hambidge, M., Cousins, R.J., Costello, R.B. (Eds.) (2000) Proceedings of the international workshop on zinc and health: current status and future directions, *J. Nutr.* **130**, 1341S–1520S.

Hellman, N.E., Gitlin, J.D. (2002) Ceruloplasmin metabolism and function, *Annu. Rev. Nutr.* **22**, 439–458.

Institute of Medicine. (2001) Zinc, in: *Dietary Reference Intakes for Vitamin A, Vitamin K, Arsenic, Boron, Chromium, Copper, Iodine, Iron, Manganese, Molybdenum, Nickel, Silicon, Vanadium, and Zinc*, National Academy Press, Washington, DC, pp. 442–501.

King, J.C. (Ed.) (2003) 11th international symposium on trace elements in man and animals, *J. Nutr.* **133**, 1429S–1587S.

Liuzzi, J.P., Cousins, R.J. (2004) Mammalian zinc transporters, *Annu. Rev. Nutr.* **24**, 151–172.

Lutsenko, S., Petris, M.J. (2003) Function and regulation of the mammalian copper-transporting ATPases: insights from biochemical and cell biological approaches, *J. Membr. Biol.* **191**, 1–12.

Miret, S., Simpson, R.J., McKie, A.T. (2003) Physiology and molecular biology of dietary iron absorption, *Annu. Rev. Nutr.* **23**, 283–301.

Rutherford, J.C., Bird, A.J. (2004) Metal-responsive transcription factors that regulate iron, zinc, and copper homeostasis in eukaryotic cells, *Eukaryot. Cell* **3**, 1–13.

Primary Literature

Beck, M.A., Levander, O.A., Handy, J. (2003) Selenium deficiency and viral infection, *J. Nutr.* **133**, 1463S–1467S.

Beguin, Y. (1992) The soluble transferrin receptor: biological aspects and clinical usefulness as quantitative measure of erythropoiesis, *Haematologica* **77**, 1–10.

Blalock, T.L., Dunn, M.A., Cousins, R.J. (1988) Metallothionein gene expression in rats: tissue-specific regulation by dietary copper and zinc, *J. Nutr.* **118**, 222–228.

Blanchard, R.K., Cousins, R.J. (1996) Differential display of intestinal mRNAs regulated by dietary zinc, *Proc. Natl. Acad. Sci. U.S.A.* **93**, 6863–6868.

Bremner, I., Morrison, J.N., Wood, A.M., Arthur, J.R. (1987) Effects of changes in dietary zinc, copper and selenium supply and of endotoxin administration on metallothionein I concentrations in blood cells and urine in the rat, *J. Nutr.* **117**, 1595–1602.

Bulteau, A.L., O'Neill, H.A., Kennedy, M.C., Ikeda-Saito, M., Isaya, G., Szweda, L.I. (2004) Frataxin acts as an iron chaperone protein to modulate mitochondrial aconitase activity, *Science* **305**, 242–245.

Burk, R.F., Hill, K.E., Motley, A.K. (2003) Selenoprotein metabolism and function: evidence for more than one function for selenoprotein P, *J. Nutr.* **133**, 1517S–1520S.

Chan, R.Y., Schulman, H.M., Ponka, P. (1993) Expression of ferrochelatase mRNA in erythroid and non-erythroid cells, *Biochem. J.* **292**, 343–349.

Cousins, R.J. (1985) Absorption, transport, and hepatic metabolism of copper and zinc: special reference to metallothionein and ceruloplasmin, *Physiol. Rev.* **65**, 238–309.

Cousins, R.J., Blanchard, R.K., Popp, M.P., Liu, L., Cao, J., Moore, J.B., Green, C.L. (2003) A global view of the selectivity of zinc deprivation and excess on genes expressed in human THP-1 mononuclear cells, *Proc. Natl. Acad. Sci. U.S.A.* **100**, 6952–6957.

Cousins, R.J., Leinart, A.S. (1988) Tissue-specific regulation of zinc metabolism and metallothionein genes by interleukin 1, *FASEB J.* **2**, 2884–2890.

Eisenstein, R.S., Garcia-Mayol, D., Pettingell, W., Munro, H.N. (1991) Regulation of ferritin and heme oxygenase synthesis in rat fibroblasts by different forms of iron, *Proc. Natl. Acad. Sci. U.S.A.* **88**, 688–692.

Failla, M.L., Cousins, R.J. (1978) Zinc accumulation and metabolism in primary cultures of adult rat liver cells. Regulation by glucocorticoids, *Biochim. Biophys. Acta* **543**, 293–304.

Fraker, P.J., DePasquale-Jardieu, P., Zwickl, C.M., Luecke, R.W. (1978) Regeneration of T-cell helper function in zinc-deficient adult mice, *Proc. Natl. Acad. Sci. U.S.A.* **75**, 5660–5664.

Granick, S. (1946) Ferritin. IX. Increase of the protein apoferritin in the gastrointestinal mucosa as a direct response to iron feeding. The function of ferritin in the regulation of iron absorption, *J. Biol. Chem.* **164**, 737–746.

Grider, A., Bailey, L.B., Cousins, R.J. (1990) Erythrocyte metallothionein as an index of zinc status in humans, *Proc. Natl. Acad. Sci. U.S.A.* **87**, 1259–1262.

Hamza, I., Faisst, A., Prohaska, J., Chen, J., Gruss, P., Gitlin, J.D. (2001) The metallochaperone Atox1 plays a critical role in perinatal copper homeostasis, *Proc. Natl. Acad. Sci. U.S.A.* **98**, 6848–6852.

Hamza, I., Prohaska, J., Gitlin, J.D. (2003) Essential role for Atox1 in the copper-mediated intracellular trafficking of the Menkes ATPase, *Proc. Natl. Acad. Sci. U.S.A.* **100**, 1215–1220.

Harris, E.D. (1976) Copper-induced activation of aortic lysyl oxidase in vivo, *Proc. Natl. Acad. Sci. U.S.A.* **73**, 371–374.

Harris, Z.L., Durley, A.P., Man, T.K., Gitlin, J.D. (1999) Targeted gene disruption reveals an essential role for ceruloplasmin in cellular iron efflux, *Proc. Natl. Acad. Sci. U.S.A.* **96**, 10812–10817.

Hershko, C., Cook, J.D., Finch, C.A. (1974) Storage iron kinetics. VI. The effect of inflammation on iron exchange in the rat, *Br. J. Haematol.* **28**, 67–75.

Huang, L., Gitschier, J. (1997) A novel gene involved in zinc transport is deficient in the lethal milk mouse, *Nat. Genet.* **17**, 292–297.

Kennedy, M.C., Mende-Mueller, L., Blondin, G.A., Beinert, H. (1992) Purification and characterization of cytosolic aconitase from beef liver and its relationship to the iron-responsive element binding protein, *Proc. Natl. Acad. Sci. U.S.A.* **89**, 11730–11734.

Klotz, L.-O., Kröncke, K.-D., Buchczyk, D.P., Sies, H. (2003) Role of copper, zinc, selenium and tellurium in the cellular defense against oxidative and nitrosative stress, *J. Nutr.* **133**, 1448S–1451S.

Klug, M., Rhodes, D. (1987) Zinc fingers: a novel protein motif for nucleic acid recognition, *Trends Biochem. Sci.* **12**, 464–469.

Knutson, M., Wessling-Resnick, M. (2003) Iron metabolism in the reticuloendothelial system, *Crit. Rev. Biochem. Mol. Biol.* **38**, 61–88.

Kuo, Y.M., Zhou, B., Cosco, D., Gitschier, J. (2001) The copper transporter CTR1 provides an essential function in mammalian embryonic development, *Proc. Natl. Acad. Sci. U.S.A.* **98**, 6836–6841.

Lee, J., Prohaska, J.R., Thiele, D.J. (2001) Essential role for mammalian copper transporter Ctr1 in copper homeostasis and embryonic development, *Proc. Natl. Acad. Sci. U.S.A.* **98**, 6842–6847.

Lönnerdal, B. (2003) Genetically modified plants for improved trace element nutrition, *J. Nutr.* **133**, 1490S–1493S.

Lyons, T.J., Liu, H., Goto, J.J., Nersissian, A., Roe, J.A., Graden, J.A., Café, C., Ellerby, L.M., Bredesen, D.E., Gralla, E.B., Valentine, J.S. (1996) Mutations in copper-zinc superoxide dismutase that cause amyotrophic lateral sclerosis alter the zinc binding site and the redox behavior of the protein, *Proc. Natl. Acad. Sci. U.S.A.* **93**, 12240–12244.

Maret, W. (2004) Zinc and sulfur: a critical biological partnership, *Biochemistry* **43**, 3301–3309.

Mason, K.E. (1979) A conspectus of research on copper metabolism and requirements of man, *J. Nutr.* **109**, 1979–2066.

Mercer, J.F.B., Llanos, R.M. (2003) Molecular and cellular aspects of copper transport in developing mammals, *J. Nutr.* **133**, 1481S–1484S.

Michalska, A.E., Choo, K.H. (1993) Targeting and germ-line transmission of a null mutation at the metallothionein I and II loci in mouse, *Proc. Natl. Acad. Sci. U.S.A.* **90**, 8088–8092.

Nemeth, E., Rivera, S., Gabayan, V., Keller, C., Taudorf, S., Pedersen, B.K., Ganz, T. (2004) IL-6 mediates hypoferremia of inflammation by inducing the synthesis of the iron regulatory hormone hepcidin, *J. Clin. Invest.* **113**, 1271–1276.

Owen, C.A. Jr. (1964) Distribution of copper in the rat, *Am. J. Physiol.* **207**, 446–448.

Osaki, S., Johnson, D.A. (1969) Mobilization of liver iron by ferroxidase (ceruloplasmin), *J. Biol. Chem.* **244**, 5757–5758.

Palmiter, R.D., Findley, S.D. (1995) Cloning and functional characterization of a mammalian zinc transporter that confers resistance to zinc, *EMBO J.* **14**, 639–649.

Pekarek, R.S., Burghen, G.A., Bartelloni, P.J., Calia, F.M., Bostian, K.A., Beisel, W.R. (1970) The effect of live attenuated Venezuelan equine encephalomyelitis virus vaccine on serum iron, zinc, and copper concentrations in man, *J. Lab. Clin. Med.* **76**, 293–303.

Pietrangelo, A. (2004) Hereditary hemochromatosis-a new look at an old disease, *N. Engl. J. Med.* **350**, 2383–2397.

Prasad, A.S. (1982) Clinical and Biochemical Spectrum of Zinc Deficiency in Human Subjects, in: Prasad, A.S. (Ed.) *Clinical, Biochemical and Nutritional Aspects of Trace Elements*, Alan R. Liss, New York, pp. 3–62.

Requena, J.R., Groth, D., Legname, G., Stadtman, E.R., Prusiner, S.B., Levine, R.L. (2001) Copper-catalyzed oxidation of the recombinant SHa(29–231) prion protein, *Proc. Natl. Acad. Sci. U.S.A.* **98**, 7170–7175.

Richards, M.P., Cousins, R.J. (1975) Mammalian zinc homeostasis: requirement for RNA and metallothionein synthesis, *Biochem. Biophys. Res. Commun.* **64**, 1215–1223.

Roesijadi, G., Bogumil, R., Vasak, M., Kagi, J.H. (1998) Modulation of DNA binding of a tramtrack zinc finger peptide by the metallothionein-thionein conjugate pair, *J. Biol. Chem.* **273**, 17425–17432.

Schroeder, J.J., Cousins, R.J. (1990) Interleukin 6 regulates metallothionein gene expression and zinc metabolism in hepatocyte monolayer cultures, *Proc. Natl. Acad. Sci. U.S.A.* **87**, 3137–3141.

Shim, H., Harris, Z.L. (2003) Genetic defects in copper metabolism, *J. Nutr.* **133**, 1527S–1531S.

Snyder, S.L., Walker, R.I. (1976) Inhibition of lethality in endotoxin-challenged mice treated with zinc chloride, *Infect. Immun.* **13**, 998–1000.

Squibb, K.S., Cousins, R.J., Feldman, S.L. (1977) Control of zinc-thionein synthesis in rat liver, *Biochem. J.* **164**, 223–228.

Suzuki, K.T., Imura, N., Kimura, M. (Eds.) (1993) *Metallothionein III: Biological Roles and Medical Implications*, Birkhäuser Verlag, Basel, Switzerland.

Tang, C.K., Chin, J., Harford, J.B., Klausner, R.D., Rouault, T.A. (1992) Iron regulates the activity of the iron-responsive element binding protein without changing its rate of synthesis or degradation, *J. Biol. Chem.* **267**, 24466–24470.

Valentine, J.S., Hart, P.J. (2003) Misfolded CuZnSOD and amyotrophic lateral sclerosis, *Proc. Natl. Acad. Sci. U.S.A.* **100**, 3617–3622.

Vallee, B.L., Galdes, A. (1984) The metallobiochemistry of zinc enzymes, *Adv. Enzymol. Relat. Areas Mol. Biol.* **56**, 283–430.

Vallee, B.L., Williams, R.J. (1968) Metalloenzymes: the entatic nature of their active sites, *Proc. Natl. Acad. Sci. U.S.A.* **59**, 498–505.

Vulpe, C., Levinson, B., Whitney, S., Packman, S., Gitschier, J. (1993) Isolation of a candidate gene for Menkes disease and evidence that it encodes a copper-transporting ATPase, *Nat. Genet.* **3**, 7–13.

Wang, K., Zhou, B., Kuo, Y.M., Zemansky, J., Gitschier, J. (2002) A novel member of a zinc transporter family is defective in acrodermatitis enteropathica, *Am. J. Hum. Genet.* **71**, 66–73.

Wang, Y., Wimmer, U., Lichtlen, P., Inderbitzin, D., Stieger, B., Meier, P.J., Hunziker, L., Stallmach, T., Forrer, R., Rulicke, T., Georgiev, O., Schaffner, W. (2004) Metal-responsive transcription factor-1 (MTF-1) is essential for embryonic liver development and heavy metal detoxification in the adult liver, *FASEB J.* **18**, 1071–1079.

Weinberg, E.D. (1984) Iron withholding: a defense against infection and neoplasia, *Physiol. Rev.* **64**, 65–102.

Zahringer, J., Baliga, B.S., Munro, H.N. (1976) Novel mechanism for translational control in regulation of ferritin synthesis by iron, *Proc. Natl. Acad. Sci. U.S.A.* **73**, 857–861.

Mitochondrial Porins, Eukaryotic

Roland Benz
Biocenter of the University of Würzburg, Würzburg, Germany

Encyclopedia of Molecular Cell Biology and Molecular Medicine, 2nd Edition. Volume 8
Edited by Robert A. Meyers.
Copyright © 2005 Wiley-VCH Verlag GmbH & Co. KGaA, Weinheim
ISBN: 3-527-30550-5

Keywords

β-Barrel Cylinder
Membrane-spanning hollow cylinder formed by antiparallel amphipathic β-strands, typical in the structure of bacterial and mitochondrial outer membrane channels formed by porins.

Mitochondria
Cell organelle responsible for oxidative phosphorylation. Mitochondria are probably descendants of strictly aerobic gram-negative ancestors.

Mitochondrial (Eukaryotic Porin)
Channel-forming protein responsible for exchange of mitochondrial metabolites across the mitochondrial outer membrane.

Mitochondrial Inner Membrane
Surrounds the mitochondrial matrix space and contains the enzymes of the respiration chain and the H^+-ATPase. The mitochondrial inner membrane has many invaginations (cristae) to increase its surface and contains many carrier proteins responsible for specific substrate transport.

Mitochondrial Outer Membrane
Outer boundary membrane of mitochondria, forms with the mitochondrial inner membrane the intermembrane space that contains many different enzymes important for mitochondrial metabolism such as the mitochondrial creatine kinase.

Peripheral Kinases
Enzymes such as hexokinase and glycerokinase that bind to porin on the surface of mitochondria and utilize mitochondrial ATP.

Porin Family
Most genomes of eukaryotic cells contain more than one gene coding for mitochondrial (eukaryotic) porins. The porins form a family of highly homologous proteins that are posttranslationally imported into mitochondria.

VDAC
Voltage-dependent anion-selective channel, synonym for mitochondrial porin.

■ The mitochondrial outer membrane contains a channel that is responsible for the passage of hydrophilic metabolites across the membrane. The channel-forming protein known for many organisms, called *mitochondrial porin* or VDAC (voltage-dependent anion-selective channel), has a length of about 280 amino acids. The genomes of eukaryotic organisms contain several porin isoforms of not well-defined function. The primary structure of mitochondrial or eukaryotic porins is not particularly hydrophobic and secondary structure predictions suggest that a β-barrel cylinder typical for the secondary structure of bacterial porins also forms the mitochondrial channel. Kinases involved in mitochondrial metabolism such as hexokinase or glycerokinase bind to porin and play an important role in compartment formation in mitochondria. Mitochondrial porins are voltage-gated and switch into ion-permeable substates at voltages higher than 20 to 30 mV. The open channel has a small preference for anion over cations of the same aqueous mobility. The closed states are cation selective and impermeable for ATP and ADP. Mitochondrial porins seem to be involved in apoptosis and the release of cytochrome c. There exists emerging evidence that the cytoplasmic membrane of eukaryotic cells also contains porins of the eukaryotic porin family with a role in cellular metabolism that is not yet understood.

1
Introduction

The matrix space of mitochondria is, similar to the situation in gram-negative bacteria, surrounded by two membranes. The mitochondrial inner membrane is especially rich in proteins. Important for mitochondrial function are the enzymes of the respiration chain and oxidative phosphorylation and the many highly specific carriers for the substrates of mitochondria. In contrast to this, the mitochondrial outer membrane is far less substrate specific and acts as a molecular filter for hydrophilic solutes since it is freely permeable to a variety of small hydrophilic solutes up to a well-defined exclusion limit of about 3000 to 5000 Da. Molecules with higher molecular masses are retained. The well-defined exclusion limit indicates that the mitochondrial outer membranes contain defined passages for substrates.

A considerable part of its permeability properties is caused by the presence of general diffusion pores, called *mitochondrial porins* or voltage-dependent anion-selective channels (VDACs). As early as 1976, a channel-forming component was discovered in mitochondria, but not in other membrane fractions, of *Paramecium aurelia*. Because of the voltage dependence and the anion selectivity of the channel, the name VDAC was used. It is obvious that the anion selectivity of mitochondrial porin was overestimated at the beginning because the selectivity ratio of the channel was estimated to be more than 7 for chloride over potassium ions, while more recent data suggest that it is close to 2 at neutral pH. Mitochondrial porins were isolated from a variety of mitochondria from different organisms and tissues, and there exists emerging evidence that similar proteins are also present in cytoplasmic membranes of eukaryotic cells with yet unknown function. They form a family of slightly basic proteins with molecular masses around 30 kDa. The genes of many mitochondrial porins are known to date. The proteins are encoded in the nucleus without a leader sequence and are synthesized on cytoplasmic ribosomes. They are targeted to mitochondria as chaperone-bound species, recognized by the outer membrane general import complex (TOM) and then inserted into the outer membrane probably from the intermembrane space, similar to the situation for porins of the putative gram-negative ancestor of mitochondria.

The eukaryotic genomes may contain several genes and pseudogenes coding for porins. Mitochondria contain mostly isoform 1 (VDAC1, porin 31HL), but the role of the other one to two isoforms in eukaryotic cells was also studied. Experiments with water-soluble porin from *Neurospora crassa* and the composition of isolated mammalian porin suggest that the channel-forming unit is a complex composed of probably only one polypeptide chain and sterols. The amino acid composition exhibits a high polarity similar to bacterial porins, although the channel-forming unit is a membrane protein that contains many membrane-spanning segments. The exact structure of the pore is still a matter of debate because it has not been crystallized to date. Secondary structure predictions, electron microscopic analyzes of two-dimensional arrays and circular dichroism (CD) spectra of mitochondrial porins suggest that they contain large amounts of β-sheet structure with no indication of hydrophobic transmembrane α-helical regions, which makes it likely that bacterial and mitochondrial porins share common structural motifs. So far it is not clear if this

has to do with the evolution of mitochondria or if it simply reflects that both membrane channels have water-soluble precursors.

Reconstitution experiments with planar lipid bilayers and liposomes define mitochondrial porins as pore-forming components, which have a high permeability for both ionic and neutral substrates. According to these studies, the mitochondrial outer membrane pore has a diameter of about 2 to 3 nm in the open state, which agrees well with the electron microscopic analysis of mitochondrial porin *in vitro* (i.e. in reconstituted systems) and *in vivo* (i.e. in the outer membrane The open channel appears to be wide and water filled. It is slightly anion selective for salts composed of cations and anions of the same aqueous mobility such as potassium and chloride, but may change selectivity when the anion has a smaller aqueous mobility than the cation. For membrane voltages larger than about 30 mV, the channel closes in distinct steps in several closed states. These closed states have a reduced permeability toward hydrophilic solutes and a completely different selectivity than the open state. In addition to increasing membrane potential, polyanions such as Konig's polyanion (a copolymer of molecular mass 10 kDa from methacrylate, maleate, and styrene in a $1:2:3$ proportion) or dextran sulfate result in channel closure *in vivo* and *in vitro*. In particular, adenosine triphosphate (ATP) and adenosine diphosphate (ADP) are not permeable through the closed states of mitochondrial porin.

The role of the mitochondrial outer membrane in the physiology and metabolism of mitochondria was underestimated for a long time. It was described as a simple barrier that only prevented the destruction of the mitochondrial inner membrane during osmotic swelling. More recent papers gave some insight into the function of the mitochondrial outer membrane and the role of the porin in mitochondrial metabolism. Porin is the hexokinase-binding protein and glycerokinase also binds to this protein. Cross-linking studies of the mitochondrial outer membrane of yeast revealed that the pore associates or interacts with a 14-kDa protein that was identified as glutathione transferase. Subfractionation of the mitochondrial membranes by sucrose density centrifugation allowed the isolation of contact sites between inner and outer membranes. These contact sites are especially enriched in hexokinase-binding capacity. A further interesting aspect of mitochondrial porin is its possible role in the mitochondrial permeability transition pore (PTP) and its part in cell signaling. Furthermore, there exists a strong indication that porin is involved in apoptosis because of its interaction with proapoptotic and antiapoptotic factors such as Bax and Bcl-2. Furthermore, it shows negative regulation by C-Raf kinase. This may indicate that mitochondrial porin pore plays a crucial role in the interaction of mitochondria with cellular components.

2
Reconstitution of Mitochondrial Porins (VDACs) in Model Membranes

2.1
Isolation and Purification of Mitochondrial Porins

Several different methods are well established for the successful isolation and purification of mitochondrial porins. All start from mitochondria that are obtained by differential centrifugation or density gradient centrifugation of homogenized cellular material. It has to be noted, however, that both preparations also contain other cellular material. For preparation of the mitochondrial outer membrane, the mitochondria are exposed to a swelling/shrinking procedure. Mitochondria are first exposed to 10 mM phosphate followed by addition of 0.45 M sucrose. Centrifugation at 10 000 g yields the mitoplasts in the pellet that also contain most of the contact sites between the inner and outer membrane. The outer membrane is in the supernatant. Porin can be isolated from the outer membrane following a standard protocol by using neutral detergents but not by ionic detergents because they destroy its channel-forming activity. A higher yield of porin may be obtained following an alternative method starting from whole mitochondria. The cell organelles are extracted with neutral detergents such as Genapol X-80, N,N-dimethyldodecylamine-N-oxide (LDAO) or Triton X-100. The extract is passed through a dry hydroxyapatite (HTP) column where most of the proteins bind. Porin appears just after the void volume of the column because it is deeply buried in lipid-detergent micelles and does not interact with the column material when Triton X-100 or Genapol X-80 are employed. The use of LDAO instead of Genapol or Triton obviously leads to smaller micelles that bind to HTP and allow a normal elution protocol. The use of ionic detergents such as cholate, desoxycholate, and sodium dodecylsulfate (SDS) during the isolation and purification procedure seems to destroy the structure of the channel-forming complex, and leads to its inactivity in reconstitution experiments.

2.2
Heterologous Expression of Eukaryotic Porins

It is possible to create a water-soluble form of mitochondrial porin from *N. crassa* and other organisms. Water-soluble porin is inactive in reconstitution experiments and does not form ion-permeable channels. However, when the porin is treated with Triton or Genapol in the presence of sterols, channel formation is regained with properties that cannot be distinguished from its detergent-solubilized form. It is possible that the detergents act as some kind of chaperone. Careful analyzes of the sterol requirements suggested that cholesterol leads to highest activation, whereas the addition of epicholesterol did not influence the low channel-forming ability of water-soluble porin. The possibility to refold mitochondrial porin allows an interesting insight into the structure–function relationship of porin mutants because the genes of many of the mitochondrial porins are known to date. For heterologous expression of mitochondrial porins in *Escherichia coli*, the porin-coding sequences have to be cloned in an expression vector that contains, for easy purification of the expressed protein, a DNA sequence coding for additional six histidines (His-tag) upstream of the polycloning site. This His-tag allows the purification of the overexpressed protein by Ni-NTA affinity chromatography. The expressed protein is found within inclusion bodies and needs to be solved in 1% SDS or a high concentration of urea. After chromatography across a Ni-NTA column in SDS or urea, the protein is renatured by an overnight dialysis against a 1% Genapol or Triton X-100 solution and by the addition of a cholesterol suspension in the same detergent.

2.3
Reconstitution Methods

The properties of the channels formed by mitochondrial porins (VDACs) were studied in different model membranes. Mitochondrial porins from mung bean and rat liver were investigated in lipid vesicles. The reconstitution of rat liver porin into the vesicles made them permeable for [^{14}C]sucrose, but not for [^{3}H]dextran, which suggested that a specific pathway and not a leak is introduced into the vesicles. The exclusion limit of the channel from the outer membrane of mung bean mitochondria was measured in similar experiments using radioactively labeled solutes of different molecular mass to be between 1 and 10 kDa, probably between 3 and 6 kDa. An alternative method was used for the study of porin from *N. crassa* mitochondria. Porin-containing liposomes were transferred into a hypertonic solution containing a test solute. First, the liposomes shrink and can only swell when the test solute is permeable through the channels. The results of experiments using carbohydrates and polyethylene glycols (PEG) of different molecular masses suggest that PEGs with molecular masses up to about 6.8 kDa are permeable through the mitochondrial pore. From these results, the diameter of the mitochondrial pore was calculated to be around 3 to 4 nm. It is noteworthy that this type of vesicle-swelling assay is not as precise as the swelling method used previously for bacterial porins. The reason for this is that the method used for the study of the channel diameter of bacterial porins is based on the relative rate of permeation of many different solutes through the channel, which depends on the channel diameter according to the Renkin equation.

Fig. 1 Stepwise increase of the membrane current (given in pA) after the addition of *N. crassa* porin to a black lipid bilayer membrane given as a function of time. The aqueous phase contained 10 ng mL^{-1} protein and 1 M KCl. The membrane was formed from diphytanoyl phosphatidylcholine/*n*-decane. The voltage applied was 10 mV; $T = 25\,^{\circ}$C.

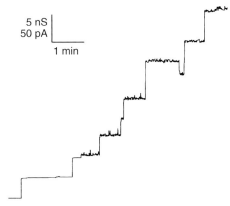

Many reconstitution studies with mitochondrial porins were performed using the planar lipid bilayer technique. The reconstitution is performed as follows. Purified porin is added at a low concentration (10 ng mL^{-1} to 1 μg mL^{-1}) to the aqueous phase bathing black lipid bilayer membranes formed according to the two different methods, which yield more or less similar results. After the addition of the porins to a preformed membrane, the membrane current starts increasing in a stepwise fashion after an initial lag of about 2 min (see Fig. 1 for the reconstitution of mitochondrial porin from *N. crassa*). This process indicates the insertion of ion-permeable channels into the membrane. The number of channels formed is dependent on time and the concentration of protein. In general, the membrane conductance increased for about 15 to 20 min. After that time, the membrane conductance (i.e. the current per unit voltage) increased at a much slower rate. When the rate of conductance increase was relatively slow (as compared with the initial one), it was shown for different mitochondrial porins that the membrane conductance was a linear function of the protein concentration up to porin concentrations of about 1 μg mL^{-1}. The results are consistent with

the assumption that the protein samples contain preformed channels.

3
Characterization of the Pore-forming Properties of Eukaryotic Porins

3.1
Single-channel Analysis of the Mitochondrial Pore in the Open State

Mitochondrial porins from different eukaryotic cells form ion-permeable channels in lipid bilayer membranes – each step in Fig. 1 reflects the insertion of one conductive unit (i.e. of one channel into the membrane). These conductance steps are caused by the reconstitution of channels since they are not observed when only the detergents Triton X-100, Genapol X-80 or LDAO are added to the aqueous phase. Figure 1 demonstrates that most of the conductance steps are directed upwards and closing steps are only rarely observed at small transmembrane potentials of about 10 to 20 mV. The most frequent value for the single-channel conductance of mitochondrial porins in 1 M KCl (the conditions of Fig. 1) was about 4 to 5 nS (see Fig. 2 for a histogram of conductance

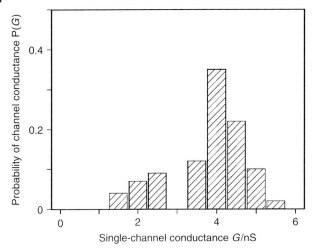

Fig. 2 Histogram of conductance fluctuations observed with membranes of diphytanoyl phosphatidylcholine/*n*-decane in the presence of *N. crassa* porin. $P(G)$ is the probability for the occurrence of a conductance step with a certain single-channel conductance (given in nS). The aqueous phase contained 1 M KCl. The voltage applied was 10 mV. The mean value of all upward directed steps was 4.2 nS for the right-side maximum and 2.5 nS for the left-side maximum (in total 187 single events); $T = 25\,^{\circ}\mathrm{C}$.

fluctuations observed with *N. crassa* porin). Only a limited number of smaller steps were observed under these conditions (about 25% of the total number of conductance steps). It should be noted that smaller steps are also found for other mitochondrial porins, including those from yeast, rat liver, and a number of cells from other tissues including mammalian cells. They may be explained as substates of the pore with the exception of porin from *Paramecium* mitochondria, where only small pores with a single-channel conductance around 2.5 nS were observed in 1 M KCl. This result may indicate that the mitochondrial outer membrane pore does not only switch to substates but that it may also exist in two different stable conformations. The channels formed by eukaryotic porins are wide and water filled. This is the result of measurements with

salts composed of different anions and cations. Even large organic anions and cations such as Tris^{+} and HEPES^{-}, were found to be permeable through the open state of mitochondrial porins (see Table 1). The single-channel conductance of these salts is a linear function of the specific conductivity of the bulk aqueous phase at small membrane potentials.

3.2
Eukaryotic Porins are Voltage-gated

The current recording shown in Fig. 1 demonstrates that closing events represent only a minor fraction of the recordings at 10 mV transmembrane potential. At larger voltages, beginning from about 20 to 25 mV, the number of closing events increases. They are always smaller than the initial on-steps, which indicate that

Tab. 1 Average single-channel conductance of porin 31HL from human T-lymphocytes in different salt solutions.

Salt	c [M]	G [nS]
KCl	0.01	0.05
	0.03	0.15
	0.1	0.45
	0.3	1.3
	1.0	4.3
	3.0	11
LiCl	1	3.2
K-acetate	1	1.5
Tris–HCl	0.5	1.5
Tris–HEPES	0.5	0.18
K-MES	0.5	0.70

The solutions contained 5–10 ng mL^{-1} porin 31HL and less than 0.1 μg mL^{-1} of the detergent NP-40; the pH was between 6.0 and 7.0. The membranes were made of diphytanoyl phosphatidylcholine/n-decane; $T = 25\,°C$; $V_m = 10$ mV. G was determined by recording at least 70 conductance steps and averaging over the right-hand maximum (see Fig. 2). c is the concentration of the salt solution.

the pores switched to ion-permeable substates at voltages higher than 20 mV. An experiment of this type is shown in Fig. 3. Transmembrane potentials of +50 and −50 mV were applied to a membrane containing first three (upper trace) and then four (lower trace) mitochondrial pores from *Paramecium*. The arrow indicates the reconstitution of the fourth channel, which closes after about 0.8 s. The membrane conductance decays in defined steps to smaller values. Similar experiments were also performed with membranes containing many mitochondrial porins, such as shown in Fig. 4. Voltages of −40 and +40 mV were applied to a diphytanoyl phosphatidylcholine/n-decane membrane containing about 150 mitochondrial porin channels from *Paramecium*. In this case, the closing steps cannot be resolved because of the large number of pores in the membrane. The current decays in a single exponential function (see Fig. 4). The

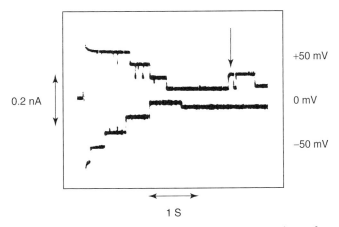

Fig. 3 Voltage-dependence of *Paramecium* porin. First a voltage of +50 mV was applied to a membrane containing three channels (upper trace). During the experiment, a fourth channel was reconstituted into the membrane (arrow), which closed after 0.8 s. Then, −50 mV was applied to the same membrane containing four channels (lower trace). The porin channels closed in defined steps. The membrane was formed from diphytanoyl phosphatidylcholine/n-decane. The aqueous phase contained 0.5 M Tris–HCl (pH 7.2); $T = 25\,°C$.

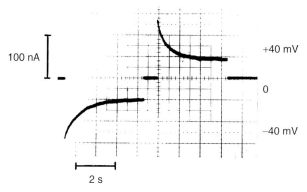

Fig. 4 Voltage dependence of *Paramecium* porin in a mutichannel experiment. About 150 porin pores were incorporated in a membrane from 1% diphytanoyl phosphatidylcholine/n-decane. The voltage across the membrane was switched to -40 mV (with respect to the cis-side, the side of the addition of 500 ng mL^{-1} protein) and then to $+40$ mV. The channels switched to substates of the open state in a single exponential curve. The aqueous phase contained 1 M KCl (pH 6); $T = 25\,^\circ$C.

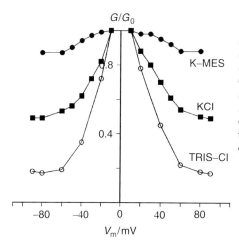

Fig. 5 Ratio of the conductance, G, at a given voltage, V_m, divided by the conductance, G_0, at 5 mV as a function of the voltage. The aqueous phase contained either 0.5 M KCl, 0.5 M K-MES or 0.5 M Tris–HCl (pH in all cases, 7.2). The cis-side contained 50 ng mL^{-1} *Paramecium* porin. The sign of the voltage is given with respect to the cis-side, the side of the addition of porin.

results of this and additional experiments at different voltages allowed the evaluation of the voltage dependence. The steady state conductance showed a bell-shaped curve as a function of the applied voltage when the conductance at a given voltage $G(V)$ divided by G_0 at zero potential is plotted as a function of membrane voltage. Figure 5 shows the results for *Paramecium* porin and three different salts (KCl, K-MES and Tris–HCl, all at pH 7.2). The results differ somewhat for different salts, although the voltage dependence is approximately the same. This has to do with the selectivity of the open channel as compared to that of the closed state (see below). The data

Fig. 6 Semilogarithmic plot of the ratio, N_o/N_c as a function of the transmembrane potential V_m. The data were taken from Fig. 5. The closed circles indicate the measurements with positive voltages, V_m, while the squares show measurements with negative potentials (with respect to the cis-side, the side of the addition of the protein). The slope of the straight lines is such that an e-fold change of N_o/N_c is produced by a change in V_m of 13 mV. The midpoint potential of the N_o/N_c distribution (i.e. $N_o = N_c$) is 35 mV for positive potentials (circles) and -28 mV for negative potentials (squares). Only data for KCl are shown.

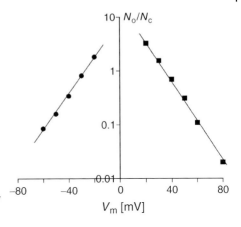

given in Fig. 5 could be analyzed as proposed previously, based on an earlier study of excitability-inducing material (EIM) by calculating the ratio of the number of open, N_o, to closed channels, N_c, from the data of the bell-shaped curves of Fig. 5:

$$N_o/N_c = (G - G_{min})/(G_0 - G) \quad (1)$$

where G is in this equation the conductance at a given membrane potential V_m, G_0 and G_{min} are the conductance at zero voltage and very high potentials, respectively. The open to closed ratio of the channels, N_o/N_c, is given by:

$$N_o/N_c = \exp[-nF(V_m - V_0)/RT] \quad (2)$$

where F (Faraday's constant), R (gas constant) and T (absolute temperature) are standard symbols, n is the number of gating charges moving through the entire transmembrane potential gradient for channel gating (i.e. a measure for the strength of the interaction between electric field and the open channel), and V_0 is the potential at which 50% of the total number of channels are in the closed configuration (i.e. $N_o/N_c = 1$).

A semilogarithmic plot of the data given in Fig. 5 shows that they could be fitted

to a straight line with a slope of 13 mV for an e-fold change of V_m (because $RT/F = 25$ mV). This result suggests that the number of charges involved in the gating process is approximately 2 in the case of *Paramecium* porin (see Fig. 6). The distribution of open and closed channels, N_o/N_c, follows a Boltzmann distribution. This means that (2) allows also the calculation of the energy difference for channel gating according to:

$$N_o/N_c = \exp[-W(V_m)/RT] \quad (3)$$

$W(V_m)$ is the voltage-dependent energy difference of one mole channels between the open and the closed state. A comparison of (2) and (3) shows that $W(V_m) = nF(V_m - V_0)$. The energy needed for channel closure, nFV_0, calculated from the data of Figs. 5 and 6 is approximately 7.7 kJ mol^{-1}, an energy that is not very high. It is considerably below the energy of one hydrogen bond, which means that channel gating is a low-energy process.

The time constant τ of the single exponential relaxation process shown in Fig. 4 decreases with increasing voltage. Interestingly, the time constants could be fitted to a similar formalism as given in (2) for the ratio N_o/N_c, which means that a semilogarithmic plot of the time constants versus voltage yields a straight line. The slope of this line corresponded to an e-fold decrease of the time constant τ for an increase of the voltage by about 13 mV under the conditions of Fig. 4. This result suggested again that the number of gating charges involved in channel gating is about 2, which agreed satisfactorily with the number of gating charges derived from the plot of N_o/N_c (see Fig. 5). The time constant of the switching of the pores from the "closed" to the "open" state could not be followed for mitochondrial porins within the time resolution of the experimental instrumentation (about 1 ms). Figure 7 shows an experiment in which the voltage across a lipid bilayer membrane containing about 30 yeast porins was first set to 70 mV. After the exponential decay of the membrane current, the voltage was decreased to 20 mV (arrow), but no relaxation was observed (see Fig. 7). This result indicated largely different reaction rates for the closing and opening of the mitochondrial porins.

All eukaryotic porins studied to date are voltage dependent. Nevertheless, the threshold voltage for complete channel closure is somewhat different for porins from different organisms. Whereas 35 mV is sufficient to decrease the initial conductance of yeast porin to 50%, about 90 mV has to be applied to membranes containing porin from rat brain to obtain the same effect. Semilogarithmic plots of the ratio N_o/N_c showed that both V_0 and n were different in these cases. These results indicated that the number of charges involved in the gating process varies for mitochondrial porins from different organisms.

3.3
Single-channel Conductance of the Closed States

The closed state of eukaryotic porins has a reduced permeability for ions. Single-channel conductance experiments allow the evaluation of the conductance of the closed state at higher membrane potentials of 30 or 40 mV. At these voltages, the open state of the channels has only a limited lifetime because of the voltage dependence of the open state. It is possible to evaluate the single-channel conductance of the closed state by subtracting the conductance of the closing events from those of the open state. Table 2 shows the results of these type of measurements obtained for three different salts and two types of porin, from yeast and VDAC1 from a human B-lymphocyte cell line (porin 31HL). The single-channel conductance

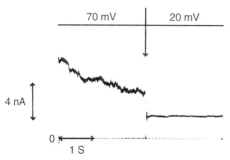

Fig. 7 Voltage dependence of the open and closed states of yeast porin. The voltage across a diphytanoyl phosphatidylcholine/*n*-decane membrane containing about 30 channels was first set to 70 mV at the cis-side of the membrane. After the relaxation of the membrane current (time constant about 1.5 s), the voltage was decreased to 20 mV (arrow). The opening of the channels was very fast (less than 50 ms) and could not be followed at the timescale of the experiment. The aqueous phase contained 1 M KCl and 20 ng mL^{-1} 1 yeast porin at the cis-side of the membrane; $T = 25\,^{\circ}$C.

Tab. 2 Average single-channel conductance of the open and closed states of yeast and human porins in different 0.5 M salt solutions.

Salt	Open state [nS]	Closed state [nS]
Yeast porin		
KCl	2.3	1.3
K-MES	0.95	0.65
Tris–HCl	1.5	0.30
Human porin (porin 31 HL)		
KCl	2.4	1.4
K-MES	0.70	0.65
Tris–HCl	1.5	0.30

The pH of the aqueous salt solutions was adjusted to 7.2. The protein concentration was between 5 and 10 ng/ml; $V_m = 30$ mV, $T = 25\,°C$. The single-channel conductance of the closed state was calculated by subtracting the conductance of the closing events from the conductance of the initial opening of the pores.

of the closed state of the pore was considerably smaller for Tris–HCl than for K-MES, despite a similar aqueous mobility of K^+ and Cl^-. This result suggested that the closed state(s) of mitochondrial porins is cation selective.

3.4

Selectivity of the Open and Closed States of Eukaryotic Porins

The data in Table 1 suggests that ions move through the open state of mitochondrial porin in the same way they move in the bulk aqueous phase. Interestingly, the pores also exhibit certain specificity for charged solutes because the single-channel conductance in potassium acetate is somewhat smaller than that in LiCl despite the same aqueous mobility of lithium ions as compared to acetate (see Table 1). This means that the mitochondrial porin is selective although it is a

wide and water-filled channel in the open state. Experiments with lipid bilayer membranes under zero-current conditions and an externally applied concentration gradient, c''/c', across the membrane allow the evaluation of the ionic selectivity. From the membrane potential, V_m, caused by the preferential movement of one sort of ions through the channel, the ratio P_{cation}/P_{anion} of the permeabilities for cations and anions can be calculated using the Goldman–Hodgkin–Katz equation.

Table 3 shows the zero-current membrane potentials and the permeability ratios for different mitochondrial and eukaryotic porins [porin 31HL (human), rat liver, yeast, and *Paramecium*] in potassium chloride, potassium acetate, and lithium chloride. It is obvious from the data of Table 3 that the asymmetry potential and

Tab. 3 Zero-current membrane potentials, V_m, of membranes from diphytanoyl phosphatidylcholine/n-decane in the presence of rat liver, yeast, and *Paramecium* mitochondrial porins measured for a 10-fold gradient of different salts[a].

Salt	V_m [mV]	P_{anion}/P_{cation}
Rat liver		
KCl (pH 6)	−11	1.7
LiCl (pH 6)	−24	3.4
K acetate(pH 7)	+14	0.50
Yeast		
KCl (pH 6)	−7	1.4
LiCl (pH 6)	−20	2.6
K acetate (pH 7)	+14	0.5
Paramecium		
KCl (pH 6)	−11	1.7
LiCl (pH 6)	−24	3.4
K acetate (pH 7)	+14	0.50

[a] V_m is defined as the potential of the dilute side (10 mM) relative to that of the concentrated side (100 mM); the temperature was 25 °C. P_{anion}/P_{cation} was calculated from the Goldman–Hodgkin–Katz equation.

the ion selectivity of mitochondrial porins are dependent on the combination of different anions and cations. The different porin channels are slightly anion selective (ratio $P_{cation}/P_{anion} = 2:1$) for potassium and chloride, which have approximately the same aqueous mobility. For the combination of the less mobile lithium ion (because of its larger hydration shell) with chloride, the anion selectivity increases. Since the ions move within the channel in a similar way as in bulk aqueous phase, the channel becomes cation selective for potassium acetate because of the smaller mobility of acetate as compared with potassium ions. This result represents another support for the assumption of the mitochondrial pore as a wide water-filled channel in the "open" state.

The open state of all mitochondrial porins characterized so far is slightly anion selective for salts composed of equally mobile cations and anions such as KCl (see above). The mitochondrial porins switch to closed states when the transmembrane voltage exceeds 20 to 30 mV. These states definitely have reduced ion permeability as the bell-shaped curves of Fig. 5 clearly indicate. The ion selectivity of the closed states cannot be measured using zero-current membrane potential measurements with pores in the closed state. However, the results of Fig. 5 and Table 2 suggest that the closed state is cation selective since for the combination K-MES (a mobile cation combined with a less mobile anion) the conductance in the open and in the closed states differs only little. The difference between open state and closed state is more substantial for Tris–HCl (a mobile anion combined with a less mobile cation). This result suggests indeed that the channel is cation selective in the closed state.

4
Inhibition of the Mitochondrial Pore

Konig and coworkers showed that an amphiphilic, synthetic polyanion (a copolymer of 10 kDa molecular mass of methacrylate, maleate, and styrene in a 1:2:3 proportion) inhibits different carriers in the mitochondrial inner membrane and the ATPase. It is possible that these effects have nothing to do with a direct interaction between the polyanion and inner membrane carriers. Moreover, it seems that the polyanion binds to mitochondrial porin and shifts its voltage dependence in a defined way, thus closing the channel when the membrane voltage has a negative sign at the cis-side, the side of the addition of the polyanion. For a positive potential at the cis-side, the channel is always in its open configuration even for voltages up to 120 mV and higher. The effect of the polyanion on mitochondrial porin from rat liver is demonstrated in Fig. 8. First the voltage dependence of reconstituted rat liver porin was studied and showed the typical bell-shaped curve. The polyanion was added at a concentration of 100 ng mL^{-1} polyanion to one side of the membrane (the cis-side). When the voltage had a negative sign on the cis-side, the membrane current started to decrease at smaller voltages than without the polyanion. A membrane potential of -10 mV was already sufficient to close the channels. Higher voltages (-40 mV; see Fig. 8) resulted in a more rapid closure of rat liver porin (relaxation time constant smaller than 100 ms), whereas $+40$ mV had no effect on porin-induced membrane conductance. Even voltages with magnitudes up to $+120$ mV did not result in any current decrease, which means that the polyanion stabilized the pore in the open state if the sign of the transmembrane

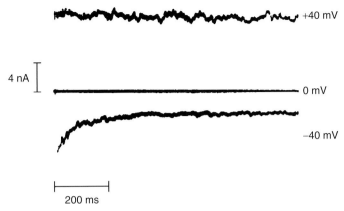

Fig. 8 Asymmetric current response of rat liver porin after application of membrane potentials of different polarity. The membrane was made of asolectin/n-decane. The concentration of the porin was 100 ng mL^{-1} in the 1 M KCl solution. The cis-side contained in addition 1μg mL^{-1} polyanion. A membrane potential of +40 mV (upper trace) and −40 mV (lower trace) was applied to the membrane (referred to as the *cis-side*) 40 min after the addition of the porin and 10 min after that of the polyanion. Note that the membrane current only decreased if the cis-side was negative; $T = 25\,°C$.

potential was positive on the cis-side. The addition of the polyanion to both sides resulted in a bell-shaped curve because the channels closed at positive and negative potentials. However, the parameters of channel closure were completely different to those without polyanion, that is, n was much larger (about 4–5) and V_0 was smaller than 10 mV. The experimental data suggest that there does not exist a "sideness" for the interaction of the polyanion with porin, which means that the polyanion can interact with the gate(s) from both sides of the channel. Furthermore, the asymmetric effect of the polyanion on rat liver porin was stable during the whole lifetime of the membranes (up to several hours), indicating that the polyanion could not penetrate the channel.

The interaction of the polyanion with mitochondrial porins from different sources also allows us to measure the conductance of the closed state when the polyanion was added at a concentration of 100 ng mL^{-1} to the side of the membrane with negative polarity (the cis-side). The single-channel conductance was about half that of the open state measured at low voltage, which means that the closed channels are still permeable for ions, which makes a direct polyanion-induced block of the channel rather unlikely. Table 4 shows the effect of polyanion on the single-channel conductance obtained with salts other than KCl. The reduction of the single-channel conductance caused by the polyanion is especially large when a mobile anion (e.g. Cl$^-$) is combined with a less mobile cation (e.g. Tris$^+$). In this case, the single-channel conductance was reduced to less than 10% of the open state value. The effect of the polyanion on combinations of mobile cations with less mobile anions (e.g. on K-MES) was very small. These results are consistent with the assumption that the slightly anion-selective channel in the

Tab. 4 Average single-channel conductance of the open and the polyanion-induced closed state of rat liver mitochondrial porin in different 0.5 M salt solutions. (The pH of the aqueous salt solutions was adjusted to 7.2. The membrane voltage was 10 mV at the cis-side; $T = 25\,°C$. The aqueous phase contained in the measurements of the closed state $0.1\ \mu g\ mL^{-1}$ polyanion added to the cis-side).

Salt	Open state [nS]	Closed state [nS]
KCl	2.2	1.2
LiCl	1.8	0.40
K acetate	1.1	0.85
K-MES	0.88	0.74
Tris–HCl	1.5	0.25

open state becomes highly cation selective in the closed state (see also Fig. 5). This was also the result of selectivity measurements in the presence of polyanion. The ratio P_{cation}/P_{anion} for rat liver porin in selectivity measurements without polyanion is approximately 0.6 for KCl. For a polyanion concentration of 15 $\mu g\ mL^{-1}$ on both sides of the membrane, P_{cation}/P_{anion} increases to about 10, which suggests that chloride has a very small permeability through the polyanion-mediated closed state of rat liver porin.

Channel blockage by the polyanion can be compared to the situation in whole mitochondria in which polyanions added to the outside (presumably the negative side of the outer membrane) are able to close the channel and inhibit the passage of anionic metabolites across the mitochondrial outer membrane, including ADP and ATP, thus blocking the mitochondrial metabolism. Other polyanions such as dextran sulfate have a similar effect on the voltage dependence of mitochondrial porin, but these compounds need a considerably larger concentration

for channel closure than the polyanion. In addition to the different polyanions, the action of a protein component from mitochondria was studied in mitochondria and in reconstituted systems. This "modulator" exhibits a similar action on the porin channel as the polyanion, which means that it disturbs nucleotide exchange across the outer membrane and shifts the voltage dependence of the channel in a dose-dependent way. Although its action is known, it has not been possible to identify the modulator. It is presumably a polypeptide since it is susceptible to pronase, that is, its action on porin is destroyed by pronase treatment.

5
Structure of the Channel Formed by Eukaryotic Porins

5.1
Primary Structure of Eukaryotic Porins

The primary structure of eukaryotic porins from many organisms, including humans, mouse, fruit fly, plants, yeast, fungi, and *Dictyostelium*, is known to date. With the exception of porin 31HL (HVDAC1 from humans), all were derived from their cDNA sequences. Porin 31HL is known from its amino acid sequence. Despite many variations of single amino acids, all sequences are related to one another and have an approximate length of about 280 amino acids. Figure 9 shows a comparison of the amino acid sequences of human (HVDAC1, porin 31HL), *Drosophila melanogaster*, potato, *Saccharomyces cerevisiae* and *N. crassa* mitochondrial porins. Despite the clear homology between the different sequences, only 15 (out of about 280) amino acid residues are strictly

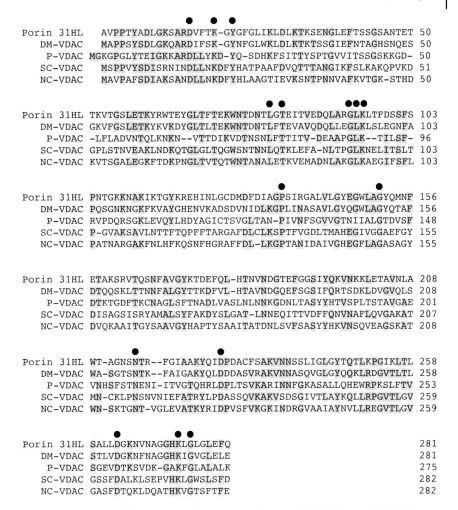

```
Porin 31HL  AVPPTYADLGKSARDVFTK-GYGFGLIKLDLKTKSENGLEFTSSGSANTET  50
   DM-VDAC  MAPPSYSDLGKQARDIFSK-GYNFGLWKLDLKTKTSSGIEFNTAGHSNQES  50
    P-VDAC  MGKGPGLYTEIGKKARDLLYKD-YQ-SDHKFSITTYSPTGVVITSSGSKKGD-  50
   SC-VDAC  MSPPVYSDISRNINDLLNKDFYHATPAAFDVQTTTANGIKFSLKAKQPVKD  51
   NC-VDAC  MAVPAFSDIAKSANDLLNKDFYHLAAGTIEVKSNTPNNVAFKVTGK-STHD  50


Porin 31HL  TKVTGSLETKYRWTEYGLTFTEKWNTDNTLGTEITVEDQLARGLKLTFDSSFS  103
   DM-VDAC  GKVFGSLETKYKVKDYGLTLTEKWNTDNTLFTEVAVQDQLLEGLKLSLEGNFA  103
    P-VDAC  -LFLADVNTQLKNKN--VTTDIKVDTNSNLFTTITV-DEAAPGLK--TILSF-  96
   SC-VDAC  GPLSTNVEAKLNDKQTGLGLTQGWSNTNNLQTKLEFA-NLTPGLKNELITSLT  103
   NC-VDAC  KVTSGALEGKFTDKPNGLTVTQTWNTANALETKVEMADNLAKGLKAEGIFSFL  103


Porin 31HL  PNTGKKNAKIKTGYKREHINLGCDMDFDIAGPSIRGALVLGYEGWLAGYQMNF  156
   DM-VDAC  PQSGNKNGKFKVAYGHENVKADSDVNIDLKGPLINASAVLGYQGWLAGYQTAF  156
    P-VDAC  RVPDQRSGKLEVQYLHDYAGICTSVGLTAN-PIVNFSGVVGTNIIALGTDVSF  148
   SC-VDAC  P-GVAKSAVLNTTFTQPFFTARGAFDLCLKSPTFVGDLTMAHEGIVGGAEFGY  155
   NC-VDAC  PATNARGAKFNLHFKQSNFHGRAFFDL-LKGPTANIDAIVGHEGFLAGASAGY  155


Porin 31HL  ETAKSRVTQSNFAVGYKTDEFQL-HTNVNDGTEFGGSIYQKVNKKLETAVNLA  208
   DM-VDAC  DTQQSKLTTNNFALGYTTKDFVL-HTAVNDGQEFSGSIFQRTSDKLDVGVQLS  208
    P-VDAC  DTKTGDFTKCNAGLSFTNADLVASLNLNNKGDNLTASYYHTVSPLTSTAVGAE  201
   SC-VDAC  DISAGSISRYAMALSYFAKDYSLGAT-LNNEQITTVDFFQNVNAFLQVGAKAT  207
   NC-VDAC  DVQKAAITGYSAAVGYHAPTYSAAITATDNLSVFSASYYHKVNSQVEAGSKAT  208


Porin 31HL  WT-AGNSNTR--FGIAAKYQIDPDACFSAKVNNSSLIGLGYTQTLKPGIKLTL  258
   DM-VDAC  WA-SGTSNTK--FAIGAKYQLDDDASVRAKVNNASQVGLGYQQKLRDGVTLTL  258
    P-VDAC  VNHSFSTNENI-ITVGTQHRLDPLTSVKARINNFGKASALLQHEWRPKSLFTV  253
   SC-VDAC  MN-CKLPNSNVNIEFATRYLPDASSQVKAKVSDSGIVTLAYKQLLRPGVTLGV  259
   NC-VDAC  WN-SKTGNT-VGLEVATKYRIDPVSFVKGKINDRGVAAIAYNVLLREGVTLGV  259


Porin 31HL  SALLDGKNVNAGGHKLGLGLEFQ                               281
   DM-VDAC  STLVDGKNFNAGGHKIGVGLELE                               281
    P-VDAC  SGEVDTKSVDK-GAKFGLALALK                               275
   SC-VDAC  GSSFDALKLSEPVHKLGWSLSFD                               282
   NC-VDAC  GASFDTQKLDQATHKVGTSFTFE                               282
```

Fig. 9 Comparison of the amino acid sequences of human (HVDAC1, porin 31HL), *D. melanogaster*, potato, *S. cerevisiae* and *N. crassa* mitochondrial porins using BLAST. Three or more identical amino acids are shaded. Points indicate identical amino acids in all five sequences.

conserved in all five porins (marked by points in Fig. 9). The most remarkable identity in all five porins is the GLK triplet near amino acid 90, which may play an essential, but unknown, role in eukaryotic porin function. However, it is not the place of ATP binding. Otherwise, the identities between the five porins are rather small in large stretches of the aligned porin sequences. These results suggest that the possible β-barrel structure of mitochondrial porins (see below) tolerates extensive amino acid variations without substantial alterations in the secondary structure and of the function of a voltage-dependent membrane channel, which was found for all the mitochondrial porins. Similarly, all porins form wide water-filled channels with an average single-channel conductance of about 4 nS in 1 M KCl, that are slightly cation selective in the open state and highly cation selective in the voltage-dependent closed state.

No identity exists between primary sequences of mitochondrial porins and those of bacterial porins, especially of phototrophic bacteria, which are related to the ancestor of mitochondria. On the other hand, the structure and function of bacterial porins as channels are clearly similar to those of the mitochondrial porins, since they both contain predominantly antiparallel β-barrel stave structures, which form hollow cylinders in the outer membrane and they both have similar molecular masses. Hydropathicity plots indicate that the primary sequence of mitochondrial porins is predominantly hydrophilic (46–50% polar residues). This clearly distinguishes them from the ion channels in nerve and muscle tissue with their typical α-helical structure, although the p-elements of potassium and similar ion channels contain an approximately 20-amino acid stretch that lines the channel and is organized similar to a β-sheet structure. Eukaryotic porins contain only about a 19-amino acid long stretch at the N-terminal end that may form an α-helical region. The function of the amphipathic α-helical structure is still a matter of debate, and it is suggested that it has to do with channel gating; protein targeting or it is even a part of the channel itself.

5.2
Secondary Structure of Mitochondrial Porins

The secondary structure of most mitochondrial porins was analyzed using computer-based prediction programs. The simplest possibility of identifying amphipathic β-strands is the method proposed by Vogel and Jahnig. It is based on the observation that every second amino acid in a sided β-strand is either on average hydrophobic or hydrophilic. This means that one side of a β-strand is hydrophilic and faces the channel interior; the other side is hydrophobic and faces the lipid side chains. In addition to this simple approach, more sophisticated methods are also possible, such as secondary structure predictions based on a neural network-based predictor, which is suitable for finding β-sheets along protein sequences and is trained to recognize the topography of known bacterial β-strands. An example of this treatment is given in Fig. 10 for *N. crassa* porin. Within its primary sequence, 16 amphipathic β-strands (indicated by shaded boxes) can be recognized by this method. All studies of porin structure agree that the eukaryotic porins contain a high degree of antiparallel β-sheets as the main structural feature. The results of the secondary structure predictions are supported by CD measurements of eukaryotic porins. Native and recombinant porin from *N. crassa* shows, in these experiments, a high degree of β-sheet structure. In general, the recombinant porins seem to be a valuable model system for the study of eukaryotic porins by spectroscopic methods, in which high amounts of protein are needed. CD spectroscopy of root plastid porin was performed to determine the secondary structure of the protein under different conditions. It has a high degree of β-sheet structure in the non-ionic detergent Genapol X-80 and in lipid vesicles. Although the presence of sterols is essential for reconstitution of eukaryotic porins in lipid bilayer membranes, the presence of sterols in the experiments with lipid vesicles does not change the CD spectra. The more polar detergent

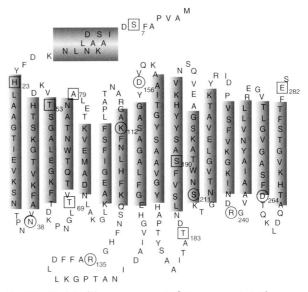

Fig. 10 Model of the arrangement of *N. crassa* porin in the membrane. The 16 putative β-strands are shaded in gray. Taken from Casadio, R., Jacoboni, I., Messina, A., De Pinto, V. (2002) A 3D model of the voltage-dependent anion channel (VDAC), *FEBS Lett.* **520**(1–3), 1–7 with permission.

SDS induces a large amount of α-helix structure in the protein. Similar results were obtained from experiments with two plant porin isoforms using attenuated total reflection Fourier-transform infrared spectroscopy (ATR-FTIR). Here, the β-sheet content of the eukaryotic porins was about 50%. Furthermore, the arrangement of the porins as detected by ATR-FTIR was very similar to that of OmpF of *E. coli*.

5.3
Structure of the Channel-forming Unit

Two different models of the channel-forming unit were derived from secondary structure predictions, site-directed mutagenesis, and deletion of part of the primary sequence of mitochondrial porins. One model assumes that the channel is composed of either 12 (yeast) or 13 (*N. crassa*) β-strands and the amphipathic N-terminal α-helix as part of the channel. This means that in the yeast model C- and N-termini of porin are on different sides of the membrane. It is, however, more likely that both are on the same side of the mitochondrial outer membrane because of the similarity to the situation in gram-negative bacteria, where they are at the periplasmic side of the outer membrane. According to the model, the amphipathic N-terminal α-helix resides in the lumen wall as part of channel gating. Its displacement by voltage should result in some sort of structural rearrangement of the pore, which leads to partial channel closure. Our own experiments with ΔN mutants of *N. crassa* porin (ΔN2–12porin and ΔN3–20porin) suggest that the N-terminus itself seems to be important for overall channel stability, rather than being directly involved in gating. The gating characteristics of the ΔN channels were only slightly different from those of the wild-type channels, indicating that the structures responsible for the gating process are still intact in these mutants. This means that the N-terminus probably does not reside in the lumen wall of the mitochondrial outer membrane channel.

The second model assumes that the channel is formed by 16 antiparallel membrane-spanning β-strands, similar to the arrangement for general diffusion pores in the bacterial outer membranes. In this case, the amphipathic α-helix at the N-terminal end is localized within the surface of the membrane and exposed to the aqueous phase because it represents a major antigenic determinant of mitochondrial porins. These results suggest that the N-terminus is indeed localized in the surface of the membrane, most probably on the side of the intermembrane space of mitochondria; however, its possible role in channel gating remains open, but it may play a major role in channel

stability (see below). Treatment of intact and broken bovine heart mitochondria with different proteases identified different cleavage sites of the pore-forming unit exposed to the aqueous phase and this was unchanged whether intact or broken mitochondria were exposed to the proteases. The cleavage sites are probably localized within surface-exposed loops of the protein. The C-terminal end may also be exposed to the surface, but it is not accessible to proteases because only a few amino acids may be localized outside the membrane. The C-terminus of *N. crassa* porin is clearly organized in antiparallel β-strands. The mutant that lacks 15 amino acids at the C-terminal (ΔC269–283porin) showed the same channel-forming activity as wild-type porin. It had a smaller single-channel conductance than wild-type protein, indicating a smaller channel size. However, its other biophysical properties, in particular, the voltage dependence, were identical to those of wild-type porin.

Figure 11 shows a model for *N. crassa* porin derived from neural network-based

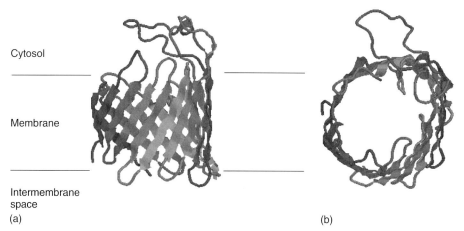

Cytosol

Membrane

Intermembrane space

(a) (b)

Fig. 11 Schematic view of the 3-D structure of the *N. crassa* porin. Note that the location of the N-terminus is not given. The coordinates were kindly provided by Rita Casadio and Vito De Pinto. (a) Side view. (b) View from the mitochondrial surface-exposed side.

predictors. The result of the prediction was compared to bacterial β-sheet proteins and used to align the primary sequence in coordinates of a three-dimensional (3-D) model. It is very similar to that proposed for yeast porin using the same formalism and has 16 antiparallel β-strands similar to the general diffusion porins of gram-negative bacteria. The model of *N. crassa* porin is shown in Fig. 11 ((a) from the side and (b) from the top). N- and C-termini are probably localized on the inner surface of the mitochondrial outer membrane (similar to the situation in gram-negative bacteria. The β-strands of *N. crassa* porin are tilted by about 40° to 60° similar to those of general diffusion pores. The loops connecting two β-strands are short on the internal surface of mitochondrial outer membrane and small at the opposite side. The hole is slightly oval and has a diameter of about 2.5 to 3 nm. The position of the amphipathic α-helical structure (not shown in Fig. 11) presumably represents a major problem for the validity of the model because it should somehow be involved in the stability of the pore as experiments with N-terminal deletion porins suggest. A simple, hollow β-stranded cylinder is presumably not stable in a membrane because of the lateral pressure in a lipid bilayer. It may result in a variation of its size as was shown for FhuA when the cork inside the β-barrel cylinder was removed. All known *β-stranded membrane channels* are somehow stabilized in their structure. External loop 3 stabilizes the bacterial porin channels. The α-hemolysin channel of *Staphylococcus aureus* is stabilized by the huge mushroom-like protrusion on the cis-side of the membrane and the receptors (for iron-siderophores or vitamin B12) are stabilized by the cork. A similar stabilizing element is also missing in both models for eukaryotic porins.

The model with 16 membrane-spanning β-strands appears to be more likely than that containing 13 (12) β-strands and the amphipathic N-terminus, although the latter model cannot be ruled out completely. The reasons for this can be summarized as follows:

- The porin channel is the primordial outer membrane channel and existed already when mitochondria evolved. Its overall structure may have been conserved, although there does not exist any homology between the primary structures of mitochondrial and bacterial porins. It has to be noted that homology between different bacterial porins can be very low, but the channel structure is well preserved.
- The amphipathic α-helix at the N-terminal end is the major antigenic determinant of eukaryotic porins, which means that it is probably not hidden inside the membrane.
- The results of experiments in which channel properties are changed by site-directed mutagenesis or by chemical modifications have to be taken with some care since the change of a single amino acid may induce a change that is related to a completely different region of the channel.

5.4
Are Sterols Involved in the Formation of the Channel-forming Unit?

Mitochondrial inner and outer membranes have different lipid compositions. The inner membrane contains cardiolipin, whereas sterols of different kind are exclusively found in the outer membrane dependent on the organism. Purified *N. crassa* porin contains the sterol ergosterol. Water-soluble mitochondrial porin from *N. crassa* that lost its associated lipids and sterols needs sterols for functional reconstitution into lipid bilayer membranes. The detergent Triton X-100 plays a critical chaperone-like role in this process, similar to what has been found in the folding of bacterial porins. The role of different sterols was studied in the functional reconstitution of water-soluble porins from *Dictyostelium discoideum*, *Paramecium*, plant, and rat liver. The water-soluble porins regain their channel-forming ability after preincubation of the polypeptides with sterols in the presence of detergents. Some of the sterols can replace one another for channel formation, but influence channel gating. Interestingly, a sterol content of five cholesterol molecules per polypeptide unit was calculated in pig heart porin purified using the detergent LDAO. Sterols are probably not tightly bound to the polypeptide. The removal of sterols from the channel-forming complex may be responsible for the substantial loss of channel-forming activity during purification of mitochondrial porins.

5.5
Electron Microscopic Analysis of Mitochondrial Porin

Three-dimensional crystals of mitochondrial porin have not been obtained to date, which means that its tertiary folding structure is still not clear. However, it was possible to study 2-D crystals of fungal and recombinant human porin using electron microscopic analysis. Porin from *N. crassa* mitochondria crystallizes in periodic polymorphs with well-defined lattice parameters when outer membranes are subjected to phospholipase A2 treatment. According to the Fourier-filtered electron microscopic images of the crystalline arrays, the mitochondrial pore is a cylinder normal to the membrane plane with an outer diameter of about 3.8 nm for the hydrocarbon cylinder (the diameter of the polypeptide backbone is about 3.8 nm) and an inner diameter of about 2.5 nm. The crystalline arrays are composed of subunits that contain six porin channels. The lattice geometry responds to a number of external compounds. In particular, polyanion binds to the porin and leads to a substantial contraction of the lattice. Besides polyanion, apocytochrome c and the targeting sequence of an imported cytochrome oxidase subunit seem to bind to the periphery of the porin channel. More recently, 2-D crystals of *N. crassa* porin were analyzed using single-particle approaches. The results look promising, but it is not known yet if these approaches will lead to an improved resolution of porin structure. Human porin expressed in *E. coli* and reconstituted using detergent dialysis also forms 2-D crystals. The data derived from electron microscopic analysis of these crystals agree basically with those derived from 2-D crystals of *N. crassa* porin, but two neighboring molecules have opposite directions, which means that the 2-D crystals of human porin are not suitable for high-resolution imaging.

6
Conclusions

Eukaryotic porins form large water-filled channels in the "open" state with diameters around 2.5 to 3 nm that are slightly anion selective. The channels appear to be voltage-gated starting at about 20 to 30 mV and switch to ion-permeable closed states that are highly cation selective. Eukaryotic porins represent the major permeability pathway of mitochondrial outer membranes, although porin-deficient mitochondria also exhibit oxidative phosphorylation. In addition to mitochondrial localization, there exists emerging evidence that other cellular compartments also contain eukaryotic porins with a yet unknown function. The first indication was the extramitochondrial location of porin 31HL, which is localized in the cytoplasmic membrane of a transformed human B-lymphocyte cell line. A similar channel is localized in astrocytic plasma membranes. Later, more evidence was found of such a location because a 36-kDa porin-like protein from rat brain copurified with the γ-aminobutyric acid/benzodiazepine receptor protein. It is possible that the channel from the astrocytic plasma membrane and the 36-kDa protein from rat brain relate to the "maxi" chloride channels present in the cell membrane of a number of eukaryotic cells. In excised patches, the channel has a maximum open probability at zero voltage, similar to the situation in lipid bilayer experiments.

The electrophysiological characterization of mitochondrial porin suggests that the permeability properties of the mitochondrial outer membrane could be voltage controlled. The generation of a voltage across the mitochondrial outer membrane is still open and it is not clear if the close apposition of inner and outer membranes *in situ* mitochondria is sufficient for electric coupling of both membranes and for channel gating. Furthermore, the exact mechanism of channel gating and the residues responsible for selectivity in the different states are completely open. In general, there exists some knowledge about the secondary structure of the eukaryotic porins. However, the precise 3-D structure, including the exact number of membrane-spanning β-strands and the diameter, are still unknown. Crystallographic data are badly needed to give clear answers to these questions and possibly also the localization of sterols in the channel-forming unit. Another open problem is the assembly of the porin channel. It is unlikely that it (i.e. the β-strands) forms inside a membrane since eukaryotic porins are polar proteins and are synthesized as water-soluble precursors outside membranes. This means that the channel is formed chaperone-promoted on the (inner?) surface of the mitochondrial outer membrane at a receptor site (Tom 20) from polypeptide and sterols. After formation, the channel inserts into the membrane using the general import pathway (TOM). So far it is not clear if it requires special insertion sites in the mitochondrial outer membrane or if the insertion process is similar to that in lipid bilayer membranes. In any case, it is likely that the channel is preformed before insertion. The preformation of channels in the aqueous phase or on the membrane surface is probably also the reason why lipid bilayer membranes represent a powerful tool for the study of mitochondrial and bacterial porins.

Acknowledgments

The generous supply of Figs. 10 and 11 by Rita Casadio and Vito De Pinto is gratefully acknowledged. I would like to thank Dieter Brdiczka, Vito De Pinto and Friedrich P. Thinnes for the fruitful and excellent collaboration during many years, and my collaborators Otto Ludwig, Birgit Popp, Elke Maier, and Angela Schmid for their excellent work with outer-membrane porins on which this review is largely based.

My own research was supported by the Deutsche Forschungsgemeinschaft (Project Be865/10) and by the Fonds der Chemischen Industrie.

See also Calcium Biochemistry.

Bibliography

Books and Reviews

Bauer, M.F., Hofmann, S., Neupert, W. (2002) Import of mitochondrial proteins, *Int. Rev. Neurobiol.* **53**, 57–90.

Bay, D.C., Court, D.A. (2002) Origami in the outer membrane: the transmembrane arrangement of mitochondrial porins, *Biochem. Cell. Biol.* **80**(5), 551–562.

Benz, R. (Ed.) (2004) *Bacterial and Eukaryotic Porins: Structure, Function, Mechanism*, VCH-Wiley, Weinheim, Germany.

Colombini, M. (2004) VDAC: the channel at the interface between mitochondria and the cytosol, *Mol. Cell Biochem.* **256–257**(1–2), 107–115.

De Pinto, V., Messina, A., Accardi, R., Aiello, R., Guarino, F., Tomasello, M.F., Tommasino, M., Tasco, G., Casadio, R., Benz, R., De Giorgi, F., Ichas, F., Baker, M., Lawen, A. (2003) New functions of an old protein: the eukaryotic porin or voltage dependent anion selective channel (VDAC), *Ital. J. Biochem.* **52**(1), 17–24.

Graham, B.H., Craigen, W.J. (2004) Genetic approaches to analyzing mitochondrial outer membrane permeability, *Curr. Top. Dev. Biol.* **59**, 87–118.

Palmieri, L., Runswick, M.J., Fiermonte, G., Walker, J.E., Palmieri, F. (2000) Yeast mitochondrial carriers: bacterial expression, biochemical identification and metabolic significance, *J. Bioenerg. Biomembr.* **32**(1), 67–77.

Shoshan-Barmatz, V., Gincel, D. (2003) The voltage-dependent anion channel: characterization, modulation, and role in mitochondrial function in cell life and death, *Cell. Biochem. Biophys.* **39**(3), 279–292.

Taylor, R.D., Pfanner, N. (2004) The protein import and assembly machinery of the mitochondrial outer membrane, *Biochim. Biophys. Acta* **1658**(1–2), 37–43.

Vyssokikh, M.Y., Brdiczka, D. (2003) The function of complexes between the outer mitochondrial membrane pore (VDAC) and the adenine nucleotide translocase in regulation of energy metabolism and apoptosis, *Acta Biochim. Pol.* **50**(2), 389–404.

Primary Literature

Aiello, R., Messina, A., Schiffler, B., Benz, R., Tasco, G., Casadio, R., De Pinto, V. (2004) Functional characterization of a second porin isoform in *Drosophila melanogaster*. DmPorin2 forms voltage-independent cation-selective pores, *J. Biol. Chem.* **279**(24), 25364–25373.

Albrecht, H., Goormaghtigh, E., Ruysschaert, J.M., Homble, F. (2000) Structure and orientation of two voltage-dependent anion-selective channel isoforms. An attenuated total reflection Fourier-transform infrared spectroscopy study, *J. Biol. Chem.* **275**(52), 40992–40999.

Albrecht, H., Wattiez, R., Ruysschaert, J.M., Homble, F. (2000) Purification and characterization of two voltage-dependent anion channel isoforms from plant seeds, *Plant. Physiol.* **124**(3), 1181–1190.

Aljamal, J.A., Genchi, G., De Pinto, V., Stefanazzi, L., Benz, R., Palmieri, F. (1993) Purification and characterization of porin from corn (*Zea mays* L.) mitochondria, *Plant Physiol.* **102**, 615–621.

Arora, K.K., Parry, D.M., Pedersen, P.L. (1992) Hexokinase receptors: preferential enzyme binding in normal cells to nonmitochondrial sites and in transformed cells to mitochondrial sites, *J. Bioenerg. Biomembr.* **24**(1), 47–53.

Babel, D., Walter, G., Gotz, H., Thinnes, F.P., Jurgens, L., Konig, U., Hilschmann, N. (1991) Studies on human porin. VI. Production and characterization of eight monoclonal mouse antibodies against the human VDAC "Porin 31HL" and their application for histotopological studies in human skeletal muscle, *Biol. Chem. Hoppe Seyler* **372**(12), 1027–1034.

Benz, R., Kottke, M., Brdiczka, D. (1990) The cationically selective state of the mitochondrial outer membrane pore: a study with intact mitochondria and reconstituted mitochondrial porin, *Biochim. Biophys. Acta* **1022**(3), 311–318.

Benz, R., Schmid, A., Dihanich, M. (1989) Pores from mitochondrial outer membranes of yeast and a porin-deficient yeast mutant: a comparison, *J. Bioenerg. Biomembr.* **21**(4), 439–450.

Benz, R., Janko, K., Boos, W., Lauger, P. (1978) Formation of large, ion-permeable membrane channels by the matrix protein (porin) of *Escherichia coli, Biochim. Biophys. Acta* **511**(3), 305–319.

Benz, R., Wojtczak, L., Bosch, W., Brdiczka, D. (1988) Inhibition of adenine nucleotide transport through the mitochondrial porin by a synthetic polyanion, *FEBS Lett.* **231**(1), 75–80.

Benz, R., Maier, E., Thinnes, F.P., Gotz, H., Hilschmann, N. (1992) Studies on human porin. VII. The channel properties of the human B-lymphocyte membrane-derived "Porin 31HL" are similar to those of mitochondrial porins, *Biol. Chem. Hoppe Seyler* **373**(6), 295–303.

Blachly-Dyson, E., Peng, S., Colombini, M., Forte, M. (1990) Selectivity changes in site-directed mutants of the VDAC ion channel: structural implications, *Science* **247**(4947), 1233–1236.

Blumenthal, K., Kahn, A., Beja, O., Galun, E., Colombini, M., Breiman, A. (1993) Purification and characterization of the voltage-dependent anion-selective channel protein from wheat mitochondrial membranes, *Plant Physiol.* **101**, 579–587.

Brdiczka, D. (1994) Function of the outer mitochondrial compartment in regulation of energy metabolism, *Biochim. Biophys. Acta* **1187**(2), 264–269.

Bureau, M.H., Khrestchatisky, M., Heeren, M.A., Zambrowicz, E.B., Kim, H., Grisar, T.M., Colombini, M., Tobin, A.J., Olsen, R.W. (1992) Isolation and cloning of a voltage-dependent anion channel-like Mr 36,000 polypeptide from mammalian brain, *J. Biol. Chem.* **267**(12), 8679–8684.

Carbonara, F., Popp, B., Schmid, A., Iacobazzi, V., Genchi, G., Palmieri, F., Benz, R. (1996) The role of sterols in the functional reconstitution of water-soluble mitochondrial porins from plants, *J. Bioenerg. Biomembr.* **28**(2), 181–189.

Casadio, R., Jacoboni, I., Messina, A., De Pinto, V. (2002) A 3D model of the voltage-dependent anion channel (VDAC), *FEBS Lett.* **520**(1–3), 1–7.

Castellan, G.W. (1983) The Ionic Current in Solutions, *Physical Chemistry*, Addison-Wesley, Reading, MA, pp. 769–784.

Cheng, E.H., Sheiko, T.V., Fisher, J.K., Craigen, W.J., Korsmeyer, S.J. (2003) VDAC2 inhibits BAK activation and mitochondrial apoptosis, *Science* **301**(5632), 513–517.

Colombini, M. (1979) A candidate for the permeability pathway of the outer mitochondrial membrane, *Nature* **279**(5714), 643–645.

Colombini, M. (1983) Purification of VDAC (voltage-dependent anion-selective channel) from rat liver mitochondria, *J. Membr. Biol.* **74**(2), 115–121.

Colombini, M., Holden, M.J., Mangan, P.S. (1989) Modulation of the Mitochondrial Channel VDAC by a Variety of Agents, in: Nalecz, K.A., Nalecz, M.J., Wojtczak, L., Azzi, A. (Eds.) *Ion Carriers of Mitochondrial Membranes*, Springer, New York, pp. 215–224.

Colombini, M., Yeung, C.L., Tung, J., Konig, T. (1987) The mitochondrial outer membrane channel, VDAC, is regulated by a synthetic polyanion, *Biochim. Biophys. Acta* **905**(2), 279–286.

Cowan, S.W., Schirmer, T., Rummel, G., Steiert, M., Ghosh, R., Pauptit, R.A., Jansonius, J.N., Rosenbusch, J.P. (1992) Crystal structures explain functional properties of two E. coli porins, *Nature* **358**(6389), 727–733.

De Pinto, V., Benz, R., Palmieri, F. (1989) Interaction of non-classical detergents with the mitochondrial porin. A new purification procedure and characterization of the pore-forming unit, *Eur. J. Biochem.* **183**(1), 179–187.

De Pinto, V., Benz, R., Caggese, C., Palmieri, F. (1989) Characterization of the mitochondrial porin from *Drosophila melanogaster, Biochim. Biophys. Acta* **987**(1), 1–7.

De Pinto, V., Ludwig, O., Krause, J., Benz, R., Palmieri, F. (1987) Porin pores of mitochondrial outer membranes from high and low eukaryotic cells: biochemical and biophysical characterization, *Biochim. Biophys. Acta* **894**(2), 109–119.

De Pinto, V., Prezioso, G., Thinnes, F., Link, T.A., Palmieri, F. (1991) Peptide-specific antibodies and proteases as probes of the transmembrane topology of the bovine heart mitochondrial porin, *Biochemistry* **30**(42), 10191–10200.

De Pinto, V., Messina, A., Schmid, A., Simonetti, S., Carnevale, F., Benz, R. (2000) Characterization of channel-forming activity in muscle biopsy from a porin-deficient human patient, *J. Bioenerg. Biomembr.* **32**(6), 585–593.

Dermietzel, R., Hwang, T.K., Buettner, R., Hofer, A., Dotzler, E., Kremer, M., Deutzmann, R., Thinnes, F.P., Fishman, G.I., Spray, D.C., Siemen, D. (1994) Cloning and in situ localization of a brain-derived porin that constitutes a large-conductance anion channel in astrocytic plasma membranes, *Proc. Natl. Acad. Sci. USA* **91**(2), 499–503.

Dihanich, M., Suda, K., Schatz, G. (1987) A yeast mutant lacking mitochondrial porin is respiratory-deficient, but can recover respiration with simultaneous accumulation of an 86-kD extramitochondrial protein, *EMBO J.* **6**(3), 723–728.

Dolder, M., Zeth, K., Tittmann, P., Gross, H., Welte, W., Wallimann, T. (1999) Crystallization of the human, mitochondrial voltage-dependent anion-selective channel in the presence of phospholipids, *J. Struct. Biol.* **127**(1), 64–71.

Ehrenstein, G., Lecar, H., Nossal, R. (1970) The nature of the negative resistance in bimolecular lipid membranes containing excitability-inducing material, *J. Gen. Physiol.* **55**(1), 119–133.

Fiek, C., Benz, R., Roos, N., Brdiczka, D. (1982) Evidence for identity between the hexokinase-binding protein and the mitochondrial porin in the outer membrane of rat liver mitochondria, *Biochim. Biophys. Acta* **688**(2), 429–440.

Freitag, H., Janes, M., Neupert, W. (1982) Biosynthesis of mitochondrial porin and insertion into the outer mitochondrial membrane of *Neurospora crassa*, *Eur. J. Biochem.* **126**(1), 197–202.

Freitag, H., Neupert, W., Benz, R. (1982) Purification and characterisation of a pore protein of the outer mitochondrial membrane from *Neurospora crassa*, *Eur. J. Biochem.* **123**(3), 629–636.

Godbole, A., Varghese, J., Sarin, A., Mathew, M.K. (2003) VDAC is a conserved element of death pathways in plant and animal systems, *Biochim. Biophys. Acta* **1642**(1–2), 87–96.

Guo, X.W., Mannella, C.A. (1993) Conformational change in the mitochondrial channel,

VDAC, detected by electron cryo-microscopy, *Biophys. J.* **64**(2), 545–549.

Guo, X.W., Smith, P.R., Cognon, B., D'Arcangelis, D., Dolginova, E., Mannella, C.A. (1995) Molecular design of the voltage-dependent, anion-selective channel in the mitochondrial outer membrane, *J. Struct. Biol.* **114**(1), 41–59.

Heins, L., Mentzel, H., Schmid, A., Benz, R., Schmitz, U.K. (1994) Biochemical, molecular, and functional characterization of porin isoforms from potato mitochondria, *J. Biol. Chem.* **269**(42), 26402–26410.

Hoch, F.L. (1992) Cardiolipins and biomembrane function, *Biochim. Biophys. Acta* **1113**(1), 71–133.

Hovius, R., Lambrechts, H., Nicolay, K., de Kruijff, B. (1990) Improved methods to isolate and subfractionate rat liver mitochondria. Lipid composition of the inner and outer membrane, *Biochim. Biophys. Acta.* **1021**(2), 217–226.

Jacoboni, I., Martelli, P.L., Fariselli, P., De Pinto, V., Casadio, R. (2001) Prediction of the transmembrane regions of beta-barrel membrane proteins with a neural network-based predictor, *Protein Sci.* **10**(4), 779–787.

Kayser, H., Kratzin, H.D., Thinnes, F.P., Gotz, H., Schmidt, W.E., Eckart, K., Hilschmann, N. (1989) Identification of human porins. II. Characterization and primary structure of a 31 kDa porin from human B lymphocytes (Porin 31HL), *Biol. Chem. Hoppe Seyler* **370**(12), 1265–1278.

Kleene, R., Pfanner, N., Pfaller, R., Link, T.A., Sebald, W., Neupert, W., Tropschug, M. (1987) Mitochondrial porin of *Neurospora crassa*: cDNA cloning, in vitro expression and import into mitochondria, *EMBO J.* **6**(9), 2627–2633.

Konig, T., Stipani, I., Horvath, I., Palmieri, F. (1982) Inhibition of mitochondrial substrate anion translocators by a synthetic amphipathic polyanion, *J. Bioenerg. Biomembr.* **14**(5–6), 297–305.

Konig, T., Kocsis, B., Meszaros, L., Nahm, K., Zoltan, S., Horvath, I. (1977) Interaction of a synthetic polyanion with rat liver mitochondria, *Biochim. Biophys. Acta* **462**(2), 380–389.

Koppel, D.A., Kinnally, K.W., Masters, P., Forte, M., Blachly-Dyson, E., Mannella, C.A. (1998) Bacterial expression and characterization of the mitochondrial outer membrane channel.

Effects of N-terminal modifications, *J. Biol. Chem.* **273**(22), 13794–13800.

Krause, J., Hay, R., Kowollik, C., Brdiczka, D. (1986) Cross-linking analysis of yeast mitochondrial outer membrane, *Biochim. Biophys. Acta* **860**(3), 690–698.

Krimmer, T., Rapaport, D., Ryan, M.T., Meisinger, C., Kassenbrock, C.K., Blachly-Dyson, E., Forte, M., Douglas, M.G., Neupert, W., Nargang, F.E., Pfanner, N. (2001) Biogenesis of porin of the outer mitochondrial membrane involves an import pathway via receptors and the general import pore of the TOM complex, *J. Cell Biol.* **152**(2), 289–300.

Kunkele, K.P., Heins, S., Dembowski, M., Nargang, F.E., Benz, R., Thieffry, M., Walz, J., Lill, R., Nussberger, S., Neupert, W. (1998) The preprotein translocation channel of the outer membrane of mitochondria, *Cell* **93**(6), 1009–1019.

Le Mellay, V., Troppmair, J., Benz, R., Rapp, U.R. (2002) Negative regulation of mitochondrial VDAC channels by C-Raf kinase, *BMC Cell Biol.* **3**(1), 14.

Linden, M., Gellerfors, P., Nelson, B.D. (1982) Purification of a protein having pore forming activity from the rat liver mitochondrial outer membrane, *Biochem. J.* **208**(1), 77–82.

Liu, M.Y., Colombini, M. (1992) A soluble mitochondrial protein increases the voltage dependence of the mitochondrial channel, VDAC, *J. Bioenerg. Biomembr.* **24**(1), 41–46.

Ludwig, O., Benz, R., Schultz, J.E. (1989) Porin of *Paramecium* mitochondria isolation, characterization and ion selectivity of the closed state, *Biochim. Biophys. Acta* **978**(2), 319–327.

Ludwig, O., De Pinto, V., Palmieri, F., Benz, R. (1986) Pore formation by the mitochondrial porin of rat brain in lipid bilayer membranes, *Biochim. Biophys. Acta* **860**(2), 268–276.

Ludwig, O., Krause, J., Hay, R., Benz, R. (1988) Purification and characterization of the pore forming protein of yeast mitochondrial outer membrane, *Eur. Biophys. J.* **15**(5), 269–276.

MacKinnon, R. (2003) Potassium channels, *FEBS Lett.* **555**(1), 62–65.

Mannella, C.A. (1984) Phospholipase-induced crystallization of channels in mitochondrial outer membranes, *Science* **224**(4645), 165–166.

Mannella, C.A. (1998) Conformational changes in the mitochondrial channel protein, VDAC, and their functional implications, *J. Struct. Biol.* **121**(2), 207–218.

Mannella, C.A., Guo, X.W. (1990) Interaction between the VDAC channel and a polyanionic effector. An electron microscopic study, *Biophys. J.* **57**(1), 23–31.

Mannella, C.A., Forte, M., Colombini, M. (1992) Toward the molecular structure of the mitochondrial channel, VDAC, *J. Bioenerg. Biomembr.* **24**(1), 7–19.

Mannella, C.A., Guo, X.W., Dias, J. (1992) Binding of a synthetic targeting peptide to a mitochondrial channel protein, *J. Bioenerg. Biomembr.* **24**(1), 55–61.

Mannella, C.A., Ribeiro, A.J., Frank, J. (1987) Cytochrome c binds to lipid domains in arrays of mitochondrial outer membrane channels, *Biophys. J.* **51**(2), 221–226.

Mannella, C.A., Ribeiro, A., Cognon, B., D'Arcangelis, D. (1992) Structure of paracrystalline arrays on outer membranes of rat-liver and rat-heart mitochondria, *J. Struct. Biol.* **108**(3), 227–237.

Messina, A., Neri, M., Perosa, F., Caggese, C., Marino, M., Caizzi, R., De Pinto, V. (1996) Cloning and chromosomal localization of a cDNA encoding a mitochondrial porin from *Drosophila melanogaster*, *FEBS Lett.* **384**(1), 9–13.

Michejda, J., Guo, X.J., Lauquin, G.J. (1990) The respiration of cells and mitochondria of porin deficient yeast mutants is coupled, *Biochem. Biophys. Res. Commun.* **171**(1), 354–361.

Mihara, K., Sato, R. (1985) Molecular cloning and sequencing of cDNA for yeast porin, an outer mitochondrial membrane protein: a search for targeting signal in the primary structure, *EMBO J.* **4**(3), 769–774.

Nikaido, H. (1983) Proteins forming large channels from bacterial and mitochondrial outer membranes: porins and phage lambda receptor protein, *Methods Enzymol.* **97**, 85–100.

Nikaido, H., Rosenberg, E.Y. (1981) Effect on solute size on diffusion rates through the transmembrane pores of the outer membrane of *Escherichia coli*, *J. Gen. Physiol.* **77**(2), 121–135.

Ostlund, A.K., Gohring, U., Krause, J., Brdiczka, D. (1983) The binding of glycerol kinase to the outer membrane of rat liver mitochondria: its importance in metabolic regulation, *Biochem. Med.* **30**(2), 231–245.

Pedersen, P.L., Greenawalt, J.W., Reynafarje, B., Hullihen, J., Decker, G.L., Soper, J.W., Bustamente, E. (1978) Preparation and characterization of mitochondria and submitochondrial particles of rat liver and liver-derived tissues, *Methods Cell Biol.* **20**, 411–481.

Peng, S., Blachly-Dyson, E., Colombini, M., Forte, M. (1992) Determination of the number of polypeptide subunits in a functional VDAC channel from *Saccharomyces cerevisiae*, *J. Bioenerg. Biomembr.* **24**(1), 27–31.

Pfaff, E., Klingenberg, M., Heldt, H.W. (1965) Unspecific permeation and specific exchange of adenine nucleotides in liver mitochondria, *Biochim. Biophys. Acta* **104**(1), 312–315.

Pfaller, R., Freitag, H., Harmey, M.A., Benz, R., Neupert, W. (1985) A water-soluble form of porin from the mitochondrial outer membrane of *Neurospora crassa*. Properties and relationship to the biosynthetic precursor form, *J. Biol. Chem.* **260**(13), 8188–8193.

Popp, B., Schmid, A., Benz, R. (1995) Role of sterols in the functional reconstitution of water-soluble mitochondrial porins from different organisms, *Biochemistry* **34**(10), 3352–3361.

Popp, B., Court, D.A., Benz, R., Neupert, W., Lill, R. (1996) The role of the N and C termini of recombinant *Neurospora* mitochondrial porin in channel formation and voltage-dependent gating, *J. Biol. Chem.* **271**(23), 13593–13599.

Popp, B., Gebauer, S., Fischer, K., Flugge, U.I., Benz, R. (1997) Study of structure and function of recombinant pea root plastid porin by biophysical methods, *Biochemistry* **36**(10), 2844–2852.

Renkin, E.M. (1954) Filtration, diffusion, and molecular sieving through porous cellulose membranes, *J. Gen. Physiol.* **38**(2), 225–243.

Roos, N., Benz, R., Brdiczka, D. (1982) Identification and characterization of the pore-forming protein in the outer membrane of rat liver mitochondria, *Biochim. Biophys. Acta* **686**(2), 204–214.

Runke, G., Maier, E., O'Neil, J.D., Benz, R., Court, D.A. (2000) Functional characterization of the conserved "GLK" motif in mitochondrial porin from *Neurospora crassa*, *J. Bioenerg. Biomembr.* **32**(6), 563–570.

Schein, S.J., Colombini, M., Finkelstein, A. (1976) Reconstitution in planar lipid bilayers of a voltage-dependent anion-selective channel obtained from *Paramecium* mitochondria, *J. Membr. Biol.* **30**(2), 99–120.

Schmid, A., Kromer, S., Heldt, H.W., Benz, R. (1992) Identification of two general diffusion channels in the outer membrane of pea mitochondria, *Biochim. Biophys. Acta* **1112**(2), 174–180.

Sen, K., Nikaido, H. (1990) In vitro trimerization of OmpF porin secreted by spheroplasts of *Escherichia coli*, *Proc. Natl. Acad. Sci. U S A* **87**(2), 743–747.

Shao, L., Kinnally, K.W., Mannella, C.A. (1996) Circular dichroism studies of the mitochondrial channel, VDAC, from *Neurospora crassa*, *Biophys. J.* **71**(2), 778–786.

Shinohara, Y., Ishida, T., Hino, M., Yamazaki, N., Baba, Y., Terada, H. (2000) Characterization of porin isoforms expressed in tumor cells, *Eur. J. Biochem.* **267**(19), 6067–6073.

Smack, D.P., Colombini, M. (1985) Voltage-dependent channels found in the membrane fraction of corn zea-mays cultivar w-64an mitochondria, *Plant Physiol.* **79**(4), 1094–1097.

Song, J., Midson, C., Blachly-Dyson, E., Forte, M., Colombini, M. (1998) The topology of VDAC as probed by biotin modification, *J. Biol. Chem.* **273**(38), 24406–24413.

Song, L., Hobaugh, M.R., Shustak, C., Cheley, S., Bayley, H., Gouaux, J.E. (1996) Structure of staphylococcal alpha-hemolysin, a heptameric transmembrane pore, *Science* **274**(5294), 1859–1866.

Sottocasa, G.L., Kuylenstierna, B., Ernster, L., Bergstrand, A. (1967) An electron-transport system associated with the outer membrane of liver mitochondria. A biochemical and morphological study, *J. Cell Biol.* **32**(2), 415–438.

Thinnes, F.P., Gotz, H., Kayser, H., Benz, R., Schmidt, W.E., Kratzin, H.D., Hilschmann, N. (1989) Identification of human porins. I. Purification of a porin from human B-lymphocytes (Porin 31HL) and the topochemical proof of its expression on the plasmalemma of the progenitor cell, *Biol. Chem. Hoppe Seyler* **370**(12), 1253–1264.

Troll, H., Malchow, D., Muller-Taubenberger, A., Humbel, B., Lottspeich, F., Ecke, M., Gerisch, G., Schmid, A., Benz, R. (1992) Purification, functional characterization, and cDNA sequencing of mitochondrial porin from *Dictyostelium discoideum*, *J. Biol. Chem.* **267**(29), 21072–21079.

Tsujimoto, Y., Shimizu, S. (2002) The voltage-dependent anion channel: an essential player in apoptosis, *Biochimie* **84**(2–3), 187–193.

Verschoor, A., Tivol, W.F., Mannella, C.A. (2001) Single-particle approaches in the analysis of small 2D crystals of the mitochondrial channel VDAC, *J. Struct. Biol.* **133**(2–3), 254–265.

Vogel, H., Jahnig, F. (1986) Models for the structure of outer-membrane proteins of *Escherichia coli* derived from Raman spectroscopy and prediction methods, *J. Mol. Biol.* **190**(2), 191–199.

Zalman, L.S., Nikaido, H., Kagawa, Y. (1980) Mitochondrial outer membrane contains a protein producing nonspecific diffusion channels, *J. Biol. Chem.* **255**(5), 1771–1774.

Mobile Structures: Cilia and Flagella

Koji Yonekura[1,2,3] *and Keiichi Namba*[2,3]
[1] *University of California, San Francisco, CA*
[2] *Osaka University, Osaka, Japan*
[3] *Dynamic NanoMachine Project, Suita, Osaka, Japan*

Encyclopedia of Molecular Cell Biology and Molecular Medicine, 2nd Edition. Volume 8
Edited by Robert A. Meyers.
Copyright © 2005 Wiley-VCH Verlag GmbH & Co. KGaA, Weinheim
ISBN: 3-527-30550-5

Keywords

Axoneme
The core structure of the eukaryotic flagella and cilia composed of nine outer doublets of microtubule and a central pair of single microtubules, called "9 + 2".

Bacterial Flagellum
A long filamentous propeller rotated by a rotary motor used for bacteria swimming, composed of ~25 different proteins. Driven by ion flows through the motor part embedded in the cytoplasmic membrane.

Cilium
A short hairlike organelle found in many cells with almost the same structure with the eukaryotic flagella. Working for the transport of fluids over epithelial cells by a whiplike stroke. Some work as a sensor.

Dynein
A processive motor protein complex. Axonemal dynein generates the force of the eukaryotic flagellar and ciliary bending.

Eukaryotic Flagellum
A long filamentous organelle used for protozoan and metazoan sperm swimming, composed of highly organized microtubules and ~250 associated proteins. Showing a wavelike oscillatory motion generated by the dynein motor within the filament using the energy of ATP hydrolysis.

Flagellin
A structural protein forming the bacterial flagellar filament.

Intraflagellar Transport (IFT)
A system for transport of eukaryotic flagellar and ciliary components to the tip or base. Necessary for their assembly, disassembly, and maintenance. Driven by a kinesin motor to the distal tip (anterograde IFT) and a cytoplasmic dynen motor to the base (retrograde IFT).

Microtubule
A cytoskeleton composed of the dimer of α- and β-tubulin. Forming the core of the axoneme. Also associated with various cell activities, for example, mitosis, and so on.

MotA/B, PomA/B
An ion-driven torque generation unit for the rotation of the bacterial flagellum. MotA/B is a proton channel found in *Escherichia coli, Salmonella*, and so on. PomA/B is a sodium channel in alkaliphilic *Bacillus* and *Vibrio* species.

Type III Protein Export Apparatus

A protein complex for export of bacterial flagellar axial proteins into the cell exterior or virulence factors into host cells, driven by the energy of ATP hydrolysis.

■ Both eukaryote and prokaryote have flagella with a similar overall look, and use them for swimming in liquids, but they are completely different from each other. Cilia look like short hairs and have almost the same structure and mechanism as the eukaryotic flagella. The long filamentous structure of the eukaryotic flagellum is composed of highly organized microtubule-based structure and huge numbers of its associate proteins, while that in the bacterial flagellum is made of a single protein called *flagellin*. The length and the diameter of the filament in eukaryote are more than 10 times larger than those in bacteria. The total number of component proteins involved in the structure of eukaryote is also ~10 times of the prokaryote. The eukaryotic flagellum makes wavelike movements generated by the dynein motor within the filament using the energy of ATP hydrolysis, while the bacterial flagellum is a rotary motor driven by ion flows through the motor part embedded in the membranes. They both grow at their distal end, but use quite different mechanism for transport of their components to the distal tip.

Although the eukaryotic and bacterial flagella are complex macromolecules, recent studies are beginning to elucidate the roles of component proteins, their complex structures and growth mechanisms.

1
Eukaryotic Flagella and Cilia

Eukaryotic flagellum is a highly conserved organelle throughout protozoan to mammalian sperm. Its length is diverged in 10 to 200 µm, and the diameter is ~250 nm or more. The flagella show a wavelike oscillatory motion for swimming in fluid environments. Cilia are hairlike projections extended from the surface of various cells. They are not only shorter versions of the flagella but also work as systems for sensory perception of environments. In this article, we will not further discuss on these nonmotile cilia. Motile cilia show a whiplike stroke at 0.1 to 0.2 s per cycle for fluid transport over epithelial cells. For example, there are huge numbers of cilia in our respiratory tract. They sweep layers of mucus to transport trapped dusts and bacteria by their harmonic motions, up to the mouth where they are drunk or spewed out.

The motions of flagella and cilia are generated by bending of their internal cytoskeletal structure called the *axoneme*. The axoneme is composed of highly organized microtubules and their associated proteins, which include a motor protein complex, axonemal dynein. It generates the force for axoneme bending by utilizing the energy of ATP hydrolysis. In the axoneme, there are nine outer doublet microtubules surrounding the two single microtubules at its center, which

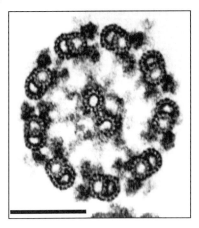

Fig. 1 Electron micrograph of a cross section of the flagellum from *Chlamydomonas*. Bar represents 100 nm (permission from Alberts, B., Johnson, A., Lewis, J., Raff, M., Roberts, K., Walter, P. (2002) *Molecular Biology of the Cell*, 4th edition, Garland Science, New York).

has been preserved for almost all eukaryotic flagella and cilia (Figs. 1 and 2). This characteristic structure is often called "9 + 2". A sensory cilium is an exception, which lacks the two central microtubules, and so-called "9 + 0". The base structure known as the basal body also shows a complicated structure, composed of nine triplets of microtubules without the central pair (Fig. 3). The basal body is critically important for constructing the axoneme. *Chlamydomonas*, a green alga, has two motile flagella at the anterior end of the cell. Recent studies especially on this alga have explored its huge number of constituent molecules and have begun to reveal the growth mechanism of the flagella and cilia.

1.1
The Axoneme

The axoneme has a huge complex structure composed of ~250 different proteins, which include cytoskeletal proteins, motor proteins, molecular chaperons, calcium-binding proteins, protein kinases, and so on. However, more than half of them have not yet been characterized. In general, the component proteins are well conserved among the flagella of

protozoan and metazoan sperm, except for some components. Figures 1 and 2 shows a typical electron micrograph and a schematic diagram of a cross section of the axoneme.

1.1.1 Microtubule

Microtubules form the core 9 + 2 structure in the axoneme. Nine outer doublet microtubules are arranged in a ring around the central two microtubules (Figs. 1 and 2). Each of the outer doublets is made of a complete tube called *A-tubule* and a partial tube called *B-tubule*, fused together. This characteristic design has been unchanged throughout protozoan to mammals. The structural unit of microtubule is a heterodimer of α- and β-tubulin, which share ~40% sequence homology. α- and β-tubulin are evolutionarily well-conserved molecules comprising ~450 residues. The atomic model of the dimer was first determined by electron crystallography on zinc-induced two-dimensional sheets, in which the dimers form a regular array.

α- and β-tubulin show very similar folds, having two β-sheets made of 4 and 6 strands, flanked by 12 α-helices. They have a typical Rossman fold for nucleotide binding at their N-terminus,

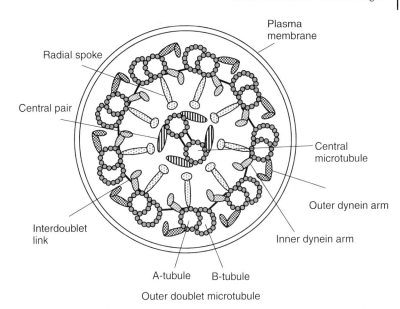

Fig. 2 Cartoon of the cross section of the flagellum or the cilium shown in Fig. 1.

where unexchangeable GTP is tightly bound in α-tubulin and exchangeable GTP or GDP is bound in β-tubulin. The dimers form a protofilament by connecting α-tubulin of one dimer to β-tubulin of another. The nucleotide binding sites of β-tubulin is placed at the interface of the dimers along the protofilament. A-tubule is a cylindrical hollow structure built of 13 protofilaments, while B-tubule is made of 11 protofilaments. Because all the protofilaments are oriented in parallel, one end of microtubule is occupied by all α-tubulin and the other end by all β-tubulin. The end made of α-tubulin is called the *minus end* and that made of β-tubulin is called the *plus end*. Microtubule assembles or disassembles $\sim$100 times faster from the plus end than the minus end. Whether the β-tubulin molecules at the plus end bind GTP or GDP affects the assembly and disassembly; GTP promotes assembly, while GDP promotes disassembly. In general, the GTP-bound filament becomes straight and stiff, while the GDP-bound one tends to be curved and unstable.

In the zinc-induced two-dimensional sheets, the protofilaments are arranged in an antiparallel fashion. Hence, details of the lateral interactions between the protofilaments in microtubule are not yet clear, although the same kinds of molecules are known to make major lateral contacts, such as α–α or β–β. α- and β-tubulin form a complete microtubule *in vitro*, while they never form a doublet microtubule, as found in the axoneme. This requires the basal body and other associated proteins described below.

1.1.2 Axonemal Dynein

Dynein is a microtubule-specific motor protein complex that consumes ATP for force generation. The dynein family is divided into two major groups: cytoplasmic dynein and axonemal dynein. Axonemal dynein is a huge molecular complex with a molecular mass of $\sim$2 MDa and is involved in the motility of eukaryotic flagella and cilia. In the axoneme, a pair of dynein "arms" with a dimension of ~25 nm are projected from A-tubule to form bridges with B-tubule of the adjacent outer doublet microtubule (Fig. 2). They are placed at a regular interval of ~24 nm along A-tubule. The arms projected inside and outside the ring of the microtubule doublet are called the *inner* and *outer dynein arms* respectively. When the motor is activated, the dynein molecules try to "walk" along the adjacent B-tubule just as myosin molecules slide along the actin filament on muscle contraction. However, owing to the interdoublet links (see below) among the outer doublets, the walking motion is converted to bending of the cilium or the flagellum.

The outer dynein arm consists of two heavy chains, three to five intermediate chains, and six light chains in metazoan sperm, and is made of three heavy chains, two intermediate chains, and eight light chains in *Chlamydomonas*. Dynein heavy chains are huge proteins with molecular masses of ~500 kDa, and act as the fundamental motor unit. They contain a globular head with ATPase activity and a microtubule-binding stalk. The globular head shows a ringlike structure made of six subdomains, each of which corresponds to AAA (ATPase Associated with various cellular activity) domains. Four of the six AAA domains have P-loop for ATP binding. There is a coiled-coil domain between the 4th and 5th AAA domains, which is thought to be essential for ATP-sensitive binding to the adjacent B-tubule. Domain between each AAA domain seems to be important for conformational changes necessary for the dynein power stroke. Dynein intermediate chains are 120- to 60-kDa proteins, and are presumably necessary for binding on

A-tubule. Dynein light chains are 30- to 8-kDa proteins, and might be a regulator for activation of the outer dynein arm.

The inner dynein arm has more complicated molecular compositions than the outer dynein arm. It contains multiple molecular species with more heavy chains.

1.1.3 The Radial Spoke and the Central Pair

The radial spoke and the central pair (Fig. 2) are composed of a large number of proteins as well, and some of them include a domain that anchors cAMP-dependent protein kinase and a calcium binding domain. Therefore, the radial spoke presumably regulates the inner dynein arm through phosphorylation/dephosphorylation and calcium-dependent changes in the flagellar waveform. The central pair probably determines the plane of bending of the flagellum by sending signals to the radial spoke.

1.1.4 Other Proteins Associated with the Axoneme Structure

The interdoublet link (Fig. 2) known as the "nexin" linkage is thought to hold the outer doublet microtubules together and prevent their mutual sliding by the power stroke of the dynein arms.

Fibrous proteins called *tektin* presumably connect with the interdoublet links and support the ninefold symmetry architecture of the axoneme. Chaperons homologous to HSP70 are associated with this fiber. Also, another chaperon, HSP40, is identified as an axonemal component. These chaperons might help the folding of newly transported proteins through the intraflagellar transport described below.

Proteins related to determine a regular 24-nm interval of the outer dynein arms have also been identified and are called the *outer dynein arm docking complex*. They are predicted to have a coiled-coil structure and may work as a molecular scale.

1.2
The Basal Body

In the root of the cilia and flagella, there is a cylindrical structure called *the basal body*. The basal body has another characteristic structure constructed from nine sets of triplet microtubules (Fig. 3). The triplets are tilted inward and are connected to the neighbors as well as to the radial spokes extended from the central core, forming a pinwheel-like structure. Each triplet comprises one complete microtubule (A-tubule) fused to two partial microtubules (B- and C-tubule). The basal body has the same structure as centrioles, which are embedded in the center of mitotic spindles known as *centrosomes*. Indeed, the basal body of sperms and flagella is converted to the centriole during fertilization in mammals and mitosis in *Chlamydomonas* respectively. The basal body and the centriole are duplicated from its origin so as to form a smaller daughter perpendicular to the origin. In *Chlamydomonas*, specialized microtubule bundles called *rootlet microtubules* are known to be involved in positioning the daughters, but little is known about the replication mechanism.

The basal body is essential for the assembly of the cilia and the flagella. It not only serves as a template for organization of the nine doublet microtubules of the axoneme but also incorporates huge numbers of proteins involved in the intraflagellar transport (see below). A recent electron tomographic study revealed various cross sections of the basal body along the central axis and transition zones to the axoneme (Fig. 4). The basal body starts from an amorphous disk at its proximal end (Fig. 4-1) and then continues to the pinwheel-like structure made of the nine triplets (2 and 3). It extends to ~40 nm in length, and then becomes the distal transitional fibers with striated structures, which end on the plasma membrane (4). There are stellate fibers in the transition zone associated with the loss of the C-tubule (5 and 6). At the end of the stellate fibers (7 and 8), the central pair of microtubules appears along with additional axonemal structures (9).

The basal body assembly also needs over 200 distinct proteins, whose roles are being characterized. A member of the tubulin

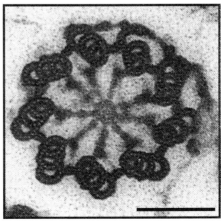

Fig. 3 Electron micrograph of a cross section of the basal body from a protozoan. Bar represents 100 nm (courtesy of Linck, R.W. and Woodrum, D.T. and permission from Alberts, B., Johnson, A., Lewis, J., Raff, M., Roberts, K., Walter, P. (2002) *Molecular Biology of the Cell*, 4th edition, Garland Science, New York).

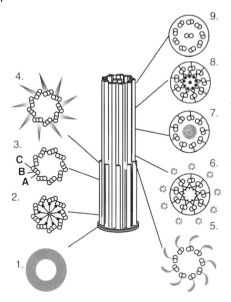

Fig. 4 Structural organization of the basal body, transition zone, and flagellum. Cross sections (1–9) are displayed from proximal to distal, respectively. Permission from O'Toole, E.T., Giddings, T.H., McIntosh, J.R., Dutcher, S.K. (2003) Three-dimensional organization of basal bodies from wild-type and δ-tubulin deletion strains of *Chlamydomonas reinhardtii*, *Mol. Biol. Cell.* **14**, 2999–3012.

super family, γ-tubulin, which is $\sim 30\%$ identical to α- and β-tubulin, is involved as an essential factor. Microtubules are nucleated at their minus end, but the concentrations of α- and β-tubulin inside the cells are below those required for spontaneous nucleation *in vitro*. Therefore, in all eukaryotes, the assembly of microtubules requires γ-tubulin. γ-tubulin is attached at the minus end of microtubules, forming a microtubule-organizing center (MTOC) together with other associated proteins. The basal body also includes other members of the tubulin super family, δ-, ε-, ζ-, and η-tubulin, which are probably required for construction of the basal body and formation of the axoneme.

1.3
The Intraflagellar Transport

The eukaryotic flagella and cilia grow at their distal tip. Components synthesized in the cytoplasm are transported through a system known as the *intraflagellar transport* (IFT). IFT was identified only a decade ago as particles rapidly moving ($2-4\,\mu m/s$) within the flagella of *Chlamydomonas*. This movement has been recognized to be essential for the growth, maintenance, and shortening of the cilia and flagella in all eukaryotes.

IFT consists of bidirectional movement of large protein particles along the outer doublet microtubules. Particles called the *IFT particles* are transported to the distal end (anterograde IFT) at $\sim 2\,\mu m\,s^{-1}$ by a motor protein, kinesin II, and back to the cell (retrograde IFT) at $\sim 4\,\mu m\,s^{-1}$ by a cytoplasmic dynein. Kinesin II is a member of the kinesin super family, which, in general, proceeds toward the plus end of microtubule, but is unrelated to dynein, which proceeds toward the minus end. The IFT particles work as a raft to carry components of the axoneme and form large linear arrays up to several hundred nanometers long. They can be divided into two complexes known as "A and B". Complex A is composed of four to six subunits with an apparent molecular mass of ~550 kDa, and complex B is made of at least nine subunits with a molecular mass of ~750 kDa. Complex A and B seem to be associated with retrograde IFT and anterograde IFT respectively.

IFT never stops and keeps moving at the same speed when the flagellum reaches its full length. Also, the IFT particles are remodeled at the base and the distal tip of the flagellum. Therefore, cells could regulate the assembly or disassembly of the flagellum by changing loads of the IFT particles at the base or the tip.

1.4
Perspectives

The eukaryotic flagella and cilia show a characteristic structure made of microtubules and ~250 associated proteins, half of which have not been characterized yet. Different from the bacterial flagella (see below), they have adopted a widely used system, the dynein motor for the force generation. Microtubule-dependent motors are also used for the process known as *IFT*, for growth, maintenance, and shortening of the cilia and flagella. Extensive studies, especially on the *Chlamydomonas* flagella, have just started to identify the roles of each protein and the growth mechanism. However, there are still too many unknowns. High-resolution structural information would be essential for further understanding of this system, and electron microscopy would be one of the most powerful tools to explore the structures and mechanisms of this complicated organelle.

2
Bacterial Flagella

Many bacteria swim by rotating helical flagellar filaments, which grow up to ~15 μm. The motor at the base of the filament drives the rotation of this helical propeller at hundreds of revolutions per second. In *Escherichia coli* and *Salmonella*, proton flow through the cytoplasmic membrane is utilized for the rotation. For chemotaxis and thermotaxis, the swimming pattern of bacteria alternates between "run" and "tumble"; a run lasts for a few seconds and a tumble for a fraction of second. During a run, the motor rotates counterclockwise (as it is viewed from outside the cell), and several flagellar filaments with a left-handed helical shape form a bundle and propel the cell. A tumble is caused by quick reversal of the motor to clockwise rotation, which produces a twisting force that transforms the left-handed helical form of the filament into a right-handed one, causing the bundle to fall apart rapidly. The separated filaments act in an uncoordinated way to generate forces that change the orientation of the cell. This leads to the tactic behavior, moving toward favorable or away from unfavorable conditions. Thus, the structure of the flagellar filament and its dynamic properties play an essential role in bacterial taxis.

The bacterial flagellum consists of several substructures. The motor structure called the *flagellar basal body* crosses both cytoplasmic and outer membrane, and continues to the extracellular structure called the *hook* and the *filament*. The largest of all is the filament, which is only ~20 nm in diameter, but grows up to a length around 15 μm by self-assembly of as many as 20 000 to 30 000 flagellin subunits. The assembly of the flagellum proceeds in a highly regulated fashion by the type III flagellar protein export apparatus, and flagellar axial proteins are always incorporated into its distal end, where capping proteins help the assembly in most cases. The bacterial flagella are best understood in terms of the structure and function for those of *E. coli* and *Salmonella typhimurium*, because much of relevant work has been carried out using them. Overall structure of the bacterial flagellum is shown in Fig. 5.

2.1
Component Structures

2.1.1 The Basal Body

The basal body of the bacterial flagellum consists of the MS ring (also called *FliF ring*), the C ring, the rod, and the L–P ring.

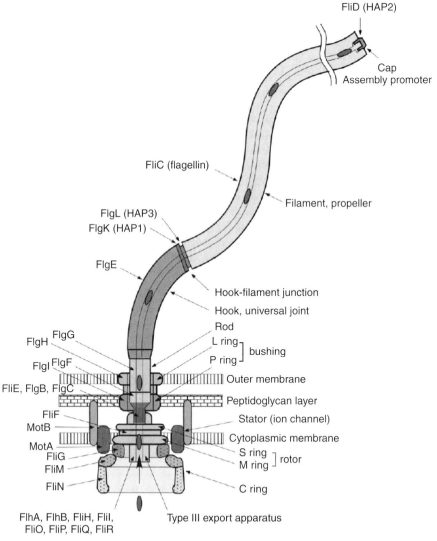

Fig. 5 Schematic diagrams showing the architecture of the bacterial flagellum. Reproduced with permission from Namba, K., Vonderviszt, F. (1997) Molecular architecture of bacterial flagellum, *Q. Rev. Biophys.* **30**, 1–65.

The MS ring with a diameter of ∼250 Å is made of ∼26 copies of FliF protein arranged cylindrically in the cytoplasmic membrane. "MS" means membrane and supramembrane. The C ring, a large cup-like structure attached on the cytoplasmic surface of the MS ring, is made of FliG, FliM, and FliN. "C" stands for cytoplasm. The rod, which traverses through the periplasmic space and the outer membrane, is made of FliE, FlgB, FlgC, FlgF, and FlgG proteins and is attached to the periplasmic face of the MS ring. The L–P ring is made of FlgH and FlgI, which are located within the outer membrane and the peptidoglycan layer respectively. L–P means "lipopolysaccharide" and "peptidoglycan." The MS ring is the base for self-assembly of all the flagellar proteins to form the flagellum and is partly involved in the type III flagellar protein export mechanism. The C ring is involved in torque generation as well as in regulation of its direction. The rod transmits the motor torque to the extracellular structures such as the hook and filament. The L–P ring is a bushing that mechanically supports the rapid rotation of the flagellum.

2.1.2 The Motor

Two cytoplasmic membrane proteins, MotA and MotB, are responsible for torque generation of the flagellar motor. MotA and MotB have four and one transmembrane-spanning helices respectively, and together form a proton channel. The MotA/B complex also functions as the stator by being attached to the peptidoglycan layer of the cell through a binding motif in the periplasmic domain of MotB. Electron micrographs of freeze-fractured samples showed that the stator complexes appear as a circular array of particles on the cytoplasmic membrane, surrounding the MS ring.

Some bacteria, such as alkaliphilic *Bacillus* and *Vibrio* species, normally have only one flagellum at their cellular pole, and use the electrochemical potential of sodium ions to generate the torque. The maximum rotational speed of the sodium-driven motor is ∼1700 Hz, while that of the proton-driven motor is ∼300 Hz. Four proteins, PomA (POlar flagellar Motility A), PomB, MotX, and MotY have been identified and are shown to be necessary for the rotation of the polar flagellar motor of *Vibrio*, such as *V. alginolyticus*, *V. parahaemolyticus*, and *V. cholerae*. PomA and PomB are homologous to MotA and MotB, while MotX and MotY are unique to the sodium-driven motor of *Vibrio*. Although the functions of MotX and MotY are not yet clear, a recent study showed that they are located in the outer membrane, and suggested that they may be essential for sodium recognition and for fast rotation of the polar flagellum.

For torque generation, MotA/B and PomA/B are thought to undergo a conformational change coupled to the proton flow and sodium flow, respectively while they interact with FliG, which is a component of the C ring. The X-ray crystal structures of the C-terminal and middle domains of FliG have revealed that important charged residues are lined up on one side of FliG so as to interact with a cytoplasmic loop of MotA or PomA. The functional unit of the motor complex is thought to be 4 MotA/2 MotB or 4 PomA/2 PomB.

2.1.3 The Hook and the Hook-filament Junction

The hook is a tubular structure, continuing directly from the rod into the outside of the cell. It is ∼55 nm long and ∼180 Å wide, and is made of ∼120 copies of

FlgE, which is also called the *hook protein*. The hook and the filament are connected with a short segment composed of two proteins, FlgK (also called *hook-associated protein 1* or HAP1) and FlgL (HAP3). Because the hook shows a relatively high flexibility in bending, it probably works as a universal joint to transmit the torque to the filament in a wide range of its orientation, thereby allowing several filaments to form a bundle behind the cell during run as well as to rotate individually during tumble.

2.1.4 The Flagellar Filament

The flagellar filament is a helical assembly of a single protein, flagellin, with roughly 5.5 subunits per one turn of the 1-start helix. The filament can also be described as a tubular structure comprising 11 protofilaments, which are nearly longitudinal helical arrays of subunits. Left- and right-handed helical forms of the filament are produced by supercoiling caused by mixture of two distinct protofilament conformations, L- and R-type. When all of the 11 protofilaments are the same type, two types of straight filaments with distinct helical symmetries are formed: The longitudinal 11-start helix is left-handed in the L-type and right-handed in the R-type straight filament.

The structures at ~10 Å resolution by earlier studies using electron microscopy and X-ray fiber diffraction revealed the subunit shape, domain organization, and subunit packing in these two types of straight filaments. However, higher resolution was needed to understand the structural mechanisms of the filament formation and supercoiling in atomic detail. Flagellin has a strong tendency to polymerize into filaments and this has prevented its crystallization. By clipping off terminal chains for ~100 residues in total, a fragment named "F41" was crystallized, and its structure was solved at 2.0 Å resolution by X-ray crystallography. Recently, the complete atomic model of the R-type straight flagellar filament was built on a density map at ~4.0 Å resolution obtained by electron cryomicroscopy and helical image reconstruction.

Flagellin from *S. typhimurium* SJW1103 strain consists of 494 amino acid residues, and its molecular mass is 51.5 kDa. Flagellin in Fig. 6 is viewed perpendicular to the filament axis. The overall shape of flagellin looks like a Greek letter "Γ" with a vertical dimension of ~140 Å and a horizontal dimension of ~110 Å. Flagellin consists of four linearly connected domains labelled D0, D1, D2, and D3, which are arranged from inside to outside of the filament (Fig. 7a). The N-terminal chain starts from D0, going through D1, D2, and reaches D3, and then comes back through D2 and D1, and the C-terminal chain ends in D0. While all three domain connections are formed by pairs of short antiparallel chains, the one that connects domains D0 and D1 is longer than the other two, and therefore it is called the *spoke region*. The top and the bottom correspond to the distal and proximal side of the flagellum respectively. We also define the distal side "up" and proximal side "down," and describe the structure accordingly.

Domain D0 consists of the N- and C-terminal α-helices forming an α-helical coiled-coil structure (labelled ND0 and CD0 respectively in Fig. 6a). The spoke region consists of two chains (NS and CS), which connect to two α-helices in domain D1 (ND1a and CD1). Helix ND1a is followed by a loop connecting to the second, shorter α-helix (ND1b), which goes down, and the chain continues to two β-turns, a β-hairpin pointing down and an extended chain going up. The chain

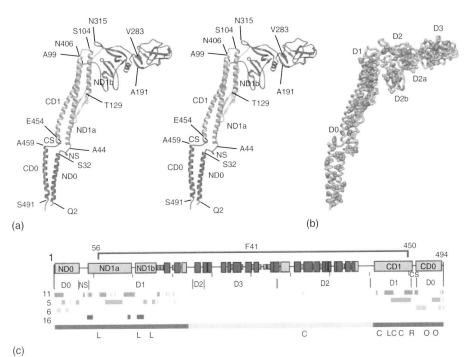

Fig. 6 The Cα backbone trace, hydrophobic side-chain distribution, and structural information of flagellin. (a) Stereo diagram of the Cα backbone. The chain is color coded as follows: residues 1 to 44, blue; 44 to 179, cyan; 179 to 406, green; 406 to 454, yellow; 454 to 494, red. (b) Distribution of hydrophobic side chains, mainly showing hydrophobic cores that define domains D0, D1, D2a, D2b, and D3. Side-chain atoms are color-coded: Ala and Met, yellow; Leu, Ile, and Val, orange; Phe, Tyr, and Pro, purple (carbon) and red (oxygen). (c) Position and region of various structural features in the amino acid sequence of flagellin. Shown are, from top to bottom, the atomic model of a major fragment of flagellin called F41 in blue; the secondary structure distribution with α-helix in yellow, β-structure in green, and β-turn in purple; tic mark at every 50th residue in blue; domains D0, D1, D2, and D3, and spoke regions NS and CS; the subunit contact regions along the 11-start in cyan, along the 5-start in orange, along the 6-start in pink, and along the 16-start in green; the well-conserved amino acid sequence in red and variable region in violet; point mutations that produce the filament of different supercoils. Letters at the bottom indicate the morphology of mutant filaments: L (F53V, D107E, R124A, R124S, G426A), L-type straight; R (A449V), R-type straight; C (D313Y, A414V, A427V, N433D) curly; O (Q472L, Q481L, Q481S) coiled. Permission from Yonekura, K., Maki-Yonekura, S., Namba, K. (2003) Complete atomic model of the bacterial flagellar filament by electron cryomicroscopy, *Nature* **424**, 643–650. (See color plate. p. xxvii)

finally goes into the outer parts, domain D2 and D3, which are mainly composed of β-structures. The vertical dimensions of domain D0 and D1 are ∼50 Å and 80 Å respectively. The extended chains in the spoke region are ∼20 Å long.

The end-on view of the R-type straight filament from the distal end of the flagellum (Fig. 7a) shows the concentric double-tubular structure made of domains D0 and D1 in the densely packed filament core. Domains D2 and D3, which project

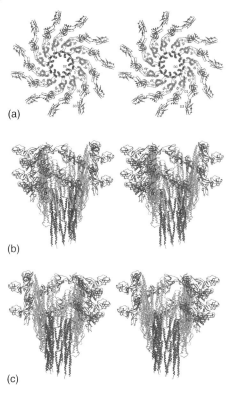

(a)

(b)

(c)

Fig. 7 Ribbon diagram of the Cα backbone of the filament model in stereo view. (a) End-on view from the distal end of the filament. Eleven subunits are displayed. (b) Side view from outside the filament. Three protofilaments on the far side have been removed for clarity. (c) Side view from inside the filament. Three protofilaments on the near side have been removed. Top and bottom of the side view images correspond to the distal and proximal ends of the filament respectively. The chain is color coded as in Fig. 6(a). Permission from Yonekura, K., Maki-Yonekura, S., Namba, K. (2003) Complete atomic model of the bacterial flagellar filament by electron cryomicroscopy, *Nature* **424**, 643–650. (See color plate. p. xxiii)

out from the filament core, are relatively well separated from one another. The diameter of the filament is ∼240 Å and that of the central channel is ∼20 Å. The N- and C-terminal α-helices (ND0 and CD0) are radially arranged with ND0 outside and CD0 inside, exposed to the central channel. The chains connecting domains D0 and D1 (NS and CS) look exactly like radial spokes in this view. As shown in side views from outside and inside the filament (Figs. 7b and 7c), domains D0 and D1 are making intimate intersubunit interactions, both axially and laterally. As a polar assembly of flagellin having a largely asymmetric structure, the filament model shows the concaved feature at the distal end (top of Fig. 7c) and the pointed tip at the proximal end (bottom), which are observed even in electron micrographs of negatively stained

filaments. The terminal chains of ∼65 N-terminal and 45 C-terminal residues are unfolded in the monomeric form of flagellin in solution. It can be inferred from this structure that, in the absence of axial and lateral packing interactions of these terminal chains in the inner tube, the two-stranded α-helical coiled coil in domain D0 with relatively less extensive hydrophobic core and a pair of extended, rather flexible looking chains connecting to domain D1 would be highly unstable. In contrast, the upper two-thirds of domain D1 can form its compact tertiary structure in the monomeric form because of its extensive hydrophobic core formed by three α-helices and one β-hairpin. The outer parts of flagellin are variable so that bacteria can escape from the attacks of host immune system, while the core part consisting of domains D0 and D1 are highly conserved (Fig. 6c).

The interactions within the inner and outer tubes are extensive, both along and between the protofilaments, while those between the inner and outer tubes are limited and are mainly in the spoke region. Most of the intersubunit interactions found within the outer tube are polar–polar or charge–polar, and contributions of hydrophobic interactions are relatively small, whereas those found within the inner tube and between the inner and outer tubes are mostly hydrophobic, contributing to the high stability of the filament structure.

There are heptad repeats of hydrophobic amino acid residues in the terminal regions of flagellin as well as other axial components, such as FlgE, HAP1, HAP3, and so on. In the filament structure, the terminal chains of neighboring subunits fold together into α-helical coiled coils. This interlocking organization would be the common motif by which the flagellar axial proteins form a continuous, mechanically stable tubular structure. The needle structures of the type III protein export system (see below) of pathogenic bacteria used to inject virulence proteins into host cells

are genetically related to the bacterial flagellum, and are thought to have α-helical coiled coil as a common motif as well.

The atomic structure of the R-type straight filament has also revealed interactions of mutation sites responsible for polymorphic supercoiling of the filament. However, the true understanding of the mechanism of polymorphic supercoiling and its dynamic transition has to wait until the atomic model of the L-type straight filament becomes available.

2.1.5 The Filament Cap

The distal end of the bacterial flagellum shows a thin, flat plate structure as identified by electron microscopy. This flat plate is a part of the filament cap made of HAP2 subunits. HAP2 from *S. typhimurium* is composed of 466 amino acid residues, and its molecular mass is 49.8 kDa. HAP2 forms a decameric complex in solution. The decamer shows a pentameric feature in its end-on view and a bipolar structure with two thin plates on both edge connected by a pair of long, slightly tapered cylindrical walls in its side view. The connecting portion appears to be relatively flexible. Figure 8

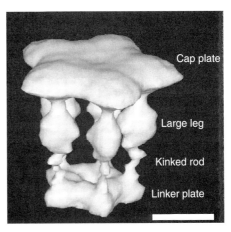

Fig. 8 Reconstructed three-dimensional image of the HAP2 pentamer. Bar represents 50 Å. Permission from Maki-Yonekura, S., Yonekura, K., Namba, K. (2003) Domain movements of HAP2 in the cap-filament complex formation and growth process of the bacterial flagellum, *Proc. Natl. Acad. Sci. U.S.A.* **100**, 15528–15533.

shows the three-dimensional structure of the HAP2 cap reconstructed from electron cryomicrographs. The HAP2 cap is composed of a pentagonal plate and five leg domains, whose lower part has no particular feature. This portion is made of both terminal regions of HAP2, because when ~40 N- and 50 C-terminal residues of HAP2 are cleaved off, the fragment does not form the decamer. These terminal chains are disordered in the pentameric or monomeric form of HAP2, suggesting that they also have the common structural motif shared by the flagellar axial proteins to form α-helical coiled coil in the inner core of the tubular structure. The heptad repeats are actually identified in the amino acid sequence of these terminal regions.

2.2
The Assembly Process of the Bacterial Flagellum

The assembly process of the bacterial flagellum starts from the formation of the MS ring complex in the cytoplasmic membrane. Then, the C ring is formed on the cytoplasmic face of the MS ring. The type III protein export apparatus, which is homologous to those that enable pathogenic bacteria to secret virulence factors, is presumably formed within the C ring. The export apparatus is composed of six membrane proteins and several cytoplasmic proteins (see below). The torque generation unit MotA/B assembles around the MS ring and is anchored to the peptidoglycan layer with the periplasmic domain of MotB. Each flagellum probably has about eight torque generation units.

The outward assembly of the flagellar axial structures proceeds at the distal end as it was initially found for the filament growth. The L–P ring assembly is an exception, where the subunit proteins FlgH and FlgI assemble around the rod in the outer membrane and peptidoglycan layer respectively. These proteins are exported by a general secretion system known as the *Sec pathway*. The formation of the P ring needs a periplasmic chaperone, FlgA. The type III protein export apparatus selectively binds and translocates flagellar axial proteins into the central channel of the growing flagellum by using the energy of ATP hydrolysis by FliI (see below). The axial proteins construct the rod, the hook, the hook-filament junction, and the long filament in this order. The proximal rod components FliE, FlgB, FlgC, and FlgF are connected with the periplasmic side of the MS ring, probably in this order, and the distal rod component FlgG follows to assemble. The rod assembly requires a help of the rod cap made of FlgJ, whose C-terminal domain has a muramidase activity to digest the peptidoglycan layer and to promote the rod growth through it. After the rod assembly is completed, the rod cap is somehow released, and FlgE assembles on the distal end of the rod in a helical manner to form the hook. A cap complex made of FlgD is attached at the distal end of the hook for its efficient assembly until it grows up to a length of ~55 nm. The length of the hook is controlled by FliK, one of the rod/hook-type export substrates, and FlhB, a member of the export apparatus (see below). A relatively short C-terminal region of FliK interacts with the cytoplasmic domain of FlhB at an appropriate timing to switch its conformation, resulting in the switching of substrate specificity of the export apparatus from the rod/hook type to the filament type (see below). This is how the hook growth stops, but the detailed mechanisms as to how the hook length information is transmitted to the cytoplasmic domains of the export apparatus is not yet well understood. Recent studies suggest that FliK acts as a ruler, but an alternative hypothesis called the *measuring cup mechanism*, in which 120 copies of FlgE are measured by the C-ring, may also be valid.

Then, the hook cap falls off the tip of the hook, probably by the binding of FlgK (HAP1), and then FlgL (HAP3) and FliD (HAP2) are transported and bound at the distal end in this order. FlgK, FlgL, and FliD are also called hook-associated proteins, HAP1, HAP3, and HAP2 respectively, because the hook-HAP1-HAP3-HAP2 complex is formed momentarily before the initiation of flagellin assembly into the long helical filament. Flagellin, also called "FliC", travels a long way through the narrow tunnel, only ~20 Å in diameter, but up to 10 to 15 µm long, to the distal end of the growing flagellum, where it assembles with a helical symmetry into the mechanically stable filamentous structure to function as a helical propeller. The filament cap made of HAP2 is always attached at the distal end, promoting the efficient assembly process.

2.2.1 The Type III Protein Export Apparatus

Most of the flagellar axial proteins are exported by the type III protein export apparatus into the central channel and to the distal end. The flagellum-specific secretion apparatus is made of six membrane proteins (FlhA, FlhB, FliO, FliP, FliQ, and FliR) and two soluble proteins, FliH and FliI. FliI has an ATPase activity and acts as the engine of export.

In addition to those, there are a general chaperon, FliJ, and substrate-specific chaperons, FlgN, FliS, and FliT, which bind to the export proteins synthesized in the cytoplasm and prevent them from undesirable aggregation or digestion, and from premature assembly within the cell. The partners of the substrate-specific chaperons are as follows: FliS–FliC; FlgN–FlgK; FlgN–FlgL; and FliT–FlgD. The structure of a complex of FliS with a short fragment of flagellin was recently solved by X-ray crystallography. This fragment corresponds to the C-terminal ~40 residues, which forms helix CD0 in the filament (Figs. 6 and 7). In the crystal structure of the complex, this helix is bent in the middle and partly extended, so as to keep the hydrophobic residues away from solution. FliS would prevent flagellin from forming the filament in this manner.

The flagellar type III protein export apparatus also specifies the order of the export substrate proteins. A recent study shows that there are two clear modes; the rod-type (FliE, FlgB, FlgC, FlgF, and FlgG) and the hook-type (FlgD, FlgE, FliK) proteins are exported in one mode, and the filament-type proteins, including FlgK, FlgL, FliD, and FliC, are exported in the other. The export of rod/hook-type proteins is followed by the export of the filament-type, and the switching of the substrate specificity occurs at a well-defined point in the assembly process to control the hook length, which is important for the proper hook function as a universal joint.

2.2.2 Growth of the Flagellar Filament

The flagellar filament is extended from the cell. Possible ways to construct such structures are either that constituent molecules are inserted at the base or they are transported to the distal end and bound there. As mentioned above, the latter is adopted by the bacterial flagellar filament assembly, probably because the bonding between the basal body, the hook, and the flagellar filament must always be tight and stable against the mechanical stress from high-speed rotation and quick reversal of the motor with the long helical propeller in viscous environments.

Flagellin subunits passing through the central channel of the flagellum polymerize just below the cap one after another to form the long helical filament. The cap stays attached at the distal end during the filament growth, and the simplest role of the cap would be to prevent flagellin from leaking out. This results in an increase in the local concentration of flagellin at the filament end above the critical concentration for polymerization. However, the cap does not completely cover the openings because HAP proteins that are occasionally transported pass through the capped end. Just in case the filament is broken in the middle, those transported HAP2 molecules can bind to the broken end of the filament and form a new cap again. Thus, the cap appears to act as a gatekeeper, which can be recovered when it is lost. The cap may also induce a conformational change of flagellin for efficient polymerization because flagellin must be unfolded to fit in the narrow channel during its transport. For insertion of flagellin subunits, the cap must move up by steps to make a room for the next flagellin monomer. Thus, the cap must play apparently contradictory roles; it must remain stably attached while permitting the insertion of flagellin subunits.

Structure analysis of the cap–filament complex by electron cryomicroscopy showed that the pentagonal plate of the cap is attached to the distal end of the filament at five positions through the anchor domains of the cap, which look just like five legs, where these anchor domains show significant deviation from the fivefold symmetry (Fig. 9). The five anchor domains are well separated from one another, forming five gaps with different sizes and shapes between the plate and the filament end (Fig. 9). One of

the gaps (Fig. 9c-1) is distinctly larger than the other four, having an inverted "L" shape and the size equivalent to domain D1 of flagellin. Because of the axial stagger in the lateral subunit packing, the end of the filament is not flat, but has five indentations around its circumference, where one of them is a double indentation. Thus, this is most likely to be the site of flagellin assembly.

Inside the filament just below the cap plate, there is a cavity ~ 30 Å wider than the central channel of the filament. Folded flagellin molecules cannot pass through the narrow central channel with a diameter of ~ 20 Å. Hence, flagellin definitely needs to be considerably unfolded until it reaches the distal end. This cavity may play some role in helping flagellin to fold up, perhaps in a similar way to Anfinsen's cage in chaperonin. The cavity appears to have the right size to accommodate only one flagellin subunit at a time, allowing its refolding without aggregation with other subunits.

During the growth of the filament, a large number of flagellin molecules, together with a small number of the hook-associated proteins, HAP1, HAP2, and HAP3, are synthesized in the cell and selectively translocated by the export apparatus into the long, narrow central channel of the flagellum, through which they are transported to the distal end by diffusion. This narrow channel would prevent unfolded flagellar proteins from undesirable aggregation.

The inner surface of the channel consists of mainly polar amino acids with one positively charged residue. The polar nature of the surface may be advantageous for fast diffusion of unfolded proteins, because unfolded proteins have many hydrophobic side chains exposed, which would be trapped on the channel surface of hydrophobic nature. This would also

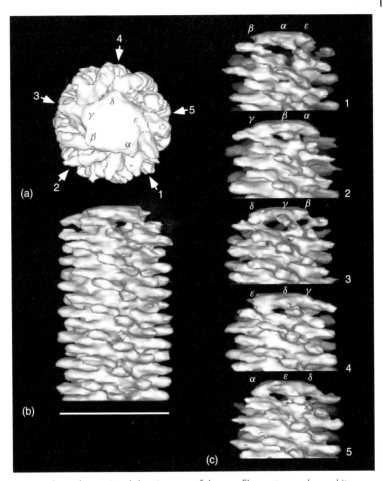

Fig. 9 Three-dimensional density map of the cap-filament complex and its central section. (a) End-on view from the top, showing a pentagonal shape of the plate domain of the cap. The five vertices of the pentagon are labeled with Greek letters, $\alpha-\varepsilon$, which are used to guide the orientation of the pentagon in the side views in (c). (b) Side view, showing a regular helical array of flagellin subunits and the plate domain of the cap. (c) Five side views, showing each of the five gaps between the plate and the filament end. The viewing directions are as indicated by arrows in (a), labeled with corresponding numbers. Greek letters labeling the vertices as in (a) also indicate the orientations of the pentagonal plate. Note that part of the gap in view one has a significantly larger axial extension than the other four gaps, and this is likely to be the site of assembly for the newly arriving flagellin subunit. Bar represents the diameter of the filament, 240 Å. Permission from Yonekura, K., Maki, S., Morgan, D.G., DeRosier, D.J., Vonderviszt, F., Imada, K., Namba, K. (2000) The bacterial flagellar cap as the rotary promoter of flagellin self-assembly, *Science* **290**, 2148–2152.

be a common feature of the type III protein export system including the needle complex of pathogenic bacteria.

Because of the symmetry mismatch between the helical subunit array of the filament with 11 protofilaments forming the tube and the pentameric annular structure of the cap, the five leg domains of the cap must have significantly different conformations from one another for interaction with flagellin. The flexible nature of the leg domain allows such polymorphic interactions, reducing conformational strains and permitting stable interactions between the two structures in spite of the symmetry mismatch. At the same time, the high conformational entropy of these cap anchor domains would counterbalance the strong binding of the cap to the filament as a whole for efficient insertion of flagellin.

Every insertion of a flagellin subunit is likely to force the cap to move into the next stable position energetically equivalent to the current position. A prediction from this structure and a previous discussion is that the cap moves up 4.7 Å and rotates 6.5° along the left-handed 5-start helix of the filament (Fig. 10). In terms of the cap conformation, the next position is approximately 65.5° away along the right-handed 1-start helix (arrangement of yellow subunits in Fig. 10). However, with permutation of the conformational states of the legs, which are 72° apart, a rotation of only 6.5° in the other direction would suffice. This would result in a roughly complete rotation of the cap by the assembly of 55 flagellin subunits. Upon every subunit incorporation, four legs of the cap rearrange their conformations and the last one changes its binding partner; namely, these legs walk along the helical steps of the filament end. These dynamic movements of the cap and its leg domains

would be required for efficient promotion of the flagellin self-assembly.

The energy for the cap rotation with the conformational rearrangements is presumably supplied by the binding energy of newly incorporated flagellin subunit to domains of multiple subunits of flagellin and HAP2, where the binding occurs by concentration of flagellin within the channel. The energy to keep the effective concentration of flagellin at the distal end of the export channel above the critical concentration for its polymerization would therefore come from ATP hydrolysis by the type III protein export apparatus located at the cytoplasmic face of the flagellar basal body. The energy transduction network formed here is unique and interesting because the energy produced in the cytoplasmic space is transmitted quite a long distance (10 ∼ 15 μm) through flagellin export to be used by the rotational movement of the cap at the end. The energy required for these movements would be relatively small because the process would involve a number of small conformational rearrangements in sequence, where the energy would propagate from one to the next.

2.3
Perspectives

The bacterial flagellum is a macromolecular complex made of ∼25 different proteins and is used for bacterial swimming. Bacteria have evolved most of those various apparatuses for this specific purpose. Biochemical and genetic studies have identified functions of each protein and have played important roles in understanding the assembly process of the flagellum. Recent electron microscopy and X-ray crystallography have also been elucidating details of the components structures,

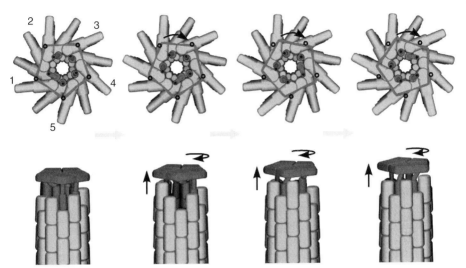

Fig. 10 Schematic diagram depicting the rotary cap mechanism promoting the flagellin assembly. This describes the rotation and axial translation of the cap plate and accompanied rearrangement of the legs upon every incorporation of a flagellin subunit (from left to right). Top view in the upper panel and an oblique view in the lower panel. In the upper panel, the cap plate is made transparent to show the different ways of leg domain binding, where black dots indicate the five positions of leg domain attachment to the plate. The plate and the black dots have strict fivefold symmetry, while the leg domains do not. In the lower panel, the outer domain of flagellin is removed for clarity. Subunits in yellow are newly incorporated flagellin molecules. Five open circles in the upper panel indicate the initial positions of the cap plate vertices as a reference for the cap rotation. The flagellin assembly proceeds along the 1-start helix, which is in the counterclockwise direction when viewed from the top, approximately at every 65.5° (360 × 2/11). This is also the angle of rotation after which the next binding site appears. However, because the legs of the cap are located every 72° (360/5), a 6.5° clockwise rotation with permutation of the leg conformations is sufficient to make the appropriate interactions between the leg and flagellin subunits. Numbers indicate the directions of views in Fig. 9(a). Animation at http://www.npn.jst.go.jp/yone.html. Permission from Yonekura, K., Maki, S., Morgan, D.G., DeRosier, D.J., Vonderviszt, F., Imada, K., Namba, K. (2000) The bacterial flagellar cap as the rotary promoter of flagellin self-assembly, *Science* **290**, 2148–2152. (See color plate. p. xxviii)

in particular, for the extracellular components. However, still little is known about molecular mechanisms of how each part of the flagellum functions. Obviously, we need high-resolution structures of not only the component proteins but also complexes made of many proteins. How the energy of ion flow is converted into the rotation of the flagellum would be the most important question to answer, and it still remains a mystery. High-resolution structure of the motor complex would be a starting point to solve the problem.

See also Bacterial Cell Culture Methods; Bacterial Growth and Division; Electron Microscopy in Cell Biology; Motor Proteins.

Bibliography

Books and Reviews

Alberts, B., Johnson, A., Lewis, J., Raff, M., Roberts, K., Walter, P. (2002) *Molecular Biology of the Cell*, 4th edition, Garland Science, New York.

Cole, D.G. (2003) The intraflagellar transport machinery of *Chlamydomonas reinhardtii*, *Traffic* **4**, 435–442.

Dutcher, S.K. (2003) Elucidation of basal body and centriole functions in *Chlamydomonas reinhardtii*, *Traffic* **4**, 443–451.

Inaba, K. (2003) Molecular architecture of the sperm flagella: molecules for motility and signaling, *Zoolog. Sci.* **20**, 1043–1056.

Minamino, T., Namba, K. (2004) Self-assembly and type III protein export of the bacterial flagellum, *J. Mol. Microbiol. Biotechnol.* **7**, 5–17.

Namba, K., Vonderviszt, F. (1997) Molecular architecture of bacterial flagellum, *Q. Rev. Biophys.* **30**, 1–65.

Snell, W.J., Pan, J., Wang, Q. (2004) Cilia and flagella revealed: from flagellar assembly in *Chlamydomonas* to human obesity disorders, *Cell* **117**, 693–697.

Yorimitsu, T., Homma, M. (2001) Na^+-driven flagellar motor of *Vibrio*, *Biochim. Biophys. Acta* **1505**, 82–93.

Primary Literature

Evdokimov, A.G., Phan, J., Tropea, J.E., Routzahn, K.M., Peters, H.K., Pokross, M.,

Waugh, D.S. (2003) Similar modes of polypeptide recognition by export chaperones in flagellar biosynthesis and type III secretion, *Nat. Struct. Biol.* **10**, 789–793.

Maki-Yonekura, S., Yonekura, K., Namba, K. (2003) Domain movements of HAP2 in the cap-filament complex formation and growth process of the bacterial flagellum, *Proc. Natl. Acad. Sci. U.S.A.* **100**, 15528–15533.

Nogales, E., Wolf, S.G., Downing, K.H. (1998) Structure of the $\alpha\beta$ tubulin dimer by electron crystallography, *Nature* **391**, 199–203.

O'Toole, E.T., Giddings, T.H., McIntosh, J.R., Dutcher, S.K. (2003) Three-dimensional organization of basal bodies from wild-type and δ-tubulin deletion strains of *Chlamydomonas reinhardtii*, *Mol. Biol. Cell.* **14**, 2999–3012.

Samatey, F.A., Imada, K., Nagashima, S., Vonderviszt, F., Kumasaka, T., Yamamoto, M., Namba, K. (2001) Structure of the bacterial flagellar protofilament and implications for a switch for supercoiling, *Nature* **410**, 331–337.

Yonekura, K., Maki, S., Morgan, D.G., DeRosier, D.J., Vonderviszt, F., Imada, K., Namba, K. (2000) The bacterial flagellar cap as the rotary promoter of flagellin self-assembly, *Science* **290**, 2148–2152.

Yonekura, K., Maki-Yonekura, S., Namba, K. (2003) Complete atomic model of the bacterial flagellar filament by electron cryomicroscopy, *Nature* **424**, 643–650.

Molecular Display Technologies

Ece Karatan, Zhaozhang Han, and Brian Kay
Argonne National Laboratory, Argonne, IL, USA

Encyclopedia of Molecular Cell Biology and Molecular Medicine, 2nd Edition. Volume 8
Edited by Robert A. Meyers.
Copyright © 2005 Wiley-VCH Verlag GmbH & Co. KGaA, Weinheim
ISBN: 3-527-30550-5

Keywords

Antibody Mimetic
A protein engineered to act as an antibody such that it binds target molecules other than its natural binding partners.

Affinity Reagent
An antibody, antibody mimetic or a peptide that has an affinity for a molecule of interest.

Bacteriophage/phage
A virus that infects bacteria.

Binder
Same as "affinity reagent."

CDR
Complementarity Determining Regions. Highly variable regions of the antibody light and heavy chain variable domains that are responsible for antigen recognition.

Fab
Fragment of antigen binding of antibodies generated by either recombinant methods or limited proteolysis.

scFv
Single-chain Fragment of variable region. A short fragment of an antibody molecule that is engineered in such a way that it contains only the variable regions of the light and heavy chains in a single polypeptide.

■ Molecular display refers to the presentation or the display of molecules – mostly proteins and peptides – on the surface of a virus, a cell, or a molecular complex such as a ribosome or messenger RNA. The coding information of the displayed molecule is carried by the entity displaying it, thereby establishing a physical link between the properties (phenotype) and DNA sequence (genotype) of the displayed molecule. All molecular display techniques have *in vivo* or *in vitro* methods for enrichment by selective propagation of the molecule with the desired properties. This, along with the genotype–phenotype link, allows large numbers (up to 10^{14}) of variations to be surveyed for the desired characteristics. These properties have made molecular display technologies a very valuable tool for the study of molecular interactions, the isolation of affinity reagents with increased affinities and stabilities, and the development of therapeutics with tissue- and cell-targeting capabilities. This article introduces different display methods, the types of random libraries most commonly used to isolate binders, and some of the interesting and creative applications of molecular display technologies in answering a variety of biological questions.

1
Display Methods

1.1
Phage display

In 1985, Dr George Smith, a professor at the University of Missouri-Columbia, published an article in which he described an experiment where he inserted a segment of the gene encoding the *Eco*RI restriction enzyme into the genome of the M13 bacteriophage. Smith found that fusions of the *Eco*RI protein fragment with one of the minor coat proteins were tolerated by the virus, as they still produced infectious particles. Interestingly, these particles incorporated the chimeric protein, and they could be fractionated from nonrecombinant virus particles with an antibody that recognized the expressed portion of the *Eco*RI protein. With sequential rounds of affinity selection, termed *biopanning*, the recombinant particles that displayed the antigen could be selectively enriched from an excess of nonrecombinant particles.

These results led to the concept of "phage display," in which coding regions for peptide or protein segments are inserted in a bacteriophage genome and the expressed peptide or protein segments are accessible for molecular interactions. The major benefit of phage display, as well as other display technologies we review here, is the physical linkage between the displayed entity (phenotype) and its genetic code (genotype), which permits a method for "bar-coding" and isolating rare peptides and proteins with desired properties from diverse libraries. In this section, we briefly discuss different molecular display methods that are in use today.

1.1.1   Bacteriophage M13
To date, bacteriophage M13 display has been the most extensively used display method. M13 phage particle are filamentous in shape (i.e. 9 nm wide, 900 nm long) and contain five types of coat proteins as well as a single-stranded DNA molecule (Fig. 1a). In the original experiments of

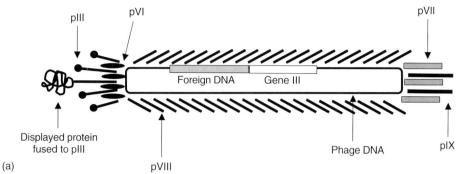

Fig. 1 Bacteriophage M13 display.
(a) Schematic representation of M13. (b) Affinity selection of phage-displayed libraries. The protein (pIII, pVI, pVII, pVIII, and pIX) and DNA (single-stranded) components of a prototypic bacteriophage M13 particle are shown. DNA fusions to the N-terminus of the coding region for mature protein III leads to the generation of a chimeric protein that is incorporated in the secreted viral particle. If the virus particle is assembled in an infected cell that carries a recombinant phagemid and a helper virus, display can be monovalent (shown); alternatively, use of a different vector can lead to pentavalent display (not shown). A population of virus particles displaying a library of antibody fragments, cDNA segments, combinatorial peptides, or engineered proteins can be fractionated by affinity selection. Phage are incubated in a microtiter plate well coated with a target protein and nonbinding particles are washed away, and retained particles are recovered by chemically denaturing the target. Infection of E. coli cells with the released particles amplifies the phage enriched for binding and the process is repeated two or more additional times. After limiting dilution of the final sample, individual clones are amplified and tested for binding to the target (and not a negative control protein) by an ELISA. DNA sequencing of the phage insert leads to identification of the affinity selected antibody fragment, cDNA segment, combinatorial peptide, or engineered protein.

Smith, protein fragments were displayed at the N-terminus of mature protein III. Fortuitously, the site chosen for display did not interfere with either virus morphogenesis or the infection process. Since then, research has shown that it is possible to display at the N-terminus of proteins III, VII, VIII, IX and at the C-terminus of proteins III, VI, and VIII. The display of proteins and peptides at the N-terminus of proteins III or VIII is most common.

Phage-displayed protein or peptide libraries are typically fractionated by affinity selection (Fig. 1b). A target protein is immobilized on a surface, incubated with an aliquot of viral particles, nonbinding particles are washed away, and bound particles are recovered. To release the bound particles without destroying their infectivity, it is possible to expose them to pH extremes (i.e. pH 2 or 12), protein denaturants (i.e. 6 M urea, 15 mM dithiothreitol), or proteases (i.e. trypsin, subtilase). Once the conditions for elution have been neutralized, the recovered phage can be used to infect *Escherichia coli* cells, thereby amplifying the selection output. Since bacteriophage M13 is not a lytic virus, infected cells secrete phage particles on a continual basis as they grow, yielding $\sim 10^{12}$ plaque-forming units (pfu) per milliliter in an overnight liquid culture.

The target proteins can be affixed to the surface of microtiter plate wells either directly by nonspecific adsorption or indirectly through a streptavidin-biotin

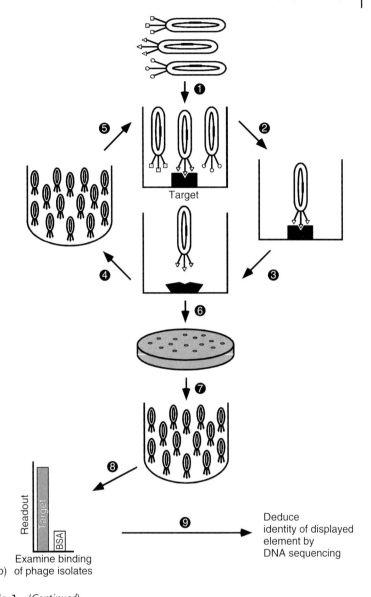

Target

Readout

Target

BSA

Examine binding
(b) of phage isolates

Deduce
identity of displayed
element by
DNA sequencing

Fig. 1 (*Continued*)

linkage (i.e. the target protein is biotiny-lated and bound to immobilized strep-tavidin). Another option is to capture complexes formed between soluble bi-otinylated protein targets and phage on streptavidin-coated microtiter plate wells or magnetic beads. Finally, it is possible to select phages that bind to the surfaces of, or are internalized by, mammalian cells or tissues. In general, three rounds of affinity selection are sufficient to enrich bind-ing phage particles from a library. After confirming that the output phage clones actually bind to the target, often through an enzyme-linked immunosorbent assay (ELISA), sequencing the DNA insert of the individual phage clones reveals the coding sequence of the displayed peptide or protein.

Many of the proteins displayed on bacteriophage M13 adopt their native structure and activity, even though they are fused to a coat protein that is assembled into a viral particle. However, in some cases, the displayed protein interferes with the function of the coat protein, viral particle morphogenesis, or secretion of the viral particle by *E. coli*. In many of these situations, these limitations can be overcome by displaying the protein on a shortened version of the coat protein and providing the wild-type form of the protein, with the aid of a helper virus. Thus, instead of being displayed on all five copies of protein III (multivalent display), the protein can be displayed in a monovalent manner, with four copies of wild-type protein III available for attaching to the pili of *E. coli* cells and infecting them. M13 has been a versatile tool for the display of many different kinds of peptides and proteins, thereby leading to many interesting biomedical and biotechnical applications, as discussed below.

1.1.2 Bacteriophage T7
It has been possible to display proteins and peptides on a number of other bacteriophages as well. One popular virus is bacteriophage T7, in which it is possible to fuse peptides and proteins to the C-terminus of the 10B capsid protein. Vectors have been developed that direct display from 1 to 415 copies of polypeptides, 10 to 1000 amino acids in length, per virion. The low-valency display vectors are suitable for the selection of high-affinity ligands, while

the higher valency display vectors enhance the recovery of low-affinity binders. Unlike the filamentous phage-display system, in which the displayed proteins need to be secreted into the periplasmic compartment of *E. coli* cells, T7 phage particles assemble in the cytoplasm, followed by lysis of the host cell. This characteristic overcomes the limited capacity of the protein secretion machinery of the *E. coli* cell and is more favorable for display of cytosolic proteins that carry reduced cysteine residues. The short life cycle (i.e. 2 h) for this lytic virus also makes it attractive for accelerating the screening process. Finally, since the protein of interest can be fused to the C-terminus of the major capsid protein, T7 phage is ideal for displaying segments of proteins encoded by cDNA fragments, which are typically biased toward containing C-terminal stop codons.

1.1.3 Bacteriophage Lambda (λ)
The λ display system has several attractive attributes. Compared to the T7 phage system, in which multivalent display of larger proteins has been reported to affect plaque formation, λ appears to tolerate display of proteins up to 300 amino acids in length on at least 90% of the copies of its capsid D protein. In addition, it is possible to display at either the N- or the C-termini of protein D. From a λ library displaying protein segments encoded by cDNA or genomic DNA fragments, one can identify interacting proteins for a variety of targets.

1.2 Cell Surface Display

In the recent years, cell surface display has emerged as an alternative to phage display. While cells can display thousands of copies of a peptide or protein, one of the biggest advantages of cell display is that one can use fluorescence-activated cell sorting (FACS) to identify and quantitate positive clones during the selection process. In addition, the carbohydrates on the cell surface offer a largely inert surface in terms of nonspecific binding to target proteins, potentially leading to lower numbers of false positives. This section discusses various bacterial and yeast cell surface display techniques.

1.2.1 Bacterial Display

A variety of expression systems for the display of either short peptides or fully folded proteins on *E. coli* and, to a lesser extent, on gram-positive bacteria have been developed. Peptides or proteins have been targeted to the cell surface of bacteria by fusing them with protein components of the outer membrane, pilus, or flagella of *E. coli*. It has also been possible to display on the surface of gram-positive bacteria, *Staphylococcus carnosus* and *S. aureus*. Through FACS, bacteria that bind a particular fluorescent target protein can be enriched over 10 000-fold from a library, and with subsequent rounds of growth and cell sorting, they are eventually purified. Bacterial display has many applications, including vaccine development, mapping epitopes, bioremediation, biocatalysis, and biosensor development.

1.2.2 Yeast Display

Surface display of proteins on baker's yeast *Saccharomyces cerevisiae* was first reported in 1997 and, since then, it has been gaining increased use in the field of protein engineering. This system utilizes the cell surface receptor a-agglutinin, which mediates cell–cell adhesion during mating of the "a" and "α" haploid forms to produce a diploid yeast cell. The large subunit of a-agglutinin, Aga1p, is anchored to the yeast cell wall by a covalent β-glycan linkage, while the small subunit Aga2p is linked to Aga1p by two disulfide bonds (Fig. 2a). Proteins of interest can be displayed as C-terminal fusions to the Aga2p protein without disturbing assembly of the agglutinin protein heterodimer on the cell surface. To tightly regulate display, expression of the *Aga2p* gene fusion is directed by the GAL1-10 promoter, which is completely repressed by growth on glucose and induced by galactose. Thus, a yeast-display library can be propagated in the presence of glucose without expression of the fusion protein, which prevents loss of recombinant clones whose Aga2p fusion protein expression is toxic. A variety of proteins have been successfully displayed on yeast, including scFv's, single-chain T-cell receptors, Fabs, a single-chain form of the class II major histocompatibility complex product, and selectins.

Affinity selection of a yeast-display library is commonly performed by mixing the yeast cells displaying the protein of interest with a fluorescently labeled target, followed by sorting of the yeast-target complexes using FACS. The amount of Aga2p fusion protein displayed on cells can be monitored in parallel through detection of an epitope added to the C-terminus of the Aga2p fusion protein (Fig.2a). With FACS, the enrichment of yeast displaying an interacting protein is evident in real time during each selection step (Fig.2b). One less-expensive alternative to FACS is the use of streptavidin magnetic beads with biotinylated targets for fractionating yeast. To date, yeast display has been applied primarily to *in vitro* evolution studies in which 10^5 to 10^7 variants of a protein are generated and screened for an

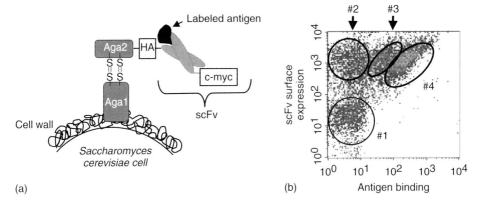

(a)   (b)   Antigen binding

Fig. 2 The yeast display method. (a) Aga2-scFv fusion displayed on the surface of *S. cerevisiae*. The protein of interest (i.e. an scFv) is expressed on the yeast surface as a C-terminal fusion to Aga2. Aga2 is attached to Aga1 by disulfide bonds, which in turn is anchored in the yeast cell wall. An HA tag C-terminal to Aga2 and a c-myc tag C-terminal to the scFv allow determination of the fraction of cells expressing the recombinant clone and those expressing Aga2 only. The antigen is labeled so that the yeast, which bind the antigen, can be fractionated and quantified using FACS. (b) Bivariate flow cytometric analysis of yeast-displayed antibody fragments. A greatly enriched antigen-binding population shows the discrimination of clones with slight differences in affinities and nonbinding populations. Cells are double labeled with biotinylated antigen/streptavidin-phycoerythrin, and anti-c-myc/anti-mouse FITC labels. Subpopulation #4 expresses higher affinity scFv's and labels more brightly with antigen at a given scFv expression level. Subpopulation #3 expresses somewhat lower affinity scFv's, and subpopulation #2 does not detectably bind antigen. Subpopulation #1 is not expressing scFv on the cell surface. (Figure 2b was reproduced with permission from Feldhaus, M.J., Siegel, R.W. et al. (2003) Flow-cytometric isolation of human antibodies from a nonimmune *Saccharomyces cerevisiae* surface display library, *Nat. Biotechnol.* **13**(2), 163–170.)

optimal variant. Even though yeast cells are not efficiently transformed with exogenous DNA, such studies generally involve smaller libraries of variants, which are easily achievable using yeast. Recently, Feldhaus and coworkers reported the construction and characterization of the first nonimmune human scFv library displayed on yeast, with a diversity of 10^9 clones. By performing a considerable number of transformations, it was possible to overcome the low transformation efficiency of yeast. With this library, scFv's to many targets have been isolated and demonstrated to have affinities similar to those obtained from phage-display libraries. Moreover, the researchers were also able to amplify this library 10^{10} fold without loss of diversity, suggesting that the original library is essentially an unlimited source for scFv's.

1.3
In vitro Display

In vitro display technologies utilize cell-free systems, and are not limited by the biology of the virus or cells. *In vitro* libraries are typically 10^{12} to 10^{14}, orders of magnitude higher than the largest phage libraries available, which allows a much larger "sequence space" to be surveyed and tighter binders to be selected.

1.3.1 Ribosome Display

Ribosome display was first applied to the display of short peptides and later modified to display proteins and protein fragments. In this technique, a linear DNA segment, containing a promoter recognized by the T7 RNA polymerase, a ribosome binding site, the coding region for the peptide or protein, and a spacer region encoding a long unstructured stretch of amino acids, is transcribed and translated *in vitro* (Fig.3a). If the translation product lacks a stop codon, the nascent polypeptide chain remains tethered to the ribosome and

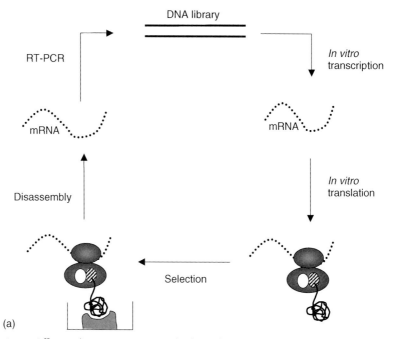

(a)

Fig. 3 Affinity selection using *in vitro* display techniques: DNA and mRNA are denoted by solid and broken lines, respectively. (a) Ribosome Display. A DNA library encoding proteins of interest is transcribed and translated *in vitro*. A long C-terminal spacer allows proteins to fold and lack of a stop codon prevents release of the protein. The mRNA-protein–ribosome complex is then used for selection on immobilized target. At the end of the selection step, the bound mRNA is released from the ribosome with EDTA, reverse transcribed, and the resulting cDNA is amplified by PCR and used either for another round of affinity selection or cloned and sequenced to determine its identity. (b) mRNA display. A DNA library encoding proteins of interest is transcribed *in vitro* and a short segment of DNA carrying a 5′ psoralen linker and a 3′ puromycin moiety (P) is hybridized to the mRNA. Photo-activation of the psoralen leads to covalent cross-linking of the mRNA and the DNA. This complex is then *in vitro* translated. The ribosomes stall at the RNA–DNA junction, allowing the puromycin to enter the A site of the ribosome (shaded circle) and accept the nascent polypeptide chain. Because of the covalent linkage of the mRNA and the protein, the ribosome can be dissociated at this point. Second-strand synthesis of the mRNA results in a DNA–RNA hybrid that is more stable and the selections are performed with this complex. At the end of the selection, the DNA is amplified by PCR and either used for additional rounds of selection or sequenced to yield the identity of the clone.

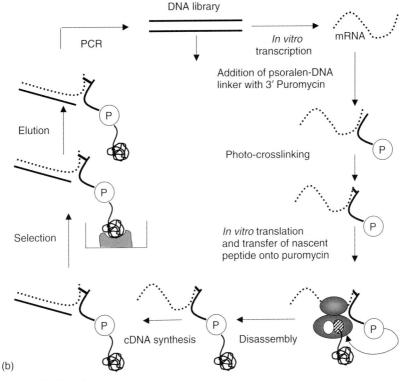

DNA library

PCR

In vitro transcription

mRNA

Addition of psoralen-DNA linker with 3′ Puromycin

Elution

Photo-crosslinking

Selection

In vitro translation and transfer of nascent peptide onto puromycin

cDNA synthesis

Disassembly

(b)

Fig. 3 (*Continued*)

mRNA, thereby creating a physical linkage between the genotype and phenotype. This RNA-protein complex, which is stabilized by high concentrations of magnesium and low temperatures, can be used in selection experiments. Sequences of interest are recovered by reverse transcribing the selected mRNA molecules, synthesizing the second DNA strand, and amplifying the resulting DNA by the polymerase chain reaction (PCR). Through additional rounds of *in vitro* transcription, translation and affinity selection, peptides or protein fragments with interesting properties are isolated from libraries.

Ribosome display has been mainly used for the display and selection of scFv's. By this technique, scFv's selected from libraries of $>10^{12}$ molecules generally have dissociation constants in the low nanomolar range. Since the PCR amplification step in the screening process can be modified to be mutagenic (i.e. 0.5–1%), this technique is well suited for molecular evolution experiments such as affinity maturation and stability engineering of scFv's. In one application, a low fidelity DNA polymerase was used for PCR amplification of scFv's selected with bovine insulin, and after six cycles of selection, scFv's were recovered with dissociation constants of 80 picomolar. In another study, scFv's that recognize the guanine quadruplex of human telomeric DNA sequences were isolated from a ribosome display library and demonstrated to have K_d values in the low nanomolar range. In immunofluorescence studies, it was possible to prove the existence of the guanine quadruplex *in vivo,* for which there had previously been no conclusive evidence, using such antibody fragments.

Ribosome display has also been used to select for proteins with catalytic activity. A beta-lactamase could be enriched over an inactive mutant by ribosome display using a suicide substrate inhibitor. Even though the active enzyme made a covalent bond with the suicide substrate, chelating the magnesium ions releases its encoding mRNA for subsequent amplification and rescreening steps.

1.3.2 mRNA Display
mRNA display was first reported in 1997 by two groups working independently. This technique is similar to ribosome display, except that the nascent polypeptide chain is linked to its mRNA covalently by use of the aminoacyl-tRNA mimic puromycin. Puromycin can be attached to the 3′ end of the mRNA by hybridizing with single-stranded DNA containing psoralen and puromycin moieties, followed by photomediated cross-linking of the hybrid through the psoralen. This RNA–DNA partial duplex is then used for *in vitro* translation: the ribosome stalls at the RNA–DNA junction, which allows time for the puromycin to enter the "A site" of the ribosome and accept the nascent polypeptide chain, thereby creating a covalent linkage between the mRNA and the protein it encodes (Fig. 3b). To stabilize the mRNA component of the complex prior to selection experiments, it is reverse transcribed or replaced by double-stranded cDNA, which is covalently attached to the protein and the selection is performed with this complex. Since proteins are covalently linked to their coding sequence, affinity selection experiments can be performed under stringent conditions.

A variety of peptides and proteins have been displayed on mRNA and have been used to isolate high-affinity reagents that bind RNA, small molecules, and other proteins. RNA-binding peptides, which bound their target with low nanomolar affinity and high specificity, were isolated from combinatorial peptide libraries of 10^{12} to 10^{14} individual members. Some of these peptides were even able to differentiate between a single nucleotide change in the loop structure of the RNA. From a combinatorial peptide library containing 80 randomized amino acids, specific binders of ATP have been isolated with high affinity ($K_d = 100$ nM). From tissue-specific cDNA libraries, interaction partners for the antiapoptotic protein Bcl-X_L, which included known partners as well as previously unidentified proteins, have been isolated. Finally, using this technology, it was possible to isolate antibody mimetics, which were based on randomizing residues within the tenth fibronectin type III domain, to tumor necrosis factor-α (TNF-α) with K_d values in the picomolar range.

mRNA display also lends itself to unique applications such as incorporation of unnatural amino acids and chemical adducts into the synthesized protein. Attachment of a penicillin moiety to a fixed cysteine in a combinatorial peptide library allowed identification of novel inhibitors of the penicillin binding protein 2a, which were 100-fold more potent than penicillin alone in inhibiting synthesis of bacterial cell walls. The same research group also introduced a biotinylated amino acid into a displayed peptide by adding a biocytin-charged amber suppressor tRNA to the *in vitro* translation reaction. The ability to use unnatural amino acids overcomes the diversity limits imposed by the 20 natural amino acids and can lead to the identification of peptides and proteins with unique characteristics.

2
Types of Libraries

Molecular display technologies allow the recovery and identification of molecules with desired properties without the need for any prior sequence information. This property has made it possible to survey very large libraries(i.e. 10^9–10^{14} unique sequences) for those rare members with the desirable biochemical traits. In this section, we briefly discuss the different kinds of libraries that are often used to accomplish this previously daunting task.

2.1
Combinatorial Peptide Libraries

One of the most common uses for bacteriophage M13 has been the display of combinatorial peptides. Oligonucleotides, which are fixed in length but degenerate in sequence, are inserted into one of the genes encoding one of the six capsid proteins, thereby generating a library of combinatorial peptides displayed on the surface of phage particles. While there are many codon schemes to choose from, the degenerate regions are commonly based on the NNK coding scheme: N is an equimolar mixture of A, C, G, and T, and K is an equimolar mixture of G and T nucleotides. In this scheme, 32 codons encode one stop codon (TAG), which is suppressed in certain *E. coli* strains, and all 20 amino acids, which are represented once (C, D, E, F, H, I, K, M, N, Q, W, Y), twice (A, G, P, V, T), or three times (L, R, S). These combinatorial peptide libraries, also called *random peptide libraries*, typically contain over one billion clones, with each viral particle displaying a different peptide sequence fused to protein III. A variety of combinatorial peptide libraries have been constructed over the years, of 6 to 42 amino acids in length, and with no or several fixed amino acids, such as cysteines. (Since the phage particles are secreted into an oxidizing environment, peptides with even numbers of cysteines will form intramolecular disulfide bonds, thereby conformationally constraining the peptides.) While such libraries are diverse enough to encode all possible six and shorter amino-acid-long peptide sequences, there is evidence for exclusion from phage-displayed combinatorial peptide libraries of certain peptide sequences that interfere with viral morphogenesis or secretion through the membrane.

2.2
scFv and Fab Libraries and Fab Libraries

Fabs or scFv's have been successfully displayed on the surface of bacteriophage. Creation of large repertoires of antibody fragments from antibody genes, as well as the development of high-throughput screening methods, reduce, and in some cases eliminate, the need to use animals for antibody production. Phage-displayed antibodies have been generated from a variety of animals (camel, chimpanzee, human, llama, mouse, and rabbit), and since each phage particle displays a single antibody fragment, they are monoclonal in nature. M13 phage display remains the most extensively used method for isolation of antibody fragments partly because it is well suited for the expression of immunoglobulin protein fragments; as the chimeric antibody-capsid proteins pass through the periplasm during viral morphogenesis, the disulfide bonds in the heavy and light chains have an opportunity to form properly.

There are three routes for generating antibody display libraries. First, it is possible to immunize an animal with an immunogen, recover the circulating B-cells, extract mRNA, and clone portions of the immunoglobulin heavy and light chains into a display vector. In this manner, monoclonal antibodies can be rescued from immunized animals *in lieu* of generating hybridomas. It is possible to use this method to obtain neutralizing antibodies from human patients that recovered from exposure to infectious agents, such as the human immunodeficiency virus, respiratory syncytial virus, and herpes simplex virus, and *Helicobacter pylori*. Second, it is possible to generate a naïve antibody library by amplifying the variable regions of the heavy and light chains from a small number of B-cell donors and then cloning and displaying a large number of pair-wise combinations. From libraries containing over one billion clones, it is possible to generate antibodies to most antigens with nanomolar dissociation constants. Third, it has been possible to randomize the codons for the hypervariable regions within the heavy and light chain variable regions, thereby generating "semisynthetic" antibody libraries. All three approaches have been extremely productive in generating antibodies for basic research, biotechnology, and medical applications.

In the postgenomic era, antibody fragments can help to define novel gene function and to measure protein expression in pathological states. In addition to these new directions, antibody fragments have begun to show success in clinical diagnosis and as immunotherapies for cancer and other diseases. Radiolabeled antibodies, derived by display technologies, have great potential for cancer imaging and treatment. Antibodies have also been exploited for prodrug therapy, cancer treatment, and gene delivery by fusing them with different molecules such as toxins, enzymes, and viruses. The list of FDA approved antibody therapeutics against cancer, viral infection, and inflammatory disease is growing rapidly (http://www.fda.gov/cber/efoi/approve. htm).

2.3
Antibody Mimetic Libraries

Although scFv's and Fabs have been used extensively as affinity reagents, they sometimes suffer from poor yields when expressed in *E. coli* and can be unstable. These drawbacks have prompted a search for alternative scaffolds, which bind targets with comparable specificity and affinity. These mimetics generally have improved stabilities, better expression yields in *E. coli*, and, in some cases, are smaller, inherently fluorescent, or lack disulfide bonds. Mimetics without disulfide bonds can be used for *in vivo* inhibition of cytoplasmic targets because they adopt stable structures in the reducing environment of the cytoplasm. Although it remains to be determined if antibody mimetics can replace therapeutic antibodies, they offer great promise in research and diagnostics.

Antibody mimetic libraries can be constructed by randomizing residues that are exposed on the scaffold surface or by grafting the complementarity determining regions (CDR) of antibodies onto the scaffold. Some of these mimetics have been termed *affibodies, monobodies,* and *anticalins,* which are based on a three-helix bundle Protein A domain from *Staphylococcus aureus*, the tenth type III domain of human fibronectin, and lipocalin, respectively. Affibodies, mono-bodies, and fluorobodies have been used to generate binders with nanomolar to picomolar affinities against protein targets, whereas anticalins have been mainly used for isolating binders with midnanomolar affinity for small molecules such as fluorescein and digoxigenin. Mono-bodies isolated against estrogen receptor complexed with different ligands have been employed as intracellular detectors for ligand-induced conformational changes of the target. Recently, repeat proteins such as ankyrins and leucine repeats have been explored as antibody mimetics where multiple randomized repeats are stacked together to create a binding surface for targets. The modular structure of this class of antibody mimetics may allow greater flexibility in designing binders with greater diversity and increased surface area for interactions with their targets. Ankyrins are also promising reagents for intracellular protein "knock-out" experiments because they do not require disulfide bond formation for their stability.

2.4
cDNA and Genomic DNA Libraries

Bacteriophages have also been used to display libraries of protein segments encoded by complementary DNA (cDNA) fragments. Libraries of plant, fungal, and dust mite cDNAs have been cloned and displayed on bacteriophage M13 for the purpose of identifying human allergens: the libraries are screened with serum IgE and phage-displaying reactive allergens are selected and subsequently identified by DNA sequencing. Bacteriophage λ and T7 have been shown effective in displaying protein fragments encoded by cDNA segments cloned at the C-terminus of

their major capsid proteins. Libraries of cDNA inserts corresponding to mRNA purified from different cell types are now widely available and such libraries have been screened extensively for interacting with specific protein, RNA, DNA promoter, carbohydrate, or small chemical molecules.

3
Applications

3.1
Protein Function and Target Validation

3.1.1 Mapping Protein–protein Interactions with Peptides

Peptide ligands selected from a phage-displayed library appear to bind preferentially to "hot spots" on the surface of protein targets. A hot spot is defined as a small region on the surface of a protein, often composed of several hydrophobic, disordered amino acids, which binds other proteins or small molecules. Peptide ligands selected from a phage library tend to home in on hot spots and to mimic proteins that normally interact with the target protein. In fact, very often, the primary structures of the selected peptides resemble regions within the interacting proteins. This phenomenon has been termed an example of *convergent evolution*, where the biological and molecular evolution has yielded similar results. Consequently, phage display has proven very useful in mapping the protein–protein interactions of extracellular and intracellular proteins (Fig. 4). Several notable examples include defining the epitopes of monoclonal and polyclonal antibodies, discovering agonists and antagonists of hormone receptors, and defining the specificity of protein interaction modules

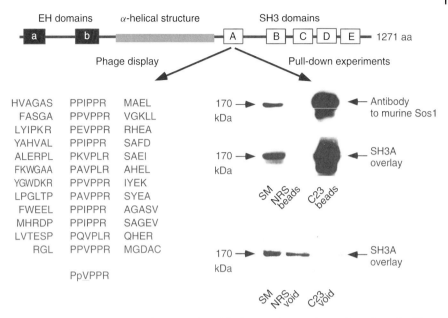

Fig. 4 Mapping protein–protein interactions. The intersectin protein consists of two Eps15 Homology (EH) domains, a central coiled-coil region, and five Src Homology 3 (SH3) domains. A protein fusion between the N-terminal SH3 domain (SH3A) and the glutathione-S-transferase (GST) can be used to affinity select phage-displayed combinatorial peptide libraries. The predicted amino acid sequences of selected phage are aligned to reveal the motif, PpVPPR, where p is commonly proline. Computer searches with the motif matched a sequence (PPVPPR) within the Son-of-Sevenless (Sos) protein, suggesting that it may interact with intersectin. Sos was demonstrated to interact with intersectin through pull-down and coimmunoprecipitation experiments. Sos is 170 kilodaltons (kDa) and reacts with C23, a commercial monoclonal antibody. SM and NRS void refer to starting material (rat brain extract) and nonreactive serum respectively.

present in signal transduction and endocytic proteins.

3.1.2 Elucidation of Enzyme Function

Peptides selected for binding to enzymes appear to preferentially bind at the active sites of the enzymes, even though they are not involved in protein–protein interactions. In a survey of ten different enzymes, peptide ligands were isolated for eight of the target proteins, and at least six were shown to inhibit the enzymes. Synthetic forms of the selected peptides acted as competitive inhibitors in activity assays. In addition, they were used in displacement assays to screen libraries of chemicals and natural products for those that prevented binding of the peptide ligands to the enzymes *in vitro*, and thus bind to the active site of the enzyme. Such peptides can also be expressed inside cells where they can bind to the enzyme and inhibit its *in vivo* activity; in some cases, inhibition leads to a cellular phenotype.

It is also possible to screen combinatorial peptide libraries to define the substrate specificity of enzymes. By tethering phage through an anchor sequence, the substrate specificity of proteases can be defined: only those viruses that carry a cleavage site in the linker composed of combinatorial peptides are released. In this manner, "substrate phages" have been identified for a variety of proteases, such as subtilisin, factor Xa, furin, matrix metalloproteinases, urokinase, granzyme B, prostate-specific antigen, and HIV-1 protease. Substrate phages have also proven useful for defining the substrate specificity of various protein tyrosine kinases.

3.1.3 Functional Knock-outs Using Intrabodies

Over the past few years, several groups have expressed antibody fragments inside cells. These intracellularly expressed antibody fragments, also named "intrabodies," can provide a way of validating targets by blocking pathways in cells. While only a small fraction of phage-displayed scFv molecules can fold properly in the reducing environment of the cytoplasm, those which do fold are still able to bind to their targets and yield interesting biological results, such as arresting HIV infection, blocking apoptosis, disturbing proper embryo development, and arresting oncogenic transformation of cells in culture. Future experiments in functional genomics will rely heavily on antibodies or mimetics that fold efficiently inside cells.

3.2 Drug Discovery and Therapeutics

3.2.1 Identification of Tissue- and Tumor-specific Markers via Whole-cell and *in vivo* Panning

Phage-displayed peptide and scFv libraries have also been used successfully for affinity selections against target mixtures present on the surface of cultured mammalian cells or tissues. Because whole-cell or tissue panning bypasses the need to purify the antigen, this method is useful for generating affinity reagents against antigens that are hard to purify, such as transmembrane proteins. Using this procedure, scFv's and peptides have been isolated for cell surface receptors such as urokinase plasminogen activated receptor, thrombin receptor, melanocortin receptor, and cell types such as colorectal tumor cells, lymphatic vessels, and *Bacillus* spores. Cell- and tumor-specific peptide ligands and antibodies have great potential usefulness in profiling the type or the stage of cancer in a manner similar to expression profiling using DNA microarrays.

In working with intact cells, scientists have realized that phage particles that promote cellular internalization can be selected from a display library. After incubating the phage particles with intact cells, the unbound phage are washed away and the cells are lysed in order to recover the internalized phage. Internalizing scFv's, which were shown to recognize the ErbB2 antigen, have been identified for SKBR3 breast cancer cells, and when fused to liposomes, target delivery into ErbB2 expressing tumor cells. Peptides selective for internalization by fibroblast cells, laryngeal carcinoma cells and ECV304 endothelial cells, and the human urothelium have been reported.

Phage-displayed combinatorial peptide libraries have also been used in *in vivo* panning. In a series of ground-breaking papers, Renata Pasqualini and Erki Ruoslahti injected phage particles displaying combinatorial peptides into mice, dissected out various tissues, amplified the bound phage particles in *E. coli*, and repeated the selection process several times. Individual clones, when tested, were confirmed to bind selectively to their respective tissue targets. Similar experiments have also been conducted in humans: phage particles have been injected into patients and peptides were identified that bind selectively to human vasculature. Such "organ homing" peptides can be linked to drugs, and adenovirus or adeno-associated virus, for the purpose of targeting the delivery of the drugs or gene therapy agents respectively.

3.2.2 Screening of Small Molecule Inhibitors Using Surrogate Peptide Ligands

Since peptide ligands selected from a phage-displayed combinatorial peptide library often bind at sites of protein–protein interactions or catalysis, assays for discovering small molecule inhibitors can be developed. For example, peptides chemically synthesized with fluorophores attached at one end can be incubated in microtiter plate wells with the target protein and compounds from natural and chemical libraries. If the compound does not bind to the target, the fluorescence polarization of the labeled peptide will become anisotropic because of binding of a small peptide to a larger target molecule. On the other hand, if the compound binds to the target, thereby blocking peptide binding, the fluorescence polarization of the peptide will be isotropic, because it is free in solution. High-end devices can scan 96-, 384-, 1536-, or 6144-well microtiter plates rapidly in a laboratory, thereby permitting research groups to survey libraries of thousands and thousands of compounds for potential inhibitors. This general approach has lead to the discovery of compounds with potential antimicrobial activity.

3.2.3 Zinc Fingers for Controlling Gene Expression

One fruitful avenue for designing affinity reagents has been with zinc fingers, which typically bind DNA. Multiple zinc fingers can be displayed on phage and can bind to specific DNA sequences, with each finger (30 amino acids) roughly binding three adjacent bases in a double-stranded DNA molecule. Thus, it has been possible to randomize residues within a phage-displayed zinc finger and select for variants that could bind to specific double-stranded DNA molecules. Through a concerted effort, zinc fingers that bind to many different DNA triplet sequences have been identified. Such fingers can be "stitched" together to make polydactyl fingers (i.e. 6 adjacent zinc fingers) that recognize unique 18-bp sequences with subnanomolar to femtomolar dissociation constants. One practical consequence of this work is that it is now possible to introduce engineered finger proteins into cells and induce or repress expression of specific genes. It will be interesting to follow how this approach matures.

3.3 Protein Engineering and Industrial Applications

3.3.1 Molecular Evolution

Molecular evolution refers to improvement of the desired characteristics of a molecule using techniques such as error-prone PCR or DNA shuffling and the appropriate selection protocols. This strategy is used extensively in protein engineering in situations where the mutations that might bring about a desired property cannot be predicted from existing biochemical or structural information, necessitating the generation and screening of random mutants. With the application of different display technologies, the number of clones that can be screened at the end of a molecular evolution experiment has increased by orders of magnitude. This allows researchers to access larger numbers of permutations for the desired protein products. By enhancing the specificity, affinity, or stability of an scFv, one can generate a reagent with improved therapeutic value. Using yeast display and a combination of DNA shuffling and error-prone mutagenesis, the dissociation constant of an scFv that binds fluorescein was improved over three orders of magnitude to 48 femtomolar.

Through phage display and targeted mutagenesis, the dissociation constant of an anti-Vascular Endothelial Growth Factor (VEGF) antibody fragment was improved 120 fold over the parent molecule. Finally, by using ribosome display, an antihemagglutinin scFv was evolved to have increased stability in DTT. Some of the clones were able to bind their antigen when expressed in the reducing environment of the *E. coli* cytoplasm. These types of antibody fragments would be invaluable as reagents for protein "knock-out" and target validation studies.

Phage-displayed proteins are amenable to selection for their resistance to extreme conditions, such as denaturants, detergents, proteases, and elevated temperatures. Thus, one can select for protein variants that maintain their shape or function under such conditions. For example, from a phage library displaying mutated ribonuclease T1 molecules, it was possible to select variants more stable than the wild-type molecule. A screen of mutagenized *Bacillus subtilis* cold shock protein CspB led to the identification of variants that had higher thermal or chemical stability, depending on whether the selection was done at elevated temperature or in the presence of a denaturant, respectively.

In another case, from a library of mutagenized alpha-amylase proteins, variants were discovered that hydrolyze starch at low pH conditions that favor industrial applications. Finally, a partially unfolded four-helix bundle protein, apocytochrome b(562), was phage displayed and challenged with a protease, to yield a stably folded form of the protein.

3.3.2 Bacterial and Yeast Surface Display as Biocatalysts

Ability to display proteins and enzymes on bacterial and yeast cell surfaces have proven to be valuable in industrial applications. For example, construction of a whole-cell biocatalyst with its sequential reaction has been achieved by the genetic immobilization of two amylolytic enzymes on the yeast cell surface. Organophosphorus hydrolase, a bacterial enzyme that has been shown to degrade a wide range of neurotoxic organophosphate nerve agents, was evolved and displayed on *E. coli* surface to hydrolyze methyl parathion, providing the possibility of creating biocatalysts that can degrade pesticides (i.e. diazinon and chlorpyrifos) and nerve agents (i.e. sarin and soman). *E. coli* and *Moraxella sp* (a bacterium known to survive in contaminated environments) have been successfully explored to display phytochelatins, which could be used for bioaccumulation of cadmium and mercury, providing a new method for treating heavy metal contamination. Finally, display of glycosyltransferases at the yeast cell surface through a cell wall anchored protein has shown potential as a simple *in vitro* method for oligosaccharide synthesis using the yeast intact cell as a biocatalyst.

3.3.3 Nanotechnology

Recently, several groups have screened phage-displayed combinatorial peptide libraries for peptides that bind to inert surfaces. Even though such peptide-surface interactions are intrinsically weak, they are detected because of the avidity resulting from the pentavalent presentation of protein III on M13. Peptides have been discovered that bind selectively to polystyrene, semiconductor surfaces, silver nanoparticles, and ZnS or CdS nanocrystals. These peptides have great promise for engineering new nanostructures.

4
Outlook and Perspectives

Scientists in both basic science and biotechnology have used display technologies creatively. While the technologies have matured over the past few years, new applications appear on a regular basis, suggesting that they will continue to impact various fields of science in a significant manner for the next 10 years. In particular, it is anticipated that these technologies will contribute significantly in the emerging areas of nanotechnology and personalized medicine, where display technologies can be used for fabricating nanostructures and treating disease respectively.

Acknowledgments

The submitted manuscript has been created by The University of Chicago as Operator of Argonne National Laboratory ("Argonne"). Argonne, a US Department of Energy Office of Science laboratory, is operated under Contract No. W-31-109-Eng-38. The US Government retains for itself, and others acting on its behalf, a paid-up nonexclusive, irrevocable worldwide license in said article to reproduce, prepare derivative works, distribute copies to the public, and perform publicly and display publicly, by or on behalf of the Government.

See also Combinatorial Phage Antibody Libraries; DNA Libraries; Peptide and Non-Peptide Combinatorial Libraries.

Bibliography

Books and Reviews

Amstutz, P., Forrer, P., Zahnd, C., Pluckthun, A. (2001) *In vitro* display technologies: novel developments and applications, *Curr. Opin. Biotechnol.* **12**(4), 400–405.

Brown, K.C. (2000) New approaches for cell-specific targeting: identification of cell-selective peptides from combinatorial libraries, *Curr. Opin. Chem. Biol* **4**(1), 16–21.

Dower, W.J., Mattheakis, L.C. (2002) *In vitro* selection as a powerful tool for the applied evolution of proteins and peptides, *Curr. Opin. Chem. Biol.* **6**(3), 390–398.

Kay, B.K., Kasanov, J., Knight, S., Kurakin, A. (2000) Convergent evolution with combinatorial peptides, *FEBS Lett.* **480**(1), 55–62.

Kretzschmar, T., von Ruden, T. (2002) Antibody discovery: phage display, *Curr. Opin. Biotechnol.* **13**(6), 598–602.

Ladner, R.C., Ley, A.C. (2001) Novel frameworks as a source of high-affinity ligands, *Curr. Opin. Biotechnol.* **12**(4), 406–410.

Ruoslahti, E. (2002) Drug targeting to specific vascular sites, *Drug Discov. Today* **7**(22), 1138–1143.

Skerra, A. (2000) Engineered protein scaffolds for molecular recognition, *J. Mol. Recognit.* **13**(4), 167–187.

Primary Literature

Abe, H., Shimma, Y., Jigami, Y. (2003) *In vitro* oligosaccharide synthesis using intact yeast cells that display glycosyltransferases at the cell surface through cell wall-anchored protein Pir, *Glycobiology* **13**(2), 87–95.

Amstutz, P., Pelletier, J.N., Guggisberg, A., Jermutus, L., Cesaro-Tadic, S., Zahnd, C., Pluckthun, A. (2002) *In vitro* selection for catalytic activity with ribosome display, *J. Am. Chem. Soc.* **124**(32), 9396–9403.

Arkin, M.R., Randal, M., DeLano, W.L., et al. (2003) Binding of small molecules to an adaptive protein-protein interface, *Proc. Natl. Acad. Sci. U.S.A.* **100**(4), 1603–1608.

Bae, W., Mulchandani, A., Chen, W. (2002) Cell surface display of synthetic phytochelatins using ice nucleation protein for enhanced heavy metal bioaccumulation, *J. Inorg. Biochem.* **88**(2), 223–227.

Barrick, J.E., Takahashi, T.T., Ren, J., Xia, T., Roberts, R.W. (2001) Large libraries reveal diverse solutions to an RNA recognition problem, *Proc. Natl. Acad. Sci. U.S.A.* **98**(22), 12374–12378.

Beerli, R.R., Barbas, C.F. III. (2002) Engineering polydactyl zinc-finger transcription factors, *Nat. Biotechnol.* **20**(2), 135–141.

Benson, R.E., Gottlin, E.B., Christensen, D.J., Hamilton, P.T. (2003) Intracellular expression of Peptide fusions for demonstration of protein essentiality in bacteria, *Antimicrob. Agents Chemother.* **47**(9), 2875–2881.

Beste, G., Schmidt, F.S., Stibora, T., Skerra, A. (1999) Small antibody-like proteins with prescribed ligand specificities derived from the lipocalin fold, *Proc. Natl. Acad. Sci. U.S.A.* **96**(5), 1898–1903.

Binz, H.K., Stumpp, M.T., Forrer, P., Amstutz, P., Pluckthun, A. (2003) Designing repeat proteins: well-expressed, soluble and stable proteins from combinatorial libraries of consensus ankyrin repeat proteins, *J. Mol. Biol.* **332**(2), 489–503.

Boder, E.T., Midelfort, K.S., Wittrup, K.D. (2000) Directed evolution of antibody fragments with monovalent femtomolar antigen-binding affinity, *Proc. Natl. Acad. Sci. U.S.A.* **97**(20), 10701–10705.

Boder, E.T., Wittrup, K.D. (1997) Yeast surface display for screening combinatorial polypeptide libraries, *Nat. Biotechnol.* **15**(6), 553–557.

Chen, Y., Wiesmann, C., Fuh, G., Li, B., Christinger, H.W., McKay, P., de Vos, A.M., Lowman, H.B. (1999) Selection and analysis of an optimized anti-VEGF antibody: crystal structure of an affinity-matured Fab in complex with antigen, *J. Mol. Biol.* **293**(4), 865–881.

Cho, C.M., Mulchandani, A., Chen, W. (2002) Bacterial cell surface display of organophosphorus hydrolase for selective screening of improved hydrolysis of organophosphate nerve agents, *Appl. Environ. Microbiol.* **68**(4), 2026–2030.

Choo, Y., Klug, A. (1994) Toward a code for the interactions of zinc fingers with DNA: selection of randomized fingers displayed on phage, *Proc. Natl. Acad. Sci. U.S.A.* **91**, 11163–11167.

Christensen, D.J., Gottlin, E.B., Benson, R.E., Hamilton, P.T. (2001) Phage display for target-based antibacterial drug discovery, *Drug Discov. Today* **6**(14), 721–727.

Chu, R., Takei, J., Knowlton, J.R., Andrykovitch, M., Pei, W., Kajava, A.V., Steinbach, P.J., Ji, X., Bai, Y. (2002) Redesign of a four-helix bundle protein by phage display coupled with proteolysis and structural characterization by NMR and X-ray crystallography, *J. Mol. Biol.* **323**(2), 253–262.

Clackson, T., Wells, J.A. (1995) A hot spot of binding energy in a hormone-receptor interface, *Science* **267**, 383–386.

Crameri, R., Kodzius, R. (2001) The powerful combination of phage surface display of cDNA libraries and high throughput screening, *Comb. Chem. High Throughput Screen.* **4**(2), 145–155.

DeLano, W.L., Ultsch, M.H., de Vos, A.M., Wells, J.A. (2000) Convergent solutions to binding at a protein-protein interface, *Science* **287**(5456), 1279–1283.

Falke, D., Juliano, R.L. (2003) Selective gene regulation with designed transcription factors: implications for therapy, *Curr. Opin. Mol. Ther.* **5**(2), 161–166.

Feldhaus, M.J., Siegel, R.W., Opresko, L.K., et al. (2003) Flow-cytometric isolation of human antibodies from a nonimmune *Saccharomyces cerevisiae* surface display library, *Nat. Biotechnol.* **21**(2), 163–170.

Goldenberg, D.M. (2002) Targeted therapy of cancer with radiolabeled antibodies, *J. Nucl. Med.* **43**(5), 693–713.

Goodson, R.J., Doyle, M.V., Kaufman, S.E., Rosenberg, S. (1994) High-affinity urokinase receptor antagonists identified with bacteriophage peptide display, *Proc. Natl. Acad. Sci. U.S.A.* **91**(15), 7129–7133.

Hammond, P.W., Alpin, J., Rise, C.E., Wright, M., Kreider, B.L. (2001) *In vitro* selection and characterization of Bcl-X(L)-binding proteins from a mix of tissue-specific mRNA display libraries, *J. Biol. Chem.* **276**(24), 20898–20906.

Hanes, J., Pluckthun, A. (1997) *In vitro* selection and evolution of functional proteins by using ribosome display, *Proc. Natl. Acad. Sci. U.S.A.* **94**(10), 4937–4942.

He, M., Taussig, M.J. (1997) Antibody-ribosome-mRNA (ARM) complexes as efficient selection particles for in vitro display and evolution of antibody combining sites, *Nucl. Acids Res.* **25**(24), 5132–5134.

Hoess, R.H. (2002) Bacteriophage lambda as a vehicle for peptide and protein display, *Curr. Pharm. Biotechnol.* **3**(1), 23–28.

Hyde-DeRuyscher, R., Paige, L.A., Christensen, D.J., et al. (2000) Detection of small-molecule enzyme inhibitors with peptides isolated from phage-displayed combinatorial peptide libraries, *Chem. Biol.* **7**(1), 17–25.

Jermutus, L., Honegger, A., Schwesinger, F., Hanes, J., Pluckthun, A. (2001) Tailoring in vitro evolution for protein affinity or stability, *Proc. Natl. Acad. Sci. U.S.A.* **98**(1), 75–80.

Keefe, A.D., Szostak, J.W. (2001) Functional proteins from a random-sequence library, *Nature* **410**(6829), 715–718.

Koide, A., Bailey, C.W., Huang, X., Koide, S. (1998) The fibronectin type III domain as a scaffold for novel binding proteins, *J. Mol. Biol.* **284**(4), 1141–1151.

Kretzschmar, T., von Ruden, T. (2002) Antibody discovery: phage display, *Curr. Opin. Biotechnol.* **13**(6), 598–602.

Lee, S.Y., Choi, J.H., Xu, Z. (2003) Microbial cell-surface display, *Trends Biotechnol.* **21**(1), 45–52.

Li, S., Millward, S., Roberts, R. (2002) *In vitro* selection of mRNA display libraries containing an unnatural amino acid, *J. Am. Chem. Soc.* **124**(34), 9972–9973.

Li, S., Roberts, R.W. (2003) A novel strategy for in vitro selection of peptide-drug conjugates, *Chem. Biol.* **10**(3), 233–239.

Martin, A., Sieber, V., Schmid, F.X. (2001) *In vitro* selection of highly stabilized protein variants with optimized surface, *J. Mol. Biol.* **309**(3), 717–726.

Mattheakis, L.C., Bhatt, R.R., Dower, W.J. (1994) An *in vitro* polysome display system for identifying ligands from very large peptide libraries, *Proc. Natl. Acad. Sci. U.S.A.* **91**(19), 9022–9026.

Murai, T., Ueda, M., Shibasaki, Y., Kamasawa, N., Osumi, M., Imanaka, T., Tanaka, A.

(1999) Development of an arming yeast strain for efficient utilization of starch by co-display of sequential amylolytic enzymes on the cell surface, *Appl. Microbiol. Biotechnol.* **51**(1), 65–70.

Nemoto, N., Miyamoto-Sato, E., Husimi, Y., Yanagawa, H. (1997) *In vitro* virus: bonding of mRNA bearing puromycin at the 3'-terminal end to the C-terminal end of its encoded protein on the ribosome in vitro, *FEBS Lett.* **414**(2), 405–408.

Nord, K., Nilsson, J., Nilsson, B., Uhlen, M., Nygren, P.A. (1995) A combinatorial library of an alpha-helical bacterial receptor domain, *Protein Eng.* **8**(6), 601–608.

Pasqualini, R., Ruoslahti, E. (1996) Organ targeting in vivo using phage display peptide libraries, *Nature* **380**, 364–366.

Roberts, R.W., Szostak, J.W. (1997) RNA-peptide fusions for the in vitro selection of peptides and proteins, *Proc. Natl. Acad. Sci. U.S.A.* **94**(23), 12297–12302.

Rodi, D.J., Makowski, L., Kay, B.K. (2002) One from column A and two from column B: the benefits of phage display in molecular-recognition studies, *Curr. Opin. Chem. Biol.* **6**(1), 92–96.

Sarikaya, M., Tamerler, C., Jen, A.K., Schulten, K., Baneyx, F. (2003) Molecular biomimetics: nanotechnology through biology, *Nat. Mater.* **2**(9), 577–585.

Schaffitzel, C., Berger, I., Postberg, J., Hanes, J., Lipps, H.J., Pluckthun, A. (2001) In vitro generated antibodies specific for telomeric guanine-quadruplex DNA react with *Stylonychia lemnae* macronuclei, *Proc. Natl. Acad. Sci. U.S.A.* **98**(15), 8572–8577.

Sieber, V., Pluckthun, A., Schmid, F.X. (1998) Selecting proteins with improved stability by a phage-based method, *Nat. Biotechnol.* **16**(10), 955–960.

Smith, G.P. (1985) Filamentous fusion phage: novel expression vectors that display cloned antigens on the virion surface, *Science* **228**(4705), 1315–1317.

Stemmer, W.P. (1994) DNA shuffling by random fragmentation and reassembly: in vitro recombination for molecular evolution, *Proc. Natl. Acad. Sci. U.S.A.* **91**(22), 10747–10751.

Stumpp, M.T., Forrer, P., Binz, H.K., Pluckthun, A. (2003) Designing repeat proteins: modular leucine-rich repeat protein libraries based on the mammalian ribonuclease inhibitor family, *J. Mol. Biol.* **332**(2), 471–487.

Tao, J., Wendler, P., Connelly, G., et al. (2000) Drug target validation: lethal infection blocked by inducible peptide, *Proc. Natl. Acad. Sci. U.S.A.* **97**(2), 783–786.

Tong, A.H., Drees, B., Nardelli, G., et al. (2002) A combined experimental and computational strategy to define protein interaction networks for peptide recognition modules, *Science* **295**(5553), 321–324.

Verhaert, R.M., Beekwilder, J., Olsthoorn, R., van Duin, J., Quax, W.J. (2002) Phage display selects for amylases with improved low pH starch-binding, *J. Biotechnol.* **96**(1), 103–118.

Winter, G., Griffiths, A.D., Hawkins, R.E., Hoogenboom, H.R. (1994) Making antibodies by phage display technology, *Annu. Rev. Immunol.* **12**, 433–455.

Xu, L., Aha, P., Gu, K., et al. (2002) Directed evolution of high-affinity antibody mimics using mRNA display, *Chem. Biol.* **9**(8), 933–942.

Yamabhai, M., Hoffman, N.G., Hardison, N.L., McPherson, P.S., Castagnoli, L., Cesareni, G., Kay, B.K. (1998) Intersectin, a novel adaptor protein with two Eps15 homology and five Src homology 3 domains, *J. Biol. Chem.* **273**(47), 31401–31407.

Molecular Immunology: *see*
Immunology

Molecular Mediators: Cytokines

Jean-Marc Cavaillon
Institut Pasteur, Paris, France

Encyclopedia of Molecular Cell Biology and Molecular Medicine, 2nd Edition. Volume 8
Edited by Robert A. Meyers.
Copyright © 2005 Wiley-VCH Verlag GmbH & Co. KGaA, Weinheim
ISBN: 3-527-30550-5

Keywords

Cytokines
Constitutively released or produced upon cell activation, cytokines are soluble
mediators – sometimes membrane bound – binding to receptors on target cells and
inducing, modulating, or inhibiting cellular functions. Most cytokines are produced by
immune cells and act on immune cells, but cytokines are produced by numerous cell
types and act on a great variety of cells, accordingly displaying a large spectrum of
activities. They are characterized by their redundancy, their synergistic action, and their
regulatory loops. Those produced constitutively contribute to the tissue and systemic
homeostasis. The cytokine are elements of a universal language used by cells to
communicate.

Chemokines
A subfamily of cytokines with structural similarities and sharing the property to recruit
cells within the extravascular space of the tissues (chemotaxis), either at homeostasis or
during inflammatory processes.

Interferons (IFN)

A subfamily of cytokines sharing an antiviral bioactivity. Interferons are characterized by different subgroups (type I and II), depending on their origin. In addition to their common property, some (e.g. IFNγ) are involved at different stages of the onset and further development of the immune responses, or play a role during gestation in mammals (e.g. IFNτ).

Interleukin (IL-)

A convenient name to classify cytokines. The numbers have been given as a reflection of their discovery through the years. This term does not define a family, since none are biochemically or biologically related. However, the most recent interleukins have been described after gene banks and are thus related to previous well-known interleukins. When the word was created (1979), there was only IL-1 and IL-2.

Receptors

The receptors of cytokines allow the cells to respond to the signal delivered by cytokines. They can involve 1, 2, or 3 different chains. Most chains are constituted with one extracellular domain, one transmembrane domain and one intracellular domain, except the chemokine receptors that are constituted by 7 transmembrane domains.

> Cytokines are proteins or glycoproteins produced by cells and acting on other cells that display on their surface the specific cytokine receptor. Cytokines are words used by the cells to communicate. Alone or within a determined sequence, these mediators lead the responding cells to modify its function (e.g. secretion, proliferation, induction, inhibition, enhanced/reduced function, migration, apoptosis, etc.). Despite cytokines having been mainly discovered by immunologists as a product of cells of the leukocyte lineages, they are now recognized as elements of a universal language used by most cells of any other lineages. Accordingly, they are essential for many events through the life (reproductive tissue remodeling, embryogenesis, steady state and adaptive hematopoiesis, surveillance and maintenance of the tissue structure, functions by the immune system, cell survival, and cell death, etc). They also allow a dialogue with the central nervous system and some cytokines may modify different behaviors (fever, anorexia, sleep, etc). They usually act within their vicinity, but they can also act via an endocrine fashion. Their productions are tightly controlled within a complex network of positive and negative loops. They are a prerequisite to control infection by invasive microorganisms, but their exacerbated production can be deleterious at the local or systemic levels. Their half angel–half devil aspect has rendered them difficult to use with therapeutic purposes, although some recombinant cytokines or some cytokine-neutralizing strategies have been successfully used in different pathological conditions.

1
Historical Record

The history of cytokines can be divided in four main steps through the years. The first step corresponds to the identification of biological activities of factors present in cell culture supernatants or within the blood stream (late 40s–early 70s). During the second step (late 70s–early 80s), the purification and the biochemical characterization of these factors allowed to reach the stage of defined molecular entities. During the third step (mid-80s–mid-90s), the cloning of the coding and noncoding sequences, the production of recombinant cytokines, as well as the production of genetically manipulated mice extended the understanding of the complexity of the cytokine-mediated language. Since the late 90s (the fourth step), the discovery of new cytokines has been relying on the availability of different genome sequences and the different gene banks. In the later case, bioactivities of newly identified cytokines remained to be discovered. Probably, the first reported bioactivity was the induction of fever and the discovery of an endogenous pyrogen in the late 40s, early 50s. The purification of human endogenous pyrogen was published in 1977 by Charles Dinarello. This allowed the identification of endogenous pyrogen as interleukin-1 (IL-1). It definitively established that fever was due to endogenous factor(s) and not directly induced by microbial products. After the cloning of IL-1 in 1984, this was confirmed the following year, when injection of recombinant IL-1 was shown to induce fever in rabbits. Later, genes encoding different IL-1 homologues have been identified by means of their significant sequence similarities to IL-1, but further studies are required to

know whether some members of the IL-1 family (IL-1F5–10) are pyrogenic. The other earlier-described bioactivity was reported in 1957 by Isaacs and Lindenmann. They showed that resistance to virus development could be conferred to a cell by another virus-infected cell. This bioactivity interfering with virus spreading was named "interferon." In the 60s, numerous bioactivities had been detected in supernatants of lymphocytes. Accordingly, identified as nonantibody mediators, D. Dumonde et al. proposed in 1969, the word "lymphokines." In the following years, the word "monokine" was employed to characterize other bioactivities found in monocyte/macrophage supernatants. In the 70s, the most famous lymphokine was the "macrophage migration inhibitory factor" (MIF). Owing its name to its *in vitro* property, MIF had been first identified in 1966 as a T-lymphocyte-derived product associated with delayed-type hypersensitivity. In the absence of cell lines producing high amounts of MIF, its biochemical characterization was delayed. The cloning of this molecule occurred in 1989, long after its discovery. In the meantime, MIF-like activity was reported in cultures of fibroblasts, and S. Cohen et al. found a MIF-like activity in the supernatants of virus-loaded cell lines. Then, it appeared evident that lymphokines were incorrectly named because many cell types other than lymphocytes could produce similar bioactivities. And, in 1974, S. Cohen coined the name "cytokine." Indeed, MIF is a good example of a cytokine produced by many cell types, including macrophages, dendritic leukocytes, endothelial cells, epithelial cells, and fibroblasts. Most fascinating was the rediscovery of MIF as a product of the pituitary gland, further reinforcing the concept of cytokine and illustrating that, sometimes, the border between cytokines and hormones may be difficult to draw.

Names of factors were traditionally reflecting their various biologic activities identified in different laboratories. This approach resulted in an imprecise and redundant nomenclature. In 1979, during the 2nd international Lymphokine Workshop in Ermatingen (Switzerland), it was decided that the name "interleukin" (IL-) be given to two different characterized factors. Under the new term IL-1 were gathered numerous names or acronyms defined according to one of the properties of this cytokine: osteoclast activating factor (OAF), lymphocyte activating factor (LAF), B-cell activating factor (BAF), epidermal thymocyte activating factor (ETAF), T-cell replacing factor III (TRFIII), leukocyte endogenous mediator, endogenous pyrogen, mitogenic protein, catabolin, hemopoietin-1, and so on. The name interleukin (between leukocytes) was obviously not the best choice since these factors were not restricted to leukocytes, but it had been mainly chosen by immunologists too self-absorbed, who were considering the immune system working independently of the other systems. When interviewed about it, Stanley Cohen specified: "It is a common misconception that the term interleukin was accepted for adoption by the nomenclature committee. When it was proposed, both Byron Waksman and I, as Co-chairs, felt it would turn out to be too restrictive (even then) and there was general agreement by the committee that it would be premature to refer to mediators as interleukins. This view was conveyed to the proposers. However, shortly thereafter, an editorial appeared in Journal of Immunology stating that the term 'interleukin' had been discussed and considered by the committee (true) and that henceforth the term would be used in scientific communications by the scientists writing the editorial. This was a true-true-unrelated kind of position to take, but it caught on in the general scientific community."

The last neologism, "chemokine," was created in 1992 to gather under one name a large family of small biochemically related cytokines, all possessing chemoattractant properties.

The availability of the first cloned molecules allowed major progresses, not only in fundamental research but also for clinical applications. While growth hormone was the first recombinant molecule used in humans in 1980, interferon-alpha was the first cytokine injected in humans in 1981, and IL-2 the second in 1985. Indeed, in 1984, 10 000 L of supernatants of activated Jurkat cell were required to prepare 30 mg of natural IL-2; the following year, 10 L of supernatants of recombinant *Escherichia coli* expressing the *IL-2* gene were sufficient to prepare 1 g of recombinant IL-2.

2
A Universal Language of Cells

2.1
Definitions

Cytokines have been defined by immunologists as cooperative factors produced by activated lymphocytes, dendritic leukocytes, macrophages that induce, modulate, favor, increase, inhibit cell proliferation, cell differentiation, or cell activities such as antibody production. The onset of the messages delivered by these soluble (or membrane-bound) molecules relies on the binding to their specific receptors displayed on the surface of the immune cells, a process leading to an intracellular signaling cascade. However, far from being limited to the leukocyte lineages, this is a universal language of cells of the whole body, independently of their lineages, their nature, and their localization. This is probably the main difference between cytokines and other soluble molecules such as neuromediators mainly produced by the brain and the neurons, or hormones mainly produced by the endocrine system (Table 1).

Like hormones, cytokines can act in an endocrine fashion, but their actions mainly occur within the cellular microenvironment in a paracrine fashion. However, since some cytokines can be expressed on the cell surface of the producer cell, cytokines can act in a juxtacrine fashion within a close contact between cells. Finally, cytokines can act as an autocrine factor for a cell that both produces a cytokine and displays its receptor (Fig. 1).

Cell interactions involve more than one cytokine. One can consider each cytokine as a word, and the full sentence that will finally emerge is the reflection of the delivery of different cytokine-mediated signals. Together, these signalings lead to orientate the activities of the target cell. Thus, the order of delivery of cytokines has major consequences on the nature of the message. Furthermore, the concentration of cytokines, their localization, the nature of the target cell, and the timing are important parameters that affect the exact nature of the cytokine-initiated process.

Tab. 1 Comparison between hormones and cytokines.

	Sources	Targets	Activities	Action
Hormones	Secreted by a unique cell lineage	Specificity rather limited to one single type of target cell	Single action	Endocrine
Cytokines	Produced by many cell types	Numerous target cells	Wide spectrum of activities (redundancy)	Paracrine Juxtacrine Autocrine Endocrine

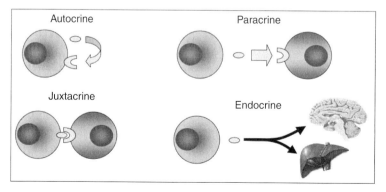

Fig. 1 Mode of action of cytokines.

In most cases, constitutive production of cytokines is limited and most cytokines are produced upon activation of cells. However, highly sensitive readout assays have allowed to determine a constitutive production at homeostasis in numerous tissues and cells.

2.2
Families

The classification of cytokines can be achieved according to (1) a common bioactivity: antiviral activity (interferons), chemoattraction (chemokines), hematopoiesis (colony-stimulating factors, CSF); (2) a common biochemical structure, suggesting a common ancestral molecule (chemokines, tumor necrosis factor (TNF) family); or (3) the sharing of a common chain of a receptor (gp130 cytokine family) (Table 2). Different subfamilies of interferons are known. Twenty-six genes of IFNα have been identified, of which at least five are pseudo genes. Interferons-alpha are produced by leukocytes and interferon-beta is produced by fibroblasts. IFNγ produced during immune response is a potent macrophage activator. IFNω is induced by virus, IFNτ is produced by trophoblast, and IFNκ is expressed by keratinocytes. Among chemokines, four different subfamilies can be identified on the basis of the highly conserved presence of the first two cysteine residues, which are either separated or are not separated by other amino acids: the CCL, CXCL, CX3CL, and the CL

Tab. 2 Families of cytokines.

Interferons	IFNα, IFNβ, IFNγ, IFNδ, IFNκ, IFNτ, IFNλ
Interleukins[1]	$\boxed{\text{IL-1}\alpha, \text{IL-1}\beta, \text{IL-1Ra, IL-18, IL-1F5-10}}$ $\boxed{\text{IL-2, IL-15, IL-21}}$
	$\boxed{\text{IL-10, IL-19, IL-20, IL-22, IL-26}}$ $\boxed{\text{IL-12, IL-23, IL-27}}$
	$\boxed{\text{IL-17A-E, IL-25}}$ $\boxed{\text{IL-28, IL-29}}$ IL-3, IL-4, IL-5, IL-6, IL-7
	IL-8 (CXCL8), IL-9, IL-11, IL-13, IL-14, IL-16
Hematopoietic factors	M-CSF, G-CSF, GM-CSF, Stem cell factor
Chemokines	CCL1 ... CCL28 ; CXCL1 ... CXCL16 ; XCL1, XCL2 ; CX3CL1
TNF superfamily	TNF, Ltα, Ltβ, NGF, CD27L, CD30L, CD40L, CD137L, APRIL, BAFF, FasL, GITRL, LIGHT, OPGL TRAIL, RANKL, TWEAK, VEGI.
Growth factors	TGF$\beta_{1,2,3}$
Gp130 cytokine family	IL-6, IL-11, CNTF, LIF, Oncostatin-M, Cardiotrophin-1

[1] Biochemically related interleukins are framed.
APRIL: apoptosis-inducing ligand; BAFF: B cell-activating factor; CCL: chemokine CC ligand; CNTF: ciliary neurotrophic factor; CSF: colony-stimulating factor; CXCL: chemokine CXC ligand; GITRL: glucocorticoid-induced TNF receptor family related gene ligand; IL-1F: IL-1 family; LIF: leukemia inhibitory factor; LIGHT : lymphotoxins, inducible expression, competes with HSVglycoprotein D for HVEM, a receptor expressed on T-lymphocyte; Lt: lymphotoxin; OPGL: osteoprotegerin ligand; RANKL: receptor activator of NF-kappaB ligand; TGFβ : transforming growth factor-β; TRAIL: TNF-related apoptosis-inducing ligand; TWEAK: TNF-like weak inducer of apoptosis; VEGI: vascular endothelial cell growth inhibitor.

chemokines. The assigned numbers of interleukins only reflect the timing of their discovery. Thus, some interleukins are also hematopoietic factors (e.g. IL-3 previously named multi-CSF), chemokines (IL-8, also named CXCL8), or even interferon (IL-28 and IL-29 are identical to IFNλ). The first interleukins have been discovered as bioactive factors, and there are no similarities in their amino acid sequences. However, there are some homologies in tertiary and quaternary structures and similar helical structures are found. The more recently discovered interleukins have been identified in gene banks according to sequence homology to previously identified interleukins. The TNF ligand superfamily gathers 18 distinct members that include soluble molecules with biochemical homology as well as membrane-bound ones. There are few other cytokines that can be displayed at the plasma membrane of the producing cells: IL-1α, IL-10, IL-15, and IFNγ. The TNF ligand superfamily also includes a growth factor (i.e. nerve growth factor). Usually, growth factors have a rather narrow spectrum of activity, are produced constitutively, are not acting on hematopoietic cells, and thus are distinct from cytokines. One exception is the transforming growth factor-β family (TGFβ), particularly because at least one member, namely TGFβ1, displays anti-inflammatory properties.

Localization of genes on chromosomes is known, and often genes of a family are gathered on the same chromosome (e.g. IL-1α, IL-1β, IL-1 receptor antagonist on human chromosome 2; most CXCL chemokines on human chromosome 4; most CCL chemokines on human chromosome 17). Most cytokines are glycosylated, their molecular weight ranging between 7 and 30 kDa. Most cytokines are monomers (e.g. IL-1, IL-2, IL-3, IL-4. . .), and few are

homodimers (e.g. IL-8) or homotrimers (e.g. TNF). Some are heterodimers like IL-12, IL-23, and IL-27: IL-12 and IL-23 sharing the same p40 chain. Some are heterotrimers like lymphotoxin-β (Ltβ) that is the association of Ltα and Ltβ chains.

3
Receptors

3.1
Families

Most cytokine receptors had been characterized once their ligands had been identified, but there were few receptors, called *orphan receptors*, that were known before their ligands (e.g. c-kit, the receptor of stem cell factor; CD40, the receptor of CD40L; IL-1 receptor-related protein-1, part of the IL-18 receptor, etc). In most cases, receptors are the association of more than one transmembrane chain. Usually, one chain is essential for the binding (α-chain), while the second chain is required to initiate the intracellular signaling (β- or γ-chain). Some chains are common for receptors of few cytokines; this is the case of the gp130 chain of the receptors of IL-6, IL-11, CNTF, LIF, Oncostatin-M, and Cardiotrophin-1, or the gamma chain of IL-2, IL-4, IL-7, IL-9, IL-15, and IL-21. Cytokine receptors are characterized by domains derived from ancestral common genes. In terms of evolution, it would be interesting to decipher which mutations in genes coding for cytokines or for cytokine receptors were first fixed. A common domain, called *hemopoietin receptor superfamily domain* characterized by the presence of four cysteine residues and a conserved sequence tryptophan-serine-x-tryptophan-serine, defines a family of numerous cytokine receptors (Fig. 2).

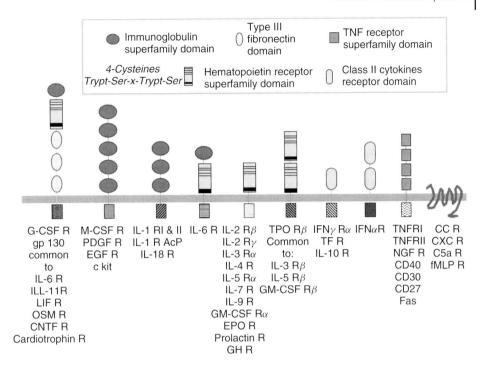

Fig. 2 Families of cytokine receptors.

Most interestingly, this family of receptors also includes receptors for hormones (e.g. growth hormone, prolactin) or the hematopoietic factor (e.g. erythropoietin). This domain can be duplicated (the common β chain of IL-3R, IL-5R, and GM-CSFR) or be associated with other domains like an immunoglobulin-like domain (IL-6Rα) or a fibronectin type III–like domain (gp130, G-CSFR). Another important family is the TNF receptor family. The activation by TNF is consecutive to the bridging of similar chains of the TNF receptor by the cytokine. Each cytokine receptor is highly specific for its ligand, except the TNFR p55 and p75 that bind both TNF and Ltα, and the chemokine receptors that can bind various chemokines. Chemokine receptors, constituted by a 7-transmembrane chain, are similar to receptors for other chemoattractant molecules such as the anaphylatoxin C5a or the bacterial fMet-Leu-Phe peptide. Cytokine receptors are often internalized after ligation to their ligands, and their expression can be induced, down- or upregulated.

3.2
Signaling

Upon binding of the cytokine to the extracellular domain of receptors, the intracellular receptor domains become associated with a variety of signaling molecules. A cascade of phosphorylation of various adaptor proteins occurs following interaction of the intracellular domain of the receptors with cytoplasmic tyrosine kinases (e.g. Janus kinases, Jak). Phosphorylation of latent cytoplasmic transcriptional activators (e.g.

signal transducers and activators of transcription, STAT) allows their dimerization and their translocation into the nucleus where they bind to specific sequences present in promoters of certain genes. Other pathways involving the mitogen-activated protein kinase (MAPK) and other kinases lead to the activation of various transcription factors (e.g. c-fos/c-jun, NF-κB, etc). Some signaling pathways are shared with other receptors. This is the case of the IL-1 and IL-18 receptors of which the intracellular domain is similar to the toll-like receptors (Toll–IL-1 Receptor, TIR domain). Thus, similar adaptors (e.g. MyD88, IRAK, etc) are involved after interaction of the respective receptors with their ligands, either members of the IL-1 family or pathogen-associated molecular patterns (PAMPs) such as endotoxin (lipopolysaccharide). Intracellular domain of some receptors (TNFR p55; Fas; TRAIL R) possesses a death domain that initiates a specific signaling cascade, leading to apoptosis.

3.3
Soluble Receptors

Receptors are integral part of the cytokine network (*cf* Sect. 7): not only can their membrane expression be modulated but they also exist as soluble molecules. Indeed, most cytokine receptors, except the chemokine receptors, can be shed from the cell surface after enzymatic cleavage following or not following a neosynthesis. As shown in Fig. 3, depending upon the nature of the receptor, its soluble form can behave as an inhibitor (e.g. soluble IL-1R, soluble TNFR) and can prevent the cytokine from reaching the membrane form. It can also be a carrier, protecting the cytokine within a complex, and increasing

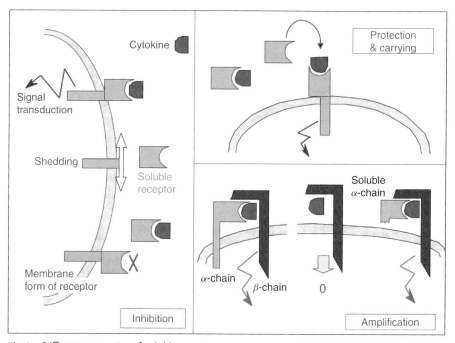

Fig. 3 Different properties of soluble receptors.

its half-life in the plasma or other biological fluids. Then, the dissociation of the cytokine-soluble receptor complex can allow the release of the cytokine close to a membrane receptor (e.g. TNFR). Lastly, in the case of a receptor constituted by one α-chain that binds the cytokine and one β-chain that recognizes the bound cytokine, the soluble α-chain receptor can enhance the responsiveness to the cytokine or can even allow the responsiveness of cells missing the α-chain, but expressing the β-chain. One example of the latter case is provided by the soluble IL-6R that, together with IL-6, controls the switch of leukocyte recruitment during inflammation.

4
Functions

4.1
Hematopoiesis

Production of leukocytes within the bone marrow is a very active and efficient process (4×10^8 white cells/leukocytes per hour in humans). Numerous cytokines, expressed at homeostasis in human bone marrow contribute to hematopoiesis, while others (like GM-CSF) are only active during infection to further increase the renewal of available leukocytes. Maturation, proliferation, and differentiation of bone marrow progenitor cells are under the control of different cytokines acting concomitantly. For example, IL-1 alone is unable to allow the differentiation of bone marrow precursors cells, but it acts in synergy with other hematopoietic cytokines. IL-3 and stem cell factor (also named *c-kit ligand*) are cytokines with a wide spectrum of activity, acting on pluripotent hematopoietic stem cells. GM-CSF favors the differentiation of myeloid

progenitors cells, while M-CSF or G-CSF are specific of monocyte/macrophage and granulocyte lineages respectively. IL-6, IL-11, and mainly thrombopoietin favor thrombocytopoiesis, leading to megakaryocyte development and platelet production. Interleukin-5 has been identified as a major regulator of eosinophil development and function. Interleukin-7 is the main cytokine of lymphopoiesis required for the development of B, alpha-beta, and gamma-delta T-lymphocytes. Most of these hematopoietic cytokines also act as activators of mature cells and as antiapoptotic ligands.

4.2
Immune Response

4.2.1 Innate Immunity
Innate immunity is characterized by a response localized within the site of infection (Fig. 4). Epithelial cells and leukocytes residing in the extravascular space of the tissue (mast cells, resident macrophages, and immature dendritic leukocytes), and endothelial cells of the microvessels are triggered by microorganisms and microbial-derived products and release inflammatory cytokines (e.g. IL-1, TNF, IL-12, IL-18, etc). These cytokines further activate the endothelial cells of the postcapillary venules, leading to the increased adherence of circulating leukocytes and to coagulation process. Other cytokines, such as IFNγ, are produced within an amplificatory loop. Local production of chemokines contributes to the recruitment of new leukocytes from the blood compartment to further prevent the infectious process. Cytokines contribute to enhance the antimicrobial activities of newly recruited neutrophils and monocytes/macrophages (e.g. production of free radicals, enhanced phagocytosis), and the

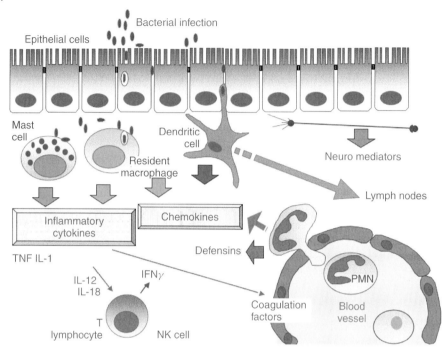

Fig. 4 Cytokines and soluble mediators as coordinators of regulated processes taking place in a bacteria-loaded sites.

antiviral activity of natural killer (NK) cells. Human leukocytes release defensins, a kind of natural antibiotics that contribute to limit bacterial dissemination. Neuromediators help control the inflammatory process. In addition to local responsiveness, a systemic response occurs, which includes fever, a reflection of the presence of pyrogenic cytokines within the plasma (e.g. IL-1, TNF, IL-6). Hyperthermia contributes to the anti-infectious process by enhancing some immune cell activities, and by reducing bacterial growth or viral replication. Finally, the production of hematopoietic cytokines (e.g. GM-CSF) further increases the number of leukocytes otherwise known to contribute to the clearance of the microorganisms and to the initiation of the healing process.

4.2.2 Adaptive Immunity

Adaptive immunity is characterized by processes relying on unique properties of the secondary lymphoid organs. Briefly, in secondary lymphoid organs such as the peripheral lymph nodes, professional antigen-presenting cells (dendritic leukocytes and mononuclear phagocytes), after processing of the native antigen, display on their membranes antigenic peptides and interact with T-lymphocytes, expressing a T-cell receptor able to specifically recognize these peptides. B-lymphocytes recognize native antigen. As a result of the cross talks between antigen-presenting cells, T- and B-lymphocytes, T-lymphocytes proliferate and differentiate as primed effector or memory T-cells, and B-lymphocytes proliferate and differentiate as antibody-secreting plasma cells. All these events are controlled by cytokines. For example, within the secondary lymphoid organs, IL-2 and IL-4 favor T-cell proliferation, IL-12 activates cytotoxic T-cells, IL-2 and IL-4 allow B-cell proliferation, IL-4, IL-5, IL-6, IL-13, and TGFβ influence plasma cell differentiation and favor different immunoglobulin classes synthesis. Within microorganisms-loaded tissues, the effector T-lymphocytes, once reactivated could release IFNγ and GM-CSF that activate professional phagocytes conferring them microbicidal functions.

4.2.3 Th1 and Th2 Cytokines

In 1986, Tim Mosmann et al. described two types of murine helper CD4$^+$ T-cell (Th) clones defined according to the profile of released cytokines. Th1 cells produce IL-2, IFNγ, and Ltα, and Th2 cells produce IL-4, IL-5, IL-6, IL-10, and IL-13. Both subpopulations produce IL-3 and GM-CSF. Their differentiation from a Th0 precursor occurs, depending upon the cytokine environment (Fig. 5). Th1 and Th2 cells control each other: IFNγ derived from Th1 neutralizes Th2 cells, and IL-4 and IL-10 inhibit Th1 cells. Th1 cells are involved in cellular immunity and activate macrophages, Th2 cells are involved in humoral immunity and activate

B-lymphocytes. Once stably differentiated as long-lived clones, some chemokine receptors allow to distinguish between these subpopulations: CXCR3 is mainly expressed on Th0 and Th1 cells, while CCR3 and CCR4 are expressed on Th2 cells. The Th1/Th2 dichotomy also exists in humans, as nicely illustrated by the early detection of Th2 cytokine mRNA expressed in response to an intradermal allergen challenge (immediate hypersensitivity) and the delayed detection of Th1 cytokine mRNA after tuberculin injection (delayed-type hypersensitivity). However, in most clinical settings, this dichotomy is not that obvious and mixed patterns are observed. More recently, other T-cell subsets with immunosuppressive properties have been characterized. Th3-type cells are a unique T-cell subset that primarily secretes TGFβ, provides help for IgA switch, and has suppressive properties for Th1 and other immune cell. Type-1 regulatory T-cells (Tr1) inhibit Th1 and Th2-type immune responses through the secretion of IL-10. They are also able to secrete TGFβ.

4.3
Cell Survival and Cell Death

Not only can cytokines be maturation and differentiation factors but they can

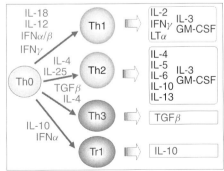

Fig. 5 T-cell subpopulations and the nature of environmental cytokines required for their differentiation.

also contribute to cell survival. For example, IL-2, IL-4, IL-6, IL-7, and IL-15 have all been shown to inhibit resting T-cell death *in vitro*. B-cell survival factors include IL-4 and B cell-activating factor (BAFF) of the TNF family. Plasma cell longevity depends on combination of IL-5, IL-6, stromal cell-derived factor-1α (SDF-1/CXCL12), and TNFα. IL-15 can sustain NK cell survival in the absence of serum. Indeed, *in vitro* serum starvation often leads to apoptosis that can be prevented by cytokines like TGFβ (e.g. survival of epithelial cell). Indeed, these properties reflect an antiapoptotic mechanism initiated by cytokines like the suppression of caspase activity or the maintenance or induction of Bcl-2 protein expression. RANKL suppresses apoptosis of primary cultured endothelial cells; SDF-1/CXCL12 enhances the survival of myeloid progenitor cells *in vitro*; stem cell factor (SCF) regulates the survival of cellular lineages by suppressing apoptosis. In addition to SCF, most hematopoietic factors allow survival of hematopoietic progenitors. In addition to sustain cell survival, cytokines can allow cell growth. For example, this is the case of IL-2 for T-cells, or IL-8 and some other chemokines that favor angiogenesis. Cytokines can also be growth factors for tumor cells; this is, for example, the case of IL-6 for malignant myeloma plasma cells. In contrast, certain cytokines can induce apoptosis of a variety of tumor cells and normal cells. Indeed, TNF was first identified in 1984 as a cytokine with antitumor effects, and other members of the TNF superfamily can induce proliferation, survival, as well as cell death: specifically, TNFα, FasL, TRAIL, and VEGI. The intracytoplasmic portion of their receptors contains a "death domain." Upon activation by their ligands, the death domain recruits intracytoplasmic adaptors expressing a death domain: TRAF (TNF receptor-associated factor) and FADD (Fas-associated death domain), which in turn recruit the proform of caspase-8. Autoactivation of caspase-8 leads to the subsequent activation of caspase-3, a proapoptotic enzyme.

4.4
Link with the Central and Peripheral Nervous System

A cross talk exists between immune leukocytes and the central nervous system. Proinflammatory cytokines, particularly IL-1 and TNF, induce fever, anorexia, and slow wave sleep. Part of their activities is conveyed by the vagal nerve. Fever involves local production of IL-6 and prostaglandins E2 (PGE2). Certain chemokines are also pyrogenic (e.g. IL-8, MIP-1α, MIP-1β, RANTES), but do not induce PGE2. IL-1 and TNF also induce a neuroendocrine loop in response to their signal, hypothalamus produces a corticotropin-releasing factor (CRF), and CRF induces the release of adrenocorticotropin hormone (ACTH) by pituitary gland. ACTH induces the release of glucocorticoid by adrenals. Glucocorticoids inhibit the production of most cytokines and antagonize many of their activities. Vagal nerve releases acetylcholine that represses inflammatory cytokine production. In addition, other neural and neuronal mediators, adrenaline, vasoactive intestinal peptide (VIP), pituitary adenylate cyclase-activating polypeptide (PACAP) are inhibitors of IL-1 and TNF production, while noradrenaline and substance P are potentiators.

4.5
Reproduction

There is compelling evidence that cytokines are involved in spermatogenesis, ovogenesis, and gestation. Cytokines play an important regulatory role in the development and normal function of the testis. They are produced by Leydig cells, Sertoli cells, and germinal cells. Proinflammatory cytokines, including IL-1 and IL-6, have direct effects on spermatogenic cell differentiation and testicular steroidogenesis. Stem cell factor and leukemia inhibitory factor, cytokines normally involved in haematopoiesis, also play a role in spermatogenesis. Anti-inflammatory cytokines of the TGFβ family are involved in testicular development. TNFα, a secretory product of round spermatids, increases endogenous androgen receptors expression in primary cultures of Sertoli cells. Given the requirement of testosterone for spermatogenesis and the importance of androgen receptors in mediating Sertoli cell responsiveness to testosterone, the stimulation of androgen receptors expression by TNFα may represent an important regulatory mechanism required to maintain efficient spermatogenesis. M-CSF is the principal growth factor regulating macrophage populations in the testis, male accessory glands, ovary, and uterus. Both male and female M-CSF deficient (op/op) mice have fertility defects. Males have low spermatozoid number and libido as a consequence of dramatically reduced circulating testosterone. Females have extended estrous cycles and poor ovulation rates.

Normal ovarian tissue is rich in cytokines. Cytokines and chemokines are important in the physiology of ovarian function and of ovulation. Actual rupture of the follicle during ovulation may be dependent on tissue remodeling that shares some features with an acute inflammatory process. TNF and IL-1 are implicated in ovarian follicular development and atresia, ovulation, steroidogenesis, and corpus luteum function (including formation, development, and regression).

Chemokines such as MCP-1 are also involved in luteolysis.

A great number of cytokines are produced by endometrium and placenta. Cytokines released at the fetomaternal interface play an important role in regulating embryo survival, controlling not only the maternal immune system but also angiogenesis and vascular remodeling. For example, LIF plays a role in the embryo pre-, peri-, and postimplantation, as does IFNτ in bovine and ovine. IFNτ is produced constitutively by embryonic trophectoderm during the period immediately prior to implantation. It acts on uterine epithelium to suppress transcription of the genes for estrogen receptor and oxytocin receptor. It blocks development of the uterine luteolytic mechanism. TGFβ is involved in regulating placental development and functions. The abortive influences of TNFα and IFNγ may terminate pregnancy during infection of uteroplacental unit. The slight dominance of proinflammatory cytokines in the fetal membranes and decidua suggest that inflammatory processes occur modestly with term labor, but much more robustly in preterm delivery, particularly in the presence of intrauterine infection.

4.6
Inflammation

"*Notae vero inflammationis sunt quatuor: rubor and tumor cum calore et dolore*" (Celsus, first century B.C.). Redness, swelling, heat, and pain are the four main parameters that characterize inflammation. They reflect the action of numerous inflammatory mediators and proinflammatory events that are orchestrated by inflammatory cytokines, mainly IL-1 and TNF, but also IL-12, IL-18, and IFNγ (Fig. 6). These cytokines are produced following

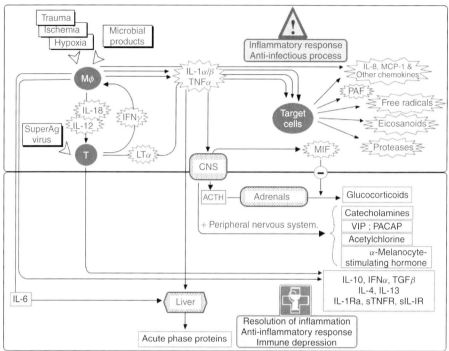

Fig. 6 Inflammation is the consequence of a cascade of events initiated by IL-1 and/or TNF. The action of these cytokines on various target cells leads to the release of numerous inflammatory mediators. Anti-inflammatory cytokines, the activation of the neuroendocrine pathway and the effects of glucocorticoids as well as the enhanced production of acute-phase proteins control negatively the inflammatory process (ACTH: adrenocorticotropic hormone; CNS: central nervous system; PACAP: pituitary adenylate cyclase-activating polypeptide; PAF: platelet activating factor; VIP: vasoactive intestinal peptide).

tissue or systemic steady state disruption like tissue loading by pathogens, ischemia, hypoxia, or trauma. Acting on target cells, they induce a cascade of mediators, including other cytokines and chemokines, proteases, lipid mediators (e.g. prostaglandins, platelet activating factor), free radicals (e.g. superoxide anion, nitric oxide) that contribute to the inflammatory process and tissue injury. Acting on endothelial cells, especially those of the postcapillary venules, they increase vascular permeability and plasma transudation, they increase circulating leukocyte adherence and margination, and they initiate the coagulation process by inducing tissue factor expression. Chemokines contribute to the recruitment of leukocytes that further maintain the inflammatory process. Some cytokines enhance the production of IL-1 and TNF (e.g. IFNγ, GM-CSF, IL-3), while others inhibit it (e.g. IL-4, IL-10, IL-13, TGFβ, IFNα). The latter contribute to attenuate the inflammatory process and to allow the tissue-repair process to occur. Other anti-inflammatory mediators include soluble IL-1 and TNF receptors, IL-1 receptor antagonist (IL-1Ra), some neuropeptides (e.g. adrenaline, acetylcholine), glucocorticoids induced after activation of the neuroendocrine loop by IL-1 and TNF, and the acute-phase proteins produced by the hepatocytes in response to IL-1, TNF, and particularly IL-6. Acute-phase proteins contribute to eliminate debris due to cell lysis, to favor phagocytosis, to neutralize toxic mediators, and to inhibit proteases. Recently, IL-1Ra was recognized as one of the acute-phase proteins. In vivo, in the endotoxin shock model, most of these molecules have been shown to be protective.

5
Life without Cytokines (What Knockout Mice Tell Us)

Deletion of genes encoding for a cytokine or a cytokine receptor has led to the generation of many different knockout (KO) mice that further allowed to decipher the contribution of cytokines to different physiological events. In most cases, viable, fertile, and clinically healthy mice could be obtained. In few cases, the deletion of the genes led to death *in utero*, illustrating a major role of certain cytokines during embryogenesis. This was the case for SDF1 (CXCL12) and its receptor CXCR4, LIF, Cardiotrophin-1, and the gp130 chain of their receptors (see Table 3). Deletions of certain signaling molecules (e.g. STAT3) are also lethal during embryonic life. Other deletions reveal the role of certain cytokines for organogenesis during embryonic life. This is the case of lymphotoxin-α, lymphotoxin-β, and LtβR for the development of lymph nodes and Peyer's patches, and RANKL for the development of lymph nodes and mammary glands.

Tab. 3 Cytokine or cytokine receptor KO mice reveal relative contribution of cytokines to life processes.

Biological events	Deleted gene	Consequences
Embryogenesis	LIF	Absence of LIF in female prevents implantation of blastocytes
	Cardiotrophin-1 & gp130 receptor chain	Heart development blocked; death *in utero* on day 16
	SDF-1 & CXCR4	Defects in heart and brain development, intestinal vascularization, and B-cell hematopoiesis. Death *in utero*
Organogenesis	Ltα, Ltβ, LtβR	Absence of lymph nodes and Peyer's patches
	TNFα	Altered spleen and lymph node organization
	RANKL	Absence of lymph nodes, mammary gland defects
	IL-8R	Splenomegaly, lymphoadenopathy, increased circulating neutrophils
Hematopoiesis	G-CSF	Chronic neutropenia
	Stem cell factor	Absence of mast cells
	M-CSF	Altered function of monocytes and osteoclast
	IL-7	Reduced number of lymphocytes

(*continued overleaf*)

Tab. 3 *(Continued)*

Biological events	Deleted gene	Consequences
Leukocyte differentiation	Ltα, Ltβ IL-15 γ-chain	Absence of follicular dendritic cells Absence of NK and NK-T cells Impaired B, T and NK cell development
Osteoclastogenesis	RANK & RANKL	Defect in the differentiation of hematopoietic osteoclast progenitor
Pulmonary homeostasis	GM-CSF & β-chain of GM-CSF R	Accumulation of surfactant lipids and proteins in the alveolar space. Lymphoid hyperplasia.
Gut homeostasis	IL-10	Intestinal mucosal hyperplasia, inflammatory cell infiltration, MHC Class II expression on colonic epithelium.
	IL-2	Colitis
	IL-15	Reduced intestinal epithelial B and T lymphocytes
Fever	IL-6	No pyrogenic response
Acute-phase response	IL-6	Profound alteration of acute-phase proteins production
Inflammation	TGFβ IL-1-Ra	Systemic lethal inflammation (day 24) Chronic arthropathy, spontaneous dermatosis, exacerbated delayed-type hypersensitivity
	IL-5	Reduced eosinophilia
Antibody production	IL-4, IL-6, IL-13, CD40L	Altered antibody production
Immune function	IL-2, IL-4, IL-6, IL-12, IL-13, IL-15, IL-18, TNF, IFNγ, Stem cell factor, CSF, Chemokines, and their receptors.	Altered anti-infectious response

Organization of hematopoietic compartments are under the control of numerous cytokines and absence of TNFα, or the γ-chain of the receptor of IL-3, -5, -7, -9, -15, and -21 have profound perturbations on tissue organization and leukocyte differentiation. This is also true for the deletion of chemokines that contribute to tissue colonization at homeostasis. While studies of KO mice often comforted the knowledge, there were some surprises. One of them was the discovery that the absence of the hematopoietic cytokine GM-CSF did not alter hematopoiesis (mice had normal numbers of peripheral leukocytes, bone marrow progenitors, and tissue hematopoietic populations), but led to a profound alteration of the lung status, demonstrating a role of GM-CSF in pulmonary homeostasis. Of major interest was the demonstration that the deletion of anti-inflammatory cytokines (IL-10, TGFβ, IL-1Ra) was deleterious at homeostasis in the absence of experimental exogenously delivered inflammatory signals, establishing their crucial role to prevent potential ongoing inflammatory processes at homeostasis. As expected, numerous deletions had major effects on the quality of the anti-infectious process. For example, in the absence of IL-6, mice produce lower levels of IgG antibodies against stomatitis vesicular virus, display a lower cytotoxic T-cell activity during vaccinia virus infection, leading to an increased number of virus in lungs, and failed to control *Listeria monocytogenes* load in the tissues they reached. In contrast, the same deficiencies may appear beneficial in certain experimental models of septic shock in response to endotoxin (e.g. TNFα, TNFR p55, IL-1β converting enzyme).

In humans, natural mutations occur that may affect cytokines, signaling molecules of receptors. In the latter case, the most famous mutation leads to the deletion of the γ-chain of the receptor of IL-3, -5, -7, -9, -15, and -21 and is associated with a severe combined immunodeficiency. This is for this clinical setting that occurred in 2000, the first successful gene therapy.

6
Cytokine Synthesis

6.1
Homeostasis

Owing to highly sensitive techniques (RT-PCR, *in situ* hybridization, ELISpot), it has been possible to show the presence of cytokine mRNA in various types of cells or to demonstrate the presence of cytokine-producing cells in the absence of activation. For example, IL-6 is spontaneously produced by 0.5% of bone marrow cells, 0.1% of spleen cells, and 0.01% of mesenteric lymph node cells. IL-6 is also produced in the absence of exogenous stimuli by enterocytes, eosinophils, neutrophils, epidermal cells, smooth muscle cells, bone marrow stromal fibroblasts, anterior pituitary cells, trophoblast cells, and so on. Furthermore, at homeostasis, certain cells contain cytokines. This is particularly the case for keratinocytes that contain large amounts of IL-1α. Mast cells contain a large panel of preformed cytokines (e.g. IL-1, IL-4, IL-5, IL-6, IL-13, TNF, etc), and MIF is preformed in numerous leukocytes. Furthermore, the presence of cytokines in biological fluids has been reported in absence of any infection or inflammatory diseases: IL-1α and IL-8 are found in sweat; IL-1α, IL-1β, IL-6, IL-8, GM-CSF, and TGFβ have been reported in tears; IL-2, IL-8, TNFα, TGFβ, and soluble TNFR are present in saliva; IL-2, IL-6, IL-8, IL-10, IL-12, TNF, and soluble IL-2R and IL-6R have been detected in seminal fluid; and a great number of different cytokines have been identified in human colostrum. In the later case, all of these cytokines probably act on the oropharyngeal and gut-associated lymphoid tissue of the newborn and favor the development and maturation of the immune system and may protect the newborn against invasive microorganisms delivered by the oral route. Finally, certain chemokines are present in tissues where they contribute to leukocyte recruitment at homeostasis (e.g. CCL17 & CCL25 in thymus, CCL21 & CXCL13 in lymph nodes, and Peyer's patches, CXCL12 in numerous tissues).

6.2
Induction

The production of cytokines by professional antigen-presenting leukocytes and activated T-cells occurs during the cellular cooperation required to initiate the adaptive immunity. This production is representative of the dialogue between immune leukocytes. In contrast, innate immunity is associated with the production of cytokines in response to exogenous activators, mainly the "pathogen associated molecular patterns" (PAMPs). Following their interaction with the "pattern recognition receptors" (PRR), various intracellular signaling cascades lead to the synthesis and the release of cytokines. The best-known PRRs are the toll-like receptors (TLR) that share with IL-1R and IL-18R a homologous Toll-IL-1R (TIR) domain. Ten different TLRs have been identified so far in humans that recognize bacterial, viral, parasitic, or fungal conserved structures expressed on the surface or within the microbes (e.g. DNA, single- or double-strand RNA). Endotoxin of Gram-negative bacteria (lipopolysaccharide, LPS) is among the most potent PAMPs and picograms of this molecule are sufficient to induce the production of a whole panel of cytokines by macrophages. LPS is trapped by the CD14 molecule on the cell surface and triggers the cell via a complex MD2/TLR4. In the case of Gram-positive bacteria, some of these bacteria release exotoxins, also known as superantigens that trigger the production of cytokines by both T-lymphocytes and macrophages.

6.3
Parameters that Affect Production

In addition to the genetic polymorphism (see Sect. 8), there are numerous parameters that influence the level of the production of cytokines. Of course, *in vitro* experiments require careful sampling of the cells. The use of medium and

agents free of endotoxin is very important. There are probably numerous wrong statements that suggest the capacity of a given molecule to induce the production of cytokines that are consecutive to their contamination by endotoxin, as it has been shown for heat shock proteins. *Ex vivo* studies of cytokine production by human cells is influenced by age, nutrition status, the use of drugs, alcohol, and smoking habit. Seasonal and circadian rhythms, physical exercise, altitude exposure, microgravity, social and psychological stress, physical stress, and gender affect cytokine production. In the later cases, neuromediators and sexual hormones modify cellular production of cytokines.

6.4
Measurements

The measurement of cytokines in humans can be performed in natural biological fluids, in induced biological fluids (e.g. broncho-alveolar or peritoneal lavages), on tissue biopsies, and with blood leukocytes. The detection of cytokines in biological fluids may represent the "tip of the iceberg," and can be detected in the case of exacerbated production since, once produced, cytokines are efficiently trapped by their specific receptors on environmental cells. Cytokine mRNA analysis can be achieved by Northern blots, *in situ* hybridization or reverse transcriptase polymerase chain reaction (RT-PCR). Cytokine-producing cells can be monitored by ELISpot or by flow cytometry. Biological assays are not performed anymore and, nowadays, the measurement of cytokines is mainly achieved by enzyme-linked immunosorbent assays (ELISA). Cytokines constitute a tightly regulated network (see below). Thus, information concerning cytokines should be achieved through the analysis of simultaneous quantification of several cytokines. Indeed, it is the cytokine milieu that influences the cellular response rather than the action of a single cytokine. Techniques recently available such as microarrays combined to real-time PCR or multiplex immunological detection of cytokines might give precious information in a near future.

7
The Cytokine Network

As illustrated in Fig. 7, the interactions between cytokine-producing cells and target cells lead to define a cytokine network. Once produced in the context of a specific immune response or following microbial activation, cytokines act on target cells and induce synthesis of new cytokines. The induction by IL-1 of the release of IL-2 by lymphocytes, IL-6 by fibroblasts, G-CSF by macrophages, and IL-8 by endothelial cells are examples of cytokine cascades. Cytokines can induce their own synthesis,

like IL-1, or further enhance the productions, leading to amplificatory loops, like IFNγ. Another key word to define this network is synergy. Owing to the great redundancy of cytokines, two cytokines can share similar activities. When acting together on the same target cell, the effect will be far greater than the only additive effect of each individual cytokine. For example, the production of complement factor C3 by endothelial cells in response to TNF and IL-1 is far higher than that induced by these cytokines alone. The synergy may be the consequence of the induction of cytokine receptors, allowing cells to respond to a second cytokine they would not do normally because they lack the specific receptor. The network also involves negative loops as a consequence of the action of cytokines, blocking the production or the effects of others. IL-10 is an example of a cytokine blocking the production and the action of other cytokines. However, the lineage of its target cell may lead to opposite observations. For example, IL-10 represses LPS-induced IL-8

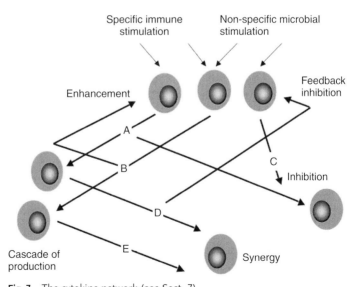

Fig. 7 The cytokine network (see Sect. 7).

production by monocytes, but enhances that produced by endothelial cells.

8
Individual Heterogeneity

Among genetic predisposition to diseases, premature death due to uncontrolled invasion and growth by microorganisms has the higher relative risk linked to heritability. The individual heterogeneity reflects in part genetic polymorphisms of cytokines. Three levels can be dissociated (Fig. 8) as follows:

1. A genetic polymorphism exists for the receptors of PAMPs. These sensors are the first essential elements in initiating an intracellular signaling cascade that leads to cytokine production. Certain single nucleotide polymorphisms (SNP) or mutations can directly modify the intensity of the responsiveness and influence disease susceptibility, as recently shown by the existence of rare mutations of TLR4 among patients with meningococcal infection.
2. SNP or mutations can affect cytokine genes, particularly when expressed within gene promoters or other regulatory sequences. Accordingly, high,

intermediate, or low producers are reported when assessing the levels of cytokines produced in response to an activator. Correlation between TNF genotypes and levels of released TNF in response to LPS has been shown.

3. Finally, depending upon the donors, target cells can react intensively, moderately, or weakly to a given amount of a cytokine. This has been nicely demonstrated with genetically distinct endothelial cells, which express various levels of adhesion molecules in response to similar amounts of IL-1 or TNF. In all these situations, the genetic polymorphism could also reflect SNP or mutations among the genes of the adaptors and signaling molecules involved in the signaling cascades.

As illustrated in Table 4, numerous cytokine gene polymorphisms have been associated with the occurrence or severity of diseases. The reported SNP or mutations can be associated with a protection against the disease or in contrast with a higher susceptibility. More recently, similar gene polymorphisms have been associated with the efficiency or the lack of efficiency of certain therapeutic approaches, particularly when the treatment targets cytokines.

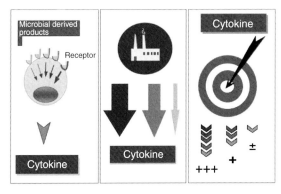

Fig. 8 Individual heterogeneity for cytokine production and responsiveness.

Tab. 4 Some examples of cytokine or cytokine receptor gene polymorphisms associated with diseases.

Disease	Gene polymorphism
Acute renal failure mortality	TNF, IL-10
Allergy	RANTES promoter; MIF promoter; IL-13 promoter
	IL-16 promoter; IL-1 gene complex; IL-18; IL-12B promoter
Asthma	IL4R α-chain; IL-10; IL-15; IL-18; TGFβ1 promoter; Eotaxin
Alzheimer	IL-1α, TNFR2, IL-6
Breast cancer	IL-6
Chronic periodontitis	IL-10 promoter,
	IL-1B +3953 and TNF-A-308 allele 2 positive
Coronary disease	IL-1Ra
Crohn's disease	IL-10, MCP-1 (CCL2); IL-16 promoter; TNFR1
HIV resistance	CCR5 deletion
Idiopathic pulmonary fibrosis	TNF-alpha (-308 A) allele
Infectious nephropathy	low CXCR1 expression
Infertility	functional mutations in the LIF gene
Lupus nephritis	MCP-1 (CCL2) promoter
Multiple sclerosis	microsatellite allele of TNF gene; IL-2 promoter
Parkinson's disease	promoter region of IL-8
Rheumatoid arthritis	noncoding region of IFNγ gene; TGFβ1; MIF promoter; IL-1 gene cluster; IL-4
Schizophrenia	IL-1 gene complex
Sepsis mortality	TNF
Systemic lupus erythematosus	IL-1α; TNFα
Sudden infant death syndrome	IL-10
Susceptibility to cerebral malaria	TNF
Transplantation	TNF, IL-10

9
Cytokines and Infections

9.1
Half Angel–Half Devil

As mentioned in Sect. 4.2.1, cytokines play a major role in innate immunity during the early processes that occur in sites of microorganisms' invasion. During viral infection, the antiviral properties of interferons have been established for more than half a century. After interaction of IFNα or IFNβ with its receptor, the activation of 2–5 A synthetase leads to the production of an endoribonuclease that degrades viral RNA, while activation of protein kinase P1 contributes to the capacity of the cell to inhibit viral protein synthesis. During bacterial infection, the beneficial effects of proinflammatory cytokines have been widely demonstrated. The injection of cytokines such as IL-1, IL-6, IL-12, IL-18, TNF, G-CSF, M-CSF, GM-CSF, and IFNγ before the injection of bacteria have been shown to promote faster clearance of bacteria, and to enhance survival. The use of cytokine-neutralizing antibodies or antagonists molecules further demonstrated that endogenous cytokines produced in the course of infections play an essential role in preventing the microorganisms to establish themselves, or to grow, and their neutralization was highly deleterious. Finally, mice rendered deficient for the expression of a cytokine or a cytokine receptor display enhanced "susceptibility" to infection such as an enhanced microbial load in tissues, and reduced survival.

Despite their undoubtedly beneficial effects to clear microorganisms, some cytokines can also contribute to death in some clinical situations like the septic shock in response to endotoxin injection or after a lethal bacterial infection. Antibodies against TNF protect mice against lethal injection of LPS and baboons against lethal injection of *E. coli*. The ambivalent role of cytokines is illustrated in TNF receptor p55 knockout mice: these mice are more sensitive to *L. monocytogenes* infection than normal mice, but they are more resistant to toxic shock induced by endotoxin or Staphylococcal enterotoxin B. Furthermore, the synergy already described as a key word to define cytokine network is also valid for lethality and none lethal dose of TNF, injected with non-lethal dose of IL-1 or IFNγ leads to lethality in mice.

9.2
The Strategies of Microbes against Cytokines

The fact that pathogenic microorganisms have developed numerous subterfuges to counteract the action of cytokines further illustrates that cytokines are crucial in the fight against the infectious process. Hijacking of genes for cytokine receptors has been perpetrated by virus to elaborated soluble molecules that neutralize cytokines of the host (e.g. IFNγR and TNFR homologs of Myxoma virus, IL-1R homolog of Vaccinia virus). Another hijacking concerns the gene of IL-10, allowing viruses (Epstein–Barr virus, equine herpesvirus, parapoxvirus orf virus) to create an immunosuppressive environment. Another strategy has been developed by Vaccinia

virus to prevent the maturation of biologically active IL-1 and IL-18. This virus produces an inhibitor (serpine) of caspase-1, the enzyme required for the cleavage of the proforms of these cytokines. Other mechanisms include the production of viral factors that downregulate cytokine mRNA expression (e.g. HTLV, adenovirus), the use of cytokines to further enhance the viral replication (e.g. HIV), the use of chemokine receptors to enter the cell (e.g. HIV), or the synthesis of viral chemokines that can behave as antagonists (e.g. Kaposi' sarcoma associated virus, human Herpesvirus 8, Stealth virus, etc.). Bacteria also display strategies to block cytokine production (e.g. *Yersinia enterocolitica*, enteropathogenic *E. coli*) and to use them as growth factors (*Staphylococcus aureus*, *Pseudomonas aeruginosa*, *Acinetobacter*).

10
Cytokines and Diseases

As the discovery of new cytokines were occurring, great hopes to cure diseases were mentioned, often followed by great disappointments. For example, while TNF was discovered for its antitumor effect, it was soon demonstrated that it was one of the most potent proinflammatory cytokines, obviously limiting its therapeutic use. One unique cytokine is rarely the unique key effector underlying the pathogenic process, and the complexity of the cytokine network has rendered difficult their use as therapeutic tools or as therapeutic target. In addition, the identification of an overexpressed cytokine at one peculiar stage of the disease process does not allow discriminating whether this cytokine is a marker or an actor. Nevertheless, there exist certain clinical situations in which the use of recombinant cytokines or the use of neutralizing antibodies has allowed significant therapeutic progresses. Few examples of diseases for which the involvement of cytokines is well established and clinical applications have been considered are summarized below.

10.1
Autoimmune Diseases

Among autoimmune diseases, systemic lupus erythematosus has been associated with an exacerbated production of IL-10. Considering the properties of IL-10, which activates B-lymphocytes and inhibits T-lymphocytes, its overproduction in lupus could explain the immune anomalies of this disease, which is characterized by anti-self antibodies production and by deficient T-responses. Thus, the use of IL-10 antagonists may be beneficial in the management of systemic lupus erythematosus, as suggested by a preliminary report. Human lupus patients also have elevated blood levels of BAFF that support differentiation of selected B-cells into mature long-lived B-cells and may be critical in generating deleterious autoimmune responses. Indeed, overexpression of BAFF in mice may lead to systemic lupus erythematosus–like disease. Thus, BAFF could be another potential cytokine to neutralize in systemic lupus erythematosus.

10.2
Allergy

Allergen specific IgE antibodies, mast cell degranulation releasing inflammatory mediators such as histamine, and influx of eosinophils in airway mucosa and airway lumen are a hallmark of atopy and allergic asthma. Thus, Th2 cytokines remain important candidates for a role

in the pathogenesis of atopy and allergic asthma: IL-4 and IL-13 are required for IgE production, IL-9 is involved in mast cell growth, and IL-5 is required for eosinopoiesis. Humanized monoclonal antibodies (hMAbs) against IL-5, anti-IL-4, a recombinant soluble human IL-4 receptor, anti-IL-9, CCR3 antagonists (which block eosinophil chemotaxis), and CXCR2 antagonists (which block neutrophil and monocyte chemotaxis) have been developed as possible therapeutic interventions.

10.3
Sepsis

Animal models of sepsis have clearly shown that TNF, IL-1, or MIF are involved in sepsis-related lethality, and antibodies against these cytokines are highly protective. Unfortunately, anti-TNF antibodies and IL-1Ra have been used in numerous placebo double-blinded controlled assays in humans without any success. This failure may reflect that antibodies in animal models and in humans are not used with similar timings, and that mice, often used in experimental models, being 10^5 less sensitive to Gram negative bacteria–derived endotoxin than humans, are not the most relevant laboratory animals to model the human settings!

10.4
Chronic Inflammatory Diseases

Fortunately for companies that had developed these anti-TNF antibodies, successful treatments have been achieved for Crohn's disease and rheumatoid arthritis. The significant improvement of patients allowed the marketing of these antibodies. However, in none of these studies, the successful improvement concerned all patients. For rheumatoid arthritis, the first successful treatment was reported in 1994. In responding patients, clinical improvements were associated with biological changes, reflecting the attenuation of the inflammatory process. There was no clinical response in 30% of the cases. While the effect of a single injection is long lasting (few months), it has to be repeated. As a consequence, the occurrence of microbial pathogenic processes (tuberculosis, *Pneumocystis*- triggered pneumonia, histoplasmosis, listeriosis, sepsis) has been deplored. Other approaches have been proposed to neutralize TNF with recombinant soluble TNF receptors. IL-1Ra has also been approved by the Food and Drug Administration (FDA) for treatment of rheumatoid arthritis. For this disease, the targeting of IL-17 has also been suggested to be of potential value. For Crohn's disease, the anti-TNF treatment was approved in 1998, and clinical response was still observed 12 weeks after the infusion. A remission could be obtained for half of the patients after 4 doses every 8 weeks.

10.5
Cancer

IL-2 has been extensively studied in cancer patients, either alone, or associated with chemotherapy, with activated NK cells (Lymphokine-activated killer cells, LAK), with tumor-infiltrating lymphocytes (TIL), with IFNα, and, more recently, with tumor-pulsed dendritic cells. In some specific cases (melanoma, metastatic renal cancer), success remained modest. The major achievement has been obtained with IFNα in hairy cell leukemia and chronic myelogenous leukemia.

10.6
Hepatitis C

IFNα has been successful in a subset of hepatitis C virus–infected patients. More recently, once associated to ribavirin, a polymerized form of IFNα, the half-life of which is prolonged, has been shown to lead to sustained virological response rates of >50% in chronic hepatitic C patients.

10.7
Multiple Sclerosis

In 1993, interferon-β was approved in the United States for relapsing–remitting multiple sclerosis. The use IFNβ and glatiramer acetate for the treatment of multiple sclerosis has, to some extent, changed the course of the disease. The annual relapse rate of patients treated with these drugs is lower than that in placebo-treated patients, and more treated patients remain relapse-free as compared to untreated patients.

10.8
Transplantation

The alpha-chain of the IL-2R is a specific peptide against which monoclonal antibodies have been raised, with the aim of blunting the immune response by means of inhibiting proliferation and inducing apoptosis in primed lymphocytes. One of such antibodies has proved to be effective in reducing the episodes of acute rejection after kidney and pancreas transplantation. The use of this antibody was associated with a significant reduction in the incidence of any treated rejection episodes after kidney transplantation in the two major randomized European and US studies.

10.9
Neutropenia

Following chemotherapy for leukemia, neutropenia renders the patients more susceptible to develop infection. Treatment of patients with G-CSF and GM-CSF accelerates the recovery of normal neutrophil counts and significantly reduces the occurrence of documented infections.

11
Conclusion

Despite the huge number of molecules that belong to the cytokine family, the complexity of the cytokine network, the great number of parameters that may affect the biological properties of cytokines, tremendous amount of work has allowed to decipher the language of cells. Despite the difficulty, new promising use of cytokines or cytokine antagonists can be expected to master certain diseases.

See also Cytokines: Interleukins; Growth Factors; Programmed Cell Death.

Bibliography

Books and Reviews

Aggarwal, B.B. (1998) *Human Cytokines: Handbook for Basic and Clinical Research*, Blackwell, Oxford, UK.

Bona, C.A., Revillard, J-P. (2000) *Cytokines and Cytokine Receptors: Physiology and Pathological Disorders*, Harwood Academic Publishers, Amsterdam, Netherlands.

Cavaillon, J.M. (1994) Cytokines and macrophages, *Biomed. Pharmacother.* **48**, 445–453.

Cavaillon, J-M. (2001) Pro- versus anti-inflammatory cytokines: myth or reality, *Cell. Mol. Biol.* **47**, 695–702.

Fitzgerald, K.A. (2001) *Cytokine Facts Book*, Academic Press, San Diego, CA.

Giamila, F., Durum, S.K. (2003) *Cytokine Knockouts*, Contemporary immunology, 2nd edition, Humana Press, Totowa, NJ.

Kotb, M., Calandra, T. (2003) *Cytokines and Chemokines in Infectious Diseases Handbook*, Humana Press, Totowa, NJ.

Oppenheim, J.J., Feldmann, M., Durum, S.K. (2000) *Cytokine Reference: A Compendium of Cytokines and Other Mediators of Host Defense*, Academic Press, London, San Diego, CA.

Santamaria, P. (2003) *Cytokines and Chemokines in Autoimmune Disease*, Landes Bioscience/Eurekah.com, Georgetown, TX.

Thomson, A.W., Lotze, M.T. (2003) *The Cytokine Handbook*, 4th edition, Academic Press, Amsterdam, Netherlands.

Primary Literature

Auron, P.E., Webb, A.C., Rosenwasser, L.J., Mucci, S.F., Rich, A., Wolff, S.M., Dinarello, C.A. (1984) Nucleotide sequence of human monocyte interleukin 1 precursor cDNA, *Proc. Natl. Acad. Sci. USA* **81**, 7907–7911.

Bartfeld, H., Atoynatan, T. (1969) Cytophilic nature of migration inhibitory factor associated with delayed hypersensitivity, *Proc. Soc. Exp. Biol. Med.* **130**, 497–501.

Bazan, J.F. (1990) Structural design and molecular evolution of a cytokine receptor superfamily, *Proc. Natl. Acad. Sci. USA.* **87**, 6934–6938.

Bender, J.R., Sadeghi, M.M., Watson, C., Pfau, S., Pardi, R. (1994) Heterogeneous activation thresholds to cytokines in genetically distinct endothelial cells: evidence for diverse transcriptional responses, *Proc. Natl. Acad. Sci. U S A.* **91**, 3994–3998.

Bernhagen, J., Calandra, T., Mitchell, R.A., Martin, S.B., Tracey, K.J., Voelter, W., Manogue, K.R., Cerami, A., Bvcala, R. (1993) MIF is a pituitary-derived cytokine that potentiates lethal endotoxaemia, *Nature* **365**, 756–759.

Bigazzi, P.E., Yoshida, T., Ward, P.A., Cohen, S. (1975) Production of lymphokine-like factors (cytokines) by simian virus 40-infected and simian virus 40-transformed cells, *Am. J. Pathol.* **80**, 69–78.

Burdin, N., Peronne, C., Banchereau, J., Rousset, F. (1993) Epstein-Barr virus transformation induces B lymphocytes to produce human interleukin 10, *J. Exp. Med.* **177**, 295–304.

Carswell, E.A., Old, L.J., Kassel, R.L., Green, S., Fiore, N., Williamson, B. (1975) An endotoxin-induced serum factor that causes necrosis of tumors, *Proc. Natl. Acad. Sci. USA* **72**, 3666–3670.

Cavazzana-Calvo, M., Hacein-Bey, S., de Saint Basile, G., Gross, F., Yvon, E., Nusbaum, P., Selz, F., Hue, C., Certain, S., Casanova, J.L., Bousso, P., Deist, F.L., Fischer, A. (2000) Gene therapy of human severe combined immunodeficiency (SCID)-X1 disease, *Science* **288**, 669–672.

Chai, Z., Gatti, S., Toniatti, C., Poli, V., Bartfai, T. (1996) Interleukin (IL)-6 gene expression in the central nervous system is necessary for fever response to lipopolysaccharide or IL-1 beta: a study on IL-6-deficient mice, *J. Exp. Med.* **183**, 311–316.

Coffman, R.L., Ohara, J., Bond, M.W., Carty, J., Zlotnik, A., Paul, W.E. (1986) B cell stimulatory factor-1 enhances the IgE response of lipopolysaccharide-activated B cells, *J. Immunol.* **136**, 4538–4541.

Cohen, S., Bigazzi, P.E., Yoshida, T. (1974) Similarities of T cell function in cell-mediated immunity and antibody production, *Cell. Immunol.* **12**, 150–159.

DeForge, L.E., Kenney, J.S., Jones, M.L., Warren, J.S., Remick, D.G. (1992) Biphasic production of IL-8 in lipopolysaccharide (LPS)-stimulated human whole blood. Separation of LPS- and cytokine-stimulated components using anti-tumor necrosis factor and anti-IL-1 antibodies, *J. Immunol.* **148**, 2133–2141.

de Waal Malefyt, R., Figdor, C.G., Huijbens, R., Mohan-Peterson, S., Bennett, B., Culpepper, J., Dang, W., Zurawski, G., de Vries, J.E. (1993) Effects of IL-13 on phenotype, cytokine production, and cytotoxic function of human monocytes. Comparison with IL-4 and modulation by IFN-gamma or IL-10, *J. Immunol.* **151**, 6370–6381.

Dinarello, C.A., Renfer, L., Wolff, S.M. (1977) Human leukocytic pyrogen: purification and development of a radioimmunoassay, *Proc. Natl. Acad. Sci. U S A.* **74**, 4624–4627.

Doherty, G.M., Lange, J.R., Langstein, H.N., Alexander, H.R., Buresh, C.M., Norton, J.A. (1992) Evidence for IFN-gamma as a mediator of the lethality of endotoxin and tumor necrosis factor-alpha, *J. Immunol.* **149**, 1666–1670.

Dranoff, G., Crawford, A.D., Sadelain, M., Ream, B., Rashid, A., Bronson, R.T., Dickersin, G.R., Bachurski, C.J., Mark, E.L., Whitsett, J.A. (1994) Involvement of granulocyte-macrophage colony-stimulating factor in pulmonary homeostasis, *Science* **264**, 713–716.

Dumonde, D.C., Wolstencroft, R.A., Panayi, G.S., Matthew, M., Morley, J., Howson, W.T. (1969) "Lymphokines": non-antibody mediators of cellular immunity generated by lymphocyte activation, *Nature* **224**, 38–44.

Echtenacher, B., Männel, D.N., Hultner, L. (1996) Critical protective role of mast cells in a model of acute septic peritonitis, *Nature* **381**, 75–77.

Elliott, M.J., Maini, R.N., Feldmann, M., Kalden, J.R., Antoni, C., Smolen, J.S., Leeb, B., Breedveld, F.C., MacFarlane, J.D., Bijl, H. (1994) Randomised double-blind comparison of chimeric monoclonal antibody to tumour necrosis factor alpha (cA2) versus placebo in rheumatoid arthritis, *Lancet* **344**, 1105–1110.

Fernandez-Botran, R. (1991) Soluble cytokine receptors: their role in immunoregulation, *FASEB J.* **5**, 2567–2574.

Frendeus, B., Godaly, G., Hang, L., Karpman, D., Lundstedt, A.C., Svanborg, C. (2000) Interleukin 8 receptor deficiency confers susceptibility to acute experimental pyelonephritis and may have a human counterpart, *J. Exp. Med.* **192**, 881–890.

Gabay, C., Smith, M.F., Eidlen, D., Arend, W.P. (1997) Interleukin 1 receptor antagonist (IL-1Ra) is an acute-phase protein, *J. Clin. Invest.* **99**, 2930–2940.

Gerard, C., Bruyns, C., Marchant, A., Abramowicz, D., Vandenabeele, P., Delvaux, A., Fiers, N., Goldman, M., Velu, T. (1993) Interleukin 10 reduces the release of tumor necrosis factor and prevents lethality in experimental endotoxemia, *J. Exp. Med.* **177**, 547–550.

Hapel, A.J., Lee, J.C., Farrar, W.L., Ihle, J.N. (1981) Establishment of continuous cultures of thy1.2+, Lyt1+, 2-T cells with purified interleukin 3, *Cell* **25**, 179–186.

Horai, R., Saijo, S., Tanioka, H., Nakae, S., Sudo, K., Okahara, A., Ikuse, T., Asano, M., Iwakura, Y. (2000) Development of chronic inflammatory arthropathy resembling rheumatoid arthritis in interleukin 1 receptor antagonist-deficient mice, *J. Exp. Med.* **191**, 313–320.

Howard, M., Farrar, J., Hilfiker, M., Johnson, B., Takatsu, K., Hamaoka, T., Paul, W.E. (1982) Identification of a T cell-derived B cell growth factor distinct from interleukin 2, *J. Exp. Med.* **155**, 914–923.

Hurst, S.M., Wilkinson, T.S., McLoughlin, R.M., Jones, S., Horiuchi, S., Yamamoto, N., Rose-John, S., Fuller, G.M., Topley, N., Jones, S.A. (2001) IL-6 and its soluble receptor orchestrate a temporal switch in the pattern of leukocyte recruitment seen during acute inflammation, *Immunity* **14**, 705–714.

Isaacs, A., Lindenmann, J. (1957) Virus interference I. The interferon, *Proc. Roy. Soc. B.* **147**, 258–267.

Jacob, C.O., Fronek, Z., Lewis, G.D., Koo, M., Hansen, J.A., McDevitt, H.O. (1990) Heritable major histocompatibility complex class II-associated differences in production of tumor necrosis factor alpha: relevance to genetic predisposition to systemic lupus erythematosus, *Proc. Natl. Acad. Sci. USA.* **87**, 1233–1237.

Kopf, M., Baumann, H., Freer, G., Freudenberg, M., Lamers, M., Kishimoto, T., Zinkernagel, R., Bluethmann, H., Kohler, G. (1994) Impaired immune and acute-phase responses in interleukin-6-deficient mice, *Nature* **368**, 339–342.

Kuhn, R., Lohler, J., Rennick, D., Rajewsky, K., Muller, W. (1993) Interleukin-10-deficient mice develop chronic enterocolitis, *Cell* **75**, 263–274.

Llorente, L., Zou, W., Levy, Y., Richaud-Patin, Y., Wijdenes, J., Alcocer-Varela, J., Morel-Fourrier, B., Brouet, J.C., Alarcon-Segovia, D., Galanaud, P. (1995) Role of interleukin 10 in the B lymphocyte hyperactivity and autoantibody production of human systemic lupus erythematosus, *J. Exp. Med.* **181**, 839–844.

Moore, K.W., Vieira, P., Fiorentino, D.F., Trounstine, M.L., Khan, T.A., Mosmann, T.R. (1990) Homology of cytokine synthesis inhibitory factor (IL-10) to the Epstein-Barr virus gene BCRFI, *Science* **248**, 1230–1234.

Mosmann, T.R., Cherwinski, H., Bond, M.W., Giedlin, M.A., Coffman, R.L. (1986) Two types of murine helper T cell clone. I. Definition according to profiles of lymphokine activities and secreted proteins, *J. Immunol.* **136**, 2348–2357.

Nakano, Y., Onozuka, K., Terada, Y., Shinomiya, H., Nakano, M. (1990) Protective effect of recombinant tumor necrosis factor-alpha in murine salmonellosis, *J. Immunol.* **144**, 1935–1941.

Ohlsson, K., Bjork, P., Bergenfeldt, M., Hageman, R., Thompson, R.C. (1990) Interleukin-1 receptor antagonist reduces mortality from endotoxin shock, *Nature* **348**, 550–552.

Pfeffer, K., Matsuyama, T., Kundig, T.M., Wakeham, A., Kishihara, K., Shahinian, A., Wiegmann, K., Ohashi, P.S., Kronke, M., Mak, T.W. (1993) Mice deficient for the 55 kDa tumor necrosis factor receptor are resistant to endotoxic shock, yet succumb to L. monocytogenes infection, *Cell* **73**, 457–467.

Pociot, F., Briant, L., Jongeneel, C.V., Mölvig, J., Worsaae, H., Abbal, M., Thomsen, M., Nerup, J., Cambon-Thomsen, A. (1993) Association of tumor necrosis factor and class II major histocompatibility complex alleles with the secretion of TNFα and TNFβ by human mononuclear cells: a possible link to insulin-dependent diabetes mellitus, *Eur. J. Immunol.* **23**, 224–231.

Porat, R., Clark, B.D., Wolff, S.M., Dinarello, C.A. (1991) Enhancement of growth of virulent strains of Escherichia coli by interleukin-1, *Science* **254**, 430–432.

Rosenberg, S.A., Lotze, M.T., Muul, L.M., Leitman, S., Chang, A.E., Ettinghausen, S.E., Matory, Y.L., Skibber, J.M., Shiloni, E., Vetto, J.T. (1985) Observations on the systemic administration of autologous lymphokine-activated killer cells and recombinant interleukin-2 to patients with metastatic cancer, *N. Engl. J. Med.* **313**, 1485–1492.

Sallusto, F., Lenig, D., Mackay, C.R., Lanzavecchia, A. (1998) Flexible programs of chemokine receptor expression on human polarized T helper 1 and 2 lymphocytes, *J. Exp. Med.* **187**, 875–883.

Shalaby, M.R., Waage, A., Aarden, L., Espevik, T. (1989) Endotoxin, tumor necrosis factor-alpha and interleukin 1 induce interleukin 6 production in vivo, *Clin. Immunol. Immunopathol.* **53**, 488–498.

Shirai, A., Holmes, K., Klinman, D. (1993) Detection and quantitation of cells secreting IL-6 under physiologic conditions in BALB/c mice, *J. Immunol.* **150**, 793–799.

Shull, M.M., Ormsby, I., Kier, A.B., Pawlowski, S., Diebold, R.J., Yin, M., Allen, R., Sidman, C., Proetzel, G., Calvin, D. (1992) Targeted disruption of the mouse transforming growth factor-beta 1 gene results in multifocal inflammatory disease, *Nature* **359**, 693–699.

Smith, C.A., Davis, T., Wignall, J.M., Din, W.S., Farrah, T., Upton, C., McFadden, G., Goodwin, R.G. (1991) T2 open reading frame from the Shope fibroma virus encodes a soluble form of the TNF receptor, *Biochem. Biophys. Res. Commun.* **176**, 335–342.

Stuber, F., Petersen, M., Bokelmann, F., Schade, U. (1996) A genomic polymorphism within the tumor necrosis factor locus influences plasma tumor necrosis factor-alpha concentrations and outcome of patients with severe sepsis, *Crit. Care Med.* **24**, 381–384.

Supajatura, V., Ushio, H., Nakao, A., Akira, S., Okumura, K., Ra, C., Ogawa, H. (2002) Differential responses of mast cell Toll-like receptors 2 and 4 in allergy and innate immunity, *J. Clin. Invest.* **109**, 1351–1359.

Tracey, K.J., Fong, Y., Hesse, D.G., Manogue, K.R., Lee, A.T., Kuo, G.C., Lowry, S.F., Cerami, A. (1987) Anti-cachectin/TNF monoclonal antibodies prevent septic shock during lethal bacteraemia, *Nature* **330**, 662–664.

Tsicopoulos, A., Hamid, Q., Haczku, A., Jacobson, M.R., Durham, S.R., North, J., Barkans, J., Corrigan, C.J., Meng, O., Moqbel, R. (1994) Kinetics of cell infiltration and cytokine messenger RNA expression after intradermal challenge with allergen and tuberculin in the same atopic individuals, *J. Allergy Clin. Immunol.* **94**, 764–772.

Upton, C., Mossman, K., McFadden, G. (1992) Encoding of a homolog of the IFN-gamma receptor by myxoma virus, *Science* **258**, 1369–1372.

van Dullemen, H.M., van Deventer, S.J., Hommes, D.W., Bijl, H.A., Jansen, J., Tytgat, G.N., Woody, J. (1995) Treatment of Crohn's disease with anti-tumor necrosis factor chimeric monoclonal antibody (cA2), *Gastroenterology* **109**, 129–135.

Wang, H., Yu, M., Ochani, M., Amella, C.A., Tanovic, M., Susarla, S., Li, J.H., Wang, H., Yang, H., Ulloa, L., Al-Abed, Y., Czura, C.J., Tracey, K.J. (2003) Nicotinic acetylcholine receptor alpha7 subunit is an essential regulator of inflammation, *Nature* **421**, 384–388.

Yoshimura, T., Matsushima, K., Oppenheim, J.J., Leonard, E.J. (1987) Neutrophil chemotactic factor produced by lipopolysaccharide (LPS)-stimulated human blood mononuclear leukocytes: partial characterization and separation from interleukin 1 (IL 1), *J. Immunol.* **139**, 788–793.

Molecular Motors in Plant Cells

Anireooy S.N. Reddy
Colorado State University, Fort Collins, CO

Encyclopedia of Molecular Cell Biology and Molecular Medicine, 2nd Edition. Volume 8
Edited by Robert A. Meyers.
Copyright © 2005 Wiley-VCH Verlag GmbH & Co. KGaA, Weinheim
ISBN: 3-527-30550-5

Keywords

Cytoskeleton
A dynamic network of protein tubules (microtubules) and filaments (microfilaments and intermediate filaments) in all eukaryotic cells that is necessary for cell shape, cell motility, and spatial and temporal organization of molecules, and organelles within the cells.

Microtubules
Long hollow cylindrical tubes with an external diameter of 24 nm that are formed by polymerization of α- and β-tubulin monomers.

Microfilaments
Filaments of about 7-nm diameter that are formed by polymerization of actin.

Kinesins
A family of motor proteins in eukaryotes that can hydrolyze ATP and use the derived energy to either transport organelles, vesicles, and RNA protein complexes on microtubules or regulate microtubule dynamics.

Myosins
A family of motor proteins that can hydrolyze ATP and move on microfilaments. These motors function in muscle contraction, cell division, and vesicle transport.

Preprophase Band
A narrow band of interconnected microtubules and microfilaments encircling the nucleus that appears in cells preparing to enter mitosis and marks the location of future cell wall.

Phragmoplast
A plant-specific structure made of two discs of microtubules (and microfilaments) that is formed between two daughter nuclei at late telophase. It constructs the new cell wall between the daughter nuclei.

Trichome
A hair-like branched or unbranched structure on the surface of a leaf or stem. These structures may be unicellular or multicellular.

Molecular motors regulate diverse cellular functions, including the organization and dynamics of the microtubule (MT) and actin cytoskeleton, cytoplasmic streaming, cell polarity, cell growth, morphogenesis, chromosome segregation, and transport of vesicles, organelles, and macromolecular complexes. In eukaryotes, there are three families of molecular motors: the kinesins, the dyneins, and the myosins. All three types of motors use ATP to move along filamentous structures. Kinesins and dyneins move on MTs, whereas the myosins translocate on actin filaments. Using biochemical, cell biological, molecular, and genetic approaches, several molecular motors have been identified in plants, and functions of some of these are beginning to be understood. Recent completion of genome sequence of several eukaryotes ranging from yeast, a simple unicellular eukaryote, to highly evolved multicellular organisms, including humans and flowering plants, has permitted comparative analysis of three families of motors in various species. Such analysis across phylogenetically divergent species has yielded some interesting insights into evolutionary and functional relationships among motor proteins from various organisms. Systematic analyses of the recently completed Arabidopsis genome sequence with the conserved motor domain of kinesins, myosins, and dyneins revealed the presence of 78 molecular motors (61 kinesins and 17 myosins) in this organism. Of the two families of MT-based motors, dyneins are absent in Arabidopsis. Surprisingly, Arabidopsis has the largest number of kinesins as compared to other multicellular organisms including humans, suggesting that the kinesin superfamily is expanded considerably in plants. Also, all plant myosins belong to two novel classes. Although the identification of molecular motors in plants is progressing rapidly due to genome and EST-sequencing projects, only a few plant motors have been characterized in any detail and the functions of many these motors are not known. Nevertheless, it is becoming obvious that plants contain novel families of kinesins and myosins with unique functions and regulatory mechanisms. In this article, the focus is primarily on plant molecular motors, their function, and regulation with emphasis on model plant Arabidopsis.

1
Microtubule-based Motors

The motors that move on MTs (microtubules) belong to two different families: kinesins that move either toward plus end or minus end of the MTs, and the dyneins, which move toward the minus end of MTs.

1.1
Kinesins

Members of the kinesin superfamily are known or implicated to play important roles in many fundamental cellular and developmental processes, including intracellular transport of vesicles and organelles, mitotic and meiotic spindle formation and elongation, chromosome segregation, germplasm aggregation, MT organization, and dynamics and intraflagellar transport. A large number of kinesins have been identified in plants and animals (Table 1). All members of the kinesin superfamily have a highly conserved catalytic region of about 350 amino acid residues, known as the *motor domain*, with ATP- and MT-binding sites. The motor domain in kinesins is located either in the N-terminus, C-terminus, or in the middle of the protein. In addition to the motor domain, most kinesins have a stalk region that forms an alpha-helical coiled-coil region, which aids in dimerization, and a highly variable tail, which is thought to interact with a specific cargo.

Although molecular motors have been implicated in a variety of cellular processes

Tab. 1 The total number of kinesins and myosins in some of the completely sequenced organisms.

	S. cerevisiae	S. pombe	C. elegans	D. melanogaster	H. sapiens	A. thaliana
Kinesins	6	9	21	24	45	61
Myosins	5	5	17	13	40	17
Total	11	14	38	37	85	78

in plants, including spindle function, cytokinesis, cell polarity, morphogenesis, organelle and vesicle transport, and cytoplasmic streaming, until recently very little was known about their molecular identity. Plant kinesins have been identified in pollen tubes of tobacco and other plants using antibodies to animal kinesins. An MT-based motor (90 kD) from tobacco pollen tubes, which induced MT gliding in motility assays, binds organelles associated with MTs in the cortical regions of the pollen tube, suggesting its involvement in organelle transport. Two kinesins (125 kD and 120 kD) that showed plus-end motor activity in *in vitro* motility assays and MT-dependent ATPase and GTPase activity were isolated from tobacco phragmoplasts. Using various approaches, cDNAs for several kinesins have been characterized from plants. Thirteen of these are from Arabidopsis (KatA, KatB, KatC, KatD, AtKCBP, AtPAKRP1, AtPAKRP1L, AtPAKRP2, AtMKRP1, AtMKRP2, At-NCAK1, KCA1, KCA2), 15 from tobacco (TKRP125, KCBP, NCAK1, NCAK2, and TBK1 to TBK11), and 1 from potato (KCBP). A kat gene family (katA, katB, katC, katD, and katE) encoding kinesins in *Arabidopsis thaliana* was characterized using primers corresponding to conserved regions of the kinesin motor domain. Using a similar approach, 12 different kinesins that belong to seven subfamilies have been isolated from tobacco BY-2 cells. It is interesting that in tobacco BY-2 cells alone 15 different kinesins that

belong to different subfamilies are expressed. KatA (89 kD) protein has been shown to be a minus-end-directed motor, and a similar protein has been found in carrot and tobacco. KatA (new name ATK1) is necessary for bipolar spindle assembly during male meiosis. The central region of KatD shares sequence similarity to the motor domain of kinesin and is followed by about 240 residues at the C-terminus. Phylogenetic analysis with the katD motor domain indicates that it belongs to the C-terminal family despite the fact that it has a 240 amino acid stretch following the motor domain. The amino-terminal region of katD showed sequence similarity to the calponin homology (CH) domain. KatD is a flower-specific motor protein, suggesting that it may function in transport or cytoskeleton organization in pollen. TKRP125 (tobacco kinesin–related polypeptide of 125 kD) is an N-terminal kinesin with strong similarity to members of the BimC (blocked in mitosis) subfamily. Two other members of BimC class kinesins (DcKRP120-1 and DcKRP120-2) were isolated from carrot suspension cells. DcKRP120-1 is a homolog of TKRP125. In Arabidopsis, there seem to be at least three TKRP-like proteins. Recently, three phragmoplasts-associated kinesin-related proteins (AtPAKRP1, AtPAKRP1L, and AtPAKRP2) have been isolated and characterized from Arabidopsis. AtPAKRP1 is an N-terminal kinesin and is most closely related to

XKlp2, an ungrouped kinesins from *Xenopus laevis*. AtPAKRP2 is also an N-terminal motor with no known homologs in non-plant systems. In phylogenetic analysis, AtPAKRP2 does not group with any of the existing families of kinesins. These motors have distinct functions in cytokinesis. The AtPAKRP1L that is closely related to AtPAKRP1 shows same localization as AtPAKRP1, suggesting that these two kinesins may have a similar role in cell division.

KCBP (kinesin-like calmodulin-binding protein), a novel kinesin with a C-terminal motor domain, was isolated from a number of plants as a calmodulin-binding protein. Although KCBP, like other kinesins, contains three distinct regions (a motor domain, a coiled-coil stalk, and a tail), it has two features that make it unique among members of the kinesin superfamily in eukaryotes. These include (1) a calmodulin-binding domain adjacent to the motor domain at the C-terminus and (2) a myosin tail homology and talin-like region in the N-terminal tail. KCBP binds calmodulin in a calcium-dependent manner at physiological calcium concentrations. KCBP is a minus-end directed MT motor and has two MT-binding domains, one located at the C-terminus and the second one located at the N-terminus. Unlike the MT-binding domain in the C-terminus, the N-terminal region of KCBP binds MTs, both in the presence and absence of ATP, indicating that the MT-binding domain in the N-terminus is insensitive to ATP. KCBP is highly conserved in phylogenetically divergent plant species, including dicots, monocots, gymnosperms, and algae. Recently, a calmodulin-binding C-terminal kinesin (kinesin C) was cloned from sea urchin. The calmodulin-binding domain of kinesin C shared 35% sequence identity with the calmodulin-binding domain in KCBP. The existence of calmodulin-binding kinesins in both plants and sea urchins suggests that the origin of this group of kinesins predates the divergence of plants and animals from a common ancestor, which is believed to have occurred about 1.5 billion years ago. If this was the case, there must have been insertion of some domains such as MyTH4 and talin-like regions in the tail of KCBP (or deletion of these domains in kinesin C) to acquire functional specialization of these kinesins. Alternatively, calmodulin-binding kinesins may have evolved independently in plant and animals after they diverged from a common ancestor. The amino-terminal tail and stalk regions of KCBPs from different plant systems are highly conserved and contain myosin tail homology (MyTH4) and talin-like regions that are not present in kinesin C. In phylogenetic analysis of kinesins, KCBPs and kinesin C are grouped with other known C-terminal kinesins. However, Arabidopsis KCBP together with its orthologs from other plants constitute a distinct group within the C-terminal subfamily of motors.

Recently, two Arabidopsis kinesins (AtMKRP1 and AtMKRP2) with mitochondrial targeting signals were found to be targeted to mitochondria, suggesting that these motors might function in this organelle. AtMKRP1 and AtMKRP2 belong to ungrouped kinesins. What functions these MKRPs perform in mitochondria remains to be seen. There are other Arabidopsis kinesins with targeting signals to other organelles such as chloroplasts. It would be interesting to see if any other Arabidopsis kinesins are targeted to organelles. Kinesins that are targeted to organelles have not been reported previously in the literature.

Mutations in the *HINKEL* (*HIK*) gene in Arabidopsis result in defective cytokinesis. These mutants have high frequency of incomplete cell walls and multinucleate cells. The *HIK* gene encodes a plant-specific N-terminal kinesin that plays an important role in cytokinesis. Two tobacco kinesins (NACK1 and NACK2) interact with NPK1, a mitogen-activated kinase kinase kinase, and stimulate the activity of the kinase. Orthologs of NACK1 and NACK2 have been identified in Arabidopsis and AtNACK1 was found to be identical to HIK. These kinesins are highly diverged N-terminal motors and do not group with any of the known kinesin subfamilies. The closest nonplant kinesin that shares significant similarity with the motor domain of NACKs is CENP-E. At least eight Arabidopsis kinesins are known to function in some aspect of cell division (see Sect. 3.1 on cell division).

Analysis of the recently completed Arabidopsis genome sequence with the conserved motor domain of kinesins has resulted in identification of 61 kinesins in the model plant. A corresponding cDNA or EST (expressed sequence tag) was found for most kinesins, suggesting that most kinesins in Arabidopsis are expressed and not likely to be pseudogenes. The genes encoding kinesins are distributed throughout the genome. Chromosome 3 has the highest number (18) of kinesin genes. In comparison, *Saccharomyces cerevisiae, Schizosaccharomyces pombe, Caenorhabditis elegans, Drosophila*, and *Homo sapiens* have 6, 9, 21, 24, and 45 kinesins respectively (Table 1). Surprisingly, the Arabidopsis genome contains the largest number of kinesins among all eukaryote genomes that have been sequenced. Arabidopsis has the highest percentage (0.24%) of the total number of genes as compared to *S. cerevisiae* and *S. pombe* with 0.1% and 0.17%

respectively, *C. elegans* with 0.11%, and *Drosophila* with 0.18%. On the basis of the number of kinesins in Arabidopsis, the number of cellular processes that is regulated by kinesins in plants is likely to match or exceed the number of processes controlled by animal kinesins.

Only 13 of the 61 Arabidopsis kinesins have been reported in the literature, whereas the 3 AtKRP125 kinesins (AtKRP125a,b,&c; Table 2) show sequence similarity to a tobacco kinesin (NtKRP125) that was isolated from phragmoplasts of tobacco. AtKRP125b has 68% identity with NtKRP125 over the 1000 residues they have in common.

The number of known and predicted introns in kinesins ranges from 3 to 34 (Table 2). In the Arabidopsis genome, the number of introns ranges from 0 to 77 with an average of about 5. More than 85% of Arabidopsis genes have 10 or less introns, while the Arabidopsis kinesin genes have an average of 16.4 with only 10 genes having 10 or less introns. The number and locations of the introns need to be verified experimentally. It is likely that the predicted sizes of the proteins in Tables 2 and 3 may change somewhat as corresponding cDNAs are characterized. In a few cases in which the cDNAs were isolated recently, there are some differences between the predicted and actual size of the protein due to inaccurate predictions of introns and exons.

All Arabidopsis kinesins except three have a coiled-coil region, suggesting that they may function as homo or heterodimers (Fig. 1 and Table 2). Interestingly, some Arabidopsis kinesins have predicted domains that are not present nonplant kinesins (Fig. 1, Table 2). Six of the Arabidopsis kinesins have a calponin homology (CH) domain, which is an actin-binding domain present in the N-termini

Tab. 2 Kinesin-like proteins in arabidopsis.

	Gene code (published name)	Motor location	#aa	Number of introns	Motility	Cellular localization	Other Domains
1	**At2g37420** (AtKRP125a)	N-terminal	1022	17	ND	ND	CC
2	**At2g36200** (AtKRP125b)	N-terminal	1056	22	ND	ND	CC
3	**At2g28620** (AtKRP125c)	N-terminal	1076	18	ND	ND	CC,
4	At3g45850	N-terminal	1058	21	ND	ND	CC
5	At4g05190	C-terminal	777	17	ND	ND	CC,
6	**At4g21270** (AtKatA)	C-terminal	793	8	Minus[a]	Mitotic MT arrays	CC
7	**At4g27180** (AtKatB)	C-terminal	745[d]	15	ND	ND	CC
8	**At5g54670** (AtKatC)	C-terminal	754[d]	15	ND	ND	CC
9	At5g27550	N-terminal[b]	425	6	ND	ND	CC
10	At2g22610	Internal[c]	1068	18	ND	ND	CC
11	At1g72250	Internal[c]	1195	17	ND	ND	CC,
12	At3g10310	Internal[c]	897	16	ND	ND	CC,CH
13	At1g18410	Internal[c]	1162[d]	17	ND	ND	CC
14	At1g73860	Internal[c]	1050	17	ND	ND	CC
15	At1g63640	Internal[c]	1056	19	ND	ND	CC,CH
16	At5g41310	Internal[c]	967	19	ND	ND	CC,CH
17	**At5g27000** (AtKatD)	Internal[c]	987	17	ND	ND	CC,CH
18	At1g09170	Internal[c]	1032[d]	18	ND	ND	CC,CH
19	At2g47500	Internal[c]	861	14	ND	ND	CC,CH
20	At3g44730	Internal[c]	767	14	ND	ND	CC
21	**At5g10470** (KCA1)	N-terminal[b]	1273	22	ND	ND	CC
22	**At5g65460** (KCA2)	N-terminal[b]	1264	22	ND	ND	CC
23	At5g27950	N-terminal[b]	640	8	ND	ND	CC
24	At1g55550	N-terminal[b]	887	9	ND	ND	CC
25	**At5g65930** (AtKCBP)	C-terminal	1259	20	Minus	PPB, spindle, spindle poles, Phragmo-plast	CC, MyTH4 Talin-like, CBD, PEST
26	At1g20060	N-terminal	923[d]	21	ND	ND	
27	At3g10180	N-terminal	459	11	ND	ND	CC
28	At1g59540	N-terminal	823	17	ND	ND	CC
29	At5g06670	N-terminal	997	22	ND	ND	CC

(continued overleaf)

Tab. 2 (*Continued*)

	Gene code (published name)	Motor location	#aa	Number of introns	Motility	Cellular localization	Other Domains
30	At3g12020	N-terminal	956	22	ND	ND	CC
31	**At1g21730 (MKRP1)**	N-terminal	909[d] (890)	22	ND	Mito-chondria	CC
32	**At4g39050 (MKRP2)**	N-terminal	1121 (1055)	22	ND	Mito-chondria	CC
33	At2g21380	N-terminal	857	18	ND	ND	CC
34	**At1g18370 (HINKEL/ AtNACK1)**	N-terminal	1003[d] (974)	11	ND	Phragmo-plast equator	CC
35	**At3g43210 (ATNACK2/ TES)**	N-terminal	932	13	ND	Interzonal Mts, Phragmo-plast	CC,
36	At4g38950	N-terminal	834	11	ND	ND	CC
37	At2g21300	N-terminal	581	9	ND	ND	CC
38	At3g51150	N-terminal	968	13	ND	ND	CC
39	At5g66310	N-terminal	1037	13	ND	ND	CC
40	At5g42490	N-terminal	1087	9	ND	ND	
41	At4g24170	N-terminal	1263	14	ND	ND	CC
42	At3g16630	Internal	799	12	ND	ND	CC
43	At3g16060	Internal	706	12	ND	ND	CC
44	At5g02370	N-terminal	664	10	ND	ND	
45	At1g18550	N-terminal	703[d]	3	ND	ND	CC
46	At3g49650	N-terminal	813	14	ND	ND	CC
47	At5g23910	N-terminal	665	17	ND	ND	HhH1
48	At3g50240	N-terminal	1075	21	ND	ND	CC
49	At5g47820	N-terminal	1032	22	ND	ND	CC
50	At5g60930	N-terminal	1335[d]	24	ND	ND	CC
51	At3g63480	N-terminal	439	11	ND	ND	CC
52	At1g01950	N-terminal	885[d]	18	ND	ND	CC,ARM,
53	At1g12430	N-terminal	895[d]	16	ND	ND	CC,ARM
54	At3g54870	N-terminal	1070	20	ND	ND	CC, ARM
55	At3g19050	N-terminal	2756	34	ND	ND	CC,
56	At3g17360	N-terminal	2158[d]	30	ND	ND	CC,
57	At3g44050	N-terminal	1229	20	ND	ND	CC
58	At3g20150	N-terminal	1103[d]	16	ND	ND	CC
59	**At3g23670 (ATPAKRP1L)**	N-terminal	1268[d] (1313)	14	ND	Spindle midzone, Phragmo-plast	CC,
60	**At4g14150 (AtPAKRP1)**	N-terminal	1292[e] (1294)	24	ND	Spindle midzone, Phragmo-plast	CC

Tab. 2   *(Continued)*

	Gene code (published name)	Motor location	#aa	Number of introns	Motility	Cellular localization	Other Domains
61	**At4g14330 (AtPAKRP2)**	N-terminal	959 (869)	9	ND	Punctate staining of Phrag-moplast	CC

[a] Motility was determined using in vitro assays.
[b] Although the motor is N-terminal, it groups into C-terminal family.
[c] Although the motor is internal, it groups into C-terminal family.
[d] Slight discrepancy in #aa between AtDB and MIPS predictions;
[e] NCBI predicted protein has 1662 amino acids.
Notes: The number of amino acids in parenthesis are based on the deduced number from cDNA.
CC: coiled coil; CH: calponin homology domain; MyTH4: domain present in the tail region of some myosins; Talin-like: talin-like domain found in some myosins and band 4.1 superfamily; CBD: calmodulin binding domain; PEST: motif rich in proline, glutamine, serine, and threonine residues; ARM: Armadillo/beta-catenin-like repeats; HhH1: helix-hairpin-helix.
(Modified from Reddy, A.S.N., Day, I.S. (2001a) Analysis of the myosins encoded in the recently completed *Arabidopsis thaliana* genome sequence, *Genome Biol.* **2**, 24.21–24.17).

of spectrin-like proteins. The CH domain is a protein module of approximately 110 residues found in cytoskeletal and signal transduction proteins, either as two domains in tandem or as a single copy. Proteins with a tandem pair of CH domains cross-link F-actin, bundle actin, or connect intermediate filaments to cytoskeleton. Proteins with a single copy are involved in signal transduction. Perhaps the kinesins containing a CH domain bind actin and are involved in signal transduction or linking of actin and MTs. Some domains that are involved in protein–protein or protein–DNA interactions are also present in some kinesins (Fig. 1 and Table 2). These include armadillo repeats, a helix-hairpin-helix DNA-binding domain. KCBP has MyTH4 and talin-like domains present in some myosins, suggesting that it has domains of both MT- and actin-based motors. Such motors may be involved in cross talk between MT and actin cytoskeleton. KCBP also has

been shown to have a calmodulin-binding domain.

1.1.1
Phylogenetic Analysis

In nonplant systems, nine subfamilies of kinesins with some ungrouped kinesins have been identified by phylogenetic analysis using the conserved motor domain. Three of the subfamilies (KHC, KRP85/95, and Unc104/KIF1) are involved in transport. Members from the other subfamilies (C-terminal, Kip3, MKLP1, BimC, chromokinesin/KIF4, and MCAK/KIF4) function in various processes associated with cell division. A phylogenetic analysis of the 61 Arabidopsis kinesins motor domain sequences with 113 other motor domain sequences from nonplant systems has revealed that seven of the nine families are represented in Arabidopsis (Figs. 2 and 3). However, several Arabidopsis kinesins do not fall into any of the nine subfamilies and are likely to represent

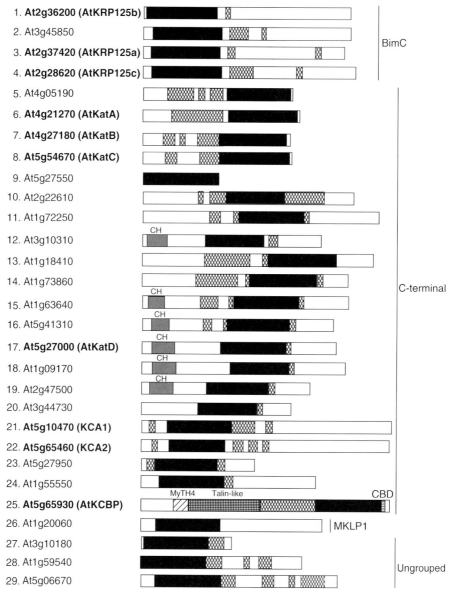

Fig. 1 Structural features of Arabidopsis kinesins. The deduced amino acid sequence of kinesins in the Arabidopsis database was analyzed for the presence of various domains. The kinesins are arranged according to their grouping in the phylogenetic tree in Fig. 3. Characterized kinesins are indicated in bold. ARM, Armadillo/beta-catenin-like repeat, CBD, calmodulin-binding domain; CC, coiled-coil region; CH, Calponin homology domain; MD, motor domain; MyTH4, myosin tail homology region; Talin-like, talin-like domain; HhH1, helix-hairpin-helix domain. Bar = 100 aa. (Modified from Reddy, A.S.N., Day, I.S. (2001b) Kinesins in the Arabidopsis genome: A comparative analysis among eukaryotes, *BMC Genomics* **2**, 2.1–2.13).

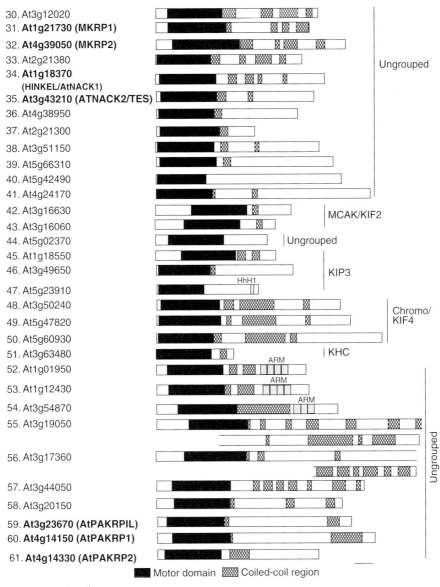

30. At3g12020
31. **At1g21730 (MKRP1)**
32. **At4g39050 (MKRP2)**
33. At2g21380
34. **At1g18370**
    (HINKEL/AtNACK1)
35. **At3g43210 (ATNACK2/TES)**
36. At4g38950
37. At2g21300
38. At3g51150
39. At5g66310
40. At5g42490
41. At4g24170
42. At3g16630
43. At3g16060
44. At5g02370
45. At1g18550
46. At3g49650
47. At5g23910
48. At3g50240
49. At5g47820
50. At5g60930
51. At3g63480
52. At1g01950
53. At1g12430
54. At3g54870
55. At3g19050
56. At3g17360
57. At3g44050
58. At3g20150
59. **At3g23670 (AtPAKRPIL)**
60. **At4g14150 (AtPAKRP1)**
61. **At4g14330 (AtPAKRP2)**

Ungrouped

MCAK/KIF2

Ungrouped

KIP3

Chromo/KIF4

KHC

Ungrouped

HhH1

ARM

ARM

ARM

■ Motor domain   ▨ Coiled-coil region

Fig. 1 (*Continued*)

additional subfamilies that are unique to plants (Figs. 2 and 3). Most of the Arabidopsis kinesins are more closely related to another Arabidopsis or another plant kinesin than to any other kinesin used in the comparison. The subfamilies that are involved in transport are underrepresented in Arabidopsis (Fig. 2). There are no members of the KRP85/95 or Unc104/KIF1 subfamilies in Arabidopsis (Figs. 2 and 3). No KHCs have been reported in plants. The phylogenetic tree indicates that there is possibly one KHC-type kinesin in Arabidopsis. At3g63480

Tab. 3 Myosins in arabidopsis.

Name	# of aa	Gene code	EST/cDNA	Class	Domains
1. **At ATM**	1166	AT3g19960 (ATM1)[a]	+	VIII	MD,CC,IQ
2. **At ATM2**	1111	AT5g54280	+	VIII	MD,CC,IQ
	1101[b]	ATM2/AtMYOS1[a]			
3. At VIIIA	1085	AT1g50360	+	VIII	MD,CC,IQ
4. At VIIIB	1126	AT4g27370		VIII	MD,CC,IQ
5. **At MYA1**	1520	AT1g17580	+	XI	MD,CC,IQ
	1599[c]	(AtMYA1)[a]			
6. **At MYA2**	1505[c]	AT5g43900	+	XI	MD,IQ
	1515	(AtMYA2)[a]			
7. At XIA	1730	AT1g04600		XI	MD,CC,IQ
8. At XIB	1519	AT1g04160	+	XI	MD,IQ
9. At XIC	1572	AT1g08730	+	XI	MD,CC,IQ
10. At XID	1611	AT2g33240	+	XI	MD,CC,IQ
11. At XIE	1529	AT1g54560		XI	MD,CC,IQ
12. At XIF	1556[d]	AT2g31900	+	XI	MD,IQ
13. At XIG	1502	AT2g20290		XI	MD,CC,IQ
14. At XIH	1452[d]	AT4g28710	+	XI	MD,CC,IQ
15. At XI-I	1374	AT4g33200	+	XI	MD,CC,IQ
16. **At XIJ**	1242	AT3g58160		XI	MD,CC,IQ
	963[b]	(AtMYOS3)[a]			
	998[b]	(AtMYA3)[a]			
17. At XIK	1544	AT5g20490		XI	MD,CC,IQ

[a] Name as reported in the literature.
[b] Number of amino acids previously reported for partial sequence.
[c] Number of amino acids predicted by NCBI.
[d] Edited for full-length sequence.
Notes: MD: motor domain; CC: Coiled-coil region; IQ: putative calmodulin-binding motif
(Modified from Reddy, A.S.N., Day, I.S. (2001a) Analysis of the myosins encoded in the recently completed *Arabidopsis thaliana* genome sequence, *Genome Biol.* **2**, 24.21–24.17).

falls into the KHC group with a closer relationship to KHCs found in fungi and, like fungal KHC, lacks binding site for kinesin light-chain proteins. However, biochemical and functional studies are needed to verify that At3g63480 codes for a KHC. One Arabidopsis kinesin groups with the MKLP1 subfamily (Fig. 3). Two internal motor Arabidopsis kinesins grouped with the MCAK/KIF2 subfamily (also called *the internal family*), whose members are involved in vesicle transport, chromosome movement, and MT catastrophe. Three Arabidopsis kinesins fall into a group with Kip3 subfamily members in which ScKip3 is involved in nuclear movement. Three Arabidopsis kinesins form a branch off of the chromokinesin/KIF4 subfamily members, some of which are involved in vesicle transport (HsKIF) and spindle organization and chromosome positioning (Xlklp1). Arabidopsis has many kinesins (24 out of 61) that did not fall into any of the known groups (Figs. 2 and 3). Members of BimC subfamily, which are N-terminal plus-end motors, are present in all five eukaryotic organisms that have been sequenced. The three NtKRP125-like

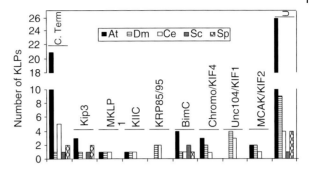

Fig. 2 The number of kinesins in each family in different organisms. C.Term, C-Terminal motor; Chromo/KIF4, chromokinesin/KIF4; U, ungrouped. At, *Arabidopsis thaliana*; Dm, *Drosophila melanogaster*; Ce, *Caenorhabditis elegans*; Sc, *Saccharomyces cerevisiae*; Sp, *Schizosaccharomyces pombe* (Adapted from Reddy, A.S.N., Day, I.S. (2001b) Kinesins in the Arabidopsis genome: A comparative analysis among eukaryotes, *BMC Genomics* **2**, 2.1–2.13).

Arabidopsis kinesins (AtKRP125a, b, and c) are grouped with the tobacco homolog in the BimC subfamily, which are involved in cross-linking and antiparallel sliding of MTs.

Twenty-one Arabidopsis kinesins were grouped into the C-terminal subfamily. This is an unusually large number compared to the other organisms. It is also unusual in that 11 kinesins in this group have internal motors and five have N-terminal motors. The internal kinesins have a motor domain that is closer to the C-terminus than the N-terminus, but each has some sequence C-terminal to the motor domain and could be called internal, depending on the parameters used to define an internal motor. It will be interesting to find out the direction of movement of these kinesins whose motor domains are either N-terminal or internal, but are most closely related to the C-terminal subfamily. Kinesins in the C-terminal subfamily translocate toward the minus end of MTs and have a conserved sequence at the neck/motor core junction. It was reported that the residues G and

N residues at the neck/motor core junction are necessary for minus-end directed movement. Analysis of the neck/motor core junction of the 21 C-terminal class Arabidopsis kinesins showed conservation of these residues in most of the C-terminal Arabidopsis kinesins.

Several Arabidopsis kinesins show some similarity to other ungrouped kinesins (Figs. 2 and 3). The ungrouped centromeric proteins (HsCENPE and UmKin1) cluster with seven Arabidopsis kinesins. CENP-E binds to the kinetochore throughout mitosis and to MTs of the spindle midzone during late stages of mitosis. One Arabidopsis kinesin is paired with HsKid, an ungrouped kinesin. HsKid is a kinesin-like DNA-binding protein that is involved in spindle formation and the movements of chromosomes during mitosis and meiosis. Arabidopsis PAKRP1 along with five other Arabidopsis kinesins are grouped with XlKlp2. XlKlp2 is required for centrosome separation and maintenance of spindle bipolarity, whereas PAKRPs associates with the phragmoplast and is

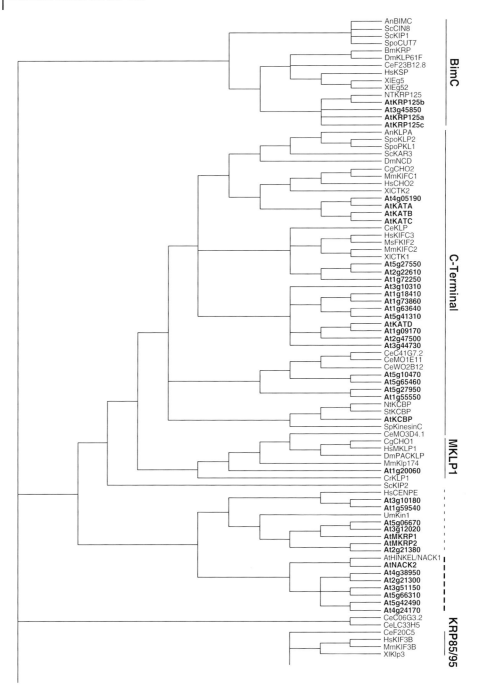

Fig. 3 *(Continued on page 476)*

expected to function in cytokinesis. Three Arabidopsis kinesins form a subgroup separate from any other kinesin, but share a branch with CeLF22F4 and DmNOD (Fig. 3). Eight other Arabidopsis kinesins form a subgroup separate from kinesins of any other organism. In many cases, a group of Arabidopsis kinesins forms a separate branch within the major subgroup in which they fall.

The large number of kinesins in plants is expected to control many diverse cellular processes including some plant-specific functions. There are many MT-associated processes that are unique to plants. For example, during cell division in plants, several plant-specific MT arrays such as the preprophase band and the phragmoplast are formed that are important in determining the future location of the cell wall and cell wall formation respectively. These unique processes are likely to require additional plant-specific motors. In addition, centrosomes play an important role in MT organization in animals, whereas plants have no well-defined centrosomes. Hence, MT organization and dynamics in plants may also require additional MT motors. The presence of unique organelles such as chloroplasts and MT-dependent processes such as cell wall synthesis may also

warrant for additional motors. Also, cell-to-cell transport of macromolecules such as RNA and viruses through plasmodesmata may require MT-based motors. It is very likely that plant kinesins may participate in functions other than transport and MT dynamics and organization. Recently, it has been shown that two plant kinesins function as activators of a protein kinase. Since Arabidopsis does not have dyneins, some kinesins may be performing functions that are carried out by dyneins in other systems.

1.2
Dyneins

Dyneins move along MTs toward the minus ends. The motor domain in dyneins (~1000 amino acids) is much longer than kinesins (~340 amino acids) and does not share sequence similarity with the kinesin motor domain. Cytoplasmic dynein is a large multisubunit complex consisting of two heavy chains (~530 kDa each), three intermediate chains (74 kDa), and four light intermediate chains (~55 kDa). The central and C-terminal regions of dynein heavy chain, which are predicted to form globular structure, interact with MTs and contain motor activity, whereas the N-terminal region is thought to bind

---

Fig. 3 Phylogenetic tree of all Arabidopsis kinesins along with 113 nonplant kinesins. The tree was generated using the motor domain of kinesins, as described. Vertical dashes indicate ungrouped kinesins (light dashes – those grouped with other kinesins, bold dashes – exclusively Arabidopsis kinesins). The Arabidopsis kinesins are in bold. Kinesins from the following organisms were used: An, *Aspergillus nidulans*; Bm, *Bombyx mori*; Ce, *Caenorhabditis elegans*; Cf, *Cylindrotheca fusiformis*; Cg, *Cricetulus griseus*; Cr, *Chlamydomonas rheinhardtii*; Dd, *Dictyostelium discoideum*; Dm, *Drosophila melanogaster*; Gg, *Gallus gallus*; Hs, *Homo sapiens*; Lc, *Leishmania chagasi*; Lm, *Leishmania major*; Lp, *Loligo pealii*; Mm, *Mus musculus*; Ms, *Morone saxatilis*; Nc, *Neurospora crassa*; Nh, *Nectria haematococca*; N, *Nicotiana tabacum*; St, *Solanum tuberosum*; Rn, *Rattus norvegicus*; Sc, *Saccharomyces cerevisiae*; Sp, *Strongylocentrotus purpuratus*; Spo, *Schizosaccharomyces pombe*; Sr, *Syncephalastrum racemosum*; Um, *Ustilago maydis*; Xl, *Xenopus laevis*. (Modified from Reddy, A.S.N., Day, I.S. (2001b) Kinesins in the Arabidopsis genome: A comparative analysis among eukaryotes, *BMC Genomics* **2**, 2.1–2.13).

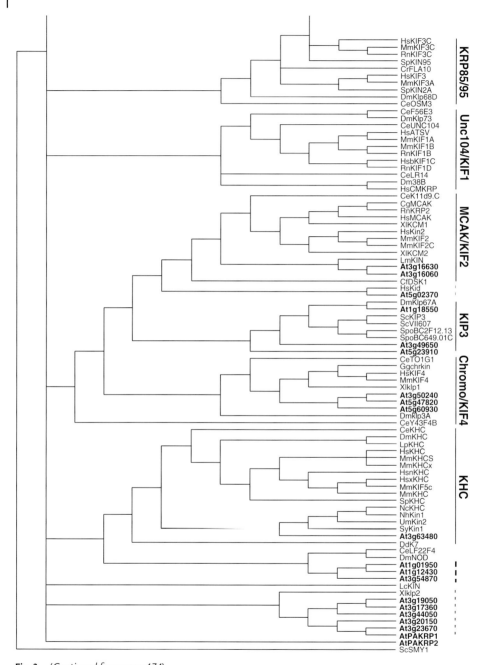

Fig. 3 *(Continued from page 474)*

cargo. The intermediate and light intermediate chains of cytoplasmic dynein associate with a protein complex called *dynactin*, which is made of 10 subunits. Although there is some immunological and biochemical data indicating the presence of dynein heavy chains in plants, sequences that are similar to dynein and dynactin subunits have not been found in the Arabidopsis genome, suggesting that they may be absent in flowering plants. The expanded family members of the kinesin superfamily may carry out the functions performed by dyneins in animals. Interestingly, dynein genes appear to be present in rice and tobacco.

2
Actin-based Motors

2.1
Myosins

Myosins, a diverse group of actin-based molecular motors, perform a broad array of cellular functions. A large number of myosins have been described in many eukaryotes. Most of these are identified on the basis of sequence similarity with the myosin motor domain. All myosin heavy chains have three common domains: a highly conserved motor domain ($\sim$850 amino acids) located at the amino-terminus in almost all myosins, which is followed by a neck region that binds to calmodulin or calmodulin-related proteins, and a C-terminal tail domain. The calmodulin-binding domain ''IQ motif'' in these myosins has a consensus sequence ''IQXXXRGXXXR'' (I, isoleucine; Q, glutamine; R, arginine; G, glycine). This motif is present in one to seven repeats, depending on the type of myosin. The only myosin that does not contain an ''IQ motif'' is

myosin XIV. The tail region of different myosins varies considerably in length and structure. The tail is characteristic to each myosin with interesting domains that are also found in nonmotor proteins. The number of myosins in an organism varies among organisms. However, not all types of myosins have been identified in a single organism.

Although there is compelling indirect evidence for the role of actin and actin-based motors in various transport processes in plants, little is known about specific plant myosins that perform these functions. Myosins have been identified in plants both biochemically and at the molecular level. Immunological detection of myosins using animal myosin antibodies identified proteins of various sizes from different plants. Immunofluorescence studies localized myosin to the surface of organelles, and the vegetative nuclei and generative cells in pollen grains and tubes, to the active streaming lanes and cortical surface in pollen tubes and, more recently, to plasmodesmata in root tissues. Motility assays and ATPase assays using myosin-like proteins isolated from plants have also demonstrated the presence of myosins in plants. Using PCR-based approaches, a few myosins have been cloned from Arabidopsis and other plants.

A 170-kDa myosin has been purified from lily pollen and the antibodies against this myosin stained particles of various sizes and the apical cytoplasm of lily and tobacco pollen tubes, suggesting that several myosins are likely to be present in pollen tubes. The 170-kDa and 175-kDa myosins that translocate F-actin at a velocity of about 9 µm/s and 3 to 4µm-s respectively were found to be associated with calmodulin. In *in vitro* motility assays, the characean myosin translocated actin at 50 µm-s. In Arabidopsis, the expression of several myosins in a given cell type implies that actin-based motors are involved in a wide range of cellular functions.

By analyzing the Arabidopsis genome sequence with the conserved motor domain of myosins, a total of 17 myosins were identified in Arabidopsis. Similar analysis in rice indicates that there are at least 14 myosins in this organism. Table 3 lists the Arabidopsis myosins by names as given in the phylogenetic tree. In comparison, *S. cerevisiae*, *S. pombe*, *C. elegans*, *Drosophila melanogaster* and *H. sapiens* have 5, 5, 17, 13, and 40 myosins respectively. Five of the 17 Arabidopsis myosins have been reported in the literature. All Arabidopsis myosins have three to six putative calmodulin-binding "IQ" motifs (Fig. 4). The IQ domains usually follow right after the motor domain, but are separated slightly from the motor domain in At XID, At XI-I, and At XIK. There are three or four IQ domains in Class VIII myosins and five or six in Class XI, except for At XIK, which has only four (Fig. 4). There are coiled-coil domains in all the myosins that differ in length and number. They often follow directly after the IQ domains, but, in some cases, there is intervening sequence. On the basis of the presence of the coiled-coil domains, it is likely that the Arabidopsis myosins either form dimers or interact with other proteins. The Class XI myosins are much longer than the Class VIII myosins with the difference being in the length of the C-terminal region following the conserved domains found in myosins. Myosins containing IQ domains are typically calmodulin-sensitive. However, the interaction with, or regulation by, calmodulin has not yet been demonstrated for Arabidopsis myosins. The information on myosin light chains is limited in plants. Calmodulin associates with two myosins from lily pollen and regulates their motor

activity. Besides the motor, IQ, and coiled-coil domains, several other domains have been identified in myosins from classes other than the plant classes VIII and XI. These include SH3 domains (Src homology 3 domains), ankyrin repeats, MYTH4 (domain of unknown function found in a few classes of myosins), zinc-binding domain, plecstrin homology, FERM/talin (band 4.1/ezrin/radixin/moesin), GPA-rich domains, and a protein kinase domain. Plant myosins do not have any of these other domains.

2.1.2 Phylogenetic Analysis

Phylogenetic analysis using the conserved motor domain grouped all known myosins into 18 distinct classes. Phylogenetic analysis of the Arabidopsis myosins motor domain with nonplant and plant myosins revealed that all the Arabidopsis myosins and other plant myosins fall into only two groups, Class VIII and Class XI (Fig. 5). Four Arabidopsis myosins are in Class VIII and 13 in Class XI. These groups contain exclusively plant or algal myosins with no animal or fungal myosins, suggesting that plants have a unique set of myosins (Fig. 5). The motor domain in all cases is in the N-terminal region (Fig. 4). The motor domain starts at about 50 to 55 residues for the Class XI myosins, while the Class VIII myosins have a longer N-terminal region prior to the motor domain (99–159 residues). An algal (*Chara corallina*) myosin, Cc ccm, does group with the plant Class XI myosins, but is on a separate branch from any other Class XI myosin (Fig. 5). A phylogenetic tree that was constructed using the full-length sequences of Arabidopsis and other plant myosins showed similar grouping of plant myosins. Among the Class XI myosins, the similarity ranges from 40 to 85% (full length) and 61 to

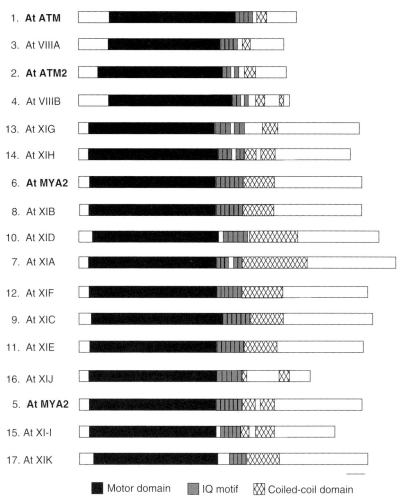

Fig. 4 Schematic diagram showing various predicted domains in Arabidopsis myosins. The Arabidopsis genome database was searched with the conserved motor domain of a myosin to identify Arabidopsis myosins. The deduced amino acid sequence of Arabidopsis myosins was analyzed for the presence of various domains. The numbers refer to the number in Table 1. The first four myosins are in Class VIII and the following 13 are in Class XI. The bar represents 100 amino acids. (Adopted from Reddy, A.S.N., Day, I.S. (2001a) Analysis of the myosins encoded in the recently completed *Arabidopsis thaliana* genome sequence, *Genome Biol.* **2**, 24.21–24.17).

91%(motor domain). The similarity between the Class VIII myosins ranges from 50 to 83% (full length) and 64 to 92% (motor domain). When Class VIII myosins are compared to Class XI myosins, the similarity only ranges from 22 to 29% (full length) and 35 to 42% (motor domain). Myosins have 131 highly conserved residues spread throughout the motor domain that define a core consensus sequence. Comparison of an alignment of Arabidopsis myosin motor domains to these conserved sequences

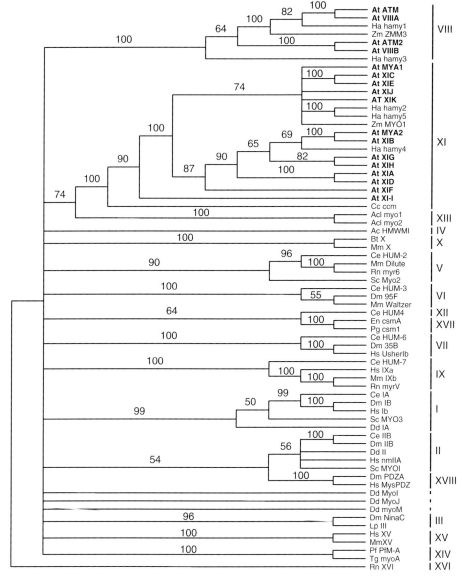

Fig. 5 Phylogenetic analysis of Arabidopsis myosins. A phylogenetic tree was generated as described earlier. The myosin groups are identified on the left in roman numerals. The Arabidopsis myosins are in bold. Myosins from the following organisms were used: Ac, *Acanthamoeba castellani;* Acl, *Acetabularia cliftoni;* Cc, *Chara corallina,* Ha, *Helianthus annuus;* Zm, *Zea mays;* Bt, *Bos taurus;* Mm, *Mus musculus;* Ce, *Caenorhabditis elegans;* Dm, *Drosophila melanogaster;* Rn, *Rattus norvegicus;* Sc, *Saccharomyces cerevisiae;* Hs, *Homo sapiens;* Dd, *Dictyostelium discoideum;* Lp, *Limulus polyphemus;* En, *Emericella nidulans;* Pg, *Pyricularia grisea;* Pf, *Plasmodium falciparum;* and Tg, *Toxoplasma gondii.* The number at the branches indicates the number of times the dichotomy was supported out of 100 bootstrap tries.

shows a great deal of conservation among them, suggesting that they are capable of motor function.

3
Cellular Roles of Motors

3.1
Cell Division

Although many of the dynamic processes of cell division such as formation of the spindle, congression of chromosomes at the equatorial region, and migration of chromatids during cell division are common to both animals and plants, there are significant differences between plant and animal cell division. For example, the dividing plant cells organize their cytoskeletal elements into unique structures that are not found in nonplant systems. Just prior to prophase, plant cortical MTs rearrange to form a band of MTs called the *preprophase band* (PPB), which accurately predicts the future location of the cell plate. Another distinctive feature of plant mitosis is the formation of a bipolar spindle in the absence of centrosomes. The structures involved in the organization of MTs are the centrosomes in animal cells and the spindle pole body (SPB) in fungal cells, which are lacking in plant cells. Finally, cytokinesis, the process that produces two daughter cells following the completion of nuclear division, in plants is accomplished quite differently from plants. Cytokinesis in plant cells occurs via the formation of a polysaccharide cell plate, which expands from the center to the periphery. Phragmoplast, a structure composed of two disks of parallel MTs, with their plus ends toward the equatorial zone, and actin filaments is involved in cell-plate formation. Vesicles carrying the cell plate materials are transported to the equatorial zone of the phragmoplast. The association of vesicles with phragmoplast MTs indicates that MT-based motors are likely to be involved in transporting the vesicles to the cell plate. The fusion of vesicles in the cell plate is mediated by phragmoplastin, a dynamin homolog, in plants. The plant spindle, which is formed in the absence of a well-defined centrosome, and several unique plant-specific MT arrays such as the preprophase band and the phragmoplast suggest that plants are likely to contain novel kinesins that are not present in animal cells.

In animals, six subfamilies of kinesins have been shown to be involved in some aspect of cell division. Several recent reports on kinesins expression and immuunolocalization indicate that at least nine plant kinesins (Kat A, KatB/C, TKRP125, AtKCBP, AtPAKRP1, AtPAKRP1L, AtPAKRP2, DcKRP120-2, HINKEL/AtNACK1, AtNACK2/TES) that belong to different families have a mitotic function. Some kinesins are expressed in a cell cycle–dependent manner. TKRP125, a member of BimC subfamily and NCAK from tobacco and two kinesins from Arabidopsis (KCBP, KatB/C) show a high level of expression in M-phase of the cell cycle. DcKRP120-1 and DcKRP120-2 are not detectable in nondividing cells. TKRP125 localizes to the cortical MTs in the S-phase, along MTs of PPB and perinuclear MTs in prophase, the equatorial plane of spindle MTs in metaphase and anaphase, and the phragmoplast MTs in telophase and cytokinesis. A carrot homolog of TKRP125 (DcKRP120-1) also localizes to the cortical MTs, the preprophase band, spindle, and the phragmoplast. Another member of BimC subfamily (DcKRP120-2) stained the spindle and phragmoplast, predominantly in the midline of the phragmoplast where plus ends of the MTs overlap. AtPAKRP1, a nonmember of BimC, which resembles an ungrouped Xklp2 from Xenopus, also localizes to the midline of the phragmoplast. KatA, a C-terminal motor, localizes near the midzone at metaphase and anaphase and the phragmoplast in Arabidopsis and tobacco BY-2 suspension cells. Anti-KCBP antibodies stain the preprophase band, the spindle, and the phragmoplast. The KCBP antibody did not stain the cell plate region of the phragmoplast, but MT bundles on either side of the cell plate are strongly labeled (Fig. 6). Localization of KCBP in *Haemanthus* also showed localization to mitotic MTs arrays. In *Haemanthus,* anti-KCBP staining is seen almost exclusively at the spindle poles in late anaphase. The assembly of the acentriolar spindle in plants may involve convergence of MT minus ends leading to the formation of spindle poles. The fact that KCBP is a minus-end-directed motor and localizes to the spindle poles suggests that it may be involved in acentriolar spindle formation in plants. Between early and late anaphase, the localization of KCBP shifts toward the spindle pole, supporting a role for KCBP in the formation of a converging bipolar spindle. Localization of minus-end (KatA and KCBP) and plus-end (TKRP125) motors in spindle suggests their involvement in counterbalancing forces generated by plus- and minus-end motors to stabilize the spindle. Constitutive activation of KCBP during late prophase caused premature breakdown of nuclear envelope and arrest of cells at prometaphase, whereas activation of KCBP at late metaphase or anaphase did not affect the progression of anaphase, but caused aberrant phragmoplasts formation and delayed cytokinesis. This study suggests that KCBP is differentially active during the various phases of cell division. Its activity is downregulated

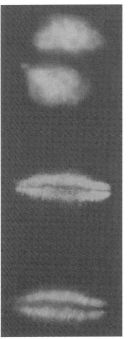

Fig. 6 Triple localization of nucleus, MTs, and KCBP in a cytokinetic cell of tobacco. Nuclei stained with DAPI (DAPI), MTs (MT) are localized with a monoclonal antibody to β-tubulin and KCBP (KCBP) was localized with affinity-purified anti-KCBP antibody. Secondary antibodies conjugated to FITC and Cy3 were used to detect MTs and KCBP respectively. (From Bowser, J., Reddy, A.S. (1997) Localization of a kinesin-like calmodulin-binding protein in dividing cells of Arabidopsis and tobacco, *Plant J.* **12**, 1429–1437).

in metaphase and telophase and upregulated in anaphase most likely by changes in cytosolic calcium level. Colocalization of some of the kinesins with mitotic MT arrays suggests that they may have a role in forming these arrays by bundling MTs. The motor domain of KCBP has been shown to bundle MTs.

Numerous forces are at play in the development, maintenance, and function of the phragmoplast. Golgi-derived vesicles are transported to the forming cell plate where the plus ends of phragmoplast MTs interdigitate. It is likely that MT motors are involved in forming the phragmoplast and powering the movement of the vesicles along the MTs. Seven plant kinesins (Kat A, TKRP125, AtKCBP, AtPAKRP1, AtPAKRP2, DcKRP120-2, HINKEL/ AtNACK1) including plus- and minus-end motors and some plant-specific kinesins are associated with the phragmoplast and so are expected to function in cytokinesis. The localization patterns with various kinesins differed considerably. Immunolocalization studies as well as phragmoplast MT gliding assay in the presence of TKRP antibodies indicates its involvement in the organization of MT arrays especially in phragmoplast function. The polarity of MTs and microfilaments in the phragmoplast indicates that plus-end motors are likely to be involved in the transport of vesicles to the cell plate. Of the eight kinesins that localize to the phragmoplast, six are N-terminal motors (TKRP125, PAKRP1, PAKRP1L, DcKRP120-2, PAKRP2, HIK/NCAK) and two are C-terminal (KatA and KCBP). TKRP125 is a plus-end motor, and the other four N-terminal kinesins are likely plus-end motors. One or more of these are likely to be involved in vesicle transport to the cell plate. The punctate-staining pattern of PAKRP2 and its association

with vesicles suggest that it is involved in transporting vesicles to the cell plate. It is unlikely that KCBP and KatA play a role in vesicle transport to the cell plate as these are known to be minus-end motors and the MTs are oriented with their plus ends facing the cell plate. The possible functions of minus-end motors are organization/stabilization of phragmoplast and/or recycling of Golgi vesicle membranes from the expanding cell plate. Functional analysis of TKRP125 indicates its role in cross-linking and sliding of antiparallel MTs in the phragmoplast. Immunofluorescence with AtPAKRP1 and AtPAKRP1L antibodies showed staining of interzonal MTs at late anaphase and later on its localization is restricted to the plus ends of interdigitating MTs in the phragmoplast. Functional studies with this kinesin indicate its involvement in maintaining the integrity of phragmoplast MTs. In order for the growth of the phragmoplast to be achieved, a rapid turnover of phragmoplast MTs must occur. The expansion of the phragmoplast from the center to the periphery involves disassembly of the MTs inside and assembly of MTs at the outside. Blocking of disassembly in the center by MT-stabilizing drugs causes cessation of phragmoplast expansion and formation of incomplete cell walls. Phragmoplast-associated kinesins may play a role in the MT assembly and disassembly associated with its growth. The loss of MTs from the center of the phragmoplast as it grows does not occur in *hik* mutants, suggesting that it controls directly or indirectly the dynamics of MTs in the phragmoplast. HIK does not appear to be involved in either vesicle transport or organization of MTs in the phragmoplast. Studies with tobacco NCAK1, a homolog of HINKEL, showed its localization to the equatorial zone of the phragmoplast, but not in the inner regions of the phragmoplast where the cell plate matures. It appears that NCAK through a protein kinase, NPK1, regulates depolymerization of MTs in the center of the phragmoplast as it expands.

There is some evidence of the presence of a myosin in the phragmoplast, suggesting that actin-based motors may also be involved in the transport of vesicles on the phragmoplast. During cell division, the anti-Class VIII myosin staining remains confined to the transverse cell walls and is strongest in the newly formed cell wall. Immunogold electron microscopy showed labeling of Class VIII myosin associated with the plasma membrane and plasmodesmata. These studies suggest that Class VIII myosins may be involved in new cell wall formation and transport in the plasmodesmata.

3.2
Cell Polarity and Morphogenesis

Both actin and MT cytoskeleton have been implicated in cell polarity and morphogenesis. The first indication that a kinesin might play a role in morphogenesis came from studies with the *zwichel* mutant. In wild-type Arabidopsis, trichomes are unicellular with a stalk and three branches. In *zwichel* (*zwi*), mutants' trichomes are abnormal with a short stalk and one or two branches, depending on the severity of the allele (Fig. 7). Cloning of *ZWICHEL* has indicated that it is a calmodulin-binding kinesin, suggesting the requirement of this motor for expansion of the stalk and branching. Using paclitaxel, a known MT stabilizer, *zwichel* mutants were induced to form similar growth points indicative of branch formation in normal trichomes. Transient stabilization of MTs could compensate for KCBP/ZWI activity, suggesting that it may be involved in stabilization of MTs. The two MT-binding sites of KCBP could be involved in stabilization or possibly a KCBP/ZWI-interacting protein(s) could be responsible for the stabilization. Several alleles of *zwi* have been characterized. All *zwi* mutants grew normally with no apparent defects in cell division, except that they contain abnormal trichomes. The lack of a phenotype in other tissues suggests that either KCBP is nonessential or another motor

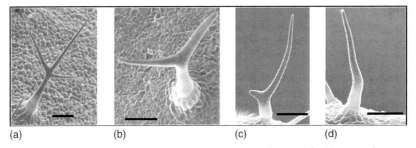

(a)   (b)   (c)   (d)

Fig. 7 Scanning electron micrographs of Arabidopsis trichomes of wild-type and *zwichel/kcbp* mutants. (a) Wild type; (b) *zwi*; (c) *zwi w2*; (d) *zwi9311-11*. In *zwi w2*, most trichomes have two branches but the length of the second branch is varied. In the *zwi9311-11* mutant, about 40% of trichomes are unbranched and the rest have two branches. Seeds of *zwichel* mutants were kindly provided by Dr. David Oppenheimer and Dr. Martin Hülskamp. Scale bar = 100 μm (From Reddy, A.S.N., Day, I.S. (2000) The role of the cytoskeleton and a molecular motor in trichome morphogenesis, *Trends Plant Sci.* **5**, 503–505).

with overlapping functions may substitute for ZWI function in other tissues. In Arabidopsis, there are 61 kinesins including several C-terminal motors. Three extragenic suppressors (suppressor of *zwichel*-3; *suz1, suz2,* and *suz3*) that rescued trichome branch number defect in a *zwichel* mutant have been isolated. All three suppressors were found to be allele-specific, indicating direct interactions between these proteins. The *suz1 zwi-3* double mutants are male sterile due to a defect in pollen germination and pollen tube growth, suggesting a role for these genes in pollen germination and tube growth. At least four proteins (two calcium-binding proteins, a plant-specific protein kinase, and ANGUSTIFOLIA, a protein required for trichome branching) have been shown to interact with KCBP/ZWI. Inactivation of one of the myosins (*MYA2*) of myosin XI family in Arabidopsis also results in abnormal (unbranched) trichomes.

3.3
Cytoplasmic Streaming

Cytoplasmic streaming in plant cells is important in intracellular transport. The force that drives the cytoplasmic streaming appears to be dependent on actin and actin-based motors. Insight into the function of plant myosins in cytoplasmic streaming has been gained primarily by studies done with algae and pollen tubes. Characean cells exhibit a very rapid cytoplasmic streaming ($\sim 70 \, \mu m \, s^{-1}$). Also, in pollen tubes, vesicles flow rapidly from the shank to the tip of the pollen tube where they fuse to support extensive tip growth of these cells. In addition to the bidirectional movement of vesicles, the generative cell and the nucleus move unidirectionally toward the tip. In *Chara,* an increase in Ca^{2+} concentration causes cytoplasmic streaming to stop. A myosin isolated from the alga *Chara corallina* was shown to be responsible for cytoplasmic streaming. The myosin was cloned and found to be a Class XI myosin related to the Arabidopsis MYA myosins. Using immunofluorescence, myosin was localized to vesicles, organelles, and generative cells and vegetative nuclei in grass pollen tubes. A myosin isolated from lily pollen has been shown to be responsible for cytoplasmic streaming in pollen tubes, and two myosins were identified in tobacco cell cultures are also thought to participate in cytoplasmic streaming. Myosin has been localized to vesicles, chloroplasts, and other organelles, indicating their involvement in transport of these organelles. The stop-and-go movements of Golgi stacks in plants also appear to be dependent on myosin. However, specific myosin(s) responsible for the movement of these organelles have not been identified. Biochemical properties and the localization data of two lily pollen myosins (175 kDa and 170 kDa) suggest that these myosins are at least partially responsible for cytoplasmic streaming, organelle transport, and calcium sensitivity of cytoplasmic streaming. It has been shown that the directional movement of mitochondria is dependent on F-Actin and myosin system. Unlike in animals, the movement of peroxisomes in plants is actin-based and actin polymerization appears to drive the movement of these organelles.

3.4
Microtubule Dynamics and Organization

Microtubules are highly dynamic structures with a half-life of an individual MT of about 10 min. Drugs that disrupt the dynamics of MTs have been shown to severely hamper the MT-dependent process. Treatment of dividing cells with colchicine leads to disappearance of the mitotic spindle, suggesting that MTs in the spindle undergo continuous polymerization and depolymerization. Taxol, in contrast to colchicine, bind to MTs and stabilizes them. Addition of taxol to dividing cells leads to arrest of dividing cells in mitosis. The dynamics of the MT cytoskeleton during cell division in plant cells have been studied extensively. The rapid turnover of plant MTs suggests that the transition between MT arrays and change in orientations may involve depolymerization/repolymerization of MTs and/or movement of MTs. The dynamic instability of MTs is important for their normal function. For example, stabilization of MTs in elongating root hairs causes loss of growth directionality and promotes branching. The MTs of the phragmoplast have a t1/2 of 60 s. However, the mechanisms that regulate the formation and dynamic instability of MT arrays in plants are not understood. Several aspects of MT dynamics such as cross-linking, zippering, bundling, sliding, and stability of MTs is regulated by motor proteins. KCBP like Ncd in *Drosophila* can cross-link and zipper MTs to focus MTs to form spindle poles. Two distinct MT-binding sites and minus-motor activity enable these proteins to perform such a function. In addition, C-terminal motors have been shown to bundle MTs. Some motors are implicated in polymerization and depolymerization of cytoskeletal elements. There is at least one example where a kinesin has no motor activity, but controls MT dynamics. Plant kinesins such as NCAK1 appear to be involved in MT dynamics. Hence, the paradigm that motors bind cargo and move along cytoskeletal tracks

may not explain the functions of all the motors.

3.5
Intercellular Transport

Plant cells are connected through plasmodesmata to form a continuous symplastic network. Recent studies show that macromolecules such as RNA and protein move from cell to cell through plasmodesmata. Cytoskeleton and molecular motors have been implicated in this regulated transport of macromolecules through plasmodesmata. The movement protein of viruses, which enables the movement of viruses from cell to cell colocalizes to MTs, suggesting a role for cytoskeleton and possibly molecular motors in the transport of protein from cell to cell through plasmodesmata. These studies hint at an interesting possibility that motors may play a role in transporting macromolecules from cell to cell and influence cell-to-cell communication. Immunolocalization studies have also detected myosin associated with plasmodesmata. A recent study using an antibody to a cloned Class VIII Arabidopsis myosin ATM1 (At ATM) localized this myosin to the plasmodesmata and the plasma membrane regions involved in the assembly of new cell walls. Earlier work suggested that actin was involved in regulation of plasmodesmal transport. Other studies using antibodies to animal myosins in root tissues of *Allium cepa*, *Zea mays*, and *Hordeum vulagare* have also indicated the presence of myosin in the plasmodesmata.

3.6
Other Functions

It appears that the certain plant motors may perform unexpected novel functions.

For example, two plant kinesins (NCAK1 and NCAK2) have been shown interact with NPK1, a MAP kinase–kinase–kinase, and stimulate its activity. Motors in plants may also be involved in regulating cell polarity, development, and differentiation by transporting the proteins or mRNA to localized regions within the cell. There is substantial indirect evidence indicating that asymmetric localization of mRNA (e.g ASH1 mRNA in yeast), protein (e.g. calmodulin in Drosophila photoreceptors, a MAPKKK in tobacco BY2 cells), and RNA/protein complexes in some eukaryotic cells is dependent on cytoskeletal motors.

4
Regulation of Motors

Although a large number of kinesins have been characterized from diverse organisms (www.blocks.fhcrc.org/~kinesin/), the mechanism that regulate the activity/function of MT-dependent motors is limited both in plants and animals.

4.1
Calcium/Calmodulin

Calcium is a key messenger in transducing many hormonal and environmental signals in plants. Calmodulin, a ubiquitous calcium-binding protein in eukaryotes, is highly conserved in eukaryotes and regulates many diverse cellular functions by modulating the activity of the proteins that interact with it. Upon binding calcium, calmodulin becomes activated via a conformational change and is able to regulate the activity of its target proteins. Cytoplasmic streaming of characean cells is regulated by the cytosolic calcium concentration. Elevated levels of calcium inhibit cytoplasmic streaming and the movement of organelles. Movement of pollen tube organelles along actin cables is inhibited by calcium, suggesting that the activity of actin-based motors is sensitive to calcium levels. The mechanisms by which calcium regulates the activity of these motors is just beginning to emerge. All myosins in Arabidopsis possess two or more putative calmodulin-binding motifs, suggesting that myosins in plants are likely to be regulated by calmodulin in response to changes in intracellular calcium. Two myosin heavy chains (170 and 175 kDa) in plants associate with calmodulin, indicating that calmodulin serves as a light chain for these myosins. Calcium inhibits the myosin motor activity as well as myosin-activated ATPase activity in plants. The inhibition of myosin activity in the presence of calcium appears to be due to dissociation of calmodulin from the heavy chain since calcium inhibition of motor activity can be restored by exogenous addition of calmodulin. The gene encoding 175-kDa myosin has been cloned. Sequence analysis indicates that it is a Class X1 myosin with six IQ motifs. This myosin has been shown to move processively on actin filaments.

KCBP and recently discovered kinesin C are the only kinesins among the kinesin superfamily that bind calmodulin. The binding of activated calmodulin to KCBP inhibits its interaction with MTs via the calmodulin-binding domain, resulting in the inhibition of MT-dependent ATPase activity and motility. These studies strongly suggest that the binding of calmodulin to KCBP affects MT-binding regions, resulting in inhibition of KCBP binding to MTs and dissociation of KCBP/MT complex. Crystal structure of

KCBP motor domain with its calmodulin-binding domain indicates that calmodulin binding to KCBP blocks MT-binding sites on the motor. The calmodulin effect on KCBP suggests that activated calmodulin can act as a molecular switch to down-regulate the activity of KCBP. On the basis of *in vitro* studies with KCBP, it is reasonable to speculate that spatial and temporal changes in free cytosolic calcium levels in response to signals are likely to regulate KCBP activity in the cell.

4.2
Protein Phosphorylation

The members of BimC family have a conserved "BimC Box" in which threonine residue is phosphorylated by cdc2-kinase during cell division. Targeting of the kinesins to spindle is regulated by phosphorylation. TKRP125, all three members of Arabidopsis TKRP-like proteins (AtKRPa, b, and c), and DcKRP120-1 and DcKRP120-2 that belong to BimC family have the conserved BimC box, suggesting that phosphorylation of these kinesins may be involved in targeting them to right cellular location. Recently, it was found that the tail region of KCBP also interacts with a protein kinase (KIPK, KCBP interacting protein kinase) in the yeast two-hybrid system. The association of KIPK with KCBP suggests regulation of KCBP or KCBP-associated proteins by phosphorylation and/or KCBP is involved in targeting KIPK to proper cellular location. The tobacco NPK1, a mitogen-activated protein kinase–kinase–kinase, NPK1, interacts with NACKs. This interaction results in activation and autophosphorylation of NPK1 and phosphorylation of NACK1. Although the significance of NACK1 phosphorylation is not known, NPK1 phosphorylation targets it

to the equatorial zone of the phragmoplasts. A cyclin-dependent kinase (CDK) has been shown to interact with two Arabidopsis kinesins (KCA1 and KCA2). This interaction results in phosphorylation of kinesins, suggesting that the activity/function of these motors may be regulated by CDK.

5
Concluding Remarks

The complete sequencing of several phylogenetically diverse model organisms ranging from a unicellular eukaryote to highly complex multicellular animals and plant has allowed comparative analysis of molecular motors in eukaryotes. Plants contain a large number of motors. About 0.4% of the predicted genes in Arabidopsis encode molecular motors. The comparative analysis provides a framework for future functional studies with plant kinesins. The number of myosins in Arabidopsis is comparable to other multicellular organisms of similar genome size. However, plant myosins are unlike myosins from any other organism except algae. Nonplant myosins contain a variety of other known domains that are lacking in plants. Furthermore, Arabidopsis has a surprisingly large number of kinesins among the completed eukaryotic genomes. The role of molecular motors in many plant cellular processes is beginning to be unraveled. Although the functions of most of the plant motors remains to be determined, phylogenetic analysis and known domains in these proteins are providing clues to their function, which can be tested empirically. Many Arabidopsis kinesins do not fall into any known subfamilies of kinesins, and several Arabidopsis kinesins are not present in other organisms and are likely to represent new plant-specific subfamilies. Analysis of expression of each motor and cellular localization using reporter fusions coupled with studies using the loss-of-function and gain-of-function mutants should help understand the functions of these motors. Identification of the proteins that are associated with motors is also important to our understanding of their roles in plants. Knockout mutants for almost all Arabidopsis motors are already available from SIGnAL (http://signal.salk.edu/). Because of the large number of motors and likely functional redundancy or overlap with other motors, it will be necessary to create not only single mutants but also double or triple mutants. Analysis of the function and regulation of plant motors will be an exciting area of research in plant cell biology and is bound to throw many surprises.

Acknowledgment

I thank members of my laboratory and all my collaborators for their contributions. I wish to thank Irene Day for preparing the figure and critically reading the manuscript. Research on motors in my laboratory is supported by grants from the National Science Foundation (MCB-9630782 and MCB-0079938).

See also E-Cell: Computer Simulation of the Cell; Metabolic Basis of Cellular Energy; Motor Proteins; Nucleic Acid and Protein Single Molecule Detection and Characterization.

Bibliography

Books and Reviews

Asada, T., Collings, D. (1997) Molecular motors in higher plants, *Trends Plant Sci.* **2**, 29–37.

Hepler, P.K., Valster, A., Molchan, T., Vos, J.W. (2002) Roles for kinesin and myosin during cytokinesis, *Philos. Trans. R. Soc. Lond. B. Biol. Sci.* **357**, 761–766.

Hirokawa, N. (1998) Kinesin and dynein superfamily proteins and the mechanism of organelle transport, *Science* **279**, 519–526.

Liu, B., Lee, Y.R.J. (2001) Kinesin-related proteins in plant cytokinesis, *J. Plant Growth Regul.* **20**, 141–150.

Oppenheimer, D.G. (1998) Genetics of plant cell shape, *Curr. Opin. Plant Biol.* **1**, 520–524.

Reddy, A.S.N. (2001) Molecular motors and their functions in plants, *Int. Rev. Cytol. & Cell Biol.* **204**, 97–178.

Reddy, A.S.N., Day, I.S. (2000) The role of the cytoskeleton and a molecular motor in trichome morphogenesis, *Trends Plant Sci.* **5**, 503–505.

Smith, L.G. (1999) Divide and Conquer: cytokinesis in plant cells, *Curr. Opin. Plant Biol.* **2**, 447–453.

Vale, R.D. (2003) The molecular motor toolbox for intracellular transport, *Cell* **112**, 467–480.

Primary Literature

Arn, E.A., MacDonald, P.M. (1998) Motors driving mRNA localization: new insights from in vivo imaging, *Cell* **95**, 150–154.

Asada, T., Collings, D. (1997) Molecular motors in higher plants, *Trends Plant Sci.* **2**, 29–37.

Asada, T., Sonobe, S., Shibaoka, H. (1991) Microtubule translocation in the cytokinetic apparatus of cultured tobacco cells, *Nature* **350**, 238–241.

Baluska, F., Wojtaszek, P., Volkmann, D., Barlow, P. (2003) The architecture of polarized cell growth: the unique status of elongating plant cells, *Bioessays* **25**, 569–576.

Barroso, C., Chan, J., Allan, V., Doonan, J., Hussey, P., Lloyd, C. (2000) Two kinesin-related proteins associated with the cold-stable cytoskeleton of carrot cells: characterization of a novel kinesin, DcKRP120-2, *Plant J* **24**, 859–868.

Baskin, T.I., Cande, W.Z. (1990) The structure and function of the mitotic spindle in flowering plants, *Annu. Rev. Plant Physiol. Plant Mol. Biol.* **41**, 277–315.

Berg, J.S., Powell, B.C., Cheney, R.E. (2001) A millennial myosin census, *Mol. Biol. Cell* **12**, 780–794.

Bezanilla, M., Horton, A.C., Sevener, H.C., Quatrano, R.S. (2003) Phylogenetic analysis of new plant myosin sequences, *J. Mol. Evol.* **57**, 229–239.

Bowser, J., Reddy, A.S.N. (1997) Localization of a kinesin-like calmodulin-binding protein in dividing cells of Arabidopsis and tobacco, *Plant J.* **12**, 1429–1437.

Cai, G., Romagnoli, S., Moscatelli, A., Ovidi, E., Gambellini, G., Tiezzi, A., Cresti, M. (2000) Identification and characterization of a novel microtubule-based motor associated with membranous organelles in tobacco pollen tubes, *Plant Cell* **12**, 1719–1736.

Cyr, R.J., Palevitz, B.A. (1995) Organization of cortical microtubules in plant cells, *Curr. Opin. Cell Biol.* **7**, 65–71.

Day, I.S., Miller, C., Golovkin, M., Reddy, A.S.N. (2000) Interaction of a kinesin-like calmodulin-binding protein with a protein kinase, *J. Biol. Chem.* **275**, 13737–13745.

Deavours, B.E., Reddy, A.S.N., Walker, R.A.N. (1998) Ca^{2+}/calmodulin regulation of the Arabidopsis kinesin-like calmodulin-binding protein, *Cell Motil. Cytoskeleton* **40**, 408–416.

Desai, A., Verma, S., Mitchison, T.J., Walczak, C.E. (1999) Kin I kinesins are microtubule-destabilizing enzymes, *Cell* **96**, 69–78.

Euteneur, U., Jackson, W.T., McIntosh, J.R. (1982) Polarity of spindle microtubules in *Haemanthus* endosperm, *J. Cell Biol.* **94**, 644–653.

Folkers, U., Kirik, V., Schobinger, U., Falk, S., Krishnakumar, S., Pollock, M.A., Oppenheimer, D.G., Day, I.S., Reddy, A.S.N., Jurgens, G., Hulskamp, M. (2002) The cell morphogenesis gene *ANGUSTIFOLIA* encodes a CtBP/BARS-like protein and is involved in the control of the microtubule cytoskeleton, *EMBO J.* **21**, 1280–1288.

Goddard, R.H., Wick, S.M., Silflow, C.D., Snustad, D.P. (1994) Microtubule components of the plant cell cytoskeleton, *Plant Physiol.* **104**, 1–6.

Goldstein, L.S.B., Philip, A.V. (1999) The road less traveled: Emerging principles of kinesin motor utilization, *Annu. Rev. Cell Dev. Biol.* **15**, 141–183.

Gunning, B.E., Wick, S.M. (1985) Preprophase bands, phragmoplasts, and spatial control of cytokinesis, *J. Cell. Sci. Suppl.* **2**, 157–179.

Hayama, T., Shimmen, T., Tazawa, M. (1979) Participation of Ca^{2+} in cessation of cytoplasmic streaming induced by membrane excitation in *Characeae* internodal cells, *Protoplasma* **99**, 305–321.

Hepler, P.K., Hush, J.M. (1996) Behavior of microtubules in living plant cells, *Plant Physiol.* **112**, 455–461.

Heslop-Harrison, J., Heslop-Harrison, Y. (1989) Myosin associated with the surface of organelles, vegetative nuclei and generative cells in angiosperm pollen grains and tubes, *J. Cell Sci.* **94**, 319–325.

Heslop-Harrison, J., Heslop-Harrison, Y. (1990) Dynamic aspects of apical zonation in the angiosperm pollen tube, *Sex. Plant Reprod.* **3**, 187–194.

Hodge, T., Cope, M.J. (2000) A myosin family tree, *J. Cell Sci.* **113**, 3353–3354.

Holweg, C., Nick, P. (2004) Arabidopsis myosin XI mutant is defective in organelle movement and polar auxin transport, *Proc. Natl. Acad. Sci. U S A* **101**, 10488–10493.

Hush, J.M., Wadsworth, P., Callaham, D.A., Hepler, P.K. (1994) Quantification of microtubule dynamics in living plant cells using fluorescence redistribution after photobleaching, *J. Cell Sci.* **107**, 775–784.

Itoh, R., Fujiwara, M., Yoshida, S. (2001) Kinesin-related proteins with a mitochondrial targeting signal, *Plant Physiol.* **127**, 724–726.

Jiang, S., Ramachandran, S. (2004) Identification and molecular characterization of myosin gene family in Oryza sativa genome, *Plant Cell Physiol.* **45**, 590–599.

Kashiyama, T., Kimura, N., Mimura, T., Yamamoto, K. (2000) Cloning and characterization of a myosin from characean alga, the fastest motor protein in the world, *J. Biochem. (Tokyo)* **127**, 1065–1070.

Kikuyama, M., Tazawa, M. (1982) Ca^{2+} ion reversibly inhibits the cytoplasmic streaming of *Nitella*, *Protoplasma* **113**, 241–243.

Kim, A.J., Endow, S.A. (2000) A kinesin family tree, *J. Cell Sci.* **113**, 3681–3682.

Kinkema, M., Wang, H., Schiefelbein, J. (1994) Molecular analysis of the myosin gene family in Arabidopsis thaliana, *Plant Mol. Biol.* **26**, 1139–1153.

Knight, A.E., Kendrick-Jones, J. (1993) A myosin-like protein from a higher plant, *J. Mol. Biol.* **231**, 148–154.

Kohno, T., Shimmen, T. (1988) Mechanism of Ca^{2+} inhibition of cytoplasmic streaming in lily pollen tubes, *J. Cell Sci* **91**, 501–509.

Kohno, T., Ishikawa, R., Nagai, T., Kohama, K., Shimmen, T. (1992) Partial purification of myosin from lily pollen tubes by monitoring with an in vitro motility assay, *Protoplasma* **170**, 77–85.

Kohno, T., Okagaki, T., Kohama, K., Shimmen, T. (1991) Pollen tube extract supports the movement of actin filaments in vitro, *Protoplasma* **161**, 75–77.

Kong, L.J., Hanley-Bowdoin, L. (2002) A geminivirus replication protein interacts with a protein kinase and a motor protein that display different expression patterns during plant development and infection, *Plant Cell* **14**, 1817–1832.

Krishnakumar, S., Oppenheimer, D.G. (1999) Extragenic suppressors of the Arabidopsis zwi-3 mutation identify new genes that function in trichome branch formation and pollen tube growth, *Development* **126**, 3079–3088.

Lawrence, C.J., Malmberg, R.L., Muszynski, M.G., Dawe, R.K. (2002) Maximum likelihood methods reveal conservation of function among closely related kinesin families, *J. Mol. Evol.* **54**, 42–53.

Lawrence, C.J., Morris, N.R., Meagher, R.B., Dawe, R.K. (2001) Dyneins have run their course in plant lineage, *Traffic* **2**, 362–363.

Lazarowitz, S.G., Beachy, R.N. (1999) Viral movement proteins as probes for intracellular and intercellular trafficking in plants, *Plant Cell* **11**, 535–548.

Lee, Y-R.J., Liu, B. (2000) Identification of a phragmoplast-associated kinesin-related protein in higher plants, *Curr. Biol.* **10**, 797–800.

Lee, Y.R., Giang, H.M., Liu, B. (2001) A novel plant kinesin-related protein specifically associates with the phragmoplast organelles, *Plant Cell* **13**, 2427–2439.

Liu, B., Lee, Y.R., Pan, R. (2003) Identification of kinesin-related proteins in the phragmoplast, *Cell Biol. Int.* **27**, 227–228.

Liu, L., Zhou, J., Pesacreta, T.C. (2001) Maize myosins: diversity, localization, and function, *Cell Motil. Cytoskeleton* **48**, 130–148.

Lloyd, C., Hussey, P. (2001) Microtubule-associated proteins in plants–why we need a MAP, *Nat. Rev. Mol. Cell Biol.* **2**, 40–47.

Lloyd, C.W. (Ed.) (1991) *The Cytoskeletal Basis of Plant Growth and Form*, Academic Press, New York.

Marc, J., Granger, C.L., Brincat, J., Fisher, D.D., Kao, T., McCubbin, A.G., Cyr, R.J. (1998) A GFP-MAP4 reporter gene for visualizing cortical microtubule rearrangements in living epidermal cells, *Plant Cell* **10**, 1927–1940.

Marcus, A.I., Li, W., Ma, H., Cyr, R.J. (2003) A kinesin mutant with an atypical bipolar spindle undergoes normal mitosis, *Mol. Biol. Cell* **14**, 1717–1726.

Marcus, A.I., Ambrose, J.C., Blickley, L., Hancock, W.O., Cyr, R.J. (2002) Arabidopsis thaliana protein, ATK1, is a minus-end directed kinesin that exhibits non-processive movement, *Cell Motil. Cytoskeleton* **52**, 144–150.

Mas, P., Beachy, R.N. (2000) Role of microtubules in the intracellular distribution of tobacco mosaic virus movement protein, *Proc. Natl. Acad. Sci. U S A* **97**, 12345–12349.

Mathur, J., Chua, N.H. (2000) Microtubule stabilization leads to growth reorientation in Arabidopsis trichomes, *Plant Cell* **12**, 465–477.

Mathur, J., Mathur, N., Hulskamp, M. (2002) Simultaneous visualization of peroxisomes and cytoskeletal elements reveals actin and not microtubule-based peroxisome motility in plants, *Plant Physiol.* **128**, 1031–1045.

Mathur, J., Spielhofer, P., Kost, B., Chua, N. (1999) The actin cytoskeleton is required to elaborate and maintain spatial patterning during trichome cell morphogenesis in *Arabidopsis thaliana*, *Development* **126**, 5559–5568.

Matsui, K., Collings, D., Asada, T. (2001) Identification of a novel plant-specific kinesin-like protein that is highly expressed in interphase tobacco BY-2 cells, *Protoplasma* **215**, 105–115.

Mitsui, H., Hasezawa, S., Nagata, T., Takahashi, H. (1996) Cell cycle-dependent accumulation of a kinesin-like protein, KatB/C, in synchronized tobacco BY-2 cells, *Plant Mol. Biol.* **30**, 177–181.

Mitsui, H., Nakatani, K., Yamaguchi-Shinozaki, K., Shinozaki, K., Nishikawa, K., Takahashi, H. (1994) Sequencing and characterization of the kinesin-related genes *katB* and *katC* of *Arabidopsis thaliana, Plant Mol. Biol.* **25**, 865–876.

Moscatelli, A., Del Casino, C., Lozzi, L., Cai, G., Scali, M., Tiezzi, A., Cresti, M. (1995) High molecular weight polypeptides related to dynein heavy chains in Nicotiana tabacum pollen tubes, *J. Cell Sci.* **108**, 1117–1125.

Nagai, R. (1993) Regulation of intracellular movements in plant cells by environmental stimuli, *Int. Rev. Cytol.* **145**, 251–310.

Narasimhulu, S.B., Reddy, A.S.N. (1998) Characterization of microtubule binding domains in the Arabidopsis kinesin-like calmodulin-binding protein, *Plant Cell* **10**, 957–965.

Narasimhulu, S.B., Kao, Y.-L., Reddy, A.S.N. (1997) Interaction of Arabidopsis kinesin-like calmodulin-binding protein with tubulin subunits: Modulation by Ca^{2+}-calmodulin, *Plant J.* **12**, 1139–1149.

Nebenfuhr, A., Gallagher, L.A., Dunahay, T.G., Frohlick, J.A., Mazurkiewicz, A.M., Meehl, J.B., Staehelin, L.A. (1999) Stop-and-Go movements of plant Golgi stacks are mediated by the acto-myosin system, *Plant Physiol.* **121**, 1127–1142.

Nishihama, R., Ishikawa, M., Araki, S., Soyano, T., Asada, T., Machida, Y. (2001) The NPK1 mitogen-activated protein kinase kinase kinase is a regulator of cell-plate formation in plant cytokinesis, *Genes Dev.* **15**, 352–363.

Nishihama, R., Soyano, T., Ishikawa, M., Araki, S., Tanaka, H., Asada, T., Irie, K., Ito, M., Terada, M., Banno, H., Yamazaki, Y., Machida, Y. (2002) Expansion of the cell plate in plant cytokinesis requires a kinesin-like protein/MAPKKK complex, *Cell* **109**, 87–99.

Oppenheimer, D.G., Pollock, M.A., Vacik, J., Szymanski, D.B., Ericson, B., Feldmann, K., Marks, D. (1997) Essential role of a kinesin-like protein in Arabidopsis trichome morphogenesis, *Proc. Natl. Acad. Sci. USA* **94**, 6261–6266.

Preuss, M.L., Delmer, D.P., Liu, B. (2003) The cotton kinesin-like calmodulin-binding protein associates with cortical microtubules in cotton fibers, *Plant Physiol.* **132**, 154–160.

Reddy, A.S.N. (2001) Calcium: silver bullet in signaling, *Plant Sci.* **160**, 381–404.

Reddy, A.S.N., Day, I.S. (2001a) Analysis of the myosins encoded in the recently completed *Arabidopsis thaliana* genome sequence, *Genome Biol.* **2**, 24.21–24.17.

Reddy, A.S.N., Day, I.S. (2001b) Kinesins in the Arabidopsis genome: A comparative analysis among eukaryotes, *BMC Genomics* **2**, 2.1–2.13.

Reddy, A.S.N., Narasimhulu, S.B., Safadi, F., Golovkin, M. (1996a) A plant kinesin heavy chain-like protein is a calmodulin-binding protein, *Plant J.* **10**, 9–21.

Reddy, A.S.N., Safadi, F., Narasimhulu, S.B., Golovkin, M., Hu, X. (1996b) A novel plant calmodulin-binding protein with a kinesin heavy chain motor domain, *J. Biol. Chem.* **271**, 7052–7060.

Reddy, V.S., Reddy, A.S.N. (2002) The calmodulin-binding domain from a plant kinesin functions as a modular domain in conferring Ca^{2+}-CaM regulation to animal plus- and minus-end kinesins, *J. Biol. Chem.* **277**, 48058–48065.

Reddy, V.S., Reddy, A.S.N. (2004) Developmental and Cell-specific expression of KCBP/ZWICHEL is regulated by the intron and exon sequences of its upstream protein-coding gene, *Plant Mol. Biol.* **54**, 273–293.

Reddy, V.S., Day, I.S., Thomas, T., Reddy, A.S. (2004) KIC, a novel Ca^{2+} binding protein with one EF-hand motif, interacts with a microtubule motor protein and regulates trichome morphogenesis, *Plant Cell* **16**, 185–200.

Reichelt, S., Knight, A.E., Hodge, T.P., Baluska, F., Samaj, J., Volkmann, D., Kendrick-Jones, J. (1999) Characterization of the unconventional myosin VIII in plant cells and its localization at the post-cytokinetic cell wall, *Plant J.* **19**, 555–567.

Reilein, A.R., Rogers, S.L., Tuma, M.C., Gelfand, V.I. (2001) Regulation of molecular motor proteins, *Int. Rev. Cytol.* **204**, 179–238.

Rogers, G.C., Hart, C.L., Wedman, K.P., Scholey, J.M. (1999) Identification of kinesin-C, a calmodulin-binding carboxy-terminal kinesin in animal (*Strongylocentrotus purpuratus*) cells, *J. Mol. Biol.* **294**, 1–8.

Romagnoli, S., Cai, G., Cresti, M. (2003) In vitro assays demonstrate that pollen tube organelles use kinesin-related motor proteins to move along microtubules, *Plant Cell* **15**, 251–269.

Schwab, B., Mathur, J., Saedler, R., Schwarz, H., Frey, B., Scheidegger, C., Hulskamp, M. (2003) Regulation of cell expansion by the DISTORTED genes in Arabidopsis thaliana: actin controls the spatial organization of microtubules, *Mol. Genet. Genomics* **269**, 350–360.

Shimmen, T., Yokota, E. (2004) Cytoplasmic streaming in plants, *Curr. Opin. Cell. Biol.* **16**, 68–72.

Siddiqui, S.S. (2002) Metazoan Motor Models: Kinesin Superfamily in C. elegans, *Traffic* **3**, 20–28.

Smirnova, E., Reddy, A.S.N., Bowser, J., Bajer, A.S. (1998) A minus end-directed kinesin-like motor protein, KCBP, localizes to anaphase spindle poles in Haemanthus endosperm, *Cell Motil. Cytoskeleton* **41**, 271–280.

Smith, L.G. (2002) Plant cytokinesis: motoring to the finish, *Curr. Biol.* **12**, R206–R208.

Song, H., Golovkin, M., Reddy, A.S.N., Endow, S.A. (1997) *In vitro* motility of AtKCBP, a calmodulin-binding kinesin-like protein of Arabidopsis, *Proc. Natl. Acad. Sci. USA.* **94**, 322–327.

Staehelin, L.A., Hepler, P.K. (1996) Cytokinesis in higher plants, *Cell* **84**, 821–824.

Staiger, C.J., Lloyd, C.W. (1991) The Plant Cytoskeleton, *Curr. Opin. Cell Biol.* **3**, 33–42.

Strompen, G., El Kasmi, F., Richter, S., Lukowitz, W., Assaad, F.F., Jurgens, G., Mayer, U. (2002) The Arabidopsis *HINKEL* gene encodes a kinesin-related protein involved in cytokinesis and is expressed in a cell cycle-dependent manner, *Curr. Biol.* **12**, 153–158.

Sylvester, A.W. (2000) Division decisions and the spatial regulation of cytokinesis, *Curr. Opin. Plant Biol.* **3**, 58–66.

Szymanski, D.B., Marks, D.M., Wick, S.M. (1999) Organized F-actin is essential for normal trichome morphogenesis in Arabidopsis, *Plant Cell* **11**, 2331–2348.

Tamura, K., Nakatani, K., Mitsui, H., Ohashi, Y., Takahashi, H. (1999) Characterization of katD, a kinesin-like protein gene specifically expressed in floral tissues of Arabidopsis thaliana, *Gene* **230**, 23–32.

Tominaga, M., Kojima, H., Yokota, E., Orii, H., Nakamori, R., Katayama, E., Anson, M., Shimmen, T., Oiwa, K. (2003) Higher plant myosin XI moves processively on actin with 35 nm steps at high velocity, *EMBO J.* **22**, 1263–1272.

Van Gestel, K., Kohler, R.H., Verbelen, J.P. (2002) Plant mitochondria move on F-actin, but their positioning in the cortical cytoplasm depends on both F-actin and microtubules, *J. Exp. Bot.* **53**, 659–667.

Vanstraelen, M., Torres Acosta, J.A., De Veylder, L., Inze, D., Geelen, D. (2004) A plant-specific subclass of C-terminal kinesins contains a conserved a-type cyclin-dependent kinase site implicated in folding and dimerization, *Plant Physiol.* **135**, 1417–1429.

Vinogradova, M., Reddy, V.S., Reddy, A.S.N. and Fletterick, R.J. (2004) Crystal structure of KCBP: Regulation by calcium/calmodulin, *J. Biol. Chem.* **279**, 23504–23509.

Vos, J.W., Safadi, F., Reddy, A.S.N., Hepler, P.K. (2000) The Kinesin-like calmodulin binding protein is differentially involved in cell division, *Plant Cell* **12**, 979–990.

Wang, D.Y., Kumar, S., Hedges, S.B. (1999) Divergence time estimates for the early history of animal phyla and the origin of plants, animals and fungi, *Proc. R. Soc. Lond. B. Biol. Sci.* **266**, 163–171.

Wang, Z., Pesacreta, T.C. (2004) A subclass of myosin XI is associated with mitochondria, plastids, and the molecular chaperone subunit TCP-1alpha in maize, *Cell Motil. Cytoskeleton* **57**, 218–232.

Wasteneys, G.O. (2002) Microtubule organization in the green kingdom: chaos or self-order? *J. Cell. Sci. Suppl* **115**, 1345–1354.

Williamson, R.E. (1976) *Actin and Motility in Plant Cells*, North-Holland Publications, Amsterdam, Netherlands.

Williamson, R.E. (1993) Organelle movements, *Annu. Rev. Plant Physiol. Plant Mol. Biol.* **44**, 181–202.

Wolenski, J.S. (1995) Regulation of calmodulin-binding myosins, *Trends Cell Biol.* **5**, 310–316.

Yamamoto, K., Hamada, S., Kashiyama, T. (1999) Myosins from plants, *Cell. Mol. Life Sci.* **56**, 227–232.

Yamashita, R.A., Sellers, J.R., Anderson, J.B. (2000) Identification and analysis of the myosin superfamily in Drosophila: a database approach, *J. Muscle Res. Cell Motil.* **21**, 491–505.

Yang, C.Y., Spielman, M., Coles, J.P., Li, Y., Ghelani, S., Bourdon, V., Brown, R.C., Lemmon, B.E., Scott, R.J., Dickinson, H.G. (2003) TETRASPORE encodes a kinesin required for male meiotic cytokinesis in Arabidopsis, *Plant J.* **34**, 229–240.

Yokota, E., Muto, S., Shimmen, T. (1999) Inhibitory regulation of higher-plant myosin by Ca^{2+} ions, *Plant Physiol.* **119**, 231–240.

Zhong, R., Burk, D.H., Morrison, W.H. 3rd, Ye, Z.H. (2002) A kinesin-like protein is essential for oriented deposition of cellulose microfibrils and cell wall strength, *Plant Cell* **14**, 3101–3117.

Molecular Nanotechnology: *see* Nanobiotechnology

Molecular Neurobiology, Single-Cell

Jennifer Spaethling, Emily Rozak, and James Eberwine
University of Pennsylvania, Philadelphia, PA, USA

Encyclopedia of Molecular Cell Biology and Molecular Medicine, 2nd Edition. Volume 8
Edited by Robert A. Meyers.
Copyright © 2005 Wiley-VCH Verlag GmbH & Co. KGaA, Weinheim
ISBN: 3-527-30550-5

Keywords

CNS
Central nervous system.

Dendrite
Postsynaptic neuronal processes branching from cell soma that receive information from presynaptic cells.

Expression Profile
Relative abundances of multiple RNAs.

Postsynaptic
Cell that receives a signal from a neuron.

Presynaptic
Cell that sends a signal to another neuron.

Proteomics
Analysis of multiple proteins simultaneously.

aRNA
Amplified antisense RNA.

RT-PCR
Reverse-transcribed PCR.

qRT-PCR
Quantitative reverse-transcribed PCR.

Synapse
Connection point between a presynaptic and postsynaptic neuron.

■ The brain is composed of hundreds of thousands of cell types, each making thousands of connections to other cells in order to communicate and to facilitate proper interactive functioning. One of the ways that cells respond to and propagate signals is to change the abundances of various signaling molecules within the cell. These molecules include RNA and proteins, which, through their unique abundances and degree of activation transmit accurate messages to responsive cells. To understand these changes, evaluation of molecular pathways in single cells is required. With the development of single-cell molecular biological analysis techniques, changes in RNA and protein abundances can be monitored. Using methods such as, aRNA amplification, qRT-PCR, IDAT and protein arrays it is possible to obtain these measurements. Such anatomical specificity has the potential to revolutionize molecular research by simplifying the complexities of a multicellular pathway to an experimentally approachable system.

1
Importance of Single-cell Molecular Experiments

With a system in charge of complex tasks like memory, senses, mood, and development, it is not surprising that the cellular composition of the brain is equally complex. Neurons and glia are the two dominant classes of cells in the central nervous system (CNS). Neurons are thought to be at the heart of all cell–cell communication; yet, recent data has highlighted the importance of glial cells in modulating neuronal function. The variety of neuronal and glial cell types and their connections allows for the distribution and integration of information involved in everyday activities. Over the last century, it has become clear that an understanding of the complexities of CNS functioning requires not only utilization of systems science but also the reductionist approach of characterization of the individual components of the system. The development of single-cell molecular techniques has enabled researchers to begin to differentiate between not only the activities of the different regions of the brain but also the different types of cells that make up each of these regions. It is estimated that there are hundreds of thousands of cell types in the brain, each expressing tens of thousands of mRNAs in differing abundances. These differences in the abundances of RNAs as well as selective gene expression in each cell enables the complexity whereby so few chemicals in the brain produce diverse cell-specific responses.

Prior to single-cell experimental procedures, experiments used whole brain or regional sections of cells to try to elucidate the molecular changes in response to outside stimuli. Whether such stimuli originate as a result of disease, behavior, or pharmacological stimulation, quantitation of the resulting cellular responses at the molecular level enables researchers to begin to decipher what underlies the subsequent response. Since an individual neuron can have thousands of connections (synapses) with other cells that feed and distribute information, it is likely that there are differences in expressed gene phenotypes in each cell depending upon the types and quantity of these connections.

Importantly, diseases such as Alzheimer's do not necessarily affect every cell in an affected brain region. Therefore, if a group of cells were evaluated to investigate the cellular response to a stimulus, it would be hard to separate the nonresponsive cells' molecular milieu from that of the responsive cells. Single-cell molecular methods now allow researchers to separate and study target genes or whole expression pathways in individual cells so that specific cellular phenotypes can be described.

2
Isolation Techniques

After determining and marking which cells are of interest in an experiment, single cells can be collected either from fixed-tissue samples or from live cells in culture.

2.1
Microaspiration of Live Cells

Live cells in primary cell culture, continuous cell culture, or slice culture can provide a source for single-cell RNA and protein analysis. These studies require the precise removal and collection of the cytoplasmic contents of individual cells. This precision is afforded through the use of electrophysiological sampling procedures, namely whole cell patch pipette harvesting (Fig. 1). In this procedure, a thin capillary tube is drawn to a very fine tip and used to impale a cell. The cytoplasmic contents (RNA and protein) can then be aspirated into the pipette and transferred to a microcentrifuge tube for further manipulation and analysis. This technique further allows for collection of electrophysiological data from the cell to be harvested. The marrying of analyses, electrophysiological and molecular, permits a detailed phenotypic portrait of the cell to be obtained. The analysis of the mRNA complement of the isolated cellular material requires amplification (discussed later in this chapter) of the RNA signal; consequently, the mRNA must initially be converted into cDNA. This is accomplished by reverse transcriptase and can be initiated either in the microcentrifuge tube or directly in the patch pipette. Very fine manipulations are possible in this approach, allowing not only for single or multiple cell collection but also for isolation of cellular subregions such as dendrites, axons, or cell bodies. Although this is a time-intensive technique, the fine control allows for very accurate analysis of the RNA content of single cells.

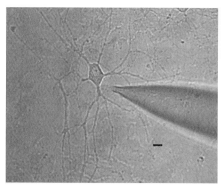

Fig. 1 Harvesting of a neuronal cell soma with a patch pipette – The manual harvesting of the cytoplasmic contents of a single rat hippocampal neuron in culture is shown. In this phase contrast image, the patch pipette used to harvest the cytoplasmic contents is in the foreground and the vast dendritic arborization associated with this neuron is clear. Aspiration through the patch pipette affords the negative pressure necessary to bring the cellular contents into the pipette. The black scale bar below the pipette corresponds to ~10 γm.

2.2
Microaspiration of Fixed Cells

Oftentimes it is difficult to obtain live cells from a particular brain region or from a particular species (e.g. human). Fixed-tissue sections offer the advantage of being able to examine cellular processes in many different cell types from essentially any organism. Fixation usually involves either chemical cross linking of proteins and nucleic acids or intracellular precipitation of these entities to preserve their localization and abundances during tissue processing. When collecting single cells from fixed-tissue sections, there are two commonly used techniques. The first technique is microaspiration, where micromanipulators and a microcontrolled vacuum are to remove the cellular constituents of the single cell. Often, the fixation protocol will dictate the success of this harvesting procedure. If the goal of these studies is to determine the RNA composition of an individual fixed cell, then as a prelude to nucleic acid amplification, cDNA synthesis is often performed *in situ* prior to microdissection of the identified cell.

2.3
Laser Capture Microdissection

The second method for single-cell collection is laser capture microdissection (LCM). This procedure uses an infrared laser beam to melt a transparent polymer transfer film (ethylene vinyl acetate polymer) so that it permeates the individual microscopically imaged cell. When the film is lifted, the desired cell is detached from all other surrounding cells and connective tissues. The film adheres to a transparent cap, which can subsequently be placed in a vial to perform any further desired manipulations. Although

a laser is used, the energy emitted by the laser is thought to not interfere with the morphology of or the chemical composition of the isolated cell. It is important to note that issues of surrounding cellular contamination and method of fixation will impact the utility of LCM. With these caveats in mind, this method promises a fully animated, fast way to obtain a single cell sample, or many cells of a similar type depending upon the goals of the experiment. Additional newer methods for laser dissection have been developed that do not rely upon film transfer. These procedures use a low-power laser to separate the cell of interest from surrounding tissue, and a higher-powered beam transfers the cell to a vial. This LCM approach offers an advantage over the cap collection methodology of limiting the amount of surrounding-cell contamination.

3
mRNA Profiling

3.1
In situ Hybridization

There are two basic questions that can be asked when evaluating mRNA in single-cell experiments. The first is "is a mRNA present within the cell" and the second is "what is the quantity of the RNA." Table 1 summarizes the benefits and disadvantages of each mRNA profiling method. The second column refers to the quantitative nature of the amplification of the technique. The third column addresses whether the method measures the presence of individual mRNAs or the mRNA population. The fourth column refers to the amount of sample required for quantification. Lastly, the final column suggests how difficult the procedure is

Tab. 1 mRNA profiling.

	Quantitative	Individual/population	Sample size	Simplicity
In situ hybridization	No	Individual	Small	
RT-PCR	Semi	Individual	Small	
qRT-PCR	Semi	Individual	Small	
aRNA amplification	Yes	Population	Small	
Microarray	Yes	Population	Large	Advanced

to use. To evaluate the existence or co-existence of mRNAs in a single cell, a commonly employed technique is *in situ* hybridization (ISH). This technique uses labeled complementary RNA probes to bind to the RNA of interest. After permeabilizing the cell membrane, the probe is hybridized to the target RNA and, in turn, the RNA tag (Fluorescence, radioactivity, DAB, etc.) is visualized to determine the cellular location of the RNA. By using more than one type of fluorescence label, it is possible to determine the colocalization of multiple RNAs in a single cell. This technique can be used to localize RNAs to subregions of cells and, more recently, as a verification of other mRNA profiling techniques. However, ISH is quite difficult to quantitate; consequently, its use as a confirmation of RNA abundances determined by other procedures such as microarray analysis is limited. ISH is quite useful in differentiating between decreased mRNA levels due to decreased cell numbers versus decreased mRNA levels.

3.2
RT-PCR

Once a single cell has been harvested, there are many methods available to evaluate differences in mRNA levels due to variables such as disease or cellular stimulation.

Since the quantity of mRNA in a single neuron or glia is quite small, often the first step in the evaluation process is amplifying the mRNA signal. This can be done by using a technique called *reverse-transcription polymerase chain reaction* (RT-PCR). After the mRNA is harvested from a cell, the first step in the amplification process is to anneal a specific primer to the mRNA of interest. Upon annealing, the reverse transcriptase will bind and, in the presence of dNTPs, create a complementary strand to the mRNA (cDNA). Next, the original mRNA is removed and a second primer binds to the cDNA strand to direct binding of a DNA polymerase that creates a copy of the first cDNA strand. This double-stranded cDNA is then used as a template for the iterative action of TAQ polymerase and the primers in the PCR amplification procedure. Such amplification is necessary since a large amount of the product is required for later quantitative analysis. During each round of PCR, the amount of DNA doubles, so in an average PCR reaction (27 rounds of amplification), the theoretical amplification yield is approximately 2^{27} or 10^8 cDNA copies (amplicons). Although this technique is used to quantify the mRNA abundance from a single cell, the PCR amplification is not linear and can lead to a skewing of amplicon abundance thereby producing aberrant results. mRNAs in the initial RNA pool that are present in lower abundances are often not amplified to the same degree as more abundant mRNAs in the reaction mixture. This misrepresentative amplification is multiplied as the PCR amplification occurs. Also, secondary structure in the RNA or cDNA may lead to the TAQ polymerases falling off the strand, producing cDNAs that are not long enough to contain the second primer binding site, resulting in erroneously low quantitative results.

3.3
qRT-PCR

A convenient method that provides rapid RNA abundance measurements is quantitative RT-PCR (qRT-PCR) and is usually used to probe individual RNAs as opposed to the population of RNAs. Using the reverse-transcription method described above, cDNA is made from the original mRNA. During the PCR step, a quenched fluorescent probe is added, which binds selectively to the cDNA. When the polymerase moves along the strand, the quenching region of the probe is cleaved and the probe fluoresces. Using a spectrofluorimeter, the level of fluorescence is measured after each round of PCR. The number of rounds required to reach a threshold point can be compared between target RNAs in the sample. Quantitative results can only be achieved through quantitation during the linear phase of amplification prior to the point of PCR saturation. The amplified products corresponding to the RNA of interest are often normalized to a housekeeping gene for further comparison purposes and to limit variability. This technique limits the need for further quantification after amplification; however, many of the nonlinear problems

described for RT-PCR also apply to this method.

3.4
aRNA Amplification

A second form of quantitative amplification is antisense RNA (aRNA) amplification. As originally described, this technique makes use of the fact that mRNA has a poly-A tail by using a complementary dT primer containing the T7 promoter to begin reverse transcription on the starting mRNA. Following the production of the first strand of cDNA, the original RNA is removed and the complement second strand cDNA is created, creating a functional T7 promoter. T7 RNA polymerase is then added to transcribe multiple antisense RNA copies from the new cDNA. Given that one round of RNA amplification produces up to 10 000-fold amplification, an important aspect of this technique is that only two rounds of amplification are required to obtain $\sim 10^6$ copies of the aRNA. This limits the problems of nonlinear amplification that are apparent from 25 to 40 rounds of RT-PCR.

3.5
Microarrays

Following a quantitative amplification protocol, microarrays are often used as a way to measure the abundance of thousands of genes simultaneously. This process usually begins with a solid support, either a glass slide or nylon membrane, with multiple oligonucleotides "printed" or deposited onto the surface in a square grid pattern. These grids can be "printed" by adding one nucleotide at a time until the complete sequence is present. Similarly, the grids can be made through direct deposition of cDNAs. The amplified RNA or cDNA, made from the aRNA, is then radioactively or fluorescently labeled and hybridized to the microarray. Any immobilized oligonucleotides or cDNAs with the complementary sequence will bind the labeled aRNAs. After hybridization occurs, either fluorescence can be measured with a spectrofluorimeter or the radioactive signal is visualized using a phosphorimager. Since the pattern of DNAs immobilized on the membrane is known, the levels of fluorescence or radioactivity represent the quantity of RNA that is bound to each particular DNA on the array. There are two main benefits of using fluorescent dyes as opposed to radiolabeling; the first is that, instead of exposing the array to a phosphor screen for up to seven days, fluorescence can be measured immediately following hybridization. The second advantage is the possibility of using multiple dyes that fluoresce with different colors allowing for competitive hybridization comparison of RNAs profiles from different samples on the same array. Radioactively labeled probes have the advantage of being more sensitive; therefore, they are better able to detect a small amount of RNA.

3.6
Experimental Examples of Profiling

There are many areas of study where the techniques of RNA profiling can be used. One is in interpreting the molecular changes associated with disease state. For example, an area of important neurological disease research is understanding the genetic, genomic, and proteomic underpinnings of schizophrenia. The hope is that by targeting the molecular deficits in affected cells, researchers may be able to create chemicals that are therapeutically effective. For example, in the prefrontal cortex (PFC) of schizophrenia patients,

there is a deficit of glutamate decarboxylase (GAD) mRNA which is necessary for the synthesis of the neurotransmitter GABA. Since the PFC is a brain region involved in working memory and is known to be impaired in schizophrenic individuals, understanding the pathways leading to these deficits might uncover novel treatments of this symptom. Single-cell mRNA profiles have shown that while some GABAergic neurons had normal expression other neurons exhibited expression far below control levels. Using molecular techniques, it is possible to investigate which subset of GABAergic neurons exhibits such deficits.

Such cellular specification highlights the selectivity of vulnerable cell types in schizophrenia. Further, recent experiments utilizing dual-label *in situ* hybridization have investigated the coincidence of GAD and parvalbumin (PV) or calretinin (CR). Since each type of GABAergic neuron in the PFC expresses only one of three calcium binding proteins, PV, CR or calbindin D-28, subdivision into three subtypes based on the presence of one of these proteins is possible. Using *in situ* hybridization, tissue samples from the PFC were labeled with probes for GAD and PV or CR, and direct correlations between GAD density and PV density were seen, whereas no correlation was seen between CR levels and GAD levels. Further insight into the relationship between GAD and PV RNA localization was achieved through dual labeling of cells with GAD and PV probes. In controls, GAD was often colocalized with PV, whereas, in schizophrenia patients, many cells expressing PV did not contain GAD mRNA. Other experiments have highlighted additional mRNAs that are downregulated in schizophrenia patients, such as those related to G-protein subunits, glutamate receptors, synapto-physin, and other proteins involved in synaptic regulation.

Diagnosis of many neurological and psychiatric diseases occurs postmortem through examination of the morphology of brain tissue. More definitive diagnoses have been troublesome because of the relative lack of biomarkers for many diseases. By using well-prepared postmortem brain samples from patients, genomics and proteomics methodologies coupled with single-cell resolution may highlight the molecules and pathways involved in diseases like Alzheimer's and Parkinson's. It is hoped that such data can be deciphered and used for diagnostic purposes and eventually for possible treatments. In recent experiments, postmortem tissue from Alzheimer's patients was used to compare mRNA profiles in affected (tangle bearing) and nonaffected cells. qRT-PCR was used to compare quantities of synaptophysin mRNA, in patients and controls. Given the downregulation of synaptophysin in patient tissue samples, this RNA may serve as a new Alzheimer's biomarker.

A hallmark of Alzheimer's is the presence of intracellular protein aggregates called *neurofibrillary tangles* (NFT). It was hypothesized that differences in mRNA profiles existed in affected versus non-affected cells. To investigate this, aRNA amplification from single tangle-bearing and nontangle bearing cells was used to generate probes that were used to screen microarrays. Large decreases in both amyloid-ß precursor as well as many dopamine (DA) receptor mRNA abundances were observed in the tangle-bearing neurons. Since these are all known to be involved in neurodegenerative diseases, this finding may be relevant to the etiology or propagation of the disease. Interestingly, lower levels of DA receptor binding

(functional DA receptor) have previously been observed in Alzheimer's tissue; consequently, there appears to be a correlation between the functional and molecular studies. When working with a neurodegenerative disorder, it is important to note that decreases in RNA abundance may result from a decrease in cell number rather than from regulation of selected genes. Consequently, data showing that the RNA abundance for α1-ACT, a protein related to inflammation, is increased in Alzheimer's disease in single cells strongly suggests a link between inflammatory pathways and the disease state.

4
mRNA Localization

4.1
Targeting and Trafficking

Not only can there be variations in mRNA levels between cells but also within regions of a single cell. In neurons, subcellular regions such as dendrites, axons, and the cell soma can have different abundances of mRNA present. Similar nucleic amplification techniques can be used for analysis of cell subregion mRNA as are used for single-cell expression profiling studies. For example, individual dendrites can be manually harvested from a live neuron using a patch pipette and aspiration. The mRNA in the dendrite can be linearly amplified to quantifiable amounts using the same aRNA amplification techniques mentioned earlier. In recent studies, investigators have shown the possible differential distribution of large numbers of selective mRNAs in dendrites as compared to the cell soma. Since synapses on the dendrite are the place where the cell initially receives information from presynaptic cells, it is reasonable to assume that the identity and abundances of dendritically localized mRNAs may be involved in the initial phase of some postsynaptic responses. This hypothesis is supported by the identification of some of the RNAs that are dendritically – targeted, such as those encoding NMDA, AMPA, and kainate classes of glutamate receptors, all of which are involved in postsynaptic responsiveness. Comparative analysis was performed on RNA that originated in the dendrites versus that of the soma. Differences in not only presence but also in quantity of some receptor subunits were observed between the two groups.

These data suggest several questions including the following: how do these differences in subcellular localization occur? Is the localization driven by differences in sequences such as the 3'-untranslated region (3'UTR) or are there RNA secondary or tertiary structural differences that help to target the mRNAs? Part of the answer comes from work investigating the somatodendritically localized microtubule-associated protein 2 (MAP2) mRNA. Normally, this RNA can be found in dendrites as well as in cell bodies (Fig. 2); however, if the 3'-UTR is removed, leaving just the coding and 5'-UTR, then the mRNA does not leave the cell soma. Also, if the 3'-UTR of the MAP2 RNA is fused to a nondendritically localized sequence, then the recombinant RNA is directed to the dendrites. It has now been shown that the 3'-UTR of many RNAs are necessary and sufficient for dendritic targeting. As our knowledge of RNA targeting sequences advances, the primary sequence as well as secondary and tertiary RNA structure will almost certainly be shown to influence RNA movement.

Advances in cell biological methodology are permitting experiments that monitor the real-time transport of RNAs into the dendrites to become feasible. An older method used pulse-chase experiments where RNA is made with incorporated radioactively labeled uridine (pulse) followed by chasing with nonlabeled uridine (chase). At different times after the chase, the movement of labeled RNA into different cellular regions can be observed. Also, visualization of the labeled RNA is done postfixing of cells, meaning that the observations are not made in real time. Further, the labeling used in pulse-chase experiments has been shown to change the stability of the mRNA. A vital dye, SYTO-14, fluoresces when it binds to nucleotides, making it possible to visualize RNA movement in live cells. RNAs were found to move at a rate of $0.1 \, \mu m \, s^{-1}$, which is similar to the rate of transport when cargoes are moved along microtubules.

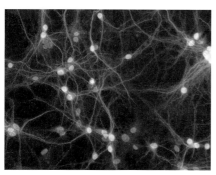

Fig. 2 Immunohistochemical localization of Map2 protein to the somatodendritic region of neurons – A rabbit polyclonal antibody raised against Map2 was applied to neurons of rat hippocampal neurons in primary culture. This first antibody was localized with a goat antirabbit secondary antibody that was fluorescein labeled (Green color). Green staining was observed in the fine hairlike structures that radiate from the cell soma. The nucleus is stained with DAPI. Glial cells in the primary cell culture system have a strong DAPI staining but no Map2 staining consistent with the absence of Map2 protein form glial cells (see color plate p. xxv).

Other types of tags can be used to visualize movement of mRNA such as a GFP-MS2 fusion protein that combines a fluorescent protein with the MS2 RNA binding protein. This is transfected into cells, and, when a fusion RNA construct is transfected into the same cells, the fusion protein binds to the engineered MS2 RNA sequence. Another *in vivo* marker of RNA is the molecular beacon probe, which is an antisense DNA oligonucleotide that is attached to a fluorophore and quencher. The structure forms a hairpin and binds with itself, quenching the fluorophore. When hybridization with the target RNA occurs, the hairpin is opened and fluorescence is emitted, allowing for very low background since only bound probes will fluoresce. This and modifications of the molecular beacon strategy are being used by several investigators to assess RNA trafficking in cells.

Different levels of mRNAs in dendrites as opposed to the cell soma have also been linked to developmental changes. For example, mRNA abundances have been monitored in dendritic growth cones through different stages of development. Amplification of RNA in growth cones at various stages of development followed by cDNA array analysis has identified RNAs that are present in the dendrite at each time point. As a function of developmental time, increasing numbers of the RNAs appeared in increased concentrations; others stayed in the soma for the entire time course of the experiment.

that suggests that dendritically localized mRNA may be translated into functioning proteins to facilitate dendritic-specific responses. Several years ago, in 1965, the ribosomal machinery necessary for protein translation was shown to exist in neuronal dendrites and to not be restricted to the cell soma. These data strongly suggested that localized translation could be a link to rapid changes in protein composition induced by synaptic modulation. This first observation of dendritically localized translational machinery was followed by several groups showing that polyribosomes localized preferentially near dendritic spines, the sites where synaptic connections with presynaptic neurons occurs. In further support of the dendritic protein synthesis hypothesis, other components of the translational machinery, including Golgi and endoplasmic reticulum, have been localized in dendrites. Currently, none of the intact neuron studies designed to prove dendritic protein synthesis have conclusively shown that protein synthesis can occur in dendrites. However, dendritic protein synthesis was confirmed using techniques to transfect isolated dendrites with exogenous RNAs encoding epitopes that when translated are immunologically or fluorescently detectable. These and subsequent studies have highlighted the complexities of dendritic protein synthesis and have further confirmed that integral membrane proteins can be synthesized and inserted into cellular membranes in dendrites.

4.2
Translational Machinery

Even if there are different levels of mRNAs in dendrites, it is likely that much of actual cellular signaling takes place through the action of proteins. It is this rationale

5
Protein Profiling

While identifying changes in mRNA levels can give researchers a beginning point suggesting molecular causes of diseases or signaling pathway responses, it is usually

Tab. 2 Protein profiling.

	Quantitative	Individual/population	Sample size	Simplicity
Immunohistochemistry	No	Individual	Small	
Protein array	Yes	Population	Large	Advanced
IDAT	Yes	Individual	Small	
Mass spectrophotometry	Yes	Population	Large	Advanced

the protein abundances and their cellular localization that provides the most functionally useful information. Given that translational control of gene expression is a well-established fact, changes in mRNA levels do not necessarily correspond to changes in protein levels. RNA abundance differences suggest protein abundance differences but this should be viewed as hypothesis-generating data rather than as functionally causative biological changes. Measurement of cellular quantities of proteins and their posttranslational modifications will provide a more accurate picture of cellular responses to stimulation. Table 2 summarizes the different types of protein profiling techniques and the benefits and disadvantages of each method. Quantitative refers to whether the technique addresses just the presence of a protein or its cellular abundance. Individual/Population addresses whether specific antibodies are used or if the method is designed to measure many of the proteins present within the sample. Sample size refers to how much sample protein is required for the method. Finally, simplicity suggests how difficult the procedure is to use.

5.1
Immunohistochemistry

Understanding which proteins are present in a cell and at what concentrations can help researchers predict what the cellular response to stimulation will be or the results of aberrant stimulation such as occurs as a result of disease. Protein profiling in single cells is considerably more difficult than mRNA profiling because of the complexities of amino acid chemistry. However, methods for the evaluation of protein profile in a single cell are developing. The simplest way is to study the presence or absence of a certain protein using immunohistochemical techniques. In this methodology, antibodies that will bind to a particular protein are applied to a thin fixed-tissue section. The primary antibodies bind to the target protein and are detected with a secondary antibody that can be fluorescently tagged. The tissue is then microscopically analyzed and the presence of the protein visualized. By using different colors of fluorescence or other detection schemes, multiple proteins can be observed simultaneously. This process enables researchers to determine colocalization; if the proteins are colocalized, possible interacting proteins can be identified.

A recent example of the application of this methodology is shown in the examination of Lewy bodies (LB), which are intraneuronal protein aggregates involved in many neurodegenerative diseases such as Parkinson's and Diffuse Lewy Body Disease (DLBD). A major protein involved in LBs is known to be α-synuclein (α-syn), which is thought to interact with other proteins that have been linked to these diseases. In other neurodegenerative diseases such as Alzheimer's, protein aggregates made of the protein tau are known as *Neurofibrillary Tangles* (NFT). Given the existence of aggregates in both diseases, it is reasonable to ask whether tau is localized to LBs. To address this question, immunohistochemistry was used to check for colocalization of α-syn and tau in LBs. Using a green fluorescent labeled (Cy3) antibody to detect one antigen and a red fluorescent labeled (Cy5) antibody to detect the second antigen, coexistence of the antigens was observed in the protein aggregates as the overlap of the fluorescence signal (yellow fluorescence). Eighty percent of the DLBD cases were found to exhibit coexistence of tau and α-syn in LBs. This high level of concurrency makes these proteins good candidates for biochemical analysis of protein interactions.

5.2
Protein Arrays

Other protein profiling techniques can provide quantitative analysis. Through the use of similar principles as microarrays, protein arrays or protein chips use protein binding as a way to separate and quantify proteins from single cells. There are two main ways for the initial protein targets to be attached to the array membrane. The first is to directly build the proteins on the arrays using a printing technique. The other way is to attach cDNAs to the membrane and run individual transcription and translation reactions, resulting in specific protein targets on the array. The latter technique eliminates worries of protein degradation and removes the need for protein purification before application onto traditional arrays.

Protein arrays work on the principle of protein–protein interactions. Similar to coimmunoprecipitation, where antibodies are attached to beads and proteins that interact with the targeted protein are precipitated out of solution, protein microarrays allow for an initial screen of antibody–protein or protein–protein interactions. By applying the proteins from a single cell to known proteins on an array, it is possible to determine which proteins readily interact. Using fluorescently labeled probe proteins from cells, the amount of binding, and therefore the quantity of protein, can be measured. This method is useful when comparing protein profiles from disease and control samples. Similar methods use immobilized antibodies in place of target proteins to probe for labeled proteins.

The latest method in single-cell protein arrays uses what are called *nanoarrays*. These miniature spots use one-thousandth of the space that a traditional microarray spot occupies. This requires a much smaller probe and target protein samples which, in theory, makes it ideal for single-cell work. The current limiting factor to this method is that of detection sensitivity. A new technique called *Immuno-Detection Amplified by T7* RNA polymerase (IDAT) uses antibodies modified to contain a covalently attached double stranded DNA with an integral T7 RNA polymerase promoter. The anitbody binds to the antigen and this interaction is detected and quantified by T7 RNA amplification of the anitbody-attached DNA. The antibody is connected to a double-stranded oligonucleotide that contains a T7 promoter. This allows for specific amplification of the targeted RNA using the linear and quantifiable T7 RNA polymerase methodology.

5.3
Mass Spectrometry

Following protein array hybridization or traditional, two-dimensional gel electrophoresis to separate proteins, mass spectrometry (MS) can be used to definitively identify the proteins and their respective concentrations in the cells. Unlike fluorescently labeled proteins, MS techniques require no labeling, which eliminates any worries of labeling induced artifacts. Using only the proteins' masses and peptide fragment masses, MS separates and identifies proteins on the basis of time of flight (TOF) of fragments through the mass spectrometer. This procedure has been used successfully to identify and quantify down to 10 proteins from a single interacting target protein. In neurons, experiments using MALDI-MS have been performed to analyze the protein profiles of dendrites. Indeed, the advent of single-cell MALDI promises to identify the most-abundant proteins in the single-cell soma and potentially in isolated regions of the neuron.

6
Protein Localization

As previously emphasized, dendrites are the first areas of the cell to receive intercellular messages. Such intercellular communication causes overall changes in the cells that facilitate and result in complex brain function. The plasticity of synapses, or the ability of a neuron to adapt to different types of stimulation, is thought to be caused by changes in the abundance of locally functional proteins. It was historically thought that these changes occurred by transport of proteins from the soma; however, given the existence of dendritic protein synthesis, it is possible that differentially stimulated regions of a dendrite might respond in a subdendritic translationally selective manner. Experiments have been performed to address this issue in which isolated dendrites (free of the cell soma) were transfected with green fluorescent protein (GFP) mRNA so that local translation could be measured, *in vivo*, though observation of fluorescence. Interestingly, dendrites that were stimulated with the metabotropic glutamate receptor agonist, DHPG (a known stimulator of protein synthesis), exhibited an increase in fluorescence that could be blocked with a translational inhibitor, anisomycin. These experiments showed that the new GFP protein was being translated locally in the isolated dendrites. Further studies showed that the newly synthesized GFP was translated in hot spots that correlated with the location of ribosomes, which previously were shown to be preferentially located near areas of synaptic input. Also, data has been generated that shows that localized dendritic translational rates not only varied from those of the cell body but also within the dendrite. These heterogeneous and specialized translational activities of dendrites give insight into the signaling processes, which are likely involved in modulating the complex biology of intercellular communication within the brain.

7
Other Applications

7.1
Electrophysiological Measurements of Molecular Changes

The hippocampus is a brain region linked to learning and memory. It is also a region where neurogenesis, the generation of new neurons, occurs. Experiments have been performed that connect these two facts, in an effort to understand the cellular events at the foundation of memory. Since new neurons are constantly being created in the hippocampus, there is always a subgroup of cells that are going through various stages of development. Investigators have determined the electrophysiological responses of neurons in each stage of development. The cells can be divided into five electrophysiologically distinct groups and single-cell RT-PCR was used to determine the developmental state of the neuron through the identification of stage-specific neuronal markers. Results showed that as development progresses, the hyperpolarity of the resting membrane potential of the cells increases and the input resistance of the cells decreases. The logical extension of these experiments is to identify the exact molecules and signaling pathways leading to these endpoints. This can be envisioned through the use of single-cell aRNA and microarray experiments coupled with proteomics.

7.2
Molecular Response to Cellular Input

One of the difficult aspects of psychiatric illness is the time course required for drug treatment to be effective. Most antipsychotic drugs take at least three weeks to begin to show their clinical efficacy. Drugs such as haloperidol are used for the treatment of diseases such as hyperactivity and mania and provide relief to the patient only if given chronically. Looking at gene expression profiles of cells that have been exposed to a drug can provide insight into the mechanisms resulting in the antipsychotic efficacy time-lag. In the rat model system, RNA was extracted from single cells in two different brain regions using LCM after administration of haloperidol for 1, 10, or 21 days. Radiolabeled aRNA was used to probe macroarrays containing 93 cDNAs of hypothesized relevance. After chronic administration, many of the mRNAs showed a steady baseline level or an increased expression level. At the intermediate time points, there were varying patterns. Some of the mRNAs spiked and later returned to a baseline level while others dropped far below the starting levels only to increase with chronic treatment. The complexities seen in the different patterns of time-dependent mRNA profiling suggest several avenues for dissection of the underlying reasons for the delayed efficacy in antipsychotic treatment.

8
Future Considerations

The last decade has seen an increase in the awareness of the importance of single-cell molecular analysis. The coordinate changes in mRNA abundance (transcription and degradation), protein abundance, posttranslational protein modifications, and the subcellular localization of RNA and proteins underlie all aspects of cellular functioning. As we progress into the twenty-first century, it is clear that we now need to not only quantitate the previously described issues *in vitro* but also to do so *in vivo*. Live cell imaging of biological processes such as transcription, translation, and subcellular movement will provide significant insight into normal cell functioning and subsequently into cell functioning as modified by disease state. To achieve these goals, it will be necessary for scientists in different disciplines to interact in an unprecedented manner. For instance, chemists making new fluorescent probes or other types of molecular detectors, physicists helping to understand the folding dynamics of RNA and protein, and biologists defining the quantifiable correlates of cellular functioning will facilitate an understanding of *in vivo* biology.

Acknowledgment

The authors gratefully acknowledge Kevin Miyashiro for supplying Fig. 2 of this manuscript. This work was funded by NIH grants to J. Eberwine.

See also Brain Development; Electron Microscopy in Cell Biology; Neuron Chemistry; Pain Transduction: Gating and Modulation of Ion Channels.

Bibliography

Books and Reviews

Eberwine, J., Yeh, H., Miyashiro, K., Cao, Y., Nair, S., Finnell, R., Zettel, M., Coleman, P. (1992) Analysis of gene expression in single live neurons, *Proc. Natl. Acad. Sci. U.S.A.* **89**(7), 3010–3014.

Eberwine, J., Miyashiro, K., Kacharmina, J.E., Job, C. (2001) Local translation of classes of mRNAs that are targeted to neuronal dendrites, *Proc. Natl. Acad. Sci. U.S.A.* **98**(13), 7080–7085.

Eberwine, J., Kacharmina, J.E., Andrews, C., Miyashiro, K., McIntosh, T., Becker, K., Barrett, T., Hinkle, D., Dent, G., Marciano, P. (2001) mRNA expression analysis of tissue sections and single cells, *J. Neurosci.* **21**(21), 8310–8314.

Ginsberg, S.D., Elarova, I., Ruben, M., Tan, F., Counts, S.E., Eberwine, J.H., Trojanowski, J.Q., Hemby, S.E., Mufson, E.J., Che, S. (2004) Single-cell gene expression analysis:

implications for neurodegenerative and neuropsychiatric disorders, *Neurochem. Res.* **29**(6), 1053–1064.

Grant, S.G., Blackstock, W.P. (2001) Proteomics in neuroscience: from protein to network, *J. Neurosci.* **21**(21), 8315–8318.

Hinkle, D., Glanzer, J., Sarabi, A., Pajunen, T., Zielinski, J., Belt, B., Miyashiro, K., McIntosh, T., Eberwine, J. (2004) Single neurons as experimental systems in molecular biology, *Prog. Neurobiol.* **72**(2), 129–142.

Job, C., Eberwine, J. (2001) Localization and translation of mRNA in dendrites and axons, *Nat. Rev. Neurosci.* **2**(12), 889–898.

Kelz, M., Therianos, S., Dent, G., Coleman, P., Eberwine, J. (2002) *Single Cell Gene Expression Analysis: Insight into Alzheimers Disease and Other Neurological Disorders*. Science's SAGE KE.

Primary Literature

Aakalu, G., Smith, W.B., Nguyen, N., Jiang, C., Schuman, E.M. (2001) Dynamic visualization of local protein synthesis in hippocampal neurons, *Neuron* **30**(2), 489–502.

Ambrogini, P., Lattanzi, D., Ciuffoli, S., Agostini, D., Bertini, L., Stocchi, V., Santi, S., Cuppini, R. (2004) Morpho-functional characterization of neuronal cells at different stages of maturation in granule cell layer of adult rat dentate gyrus, *Brain Res.* **1017**(1–2), 21–31.

Angenstein, F., Greenough, W.T., Weiler, I.J. (1998) Metabotropic glutamate receptor-initiated translocation of protein kinase p90rsk to polyribosomes: a possible factor regulating synaptic protein synthesis, *Proc. Natl. Acad. Sci. U.S.A.* **95**(25), 15078–15083.

Bassell, G.J., Zhang, H., Byrd, A.L., Femino, A.M., Singer, R.H., Taneja, K.L., Lifshitz, L.M., Herman, I.M., Kosik, K.S. (1998) Sorting of beta-actin mRNA and protein to neurites and growth cones in culture, *J. Neurosci.* **18**(1), 251–265.

Bassell, G.J., Singer, R.H. (2001) Neuronal RNA localization and the cytoskeleton, *Results Probl. Cell Differ.* **34**, 41–56.

Becker, A.J., Chen, J., Paus, S., Normann, S., Beck, H., Elger, C.E., Wiestler, O.D., Blumcke, I. (2002) Transcriptional profiling in human epilepsy: expression array and single cell real-time qRT-PCR analysis reveal distinct cellular gene regulation, *Neuroreport* **13**(10), 1327–1333.

Behar, L., Marx, R., Sadot, E., Barg, J., Ginzburg, I. (1995) cis-acting signals and trans-acting proteins are involved in tau mRNA targeting into neurites of differentiating neuronal cells, *Int. J. Dev. Neurosci.* **13**(2), 113–127.

Benson, D.L. (1997) Dendritic compartmentation of NMDA receptor mRNA in cultured hippocampal neurons, *Neuroreport* **8**(4), 823–828.

Bernhardt, R., Matus, A. (1984) Light and electron microscopic studies of the distribution of microtubule-associated protein 2 in rat brain: a difference between dendritic and axonal cytoskeletons, *J. Comp. Neurol.* **226**(2), 203–221.

Betsuyaku, T., Senior, R.M. (2004) Laser capture microdissection and mRNA characterization of mouse airway epithelium: methodological considerations, *Micron* **35**(4), 229–234.

Blichenberg, A., Schwanke, B., Rehbein, M., Garner, C.C., Richter, D., Kindler, S. (1999) Identification of a cis-acting dendritic targeting element in MAP2 mRNAs, *J. Neurosci.* **19**(20), 8818–8829.

Blichenberg, A., Rehbein, M., Muller, R., Garner, C.C., Richter, D., Kindler, S. (2001) Identification of a cis-acting dendritic targeting element in the mRNA encoding the alpha subunit of Ca2+/calmodulin-dependent protein kinase II, *Eur. J. Neurosci.* **13**(10), 1881–1888.

Bodian, D. (1965) A suggestive relationship of nerve cell RNA with specific synaptic sites, *Proc. Natl. Acad. Sci. U.S.A.* **53**, 418–425.

Chow, N., Cox, C., Callahan, L.M., Weimer, J.M., Guo, L., Coleman, P.D. (1998) Expression profiles of multiple genes in single neurons of Alzheimer's disease, *Proc. Natl. Acad. Sci. U.S.A.* **95**(16), 9620–9625.

Coss, R.G., Perkel, D.H. (1985) The function of dendritic spines: a review of theoretical issues, *Behav. Neural Biol.* **44**(2), 151–185.

Crair, M.C. (1999) Neuronal activity during development: permissive or instructive? *Curr. Opin. Neurobiol.* **9**(1), 88–93.

Crino, P.B., Eberwine, J. (1996) Molecular characterization of the dendritic growth cone: regulated mRNA transport and local protein synthesis, *Neuron* **17**(6), 1173–1187.

Crino, P., Khodakhah, K., Becker, K., Ginsberg, S., Hemby, S., Eberwine, J. (1998) Presence and phosphorylation of transcription factors in developing dendrites, *Proc. Natl. Acad. Sci. U.S.A.* **95**(5), 2313–2318.

Curran, S., McKay, J.A., McLeod, H.L., Murray, G.I. (2000) Laser capture microscopy, *Mol. Pathol.* **53**(2), 64–68.

Davis, L., Banker, G.A., Steward, O. (1987) Selective dendritic transport of RNA in hippocampal neurons in culture, *Nature* **330**(6147), 477–479.

de Moor, C.H., Richter, J.D. (2001) Translational control in vertebrate development, *Int. Rev. Cytol.* **203**, 567–608.

Denk, W., Strickler, J.H., Webb, W.W. (1990) Two-photon laser scanning fluorescence microscopy, *Science* **248**(4951), 73–76.

Eberwine, J., Cao, Y., Nair, S., Miyashiro, K., Mackler, S., Finnell, R., Surmeier, J., Dichter, M. (1995) Mechanisms of neuronal plasticity as analyzed at the single cell level, *Prog. Brain. Res.* **105**, 117–126.

Emmert-Buck, M.R., Bonner, R.F., Smith, P.D., Chuaqui, R.F., Zhuang, Z., Goldstein, S.R., Weiss, R.A., Liotta, L.A. (1996) Laser capture microdissection, *Science* **274**(5289), 998–1001.

Estee Kacharmina, J., Crino, P.B., Eberwine, J. (1999) Preparation of cDNA from single cells and subcellular regions, *Methods Enzymol.* **303**, 3–18.

Estee Kacharmina, J., Job, C., Crino, P., Eberwine, J. (2000) Stimulation of glutamate receptor protein synthesis and membrane insertion within isolated neuronal dendrites, *Proc. Natl. Acad. Sci. U.S.A.* **97**, 11545–11550.

Evanko, D.S., Zhang, Q., Zorec, R., Haydon, P.G. (2004) Defining pathways of loss and secretion of chemical messengers from astrocytes, *Glia* **47**(3), 233–240.

Fasulo, W.H., Hemby, S.E. (2003) Time-dependent changes in gene expression profiles of midbrain dopamine neurons following haloperidol administration, *J. Neurochem.* **87**(1), 205–219.

Femino, A.M., Fay, F.S., Fogarty, K., Singer, R.H. (1998) Visualization of single RNA transcripts in situ, *Science* **280**(5363), 585–590.

Gardiol, A., Racca, C., Triller, A. (1999) Dendritic and postsynaptic protein synthetic machinery, *J. Neurosci.* **19**(1), 168–179.

Garner, C.C., Tucker, R.P., Matus, A. (1988) Selective localization of messenger RNA for cytoskeletal protein MAP2 in dendrites, *Nature* **336**(6200), 674–677.

Gazzaley, A.H., Benson, D.L., Huntley, G.W., Morrison, J.H. (1997) Differential subcellular regulation of NMDAR1 protein and mRNA in dendrites of dentate gyrus granule cells after perforant path transection, *J. Neurosci.* **17**(6), 2006–2017.

Geiger, J.R., Melcher, T., Koh, D.S., Sakmann, B., Seeburg, P.H., Jonas, P., Monyer, H. (1995) Relative abundance of subunit mRNAs determines gating and Ca2+ permeability of AMPA receptors in principal neurons and interneurons in rat CNS, *Neuron* **15**(1), 193–204.

Grant, S.G., Husi, H. (2001) Proteomics of multiprotein complexes: answering fundamental questions in neuroscience, *Trends Biotechnol.* **19**(Suppl. 10), S49–S54.

Gutala, R.V., Reddy, P.H. (2004) The use of real-time PCR analysis in a gene expression study of Alzheimer's disease post-mortem brains, *J. Neurosci. Methods* **132**(1), 101–107.

Haab, B.B. (2003) Methods and applications of antibody microarrays in cancer research, *Proteomics* **3**(11), 2116–2122.

Hashimoto, T., Volk, D.W., Eggan, S.M., Mirnics, K., Pierri, J.N., Sun, Z., Sampson, A.R., Lewis, D.A. (2003) Gene expression deficits in a subclass of GABA neurons in the prefrontal cortex of subjects with schizophrenia, *J. Neurosci.* **23**(15), 6315–6326.

Haydon, P.G. (2003) Biological near-field microscopy, *Methods Enzymol.* **360**, 501–508.

Hemby, S.E., Ginsberg, S.D., Brunk, B., Arnold, S.E., Trojanowski, J.Q., Eberwine, J.H. (2002) Gene expression profile for schizophrenia: discrete neuron transcription patterns in the entorhinal cortex, *Arch. Gen. Psychiatry* **59**(7), 631–640.

Huang, E.P. (1999) Synaptic plasticity: regulated translation in dendrites, *Curr. Biol.* **9**, R168–R170.

Huber, K.M., Kayser, M.S., Bear, M.F. (2000) Role for rapid dendritic protein synthesis in hippocampal mGluR-dependent long-term depression, *Science* **288**(5469), 1254–1257.

Husi, H., Grant, S.G. (2001) Proteomics of the nervous system, *Trends Neurosci.* **24**(5), 259–266.

Ishizawa, T., Mattila, P., Davies, P., Wang, D., Dickson, D.W. (2003) Colocalization of tau and alpha-synuclein epitopes in Lewy bodies, *J. Neuropathol. Exp. Neurol.* **62**(4), 389–397.

Jakoi, E.R., Severt, W.L. (2000) Disruption of mRNA-RNP formation and sorting to dendritic synapses by antisense oligodeoxynucleotides, *Methods Enzymol.* **313**, 456–466.

Job, C., Eberwine, J. (2001) Identification of sites for exponential translation in living dendrites, *Proc. Natl. Acad. Sci. U.S.A.* **98**(23), 13037–13042.

Johansen, F.F., Lambolez, B., Audinat, E., Bochet, P., Rossier, J. (1995) Single cell RT-PCR proceeds without the risk of genomic DNA amplification, *Neurochem. Int.* **26**(3), 239–243.

Kamme, F., Salunga, R., Yu, J., Tran, D.T., Zhu, J., Luo, L., Bittner, A., Guo, H.Q., Miller, N., Wan, J., Erlander, M. (2003) Single-cell microarray analysis in hippocampus CA1: demonstration and validation of cellular heterogeneity, *J. Neurosci.* **23**(9), 3607–3615.

Kanai, Y., Hirokawa, N. (1995) Sorting mechanisms of tau and MAP2 in neurons: suppressed axonal transit of MAP2 and locally regulated microtubule binding, *Neuron* **14**(2), 421–432.

Knowles, R.B., Sabry, J.H., Martone, M.E., Deerinck, T.J., Ellisman, M.H., Bassell, G.J., Kosik, K.S. (1996) Translocation of RNA granules in living neurons, *J. Neurosci.* **16**(24), 7812–7820.

Lambolez, B., Audinat, E., Bochet, P., Crepel, F., Rossier, J. (1992) AMPA receptor subunits expressed by single Purkinje cells, *Neuron* **9**(2), 247–258.

Lynch, M., Mosher, C., Huff, J., Nettikadan, S., Johnson, J., Henderson, E. (2004) Functional protein nanoarrays for biomarker profiling, *Proteomics* **4**(6), 1695–1702.

Mackler, S.A., Brooks, B.P., Eberwine, J.H. (1992) Stimulus-induced coordinate changes in mRNA abundance in single postsynaptic hippocampal CA1 neurons, *Neuron* **9**(3), 539–548.

Mackler, S.A., Eberwine, J.H. (1993) Diversity of glutamate receptor subunit mRNA expression within live hippocampal CA1 neurons, *Mol. Pharmacol.* **44**(2), 308–315.

Mazumder, B., Seshadri, V., Fox, P.L. (2003) Translational control by the 3'-UTR: the ends specify the means, *Trends Biochem. Sci.* **28**(2), 91–98.

Mirnics, K., Middleton, F.A., Marquez, A., Lewis, D.A., Levitt, P. (2000) Molecular characterization of schizophrenia viewed by microarray analysis of gene expression in prefrontal cortex, *Neuron* **28**(1), 53–67.

Miyashiro, K., Dichter, M., Eberwine, J. (1994) On the nature and differential distribution of mRNAs in hippocampal neurites: implications for neuronal functioning, *Proc. Natl. Acad. Sci. U.S.A.* **91**(23), 10800–10804.

Miyashiro, K.Y., Beckel-Mitchener, A., Purk, T.P., Becker, K.G., Barret, T., Liu, L., Carbonetto, S., Weiler, I.J., Greenough, W.T., Eberwine, J. (2003) RNA cargoes associating with FMRP reveal deficits in cellular functioning in Fmr1 null mice, *Neuron* **37**(3), 417–431.

Mori, Y., Imaizumi, K., Katayama, T., Yoneda, T., Tohyama, M. (2000) Two cis-acting elements in the 3′ untranslated region of alpha-CaMKII regulate its dendritic targeting. [see comments], *Nat. Neurosci.* **3**(11), 1079–1084.

Morozov, V.N., Morozova, T.Y., Johnson, K.L., Naylor, S. (2003) Parallel determination of multiple protein metabolite interactions using cell extract, protein microarrays and mass spectrometric detection, *Rapid. Commun. Mass Spectrom.* **17**(21), 2430–2438.

Nitin, N., Santangelo, P.J., Kim, G., Nie, S., Bao, G. (2004) Peptide-linked molecular beacons for efficient delivery and rapid mRNA detection in living cells, *Nucleic Acids Res.* **32**(6), e58.

Pascual, O., Haydon, P.G. (2003) Synaptic inhibition mediated by glia, *Neuron* **40**(5), 873–875.

Pierce, J.P., van Leyen, K., McCarthy, J.B. (2000) Translocation machinery for synthesis of integral membrane and secretory proteins in dendritic spines, *Nat. Neurosci.* **3**(4), 311–313.

Racca, C., Gardiol, A., Triller, A. (1998) Cell-specific dendritic localization of glycine receptor alpha subunit messenger RNAs, *Neuroscience* **84**(4), 997–1012.

Ramachandran, N., Hainsworth, E., Bhullar, B., Eisenstein, S., Rosen, B., Lau, A.Y., Walter, J.C., LaBaer, J. (2004) Self-assembling protein microarrays, *Science* **305**(5680), 86–90.

Rubakhin, S.S., Garden, R.W., Fuller, R.R., Sweedler, J.V. (2000) Measuring the peptides in individual organelles with mass spectrometry, *Nat. Biotechnol.* **18**(2), 172–175.

Rusakov, D.A., Richter-Levin, G., Stewart, M.G., Bliss, T.V. (1997) Reduction in spine density associated with long-term potentiation in the dentate gyrus suggests a spine fusion-and-branching model of potentiation, *Hippocampus* **7**(5), 489–500.

Service, R.F. (2001) Proteomics. Searching for recipes for protein chips, *Science* **294**(5549), 2080–2082.

Shi, S., Hayashi, Y., Esteban, J.A., Malinow, R. (2001) Subunit-specific rules governing AMPA receptor trafficking to synapses in hippocampal pyramidal neurons, *Cell* **105**(3), 331–343.

Sokol, D.L., Zhang, X., Lu, P., Gewirtz, A.M. (1998) Real time detection of DNA. RNA hybridization in living cells, *Proc. Natl. Acad. Sci. U.S.A.* **95**(20), 11538–11543.

Steward, O., Levy, W.B. (1982) Preferential localization of polyribosomes under the base of dendritic spines in granule cells of the dentate gyrus, *J. Neurosci.* **2**(3), 284–291.

Surmeier, D.J., Eberwine, J., Wilson, C.J., Cao, Y., Stefani, A., Kitai, S.T. (1992) Dopamine receptor subtypes colocalize in rat striatonigral neurons, *Proc. Natl. Acad. Sci. U.S.A.* **89**(21), 10178–10182.

Tecott, L.H., Barchas, J.D., Eberwine, J.H. (1988) In situ transcription: specific synthesis of complementary DNA in fixed tissue sections, *Science* **240**(4859), 1661–1664.

Tiedge, H., Brosius, J. (1996) Translational machinery in dendrites of hippocampal neurons in culture, *J. Neurosci.* **16**(22), 7171–7181.

Tongiorgi, E., Righi, M., Cattaneo, A. (1998) A non-radioactive in situ hybridization method that does not require RNAse-free conditions, *J. Neurosci. Methods* **85**(2), 129–139.

Torre, E.R., Steward, O. (1996) Protein synthesis within dendrites: glycosylation of newly synthesized proteins in dendrites of hippocampal neurons in culture, *J. Neurosci.* **16**(19), 5967–5978.

Van Gelder, R.N., von Zastrow, M.E., Yool, A., Dement, W.C., Barchas, J.D., Eberwine, J.H. (1990) Amplified RNA synthesized from limited quantities of heterogeneous cDNA, *Proc. Natl. Acad. Sci. U.S.A.* **87**(5), 1663–1667.

Weiler, I.J., Irwin, S.A., Klintsova, A.Y., Spencer, C.M., Brazelton, A.D., Miyashiro, K., Comery, T.A., Patel, B., Eberwine, J., Greenough, W.T. (1997) Fragile X mental retardation protein is translated near synapses in response to neurotransmitter activation, *Proc. Natl. Acad. Sci. U.S.A.* **94**(10), 5395–5400.

Zhang, H.T., Kacharmina, J.E., Miyashiro, K., Greene, M.I., Eberwine, J. (2001) Protein quantification from complex protein mixtures using a proteomics methodology with single-cell resolution, *Proc. Natl. Acad. Sci. U.S.A.* **98**(10), 5497–5502.

Molecular Systematics and Evolution

Jeffrey H. Schwartz
University of Pittsburgh, Pittsburgh, PA, USA

Encyclopedia of Molecular Cell Biology and Molecular Medicine, 2nd Edition. Volume 8
Edited by Robert A. Meyers.
Copyright © 2005 Wiley-VCH Verlag GmbH & Co. KGaA, Weinheim
ISBN: 3-527-30550-5

Keywords

African Apes
Chimpanzees, bonobos, and gorillas.

Anthropoidea
The so-called higher primates; the subordinal rank that subsumes New and Old World monkeys plus the hominoids.

Clade
A group of related organisms; a group of organisms united by common ancestry.

Derived Feature
A relative character state determined by how restrictively shared a feature under study is.

Hominids
Humans and their fossil relatives.

Hominoids
The group of anthropoid primates that includes gibbons and siamangs, chimpanzees, gorillas, orangutans, and humans.

Homology
Similarity due to common ancestry.

Large-bodied Hominoids
Orangutans, humans, and African apes.

Lesser (Small-bodied) Hominoids
The gibbons and siamangs.

Monophyletic Group
A clade.

Outgroup
A taxon that is outside of (a sister taxon of) the focal group of phylogenetic reconstruction.

Phylogeny Reconstruction
The generation of a theories of evolutionary relationships among taxa.

Primitive Feature
A relative character state determined by how broadly shared a feature under study is.

Primitive Retention
A feature that is shared by the members of a group by inheritance from a common ancestor.

Sister Taxon
A taxon that is the closest relative of another.

Systematics
The study of the pattern and relationships of organisms, including the identification of species and delineation of groups of taxa.

Taxic
Relating to taxa.

Taxon (Plural: Taxa)
Any taxonomic rank, for example, species, genus, family, order, and class.

Studies that rely on genetic and molecular information to address evolutionary questions fall roughly into two different categories: the reconstruction of the evolutionary relationships of organisms (including the times of divergence of groups or lineages), and the formation and emergence of morphological novelties that distinguish or characterize different organisms. The former endeavor is sometimes referred to as *molecular systematics*, and when it is applied to primates and especially the relatedness of humans and apes, "molecular anthropology." The debate over the regularity of molecular change, resulting in a molecular clock, lies within the realm of molecular systematics, and especially that of molecular anthropology. In the past decade, in particular, the primary focus of molecular systematists has been DNA sequences, both nuclear and mitochondrial. Throughout, the assumption has been that the degree of similarity reflects the degree of evolutionary relatedness, because differences accrue once lineages have diverged from a common ancestor. The popularity of molecular systematics in recent years is also predicated on the notion of "a law of large numbers," that is, the thousands of bases that produce DNA sequences. The question is whether these assumptions are sustainable.

The rather newly defined discipline of evolution and development, or "evo-devo" as it has been nicknamed, is less involved in the interpretation of molecular data for purposes of reconstructing evolutionary relationships than it is in trying to identify molecular elements – whether they be transcription factors or other kinds or classes of proteins, as well as "genes" themselves – that are relevant to, or participate in, the processes of development of structure and form. Here, what is important is not a particular RNA, DNA, or amino acid sequence, but the signal transduction pathways or sequences of communication between different molecules in the regulation of development and the origin of structure. Sometimes, hypotheses of when and at what level within groups of organisms specific features emerged are generated as a result of overlaying this information on a presumed theory of relationships of the organisms under consideration. It appears that insights from developmental genetics will prove to be more useful than sequence data alone for systematic and phylogenetic inquiry.

This entry will attempt to summarize the main aspects of these fields of inquiry, including their underlying assumptions, and suggest possible avenues for future research that might bring these disciplines together in light of new ways of thinking about evolution.

1
The Beginning of Molecular Systematics

The first inquiries into systematics and phylogenetic reconstruction by investigation of the "blood relationship" of organisms can be traced to a few individuals, of whom the best known is George H. F. Nuttall. In the research that preceded his monograph of 1904, *Blood Immunity and Blood Relationship: A Demonstration of Certain Blood-Relationships Amongst Animals by Means of the Precipitin Test for Blood*, Nuttall sought to demonstrate that the degree of similarity between animals in their blood serum proteins was a reflection of their evolutionary closeness. The idea was simple: produce an antiserum (antibodies or antigens) to blood serum proteins of one animal, mix the antiserum with blood serum proteins of another animal, and measure the amount of precipitin that settled out. The more profound the precipitation, the greater the similarity (because of a greater number of antibody–antigen binding sites), and, consequently, the closer the evolutionary relationship between the two organisms being compared.

The strength of the antiserum played an important role in determining just how many species might belong to the same group. A weak antiserum would provoke reactivity in the sera of only the most closely related of organisms, while a stronger solution would produce

a precipitate when combined with the sera of a greater number of related animals. The more distantly related the groups being compared, the less the reactivity between serum and antiserum. In addition to strength of reactivity, Nuttall thought that the reaction rate was also an indicator of closeness of relatedness. In the end, he decided that "the zoological relationships between animals are best demonstrated by means of powerful antisera." Consequently, he concluded, "if we accept *the degree of blood reaction as an index of the degree of blood relationship* within the Anthropoidea [the so-called higher primates, New and Old World monkeys plus the hominoids, which includes gibbons, chimpanzees, gorillas, orangutans, and humans], then we find that the Old World apes are more closely allied to man than are the New World apes, and this is exactly in accordance with the opinion expressed by Darwin."

Although Nuttall believed that his approach accurately revealed the evolutionary relationships of animals, with the exception of reference in 1922 by the paleontologist, W. K. Gregory, to Nuttall having demonstrated "the anthropoid heritage of man," his work went largely unacknowledged until the 1960s, when various publications claimed to have demonstrated the evolutionary relationships of organisms through study of elements of their biochemistry.

2
The Molecular Assumption

Perhaps the most influential of these publications was by Emile Zuckerkandl and Linus Pauling on fetal and adult hemoglobin, which included a comparison of sequences between human, gorilla, horse, and fish. Although one might question the comparison of gill- with lung-bearing animals, the endeavor demonstrated that the fish was more dissimilar to humans than to the horse, while the gorilla was the most similar to humans in hemoglobin sequence. Since this pattern of taxic distance mirrored the morphologically accepted scheme of evolutionary relatedness of these animals, Zuckerkandl and Pauling suggested that their observations "can be understood at once if it is *assumed* [emphasis added] that in the course of time the hemoglobin-chain genes duplicate, [and] that the descendants of the duplicate genes 'mutate away' from each other." From this assumption, they felt justified in proposing the following: "[O]ver-all similarity must be an expression of evolutionary history," with descendants "mutating away" and becoming "gradually more different from each other." In other words, the more time that elapses after a lineage's divergence, the greater the molecular difference its succession of descendants would accumulate. Consequently, evolutionary closeness became synonymous with molecular difference. Also assumed in this nexus of assumption is the notion of constant and gradual unilinear change.

There is, however, nothing in Zuckerkandl and Pauling's work that justifies the notion that "overall similarity" is "an expression of evolutionary history." They merely assumed such a correspondence, just as they assumed that difference also meant change and the same kind of change, which, in turn, was achieved at a constant and gradual rate. However, the discovery of a hundred nucleotide differences between two taxa does not translate into the rate at which presumed substitutions occurred. One needs additional information, such as a suggestion of the time at which these differences began to accumulate. And then one has to assume that the rate of substitution over this period of time was gradual and that differences did not arise during a few concentrated phases of replacement.

Nevertheless, Zuckerkandl and Pauling's effort resulted in the "molecular assumption" – continual molecular change and its continual accumulation over evolutionary time – which would thereafter become the foundation of molecular systematics. As Adalgisa Caccone and Jeffrey Powell would write in 1989: "Virtually all molecular phylogenetic studies...have a major underlying assumption: the genetic similarity or difference among taxa is an indication of phylogenetic relatedness. Lineages that diverged more recently will be genetically more similar to one another than will be lineages with more ancient splits." If one embraces the assumption, the rest may follow logically, but not necessarily because of biological demonstration. One of the major extensions of the "molecular assumption" is the notion of a "molecular clock," the existence of which was promoted initially by Vincent Sarich and Allan Wilson.

Beginning with their first major paper in 1966, Sarich and Wilson sought to elucidate the evolutionary relationships of primates, and especially of humans and the large-bodied apes, using the technique of microcomplement fixation (MC'F), which requires only minute amounts of serum and antiserum for study of immunological reactivity. Their molecule of choice was the larger blood serum protein, albumin. The degree of reactivity achieved between albumin and anti-albumin was translated into an "index of dissimilarity," which was subsequently referred to as *immunological distance* (ID). The closer an ID value was to 1.0, the greater was the overall molecular similarity and thus, given an assumed equivalence between ID and evolutionary relationship, the presumption of closeness of relatedness between the organisms under study. As an apparent check on the validity of this approach, Sarich and Wilson conceived of the "test of reciprocity." For example, if chimpanzee serum was cross-reacted with antihuman serum (produced by injecting another animal, for instance, a rabbit or chicken, with human serum) the first time around, human serum was cross-reacted with antichimpanzee serum the next time to see if the resultant ID values were reasonably similar. Predictably, the test of reciprocity usually confirmed the initial immunological finding.

The supposed evolutionary arrangement of the primates that Sarich and Wilson achieved was more consistent with the general pattern of primate relationships based on comparative morphology than it was in disagreement. Because of this concordance, they concluded that "the MC'F data are in qualitative agreement with the anatomical evidence, on the basis of which the apes, Old World monkeys, New World monkeys, prosimians, and nonprimates are placed in taxa which form a series of decreasing genetic relationship to man." In 1967, in their second article on this topic, Sarich and Wilson argued that molecular change must have occurred at a constant rate among all major groups of primates. In the third and last article of this series, which was also published in 1967,

they concluded that the small amount of difference they detected between hominoids in their albumin reflected little molecular change and that this, in turn, implied that little time had elapsed since the separation of the hominoid lineages. Since, as they claimed, molecular change ticked away at a constant rate, IDs represented a "molecular clock" that could reveal the times at which the various hominoids – actually any species – diverged and went off on their own evolutionary paths.

After calibrating the molecular clock on the basis of paleontologists' interpretation, based on fossils, of when the Old World monkey and hominoid lineages might have separated, Sarich and Wilson calculated divergence dates of about 10 million years ago for the gibbon lineage, eight million years ago for the orangutan lineage, and a mere five million years ago for the human and African ape lineages. The major implication of this calculation was that fossils older than five million years, such as *Ramapithecus* (now referred to the genus *Sivapithecus*) from c. 12- to 14-million-year old deposits in Indo-Pakistan, could not, as paleontologists concluded from study of morphology, be hominid.

However, while molecular systematists were in general agreement on the premise of the molecular assumption – that degrees of similarity reflect degrees of evolutionary closeness – not all of them embraced the notion of a molecular clock that "ticked" at a constant rate. Among the most vocal dissenters was Morris Goodman, who in the early 1960s was the leading proponent of Nuttall's work. Although Goodman went beyond claiming that molecular similarity indicated a very close relationship between humans and the African apes to advocating that these three hominoids should also be placed in the same taxonomic family, Hominidae (a position that for years caused many comparative primate anatomists and paleontologists to be skeptical of molecular systematics in general), he nonetheless tried to accommodate the paleontologists' identification of pre-five-million-year old fossils as being hominid.

In contrast to Sarich (who in 1971 was the first molecular systematist to reject morphology as being phylogenetically informative), Goodman accepted the paleontologically derived date of at least 14 million years as the time of divergence between hominids and their potential ape relatives. As a consequence, though, Goodman then had to interpret the immunological and biochemical data in a way that would make them compatible with a deep timescale of hominoid diversification. He did so by constructing a selectionist argument to explain the apparent acceleration or deceleration in rates of molecular change that *de facto* must have occurred, given the paleontologically established dates for the earliest appearances of each of the large-bodied hominoids. Thus, for instance, if the various hominoid lineages had indeed originated between 18 and 14 million years ago, the extreme similarity between the various extant hominoids in their albumin had to have resulted from a slowdown in the rate of molecular change of that particular blood serum protein.

Goodman proposed that the inferred deceleration in rate of molecular change in large-bodied hominoid albumin was related to the fact that, although they are similar to other anthropoid primates in developing hemochorial placentation (the most intimate of placental modes in the approximation of maternal and fetal blood systems), large-bodied hominoids differ from other anthropoids in having very long gestation periods. He argued that

in animals with hemochorial placentation but fairly short gestation periods, there would be enough slack in the system to allow unimpeded molecular change to occur and accrue. The fetus would be born before the mother's immune system could produce antibodies to it. A large-bodied hominoid's gestation period, however, would be long enough not only for maternal-fetal immunological incompatibility to build up but also for maternally produced antibodies to diffuse through the placenta, with deleterious consequences to the fetus.

The dilemma which then had to be confronted was: How could an animal have both hemochorial placentation and a long gestation period? Goodman's answer was that natural selection must have acted to reduce the possibility of immune responses from the mother toward her fetus's proteins. Consequently, there had to have been a slowdown of molecular change that was commensurate with a prolongation of the gestation period. Although albumin had been the basis of Sarich and Wilson's constant-rate molecular clock, Goodman used the same molecule as the exemplar of how the molecular clock could run at different rates.

Concurrent with the various approaches – immunological reactivity as well as gel-column electrophoretic separation – that were brought to bear during the 1960s and 1970s on the determination of evolutionary relationships was an increased interest in protein sequences. The first major study was Zuckerkandl and Pauling's 1962 analysis of hemoglobin, and, for some years thereafter, hemoglobin was the most intensely studied molecule. Beginning in 1973, however, a research group headed by A. E. Romero-Herrera analyzed myoglobin sequences across a number of primarily mammalian taxa. As more taxa were added to the study, the resultant, most "parsimonious" arrangement of the hominoids placed the gibbon rather than the orangutan closer to a human-African clade. Only by arguing for a more complicated scheme of myoglobin "evolution" in hominoids – including, for instance, the condition that there had to have been various "back-mutations" to a state similar to the presumed unchanged state – could one arrive at the more commonly accepted arrangement of divergences: gibbon first, then the orangutan, and then a human-African ape group.

But while protein sequencing was seen by some, especially Zuckerkandl as late as 1987, as the molecular level on which to focus for purposes of determining evolutionary relationships (primarily because amino acid sequences are less subject to the problems with DNA, such as insertions, deletions, and back-mutations), there was a growing consensus that the best systematic information lay at an even deeper "genetic" level. Since base differences at the third position of a codon (a nucleotide triplet) did not yield different resultant amino acids, the concern of DNA sequence advocates was that demonstration of similarity in protein sequences could produce "false" phylogenies because the underlying nucleotide sequences themselves may be different. Consequently, it became imperative to get to the level of DNA – nuclear DNA – itself, especially because the general belief or at least expectation (which was based on bacteria and inferred for metazoans) was the existence of a direct correlation between specific DNA sequences and specific genes. And, as genes were conceptualized, there was also supposed to be correspondence between one or perhaps a few genes and a specific feature or structure. In addition, because DNA sequences were

composed of thousands upon thousands of bases, the apparent massive scale of the comparison had the appeal of providing an overwhelming amount of phylogenetically relevant information.

3
DNA Hybridization

Although DNA sequencing was possible by the late 1970s, it was an expensive and laborious procedure, which made the endeavor itself, much less the comparison of DNA sequences for phylogenetic purposes, prohibitive. There was, however, another way in which DNA sequence information could be achieved: DNA–DNA hybridization.

The theory behind using DNA–DNA hybridization (or just DNA hybridization) as an approximation of overall similarity in DNA sequences between different taxa lay in the fact that if the two strands that form the helical structure of nuclear DNA were dissociated (which could be accomplished by subjecting it to heat), their complementary bases would rebond and the two strands would reform their original helical organization. However, it is not only two cleaved but also originally helically arranged DNA strands from the same individual that will reassociate (reanneal). Analogous to the geneticist Theodosius Dobzhansky's discovery earlier in the twentieth century with chromosomes of different lengths from different species of fruit fly, any two strands of DNA, from different individuals of even different species, will attempt to anneal.

As in Dobzhansky's fruit-fly experiments, in which the larger chromosome would loop and fold so that whichever of its loci were also present in the shorter one would match up, single strands of DNA (derived from heat splitting or melting of a double helix) from different organisms would attempt to recombine or hybridize at complementary base positions. The more complementarity there was between hybridized DNA strands, the greater was the intensity of heat needed to break down the bonds holding them together. Consequently, it seemed logical to conclude that the heat (ΔT) it took to disassociate a DNA–DNA hybrid (that is, the more thermally stable the hybrid was), the more similar the annealed strands were in their nucleotide sequences. From this apparent demonstration, one could then invoke the molecular assumption to explain how higher ΔT's, reflecting greater molecular similarity, were also a reflection of closer evolutionary relatedness. As logical, though, as this thought experiment might appear to be, it relies on an assessment of overall similarity, which does not identify the bases or sequence positions that underlie differences in ΔT's. Conversely, similar ΔT's in different taxa might not be the result of the same regions hybridizing.

One of the first applications of the technique of DNA hybridization to evolutionary questions was published in 1968 by R. J. Britten and D. E. Kohne, who were primarily concerned with learning more about the genome. They discovered that the genomes of higher organisms – but not of bacteria or viruses – contained "hundreds of thousands of copies of DNA sequences." However, not only did repeated DNA sequences represent a considerable portion of a genome, Britten and Kohne also found that they were "trivial and permanently inert." Only a small fraction of a genome was composed of unrepeated, unique, single-copy DNA sequences, which apparently constituted the active or functional elements of that genome.

Britten and Kohne also speculated on pathways or mechanisms that might produce genomic change, which they argued must be considered on two different levels: change in nucleotide sequence, and the origin of new families of nucleotide sequences. As Zuckerkandl and Pauling had assumed for differences in hemoglobin, Britten and Kohne proposed that changes in nucleotide sequences – which would be identified as point mutations – occur slowly over time. However, this model would explain only the "divergence of pre-existing families" of nucleotide sequences, not the introduction of new families of sequences, as was assumed would eventually happen in the gradual-accumulation-of-change model that dominated molecular systematics. The introduction of new families of sequences, Britten and Kohne alternatively suggested, must "result from relatively sudden events," called *saltatory replications*. Accordingly, "saltatory replications of genes or gene fragments occurring at infrequent intervals during geologic history might have profound and perhaps delayed results on the course of evolution." Although Britten and Kohne's distinctions between types of molecular change were not at the time embraced by molecular systematists, their suggestion of "saltatory replication" would certainly appear to have been borne out with the subsequent identification of vertebrate regulatory genes involved in segmentation – *Hox-a*, *Hox-b*, *Hox-c*, and *Hox-d* – as replicates of the orthologous *Antennapedia* gene identified in fruit flies.

One of the early attempts at using DNA hybridization to reconstruct relationships among the primates was published in 1976 by Raoul Benveniste and George Todaro, whose primary focus was actually on the distribution among mammals of the type C viral gene, the presence of which, they discovered, was particularly characteristic of animals found in Asia, including the orangutan and the gibbon, but which, unexpectedly, was also present in humans. After reviewing these data, Benveniste and Todaro turned to DNA hybridization for sorting out the relationships of various primates. Relying on previous studies, they estimated "the effect of mismatched base-pairs on thermal stability" as being "between 0.7 and 1.7 °C per 1% altered pairs." Since, depending on the animal, nuclear DNA consists of anywhere from 10 to 100 million nucleotide pairs, every 1% of difference in nucleotide sequence between DNA hybrids would result in a drop of about 1.0 °C of the heat necessary to melt the hybrid molecule.

Although Benveniste and Todaro calculated a 1.1% sequence difference between humans and both African apes and a 2.4% difference between humans and the orangutan, their published ranges of ΔT's actually demonstrate overlap between all large-bodied hominoids (although one does not know the identity of nucleotide difference and similarity). Taking into consideration the fact that presumed-to-be-homologous DNA sequences are not the same length (that is, they do not consist of the same number of nucleotides) in all animals, including the large-bodied hominoids (see below for human and orangutan), one also does not know from DNA hybridization experiments the details of how a short sequence from one species aligns with a longer sequence of another species.

Probably because DNA hybridization seemingly provided a way of getting closer to the supposedly most "informative" genetic level, questions about what, exactly, similarity in ΔT's meant were not explored. But other questions were.

In the 1980s, Charles Sibley and Jon Ahlquist reviewed earlier DNA hybridization experiments and concluded that a major problem with them all was that only a limited number of cross hybridizations had been tested. This, Sibley and Ahlquist claimed, was the reason previous studies had such difficulty in refining certain evolutionary relationships, particularly those between humans and the African apes. Although there was not particular morphological support for this theory of relationship, molecular systematists often grouped these three hominoids together. However, the level of resolution in the majority of the molecular work was never fine enough to determine which two of the three hominoids were the most similar (and thus presumably the most closely related).

Sibley and Ahlquist felt another source of error in these earlier studies resulted from using all of an animal's DNA, repeated as well as single-copy, to form hybrids. They argued that it was important to remove all repeated DNA – since it was redundant – and to work only with single-copy DNA.

Before Sibley and Ahlquist decided to tackle the evolutionary relationships of the large-bodied hominoids, they had spent years applying their DNA hybridization technique to the phylogeny of birds, through which they also developed the notion of a uniform average rate of genomic change (UAR). UARs, they believed, characterized the nature of molecular, and thus evolutionary, change in all taxa. The notion of a UAR also seemed to negate the need for debates over constant versus irregularly running molecular clocks, which were directed at individual molecules. As they saw it, the entire genome could change at a uniform average rate, even though the rates at which individual molecules change may be quite different.

Sibley and Ahlquist argued that the power of DNA hybridization in "discovering" the phylogenetic relationships of organisms came from sampling the entire genome of an animal. They claimed that, since an organism's genome is composed of millions of nucleotides, the "law of large numbers" provides the checks and balances necessary to rule out "false" similarities or parallelisms. Accordingly, since DNA hybrid strands will only link up along those stretches of sequences that are complementary, there should be no question about homology. Conversely, sequences that did not line up were not homologous. At least that was the argument.

Sibley and Ahlquist's melting temperatures for their DNA hybrids produced numbers, which they then converted into phylogenetic distances using a procedure called *average linkage*. "Average linkage" begins "by clustering the closest pair or pairs of taxa," after which "one links the taxa which have the smallest average distance to any existing cluster," and on it goes until "all taxa are linked." The underlying assumption, which permits the linking of the most similar pairs, is that the DNA hybridization technique "measures the net divergence between the homologous nucleotide sequences of the species being compared."

Many aspects of Sibley and Ahlquist's phylogenetic reconstructions of birds were consistent with theories of relatedness derived from study of morphology. However, there were some significant differences, for instance, with regard not only to the broader phylogenetic relationships of, but also to the details of relatedness among, the flycatchers. By the 1980s, however, it was common practice, when molecular and the morphological phylogenies were

in discord, to opt for the former. As Caccone and Powel argued in 1989, once the molecular assumption is accepted, overall similarity becomes the yardstick for determining closeness of relatedness. Indeed, whether it is a presumed albumin clock or a UAR, the assumption of ongoing molecular change validates the molecular assumption, which, in turn, demands that a molecular phylogeny is correct in its entirety, even if other sources of information contradict it.

Sibley and Ahlquist's "law of large numbers" and UAR appealed to most molecular systematists. This appeal was probably the primary reason paleontologists finally succumbed to the notion that molecular phylogenies had greater authority in deciphering evolutionary relationships when morphologically based theories were in conflict with them. But not all molecular systematists were convinced. Among them was Alan Templeton, who objected to the use of DNA hybridization because it did not allow one to determine the polarity of the similarity, which is necessary to address the question: "Is similarity due to distant or recent common ancestry?" For, only by studying the actual sequences of nucleotides can one determine the identity and from this attempt to infer the significance of similarity and dissimilarity. Sibley and Ahlquist countered, however, that there was "no reason to expect data derived from base sequences to improve on those from amino acid sequences, which have produced contradictory results." In their support, they also quoted A. E. Friday (who had collaborated with Romero–Herrera on myoglobin sequencing): "Phylogenetic conclusions derived from a study of nucleotide sequences will be subject to the same suspicions as those derived from amino acid sequences."

4
Mitochondrial DNA

As the field of molecular systematics was expanding its sphere of investigation during the 1970s, DNA located outside the nucleus, in a cell's mitochondria, was attracting attention. In a study published in 1979, W. Brown, M. George, and A. Wilson noted that there was more difference between humans and a sampling of Old World monkeys in their mitochondrial (mt) DNA than there was among these same primates in their nuclear DNA. By assuming that the divergence between the human and Old World monkey lineages occurred over 20 million years ago (based on morphological studies of the primate fossil record), Brown et al. concluded that the differences between these humans and Old World monkeys could be explained if nucleotide substitutions occurred at a slower pace −5 to 10 times slower – in nuclear than in mtDNA (in other words, if mtDNA "evolved" 5 to 10 times faster than nuclear DNA). On the basis of this interpretation, Brown et al. calculated that mtDNA data are the most accurate for studying evolutionary events that occurred "within the past 3–10 million years."

In subsequent publications, they proceeded to analyze the mtDNA of the large-bodied hominoids, which, largely because of Sarich and Wilson's molecular clock, were assumed to have diverged within this time period and which, because of this assumption, were thus amenable to such analysis. Although their data showed the fewest differences (interpreted as substitutions) between humans and chimps, Brown et al. concluded that humans were related to an African ape group. Given the molecular assumption, however, the data should have yielded a human–chimpanzee sister group, as

Maryellen Ruvolo and a number of collaborators concluded in 1991 from their analysis of mtDNA.

In addition to the fact that mtDNA is single-, not double-stranded, the allure of using mtDNA was and continues to be twofold. One assumption is that, unlike nuclear DNA, mtDNA is supposedly inherited only through the maternal line and thus not subject to the complexities that occur through recombination of maternal and paternal DNA. The second assumption is the existence of a "hotspot" in the D-looped configuration of mtDNA and that this is the primary site of molecular activity. This region is called the *hypervariable zone* and its study has been interpreted as providing evidence of evolutionary change. Although crucial to the use of mtDNA for purposes of phylogenetic reconstruction, these assumptions are probably incorrect.

The introduction of paternal mtDNA into fertilized eggs has been reported in the literature from time to time. For example, in 1992, Allan Wilson and coworkers demonstrated that this happens in mice, as part of the sperm's tail, which does contain mtDNA, penetrates the ovum's membrane. More recently, John Maynard-Smith and coworkers calculated that there had to have been a recombination of maternal and paternal mtDNA during human evolution, and concluded that this possibility should cause systematists to reconsider the seemingly inviolable "fact" that humans and chimpanzees are closely related.

As Erika Hagelberg reviewed at great length, there is increasing indication that mtDNA is not as exempt from paternal inheritance and recombination as was initially believed. With regard to so-called hypervariable sites being regions of preferentially high mutation rates, which therefore lend themselves to phylogenetic analysis,

Hagelberg pointed out that "there is no direct evidence of hypervariability," although "most researchers believe that anomalous patterns of DNA substitution are best explained by mutation." Indeed, she writes, "because the notion of hypervariability fits with the received view of mtDNA clonality [maternal inheritance only], anomalies are seldom questioned." Hagelberg gives an interesting example. Depending on which subject's mtDNA is used, one can reconstruct "our most-recent female common ancestor...[as having] lived just 6000 years ago, a date more consistent with Biblical Eve than Mitochondrial Eve." As she cautions: "The picture is far from simple, and it is clear that extreme care must be taken in the interpretation of mtDNA phylogenetic trees in the face of possible recombination...[T]here are enough unexplained patterns in mtDNA data to warrant reassessment of the conclusions of many mtDNA studies."

5
DNA Sequences

In spite of Sibley and Ahlquist's and Friday's warnings about nucleotide sequence data not being any more reliable than protein sequence data for reconstructing phylogenetic relationships, the increasing ease with which DNA could be sequenced could not be ignored by molecular systematists. Nuclear as well as mtDNA were sequenced. Regardless of the genetic "level" under scrutiny, the interpretation of similarity or difference in DNA sequences under comparison was still predicated on the molecular assumption. The weight of DNA sequence data as being key to deciphering evolutionary relationships assumed special significance in some areas of molecular systematics because of the supposed information content of nuclear DNA in general. As Elizabeth Bruce and Francisco Ayala wrote as early as 1979 in an article on blood serum proteins: "Information macromolecules – that is, nucleic acids and proteins – document evolutionary history...[Thus] degrees of similarity in such macromolecules reflect, on the whole, degrees of phylogenetic propinquity."

Almost coincident with the rise in popularity of comparing DNA sequences for purposes of inferring evolutionary relationships came the cautionary notes. Of particular importance in this regard is the question of alignment, which, in 1991, J. A. Lake was among the first to address. As mentioned above, not all DNA sequences chosen for comparison, if homologous (e.g. representing the same "gene" or segment) are the same length. Therefore, decisions must be made with regard to how to subdivide the shorter sequence in order to align its nucleotides with those of the longer sequence. Typically, the alignment of compared sequences is presented in the literature without justification of the assumptions and decisions that produced the alignment, which, in turn, was then used as the basis for the phylogenetic analysis. But, as Lake warned in the title of his article, "[t]he order of sequence alignment can bias the selection of tree topology." In addition to assumptions that inform the decision to break up a short sequence so that its bases align with those of a longer sequence is the issue of whether, for one or another sequence, bases may have been added or inserted, removed or deleted, or one base substituted for another.

In *what it means to be 98% chimpanzee*, Jonathan Marks provides an example of these problems with DNA sequences from a human and from an orangutan to show three different ways in which their bases (C,T,G,A) can be aligned and the interpretive consequences. Even beforehand, the assumption must be made that the 40 bases in the human sequence actually have homologous counterparts in the 54 bases in the orangutan sequence.

There may be seven differences or there may be eleven differences, depending on how we decide the bases correspond to each other across the species – and that is, of course, assuming that a one-base gap is

HUMAN CCTCCGCCGCGCCG   CTCCGC GCCGCCGGGCA   CGGCC   CCGC

ORANG CC   GTCGCCTCCGCCACGCCGCGCCACCGGGCCGGGCCGGCCCGGCCCGCCCCGC

HUMAN   CCTCCGCCGCGCCGCT   CCGCGCCGCCGGGCACGGCCCCGC

ORANG CCGTCGCCTCCGCCACGCCGCGCCACCGGGCCGGGCCG GCCCGGCCCGCCCCGC

HUMAN CCTCCGCCGCGCCG   CT CCGCGCCGCCGGG CAC   GGCC   CCGC

ORANG   CCGTCGCCTCCGCCACGCCGCGCCACCGGGCCGGGCCGGCCCGGCCCGCCCCGC

As Marks comments:

"Tabulate the differences. The top one invokes five gaps and six base substitutions; the middle has only two gaps but nine base substitutions. And the bottom one has five gaps and only three base substitutions. The three pairs of sequences differ in the assumptions about which base in one species corresponds to which base in the other. While we might, by Occam's Razor, choose the alignment that invokes the fewest inferred hypothetical evolutionary events, we still have to decide whether a gap "cquals" a substitution. Does the bottom one win because it has a total of only eight differences? Or might the middle one win because a gap should be considered rare and thereby "worth," say, five base substitutions?

The problem is that we cannot know which is "right," and the one we choose will contain implicit information about what evolutionary events have occurred, which will in turn affect the amount of similarity we tally. How similar is this stretch of DNA between human and orangutan?

also equivalent to a five-base gap and to a base substitution.

In a more general sense, however, the problem of taking *quantitative* estimates of difference between entities that differ in *quality* is prevalent throughout the genetic comparison of human and ape. The comparison of DNA sequences presupposes that there are corresponding, homologous sequences in both species, which of course there must be if such a comparison is actually being undertaken. But other measurements have shown that a chimpanzee cell has 10% more DNA than a human ccll. (this doesn't mean anything functionally, since most DNA is functionless.) But how do you work that information into the comparison, or into the 99.44% similarity [between human and chimp]?" [comment added].

These concerns have not, however, been widely appreciated by molecular systematists, especially molecular anthropologists, who not only portray the analysis of DNA sequences as neutral and objective but also use the assumption of relatedness to inform the way in which they analyze the sequences they have aligned according to certain assumptions. Exemplary in this regard is the multiple DNA sequence analysis Maryellen Ruvolo and collaborators published in 1997, in which they sought to resolve the supposed dilemma of to which African ape humans are more closely related. In order to pursue this question, they assumed first that the orangutan was the sister taxon of a clade or evolutionary group consisting of humans, the chimpanzee, and the gorilla. Consequently, the differences in the orangutan had to be considered primitive relative to any similarities that were delineated between humans and one or the other of the African apes.

With the ever-growing popularity of the parsimony-based phylogenetic computer program PAUP (phylogenetic analysis using parsimony), it is common practice to "root" a phylogenetic analysis in a taxon that is chosen as the primitive outgroup – that is, the taxon that diverged earlier than the others – prior to the analysis taking place. Rooting parsimony or any of the other available clustering analyses (for example, nearest-neighbor joining or maximum likelihood, which are essentially similar to Sibley and Ahlquist's linking technique) in a particular taxon may be necessary for the algorithm to "work." Nevertheless, this procedure artificially determines character polarity since, by definition, the outgroup (the taxon in which the tree is rooted) is defined from the outset as being primitive in its entirety. In turn, the taxa to which the outgroup is the supposed primitive sister taxon are predetermined as being derived in whatever ways they differ from it.

The widespread use of this algorithm-based approach to analyzing nuclear and mtDNA as well as protein sequences presents its own set of problems and assumptions. Consider the molecular assumption: Since molecular change is supposedly continually occurring and being accumulated as a lineage proceeds through time, the degree of molecular similarity reflects the antiquity or recency of lineage divergence. Accordingly, each lineage accumulates it own unique array of molecular changes, which should make a lineage more distinctive (that is, different) the longer it is in existence. Although tautological, this assumption explains why more recently diverged taxa are more similar than more anciently divergent lineages. On the other hand, in order to root an algorithm for purposes of generating presumed phylogenetic relationships, one must assume that the taxon chosen as the earlier-divergent outgroup is totally primitive relative to the taxa to which it is supposed to be the sister taxon. Yet, it is the molecular assumption that validates the use of overall similarity as the key to resolving phylogenetic relationships by contrasting it with the unique differences that earlier-divergent lineages accumulated along their own, unique evolutionary trajectories. Clearly, both assumptions cannot be correct at the same time. Either the earlier divergent-taxa or lineages did not change, but remained primitive (which is the logical extension of identifying a taxon as the outgroup in which to root a computer analysis), or they did change by accumulating their own suites of molecular difference (the basis of the molecular assumption), in which case they are at least in some aspects derived (and uniquely so, for that matter, because of their unique molecular histories) and not primitive relative to the taxa to which they are being compared.

In the realm of morphological systematics, according to Hennigian or cladistic principles, overall similarity is not *de facto* a clue to evolutionary relatedness. Similarity must be sorted out into features that reflect a hierarchy of inheritance: primitive features from ancient ancestors, and derived features from recent ancestors.

Since the pattern of life is one of a hierarchy of nested sets of smaller and smaller clades (groups of related taxa), that which is considered primitive versus that which is considered derived depends on the level in the hierarchy of nested clades one is investigating. Primitive features – features retained in descendants – do not elucidate the relationships of these taxa. Only derived features can. It is also important conceptually to recognize that a derived feature at one level of the hierarchy is a primitive retention at another. There is no theoretical reason why this approach to systematics cannot be applied to molecular data. The major difficulty is that molecular systematists would have to sample and compare a wide range of taxa. This is the only way in which relative primitiveness and derivedness can be determined. It cannot be justified by *a priori* assumptions of directionality, as underlies the molecular assumption, or by choosing an outgroup on the basis of its presumed evolutionary relationship to other taxa. However, even from the beginning, it is also crucial to realize that shared similarity does not translate directly into a demonstration of relatedness. Taxa may be similar, not because they inherited changes that distinguished a recent common ancestor, but because they share primitive retentions, that is, features that have not changed in a succession of ancestors.

Nevertheless, it is becoming increasingly popular in the literature for molecular studies on the relatedness of taxa to be identified as being "cladistic." One argument is that nucleotide bases – C, G, T,

A – represent alternative character states. On one level this may appear logical, but it is actually misleading since none of them represents a character. Phylogenetically relevant alternative molecular character states would be better represented by comparison, for example, of arrangements of "gene" sequences with regard to *cis* and *trans* elements, patterns of introns and exons and of methylation of transposons or other elements, and pathways of molecular communication. Another argument in support of molecular studies being cladistic derives from the claim that molecular similarity is equivalent to synapomorphy; that is, shared similarity represents shared derived character states. This conclusion is, of course, only a restatement of the molecular assumption: the most recently diverged taxa share more recently accumulated (equate with derived) molecular states. Thus, the supposedly shared derived molecular states are delineated *a posteriori*; in other words, after the algorithm of choice has clustered taxa on the basis of their greater or lesser degrees of similarity (depending on the algorithm), typically after rooting the tree in a particular taxon (which, as pointed out above, at once defines it as being primitive and the taxa being compared to it as derived in their shared similarities). This, however, is not how a cladistic analysis proceeds. The endeavor of hypothesizing primitive versus derived character states occurs prior to hypothesizing relationships – which is the only way in which such a methodology can actually be employed.

The assumption of continual molecular change – whether through point mutations affecting nuclear or mtDNA, or altering amino acids in protein sequences – is also of interest. Recall that this idea was initially framed by Zuckerkandl and Pauling as a way of explaining their data: "overall similarity *must* [emphasis added] be an expression of evolutionary history," with descendants "mutating away" from each other, becoming "gradually more different from each other." It is this assumption that proposes that earlier-divergent taxa accumulate their own molecular differences, while the most recently divergent taxa are similar because of the longer shared history of accumulated molecular changes and shorter time of independent molecular change. The existence of molecular clocks and UAR is predicated on this notion. Nevertheless, it must be recognized that this is an extrapolation – an explanation of how something might come to be. It has not been demonstrated.

The contradiction is that while constant molecular change is predicted through the molecular assumption as occurring during gametogenesis, or in some way as to be passed on to offspring, in molecular biology, it is well known that the only source of constant molecular change is ultraviolet radiation, which produces a mutation rate of $10^{-8}-10^{-9}$. But the other element of UV-derived mutation is that it is random, with the potential of affecting either somatic or sex cells and also with regard to the molecule that is affected. Thus, while there might appear to be concordance between the reality of the physical world in which there is a relatively constant UV-provoked mutation and the concept of a constantly "evolving" molecular world, this is an illusion.

The notion of constant and accumulative mutation affecting sex cells is of further interest because it also contradicts the basic tendency of cells to remain in homeostasis. As seen, for example, in the roles of heat shock proteins (HSP) – maintaining cell membrane physical states through lipid transport, eliminating reading errors that occur during transcription or translation, DNA repair, chaperoning other proteins, and ensuring proper folding of proteins as they emerge from the ribosomes – the basic propensity of a cell is to resist change. Intuitively, this should make sense inasmuch as unabated molecular change would undermine the integrity of cell function, as would also be the case with a constant accumulation of point mutations, and more probably lead to the death of organisms than to change.

6
"Evo-devo"

In 1975, Mary-Claire King and Allan Wilson surveyed all available data on blood serum proteins, as well as the results of DNA hybridization, with regard to humans and chimpanzees. Although their publication is cited as having demonstrated the relatedness of these two hominoids, this was not their intention. As they stated: "the only two species which have been compared by all of these methods are chimpanzees...and humans," and thus "a good opportunity is...presented for finding out whether the molecular and organismal estimates of distance agree." The result was that humans and chimpanzees differed in their genetic makeup only by about one percent. King and Wilson concluded that "all the biochemical methods agree in showing that the genetic distance between humans and the chimpanzee is probably too small to account for their substantial organismal differences." In order to explain how humans and chimpanzees could be virtually identical in their genes but markedly different animals, King and Wilson suggested that humans and chimpanzees must be different in those genes that regulate development.

Since then, studies on the regulation of development have expanded exponentially, not only with regard to distinguishing between regulatory and structural genes, but especially with regard to the array of molecules that induce gene transcription and communicate via signal transduction pathways to produce structure. Interestingly, those animals that have been studied in depth – such as the fruit fly, zebra fish, frog, chick, mouse, human – demonstrate a commonality of "homeotic genes" (which contribute to segmental patterning and segment identify). In turn, through their protein products (transcription factors), homeotic genes control or at least affect gene expression. Time and time again during development, the same proteins (e.g. various growth factors, trans-inducing and bone-modifying proteins) and regulatory genes (and their products) are coopted to produce what in adult organisms are different morphologies.

In 1994, Lewis Wolpert summarized the situation: "During development, differences are generated between cells in the embryo that then lead to spatial organization (pattern formation), changes in form, and the generation of different cell types. Genes control development by controlling cell behavior." But one should not be too gene-centric in envisioning the emergence of form from genes and gene products alone. For, while there might be genetic regulation of some cells' activities, the results (e.g. cellular asymmetry, cell membrane elasticity or rigidity, compressive forces) might produce physical or mechanical responses, which may not themselves be genetically based, but which, nonetheless, greatly affect cell geometry and ultimately organismal form.

"Cell behavior," to return to Wolpert, has many different levels of meaning.

In the 1970s, Søren Løvtrup argued that one must recognize the importance of epigenesis in development: especially that changes in properties of the fertilized egg can alter the chronology and spatial organization of patterns of cellular diversification. Since the larger clades of multicellular organisms possess the same kinds of cells, as well as the same chemical substances that form the immediate environment of the cells, variation in the spatial and chronological organization of cellular differentiation must be at least one of the keys to the emergence of evolutionary novelty. For instance, whether a cell divides symmetrically or asymmetrically (which can be affected even by the positions of the chromosomes relative to the center or periphery of the cell) can greatly impact the spatial relationships of cells and, therefore, eventually have an effect on organismal shape. As Pere Alberch has emphasized, the development of organismal form and structure is also a function of the physical and mechanical properties of cells' sizes, shapes, and spatial relationships.

The application to evolutionary questions of discoveries in the regulation of development has given rise to the field of "evo-devo" (evolution and development), in which the interrelationship between the "genetics" and the "epigenetics" of development has become a primary focus. Indeed, Løvtrup's concern with the influence on metazoan development of differences in cellular differentiation during gastrulation appears to be even more germane to an understanding of the emergence of form and its conservation across taxa as well as of the emergence of differences in form, whether their expression constitutes variation (individual differences) or diversity (species differences).

7
Positional Information and Shape

One of the ongoing questions in developmental biology is how cells acquire positional information, not only in terms of entire structures themselves (e.g. where limbs will grow) but also with regard to how cells acquire information to contribute to regional shapes of a structure (e.g. the different segments of a limb).

In invertebrates, wing (e.g. as in a fruit fly) and limb (e.g. as in the brine shrimp-like crustacean, *Artemia*) positioning involves activation of the regulatory genes *nubbin* and *apterous*. In vertebrates, various regulatory genes, especially *Hox* genes, and also *dlx* (distal-less), are known to be involved in segmentation and limb positioning. In fish and tetrapods, the *Hoxd11-13* genes are expressed along the posterior margin of the enlarging limb bud; however, in tetrapods, this homeodomain expands anteriorly across the distal (lower) end of the limb bud. Additionally, in vertebrates, *Hox* genes are not only recruited in the formation of a segmented trunk but also through regional activation, in the segmentation of the hindbrain. Regional activation of *sonic hedgehog*, however, in part, underlies forebrain segmentation.

Eye development is also of interest in this regard. For although what used to be thought of as "master-control genes," such as *Pax-6*, were found to participate in signal transduction pathways leading to eye formation in invertebrates (typically multilensed, rigid) as well as in vertebrates (single lensed, deformable), there is at least one element that vertebrates have in common: Even though there are differences among vertebrates so far studied with regard to when in ontogeny and how often and in how many different regions the *Rx*

gene (which is also recruited in fruit-fly eye development) is activated, it is always expressed in the vertebrate forebrain. Thus, at one level, one can hypothesize that the last common ancestor of vertebrates was characterized by early activation of the *Rx* gene in the presumptive forebrain and that differences between taxa are the result of differences in other aspects of *Rx* gene expression: for example, the different proteins in the lenses of amphibians and mammals may be due in part to down- or upstream effects of the *Rx* gene being expressed later on in development in the amphibian (frogs) retina, whereas in mammals (mice), the *Rx* gene is expressed early on in the presumptive eye itself.

In addition to considering morphological differences in light of differences in regional (as well as in overlapping) domains of regulatory gene expression, it is becoming increasingly clear that differences in fields of molecular gradients (morphogenetic fields) also play a role. As C. Owen Lovejoy, M. J. Cohn, and T. D. White hypothesized in 1999 in their discussion of the evolution of human pelvic form, "if a particular PI [positional information] gradient were to span *n* cell diameters, and those cells defined the ultimate anteroposterior dimension of the presumptive ilium (superoinferior in the adult human), then a slight increase in the steepness of its slope would cause that signal to span fewer cells, 'distorting' the presumptive anlagen and substantially altering downstream adult morphology." In other words, although it would seem to be a process involving myriad steps, "the transformation of the common ancestral pelvis [in its entirety] into that of early hominids may have been as 'simple' as a slight modification of a gradient" [comment added]. Thus, in addition to differences in gene expression and pathways of molecular communication, as well as to the physical and mechanical consequences of cellular organization, morphological novelty in metazoans (and presumably plants as well) may also be affected by altering the domains of morphogenetic fields.

But what is the source of differences in gene or molecular gradient expression? Lovejoy et al. suggest that one need not seek the answer in mutation, which is a position that Sean Carroll has recently also strongly argued.

8
"Mutation"

The concept of "mutation" is about as slippery as that of a "gene." It means different things to different researchers, and, interestingly, the differing concepts seem to "work" in their disparate intellectual contexts. With regard to mutation, the "textbook" notions of preceding decades included point mutation, gene duplication, and chromosomal rearrangement. The latter was basic to the earlier experimental studies and theoretical considerations of the fruit-fly geneticist, Theodosius Dobzhansky. Dobzhansky's emphasis on chromosomal rearrangement as a potential source of evolutionary novelty was subsequently adopted by the developmental biologist, Richard Goldschmidt, in his theory of systemic mutation, which he argued would lead to the abrupt appearance of novel form. Unfortunately, Goldschmidt is best remembered, and consequently criticized, for suggesting that such novelties would emerge in individuals he identified as "hopeful monsters."

One of Goldschmidt's major theoretical thrusts, however, was distinguishing between what he identified as micromutation (leading to variation and microevolution) and macromutation (leading to the origin of species or evolution). The small mutations that fruit-fly population geneticists, such as Thomas Hunt Morgan, inferred lay behind small phenotypic changes, Goldschmidt identified as micromutations, which, he argued, led only to the survival of species, not to their origin. The latter required a larger source of genetic disturbance, and for that he turned to chromosomal rearrangement. The logic is understandable: If the chromosome theory was correct (that, indeed, units of heredity or genes were contained on chromosomes – as Morgan presented it, like beads in a necklace), then manipulating them on a grand scale (producing a systemic mutation) might yield evolutionarily significant novelty, that is, new species. A major problem with Goldschmidt's theory, however, was that he did not provide a mechanism by which more than one individual would be the bearer of the novelty and, thus, of the systemic mutation underlying it.

Point mutations, commonly the result of UV radiation, are random with regard to affecting somatic or sex cells. In addition, if they do not interfere with cell function, such point mutations are not a reliable source of potential genetic and subsequently morphological novelty. Indeed, it appears that point mutations do not often cause any noticeable effect, either genetically or phenotypically. Gene duplication – as seen, for instance, in the emergence of *Hoxa-d* – does sometimes occur, but knockout experiments have demonstrated that duplication typically reflects redundancy, not a source of phenotypic novelty.

It may be true that manipulation of levels of thyroxin or retinoic acid during ontogeny can affect the size of an organism, or the shapes of some of its features, just as a mother's diet can affect methylation in its fetus or fetuses, and thus aspects of her progeny's postnatal growth. However, these disturbances only affect an individual during its lifetime, and should not be expected to be repeated across generations and under the influence of fluctuating environmental stimuli (e.g. diet, temperature, amount of daylight). The problems, then, that still must be addressed are as follows: How does a genetic or cellular change remain "fixed" or constant, and how do many individuals come to bear it?

It is important to realize that there is a difference between change at the genetic level and what is perceived as phenotypic change. Common in the literature on the genetics of evolution is the mistake of conflating the two as constituting macromutation – a misconception that derives from the confusion Dobzhansky introduced with regard to the terms micromutation and macromutation when, in 1941, he sought to discredit Goldschmidt. Nevertheless, especially with the increasing awareness from molecular biology that there are not "genes for" features, we must be vigilant in making a distinction between what appears morphologically to have been the result of a macromutation (e.g. developing feathers instead of scales) and the underlying genetic–epigenetic interactive pathway.

9
Toward a Theory of Evolutionary Change

The question at hand, then, is the articulation of a mechanism that can first provide the potential for genetic novelty. Building on my original theory for the sudden appearance of morphological novelty (through the silent spread of recessive "mutations"), the molecular biologist,

Bruno Maresca, and I have proposed that the opportunity for genetic novelty may lie in overstressing cells to the extent that their HSPs can no longer maintain genetic homeostasis; that is, they cannot fulfill their roles as chaperones, respond to the needs of the cell membrane, and, perhaps most importantly, for this discussion, properly fold other proteins and assist in DNA repair. Although first identified in heat shock experiments, HSPs can be affected by a variety of stresses, including diet (saturated vs unsaturated fatty acids), wind, aridity, and cold. Since most multicellular plants and animals possess HSPs, the theory is more widely applicable than metazoan-centric Darwinian and neo-Darwinians models of evolutionary change, the latter of which relies on unwarranted extrapolations from fruit-fly population genetics.

Since most multicellular organisms have a window of heat shock response, they can "adapt" to normal fluctuations in their environmental circumstances. If environmental change exceeds this window (as when seasons change), most organisms can "reset" it, often in less than two months. Until this window is reset, the stress induces an increase in HSP production. If, however, there is a spike in environmental stress that exceeds an organism's ability to reset its HSP response, HSP function will fail, and opportunities for introducing genetic novelty will emerge – especially as a result of improper protein folding and inefficient DNA repair. In the former situation, improperly folded proteins may, for example, no longer recognize (or be recognized by) promoter or enhancer regions to which they would normally bind, but they may now be capable of binding to different sites. An obvious result could be the activation or deactivation of a "gene" or "genes," and, thus, the creation of one or more new developmentally significant signal transduction pathways. With regard to inefficient DNA repair, genetic novelty of a different sort can be introduced, with obvious potential consequences. In both cases, however, the fact that the environmental stress will be at least regional (if not global), the circumstances exist for more, perhaps many more, than one individual of a species to be affected (not, however, necessarily in the same way).

However, while it might be tempting to extrapolate immediately from these possible sources of genetic novelty to the emergence of evolutionarily relevant morphological novelty, one must be cautious. First, the effects of extreme environmental spikes on HSPs must be actualized during gametogenesis. If they are not, offspring will not be affected. Second, if the effects do not kill the individuals that inherit any of these genetically based novelties, they will probably not be expressed; that is, these genetic novelties will be in the "recessive" state. Consequently, there will not be an immediate phenotypic reflection of these genetic changes. In the recessive state, however, they can spread "silently" throughout the population, until it becomes sufficiently saturated with heterozygotes that homozygotes for the genetic novelty will be produced. If the resultant phenotypic expression – cellular or greater – does not kill its bearers, they may continue to reproduce themselves, as heterozygotes will also contribute to the numbers of individuals bearing the phenotypic novelty. Thus, the spread of a genetic basis for potential phenotypic novelty may take numerous generations before there is any statistical possibility of the phenotype being expressed. In other words, there will be a temporal disjunction between the disruption of cellular and genetic homeostasis, and what will be seen as the abrupt or sudden appearance of phenotypic novelty, and in some number of individuals. In addition, one must bear in mind that, during periods of "silently spreading" genetic novelty, there could be other environmental spikes that would contribute to the pool of potential for genetic novelty and also then phenotypic novelty (however defined). Superficially, this process – or at least the sudden appearance of phenotypic novelty – may seem macromutational, but, clearly, it is not, at least in the original sense of Goldschmidt or even that of Dobzhansky. Indeed, something as simple as slight changes in protein folding could have major cascading morphological effects.

10
Molecules and Systematics: Looking Toward the Future

It may be widely believed, and even true at some level, that, as Sean Carroll has recently reiterated, "genomes diverge as a function of time." However, the observation that genomes may be different (in whatever ways difference, and similarity, may be identified and defined) does not in and of itself provide clues to how this difference was achieved. No doubt, some difference is due to the rare and random effects of UV radiation. In addition, genomic difference may be due to failures in DNA repair. There may be something intuitively appealing about Sibley and Ahlquist's the "law of large numbers" – the idea that organisms are closely related because they share "lots" of their genome. However, as Jonathan Marks points out, humans share about 25% of their genome with bananas. Essentially, there is nothing in an observation of genomic difference or similarity that directly translates into the "molecular assumption" and, consequently, a theory of evolutionary relatedness.

Can, then, molecular information be useful in systematics and phylogenetic reconstruction?

The answer is yes, but it will have to be at the level of cell biology and pathways of molecular communication. As King and Wilson came close to predicting many decades ago, it is not through the study of molecular or genomic similarity of organisms that we will come to understand their biology, but through the investigation of those elements that underlie the development of their biology. This makes sense. For, if something as simple as the inactivation or deletion of a transcriptional enhancer can result in a more caudal repositioning of the sacrum, or if the expansion of a morphogenetic gradient can transform in its entirety a narrow pelvic girdle with tall, thin ilial blades into a broad, deep, and squat structure, then it is by seeking to identify the similarity or difference in these molecular events that we may more profitably explore the molecular basis of morphology and, consequently, the evolutionary relationships of complex organisms.

This is, perhaps, a timely occasion to both question and expand our perceptions of what is or will be evolutionarily revealing at the molecular level. There has been a steady increase in the number of studies that demonstrate virtual molecular identity between taxa that are morphologically very different and then express amazement at this apparent contradiction. As Sean Carroll pointed out with regard to the importance placed on the human and chimpanzee-genome projects – especially since so much money has been poured into them in the hope that forthcoming comparisons will instantaneously provide answers to any questions – demonstrating molecular similarity does not translate into deciphering the pathways that make these organisms so different in hard- and soft-tissue anatomy, physiology, reproductive biology, cognitive abilities, and behavior. Here, the "law of large numbers" fails to be enlightening. For, in contrast to the bacterial world, in the metazoan world, a one-to-one correspondence between a "gene" (a sequence of nucleotides bound by start and stop codons) and a "gene product" (a protein or amino acid sequence) is not there. In multicellular animals, RNA essentially directs the "show," for example, in reading select bases and splicing specific introns, as it composes different proteins from the same stretches of DNA. The surprise "The International Chimpanzee Chromosome 22 Consortium" had at finding upon comparing human chromosome 21 with its apparent orthologue in the chimpanzee, chromosome 22 – not only that these hominoids differ by 83% in their amino acid sequences but also that this large difference is produced from very similar DNA sequences – should serve as a lesson: While there may be appeal to the "law of large numbers" that comparison of chromosomes and especially of entire genomes purportedly represents, in the end, this molecular level may not be the evolutionarily informative hotspot everyone has been seeking.

See also Gene Mapping and Chromosome Evolution by Fluorescence–Activated Chromosome Sorting; Genetic Intelligence, Evolution of; Genetic Variation and Molecular Evolution; Immunoassays.

Bibliography

Books and Reviews

Frontiers in biology: development, *Science* (1994) **266**, 561–614.

Gerhart, J.C., Kirschner, M.W. (1997) *Cells, Embryos, and Evolution: Toward a Cellular and Developmental Understanding of Phenotypic Variation and Evolutionary Adaptability*, Blackwell Science, New York.

Marks, J. (2003) *What it Means to be 98% Chimpanzee*, (revised paperback edition). University of California Press, Berkeley, CA.

Müller, G.B., Newman, S.A. (Eds.) (2000) *Origination of Organismal Form*, MIT Press, Cambridge, MA.

Raff, R.A. (1996) *The Shape of Life: Genes, Development, and the Evolution of Animal Form*, University of Chicago Press, Chicago, IL.

Schwartz, J.H. (1987) *The Red Ape: Orangutans and Humans Origins*, Basic Books, New York; in press.

Schwartz, J.H. (1999) *Sudden Origins: Fossils, Genes, and the Emergence of Species*, John Wiley, New York.

Primary Sources

Averoff, M., Cohen, S.M. (1997) Evolutionary origin of insect wings from ancestral gills, *Nature* **385**, 627–630.

Awadella, P., Eyre-Walker, A., Maynard Smith, J. (1999) Linkage disequilibrium and recombination in hominid mitochondrial DNA, *Science* **286**, 2524–2525.

Baker, R.H., Xiaobo, Y., DeSalle, R. (1998) Assessing the relative contribution of molecular and morphological characters in simultaneous analysis trees, *Mol. Phylogenet. Evol.* **9**, 427–436.

Benveniste, R.E., Todaro, G.J. (1976) Evolution of type C viral genes: evidence for an Asian origin of man, *Nature* **261**, 101–8.

Britten, R. (2002) Divergence between samples of chimpanzee and human DNA sequences is 5%, counting indels, *Proc. Natl. Acad. Sci. U.S.A.* **99**, 13633–13635.

Britten, R.J., Kohne, D.E. (1968) Repeated sequences in DNA, *Science* **161**, 529–40.

Brown, W.M., George, M. Jr., Wilson, A.C. (1979) Rapid evolution of mitochondrial DNA, *Proc. Natl. Acad. Sci. U.S.A.* **76**, 1967–71.

Brown, W.M., Prager, E.M., Wang, A., Wilson, A.C. (1982) Mitochondrial DNA sequences of primates: tempo and mode of evolution, *J. Mol. Evol.* **18**, 225–39.

Bruce, E.J., Ayala, F.J. (1979) Phylogenetic relationships between man and the apes: electrophoretic evidence, *Evolution* **33**, 1040–1056.

Caccone, A., Powell, J.R. (1989) DNA divergence among hominoids, *Evolution* **43**, 925–942.

Carroll, S.R. (2003) Genetics and the making of *Homo sapiens*, *Nature* **422**, 849–857.

Cohn, M.J., Patel, K., Krumlauf, R., Wilkinson, D.G., Clarke, J.D.W., Tickle, C. (1997) Hox9 genes and vertebrate limb specification, *Nature* **387**, 97–101.

Crockford, S.J. (2003) Thyroid rhythm phenotypes and hominid evolution: a new paradigm implicates pulsatile hormone selection in speciation and adaptation changes, *Comp. Biochem. Physiol., A* **135**, 105–129.

Cronin, J.E., Sarich, V.M. (1980) Tupaiid and Archonta phylogeny: the Macromolecular Evidence, in: Luckett, W.P. (Ed.) *Comparative Biology and Evolutionary Relationships of Tree Shrews*, Plenum Press, New York, pp. 293–312.

Czelusniak, J., Goodman, M., Moncrief, N.D., Kehoe, S.M. (1990) Maximum parsimony approach to construction of evolutionary trees from aligned homologous sequences, *Methods Enzymol.* **183**, 601–615.

Dobzhansky, T. (1935) *Drosophila miranda*, a new species, *Genetics* **20**, 377–391.

Dobzhansky, T. (1941) *Genetics and the Origin of Species*, 2nd edition, Columbia University Press, New York.

Felsenstein, J. (1988) Phylogenies from molecular sequences: inference and reliability, *Annu. Rev. Genet.* **22**, 521–565.

Ferris, S.D., Wilson, A.C., Brown, W.M. (1981) Evolutionary tree for apes and humans based on cleavage maps of mitochondrial DNA, *Proc. Natl. Acad. Sci. U.S.A.* **78**, 2431–2436.

Figdor, M., Stern, C. (1993) Segmental organization of embryonic diencephalon, *Nature* **363**, 630–634.

Goldschmidt, R. (1940) *The Material Basis for Evolution*, Yale University Press, New Haven, CT, (reprinted 1982).

Goodman, M. (1962) Immunochemistry of the primates and primate evolution, *Ann. N.Y. Acad. Sci.* **102**, 219–234.

Goodman, M., Tashian, R.E. (Eds.) (1976) *Molecular Anthropology*, Plenum Press, New York.

Goodman, M., Braunitzer, G., Stangl, A., Schrank, B. (1983) Evidence on human origins from haemoglobins of African apes, *Nature* **303**, 546–48.

Goodman, M., Olson, C.B., Beeber, J.E., Czelusniak, J. (1982) New perspectives in the molecular biological analysis of mammalian phylogeny, *Acta Zoolo. Fennica* **169**, 1–73.

Graur, D., Martin, W. (2004) Reading the entrails of chickens: molecular timescales of evolution and the illusion of precision, *Trends Genet.* **20**, 80–86.

Gregory, W.K. (1922) *The Origin and Evolution of the Human Dentition*, Williams and Wilkins, Baltimore, MD.

Hagelberg, E. (2003) Recombination or mutation rate heterogeneity? Implications for mitochondrial Eve, *Trends Genet.* **19**, 84–90.

Hasegawa, M., Yano, T. (1984) Maximum likelihood method of phylogenetic inference from DNA sequence data, *Bull. Biometri. Soc. Jpn.* **5**, 1–7.

Hasegawa, M., Kishino, H., Yano, T. (1985) Dating of the human-ape splitting by a molecular clock of mitochondrial DNA, *J. Mol. Evol.* **22**, 160–174.

Hedges, S.B., Kumar, S. (2002) Vertebrate genomes compared, *Science* **297**, 1283–1285.

Hedges, S.B., Kumar, S. (2003) Genomic clocks and evolutionary timescales, *Trends Genet.* **19**, 200–206.

Hedges, S.B., Kumar, S., Tamura, K., Stoneking, M. (1992) Human origins and analysis of mitochondrial DNA sequences, *Science* **255**, 737–739.

Horai, S., Hayasaka, K., Kondo, R., Tsugane, K., Takahata, N. (1995) Recent African origin of modern humans revealed by complete sequence of hominid mitochondrial DNAs, *Proc. Natl. Acad. Sci. U.S.A.* **92**, 532–536.

Hunt, P., Gulisano, M., Cook, M., Sham, M.H., Faiella, A., Wilkinson, D., Boncinelli, E., Krumlauf, R. (1991) A distinct Hox code for the branchial region of the vertebrae head, *Nature* **353**, 861–864.

King, M.-C., Wilson, A.C. (1975) Evolution at two levels in humans and chimpanzees, *Science* **188**, 107–88.

Lake, J.A. (1991) The order of sequence alignment can bias the selection of tree topology, *Mol. Biol. Evol.* **8**, 378–385.

Lovejoy, C.O., Cohn, M.J., White, T.D. (1999) Morphological analysis of the mammalian postcranium: a developmental perspective, *Proc. Natl. Acad. Sci. U.S.A.* **96**, 13247–13252.

Lovejoy, C.O., McCollum, M.A., Reno, P.I., Rosenman, B.A. (2003) Developmental biology and human evolution, *Annu. Rev. Anthropol.* **32**, 85–109.

Løvtrup, S. (1974) *Epigenetics: A Treatise on Theoretical Biology*, Wiley, New York.

Lowenstein, J.M., Molleson, T., Washburn, S.L. (1982) Piltdown jaw confirmed as orang, *Nature* **299**, 294.

Lowenstein, J.M., Sarich, V.M., Richardson, B.J. (1981) Albumin systematics of the extinct mammoth and Tasmanian wolf, *Nature* **291**, 409–411.

Lumsden, A., Krumlauf, R. (1996) Patterning the vertebrate neuraxis, *Science* **274**, 1109–1115.

Maresca, B., Schwartz, J.H. Environmental change and stress protein concentration as a source of morphological novelty: sudden origins, a general theory on a mechanism of evolution, in manuscript.

Marshall, C.R. (1991) Statistical tests and bootstrapping: assessing the reliability of phylogenetics based on distance data, *Mol. Biol. Evol.* **8**, 386–391.

Mathers, P., Grinberg, A., Mahon, K., Jamrich, M. (1997) The Rx homeobox gene is essential for vertebrate eye development, *Nature* **387**, 604–607.

Morgan, T.H. (1916) *A Critique of the Theory of Evolution*, Princeton University press, Princeton, NJ.

Muragaki, Y., Mundlos, S., Upton, J., Olsen, B.R. (1996) Altered growth and branching patterns in synpolydactyly caused by mutations in HOXD13, *Science* **272**, 548–551.

Nuttall, G.H.F. (1904) *Blood Immunity and Blood Relationship*, Cambridge University Press, Cambridge, MA.

O'hUigin, C.O., Satta, Y., Takahata, N., Klein, J. (2002) Contribution of homoplasy and of ancestral polymorphism to the evolution of genes in anthropoid primates, *Mol. Biol. Evol.* **19**, 1501–1513.

Osborn, J.F. (1978) Morphogenetic Gradients: Fields Versus Clones, in: Butler, P.M., Joysey, K.A. (Eds.) *Development, Function, and*

Evolution of Teeth, Academic Press, New York, pp. 171–201.

Oster, G., Alberch, P. (1982) Evolution and bifurcation of developmental programs, *Evolution* **36**, 444–459.

Pennisi, E. (1999) Genetic study shakes up out of Africa theory, *Science* **283**, 1828.

Romero-Herrera, A.E., Lehmann, H., Castillo, O., Joysey, K.A., Friday, A.E. (1976) Myoglobin of the orangutan as a phylogenetic enigma, *Nature* **261**, 162–64.

Romero-Herrera, A.E., Lehmann, H., Joysey, K.A., Friday, A.E. (1978) On the evolution of myoglobin, *Philos. Trans. R. Soc. Lond. B Biol. Sci.* **283**, 61–183.

Rubenstein, J.L.R., Martinex, S., Shimamura, K., Puelles, L. (1994) The embryonic vertebrate forebrain: the prosomeric model, *Science* **266**, 578–580.

Ruvolo, M. (1997) Molecular phylogeny of the hominoids: inferences from multiple independent DNA sequence data sets, *Mol. Biol. Evol.* **14**, 248–265.

Ruvolo, M., Disotell, T.R., Allard, M.W., Brown, W.M., Honeycutt, R.L. (1991) Resolution of the African hominoid trichotomy by use of a mitochondrial gene sequence, *Proc. Natl. Acad. Sci. U.S.A.* **88**, 1570–1574.

Samollow, P.B., Cherry, L.M., Whitte, S.M., Rogers, J. (1996) Interspecific variation at the Y-linked *RPS4Y* locus in hominoids: implications for phylogeny, *Am. J. Phys. Anthropol.* **101**, 333–343.

Sarich, V.M. (1971) A Molecular Approach to the Question of Human Origins, in: Dolhinow, P., Sarich, V.M. (Eds.) *Background for Man*, Little, Brown, Boston, MA, pp. 60–81.

Sarich, V.M., Wilson, A.C. (1966) Quantitative immunochemistry and the evolution of primate albumins: micro-complement fixation, *Science* **154**, 1563–66.

Sarich, V.M., Wilson, A.C. (1967a) Rates of albumin evolution in primates, *Proc. Natl. Acad. Sci. U.S.A.* **58**, 142–148.

Sarich, V.M., Wilson, A.C. (1967b) Immunological time scale for hominid evolution, *Science* **158**, 1200–1203.

Schwartz, J.H. (2001) A review of the systematics and taxonomy of Hominoidea: history, morphology, molecules, and fossils, *Ludus Vitalis* **IX**, 15–45.

Shubin, N., Alberch, P. (1986) A morphogenetic approach to the origin and basic organization of the tetrapod limb, *Evol. Biol.* **20**, 319–387.

Shubin, N., Tabin, C., Carroll, S. (1997) Fossils, genes and the evolution of animal limbs, *Nature* **388**, 639–648.

Sibley, C.G., Ahlquist, J.E. (1983) Phylogeny and Classification of Birds Based on the Data of DNA-DNA Hybridization, in: Johnston, R.F. (Ed.) *Current Ornithology*, Vol. 1, Plenum Press, New York, pp. 245–92.

Sibley, C.G., Ahlquist, J.E. (1984) The phylogeny of the hominoid primates, as indicated by DNA-DNA hybridization, *J. Mol. Evol.* **20**, 2–15.

Sordino, P., van der Hoeven, F., Duboule, D. (1995) Hox gene expression in teleost fins and the origin of vertebrate digits, *Nature* **375**, 678–681.

Stauffer, R.L., Walker, A., Ryder, O.A., Lyons-Weiler, M., Hedges, S.B. (2001) Human and ape molecular clocks and constraints on paleontological hypotheses, *J. Hered.* **92**, 469–474.

Summerbell, D. (1981) Evidence for regulation of growth, size and pattern in the developing chick limb bud, *J. Embryol. Exp. Morphol.* **65**, 129–150.

The International Chimpanzee Chromosome 22 Consortium. (2004) DNA sequence and comparative analysis of chimpanzee chromosome 22, *Nature* **429**, 382–388.

Tickle, C. (1992) A tool for transgenesis, *Nature* **358**, 188–189.

Wolpert, L. (1980) Positional Information and Pattern Formation in Limb Development, in: Pratt, R.M., Christiansen, R.L. (Eds.) *Current Research Trends in Prenatal Craniofacial Development*, Elsevier/North Holland, New York, pp. 89–101.

Zuckerkandl, E. (1987) On the molecular evolutionary clock, *J. Mol. Evol.* **26**, 34–46.

Zuckerkandl, E., Pauling, L. (1962) Molecular Disease, Evolution, and Genic Heterogeneity, in: Kasha, M., Pullman, B. (Eds.) *Horizons in Biochemistry*, Academic Press, New York, pp. 189–225.

Motor Neuron Diseases: Cellular and Animal Models

Georg Haase
Institut de Neurobiologie de la Méditerranée, Marseille, France

Encyclopedia of Molecular Cell Biology and Molecular Medicine, 2nd Edition. Volume 8
Edited by Robert A. Meyers.
Copyright © 2005 Wiley-VCH Verlag GmbH & Co. KGaA, Weinheim
ISBN: 3-527-30550-5

Keywords

Amyotrophic Lateral Sclerosis (ALS)
A neurodegenerative disease involving the selective degeneration of motor neurons in cerebral cortex, brainstem, and spinal cord. ALS was first described in 1869 by the French neuropathologist Jean-Marie Charcot.

Chaperone
A protein that assists in the folding or assembly of another protein, without becoming part of the completed structure.

Guanine Nucleotide Exchange Factor (GEF)
A protein that facilitates the exchange of GDP for GTP in a GTP-binding protein.

Superoxide Dismutase 1 (SOD1)
An enzyme that catalyzes the conversion of superoxide anion to water and hydrogen peroxide.

◼ Motor neuron diseases are human disorders characterized by the progressive degeneration and death of motor neurons and the denervation of skeletal muscles. Their course is always fatal and there are currently no really effective therapies. While most adult motor neuron diseases appear sporadically, others are inherited, because of genetic mutations in Superoxide Dismutase 1, Alsin, or Dynactin, or associated with risk factors such as genetic polymorphisms in Neurofilaments or Vascular Endothelial Growth Factor (VEGF). Multiple molecular mechanisms might therefore cause – or contribute to – disease. Some of the proposed disease mechanisms might operate cell-autonomously in motor neurons, for example, by perturbing energy metabolism, endosome trafficking, or axonal transport, whereas others seem to act indirectly, involving nonneuronal cells and diffusible factors such as nitric oxide and excess glutamate. Relevant cellular and animal models are now available to test these hypotheses and to investigate new therapeutic strategies.

1
Human Motor Neuron Diseases

Motor neuron diseases represent a spectrum of disorders in which neurons controlling voluntary movement progressively degenerate while other types of neurons are spared. Degeneration of upper motor neurons in cerebral cortex results in spasticity and hyperreflexia. Degeneration of lower motor neurons in brainstem and spinal cord leads to muscle paralysis and atrophy. According to these clinical and pathological criteria, motor neuron diseases can be further classified: Amyotrophic lateral sclerosis (ALS, Charcot disease or Lou Gehrig's disease) affects both upper and lower motor neurons. An ALS form with predominant brainstem symptoms is called progressive bulbar palsy. Progressive muscular atrophies represent conditions in which only lower motor neurons degenerate, whereas primary lateral sclerosis involves only upper motor neurons. ALS is the most frequent motor neuron disease in the adult, with a lifetime risk of 1 in 2000 and a mean age of onset around 55 years. ALS is inevitably fatal but its course is variable between patients. Death is most often caused by failure of respiratory muscles appearing within one to five years after diagnosis.

Histopathological studies have documented a loss of motor neurons in the cervical and lumbar spinal cord of ALS patients. The number of myelinated large caliber axons in the corticospinal tract and in the ventral roots was also found to be reduced and signs of Wallerian degeneration and atrophy have been noted. Other microscopic abnormalities such as reactive gliosis, neuronal inclusion bodies, axonal spheroids, and the loss of dendrites have also been described. In skeletal muscle, denervated fibers appear atrophic and angulated. Grouping of muscle fiber types positive for esterase, ATPase, or myosin isoforms indicates muscle reinnervation from the remaining motor neurons.

2
Genetics of Motor Neuron Diseases

While most ALS cases appear spontaneously, some (~10%) are familial (FALS). These can show autosomal dominant, recessive, or X-linked inheritance. Genetic studies in FALS (familial amyotrophic lateral sclerosis) are rendered difficult by the late onset of disease, its incomplete penetrance, and the short survival of affected family members. Nevertheless, several FALS genes have been identified (Table 1) and additional FALS loci have been mapped to chromosomes 9q21-22, 15q15-22, 16, 18, and 20.

Tab. 1 Gene mutations responsible for motor neuron diseases.

Gene	Disease	Locus	Inheritance	Mutation type
SOD1	ALS1	21q22.1	Dominant[a]	Missense
Alsin	Juvenile ALS (ALS2) progressive lateral sclerosis	2q33	Recessive	Frameshift
Tau	ALS/dementia/ Parkinson complex	17q21	Dominant	Missense or intronic
Dynactin (p150 subunit)		2p13	Dominant	Missense
Androgen receptor	Spinal and bulbar muscular atrophy (SBMA)	Xq12	Recessive	CAG triplet expansion
Survival motor neuron (SMN)	SMA	5q13	Recessive	Deletion or missense

[a]With the exception of SOD1 D90A.

2.1
SOD1 Mutations

About 10 to 20% of familial ALS cases (1–2% of total) are linked to mutations in the superoxide dismutase 1 (SOD1) gene. The *SOD1* gene is ubiquitously expressed and encodes a dimeric enzyme containing 153 amino acids, one copper, and one zinc ion per monomer. The normal function of this enzyme is to catalyze the conversion of superoxide to hydrogen peroxide. Over 100 different missense mutations scattered throughout the protein have now been shown to produce motor neuron disease. All SOD1 mutations except one (D90A) show autosomal dominant transmission. It has been proposed that SOD1 mutations cause motor neuron degeneration by catalyzing aberrant free radical chemistry, by perturbing energy homeostasis, by initiating protein aggregation, by causing excessive glutamate levels, or through a combination of these mechanisms.

2.2
Alsin Mutations

Mutations in a gene termed *ALS2 or Alsin*, located on chromosome 2q33-35, have been detected in familial ALS cases with juvenile onset and slow disease progression. The gene mutations are autosomal-recessively inherited and most of them represent small deletions. Computer algorithms have predicted that the Alsin protein contains guanine nucleotide exchange factor (GEF) domains for small GTPases. Small GTPases are molecular switches that cycle between an inactive GDP-bound and an active GTP-bound state; they are activated by GEFs, which stimulate the exchange of GDP for GTP. The 183-kDa alsin protein acts as guanine nucleotide exchange factor for the small GTPase Rab5, localizes to early endosomes, and enhances their fusion. Reported alsin mutations are predicted to result in a failure of Rab5 activation, suggesting that normal endosome fusion and trafficking are essential for motor neuron maintenance.

3
Culture Models of Motor Neuron Degeneration

In an attempt to unravel disease mechanisms and to identify or validate new therapeutic candidates, various motor neuron culture models have been developed. These include neuronal cell lines, primary motor neurons, and spinal cord slices.

3.1
Neuronal Cell Lines

Neuronal cell lines, derived from tumors or *in vitro* transformed cells, only rarely resemble motor neurons. One of these exceptions is the NSC-34 cell line, which was generated by fusion of mouse neuroblastoma cells with embryonic spinal motor neurons. Like motor neurons, NSC-34 cells display a multipolar phenotype and express choline acetyltransferase (ChAT), different ion channels, and neurofilaments. These cells can also generate action potentials and induce acetylcholine receptor clusters on cocultured myotubes. NSC-34 cells transfected with mutant SOD1 plasmids display various abnormalities including mitochondrial dysfunction, increased cytochrome c release, and decreased cell survival following oxidative stress. Gene expression profiling and proteome analysis revealed underlying molecular changes such as downregulation of neurofilaments and glutathion S-transferases and up-regulation of proteins involved in NO metabolism (argininosuccinate synthase, argininosuccinate lyase, and neuronal NO synthase). Some of these changes could be confirmed *in vivo* in individual spinal cord motor neurons.

3.2
Motor Neuron Cultures

Motor neurons are notoriously difficult to purify and to maintain in primary culture. Current methods rely on three properties of these cells: their characteristic size, their expression of specific surface markers, and their dependence on trophic factors for survival and neurite outgrowth. Since motor neurons are larger in size and lower in density than most other types of neurons and glial cells, they can be enriched by differential centrifugation. As a further purification step, immuno-panning techniques for chicken and rat spinal motor neurons have been developed using monoclonal antibodies that recognize specific cell surface antigens such as SC1, BEN, or the p75 neurotrophin receptor. Spinal motor neurons can also be isolated using antibody-coated microbeads and subsequent magnetic cell sorting. Once purified, motor neurons are seeded in culture dishes coated with appropriate substrates such as polyornithine and laminin. While many motor neurons die during the first hours of culture, the remaining cells develop typical morphologies and are able to survive for several weeks when appropriate trophic factors are added. These procedures are currently limited to embryonic motor neurons from spinal cord and brainstem and not yet available for isolating cortical motor neurons.

One of the first motor neuron culture systems relevant to human ALS involved microinjection of mutant and wild-type SOD1 plasmid expression vectors into cultured motor neurons. In this system, abnormal protein aggregates appeared specifically in mutant SOD1-expressing cells. Newer studies suggest that cell death of SOD1 expressing motor neurons might also involve external triggers: In normal embryonic motor neurons, activation of the cell surface receptor Fas triggers death via two parallel pathways, the classical FADD/caspase-8 cascade and a second cascade involving Daxx, ASK1, p38, and transcriptional upregulation of neuronal NO synthase; see Fig. 1. Interestingly, motor neurons from transgenic mice for the SOD1 mutations G37R, G85R, or G93A are much more sensitive to Fas- or NO-triggered cell death than motor neurons expressing wild-type human SOD1. Cerebellar neurons, DRG neurons, or astrocytes do not display such increased sensitivity, reflecting the selective vulnerability of motor neurons in living mutant SOD1 mice. Despite their inherent technical difficulties, mutant SOD1 motor neuron cultures thus emerge as a promising tool to screen pharmacological compounds for their effects on disease-related molecular targets or cellular phenotypes.

3.3
Spinal Cord Slice Cultures

In an attempt to study motor neurons in their normal cellular environment, several groups have prepared spinal cord slice cultures from rat and mouse. These cultures can be maintained for up to 2 weeks (adult mouse) or even several months (perinatal rat). Individual motor neurons in living slice cultures are identified by their large size, the presence of action potentials, or retrograde DiI labeling prior to culturing. In fixed slices, motor neuron cell bodies

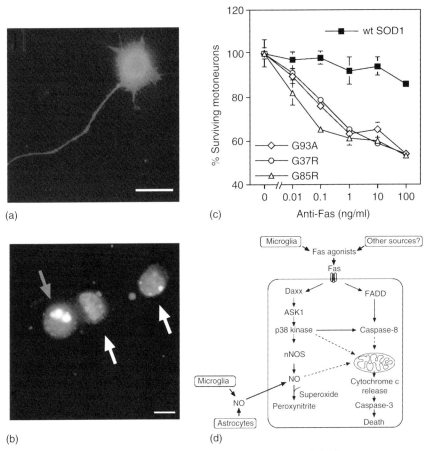

(a)

(c)

(b)

(d)

Fig. 1 Isolated motor neurons can be used to study mutant SOD1-linked motor neuron death. (a) Human superoxide SOD1 is expressed in cultured embryonic motor neurons from transgenic SOD1 G93A mice. (b) Treatment of SOD1 G85R motor neuron cultures with agonistic antibodies to the cell surface receptor Fas leads to increased apoptosis, as detected by staining with DAPI (in blue) and immunolabeling for activated caspase 3 (in red). An apoptotic motor neuron with strong caspase-3 activation and nuclear condensation c (red arrow) can be distinguished from healthy motor neurons displaying only weak caspase-3 activation and normal chromatin structure (white arrows). Scale bars: 25 μm. (c) Motor neurons from transgenic mutant SOD1 G93A, G85R, and G37R mice show higher susceptibility to Fas-triggered cell death than motor neurons from mice expressing wild-type SOD1. (d) Model of Fas-triggered motor neuron death. Cell death involves the classical FADD/caspase-8 pathway and a parallel pathway leading from Daxx, ASK1, and p38 activation to transcriptional upregulation of neuronal NO synthase (nNOS) and NO production. The presence of mutant SOD1 sensitizes motor neurons to Fas agonists and NO. Potential sources for these cell death triggers are astrocytes and microglia. (See color plate p. xxvi); see Fig. 2.

can be identified and counted by Nissl staining or immunocytochemical labeling for the neurotransmitter-synthesizing enzyme ChAT, the transcription factors Islet 1/2, or the surface receptor Ret. Slice culture systems have also been developed for the study of cortical motor neurons.

In the context of ALS research, organotypic culture systems have been used to show that chronic inhibition of glutamate uptake can cause slow degeneration of motor neurons. In this model, antiexcitotoxic agents such as CNQX, riluzole, and topiramate, nitric oxide synthase inhibitors, and neurotrophic factors such as IGF-1 and GDNF protected against motor neuron degeneration. When organotypic slice cultures were derived from transgenic mutant SOD1 mice and exposed to excitotoxic stimuli, mutant motor neuron death could be blocked with a cyclo-oxygenase-2 inhibitor. In another spinal cord slice model, chronic administration of malonate, an inhibitor of mitochondrial electron transport, was found to induce motor neuron death, which could be prevented by a number of antiexcitotoxic agents, antioxidants, and caspase inhibitors.

4
Animal Models of Motor Neuron Degeneration

4.1
Axotomy Models

In neonate animals, motor neuron death can be experimentally induced by a peripheral nerve lesion, since developing motor neurons depend for their survival on contact with their target muscle and their axonal environment. In the most widely used models, the sciatic, facial, or hypoglossal nerve is crushed or sectioned,

and the survival of the lesioned motor neurons is eventually after prior retrograde labeling. The extent of motor neuron death depends on the age of the animal and the type of lesion. When a sciatic nerve lesion is performed in 1-day old rats, 90 to 100% of the corresponding motor neurons are lost. When the same type of lesion is performed at 4 or 5 days of age, between 80 and 100% of the motor neurons survive. Lesioning the peripheral nerve by section or close to the cell body both results in slower target reinnervation and reduced motor neuron survival, as compared to lesioning the nerve by section or distally.

Axotomy-induced motor neuron death has many apoptotic features. Morphological studies on lesioned motor neurons have provided evidence for DNA fragmentation, nuclear condensation, and mitochondrial disruption. Pharmacological inhibition of caspases or mitochondrial pore proteins or overexpression of apoptosis-inhibitory proteins (IAPs) reduced the extent of axotomy-induced motor neuron death. Finally, genetic studies in transgenic mice showed that overexpression of the anti-apoptotic gene *Bcl-2* or deletion of the proapoptotic gene *Bax* protected against axotomy-induced cell death. It has also been recognized that apoptotic death of axotomized motor neurons involves activation of cell surface receptors (TNF-R and Fas) and intracellular production of nitric oxide.

4.2
Transgenic Mutant SOD1 Mice and Rats

One breakthrough in ALS research has been the generation of transgenic mice and rats expressing human ALS-linked SOD1 mutations. Several transgenic lines show clinical, histopathological, and electrophysiological signs closely resembling human motor neuron disease. Transgenic mice for the SOD1 mutants G37R, G85R, or G93A display progressive muscle atrophy and paresis, gait instability, and premature death. Similar symptoms were observed in transgenic rats expressing the SOD1 mutants G93A or H46R. In these animals, disease severity depends on the type of mutation, the number of transgenes, and the genetic background. Affected mice display a reduction in the number of spinal motor neuron cell bodies and motor axons and, to a more variable extent, astrogliosis. Other histopathological abnormalities include a fragmented Golgi apparatus, enlarged mitochondria, and various types of inclusions. Electrophysiological studies in mutant mice revealed the presence of fasciculations and fibrillations, and a decreased number of functional motor units.

Some of the molecular consequences of mutant SOD1 expression in mouse spinal cord have also been recognized: sequential activation of caspases, upregulation of apoptotic proteins such as Bad and Bax, release of cytochrome c from mitochondria, and deregulation of the cyclin-dependent kinase cdk5. How the SOD1 mutants cause these changes is still obscure. For the following reasons, altered SOD1 enzymatic activity does not seem to be causal: first, only some disease-causing SOD1 mutations are enzymatically inactive (e.g. G85R), while others retain activity (G93A or G37R). Second, no motor neuron disease is observed when SOD1 activity in mice is lowered (by knocking out the *SOD1* gene or the *CCS* gene required for SOD1 copper loading) or when SOD1 activity is elevated (by transgenic expression of wild-type SOD1). Finally, crossing transgenic mutant SOD1 mice with SOD1 knockout mice, transgenic mice overexpressing wildtype SOD1, or CCS knockout mice, has no influence on disease.

The cellular basis of mutant SOD1 toxicity also remains enigmatic: In transgenic mice, pathology is only observed when mutant SOD1 expression is ubiquitous in all cell types but not when it is restricted to astrocytes or neurons (Fig. 2). To rule out insufficient levels of transgene expression, chimeric mice have been generated that carry mixtures of wild-type and mutant

Fig. 2 Motor neuron degeneration in transgenic ALS mice is influenced by nonneuronal cells. The consequences of mutant SOD1 expression in different cell types are schematically illustrated on spinal cord cross sections. Genotypes of transgenic mice are indicated. Motor neurons are depicted as multipolar cells and glial cells as round cells. Mutant SOD1 expression in these cell types is indicated in gray or black, normal (endogenous) SOD1 expression is shown in white. (a) Normal situation. (b) In transgenic mice, ubiquitous expression of mutant (m) SOD1 causes typical motor neuron disease. (c, d) No disease is observed in transgenic mice expressing mutant SOD1, specifically in glial cells (c) or in neurons (d). (e) Crossing of Thy1: SOD1 mice with transgenic SOD1 G93A mice elevates mutant SOD1 levels (in black) in motor neurons above the ubiquitous expression levels but does not exacerbate disease. (f) Chimeric mice that carry a mixture of wild-type (wt) and mutant SOD1-expressing cells in their spinal cord show later disease onset and longer lifespan than their mutant littermates. These improvements correlate with the proportion of wild-type cells within individual spinal cords. Ubiquitin inclusions are found not only in mutant but also in wild-type motor neurons. Interestingly, mutant SOD1 motor neurons in the chimeras seem to be protected by wild-type nonneuronal cells in their environment. Taken together, these data suggest that motor neuron degeneration is the result of a complex interplay between motor neurons and surrounding nonneuronal cells.

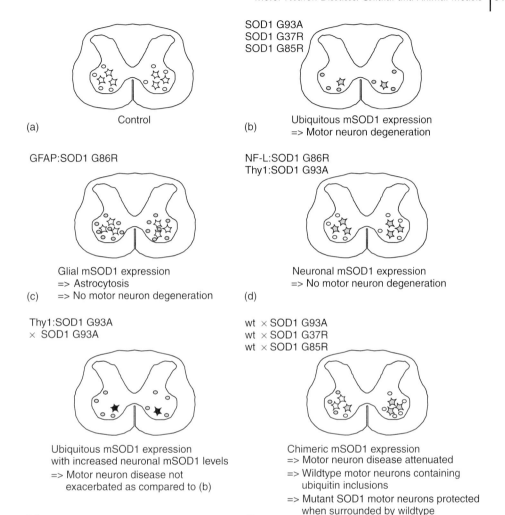

(a) Control

(b) SOD1 G93A
SOD1 G37R
SOD1 G85R

Ubiquitous mSOD1 expression
=> Motor neuron degeneration

(c) GFAP:SOD1 G86R

Glial mSOD1 expression
=> Astrocytosis
=> No motor neuron degeneration

(d) NF-L:SOD1 G86R
Thy1:SOD1 G93A

Neuronal mSOD1 expression
=> No motor neuron degeneration

(e) Thy1:SOD1 G93A
× SOD1 G93A

Ubiquitous mSOD1 expression
with increased neuronal mSOD1 levels
=> Motor neuron disease not
 exacerbated as compared to (b)

(f) wt × SOD1 G93A
wt × SOD1 G37R
wt × SOD1 G85R

Chimeric mSOD1 expression
=> Motor neuron disease attenuated
=> Wildtype motor neurons containing
 ubiquitin inclusions
=> Mutant SOD1 motor neurons protected
 when surrounded by wildtype
 non-neuronal cells

SOD1 expressing cells in the spinal cord. Interestingly, in these chimeras, degenerative signs such as ubiquitin inclusions are found not only in mutant SOD1 expressing motor neurons but also in wild-type motor neurons. Moreover, survival of mutant motor neurons seems to be enhanced by the presence of nonneuronal cells. These observations suggest that motor neuron degeneration can be triggered by non-cell-autonomous mechanisms.

4.3
pmn Mice

Mice with progressive motor neuronopathy (*pmn*) represent another popular motor neuron disease model. Homozygous *pmn* mice develop first neuromuscular symptoms at 2 weeks of age, i.e. much earlier than all reported mutant SOD1 mice. The *pmn* disease manifests with hindlimb atrophy and paresis and rapidly extends to the forelimbs and other muscle groups. Affected *pmn* mice die around 40 days of age, most probably from respiratory failure. They display fibrillations, a loss of axons in motor nerves and ventral roots and a reduced number of motor neurons in some brainstem nuclei. The genetic defect was identified as a point mutation in the *Tbce* gene, which encodes a chaperone protein for tubulin folding and microtubule assembly. The *pmn* mutation results in decreased Tbce protein stability, diminished concentrations of tubulins in peripheral nerves, and progressive loss of microtubules. Interestingly, Tbce gene mutations have also been discovered in human patients with Sanjad Sakati/Kenny Caffey disease, a rare hereditary syndrome comprising hypoparathyroidism, mental retardation, facial dysmorphism, osteosclerosis, and fulgurant infections. It is thus conceivable that the dramatic human disease course obscures motor neuron disease symptoms or that the human Tbce mutations impair another Tbce protein function than the mouse mutation.

4.4
Mutant VEGF Mice

Vascular endothelial growth factor (VEGF) is a secreted 165 amino acid protein that controls the growth and permeability of blood vessels. Under conditions of low oxygen tension, VEGF is upregulated by specific transcription factors that act on a hypoxia-responsive element located in the VEGF promoter. Mice with a targeted deletion of this hypoxia-responsive element are unable to induce VEGF in response to hypoxia. Surprisingly, these mice manifest typical signs of motor neuron degeneration at an age of 5 to 7 months, with progressive muscle atrophy, impaired motor performance, and loss of motor axons

and endplates. Interestingly, three polymorphisms in the human VEGF seem associated with reduced levels of plasma VEGF and an increased risk for ALS. These studies incriminate VEGF as an ALS-modifier gene. Whether VEGF acts directly as a trophic factor for motor neurons or more indirectly, by modifying spinal cord vasculature or perfusion, is currently under investigation.

5
Therapeutic Approaches to Motor Neuron Diseases

5.1
Studies in Animal Models

In the past decade, more than 100 different therapeutic approaches have been tested in animal models of motor neuron degeneration. Lesion models have been instrumental in validating neurotrophic factors as therapeutic candidates: CNTF, BDNF, NT-4/5, GDNF, and CT-1 all enhanced survival of motor neurons after axotomy. More recently, several pharmacological compounds such as deprenyl, a monoamino-oxidase B inhibitor, MK-801, a glutamate receptor antagonist, CGP3466B, a GAPDH antagonist, and riluzole were also shown to be neuroprotective in these models.

In mutant SOD1 mice, numerous compounds have been tested but only few have resulted in robust positive effects (Table 2). Following treatment with riluzole, currently the only FDA-approved drug for ALS, the mean survival of SOD1 G93A mice was improved by 11% but histological or functional effects could not be documented. Feeding the animals with creatine also increased survival and provided some motor neuron protection. Numerous ALS

Tab. 2 Selected therapeutic trials in mouse motor neuron disease models.

Treatment	*Model*	*Delivery*		*Outcome*		
		Begin[a]	*Mode*	*Lifespan increase [%]*	*Neuromuscular function*	*Motor neuron protection*
Riluzole	SOD1 G93A	Early	Oral	11	Not reported	Not reported
	pmn	Early	Oral	<10	Yes	Not reported
Creatine	SOD1 G93A	Early	Oral	18	Yes	Yes
Minocyclin	SOD1 G37R	Early	Oral	≈6	Yes	Yes
Minocyclin+ Riluzole+ Nimodipin	SOD1 G37R	Early	Oral	13	Yes	Yes
zVAD-fmk	SOD1 G93A	Early	Intrathecal	21	Yes	Yes
NT-3 Adeno	*pmn*	Early	i.m.	50	yes	Yes
GDNF AAV	SOD1 G93A	Early	i.m.	14	yes	Yes
CT-1Adeno	SOD1 G93A	Early	i.m	8	yes	Yes
	pmn	Early	i.m.	18	yes	Yes
IGF AAV	SOD1 G93A	Late	i.m.	18	yes	Yes

[a]Treatment begin is classified with respect to disease onset: "early" treatments are initiated in the presymptomatic period, "late" treatments are started at or after disease onset.

patients have since then taken creatine, but a recent placebo-controlled clinical trial disclosed any beneficial effect of oral creatine at 10 g per day. Encouraging results were obtained by intrathecal infusion of the broad caspase inhibitor zVAD-fmk. Recently, several studies have also reported positive effects of minocyclin, which might be related to the inhibitory actions of this tetracyclin analog on caspase activation, mitochondrial cytochrome c release, cdk5 deregulation, or microglial cell activation. When minocyclin was administered in combination with riluzole and the calcium channel blocker nimodipin, the therapeutic benefit was further enhanced.

Gene therapy strategies have been developed to overcome the side effects of systemically delivered neurotrophic factors. These strategies have been tested since 1992 in *pmn* mice and since 2001 in transgenic mutant SOD1 mice (Table 2). In *pmn* mice, the implantation of genetically engineered tumor cells or subcutaneous implantation of encapsulated fibroblasts secreting CNTF reduced degeneration of phrenic and facial motor neurons. When injected intramuscularly in newborn *pmn* mice, adenovirus vectors encoding the neurotrophic factors NT-3, CNTF, and CT-1 were shown to improve neuromuscular function, reduce motor neuron degeneration, and increase the animal's lifespan. Recently, a nonviral gene transfer method, intramuscular injection and electroporation of plasmid DNA, proved to be as efficient as the adenovirus-based strategy to attenuate disease symptoms.

In transgenic mutant SOD1 mice, adenovirus, AAV, and lentivirus vectors coding for CT-1, GDNF, and more recently IGF-1 and VEGF, administered by intramuscular injection, resulted in therapeutic benefit. Some of these therapies were also capable to slow down motor neuron degeneration when started late, after the onset of disease. In order to translate these encouraging findings into clinical therapies, a number of fundamental questions now needs to be addressed: Are the observed effects due to release of the neurotrophic factors into the bloodstream or due to retrograde vector transport from muscles to motor neuron cell bodies? What are the best factors, alone or in combination? What is the safest and most efficient vector system?

5.2
Future Concepts

Since motor axons degenerate weeks to months before cell bodies in human ALS and in corresponding mouse models, therapies aimed at axonal protection appear particularly attractive. This concept is further illustrated by studies in the mouse mutant *Wallerian degeneration slow (Wld^S, Ola)*: In normal mice, a peripheral nerve lesion leads to axonal disintegration and reactive proliferation of Schwann cells in the distal nerve stump, a process called *Wallerian degeneration*. In Wld^S mice, however, the distal nerve stump remains intact following a lesion and remains able to conduct action potentials during several weeks. The Wld^S mutation was identified as a triplication on chromosome 4 that gives rise to a fusion protein between a truncated ubiquitination factor (Ube4b) and an enzyme involved in NAD synthesis (nicotinamide mononucleotide adenylyltransferase, Nmnat). Interestingly, Wld^S protects axons not only against lesion but also in various other paradigms such as Taxol- and Vincristin-induced neurotoxicity and in the inherited motor neuron disease *pmn*. Because of this broad therapeutic potential, Wld^S and its molecular effectors have become the focus of intense research efforts.

Another innovative approach consists in knocking down the expression of pathological SOD1 alleles by using the RNA interference (RNAi) technique. RNAi has been discovered as a natural gene-silencing mechanism in which double-stranded RNA triggers the degradation of homologous messenger RNAs with extraordinary sequence specificity. The mediators of mRNA degradation are small, 21- to 23-nucleotide long, RNAs (siRNAs). Since almost most ALS-linked SOD1 mutations represent single nucleotide substitutions, it is possible to design RNA sequences that specifically silence mutant SOD1 alleles without affecting expression of the wild-type allele. These RNAs are transfected as double-stranded oligonucleotides into the target cell or expressed as small hairpin RNAs from plasmid DNA vectors or lentiviral vectors. The appropriate target cell and the level of SOD1 gene silencing necessary to achieve therapeutic effects remain to be determined.

6
Conclusions and Perspectives

In the last decade, advances in genetics, molecular and cell biology, and gene therapy have boosted research on motor neuron diseases. Several gene mutations responsible for motor neuron diseases in humans (SOD1, Alsin) and mice (Dynein, Chaperones, etc.) have been identified. These discoveries led to new pathogenic hypotheses, implicating excessive glutamate levels, toxic protein aggregates, mitochondrial damage, or axonal dysfunction. It is becoming clear that multiple disease pathways – triggered in the motor neuron or in neighboring cells – can converge and cause degeneration. Experimental therapies targeting these pathways at various levels are investigated with increasing success in relevant cellular and animal models. Translating these approaches into clinically relevant therapies represents the major challenge of the future.

See also Motor Neuron Diseases: Molecular Mechanism, Pathophysiology, and Treatments; Neuron Chemistry.

Bibliography

Books and Reviews

Beckman, J.S., Estevez, A.G., Crow, J.P., Barbeito, L. (2001) Superoxide dismutases and the death of motoneurons in ALS, *Trends Neurosci.* **24**, S15–S20.

Bruijn, L.I., Miller, T.M., Cleveland, D.W. (2004) Unraveling the mechanisms involved in motor neuron degeneration in ALS, *Annu. Rev. Neurosci.* **27**, 723–749.

Coleman, M.P., Perry, V.H. (2002) Axon pathology in neurological disease: a neglected therapeutic target, *Trends Neurosci.* **25**, 532–537.

Cleveland, D.W., Rothstein, J.D. (2001) From Charcot to Lou Gehrig: deciphering selective motor neuron death in ALS, *Nat. Rev. Neurosci.* **2**, 806–819.

Hand, C.K., Rouleau, G.A. (2002) Familial amyotrophic lateral sclerosis, *Muscle Nerve* **25**, 135–159.

Henderson, C.E., et al. (1995) in *Nerve Cell Culture: A Practical Approach*, Cohen J., Wilkin G. (Eds.) Oxford University Press, London, UK, pp. 69–81.

Hirano, A. (1982) in: Rowland L.P. (Ed.) *Human Motor Neuron Diseases*, Raven Press, New York, pp. 75–87.

Julien, J.P. (2001) Amyotrophic lateral sclerosis. unfolding the toxicity of the misfolded, *Cell* **104**, 581–591.

Nicole, S., Cifuentes-Diaz, C., Frugier, T., Melki, J. (2002) Spinal muscular atrophy: recent advances and future prospects, *Muscle Nerve* **26**, 4–13.

Orrell, R.W. (2000) Amyotrophic lateral sclerosis: copper/zinc superoxide dismutase (SOD1) gene mutations, *Neuromuscul. Disord.* **10**, 63–68.

Valentine, J.S., Hart, P.J. (2003) Misfolded CuZnSOD and amyotrophic lateral sclerosis, *Proc. Natl. Acad. Sci. U. S. A.* **100**, 3617–3622.

Primary Literature

Allen, S., et al. (2003) Analysis of the cytosolic proteome in a cell culture model of familial amyotrophic lateral sclerosis reveals alterations to the proteasome, antioxidant defenses, and nitric oxide synthetic pathways, *J. Biol. Chem.* **278**(8), 6371–6383.

Arce, V., et al. (1999) Cardiotrophin-1 requires LIFRbeta to promote survival of mouse motoneurons purified by a novel technique, *J. Neurosci. Res.* **55**(1), 119–126.

Bar, P.R. (2000) Motor neuron disease in vitro: the use of cultured motor neurons to study amyotrophic lateral sclerosis, *Eur. J. Pharmacol.* **405**(1–3), 285–295.

Beckman, J.S., et al. (2001) Superoxide dismutases and the death of motoneurons in ALS, *Trends Neurosci.* **24**(11 (Suppl.)), S15–S20.

Bommel, H., et al. (2002) Missense mutation in the tubulin-specific chaperone E (Tbce) gene in the mouse mutant progressive motor neuronopathy, a model of human motoneuron disease, *J. Cell. Biol.* **159**(4), 563–569.

Bordet, T., et al. (1999) Adenoviral cardiotrophin-1 gene transfer protects pmn mice from progressive motor neuronopathy, *J. Clin. Invest.* **104**(8), 1077–1085.

Bordet, T., et al. (2001) Protective effects of cardiotrophin-1 adenoviral gene transfer on neuromuscular degeneration in transgenic ALS mice, *Hum. Mol. Genet.* **10**(18), 1925–1933.

Bruijn, L.I., et al. (1997) ALS-linked SOD1 mutant G85R mediates damage to astrocytes and promotes rapidly progressive disease with SOD1-containing inclusions, *Neuron* **18**(2), 327–338.

Bruijn, L.I., et al. (1998) Aggregation and motor neuron toxicity of an ALS-linked SOD1 mutant independent from wild-type SOD1, *Science* **281**(5384), 1851–1854.

Carlin, K.P., et al. (2000) Dendritic L-type calcium currents in mouse spinal motoneurons: implications for bistability, *Eur. J. Neurosci.* **12**(5), 1635–1646.

Casanovas, A., et al. (1996) Prevention by lamotrigine, MK-801 and N omega-nitro-L-arginine methyl ester of motoneuron cell death after neonatal axotomy, *Neuroscience* **71**(2), 313–325.

Cashman, N.R., et al. (1992) Neuroblastoma x spinal cord (NSC) hybrid cell lines resemble developing motor neurons, *Dev. Dyn.* **194**(3), 209–221.

Chan, Y.M., et al. (2003) Inhibition of caspases promotes long-term survival and reinnervation by axotomized spinal motoneurons of denervated muscle in newborn rats, *Exp. Neurol.* **181**(2), 190–203.

Charcot, J.-M., Joffroy, A. (1869) Deux cas d'atrophie musculaire progressive avec lésion de la substance grise et des fisceaux de la moelle épinière, *Arch. Physiol.* **2**, 354, 629, 744.

Chiu, A.Y., et al. (1995) Age-dependent penetrance of disease in a transgenic mouse model of familial amyotrophic lateral sclerosis, *Mol. Cell. Neurosci.* **6**(4), 349–362.

Chou, S.M. (1992) in: Smith R.A. (Ed.) *Handbook of Amyotrophic Lateral Sclerosis*, Marcel Dekker, New York, pp. 133–181.

Cifuentes-Diaz, C., et al. (2002) Neurofilament accumulation at the motor endplate and lack of axonal sprouting in a spinal muscular atrophy mouse model, *Hum. Mol. Genet.* **11**(12), 1439–1447.

Clement, A.M., et al. (2003) Wild-type nonneuronal cells extend survival of SOD1 mutant motor neurons in ALS mice, *Science* **302**(5642), 113–117.

Coleman, M.P., et al. (1998) An 85-kb tandem triplication in the slow Wallerian degeneration (Wlds) mouse, *Proc. Natl. Acad. Sci. U. S. A.* **95**(17), 9985–9990.

Conforti, L., et al. (2000) A Ufd2/D4Cole1e chimeric protein and overexpression of Rbp7 in the slow Wallerian degeneration (WldS) mouse, *Proc. Natl. Acad. Sci. U. S. A.* **97**(21), 11377–11382.

Connelly, C.A., et al. (2000) Metabolic activity of cultured rat brainstem, hippocampal and spinal cord slices, *J. Neurosci. Methods* **99**(1–2), 1–7.

Corse, A.M., et al. (1999) Preclinical testing of neuroprotective neurotrophic factors in a model of chronic motor neuron degeneration, *Neurobiol. Dis.* **6**(5), 335–346.

Dalcanto, M.C., Gurney, M.E. (1995) Neuropathological changes in two lines of mice carrying a transgene for mutant human Cu,Zn SOD, and in mice overexpressing wild type human SOD: A model of familial amyotrophic lateral sclerosis (FALS), *Brain Res.* **676**(1), 25–40.

de Bilbao, F., Dubois-Dauphin, M. (1996) Acute application of an interleukin-1 beta-converting enzyme-specific inhibitor delays axotomy-induced motoneurone death, *Neuroreport* **7**(18), 3051–3054.

de Bilbao, F., Dubois-Dauphin, M. (1996) Time course of axotomy-induced apoptotic cell death in facial motoneurons of neonatal wild type and bcl-2 transgenic mice, *Neuroscience* **71**(4), 1111–1119.

Deckwerth, T.L., et al. (1996) BAX is required for neuronal death after trophic factor deprivation and during development, *Neuron* **17**(3), 401–411.

Drachman, D.B., Rothstein, J.D. (2000) Inhibition of cyclooxygenase-2 protects motor neurons in an organotypic model of amyotrophic lateral sclerosis, *Ann. Neurol.* **48**(5), 792–795.

Dubois-Dauphin, M., et al. (1994) Neonatal motoneurons overexpressing the bcl-2 protooncogene in transgenic mice are protected from axotomy-induced cell death, *Proc. Natl. Acad. Sci. U. S. A.* **91**(8), 3309–3313.

Durham, H.D., et al. (1993) Evaluation of the spinal cord neuron X neuroblastoma hybrid cell line NSC-34 as a model for neurotoxicity testing, *Neurotoxicology* **14**(4), 387–395.

Durham, H.D., et al. (1997) Aggregation of mutant Cu/Zn superoxide dismutase proteins in a culture model of ALS, *J. Neuropathol. Exp. Neurol.* **56**(5), 523–530.

Ferri, A., et al. (2003) Inhibiting axon degeneration and synapse loss attenuates apoptosis and disease progression in a mouse model of motoneuron disease, *Curr. Biol.* **13**(8), 669–673.

Frey, D., et al. (2000) Early and selective loss of neuromuscular synapse subtypes with low sprouting competence in motoneuron diseases, *J. Neurosci.* **20**(7), 2534–2542.

Garcia, M.L., Cleveland, D.W. (2001) Going new places using an old MAP: tau, microtubules and human neurodegenerative disease, *Curr. Opin. Cell Biol.* **13**(1), 41–48.

Gong, Y.H., et al. (2000) Restricted expression of G86R Cu/Zn superoxide dismutase in astrocytes results in astrocytosis but does not cause motoneuron degeneration, *J. Neurosci.* **20**(2), 660–665.

Greensmith, L., Vrbova, G. (1996) Motoneurone survival: A functional approach, *Trends Neurosci.* **19**(11), 450–455.

Groeneveld, G.J., et al. (2003) A randomized sequential trial of creatine in amyotrophic lateral sclerosis, *Ann. Neurol.* **53**(4), 437–445.

Guegan, C., et al. (2001) Recruitment of the mitochondrial-dependent apoptotic pathway in amyotrophic lateral sclerosis, *J. Neurosci.* **21**(17), 6569–6576.

Gurney, M.E., et al. (1994) Motor neuron degeneration in mice that express a human Cu,Zn superoxide dismutase mutation, *Science* **264**(5166), 1772–1775.

Gurney, M.E., et al. (1996) Benefit of vitamin E, riluzole, and gabapentin in a transgenic model of familiar amyotrophic lateral sclerosis, *Ann. Neurol.* **39**(2), 147–157.

Haase, G., et al. (1997) Gene therapy of murine motor neuron disease using adenoviral vectors for neurotrophic factors, *Nat. Med.* **3**(4), 429–436.

Haase, G., et al. (1999) Therapeutic benefit of CNTF in progressive motor neuronopathy depends on the route of delivery, *Ann. Neurol.* **45**(3), 296–304.

Hadano, S., et al. (2001) A gene encoding a putative GTPase regulator is mutated in familial amyotrophic lateral sclerosis 2, *Nat. Genet.* **29**(2), 166–173.

Haenggeli, C., Kato, A.C. (2002) Differential vulnerability of cranial motoneurons in mouse models with motor neuron degeneration, *Neurosci. Lett.* **335**(1), 39–43.

Hannon, G.J. (2002) RNA interference, *Nature* **418**(6894), 244–251.

Henderson, C.E., et al. (1994) GDNF: A potent survival factor for motoneurons present in peripheral nerve and muscle, *Science* **266**(5187), 1062–1064.

Hentati, A., et al. (1994) Linkage of recessive familial amyotrophic lateral sclerosis to chromosome 2q33-q35, *Nat. Genet.* **7**(3), 425–428.

Ho, T.W., et al. (2000) TGFbeta trophic factors differentially modulate motor axon outgrowth and protection from excitotoxicity, *Exp. Neurol.* **161**(2), 664–675.

Hori, N., et al. (2001) Intracellular activity of rat spinal cord motoneurons in slices, *J. Neurosci. Methods* **112**(2), 185–191.

Howland, D.S., et al. (2002) Focal loss of the glutamate transporter EAAT2 in a transgenic rat model of SOD1 mutant-mediated amyotrophic lateral sclerosis (ALS), *Proc. Natl. Acad. Sci. U. S. A.* **99**(3), 1604–1609.

Ince, P.G., et al. (1998) Amyotrophic lateral sclerosis: current issues in classification, pathogenesis and molecular pathology, *Neuropathol. Appl. Neurobiol.* **24**(2), 104–117.

Iwasaki, Y., Ikeda, K. (1999) Prevention by insulin-like growth factor-I and riluzole in motor neuron death after neonatal axotomy, *J. Neurol. Sci.* **169**(1–2), 148–155.

Iwasaki, Y., et al. (1996) Deprenyl and pergolide rescue spinal motor neurons from axotomy-induced neuronal death in the neonatal rat, *Neurol. Res.* **18**(2), 168–170.

Kaal, E.C., et al. (2000) Chronic mitochondrial inhibition induces selective motoneuron death in vitro: a new model for amyotrophic lateral sclerosis, *J. Neurochem.* **74**(3), 1158–1165.

Kashihara, Y., et al. (1987) Cell death of axotomized motoneurones in neonatal rats, and its prevention by peripheral reinnervation, *J. Physiol.* **386**, 135–148.

Kaspar, B.K., et al. (2003) Retrograde viral delivery of IGF-1 prolongs survival in a mouse ALS model, *Science* **301**(5634), 839–842.

Kennedy, W.R., et al. (1968) Progressive proximal spinal and bulbar muscular atrophy of late onset. A sex-linked recessive trait, *Neurology* **18**(7), 671–680.

Kennel, P., et al. (2000) Riluzole prolongs survival and delays muscle strength deterioration in mice with progressive motor neuronopathy (pmn), *J. Neurol. Sci.* **180**(1–2), 55–61.

Kennel, P.F., et al. (1996) Neuromuscular function impairment is not caused by motor neurone loss in FALS mice: an electromyographic study, *Neuroreport* **7**(8), 1427–1431.

Kennel, P.F., et al. (1996) Electromyographical and motor performance studies in the pmn mouse model of neurodegenerative disease, *Neurobiol. Dis.* **3**(2), 137–147.

Kirby, J., et al. (2002) Differential gene expression in a cell culture model of SOD1-related familial motor neurone disease, *Hum. Mol. Genet.* **11**(17), 2061–2075.

Klivenyi, P., et al. (1999) Neuroprotective effects of creatine in a transgenic animal model of amyotrophic lateral sclerosis, *Nat. Med.* **5**(3), 347–350.

Kong, J.M., Xu, Z.S. (1998) Massive mitochondrial degeneration in motor neurons triggers the onset of amyotrophic lateral sclerosis in mice expressing a mutant SOD1, *J. Neurosci.* **18**(9), 3241–3250.

Korinthenberg, R., et al. (1997) Congenital axonal neuropathy caused by deletions in the spinal muscular atrophy region, *Ann. Neurol.* **42**(3), 364–368.

Krassioukov, A.V., et al. (2002) An in vitro model of neurotrauma in organotypic spinal cord cultures from adult mice, *Brain Res. Brain Res. Protoc.* **10**(2), 60–68.

Kriz, J., et al. (2002) Minocycline slows disease progression in a mouse model of amyotrophic lateral sclerosis, *Neurobiol. Dis.* **10**(3), 268–278.

Kriz, J., et al. (2003) Efficient three-drug cocktail for disease induced by mutant superoxide dismutase, *Ann. Neurol.* **53**(4), 429–436.

Kunst, C.B., et al. (2000) Genetic mapping of a mouse modifier gene that can prevent ALS onset, *Genomics* **70**(2), 181–189.

La Spada, A.R., et al. (1991) Androgen receptor gene mutations in X-linked spinal and bulbar muscular atrophy, *Nature* **352**(6330), 77–79.

Lambrechts, D., et al. (2003) VEGF is a modifier of amyotrophic lateral sclerosis in mice and humans and protects motoneurons against ischemic death, *Nat. Genet.* **34**(4), 383–394.

Lefebvre, S., et al. (1995) Identification and characterization of a spinal muscular atrophy-determining gene [see comments], *Cell* **80**(1), 155–165.

Leigh, P.N., et al. (1991) Ubiquitin-immunoreactive intraneuronal inclusions in amyotrophic lateral sclerosis. Morphology, distribution, and specificity, *Brain* **114**(Pt 2), 775–788.

Lesbordes, J.C., et al. (2002) In vivo electrotransfer of the cardiotrophin-1 gene into skeletal muscle slows down progression of motor neuron degeneration in pmn mice, *Hum. Mol. Genet.* **11**(14), 1615–1625.

Li, L., et al. (1998) Characterization of spinal motoneuron degeneration following different types of peripheral nerve injury in neonatal and adult mice, *J. Comp. Neurol.* **396**(2), 158–168.

Li, M., et al. (2000) Functional role of caspase-1 and caspase-3 in an ALS transgenic mouse model, *Science* **288**(5464), 335–339.

Li, M., et al. (2000) Functional role of caspase-1 and caspase-3 in an ALS transgenic mouse model, *Science* **288**(5464), 335–339.

Lino, M.M., et al. (2002) Accumulation of SOD1 mutants in postnatal motoneurons does not cause motoneuron pathology or motoneuron disease, *J. Neurosci.* **22**(12), 4825–4832.

Liu, R., et al. (2002) Increased mitochondrial antioxidative activity or decreased oxygen free radical propagation prevent mutant SOD1-mediated motor neuron cell death and increase amyotrophic lateral sclerosis-like transgenic mouse survival, *J. Neurochem.* **80**(3), 488–500.

Lowrie, M.B., et al. (1982) Recovery of slow and fast muscles following nerve injury during early postnatal development in the rat, *J. Physiol.* **331**, 51–66.

Lyon, M.F., et al. (1993) A gene affecting Wallerian nerve degeneration maps distally on mouse chromosome 4, *Proc. Natl. Acad. Sci. U. S. A.* **90**(20), 9717–9720.

Mack, T.G., et al. (2001) Wallerian degeneration of injured axons and synapses is delayed by a Ube4b/Nmnat chimeric gene, *Nat. Neurosci.* **4**(12), 1199–1206.

Maragakis, N.J., et al. (2003) Topiramate protects against motor neuron degeneration in organotypic spinal cord cultures but not in G93A SOD1 transgenic mice, *Neurosci. Lett.* **338**(2), 107–110.

Mariotti, R., et al. (1997) Age-dependent induction of nitric oxide synthase activity in facial motoneurons after axotomy, *Exp. Neurol.* **145**(2 Pt 1), 361–370.

Martin, N., et al. (2002) A missense mutation in Tbce causes progressive motor neuronopathy in mice, *Nat. Genet.* **32**(3), 443–447.

Menzies, F.M., et al. (2002) Mitochondrial dysfunction in a cell culture model of familial amyotrophic lateral sclerosis, *Brain* **125**(Pt 7), 1522–1533.

Menzies, F.M., et al. (2002) Selective loss of neurofilament expression in Cu/Zn superoxide dismutase (SOD1) linked amyotrophic lateral sclerosis, *J. Neurochem.* **82**(5), 1118–1128.

Mourelatos, Z., et al. (1996) The Golgi apparatus of spinal cord motor neurons in transgenic mice expressing mutant Cu,Zn superoxide dismutase becomes fragmented in early, preclinical stages of the disease, *Proc. Natl. Acad. Sci. U. S. A.* **93**(11), 5472–5477.

Nagai, M., et al. (2001) Rats expressing human cytosolic copper-zinc superoxide dismutase transgenes with amyotrophic lateral sclerosis: associated mutations develop motor neuron disease, *J. Neurosci.* **21**(23), 9246–9254.

Nguyen, M.D., et al. (2001) Deregulation of Cdk5 in a mouse model of ALS: toxicity alleviated by perikaryal neurofilament inclusions, *Neuron* **30**(1), 135–147.

Oosthuyse, B., et al. (2001) Deletion of the hypoxia-response element in the vascular endothelial growth factor promoter causes motor neuron degeneration, *Nat. Genet.* **28**(2), 131–138.

Otomo, A., et al. (2003) ALS2, a novel guanine nucleotide exchange factor for the small GTPase Rab5, is implicated in endosomal dynamics, *Hum. Mol. Genet.* **12**(14), 1671–1687.

Parvari, R., et al. (2002) Mutation of TBCE causes hypoparathyroidism-retardation-dysmorphism and autosomal recessive Kenny-Caffey syndrome, *Nat. Genet.* **32**(3), 448–452.

Pasinelli, P., et al. (2000) Caspase-1 and -3 are sequentially activated in motor neuron death in Cu,Zn superoxide dismutase-mediated familial amyotrophic lateral sclerosis, *Proc. Natl. Acad. Sci. U. S. A.* **97**(25), 13901–13906.

Pennica, D., et al. (1996) Cardiotrophin-1, a cytokine present in embryonic muscle, supports long-term survival of spinal motoneurons, *Neuron* **17**(1), 63–74.

Perrelet, D., et al. (2000) IAP family proteins delay motoneuron cell death in vivo, *Eur. J. Neurosci.* **12**(6), 2059–2067.

Pramatarova, A., et al. (2001) Neuron-specific expression of mutant superoxide dismutase 1 in transgenic mice does not lead to motor impairment, *J. Neurosci.* **21**(10), 3369–3374.

Puls, I., et al. (2003) Mutant dynactin in motor neuron disease, *Nat. Genet.* **33**(4), 455–456.

Raivich, G., et al. (2002) Cytotoxic potential of proinflammatory cytokines: combined deletion of TNF receptors TNFR1 and TNFR2 prevents motoneuron cell death after facial axotomy in adult mouse, *Exp. Neurol.* **178**(2), 186–193.

Rakowicz, W.P., et al. (2002) Glial cell line-derived neurotrophic factor promotes the survival of early postnatal spinal motor neurons in the lateral and medial motor columns in slice culture, *J. Neurosci.* **22**(10), 3953–3962.

Raoul, C., et al. (1999) Programmed cell death of embryonic motoneurons triggered through the fas death receptor, *J. Cell. Biol.* **147**(5), 1049–1062.

Raoul, C., et al. (2002) Motoneuron death triggered by a specific pathway downstream of Fas. potentiation by ALS-linked SOD1 mutations, *Neuron* **35**(6), 1067–1083.

Reaume, A.G., et al. (1996) Motor neurons in Cu/Zn superoxide dismutase-deficient mice develop normally but exhibit enhanced cell death after axonal injury, *Nat. Genet.* **13**(1), 43–47.

Ripps, M.E., et al. (1995) Transgenic mice expressing an altered murine superoxide dismutase gene provide an animal model of amyotrophic lateral sclerosis, *Proc. Natl. Acad. Sci. U. S. A.* **92**(3), 689–693.

Rossiter, J.P., et al. (1996) Axotomy-induced apoptotic cell death of neonatal rat facial motoneurons: time course analysis and relation to NADPH-diaphorase activity, *Exp. Neurol.* **138**(1), 33–44.

Rothstein, J.D., et al. (1993) Chronic inhibition of glutamate uptake produces a model of slow neurotoxicity, *Proc. Natl. Acad. Sci. U. S. A.* **90**(14), 6591–6595.

Rothstein, J.D., Kuncl, R.W. (1995) Neuroprotective strategies in a model of chronic glutamate-mediated motor neuron toxicity, *J. Neurochem.* **65**(2), 643–651.

Sagot, Y., et al. (1995) Polymer encapsulated cell lines genetically engineered to release ciliary neurotrophic factor can slow down progressive motor neuronopathy in the mouse, *Eur. J. Neurosci.* **7**(6), 1313–1322.

Schmalbruch, H. (1984) Motoneuron death after sciatic nerve section in newborn rats, *J. Comp. Neurol.* **224**, 252–258.

Schmalbruch, H., et al. (1991) A new mouse mutant with progressive motor neuronopathy, *J. Neuropathol. Exp. Neurol.* **50**, 192–204.

Schmalbruch, H., Rosenthal, A. (1995) Neurotrophin-4/5 postpones the death of injured spinal motoneurons in newborn rats, *Brain Res.* **700**(1–2), 254–260.

Schnaar, R.I., Schaffner, A.E. (1981) Separation of cell types from embryonic chicken and rat spinal cord: characterization of motoneuron-enriched fractions, *J. Neurosci.* **1**(2), 204–217.

Sendtner, M., et al. (1990) Ciliary neurotrophic factor prevents the degeneration of motor neurons after axotomy, *Nature* **345**(6274), 440–441.

Sendtner, M., et al. (1992) Brain-derived neurotrophic factor prevents the death of motoneurons in newborn rats after nerve section, *Nature* **360**, 757–759.

Sendtner, M., et al. (1992) Ciliary neurotrophic factor prevents degeneration of motor neurons in mouse mutant progressive motor neuronopathy, *Nature* **358**(6386), 502–504.

Subramaniam, J.R., et al. (2002) Mutant SOD1 causes motor neuron disease independent of copper chaperone-mediated copper loading, *Nat. Neurosci.* **5**(4), 301–307.

Tandan, R., Bradley, W.G. (1985) Amyotrophic lateral sclerosis: Part 1. Clinical features, pathology, and ethical issues in management, *Ann. Neurol.* **18**(3), 271–280.

Terrado, J., et al. (2000) Soluble TNF receptors partially protect injured motoneurons in the postnatal CNS, *Eur. J. Neurosci.* **12**(9), 3443–3447.

Tian, G., et al. (1996) Pathway leading to correctly folded beta-tubulin, *Cell* **86**(2), 287–296.

Tu, P.H., et al. (1996) Transgenic mice carrying a human mutant superoxide dismutase transgene develop neuronal cytoskeletal pathology resembling human amyotrophic lateral sclerosis lesions, *Proc. Natl. Acad. Sci. U. S. A.* **93**(7), 3155–3160.

Ugolini, G., et al. (2003) Fas/tumor necrosis factor receptor death signaling is required for axotomy-induced death of motoneurons in vivo, *J. Neurosci.* **23**(24), 8526–8531.

Van Westerlaak, M.G., et al. (2001) Chronic mitochondrial inhibition induces glutamate-mediated corticomotoneuron death in an organotypic culture model, *Exp. Neurol.* **167**(2), 393–400.

Vanderluit, J.L., et al. (2003) In vivo application of mitochondrial pore inhibitors blocks the induction of apoptosis in axotomized neonatal facial motoneurons, *Cell. Death Differ.* **10**(9), 969–976.

Vukosavic, S., et al. (1999) Bax and Bcl-2 interaction in a transgenic mouse model of familial amyotrophic lateral sclerosis, *J. Neurochem.* **73**(6), 2460–2468.

Waldmeier, P.C., et al. (2000) Neurorescuing effects of the GAPDH ligand CGP 3466B, *J. Neural. Transm. Suppl.* **60**, 197–214.

Wang, L.J., et al. (2002) Neuroprotective effects of glial cell line-derived neurotrophic factor mediated by an adeno-associated virus vector in a transgenic animal model of amyotrophic lateral sclerosis, *J. Neurosci.* **22**(16), 6920–6928.

Wang, M.S., et al. (2001) The WldS protein protects against axonal degeneration: a model of gene therapy for peripheral neuropathy, *Ann. Neurol.* **50**(6), 773–779.

Wang, M.S., et al. (2002) WldS mice are resistant to paclitaxel (taxol) neuropathy, *Ann. Neurol.* **52**(4), 442–447.

Wong, P.C., et al. (1995) An adverse property of a familial ALS-linked SOD1 mutation causes motor neuron disease characterized by vacuolar degeneration of mitochondria, *Neuron* **14**(6), 1105–1116.

Xia, X.G., et al. (2003) An enhanced U6 promoter for synthesis of short hairpin RNA, *Nucleic Acids Res.* **31**(17), e100.

Yan, Q., et al. (1992) Brain-derived neurotrophic factor rescues spinal motor neurons from axotomy-induced cell death, *Nature* **360**(6406), 753–755.

Yang, Y., et al. (2001) The gene encoding alsin, a protein with three guanine-nucleotide exchange factor domains, is mutated in a form of recessive amyotrophic lateral sclerosis, *Nat. Genet.* **29**(2), 160–165.

Zhu, S., et al. (2002) Minocycline inhibits cytochrome c release and delays progression of amyotrophic lateral sclerosis in mice, *Nature* **417**(6884), 74–78.

Motor Neuron Diseases: Molecular Mechanism, Pathophysiology, and Treatments

Michael Sendtner
Institute of Clinical Neurobiology, Wuerzburg, Germany

Encyclopedia of Molecular Cell Biology and Molecular Medicine, 2nd Edition. Volume 8
Edited by Robert A. Meyers.
Copyright © 2005 Wiley-VCH Verlag GmbH & Co. KGaA, Weinheim
ISBN: 3-527-30550-5

Keywords

Amyotrophic Lateral Sclerosis (ALS)
A clinical syndrome defined by degenerative changes of lower and upper motor neurons.

Motor Neuron Disease (MND)
A variety of sporadic and inherited diseases characterized by muscle weakness due to primary degeneration of motor neurons of the spinal cord, brain stem motor nuclei, and populations of central motor neurons.

Neurotrophic Factors
Polypeptide molecules that support the survival of specific populations of neuronal cells, such as spinal motor neurons in culture and *in vivo*, via specific signal-transducing receptors.

Spinal Muscular Atrophy (SMA)
A group of disorders that are characterized by degeneration of lower (spinal and bulbar) motor neurons.

■ The term "motor neuron disease" refers to a variety of devastating neurological disorders characterized by a generalized degeneration of motoneurons. The common clinical manifestations of such disorders are progressive muscle weakness and often death by respiratory failure. Inherited and sporadic forms occur. Further clinical distinctions are made according to the onset of the disorder and whether the disease involves only spinal and bulbar motor neurons (spinal muscular atrophy), or both the lower and upper motor neurons. The most common forms of motor neuron disease are spinal muscular atrophy with an incidence of 1 : 6000 to 10 000 newborns, and sporadic amyotrophic lateral sclerosis with a prevalence of 1 : 100 000, which corresponds to an incidence of 1 : 10 000. About 90% of all cases of amyotrophic lateral sclerosis are sporadic. This disorder affects mostly adults, has a typical clinical appearance, is noninherited, still incurable, and invariably fatal. Recent years have seen the identification of a variety of disease mechanisms, which all lead to the degeneration of motor neurons. Thus, pharmacological approaches that can interfere with degeneration and cell death can now be developed but are complicated by the fact that such therapeutic strategies might work only in subpopulations and not in all patients with this disease.

1
Neuronal Cell Death: Implications for Motoneuron Disease

Motor neurons were among the first populations of neurons that were seen to undergo cell death to a significant extent during development. Such observations, made in the first part of the twentieth century, led to the concept of programmed cell death in the nervous system and to the proposal that target-derived neurotrophic factors play a key role in the maintenance of motor neuron survival and function during the development and in the adult. In the meantime, a variety of such neurotrophic factors have been characterized on the molecular and functional levels, and genetic analysis of patients with motor neuron disease has revealed that only very few patients show mutations or deletions in the genes for these factors. In such cases, it has been found that mutations in the genes for CNTF (ciliary neurotrophic factor) and LIF (Leukaemia inhibitory factor) might act as modifiers, but not as the cause for the disease. Both in patients and in a mouse model of familial ALS, homozygous inactivation of the CNTF gene leads to earlier disease onset. However, the cause of the disease might lie in other functional defects that are not related to the expression of neurotrophic factors or their effects on motor neurons. Studies with mouse models have shown that overexpression of mutant forms of Superoxide dismutase-I (SOD-I) leads to motor neuron disease and that mutations in genes that are relevant for axonal transport and maintenance of axonal processes seem to be more relevant than mechanisms that regulate survival and/or cell death of the motor neuron cell bodies. This indicates that the cell death of motor neuron cell bodies in motor neuron disease might be a secondary phenomenon that occurs as a consequence of axonal defects. On the other hand, a mouse model in which the gene for bcl-2, a key regulator of cellular survival, has been inactivated shows postnatal degeneration of motor neurons. Moreover, cell culture models have revealed that motor neurons derived from SOD-I transgenic mice are much more sensitive to pro-apoptotic stimuli such as Fas ligand and others. In addition, the induction of cell death after binding of excess glutamate to AMPA receptors is also discussed as a cause of motor neuron cell death in motor neuron disease (MND).

Movement of higher organisms requires nerve cells for coordination. The anatomy and physiology of the motor neurons innervating the skeletal musculature is highly conserved from lower to higher vertebrate species – much more than in other parts of the nervous system that are responsible for higher brain functions which are, for example, defective in Alzheimer's disease and other form of dementia. Thus, motor neurons appear as an ideal subject to study neurodegenerative mechanisms in cell culture models, animal models, and to compare these findings with clinical data.

2
Motor Neurons: Anatomical Aspects

The skeletal musculature is innervated by motor neurons located either in the ventral part (the ventrolateral horn) of the spinal cord or in the brain stem motor nuclei. The cell bodies of these motor neurons reside within the central nervous system, and the axons project into the nerves of the peripheral nervous system to the muscle, which they innervate. Many of these axons reach considerable length. For example, the cell bodies of motor neurons innervating the foot muscles are located in the lumbar spinal cord; thus, the total length of these cells can exceed 1 m in humans. This specific property identifies motor neurons as being among the most highly organized cell types of higher organisms.

The transmitter synthesized by motor neurons and released at the motor end plate, the specific synapses in skeletal muscle, is acetylcholine. Cholinergic neurotransmission to muscle cells is highly conserved from lower organisms to man.

Spinal motor neurons are in contact with many other cell types, including other neurons, in particular, interneurons of the spinal cord, astrocytes, oligodendrocytes in the central nervous system, and Schwann and muscle cells in the periphery. They receive synaptic inputs both from proprioceptive sensory neurons and from spinal interneurons. These interneurons receive input from upper motor neurons. The upper motor neurons can also project directly to spinal motor neurons. Interneurons in the spinal cord are responsible for spinal reflexes, and also coordinate patterns for coordinated motor neuron innervation, which are responsible for movement. Neurons of the cortical spinal tract (upper motor neurons), projecting to spinal and brain stem motor neurons either directly or via interneurons, are important for voluntary movement. The cell bodies of these upper motor neurons are located in the brain in specific regions of the precentral cortex. These cells are also affected by degenerative changes in various forms of human MND, in particular, in amyotrophic lateral sclerosis (ALS).

3
Clinical Appearance and Pathology

3.1
Inherited forms of MND: the Identification of Gene Defects, Animal Models, and Disease Mechanisms

In the case of inherited MND, more than 50 different forms have been distinguished on the basis of clinical parameters. The underlying gene defects for the major forms of MND in childhood, classical SMA, and spinal muscular atrophy with respiratory distress (SMARD) have been identified, and the gene defects for three out of five forms of familial amyotrophic lateral sclerosis are known. Furthermore, an increased size of a polymorphic triplet (CRG repeat) in the coding region of the androgen receptor gene on the X chromosome has been identified as the cause of X-linked spinal muscular atrophy. This defect leads to progressive muscle weakness and atrophy, as well as to gynecomastia and reduced fertility in affected males. Although only about 10% of adult onset ALS is familial, this group appears to be heterogeneous. The first identified gene defect that accounts for about 10 to 20% of familial ALS are point mutations in the gene for Cu^{2+}/Cn^{2+}-dependent superoxide dismutase (SOD-I). More than 50 different mutations in this gene have been identified most of them leading to autosomal dominant ALS, but there are also mutations in this gene that cause an autosomal recessive form of the disease. Clinically, there seems to be no clear correlation between disease onset or severity with specific mutations in the *SOD-I* gene, suggesting that modifiers exist.

However, some types of mutations in the *SOD-I* gene are prone to cause severe and rapid course of the disease, for example, the A4V mutation, whereas others, that is, E21G, G37R, D90A, G93C, G93V, I104F, L144S, and I151T are generally associated with survival times of more than 10 to 15 years. The D90A mutation is generally inherited as a recessive trait, and this mutation exists also as a polymorphism in the Scandinavian population. There are studies suggesting that about 100 000 heterozygous carriers for the D90A polymorphism exist in Finland alone.

The gene for another autosomal recessive form of ALS, which is coupled to mutations in the Chr2q33 region, has been identified as a mutation in *Alsin2*. This gene encodes for a GTP exchange factor, which resembles other RanGEF or RhoGEF molecules. This finding appears interesting, as such factors appear as the major signal mediators for extracellular regulators of axon growth and possibly also for axon maintenance. The gene for another rare autosomal dominant form of juvenile ALS (ALS 4) has been recently identified on chromosome 9q34 as missense mutation (T3I, L389S, and R2136H) in the senataxin gene. This gene encodes for a protein of more than 300 kD. The function of this protein is still unknown, but it contains a DNA/RNA helicase domain with homology to human IGHMBP2 that has been shown to be mutant in a specific form of spinal muscular atrophy with respiratory distress (SMARD 1). Interestingly, mutations in senataxin lead to an early form of ALS and patients normally fall ill at an age >25 years. This suggests that some forms of SMA and juvenile ALS are pathophysiologically related. Rapid progress is currently made in the identification of other forms of familial ALS.

A major breakthrough in understanding the cause of motor neuron disease in children was the identification of the underlying gene defect in classical SMA. Familial forms of lower motor neuron degeneration are among the most common genetic causes of death in children, leading to muscle weakness and respiratory failure within a few years after birth in the most severe forms. According to their severity, the autosomal inherited disorders are classified as spinal muscular atrophy type I (most severe form, also named Werdnig–Hoffmann disease), type

II and type III (Kugelberg–Welander disease). All three forms map to chromosome 5 (5q13). The responsible gene defect for all three forms of SMA type is located in the survival motoneuron (*SMN*1) gene, which is positioned within a duplicated region on human chromosome 5q13. Homozygous mutation or lack of the telomeric *SMN* gene is associated with SMA, whereas mutations within the centromeric copy are not responsible for SMA. The number of centromeric copies of the *SMN* gene varies, and various lines of evidence suggest that the number of centromeric *SMN* copies (*SMN2*) and the levels of *SMN2* gene expression correlate with the severity of the disease. In general, patients with less severe forms (SMA type III, Kugelberg–Welander) have more copies of the centromeric *SMN2* gene than patients with type I SMA (Werdnig–Hoffmann's disease).

The *SMN* gene is highly conserved among higher organisms. Orthologous genes have been identified in mouse, *drosophila*, *Caenorhabditis elegans*, and the fission yeast *Schizosaccharomyces pombe*. The *SMN* gene duplication is an event that took place just after the divergence of rodents and primates. In humans, both *SMN* copies differ in that full-length protein is predominantly produced from the telomeric *SMN* gene copy. The expression of the truncated SMN2 protein is caused by a C > T exchange at position +6 in exon 7, which leads to skipping of exon 7 encoded regions in about 70% of the protein product derived from the *SMN2* gene. This specific gene constellation has not been observed in primates, indicating that differential splicing for exon 7 with consequent development of SMA is restricted to humans. Point mutations in the telomeric *SMN* gene are only observed in 5% of

SMA patients. Hot spots of mutational events are found in exon 6 and the 5′ part of exon 7. These regions code for the *SMN* homodimerization domain.

In the past two years, many efforts have been made both in the identification of SMN-associated proteins and the analysis of cellular functions of SMN and SMN-associated proteins. The *SMN* gene is ubiquitously expressed and encodes for a protein of 294 amino acids, which is found both in the nucleus within specific structures called *gemini of coiled bodies* (gems) and in the cytoplasm, both in the cell body and in axons. A group of interacting proteins, which are colocalized in the nucleus includes gemin-2, gemin-3/gp103, a putative DEAD box RNA helicase, gemin-4, which interacts with gemin-3, but not directly with SMN, profilin, fibrillarin, and spliceosomal U snRNPs of the Sm class. The so-called SMN complex also includes other proteins called gemin-5 – 7. The SMN complex is involved in the biogenesis of spliceosomal U snRNPs. U snRNPs are involved in pre-mRNA splicing in the nucleus. Antibodies raised against gemin-2 and SMN interfere with binding of SMN to U snRNAs. This indicates that the SMN complex is essential for the assembly of U snRNPs. Indeed, gene knockout of *Smn* and *gemin-2* in mice is embryonic lethal at early stages of development before blastocysts implant in the uterus. In contrast to humans, the *SMN* gene is not duplicated in mice.

U snRNP assembly is essential for all types of cells and not only for motor neurons. Therefore, the specificity of a disease process for motor neurons remains obscure. The question of why reduced amounts of ubiquitously expressed SMN protein specifically cause degeneration of motor neurons without affecting other somatic cell types is still not fully answered. One explanation could be that SMN, in addition to its reported function, exhibits motor neuron–specific functions that are not relevant for other cells. It is not clear whether all SMN protein is bound to gemin-2 *in vivo*, and evidence from immunohistochemistry indicates that Smn – but not other proteins of the Smn complex – is enriched in axon terminals. In this specific part of motor neurons, the Smn protein seems associated with the hnRNP R protein. The interaction of SMN with hnRNP R requires the exon 7 encoded domains. A complex of SMN and hnRNP R is associated with the 3′-untranslated region of the β-actin mRNA, and mouse models of SMA, which exhibit reduced levels of Smn protein show reduced accumulation of β-actin mRNA in axon terminals. This correlates with β-actin protein levels in axon terminals and reduced axon growth in cultured motor neurons from Smn-deficient mice. Surprisingly, survival of isolated motor neurons with reduced SMN levels is not altered in comparison to wild-type controls. This indicates that the degeneration of motoneurons in the mouse model for spinal muscular atrophy might be a consequence of axonal defects. Lack of motor neuron function and degeneration of motor endplates and terminal axon also affects the capacity of these motor neurons to take up neurotrophic factors. Retrograde transport of these ligands together with their receptors to the cell bodies is reduced, and this should result in degeneration of motor neurons, which is similar to that in mouse models that lack receptors for neurotrophic factors such as gp130, LIFR-β, or CNTFR-α.

In patients with sporadic motor neuron disease, mutations were identified in the gene coding for heavy neurofilaments sub-unit. These mutations lead to alterations in the KSP-region, a domain with lysine-serine-proline-repeats, which is known to form the side-arm structures that cross-link the neurofilaments. This finding is particularly interesting in the light of neu-rofilaments pathology observed in ALS patients. However, such mutations are rare and only found in about 1% of patients with sporadic ALS.

Mutations were also found in the gene for dynactin, which points to a role by ax-onal transport in the pathophysiology of MND. Transgenic mice in which individ-ual components of the axonal transport machinery are inactivated show classi-cal signs of motor neuron disease. Also, the identification of the underlying gene defect in the mouse mutant progressive motoneuron disease (pmn) points to the relevance of axonal transport for the patho-physiology of MND. A point mutation in the tubulin-specific chaperone E gene is re-sponsible for the motor neuron disease in this mouse model. Interestingly, disease onset and progression can significantly be modulated by pharmacological treat-ment with the neurotrophic factor ciliary neurotrophic factor (CNTF). This indicates that the role of neurotrophic factors is not restricted to the promotion of neuronal survival, but extends to axonal mainte-nance, regulation of microtubule synthe-sis, and regulation of axonal transport.

Interestingly, the androgen receptor, the Cu^{2+}/Cn^{2+}-dependent superoxide dismu-tase, and neurofilament genes are widely expressed and little is known about what makes the disease process specific to motor neurons. It can be speculated that generalized disorders might mani-fest themselves as MND, possibly because motor neurons are more vulnerable or sen-sitive to any of these defects than other cell types. Moreover, these examples demon-strate that different, unrelated gene defects have a similar clinical appearance resem-bling typical signs of MND indicating that MND cannot be considered to be a homo-geneous disorder.

3.2
Basic Pathological Findings of ALS

As a characteristic feature of ALS (albeit with significant variation from case to case), loss of motor neurons occurs. The re-maining cells appear shrunken, dark, with basophilic cytoplasm and pyknosis of the nuclei. Shrinkage of the cells, which pre-cedes neuronal death, occurs in both upper and lower motor neurons. It has been suggested that the degeneration of lower motor neurons is a consequence of the loss of corticobulbar neurons. However, mor-phometric comparison of the loss of upper and lower motor neurons does not support such hypotheses. The morphological char-acteristics of motor neuron cell death such as disintegration of the Golgi complex, loss of Nissl structure, shrinkage, any par-ticular pyknosis of the nuclei, very much resemble the changes occurring during apoptosis. However, experimental proof is poor that motor neuron cell death in all forms of MND is apoptotic. Several stud-ies with postmortem ALS spinal cord have identified TUNEL-positive motor neurons. However, it is not clear whether classical apoptotic mechanisms are responsible for cell death, and when they are activated dur-ing the course of the disease. In particular, it remains to be demonstrated that apop-totic mechanisms are disturbed at early stages before first symptoms of the disease become clinically apparent.

Increasing evidence suggests that the disease process in ALS, the most common form of MND, might not be limited to motor neurons. There is also evidence of abnormal rates of degeneration in sensory neurons, but there is little doubt that the motor system is the site of predominant manifestation of the disease. Impairment of axonal transport might underlie many of the pathological characteristics. Inclusion bodies of several types have been described in the perikarya of motor neurons, such as basophilic inclusion, which are thought to consist mainly of RNA, eosinophilic Bunina bodies that resemble autophagosomes, and hyaline-inclusion, originally thought to be characteristic of familial ALS. This points to a role of abnormal protein degradation via the proteasome system, and experimental evidence mainly from cultured neurons suggests that increased proteasome activity can interfere with the pathological mechanisms in some models of MND, such as cell lines overexpressing mutant forms of SOD-I. Hyaline inclusions consist of densely packed phosphorylated neurofilaments, and it is believed that an impairment of slow anterograde axonal transport is responsible for their accumulation in the perikarya of motor neurons. Thus, defects in axonal transport and in protein degradation could play together in the pathophysiology of MND. The concept of impaired axonal maintenance is further supported by the finding that more myelinated nerve fibers are detectable in proximal than in distal parts of motor nerves of ALS patients. Failure of motor neurons to maintain distal structures such as the endplates appears as an early stage in the disease process. Evidence of the degeneration of endplates in early stages of the disease has been derived both from electrophysiological (spontaneous muscle fiber activity) and morphological studies. The disease then progresses to degeneration of the axons in a dying-back fashion, finally leading to degeneration of the cell

bodies in the spinal cord and brain stem. Clinical symptoms are expected to become apparent only after it has become impossible to compensate for the loss of functional motor neurons by collateral sprouting and rearrangement of motor units. This coincides with the stage at which motor neuron nerve terminals are affected. Thus, clinical symptoms can precede the degeneration of cell bodies.

4
Sporadic MND: The Pathophysiological Mechanisms Are Still Obscure

The majority of cases (>90%) of MND occur sporadically, precluding the analysis of underlying gene defects by classical molecular genetic analysis. Candidate gene approaches have been conducted in only a few places, and it is premature to state whether the identified mutations may be responsible for the development of the disease in general or only in small subsets of patients. For example, gene defects of heavy neurofilament subunits have been observed only in a few patients (see Sect. 3.1). However, these data do not tell us whether spontaneous gene defects are pathogenic, or whether epigenetic influences (such as intoxication, infection, autoimmune disorders, or any combination of such mechanisms) might be responsible for the disorder. Moreover, defects in the genes coding for neurotrophic factors or their receptors such as CNTF or CNTFR-α have been observed in healthy individuals and in patients with neurological disorders. Targeted disruption of such genes in mice leads to slow degeneration of motor neurons and changes in motor axons in the case of ciliary neurotrophic factor gene knockout. However, symptoms of muscle weakness are only very mild. Mice with gene defects in the CNTFR-α or LIFR-β gene die around birth and thus cannot be considered as a model for human MND. The severe discrepancy between gene activation of ligands such as CNTF or LIF and for receptor components such as CNTFR-α or LIFR-β suggests that defects in the genes for neurotrophic factors are compensated for by other members of the same gene family. In this case, MND would only occur when such compensatory mechanisms fail or when several kinds of degenerative influences act together, leading to such severe cellular dysfunction of motor neurons that compensatory mechanisms are not sufficient to prevent functional deficits.

There have been suggestions that many pathological mechanisms, some of them listed in Sect. 3, may be responsible for MND, but convincing evidence that interference with any of them results in an effective treatment of MND is still lacking. The only effective drug currently used for treatment of MND, Riluzole, interferes with a variety of mechanisms including glutamate toxicity, function of ion channels, and possibly additional intracellular signaling pathways, which modulate the function and structural maintenance of motor neurons. The conclusion that can be drawn from this situation is that basic mechanisms of MND are far from being understood.

4.1
The Autoimmune Hypothesis of Motoneuron Disease

There are many reports of abnormal immunological findings in MND patients. One of the most intriguing is the toxicity of serum from ALS patients to cultures of isolated motor neurons. Moreover,

abnormalities in the composition of subgroups of peripheral blood lymphocyte subsets have been described; for example, there is a decrease in mature T lymphocytes (CD3-positive cells) and a shift in the CD4/CD8-ratio in favor of CD8-positive cells, as well as an increase in natural killer cells (CD57 and CD16 positive cells). However, these findings have not been confirmed by recent studies. Another line of evidence has been derived from studies searching for autoantibodies directed at nerve antigens. Such studies have revealed antibodies against gangliosides (GM1), acetylcholine esterase, and L-type or P-type Ca^{2+}-channels. Such findings are in apparent contrast to many unsuccessful efforts to treat ALS by immunosuppressive strategies. Even total body lymph node irradiation was without effect. Some patients with high titers of anti-GM1-antibodies diagnosed as ALS exhibit peripheral nerve multifocal conduction block and are now considered to reflect a subgroup of chronic multifocal inflammatory neuropathy rather than degenerative MND. It now appears conceivable that in the course of MND epitopes on motor neurons that usually are hidden to the immune system become exposed, they allow secondary autoimmune mechanisms to come into play and add to the degenerative changes in motor neurons. Even if autoimmune mechanisms are not primary events, they could contribute to the clinical picture of MND by accelerating the functional impairment of the remaining motor neurons.

4.2
Toxic Lesions of Motor Neurons

Intoxication with ingredients of cycad plants is thought to be responsible for the ALS-Parkinsonism-dementia complex found in Chamorro populations in the

Western Pacific Islands. Since, however, the disease has been reported to occur up to several decades after exposure to the toxic agents, it is not clear whether the intoxication itself is responsible for the disease or whether it induces a long-lasting increase in the vulnerability of motor neurons to other agents. It has been suggested that β-N-methylamino-L-alanine (LBMAA) or the increased concentrations of glutamate and abnormal excitatory amino acids detectable in the cerebrospinal fluid of ALS patients could induce motor neuron degeneration via glutamate receptors. Glutamate toxicity via N-methyl-D-aspartate (NMDA) receptors is a common degenerative mechanism in other disorders of the nervous system such as stroke and CNS-trauma. However, clinical trials with NMDA-receptor antagonists such as dextromethorphan have not shown efficiency in ALS patients. However, there is evidence *in vitro* that AMPA receptors, in particular, Ca^{2+} influx through this receptor, are responsible for glutamate toxicity in motor neurons. EAAT2 variant transcripts have been found in the brain of ALS patients in combination with reduced EAAT2 protein levels. This is thought to cause elevated glutamate levels at the synapses that are located on the dendrites of bulbar and spinal motor neurons. Further evidence of glutamate toxicity in ALS is based on the efficacy of Riluzole, the only available drug for treatment of ALS up to now. This drug candidate is thought to act as a presynaptic inhibitor of glutamate release, but there is also evidence for other mechanisms of action. The role of glutamate toxicity in motor neurons has also been demonstrated after viral infection. Neuroadapted Sindbis virus mediates motor neuron death indirectly via an excitotoxic mechanism involving glutamate and calcium-permeable AMPA receptors.

5
Oxidative Stress: Is Motoneuron Degeneration in Familial ALS Due to Loss or Gain of Function of the Enzyme Superoxide Dismutase?

Mutations within the Cu/Zn-dependent superoxide dismutase (SOD-I) gene on chromosome 21 in about 2 to 5% of ALS patients has led to reinvestigations of all the observations and speculations that oxidative damage may play a pathogenic role in MND. SOD-I had already been the focus of interest in the context of the Alzheimer-like neurodegenerative changes associated with Down Syndrome. In the search for genes responsible for degenerative changes, patients with partial duplication of chromosome 21 translocated to other chromosomes were analyzed, and the 21q22 region has been found to be critical. Among the roughly 100 genes encoded in this region, the SOD-I gene was considered to be a good candidate, the duplication and consequent overexpression of which might be responsible for the degenerative changes observed in Down Syndrome: Indeed, overexpression of SOD-I in neural cell lines results in impaired neurotransmitter metabolism.

To obtain information on a possible role of increased SOD-I activity in neurodegenerative processes in more detail, transgenic mice overexpressing the human SOD-I gene under its homologous promoter were produced. These transgenic mice show pathological changes in motor neurons, such as withdrawal and destruction of terminal axons at the neuromuscular endplates and development of multiple small terminals in comparison to normal age-matched mice. Surprisingly, gene knockout of the SOD-I gene did not lead to MND, indicating that this enzyme is dispensable for maintenance of motor neurons, and its function is probably fully compensated by other enzymes such as SOD-II.

In the meantime, more than 60 mutations have been reported in patients with autosomal dominant ALS (ALS-1). These are widely scattered over the gene with hot spots in exon 4 and 5. Mutations of the conserved glycine at position 93 are quite abundant, and the A4V mutation, which is in particular found in North American families, generally leads to a severe course of ALS. The D90A mutation is quite abundant in Sweden and Finland, and in contrast to the other mutations, it is inherited as a recessive trait. 1 to 2% of the population in Scandinavia is heterozygous for the D90A allele. In Finland, about 100 000 persons are estimated to be carriers of this mutation. In general, patients with the D90A homozygous mutation show a relatively mild form of ALS with an average survival time of more than 10 years. Surprisingly, the age of onset does not differ from very severe forms such as the A4V form, indicating that other factors determine the age of onset, which act independently from those that determine survival time. Enzyme activity of SOD-I is only reduced by about 50% in most patients with familial ALS, as expected for autosomal dominant inheritance, leaving one allele of the *SOD-I* gene intact. It was suggested that the mutations lead to decreased enzyme activity either by causing an allosteric effect on the active center of the enzyme or by interfering with the formation of dimers, which is the form in which the enzyme naturally occurs in cells.

Mice with transgenic overexpression of mutant SOD-I generally show protein aggregates and other alterations in motor neurons, which become clearly detectable when the disease is clinically apparent. However, it is not fully understood whether such aggregates are responsible for the disease. Alternatively, it has been suggested that the inability of mutant SOD-I to bind copper might lead to accumulation of protein complexes and toxicity of free copper in cells and thus to neurodegeneration. Acquisition of copper by SOD-I requires a specific copper chaperone, a member of the family of metallochaperones, which regulate intracellular trafficking of metal ions. However, inactivation of this chaperone in each of the three most frequently used mouse models for ALS indicates that the disease is unaffected by elimination of copper loading. This argues against the possibility that primary toxicity arises from aberrant copper within the neurons. Morphological changes in mitochondria within motor neurons were observed as an early feature both in SOD-I^{G37R} and SOD-I^{G93A}–transgenic mice. This correlates with reduced mitochondrial function at least in SOD-I^{G93A}. Interestingly, mutant SOD-I^{G93A} has been detected within spinal cord mitochondria, and recent studies have shown that the accumulation of mutant SOD-I in mitochondria is also found in other ALS-linked mutations, and that these changes lead to damage of mitochondria-bound proteins. Therefore, it has been suggested that motor neuron–specific damage of mitochondria might be the reason for the selective toxicity of mutant SOD-I in MND.

6
Disintegration of Axonal Transport: Implications for MND

Several mouse mutants have been identified or generated in which disturbance of components of the cytoskeleton, of microtubuli, or of components of the axonal transport machinery lead to typical signs of MND. Recently, missense point mutations in the cytoplasmic dynein heavy chain have been identified as the cause for motor neuron disease in *Loa-* and *C1ra1-*mutant mice. Dynein is part of the motor complex for protein and vesicle movement along microtubules in the minus-end direction and thus an important component for retrograde axonal transport. Interestingly, the defect does not only lead to reduced retrograde transport but also to progressive postnatal loss of motor neurons. Whereas motor neuron loss is not detectable at an age of 3 months, about 50% of the motor neurons are lost at 16 to 19 months. Mutations in the p150 subunit of the transporter protein dynactin have been identified in a family with autosomal dominant progressive MND. Dynactin binds both to microtubules and to dynein, and thus acts as an essential component of the axonal transport machinery, which mediates binding of transport vesicle and transport proteins to microtubules. Mice in which dynamitin, the p50 inhibitory subunit of the dynactin complex, is overexpressed also show late onset MND.

Other widely used mouse models are *Pmn* mice, which develop a rapidly progressive form of MND between the third and fifth postnatal weeks. Histologically, the mice exhibit atrophy and loss of motor axons and at a later stage loss of motor neuron cell bodies in the spinal cord and brain stem motor nuclei. In addition, neurons in the red nucleus degenerate. The axonal projections from this nucleus to the spinal motor neurons – the corticospinal tract – are functionally linked to the corticospinal tract in rodents and contribute significantly to the innervation of spinal motor neurons. It thus functionally resembles the upper motor neuron input to lower motor neurons in higher vertebrates. As the cause of the disease, a single amino

acid exchange in the most C-terminal part of the tubulin-specific chaperone E (TBCE) protein was identified. The resulting protein appears less stable, but it could also be that translocation of the protein to axons, the function of the protein or interactions with other functionally relevant proteins are affected. These hypotheses need additional experimental data.

In summary, these data suggest that axonal transport plays an essential role in motor neuron maintenance, and future research on the pathophysiology of MND in humans will focus on the question whether defects in axonal transport are an early or late phenomenon in the disease.

The cytoskeleton of most animal cells contains a variety of elements also including thin actin filaments and intermediate filaments in addition to the microtubules. The β-actin mRNA is one among a small population of mRNAs, which are axonally transported. The functional relevance of axonal transport has been a point of debate for a long time. In contrast to dendritic transport, it has been argued that it is not clear whether the translocalized mRNAs are translated within the axonal compartment. Axonal transport of β-actin mRNA has been studied in much detail. A specific 3'-UTR region called *zip code* has been identified within β-actin mRNA that confers axonal localization. Proteins that bind to the zip-code region of β-actin mRNA have been named zip-code binding protein 1 (ZBP1) and ZBP2. It has been found that the formation of a particle involving β-actin mRNA and ZBP1 is necessary for β-actin translocation to growth cones. The translocation of the mRNA is increased by neurotrophic factors such as neurotrophin-3. In motoneurons, axonal terminals contain polyribosomes, which can be detected with specific antibodies against large subunit ribosomal RNA or by electron microscopy. The SMN protein, which is deficient in spinal muscular atrophy, binds to hnRNP R in the axonal compartment, and it has been shown that a complex of SMN and hnRNP R also interacts with the zip-code domain of β-actin mRNA. Motor neurons that lack Smn show reduced accumulation of β-actin mRNA in axon terminals. This correlates with highly reduced levels of β-actin protein in the axon terminals and reduced axon growth of isolated motor neurons in cell culture. In contrast, overexpression of SMN and/or hnRNP R in neuronal cell lines enhances neurite outgrowth, thus indicating that the translocation of β-actin mRNA and the local translation of this protein in axon terminals play a major role in axon growth and maintenance. Also in other animal models such as zebrafish, knockdown of SMN primarily leads to axonal disturbance and not to motor neuron cell death. Thus, spinal muscular atrophy might be considered as a disease that is caused by defective axonal translocation of β-actin mRNA. It will be interesting to know whether similar mechanisms also account for subtypes of adult onset MND. First indications have been obtained by the identification of mutations in senataxin, a protein with a DNA-helicase domain that is mutated in a rare form of inherited amyotrophic lateral sclerosis.

7
Identification of Neurotrophic Factors and Cytokines that can Regulate Survival and Function of Motor Neurons

Survival of embryonic spinal motor neurons depends on neurotrophic factors obtained from skeletal muscle and glial cells. Such factors are taken up by motor neurons by means of specific receptors and transported to motor neuron cell bodies where they act on specific parameters. In this way, these factors regulate apoptosis, transmitter synthesis, and other aspects of motor neuron function. Developing embryos and cultured embryonic spinal motor neurons have been used to identify neurotrophic factors and cytokines capable of supporting the survival of developing motor neurons. Positive effects have been observed for members of the ciliary neurotrophic factor (CNTF) family, for brain-derived neurotrophic factor (BDNF), neurotrophin (NT)-3 and -4, insulin-like growth factor (IGF) and members of the glial-derived neurotrophic factor (GDNF) family. These factors differ in their tissue distribution and temporal expression patterns, and gene knockout experiments have indicated that their modes of action might overlap with other members of the same gene family and probably of others. Interestingly, neurotrophic factors can interfere with the pathophysiology of motor neuron disease caused by different mechanisms. Positive effects are observed with IGF-I in mice that overexpress SOD-I. Pmn mutant mice show improved survival and motor performance after treatment with CNTF. This indicates that these factors can interfere with distinct pathophysiological mechanisms that have been identified for MND. The molecular mechanisms of these actions appear to be important for development of therapy and could lead the way to the identification of new small molecules that are active on the same targets.

8
Future Prospects for Treatment of MND

A large number of clinical trials with a variety of drugs and therapeutic strategies have been carried out in the past in the search for an effective treatment of MND. Among them were plasmapheresis (with the intention of removing toxic or autoaggressive molecules), drugs such as corticosteroids, corticotropin, cyclosporine A, and cyclophosphamide, which are known to be effective in autoimmune diseases. None of these trials were successful. Similar negative effects were observed in trials with growth hormone and thyrotropin-releasing factor.

A driving force behind some of the current therapeutic trials for the treatment of MND is the assumption that degeneration of motor neurons might be caused by mechanisms similar to those responsible for the death of other neurons of the central nervous system after glutamate exposure. Therefore, considerable attention has focused on clinical trials with glutamate antagonists and glutamate release inhibitors such as Riluzole and other drugs. Trials with free radical scavengers such as acetylcysteine, which are thought to reduce the generation of free radicals, a pathomechanism that has been implicated by mutations in the *SOD-I* gene in familial ALS, have not been successful so far.

Clinical trials with neurotrophic factors such as CNTF and BDNF have also failed. Among many potential reasons are, in particular, adverse pharmacokinetics. Thus, the factors are not available to the motor neuron cell bodies residing in the spinal cord. Alternatively, it could be that the receptors desensitize within a short time to neurotrophic factors, so that only short-time effects occur, which are not sufficient for a therapeutic effect of these drug candidates. Indeed, even promising trials with mouse models such as *Pmn* mice or SOD-I transgenic mice generally focus on treatment periods of only a few weeks, whereas treatment

of patients would require positive drug effects over several years. Nevertheless, the positive effects of IGF-I, BDNF, and CNTF in various mouse models of motor neuron disease are a good basis for further research and the development of new therapeutic strategies that might help promote long-term effects in the degenerating motor neuron.

Another problem that needs to be solved is the apparent heterogeneity of pathomechanisms, in particular in familial and probably also sporadic ALS. It could well be that drug candidates that interfere with some pathomechanisms are without any effect in patients suffering from MND due to another defect. Therefore, the genetic and molecular basis of disease needs to be better understood, and techniques for genotyping and molecular diagnosis have to be developed so that the pathomechanism of MND can be characterized for individual patients, even in patients with sporadic ALS. Potential positive or negative effects for the various drug candidates can only be tested on this basis.

This problem seems to be easier for patients with various forms of spinal muscular atrophy. The molecular basis of classical autosomal recessive SMA, which is caused by a mutation of the *SMN* gene, is known and can be easily tested and characterized. Thus, prerequisites exist to develop drugs on a rational basis. The same is also true for other forms of childhood spinal muscular atrophy such as SMARD and for adult onset X-linked spinal muscular atrophy (Kennedy's disease), which is caused by a polyglutamine-expansion in the androgen receptor gene. Multiple lines of evidence indicate that this mutation leads to a gain of pathogenic function, which requires the translocation of the receptor after binding of testosterone. Thus, anti-androgens have been suggested as drug candidates, and first trials with candidates such as leuprorelin showed promising results by reversing the behavioral and histopathological phenotypes in a mouse model. It is to be expected that clinical trials based on such results will be initiated soon, and new drug candidates will be developed with cell culture and mouse models, which can be developed once the molecular and genetic basis of the disease is known.

See also Motor Neuron Diseases: Cellular and Animal Models.

Bibliography

Books and Reviews

Bruijn, L.I., Miller, T.M., Cleveland, D.W. (2004) Unraveling the mechanisms involved in motor neuron degeneration in ALS, *Annu. Rev. Neurosci.* **27**, 723–749.

Chou, S.M., Norris, F.H. (1993) Amyotrophic lateral sclerosis: Lower motor neuron disease spreading to upper motor neurons, *Muscle Nerve* **16**, 864–869.

Cleveland, D.W., Rothstein, J.D. (2001) From Charcot to Lou Gehrig: deciphering selective motor neuron death in ALS, *Nat. Rev. Neurosci.* **2**, 806–819.

Fischbeck, K.H., Lieberman, A., Bailey, C.K., Abel, A., Merry, D.E. (1999) Androgen receptor mutation in Kennedy's disease, *Philos. Trans. R. Soc. Lond B Biol. Sci.* **354**, 1075–1078.

Jablonka, S., Wiese, S., Sendtner, M. (2004) Axonal defects in mouse models of motoneuron disease, *J. Neurobiol.* **58**, 272–286.

Katsuno, M., Adachi, H., Tanaka, F., Sobue, G. (2004) Spinal and bulbar muscular atrophy: ligand-dependent pathogenesis and therapeutic perspectives, *J. Mol. Med.* **82**, 298–307.

Smith, R.A. (1992) *Handbook of Amyotrophic Lateral Sclerosis*, Marcel Dekker, Inc., New York.

Thoenen, H., Sendtner, M. (2002) Neurotrophins: from enthusiastic expectations through sobering experiences to rational therapeutic approaches, *Nat. Neurosci.* **5**(Suppl.), 1046–1050.

Primary Literature

Bommel, H., Xie, G., Rossoll, W., Wiese, S., Jablonka, S., Boehm, T., Sendtner, M. (2002) Missense mutation in the tubulin-specific chaperone E (Tbce) gene in the mouse mutant progressive motor neuronopathy, a model of human motoneuron disease, *J. Cell Biol.* **159**, 563–569.

Chen, Y.Z., Bennett, C.L., Huynh, H.M., Blair, I.P., Puls, I., Irobi, J., Dierick, I., Abel, A., Kennerson, M.L., Rabin, B.A., Nicholson, G.A., Auer-Grumbach, M., Wagner, K., De Jonghe, P., Griffin, J.W., Fischbeck, K.H., Timmerman, V., Cornblath, D.R., Chance, P.F. (2004) DNA/RNA helicase gene mutations in a form of juvenile amyotrophic lateral sclerosis (ALS4), *Am. J Hum. Genet.* **74**, 1128–1135.

Hafezparast, M., Klocke, R., Ruhrberg, C., Marquardt, A., Ahmad-Annuar, A., Bowen, S., Lalli, G., Witherden, A.S., Hummerich, H., Nicholson, S., Morgan, P.J., Oozageer, R., Priestley, J.V., Averill, S., King, V.R., Ball, S., Peters, J., Toda, T., Yamamoto, A., Hiraoka, Y., Augustin, M., Korthaus, D., Wattler, S., Wabnitz, P., Dickneite, C., Lampel, S., Boehme, F., Peraus, G., Popp, A., Rudelius, M., Schlegel, J., Fuchs, H., Hrabe, d.A., Schiavo, G., Shima, D.T., Russ, A.P., Stumm, G., Martin, J.E., Fisher, E.M. (2003) Mutations in dynein link motor neuron degeneration to defects in retrograde transport, *Science* **300**, 808–812.

Howland, D.S., Liu, J., She, Y., Goad, B., Maragakis, N.J., Kim, B., Erickson, J., Kulik, J., DeVito, L., Psaltis, G., DeGennaro, L.J., Cleveland, D.W., Rothstein, J.D. (2002) Focal loss of the glutamate transporter EAAT2 in a transgenic rat model of SOD1 mutant-mediated amyotrophic lateral sclerosis (ALS), *Proc. Natl. Acad. Sci. U.S.A.* **99**, 1604–1609.

Kaspar, B.K., Llado, J., Sherkat, N., Rothstein, J.D., Gage, F.H. (2003) Retrograde viral delivery of IGF-1 prolongs survival in a mouse ALS model, *Science* **301**, 839–842.

Liu, J., Lillo, C., Jonsson, P.A., Velde, C.V., Ward, C.M., Miller, T.M., Subramaniam, J.R., Rothstein, J.D., Marklund, S., Andersen, P.M., Brannstrom, T., Gredal, O., Wong, P.C.,

Williams, D.S., Cleveland, D.W. (2004) Toxicity of familial ALS-linked SOD1 mutants from selective recruitment to spinal mitochondria, *Neuron* **43**, 5–17.

Nicole, S., Diaz, C.C., Frugier, T., Melki, J. (2002) Spinal muscular atrophy: recent advances and future prospects, *Muscle Nerve* **26**, 4–13.

Puls, I., Jonnakuty, C., LaMonte, B.H., Holzbaur, E.L., Tokito, M., Mann, E., Floeter, M.K., Bidus, K., Drayna, D., Oh, S.J., Brown, R.H. Jr., Ludlow, C.L., Fischbeck, K.H. (2003) Mutant dynactin in motor neuron disease, *Nat. Genet.* **33**, 455–456.

Rossoll, W., Jablonka, S., Andreassi, C., Kroning, A.K., Karle, K., Monani, U.R., Sendtner, M. (2003) Smn, the spinal muscular atrophy-determining gene product, modulates axon growth and localization of beta-actin mRNA in growth cones of motoneurons, *J. Cell Biol.* **163**, 801–812.

Subramaniam, J.R., Lyons, W.E., Liu, J., Bartnikas, T.B., Rothstein, J., Price, D.L., Cleveland, D.W., Gitlin, J.D., Wong, P.C. (2002) Mutant SOD1 causes motor neuron disease independent of copper chaperone-mediated copper loading, *Nat. Neurosci.* **5**, 301–307.

Xu, Z., Cork, L.C., Griffin, J.W., Cleveland, D.W. (1993) Increased expression of neurofilament subunit-L produces morphological alterations that resemble the pathology of human motor neuron disease, *Cell* **73**, 23–33.

Motor Proteins

Charles L. Asbury[1,3] *and Steven M. Block*[1,2]
[1] *Department of Biological Sciences, Stanford University, Stanford, California, USA*
[2] *Department of Applied Physics, Stanford University, Stanford, California, USA*
[3] *Present Address: Department of Physiology and Biophysics, University of Washington, Seattle, Washington, SA*

Encyclopedia of Molecular Cell Biology and Molecular Medicine, 2nd Edition. Volume 8
Edited by Robert A. Meyers.
Copyright © 2005 Wiley-VCH Verlag GmbH & Co. KGaA, Weinheim
ISBN: 3-527-30550-5

Keywords

Allostery
Pertaining to or involving a change in conformation caused by the attachment of a ligand or substrate.

Mechanoenzyme
A catalytic enzyme that produces motion and force.

Processivity
A measure of the number of catalytic cycles an enzyme undergoes before detaching from the substrate. In the context of motors, processivity is proportional to the average distance over which the enzyme translocates on its filamentous substrate before detaching.

Substrate
A molecule on which an enzyme acts. In the context of molecular motors, either the fuel source (e.g. NTPs), or the force-generating partner (e.g. actin filament).

Working Stroke
A conformational change that occurs during a single round of catalysis, and which drives motion and force production. The working stroke length, or working distance, is the maximal distance associated with the stroke. The working distance can be different from the distance between consecutive attachment sites on the partner filament.

Cellular motions have fascinated biologists during the 400 years since the invention of the optical microscope first allowed them to be seen. Today, we know that motions underlying the most essential processes of life – such as cell division, energy transduction, muscle contraction, DNA replication, transcription, and translation – are generated by molecular motors. A molecular motor is a protein, or a complex of proteins and nucleic acids, that produces motion and force. For fuel, many molecular motors consume nucleotide triphosphates, breaking an energy-rich phosphate bond to release chemical energy, and then converting this into mechanical work. Other motors tap electrochemical gradients that exist across membranes within bacteria, mitochondria, and chloroplasts. Motor proteins are Nature's nanomachines, and they often function with efficiency that far exceeds the best human-engineered machines.

1
Introduction

Producing motion and force is the primary role of the "classic" molecular motors, myosin, kinesin, and dynein. These *mechanoenzymes* all hydrolyze ATP as a source of energy and drive motion along protein filaments. Myosin generates motion along filamentous actin, and is well known for its role in muscle contraction. The seemingly simple act of flexing ones arm requires $\sim 10^{17}$ myosins working together to slide $\sim 10^{15}$ actin filaments toward one another. Kinesin and dynein move along microtubule filaments. An essential role of kinesin is to haul vesicles across neurons. This can be a 6-day haul since the longest neurons are more than a meter, and vesicle transport proceeds at only $2\,\mu m\,s^{-1}$. Dynein causes the beating of flagella and cilia, such as those lining the lungs, by sliding microtubules past one another.

Besides the classic motors, there are many "nontraditional" motor proteins. In some cases, motion and force production are byproducts rather than a primary function. The main role of DNA and RNA polymerases is to copy and transcribe the genetic code. In order to do so, they move along their nucleic acid templates, sometimes with amazing endurance. A single RNA polymerase molecule can transcribe all 2.5 million bases of the human dystrophin gene in 14 h, at roughly 50 bases per second. Another nontraditional motor is $F_1 F_0$-ATP synthase, which is responsible for replenishing the entire pool of ATP in all cells using energy derived from metabolism. As this enzyme toils, it also spins – it is a rotary motor. A flow of protons causes a shaft within the motor to rotate continuously, and shaft rotation is then coupled to the synthesis of ATP from ADP and phosphate. A third type of nontraditional motor activity is driven by the cytoskeletal polymers, actin, and tubulin. In addition to their roles as structural cables and girders for maintaining cell shape, and as highways for motor proteins to move along, these polymers are themselves dynamic machines that produce force. The leading edges of macrophages and other crawling cells are pushed outward by polymerizing actin filaments. Microtubule depolymerization generates tension that pulls chromosomes apart prior to cell division.

This chapter is a survey of the main classes of molecular motors. It begins with a discussion of the classic mechanoenzymes, myosin, kinesin, and dynein. These motors have been the subject of biophysical research for decades, and our understanding of their function serves as a foundation for the study of other motors. The chapter then turns to nontraditional motors, focusing on a handful of key examples, including nucleic acid enzymes (RNA polymerase), rotary motors

(F_1F_0-ATP synthase, and the bacterial flagellar motor), and protein polymers (actin and tubulin). A central goal of research on motor proteins is to determine how underlying biochemical events, such as ATP hydrolysis, are coupled to mechanical action. Progress toward this goal is chronicled throughout the chapter through description of experiments with classic and nontraditional motors.

2
Classic Molecular Motors

Myosin, kinesin, and dynein are founding members of large families of proteins whose primary function is to generate motion and force. Owing to their structural resemblance, the motors of each family operate in a manner similar to the founding proteins. However, they drive a wide variety of different cellular motions beyond the stereotypical roles of muscle contraction, vesicle transport, and the beating of cilia. There are at least 15 classes of myosin (traditionally denoted with roman numerals I through XV), and only a handful are involved in muscle contraction. Some of the other types are implicated in vesicle budding, cytokinesis, and organelle transport along actin cables. Likewise, kinesin-like proteins and cytoplasmic dyneins are essential for the formation and positioning of the mitotic spindle, chromosome separation prior to cell division, and organelle

Fig. 1 Structures of some classic and nontraditional molecular motors. (a) Muscle myosin consists of two heads connected to a common coiled-coil tail. The heads bind actin and carry ATP hydrolysis activity. A rodlike portion of each head (light gray) functions as a lever-arm, tilting $\sim70°$ relative to the remainder of the head (dark gray) upon attachment to an actin filament. The tail promotes bundling of myosin molecules into thick filaments. (b) Kinesin also has two heads, connected through short polypeptides called *neck linkers* to a common coiled-coil stalk. The heads bind microtubules and carry ATP hydrolysis activity. A conformational change of the neck linkers may drive kinesin motion. The tail binds cargo. (c) Each dynein molecule consists of a donut-shaped head with two rodlike structures, the stem and stalk, emanating from the head. The head contains four ATP-binding sites, only one of which is catalytically active. The tip of the stalk binds microtubules. To drive motion, the head and stalk rotate relative to the stem, which attaches to cargo and can also bundle two or three dynein heads together. (d) RNAP is shaped like a claw, which opens to allow a DNA template to enter, and then wraps completely around the DNA during transcription. While transcribing, RNAP separates a portion of the DNA duplex called the *transcription bubble*, and maintains registration of a short section of hybrid RNA:DNA duplex. Nucleotides enter through a channel, leading to the active site, where they are incorporated into the nascent mRNA chain. (e) F_1F_0-ATP synthase consists of two rotary motors, connected to a common shaft, which act as a motor-generator pair. The F_0 portion taps a proton gradient across the inner mitochondrial membrane to drive spinning of the rotor (c_{12}) relative to the stator (ab_2). The F_1 portion sits directly above F_0, and contains a shaft ($\gamma\varepsilon$) that is rigidly fixed to the rotor of F_0, and also a ring $(\alpha\beta)_3$ that is rigidly fixed to the stator of F_0. Spinning of the shaft relative to the ring drives ATP synthesis. The isolated F_1 portion is known as F_1-ATPase because in the presence of ATP it will spin in reverse, catalyzing ATP hydrolysis. (e) Cross-sectional view of the rotary motor of bacterial flagella. The motor core contains a stack of rings (rotor), embedded in the multilayered cell wall that rotates as a single unit about an axis (dashed line) perpendicular to the surface of the bacteria. Rotation is driven by torque-generating units composed of MotA and MotB proteins. The MotA/B complex is anchored to fixed structures (peptidoglycan) within the cell wall. Protons flow through a channel within MotA/B, where protonation and deprotonation of MotB induces conformational changes in MotA, which attaches and detaches from the base of the rotor and drives its rotation.

movement along microtubules. The discussion here centers on the founding proteins, their functional properties, and some of the experiments that uncovered these properties. Particular attention is given to *in vitro* work with single motor molecules.

2.1
Muscle Myosin: Tilting Cross-bridges Drive Contraction

The motor activity of myosin was discovered more than 50 years ago. Electron microscopy revealed that muscle fibers consist of parallel thick and thin filaments that slide past one another during contraction. Tiny structures termed "cross-bridges," connecting laterally between the filaments, were suspected to

drive filament sliding. In some images, the cross-bridges projected from the thick filaments at right angles, but in others, they were tilted, depending on tissue preparation conditions. These observations led to the theory, now well established, that cross-bridges drive filament sliding by cyclically attaching to the thin filaments, tilting, detaching, and untilting.

The thick filaments are now known to be bundles of myosin molecules. Each myosin consists of two identical 200-kDa polypeptides, plus two pairs of light chains (20 kDa). The heavy chains fold into twin globular heads connected to a common coiled-coil tail (Fig. 1a). Two light chains bind each head near the head–tail junction. Myosins bundle together by their long tails, and their heads project from the bundles, forming the cross-bridges that drive

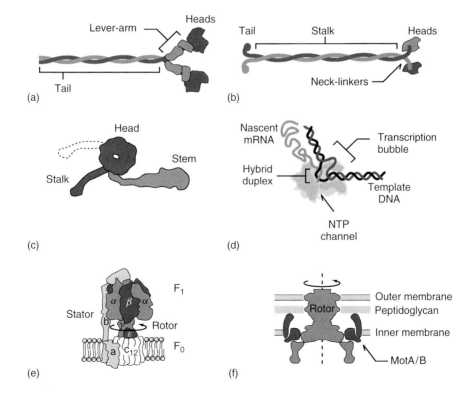

filament sliding. The heads can bind and hydrolyze ATP, and also carry a site that attaches to actin, the main component of the thin filaments, with ATP-dependent affinity. In high-resolution structures, a rodlike portion of the myosin head is found in several different orientations relative to the remainder of the head. Tilting of this "lever-arm," not the entire head, probably drives filament sliding.

The thin filaments of muscle are composed mainly of actin. Actin is a roughly spherical protein that polymerizes into a ropelike structure, with two strands, called *protofilaments*, that twist around one another. The filaments are polar, with a "plus" or "barbed" end, and a "minus" or "pointed" end, which are structurally different. During muscle contraction, the thick filaments slide toward the plus ends of the thin filaments.

2.2
Seeing is Believing: Motility Assays Demonstrate Motor Activity

A wealth of biochemical and structural information supports the tilting cross-bridge model of muscle contraction. However, the most compelling evidence for myosin motility comes from direct observation of motion generated *in vitro*. Myosin and actin are too small to see in an optical microscope. So, *in vitro* motility assays depend on various labeling schemes to render the motion visible. In the earliest assays, micron-sized beads were coated with myosin, and the beads were then observed to move along actin cables in the cytoplasm of the alga *Nitella*, and later along purified actin filaments bound to a glass surface. In an alternate strategy, actin filaments were made visible by fluorescent labeling, and gliding of these labeled filaments on myosin-coated glass surfaces

was observed in a fluorescence microscope (Fig. 2a). These important experiments established beyond doubt that actin and myosin alone, without any additional components from muscle cells, were sufficient to generate motion and force. Consistent with the rotating cross-bridge theory, ATP was required for the motility, and the myosins moved toward the plus ends of the actin filaments. Filaments glided *in vitro* at 6000 nm s^{-1} (Table 1), similar to the speed at which thick and thin filaments slide past one another during muscle contraction. As discussed below, these two basic tests – the "bead assay" and the "gliding filament assay" – have been adapted and refined to study a variety of other motors in addition to myosin (Fig. 2b through d).

2.3
Kinesin: Intracellular Porter

Motility assays were instrumental in the discovery of kinesin. Observations of the squid giant axon suggested the existence of motors that consume ATP and haul vesicles at speeds of 1 to 2 μm s^{-1} along the dense array of microtubule filaments within the axon. A putative motor was first isolated by locking the vesicles onto the microtubules using a nonhydrolyzable ATP analog, AMPPNP, followed by purification of the microtubules, and then release of the motor with ATP. A gliding filament assay identical to that developed for myosin confirmed that the purified protein, kinesin, was indeed a motor: glass surfaces coated with kinesin supported the ATP-dependent gliding of microtubules. The gliding velocity, 800 nm s^{-1} (Table 1), closely matched the speed of vesicle transport.

Additional assays, using microtubules marked to reveal their intrinsic polarity, revealed the direction of kinesin-driven

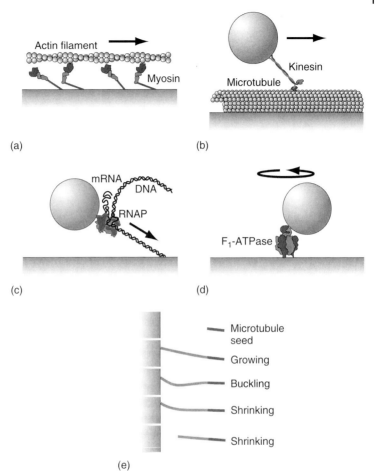

Fig. 2 Motility assays adapted for various molecular motors. (a) In the gliding filament assay, coverslip-bound motors drive filaments to move in a direction parallel to the filament long axis. Here, many myosin heads are shown interacting with a single actin filament. (b) In the bead motility assay, motors are attached to microscopic glass or plastic beads, which are pulled by the motors along coverslip-bound filaments. Kinesin is shown moving along a microtubule. (c) The tethered particle assay was developed to study the motion of nucleic acid enzymes along DNA filaments. Here, a microscopic bead is attached to RNA polymerase, which is transcribing a DNA filament that is attached at one end to the coverslip. Transcription results in a shortening (or lengthening) of the tether (depending on which end of the DNA is surface-bound). (d) In the rotation assay for F_1-ATPase, the motor is attached to a coverslip, and the orientation of the shaft is marked by off-axis attachment of a microscopic bead or filament. Spinning of the shaft causes the bead or filament to rotate. (e) Force generation by polymerization has been demonstrated by growing dynamic microtubule extensions from coverslip-bound seeds. The microtubules continue to elongate even after their growing ends encounter a barricade, which generates enough compressive force to buckle the filaments.

Tab. 1 Properties of selected molecular motors.

Protein Machine	Molecular Weight (kDa)	Force-generating Partner	Energy Source	Maximum Speed (nm s^{-1})[a]	Maximum Force (pN)[a]	Step Size (nm)[a]	Processivity (cycles)
Kinesin, native heterotetramer	340	Microtubule	ATP	1800			
Kinesin, truncated active homodimer	90	Microtubule	ATP	800	6	8	100
Myosin II, native heterohexamer	500	Actin filament	ATP	6000	1.5		
Myosin II, active HMM fragment	110	Actin filament	ATP	8000		6	1
Dynein, inner arm subspecies c	500	Microtubule	ATP	700	1.1	$8n$ ($n = 1, 2, 3 \ldots$)	~10
RNA polymerase, *E. coli* core enzyme	380	dsDNA	NTPs	5	27	0.34	>10 000
F$_1$F$_0$-ATP synthase	540	n/a	Protonmotive				n/a
F$_1$-ATPase, active rotary motor	350	n/a	ATP	150 Hz	40 pN nm	120°	n/a
Bacterial flagellar motor, basal body	9500	n/a	Protonmotive	300 Hz	4600 pN nm		n/a
Microtubule, growing	110 (tubulin dimer)	n/a	Binding	~50	4		n/a
Microtubule, depolymerizing		n/a	GTP	~500			
Actin filament, growing	40 (G-actin monomer)	n/a	Binding	~15			n/a

[a] Different units apply to the rotary motors, F$_1$F$_0$-ATP synthase, F$_1$-ATPase, and the bacterial flagellar motor. For these, the maximum rotation rate, torque, and angular step size are reported in units of Hz, pN nm, and degrees, as noted.

motion. Microtubules are rigid, tube-shaped polymers, composed of tubulin proteins arranged in a lattice, resembling a miniature drinking straw. Like actin filaments, microtubules have two structurally distinct ends, called "plus" and "minus". Polarity-marked filaments driven by kinesin glided with their minus ends leading, implying that kinesin was moving toward the plus ends.

Kinesin and myosin share many structural and functional similarities. Like myosin, each kinesin molecule consists of two identical polypeptides that form twin heads connected to a common coiled-coil stalk (Fig. 1b). The motor activity is carried by the heads. Each head hydrolyzes ATP and attaches to microtubule filaments with nucleotide-dependent affinity. But unlike myosin, the tail of kinesin does not cause bundling, it binds the motor to its cargo. Furthermore, atomic structures of kinesin heads are devoid of any rodlike structure resembling the lever-arm of myosin. Lacking a lever, kinesin's working stroke is likely to be very different from that of myosin.

2.4
Ciliary Dynein: The Dark Horse

Comparatively little functional information is available for the third classic motor, dynein, even though its activity was discovered around the same time as that of myosin. Electron micrographs revealed lateral connections between the parallel microtubules in cilia and flagella, similar to the myosin cross-bridges in muscle. The cross-bridges in cilia and flagella are dynein motors that drive bending motions by sliding microtubules past one another.

Dynein is larger and structurally more complex than kinesin or myosin, but it has many features common to all the classic motors. Depending on the source, dynein consists of one, two, or three large (500 kDa) polypeptides. Each of these forms a donut-shaped head, with two rodlike structures emanating from it, the "stem" and the "stalk" (Fig. 1c). The stem functions similar to the tails of kinesin and myosin, bundling the heads together, and also anchoring them tightly to their cargo. The tip of the stalk binds microtubules in an ATP-dependent manner, like the microtubule binding site within each of kinesin's heads. No atomic resolution structures are available for dynein, but electron microscopy of single dynein particles revealed a conformational change akin to the lever-arm tilting of myosin: under different nucleotide conditions, the stem adopts two different orientations relative to the head and stalk. Thus, dynein may move its stem-bound cargo by cyclically attaching via the stalk to a microtubule, rotating the stalk and head, detaching from the microtubule, and then unrotating. Dynein supports microtubule gliding and bead motion in *in vitro* assays. The direction of dynein-driven motion is toward the minus end of the microtubule, opposite that of kinesin-driven motion.

Dynein's complex structure contains a number of features with unknown functional significance. Dynein motors from different sources have different numbers of heads. Each donut-shaped head consists of six different subdomains arranged in a hexameric ring. Four of these subdomains bind ATP, but only one catalyzes hydrolysis. Nucleotide binding, but not hydrolysis, at one of the other subdomains is essential for motor activity. Uncovering the reasons for this complexity and elucidating dynein's mechanism of action are important frontiers for future research.

2.5
Processivity Allows Kinesin to Work Alone

Soon after its discovery, kinesin was found to possess a tenacity that set it apart from myosin and dynein. Kinesin is highly *processive*, staying attached to the microtubule as it undergoes many catalytic cycles, and translocating over relatively long distances before detaching. This processivity was first demonstrated when gliding filaments or moving beads were found to move long distances (1–2 µm), even when the surface density of motors on the slide or bead was extremely low, ensuring that single kinesin–microtubule interactions were very likely. Several independent lines of evidence now provide very strong evidence of kinesin's processivity (see Fig. 3).

The processivity of kinesin probably evolved as a means to conserve cellular resources. Kinesin's role of transporting vesicles across neurons is a critical task that must be accomplished repeatedly and with high reliability in order for these cells to function. The longest neurons contain millions of vesicles that each take weeks to make the journey from one end to the other. Kinesin's high processivity allows this Herculean task to be completed by just a few motors bound to each vesicle. In principle, the job could also be accomplished by nonprocessive motors, but many more motors would be required. The cell, in turn, would have to devote more energy and resources into producing these additional molecules.

Apart from its biological significance, kinesin's processivity has been a great advantage for experimentalists, allowing the first studies of single motor molecules. Motility assays for myosin relied on hundreds of motors acting together because the tiny tilting motions, or *working strokes*, of the individual heads are too small to see in a conventional optical microscope. However, owing to their high degree of processivity, kinesins generate hundreds of working strokes during each encounter with a microtubule moving distances of ~1 µm. The summation of many strokes renders the motion of individual kinesin motors easily visible.

Results from single molecule motility assays revealed a number of insights about how kinesin moves. The motion of kinesin-driven beads *in vitro* was not random over the microtubule surface, but appeared to follow a path parallel to the protofilaments. Gliding filament assays supplied strong evidence for protofilament tracking, when abnormal microtubules with helical protofilaments were shown to rotate about their long axis as they moved. In bead assays, engineered kinesin proteins with only one head failed to generate highly processive motion, indicating that two heads are, in fact, better than one.

2.6
Rowers Versus Porters: Duty Ratio Makes a Difference

The head domains of myosin and kinesin differ markedly in their duty ratio, the fraction of time during each biochemical cycle that they remain attached to their partner filament. Myosin possesses a low duty ratio that allows groups of molecules to work together efficiently, like the rowers of a large canoe, while kinesin has a high ratio, befitting its role as a lone porter. A myosin head has high affinity for actin just after hydrolysis, when the nucleotide-binding pocket contains either ADP and phosphate, or ADP alone, and low affinity when the pocket is empty or contains ATP. The timing of transitions between these states ensures that the high-affinity

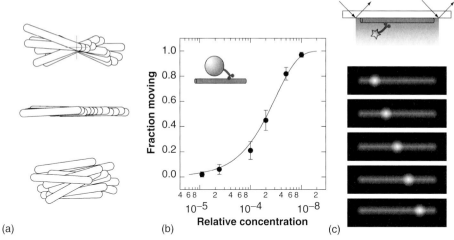

(a)

(b)

(c)

Fig. 3 Strong evidence for the processivity of kinesin. The first evidence for processivity was that kinesin-driven motion persisted *in vitro* even when the surface density of motors was extremely low. (a) Nodal point pivoting in the gliding filament assay also indicated processivity. Microtubules gliding on surfaces decorated sparsely with kinesin rotated erratically about a fixed location, even as they moved through this nodal point. When the trailing end of the microtubule reached the nodal point, it dissociated from the surface and diffused back into solution. A single motor at the nodal point presumably drove the motion (top). Negligible rotation occurred at high motor densities, when multiple motor–filament interactions constrained the filament orientation (middle). Both types of motion were distinct from thermal motion of free filaments in the absence of motor (bottom). [Adapted from Howard, J., Hudspeth, A.J., Vale, R.D. (1989) Movement of microtubules by single kinesin molecules, *Nature* **342**, 154–158.] (b) In the kinesin bead assay, the fraction of moving beads, f, decreased gradually as the relative motor concentration, C, was lowered, as expected if one molecule is sufficient to produce movement. The curve shows a one parameter (λ) fit to Poisson statistics, $f = 1 - \exp(-\lambda C)$. [Adapted from Svoboda, K., Block, S.M. (1994) Force and velocity measured for single kinesin molecules, *Cell* **77**, 773–784.] In the low-density regime, moving beads continued to translocate at normal speeds over distances that were independent of motor concentration (data not shown). (c) A third method used a microscope capable of imaging single fluorophores bound to kinesin (upper panel). The movement of labeled motors along coverslip-bound filaments was directly observed (shown schematically in the lower five panels). [Adapted from Vale, R.D., Funatsu, T., Pierce, D.W., Romberg, L., Harada, Y., Yanagida, T. (1996) Direct observation of single kinesin molecules moving along microtubules, *Nature* **380**, 451–453.] Labeling the motor by fusion to green fluorescent protein avoided chemical modification with reactive dyes, which can damage the motors, and ensured that every motor was labeled with the same number of fluorophores.

states represent <2% of the total cycle time. This low duty ratio is an adaptation that allows the myosin heads to avoid interfering with each other when many are acting on the same actin filament. They detach very quickly after undergoing a working stroke, so the speed of filament sliding is not limited by the hydrolysis rate of the individual heads. In contrast, kinesin heads have a duty ratio >50%, which partially explains the processivity of the motor. Even if the cycles of the two heads were completely uncorrelated, their high duty ratio would ensure that, on average, at least one was always bound to the microtubule.

2.7
Molecular Tug-of-war: Applying Force to Individual Motors

With the development of single molecule assays, it became possible to directly measure the forces generated by individual motors. One method for measuring force production by kinesin was to attach a microtubule filament to a flexible glass fiber, and then hold the fiber near a surface sparsely coated with kinesin. As individual kinesins on the surface bound and moved along the microtubule, they pulled against the glass fiber and caused it to bend. By measuring the amount of bending, the maximum force against which a kinesin motor could move was estimated to be 5 or 6 pN.

Another method for applying force to individual kinesin molecules, which gave a similar estimate of the stall force and also led to a number of other discoveries, was to use an optical trap. An optical trap is made by focusing a laser through the objective lens of a high-magnification microscope, creating a very bright light spot at the specimen. The focused light traps small objects such as micron-sized beads. When the trapped object is moved away from the center of focus, it feels a restoring force pulling it back that is proportional to the distance from the center, as if the trap was a stretched spring pulling on the object. To apply force to kinesin, an optical trap was used to grab beads with single kinesin molecules attached, and to place them near microtubules stuck onto to a glass surface. When the kinesin began moving along the microtubule, it pulled the bead from the trap center, and the trap supplied a restoring force that placed tension on the kinesin. As the bead was pulled gradually away from the trap center, the force increased and the motor speed decreased, halting when the force reached 6 pN.

2.8
Motors Move in Discrete Steps

Kinesin molecules move discontinuously over the microtubule surface, advancing in discrete 8-nm increments and dwelling at well-defined positions between advancements (Fig. 4). Two key innovations allowed the first observation of steps in the motion of kinesin-driven beads. First, tension supplied by an optical trap suppressed the random, thermally driven ("Brownian") motion that would otherwise dominate. Second, the bead position was measured with very high spatial and temporal resolution by monitoring the distribution of scattered light with a photodetector. The 8-nm step size (Fig. 4b) matches the spacing of tubulin dimers in the microtubule lattice. Similar experiments with dynein, which is processive under some conditions, suggest that it also moves stepwise, advancing by multiples of 8 nm.

Optical trapping has been applied to measure the motion of single myosin motors. A different technique than that used for kinesin was required because the interactions between a myosin molecule and an actin filament are fleeting, lasting only a few tens of milliseconds. To resolve these quick attachments, an assay was developed in which an actin filament with beads attached at both ends was suspended between two optical traps and held near a surface sparsely coated with myosin (Fig. 4c). Thermal motion of the beads decreased when a motor became attached to the filament, and the average position of the beads shifted abruptly, due to the tilting of the myosin head, by 6 nm. This tiny

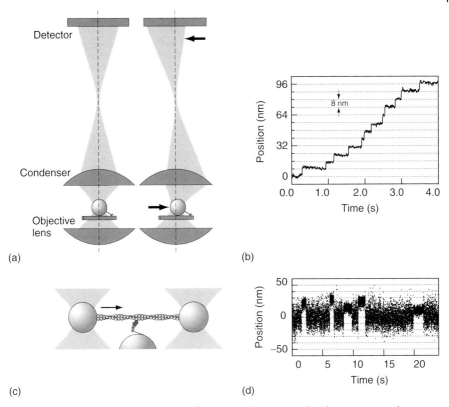

Fig. 4 Measuring the discrete motions of kinesin and myosin molecules using optical traps. Both methods use microscopic beads as handles to apply force, and as markers for the position of the motor or filament. (a) An optical trap applies tension during kinesin-driven movement of a bead along a coverslip-bound microtubule, and this tension reduces thermal motion of the bead so that individual 8-nm steps can be resolved. Bead position is detected with very high spatial resolution by monitoring the distribution of scattered light with a photodetector. (b) Example trace showing 8-nm steps generated by a single kinesin molecule. (c) The three-bead assay developed for measuring working strokes of muscle myosin. An actin filament is pulled taut between two microscopic beads held in optical traps. Binding of a single myosin head to the filament reduces thermal motion of the beads, and also induces a working stroke in the myosin, which causes the beads to deflect by 5 to 15 nm. (d) Example trace showing interactions between a single myosin head and an actin filament. [Data reprinted with permission from Lister, I., Schmitz, S., Walker, M., Trinick, J., Buss, F., Veigel, C., Kendrick-Jones, J. (2004) A monomeric myosin VI with a large working stroke, *EMBO J.* **23**, 1729–1738.]

distance is the maximum sliding distance that a myosin molecule can generate during a single interaction with actin during muscle contraction. More than a million of these interactions are evidently required just to lift a finger.

2.9
Different Strokes: Variation Within and Across Motor Families

In accordance with their diverse roles, motors of the myosin, kinesin, and dynein families often have important differences in the way they move. The cellular role of type V myosin, for example, is more similar to that of kinesin than to muscle myosin. Myosin V acts alone or in small numbers to transport vesicles and organelles along actin cables. It was therefore not surprising to find that this myosin, like kinesin, exhibits processive movement, taking several steps along an actin filament before detaching. The step size of myosin V, 36 nm, is much larger than that of kinesin, and matches the wide spacing of binding sites that occur every half-period in the actin helix. The structure of myosin V explains how it can generate such large movements. Each head contains a lever-arm that is 24 nm long, three times longer than the lever-arm of muscle myosin.

Diversity within motor families invites comparison, which can illuminate important aspects of motor function. Strong evidence supporting the tilting lever-arm model came from comparisons of gliding speeds and stroke lengths generated *in vitro* by myosin-family motors with different lever-arm lengths. The speeds and stroke lengths varied in proportion with the lever-arm length, as predicted by the model. Structural differences between kinesin and Ncd, a related motor that moves toward the opposite end of microtubules, suggested that a "gearbox" region just outside the head domain controls the direction of motion of these motors. This hypothesis was confirmed by testing chimeric motors, made by swapping the gearbox regions of the two proteins, in gliding filament assays.

2.10
Fuel Economy and Energy Efficiency

How many ATP molecules does a motor require to generate a working stroke or step? Is this "fuel economy" always the same? For kinesin, the coupling ratio – the number of ATPs consumed per step – has been measured by comparing the stepping rate in single molecule assays to the rate of ATP hydrolysis. Over a wide range of ATP concentrations and loads, one ATP is hydrolyzed per 8-nm step. This tight, 1:1 coupling implies that the energy efficiency of kinesin can be very high. ATP hydrolysis under physiological conditions is worth ~ 80 pN nm (or $2 \cdot 10^{-23}$ kcal). When kinesin generates 8-nm steps under 5 pN of load, it produces as much as 40 pN nm (10^{-23} kcal), or 50% of the total chemical energy available. So kinesin is more than twice as efficient as the best man-made gasoline engines, which are 24% efficient at full power, and typically achieve only 10 to 15% on the road.

The energy efficiency of cytoplasmic dynein, 10%, is considerably lower than that of kinesin. However, dynein's complex structure may act like an automobile transmission, allowing it to maximize fuel economy. Near dynein's stall force, 1 pN, the motor takes 8-nm steps. At <0.4 pN, however, it seems to advance in larger increments of 24 or 32 nm. Thus, dynein can apparently shift into high gear when carrying a light load. Assuming that the coupling ratio under both conditions is equal, the larger step size will result in proportionally better fuel economy.

2.11
Walk This Way: Processive Mechanoenzymes Move Hand-over-hand

The fact that single kinesin molecules generate hundreds of steps, even under load, together with the earlier findings of motion parallel to the protofilaments and the requirement for two heads, suggested that it might walk – or waddle – from one tubulin dimer to the next. An attractive hypothesis was that the twin heads each take turns, alternately detaching and moving past one another, in a "hand-over-hand" motion resembling that of a person swinging along monkey bars. The same model was also thought to apply to the processive myosin-family motor, myosin V.

Single molecule experiments with myosin V and kinesin have confirmed that both walk hand-over-hand. First, the stride length of myosin V was measured by labeling one of its two heads with a fluorophore, and tracking the label with nanometer resolution. This is like watching a person walking across a field on a moonless night with a flashlight attached to one foot: The person is invisible, but the light moves visibly with every other step. The heads of myosin V took turns making strides that were twice as long as the distance moved by the tail, evidence that strongly supported a hand-over-hand model. Next, optical trapping experiments showed that some kinesin molecules limp along the microtubule, exhibiting a difference in the timing of every other step. Limping implied that kinesin switched between two different configurations as it stepped. The most severe limpers were mutants in which one head hydrolyzed ATP more slowly than the other, arguing for a mechanism in which the two heads swap both mechanical and catalytic activities with each step. Finally, the stride length for one of kinesin's two heads was measured to be 16 nm, double the step size. Taken together, these results make a very strong case for a hand-over-hand mechanism for kinesin and myosin V.

2.12
Coordination is Required

To fully account for the hand-over-hand walking of kinesin and myosin V, coordination between the two heads is essential, and it may be achieved through mechanical tension between the heads. The mechanical cycle of both motors includes a transient state in which both heads are attached to the filament. The motors are probably stretched in this doubly attached state, owing to the relatively large distance between the heads, and the resulting intermolecular strain between the heads could bias their kinetics so that the trailing head nearly always detaches before the leading head. In support of this hypothesis, external loads have been shown to strongly affect the rate of detachment of single myosin V heads, with forward loads accelerating detachment and backward loads slowing detachment.

Coordinated action at distinct sites is a universal requirement for all motors, not just those with high processivity. Automobile engines rely on the carefully timed actions of pistons, valves, and spark plugs. Mechanoenzymes also have critical interacting parts. Consider the mechanochemical cycle of myosin: Within a few milliseconds after attachment of a myosin head to actin, small motions occur in the nucleotide-binding pocket, allowing phosphate release. This triggers lever-arm tilting, which moves the actin filament. Subsequent ADP release triggers a second, smaller motion in certain myosin types (e.g. myosin I and single myosin V heads), but it is not clear if this additional stroke occurs for muscle myosin. ATP binding triggers detachment from actin, and hydrolysis "primes" the motor for the next cycle. For other motors, the specifics and timing are different than for myosin, but this coordinated action is a universal requirement.

2.13
The Kinesin Cycle and Working Stroke

The mechanochemical cycle of kinesin is not as thoroughly understood as that of myosin. Because two heads are involved in the stepping process, the kinesin cycle is necessarily more complex than that of muscle myosin. The specific conformational changes driving motion are poorly defined, and there is uncertainty about which biochemical events are associated with motion. A working hypothesis for kinesin is that all the mechanical action is associated with just one biochemical event. In support of this "one-stroke" hypothesis, reaction schemes with just one force-dependent rate can account for force-velocity and [ATP]-velocity curves measured in single molecule assays. These schemes predict a working stroke after ATP binding, possibly upon ATP hydrolysis. By contrast, a working stroke concomitant with ATP binding is suggested by kinetic measurements, showing that ADP release from one head is stimulated by binding of a nonhydrolyzable ATP analog (e.g. AMPPNP) to the other head.

A conformational change in the structure of kinesin has been discovered that may drive its motion. This putative working stroke is quite different from either the tilting lever-arm of myosin, or the stem reorientation of dynein. In single-headed kinesin constructs, a 15-amino acid peptide known as the "neck linker," which connects each kinesin head to the coiled-coil stalk, undergoes a nucleotide-dependent transition. In the presence of ADP, or when no nucleotide is present, the neck linker is disordered. In this state, it acts like a flexible tether, pivoting about a point on the backside of the head. When the head is attached to a microtubule in the presence of ATP analogs (AMPPNP, ADP-AlF$_4^-$), the neck linker "zips" onto to the surface of the head, and its end points toward the microtubule plus end. In the full two-headed motor, zipping of the neck linker on one head could drive stepping by moving the stalk, and therefore the other head, toward the next attachment site on the microtubule lattice.

2.14
Under the Hood, Motors are Still a Mystery

Even for the best-understood motors, where high-resolution structures are available, and major mechanical steps can be identified with specific biochemical transitions, there are very fundamental questions that remain unanswered. In particular, the atomic-scale motions that transduce small chemical events in the nucleotide pocket and convert them into larger motions elsewhere are poorly understood. Important residues have been identified by comparing sequences of related motors and their precursors (e.g. myosin, kinesin, and G-proteins), and subdomains that move relative to others have been suggested by structural comparison. However, ultimately, these static structures cannot elucidate the timing of movements and the cause-and-effect relationships between the various parts. A myriad of physical and biological techniques will no doubt be essential in this effort, but single molecule techniques that simultaneously record motion and biochemical changes (e.g. through fluorescence) seem particularly valuable in this regard.

3
Nontraditional Molecular Motors

There are many other protein machines that generate force and motion, but which do not fit the classic motor paradigm. A growing number are being studied using *in vitro* assays, like those discussed above, which allow the mechanical output of single motors to be measured.

3.1
Nucleic Acid Enzymes

Some of the most important processes of life are carried out by nucleic acid enzymes, many of which are processive motors that move along DNA. Every human cell stores genetic information in the form of 23 strands of DNA, totaling three billion base pairs, and measuring 1 m in total length. All 23 strands are copied by DNA polymerase enzymes, untangled by topoisomerase enzymes, and packaged into chromosomes by condensins, before cell division. The genetic information contained in the DNA is transcribed into mRNA by RNA polymerase (RNAP), and translated into protein by the ribosome. Each of these nucleic acid enzymes is a protein machine capable of generating force and motion, and each is fascinating in its own right. A discussion of all of them is beyond the scope of this chapter, which will focus on one important example, RNAP.

3.2
More Than a Motor: Multitasking by RNA Polymerase

Even though the size of an RNAP enzyme, by comparison of total molecular weight, is not so different from that of the classic mechanoenzymes, its function is much more complex. Motion is merely a by-product of the biological role of this protein machine, transcribing the genetic code. While moving along a DNA template, RNAP separates a short section of the DNA duplex, the "transcription bubble," and builds a copy of one strand by selecting complementary nucleotides from the surrounding solution and attaching them, one at a time, to the end of the nascent mRNA chain (Fig. 1d). Along the way, it must maintain registration of a short section of "hybrid" RNA:DNA duplex, and also respond to a number of different signals that control the initiation, termination, and elongation rate of transcription. Like other mechanoenzymes, RNAP derives energy from nucleotide hydrolysis, but, in this case, each nucleotide serves a dual role. After hydrolysis, the nucleotide becomes an information-containing subunit incorporated into the growing mRNA. (If automobile engines could make such efficient use of their exhaust, urban air quality would be much improved!)

Motion may not be its primary function, but RNAP is no slouch of a motor. Its motion can be directly observed by attaching a micron-sized bead, and recording bead motion as the enzyme transcribes a DNA template bound at one end to a glass surface (Fig. 2c). Using this "tethered particle" assay in conjunction with optical trapping, the motion of a single RNAP can be tracked with high spatial resolution, and the effect of applied load can be measured. RNAP is slow, moving in these assays at 10 to 15 bp s^{-1}, or just 5 nm s^{-1} (Table 1). This speed is roughly equivalent to the rate of human hair growth, and 160-fold slower than the speed of kinesin. However, RNAP is much more processive than kinesin. In cells, RNAP molecules synthesize mRNA chains of 10^4 (in bacteria) to 10^6 (in mammals) nucleotides. *In vitro*, its processivity is reduced, but the enzyme typically moves across several thousand bases or more before detaching from the template. Movement continues, unhindered at 10 to 15 bp s^{-1}, even when backward loads as high as 27 pN are applied. A high stall force (>5-fold higher than that of kinesin) may be necessary for RNAP to function *in vivo*, perhaps allowing the enzyme to push through "road-blocks" formed by other DNA-binding proteins. RNAP is presumed to move in discrete steps corresponding to the distance between individual bases along the DNA helix, 0.34 nm. This distance is extraordinarily small (20 times smaller than kinesin's 8-nm steps), and steps of this size have not yet been directly observed. However, optical trapping technology is rapidly advancing, and such tiny motions may soon be resolvable.

3.3
RNA Polymerase Structure

RNAP is shaped like a claw. The claw opens to allow a DNA strand to enter, and during transcription it closes, wrapping completely around the DNA. Besides the DNA entry and exit channels, there is also a channel through which the newly synthesized RNA exits, and a pore for nucleotide entry (Fig. 1d). Inside the closed structure, the enzyme makes numerous contacts with the hybrid duplex and the DNA. It is unknown, which parts of RNAP are responsible for generating motion and force production, but many candidate features are apparent in the high-resolution structures. For example, a "bridge helix" located near the site of nucleotide addition may undergo a conformational change that pushes the enzyme to the next site. Determining which portions of RNAP are responsible for its motion is a great challenge for future research.

3.4
What Causes Pauses?

The motion of RNAP along the DNA template is interrupted by pauses, lasting from a few seconds to many minutes. The short-duration pauses (those with lifetimes of seconds) are very frequent, and may result from the enzyme encountering a few GC-rich base pairs of DNA that are tougher-than-average to separate. Occasionally, the enzyme pauses for a much longer duration (20 s to >30 min). These infrequent but long-lived pauses may occur for two reasons, both of which illustrate the sophisticated behavior that RNAP is capable of. First, long pauses can be induced when RNAP encounters a specific sequence in the DNA template it is transcribing. Such sequence-dependent pauses are an important mechanism for regulation of gene expression. By relieving these long pauses, a cell can greatly alter the rate of expression of a pause-containing gene. In some cases, these sequence-dependent pauses occur when the nascent mRNA chain folds into a hairpin structure, which then interacts directly with the RNAP enzyme. In other cases, these pauses occur when the RNAP transcribes a slippery AT-rich sequence, resulting in an unstable RNA:DNA hybrid duplex that allows the enzyme to backtrack. There is a second class of long-duration pauses that are not sequence-dependent, and these probably occur when the enzyme makes a copying error, misincorporating a noncomplementary nucleotide into the mRNA chain. These pauses are also associated with backtracking, and may reflect a "proofreading" activity whereby the enzyme slides backward and then (with the help of accessory factors) cleaves a short section from the end of the mRNA, removing the mistake before resuming elongation. Determining the reasons why RNAP pauses is critical for understanding how gene transcription is controlled. Single molecule experiments will be useful in this effort because the motion of unsynchronized molecules can be followed with very high resolution, in real time.

3.5
ATP Synthase

A rotary machine, F_1F_0-ATP synthase, is at the heart of energy metabolism in plants, animals, and photosynthetic bacteria, where its role is to replenish the cellular store of ATP. The importance of this job is obvious when one considers that a human body contains 100 g of ATP (0.25 moles), each molecule of which gets hydrolyzed 400 times a day to power various cellular tasks. An army of ATP synthase enzymes performs the $\sim10^{26}$ synthesis reactions required to regenerate the spent ATP. To do so, the enzymes tap into an electrochemical gradient, the protonmotive force, that exists across the inner mitochondrial membrane of animal cells, or across the thylakoid and plasma membranes of plants and eubacteria, respectively.

ATP synthase consists of two separate rotary motors, F_0 and F_1, that work together as a motor–generator pair. Normally, F_0 is the driving motor of the pair. As protons flow through it, down the electrochemical gradient, a portion of F_0 spins like a water wheel. The spinning wheel of F_0 supplies torque that rotates a shaft within the other motor, F_1, causing it to regenerate ATP from ADP and phosphate. The

motor–generator pair can also operate in reverse. In this case, ATP hydrolysis by the F_1 motor causes the shaft to rotate backward, which supplies a torque that spins the wheel of F_0 and pumps protons back up the electrochemical gradient.

F_1F_0-ATP synthase is comprised of eight different types of protein subunits (Fig. 1e). The water wheel, or "rotor" portion of F_0 is a ring of 12 identical subunits, c_{12}, that spin in the plane of the membrane relative to a "stator" composed of three other subunits, ab_2. F_1, is a donut-shaped structure made of three pairs of proteins, $(\alpha\beta)_3$, plus two additional proteins, γ and ε, which form the shaft that fits into the center of the donut. Clockwise (CW) rotation of the $\gamma\varepsilon$ shaft (as seen from the F_0 or membrane side) causes ATP synthesis to occur sequentially at three catalytic sites located symmetrically around the $(\alpha\beta)_3$ ring. Because isolated F_1 can function in reverse, catalyzing ATP hydrolysis and counter clockwise (CCW) rotation of the shaft, it is often referred to as F_1-ATPase. In the full, F_1F_0-ATP synthase enzyme, F_1 sits directly above F_0, with the $\gamma\varepsilon$ shaft of F_1 making a rigid connection to the c_{12} ring of F_0, and with the ab_2 stator of F_0 connecting to the $(\alpha\beta)_3$ donut of F_1. Normally, spinning of c_{12} drives rotation of $\gamma\varepsilon$ within $(\alpha\beta)_3$ and hence ATP synthesis.

The ring-shaped structures within F_1F_0-ATP synthase provided the first clues that rotation might be important to its function. As for the classic mechanoenzymes, proof of motion came when an *in vitro* assay was developed, allowing the rotation to be directly observed. F_1-ATPase molecules were attached sparsely to a surface, and the orientations of their $\gamma\varepsilon$ shafts were marked by attaching micron-long, fluorescent-labeled actin filaments. In the presence of ATP, the filaments rotated CCW, indicating shaft rotation. At very low concentrations of ATP, the filaments rotated in discrete steps, dwelling at well-defined orientations in between rapid, 120° reorientations. Rotation rates were one-third of the rate of ATP hydrolysis, implying that each 120° reorientation corresponds to hydrolysis of a single ATP. The filaments used to mark shaft orientation also supplied a drag force acting against the rotation. By calculating the drag on the filaments, F_1 was estimated to deliver 40 pN nm of torque during each 120° reorientation, giving 80 pN nm of mechanical work output per ATP hydrolysis. This is 100% of the available chemical energy, making F_1-ATPase one of the most efficient motors known.

F_1-ATPase is a relative newcomer to the molecular motor scene, but it is quickly becoming one of the best-understood examples of mechanochemical coupling. A series of experiments with single F_1 molecules has revealed that each 120° step occurs in two phases, or substeps, that correspond to particular biochemical transitions in the ATP synthesis reactions occurring at each of the three catalytic sites. Substeps were observed by marking the shaft orientation with 40-nm gold particles (Fig. 2d), rather than the much larger filaments used in earlier work. The small beads resulted in a lower drag force acting on the motor, allowing full speed rotation at 160 revolutions per second (Hz) (Table 1). Capturing the motion with a high-speed video camera showed a substep of 80 to 90° followed by one of 30 to 40°, underlying each of the 120° steps previously observed. Finally, simultaneous observation of shaft orientation and binding and release of a fluorescent nucleotide allowed these biochemical steps to be temporally correlated with the substeps. ATP binding to one of the catalytic sites (site 0) is concurrent with the 80 to 90° substep,

and the remaining 30 to 40° substep requires hydrolysis at the site that previously bound ATP (site − 1), and release of ADP from the remaining site (−2).

It is unknown whether the events seen during ATP hydrolysis by F_1 are simply the reverse of those occurring during synthesis, but several experiments confirm, at least, that the $\gamma\varepsilon$ shaft rotates in the opposite direction during synthesis. In one experiment, magnetic beads attached to the shaft were used to drive CW rotation in F_1-ATPase, in the presence of ADP and phosphate, and a luciferin–luciferase system that emits a photon upon reacting with ATP was used to verify synthesis. In another experiment, individual, fluorescent-labeled F_1F_0-ATP synthase complexes were embedded in liposomes. A pH difference was created across the membrane by rapid dilution, and rotation was recorded by fluorescence resonance energy transfer (FRET).

3.6
The Rotary Motor of Bacterial Flagella

Bacterial flagella are very different from the flagella and cilia of eukaryotic cells. Flagellated bacteria, such as *Escherichia coli*, swim by rotating a set of four corkscrew-shaped filaments (four, on average) that extend from the cell surface out into the surrounding medium. At the base of each filament is a large protein machine that drives filament rotation. This rotary motor, like F_1F_0-ATP synthase, is powered by the protonmotive force, but it is much larger and structurally more complex than F_1F_0-ATP synthase.

Bacteria swim to find food. They control their swimming behavior by altering the direction of rotation of their flagellar motors. When all four motors rotate CCW (as seen by an observer outside the cell), the cell swims steadily, or "runs," in a relatively straight line parallel to its long axis. When one or more motors rotates CW, the cell "tumbles," erratically moving in place and reorienting itself. The motors switch from CCW to CW at random, so that typical swimming involves runs that last ~1 s, interspersed with tumbles that last a few milliseconds. When the bacteria senses rising nutrient concentrations, it lengthens the runs by increasing the probability of CCW rotation. In this way, the cell moves, on average, toward the food.

The core of the bacterial flagellar motor is a stack of ring-shaped structures, 45 nm in diameter, embedded in the multilayered cell envelope (see Fig 1f). The rings are composed of 20 different types of proteins, but they are all thought to rotate together as a single unit, the "rotor." Rotation of the stack of rings is driven by a circular array of ≤ 16 "studs" that surround the base of the stack, and which are anchored to the framework of the cell wall. Each stud is composed of two MotA proteins (32 kDa), and one MotB protein (34 kDa). No high-resolution structures are available for the MotA/MotB complex, but both proteins span the cytoplasmic membrane, forming a transmembrane proton channel. MotB has a proton-acceptor site, and MotA contains a site that interacts with the base of the rotor. Protonation and deprotonation of MotB is thought to cause conformational changes in MotA, which probably binds and unbinds from the rotor, driving rotation and torque generation. The minimal torque-generating unit may be composed of two studs (i.e. four MotA subunits plus two MotB subunits). More detailed descriptions of the structure can be found in review articles cited in the bibliography.

Individual flagellar motors can be studied by attaching bacteria to a glass surface by one flagellum. The tail wags the dog in this tethered cell assay – with the flagellum anchored, rotation of the motor causes the whole cell body to spin. To spin the entire cell, the motor must overcome a large viscous drag, so it spins relatively slowly in this assay, at 10 Hz. But it produces an impressive 4600 pN nm of torque. When mutant cells lacking MotB are tethered, they are paralyzed and do not spin. Amazingly, these paralyzed cells can be "resurrected" by expression of MotB from an inducible gene. Resurrected cells begin to rotate within several minutes after induction of MotB expression, and their speed increases in a series of discrete jumps. Each jump in speed represents the incorporation of one additional torque-generating unit into the motor. As many as eight jumps can be seen, implying that the maximum number of torque generators is eight.

Each torque generator is itself a processive, high-duty ratio motor that moves along the surface of the rotor without detaching. The best evidence for processivity is that tethered cells in the resurrection experiment with just one torque-generating unit spin relatively smoothly, and do not freely undergo rotational Brownian motion. Evidently, a single unit is sufficient to prevent the motor from slipping, and each unit remains attached to the rotor during most, or all, of its mechanical cycle. There are twice as many studs (16) as torque-generating units (8), and one hypothesis is that each unit is a co-ordinated pair of studs, possibly moving in a hand-over-hand motion like that of kinesin or myosin V. A single unit is expected to move stepwise over the surface of the rotor, perhaps taking ~26 steps per revolution (the approximate number of subunits composing the base of the rotor), but such steps have not been directly observed. Dividing the maximum torque (4600 pN) by the number of torque-generating units (8), and by their distance from the axis of rotation (20 nm), shows that each unit generates considerable force, up to 29 pN, which is comparable to the stall force of RNAP. Their speed of motion over the rotor surface is also quite high. When the viscous load is minimal, the motor can rotate as fast as 300 Hz (Table 1). This translates into motion of the torque-generators at $38\,000$ nm s^{-1} over the rotor surface, which is >6-fold faster than muscle myosin, and similar to the speed of the fastest myosins (e.g. type XI, responsible for cytoplasmic streaming in algae).

The speed of rotation is proportional to the protonmotive force, as shown by wiring a cell to an external voltage source and watching an inert marker on the motor. To apply voltage, the cell body was drawn halfway into a micropipette, and the membrane permeabilized by chemical treatment. Estimates of the proton flux through the motor suggest that the motor is tightly coupled. Roughly 1200 protons flow through the motor during each complete revolution. By attaching a variety of different-sized latex beads to the filaments and adjusting the viscosity of the surrounding fluid, torque-speed relations have been measured over a wide range of speeds. Forward rotation under assisting torques, and backward rotation under torques above stall (>4600 pN), has been explored by using rotating electric fields or optical traps to apply torque in the tethered cell assay.

3.7
Polymers that Push and Pull

Actin filaments and microtubules are not just static polymers. In addition to their roles as structural cables and girders for maintaining cell shape, and as superhighways for mechanoenzymes to move along, these polymers are also dynamic machines that can produce force. In living cells, the cytoskeletal polymers are in a constant state of flux, and their growth and shrinkage is harnessed to drive many organelle and whole-cell movements. Crawling cells have a dense array of polymerizing actin filaments beneath their leading edge that pushes outward on the plasma membrane and causes protrusion. Similarly, the bacterial pathogen, *Listeria monocytogenes* is pushed by actin polymerization. The bacteria move in graceful arcs through the cytoplasm of a host cell, leaving "comet tails" of polymerized actin in their wake. During mitosis, chromosomes are pushed and pulled by dynamic microtubules whose ends are linked to specialized sites on the chromosomes, the kinetochores. Just before cell division, kinetochore-attached microtubules depolymerize, generating tension that pulls sister chromatids apart.

Both actin and microtubule filaments are composed of protein subunits arranged in a regular lattice. The monomeric form of actin, "G-actin," is a roughly spherical protein, 5 nm in diameter (45 kDa). Like a LEGO block, the surface of an actin monomer has several sites for attachment to other monomers. Each also has a cleft that binds an ATP molecule. Monomers assemble into a ropelike structure, "F-actin," with two strands, called *protofilaments*, that wind around each other with a helical period of 72 nm. The building blocks for microtubules are made of tubulin, a molecule that consists of two nearly identical 50-kDa proteins fused tightly to form a dimer, 8 nm in length. Each dimer has two sites that bind GTP. The dimers assemble into a hollow, tube-shaped structure, 25 nm in diameter, with 13 protofilaments that run parallel to the long axis of the tube. Both types of filaments have fast-growing "plus" ends, and slow-growing "minus" ends.

Nucleotide hydrolysis supplies energy that makes actin and tubulin polymers very dynamic. Actin monomers in solution bind ATP, and have high affinity for one another. After polymerization, the ATP is hydrolyzed and phosphate is released, leaving ADP trapped in the binding clefts of the monomers within a filament. The ADP-containing monomers have reduced affinity, so hydrolysis destabilizes the actin filament, promoting depolymerization. GTP hydrolysis by tubulin has a similar effect. Each tubulin dimer binds two molecules of GTP, one of which is hydrolyzed upon incorporation of the dimer into a microtubule filament. The GDP-containing tubulin dimers have reduced affinity for one another, which destabilizes the lattice and promotes depolymerization. Without hydrolysis, both polymers would simply grow until equilibrium was reached, when the subunit pool was spent. Hydrolysis keeps the filaments out of equilibrium, allowing coexistence of growing and depolymerizing filaments.

3.8
Microtubule Ends and Dynamic Instability

The dynamic behavior of microtubules can be directly observed *in vitro*. In the presence of GTP and pure tubulin, microtubule growth is interrupted by periods of rapid depolymerization. This "dynamic instability" requires GTP hydrolysis. Growth rates

are normal in the presence of the nonhydrolyzable GTP analog, GMPCPP, but the growth is uninterrupted. Subunit addition and removal occurs only at the ends of the filaments, which adopt different structures, depending on whether they are in a state of growth or depolymerization. The protofilaments that extend from growing ends are straight, forming sheets that are sometimes hundreds of subunits long. In contrast, the protofilaments at depolymerizing ends become highly curved, peeling away from the lattice. The ends of growing filaments are temporarily stabilized by a "cap" of GTP-containing subunits in which hydrolysis has not yet taken place. The transition between growth and depolymerization, called "catastrophe," is probably triggered when hydrolysis of the cap occurs before more GTP-containing subunits are added.

The curvature of protofilaments at the ends of depolymerizing microtubules, and the fact that the products of depolymerization are often curved oligomers, suggests a structural basis for the coupling of nucleotide hydrolysis and polymerization. Before hydrolysis, the GTP-containing dimers are probably straight, fitting snugly into the growing microtubule lattice. If the GDP-containing subunits are naturally curved, then they would be strained when trapped within the lattice. In this way, energy from hydrolysis could be stored within the lattice as mechanical strain.

3.9
Motility Assays with Cytoskeletal Filaments

Several *in vitro* experiments show that polymerization of pure actin or tubulin, without any additional proteins, can generate pushing force and do mechanical work. Polymerizing actin filaments inside liposomes causes distension of the liposomes, demonstrating that the growing filaments can push outward on the lipid bilayer. Likewise, microtubule filaments grown inside a small chamber can push against the chamber walls with enough force to buckle themselves (Fig. 1e). By analyzing the shapes of buckled filaments, the maximum pushing force of a single microtubule has been estimated at 4 pN.

These important experiments prove that growing filaments can push against an object, but they are incomplete models for the polymer-driven motility that occurs in cells. In cells, a variety of accessory proteins provide spatial and temporal control of filament dynamics, and couple the ends of growing and shrinking filaments to other structures to apply force. *Listeria* promote spatially localized actin polymerization with a nucleation factor, ActA. Likewise, kinetochores contain a host of proteins that modulate microtubule dynamics and maintain attachment to microtubule ends. Understanding the mechanisms of these accessory factors will be key to understanding how cells harness cytoskeletal filaments to produce motion and force.

In vitro motility assays that reconstitute force generation using dynamic filaments coupled to accessory proteins provide more realistic models for filament-based motility in cells. Shrinking microtubules can pull against microscopic beads when the beads are coated with proteins that maintain attachment to the depolymerizing filament ends. This motion is similar to the way chromosomes are pulled apart before cell division, and also to the way the mitotic spindle is positioned inside asymmetrically dividing yeast cells. In a reconstituted assay that closely mimics the motion of *Listeria*, beads coated with

ActA protein are pushed around by actin polymerization. The beads follow curved trajectories and leave comet tails of polymerized actin in their wake, just like the bacteria.

4
Conclusion

Motion is fundamental to life. Everyone is familiar with the macroscopic motion of muscle contraction. There are also exquisite motions taking place at the level of cells and molecules. The cells in our immune system crawl around our bodies and engulf invading bacteria. Cilia in our lungs beat to remove inhaled debris. In all these cases, the motion is generated by tiny protein machines, the molecular motors. Molecular motors are ubiquitous, and the list of known motors is growing. Besides the classic motors, myosin, kinesin, and dynein, and the cytoskeletal polymers, filamentous actin and microtubules, there are also protein machines at the heart of energy metabolism, and reading the genetic code. Studies of molecular motors, particularly *in vitro* work with single molecules, have revealed fascinating details about how they convert chemical energy into mechanical work.

While motors are arguably the most machinelike of the biological molecules, they are certainly not the only things inside living cells that remind us of man-made apparatus. The action at a distance that occurs within an allosteric enzyme, for example, is reminiscent of the push rods or levers inside an internal combustion engine. The large, multienzyme complexes that cells use to carry out sequences of reactions remind us of assembly lines. However, molecular motors are an important special case because the motions they produce are large enough to be directly measured. The study of motor proteins offers rare, direct access to address general questions about how a protein's structure dictates it's dynamics and function.

See also Nucleic Acid and Protein Single Molecule Detection and Characterization.

Bibliography

Books and Reviews

Berg, H.C. (2003) The rotary motor of bacterial flagella, *Annu. Rev. Biochem.* **72**, 19–54.

Block, S.M. (1995) Nanometres and piconewtons: the macromolecular mechanics of kinesin, *Trends Cell Biol.* **5**, 169–175.

Cameron, L.A., Giardini, P.A., Soo, F.S., Theriot, J.A. (2000) Secrets of actin-based motility revealed by a bacterial pathogen, *Nat. Rev. Mol. Cell Biol.* **1**, 110–119.

Howard, J. (2001) *Mechanics of Motor Proteins and the Cytoskeleton*, Sinauer Associates, Publishers, Sunderland, MA.

Inoue, S., Salmon, E.D. (1995) Force generation by microtubule assembly/disassembly in mitosis and related movements, *Mol. Biol. Cell* **6**, 1619–1640.

Salmon, E.D. (1995) VE-DIC light microscopy and the discovery of kinesin, *Trends Cell Biol.* **5**, 154–158.

Schliwa, M. (2003) *Molecular Motors*, Wiley-VCH, Weinheim.

Schliwa, M., Woehlke, G. (2003) Molecular motors, *Nature* **422**, 759–765.

Spudich, J.A. (2001) The myosin swinging cross-bridge model, *Nat. Rev. Mol. Cell Biol.* **2**, 387–392.

Vale, R.D., Milligan, R.A. (2000) The way things move: looking under the hood of molecular motor proteins, *Science* **288**, 88–95.

Yoshida, M., Muneyuki, E., Hisabori, T. (2001) ATP synthase–a marvellous rotary engine of the cell, *Nat. Rev. Mol. Cell Biol.* **2**, 669–677.

Primary Literature

Asbury, C.L., Fehr, A.N., Block, S.M. (2003) Kinesin moves by an asymmetric hand-over-hand mechanism, *Science* **302**, 2130–2134.

Berliner, E., Young, E.C., Anderson, K., Mahtani, H.K., Gelles, J. (1995) Failure of a single-headed kinesin to track parallel to microtubule protofilaments, *Nature* **373**, 718–721.

Block, S.M., Berg, H.C. (1984) Successive incorporation of force-generating units in the bacterial rotary motor, *Nature* **309**, 470–472.

Burgess, S.A., Walker, M.L., Sakakibara, H., Knight, P.J., Oiwa, K. (2003) Dynein structure and power stroke, *Nature* **421**, 715–718.

Case, R.B., Pierce, D.W., Hom-Booher, N., Hart, C.L., Vale, R.D. (1997) The directional preference of kinesin motors is specified by an element outside of the motor catalytic domain, *Cell* **90**, 959–966.

Diez, M., Zimmermann, B., Borsch, M., Konig, M., Schweinberger, E., Steigmiller, S., Reuter, R., Felekyan, S., Kudryavtsev, V., Seidel, C.A., Graber, P. (2004) Proton-powered subunit rotation in single membrane-bound F0F1-ATP synthase, *Nat. Struct. Mol. Biol.* **11**, 135–141.

Dogterom, M., Yurke, B. (1997) Measurement of the force-velocity relation for growing microtubules, *Science* **278**, 856–860.

Finer, J.T., Simmons, R.M., Spudich, J.A. (1994) Single myosin molecule mechanics: piconewton forces and nanometre steps, *Nature* **368**, 113–119.

Hancock, W.O., Howard, J. (1998) Processivity of the motor protein kinesin requires two heads, *J. Cell Biol.* **140**, 1395–1405.

Henningsen, U., Schliwa, M. (1997) Reversal in the direction of movement of a molecular motor, *Nature* **389**, 93–96.

Howard, J., Hudspeth, A.J., Vale, R.D. (1989) Movement of microtubules by single kinesin molecules, *Nature* **342**, 154–158.

Hua, W., Young, E.C., Fleming, M.L., Gelles, J. (1997) Coupling of kinesin steps to ATP hydrolysis, *Nature* **388**, 390–393.

Ishijima, A., Kojima, H., Funatsu, T., Tokunaga, M., Higuchi, H., Tanaka, H., Yanagida, T. (1998) Simultaneous observation of individual ATPase and mechanical events by a single myosin molecule during interaction with actin, *Cell* **92**, 161–171.

Lang, M.J., Fordyce, P.M., Block, S.M. (2003) Combined optical trapping and single-molecule fluorescence, *J. Biol.* **2**, 6.

Lister, I., Schmitz, S., Walker, M., Trinick, J., Buss, F., Veigel, C., Kendrick-Jones, J. (2004) A monomeric myosin VI with a large working stroke, *EMBO J.* **23**, 1729–1738.

Loisel, T.P., Boujemaa, R., Pantaloni, D., Carlier, M.F. (1999) Reconstitution of actin-based motility of Listeria and Shigella using pure proteins, *Nature* **401**, 613–616.

Lombillo, V.A., Stewart, R.J., McIntosh, J.R. (1995) Minus-end-directed motion of kinesin-coated microspheres driven by microtubule depolymerization, *Nature* **373**, 161–164.

Mallik, R., Carter, B.C., Lex, S.A., King, S.J., Gross, S.P. (2004) Cytoplasmic dynein functions as a gear in response to load, *Nature* **427**, 649–652.

Mehta, A.D., Rock, R.S., Rief, M., Spudich, J.A., Mooseker, M.S., Cheney, R.E. (1999) Myosin-V is a processive actin-based motor, *Nature* **400**, 590–593.

Meyhofer, E., Howard, J. (1995) The force generated by a single kinesin molecule against an elastic load, *Proc. Natl. Acad. Sci. U S A* **92**, 574–578.

Nishizaka, T., Oiwa, K., Noji, H., Kimura, S., Muneyuki, E., Yoshida, M., Kinosita, K. Jr. (2004) Chemomechanical coupling in F1-ATPase revealed by simultaneous observation of nucleotide kinetics and rotation, *Nat. Struct. Mol. Biol.* **11**, 142–148.

Noji, H., Yasuda, R., Yoshida, M., Kinosita, K. Jr. (1997) Direct observation of the rotation of F1-ATPase, *Nature* **386**, 299–302.

Ray, S., Meyhofer, E., Milligan, R.A., Howard, J. (1993) Kinesin follows the microtubule's protofilament axis, *J. Cell Biol.* **121**, 1083–1093.

Rice, S., Lin, A.W., Safer, D., Hart, C.L., Naber, N., Carragher, B.O., Cain, S.M., Pechatnikova, E., Wilson-Kubalek, E.M., Whittaker, M., Pate, E., Cooke, R., Taylor, E.W., Milligan, R.A., Vale, R.D. (1999) A structural change in the kinesin motor protein that drives motility, *Nature* **402**, 778–784.

Ryu, W.S., Berry, R.M., Berg, H.C. (2000) Torque-generating units of the flagellar motor of Escherichia coli have a high duty ratio, *Nature* **403**, 444–447.

Sakakibara, H., Kojima, H., Sakai, Y., Katayama, E., Oiwa, K. (1999) Inner-arm dynein c

of Chlamydomonas flagella is a single-headed processive motor, *Nature* **400**, 586–590.

Schafer, D.A., Gelles, J., Sheetz, M.P., Landick, R. (1991) Transcription by single molecules of RNA polymerase observed by light microscopy, *Nature* **352**, 444–448.

Schnitzer, M.J., Block, S.M. (1997) Kinesin hydrolyses one ATP per 8-nm step, *Nature* **388**, 386–390.

Shaevitz, J.W., Abbondanzieri, E.A., Landick, R., Block, S.M. (2003) Backtracking by single RNA polymerase molecules observed at near-base-pair resolution, *Nature* **426**, 684–687.

Svoboda, K., Block, S.M. (1994) Force and velocity measured for single kinesin molecules, *Cell* **77**, 773–784.

Svoboda, K., Schmidt, C.F., Schnapp, B.J., Block, S.M. (1993) Direct observation of kinesin stepping by optical trapping interferometry, *Nature* **365**, 721–727.

Vale, R.D., Reese, T.S., Sheetz, M.P. (1985) Identification of a novel force-generating protein, kinesin, involved in microtubule-based motility, *Cell* **42**, 39–50.

Vale, R.D., Funatsu, T., Pierce, D.W., Romberg, L., Harada, Y., Yanagida, T. (1996) Direct observation of single kinesin molecules moving along microtubules, *Nature* **380**, 451–453.

Veigel, C., Wang, F., Bartoo, M.L., Sellers, J.R., Molloy, J.E. (2002) The gated gait of the processive molecular motor, myosin V, *Nat. Cell Biol.* **4**, 59–65.

Yasuda, R., Noji, H., Kinosita, K. Jr., Yoshida, M. (1998) F1-ATPase is a highly efficient molecular motor that rotates with discrete 120 degree steps, *Cell* **93**, 1117–1124.

Yasuda, R., Noji, H., Yoshida, M., Kinosita, K. Jr., Itoh, H. (2001) Resolution of distinct rotational substeps by submillisecond kinetic analysis of F1-ATPase, *Nature* **410**, 898–904.

Yildiz, A., Tomishige, M., Vale, R.D., Selvin, P.R. (2004) Kinesin walks hand-over-hand, *Science* **303**, 676–678.

Yildiz, A., Forkey, J.N., McKinney, S.A., Ha, T., Goldman, Y.E., Selvin, P.R. (2003) Myosin V walks hand-over-hand: single fluorophore imaging with 1.5-nm localization, *Science* **300**, 2061–2065.

Yin, H., Wang, M.D., Svoboda, K., Landick, R., Block, S.M., Gelles, J. (1995) Transcription against an applied force, *Science* **270**, 1653–1657.

Mucosal Vaccination

W. Olszewska and Peter J. M. Openshaw
Imperial College, Paddington, London, UK

Encyclopedia of Molecular Cell Biology and Molecular Medicine, 2nd Edition. Volume 8
Edited by Robert A. Meyers.
Copyright © 2005 Wiley-VCH Verlag GmbH & Co. KGaA, Weinheim
ISBN: 3-527-30550-5

Keywords

Adjuvant
(Latin: *adjuvere* to help) substances added to vaccines to increase immune responses.

Antigen
Foreign substance recognized by immune system.

B Cells
Lymphocytes that make immunoglobulin (antibody); can also present antigens.

T Cells
Lymphocytes that require the thymus for development.

Th1 Cells
Typically make interferon (IFN)-gamma and tumor necrosis factor (TNF) and activate macrophages to kill intracellular pathogens and switch on IgG1 production from B cells.

Th2 Cells
Make cytokines such as IL-4, IL-5, and IL-13, which promote growth, activity, and survival of eosinophils and switch B cells to make IgE.

Chemokines
Chemoattractive proteins made by many cell types that cause cell recruitment to inflamed tissues.

Cytokines
Small proteins produced by T cells that act as signals to other cells of the immune system or structural cells (e.g. IL-1, IL-3, IL-4, IL-5, IL-10, and IFN).

Lymphocytes
White blood cells that concentrate in lymph nodes, spleen, and so on.

Vaccine
A substance administered to trigger memory immune responses that protect against disease.

The immune system faces a dilemma. It must tolerate benign commensal organisms and antigens present in food and air, while rapidly mounting vigorous responses to the plethora of pathogens that enter via mucosal surfaces. The existing mucosal vaccines against poliomyelitis, influenza, and measles were developed on a largely empirical basis against transient self-limiting infections that themselves induce lifelong immunity.

In this article, we review the successes and failures of mucosal immunization and the opportunities for exploiting developments in immunology to create new and effective mucosal vaccines. The challenge now is to exploit new information about immunoregulation to develop effective vaccines against more subtle agents that circumvent lifelong immunity, recur, or persist.

1
Introduction

The mucous membranes lining the lung, gut, and urogenital tract present unique problems to the host defense systems. They have specialized functions that necessitate close contact with the environment, yet most common infections enter by these routes. Immune defenses at mucosal surfaces therefore have to defend against pathogens while being tolerant of nonthreatening substances in food and inhaled air. Both innate and acquired mechanisms of immunity are involved in preventing microbial entrance and spread. Mechanisms of acquired immunity combine humoral (production of antigen-specific secretory IgA, S-IgA) and cell-mediated immune responses (Fig. 1).

Mucosal immunization has the potential to induce protective immunity against infectious diseases or to elicit antigen-specific tolerance. The mucosal immune system has to be appropriately activated to achieve effective protection against colonization and invasion by infectious agents at mucosal surfaces. Although administration of vaccines directly to mucosal sites has many advantages, mucosal immunity can be also achieved by other routes of antigen delivery. However, in this review we focus primarily on the mucosal route of vaccination.

Because of the low absorption efficiency of mucosally delivered vaccines, almost all current vaccines are administered parenterally. In addition, suboptimal immune responses are frequently induced by mucosal immunization and the use of mucosal adjuvants is required. As a result, development of successful mucosal vaccines depends largely on improvements to mucosal delivery systems and on the discovery of new and effective mucosal adjuvants (Fig 2).

2
Goals of Mucosal Vaccination

Most environmental pathogens enter the body through the mucosal membranes of the intestinal, respiratory, or genital tract. Some only replicate in the mucosa, but others use the mucosa to gain entry before dissemination to other sites. For purely mucosal infections (e.g. those caused by most common cold agents and infective diarrheal organisms), prevention of initial

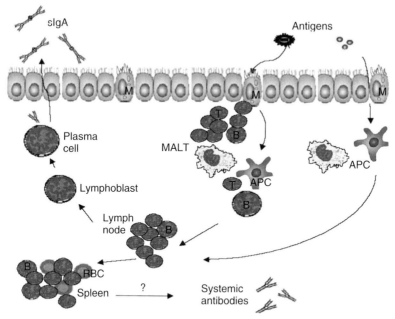

Fig. 1 Schematic representation of cells involved in immune reactions after mucosal challenge. APC, antigen-presenting cell; M, microfold epithelial cell; T, T cell; B, B cell; RBC, red blood cell.

colonization is an essential but difficult aim of vaccination; when dissemination or toxin production is essential to disease pathogenesis, prevention of surface invasion may not be necessary.

Mucosal vaccination is an attractive administration route for mass vaccination, as it does not require trained medical personnel and does not involve needles and syringes These factors affect the costs of vaccination and the risk of transmitting blood-borne infections (HIV or HepB) making mucosal vaccine particularly attractive for use in developing countries (Table 1).

New or improved vaccines are needed for a wide range of mucosal infections, including respiratory tract infections caused by *Mycobacterium* spp, *Mycoplasma pneumoniae*, influenza virus, rhinovirus, coronavirus, adenovirus, human metapneumovirus, and respiratory syncytial virus (RSV); urogenital tract infections caused by *Chlamydia*, HIV, *Neisseria gonorrhoeae*, *Treponema pallidum*, and herpes simplex virus (HSV); and gastrointestinal infections caused by *Escherichia coli*, *Salmonella*, *Shigella* spp, *Helicobacter pylori*, *Vibrio cholerae*, *Campylobacter jejuni*, *Clostridium difficile*, and rotaviruses. The challenge is to design vaccine preparations that induce neutralizing immunity to the pathogen or their toxins, preventing their attachment to mucosal surfaces, tissue invasion, and spread (Fig 3).

Induction of specific secretory IgA (S-IgA) is an essential aim of mucosal vaccination. Locally produced S-IgA is considered to be among the most important protective humoral immune factors and constitutes over 80% of all antibodies produced in mucosa. In humans, S-IgA

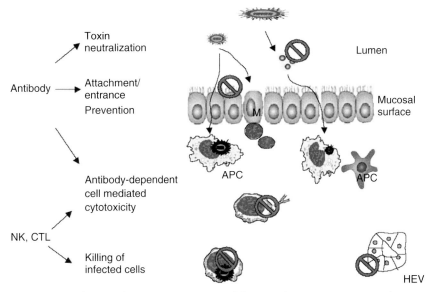

Fig. 2 Goals of mucosal immunization. Successful mucosal immunization may induce humoral and cellular immune responses. Possible roles of both are shown. M, microfold epithelial cell; HEV, high endothelial venule; APC, antigen-presenting cell; the Ø symbol indicates possible sites of immune intervention.

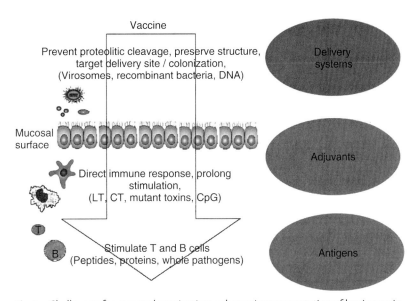

Fig. 3 Challenges for mucosal vaccination: schematic representation of key issues in mucosal vaccine development.

Tab. 1 Comparison of mucosal and parenteral delivery of vaccines.

Vaccination	Mucosal	Parenteral
Administration	Easy, self administration	Need of trained personnel
Risks	Nonaccurate dosing	Infection (HIV, Hepatitis B)
Antigen dose	Usually high	Low
Formulation	Need of mucosal adjuvant	Mixed with Alum
Immune responses	Stimulates mucosal immunity but may induce tolerance	Rarely induces mucosal responses but good systemic antibody and T cell responses
Use in human	Only few vaccines	Most of the vaccines

antibodies are usually dimeric whereas serum IgA is mainly monomeric. Antibodies secreted by mature plasma cells in various compartments of the common mucosal immune system are able to interact with invading pathogens, inhibit their attachment, and form immune complexes with potentially harmful molecules (immune exclusion). S-IgA may neutralize viruses and directly participate in antibody-dependent cell-mediated cytotoxicity (ADCC) in collaboration with macrophages and lymphocytes. However, natural killer cells and specific cytotoxic T cells are also key defenses in preventing infection with intracellular pathogens (Fig. 2). In testing vaccine efficacy, both antibody levels and cellular immunity should therefore be monitored.

3
Benefits of Mucosal Vaccination

3.1
Main Features of the Common Mucosal Immune System

A key issue in mucosal immunity is the necessity to distinguish between normal antigens, which are not harmful, and those that belong to dangerous pathogens.

The discrimination between hazardous and nonhazardous antigens is determined by activation by pattern-recognition receptors (including Toll-like receptors), which may be displayed in intestinal crypts or in the cytoplasm of mucosal cells, and are therefore inaccessible to nonhazardous commensals. The anatomical placement of pattern-recognition receptors in inaccessible sites appears to be an important factor in allowing discrimination between hazardous and nonhazardous antigens.

The mucosal immune system is highly adapted toward tolerance, the breakdown of which can result in disease (e.g. celiac disease). Mucosal exposure frequently induces regulatory T cells, of which there are several distinct types. First, Th3 cells produce transforming growth factor beta (TGF-beta); these are frequently induced by repeated feeding of low doses of antigen orally. Second, Tr1 cells are a subset of CD4 T cells, which produce IL-10, the production of which is promoted by IL-15 and Type 1 interferon. Third, CD4+ CD25+ cells are potent regulators of autoreactivity *in vivo* and can also be induced by tolerogenic oral feeding. Fourth, CD8 T cells often have a regulatory role in the gut, but their exact functions have not been clearly defined. In addition, there is mounting evidence that gamma delta T cells may, under some conditions, also promote a tolerogenic state after oral feeding. All these subsets are readily induced by oral administration of antigen, but induction must be avoided if mucosal vaccination is to be successful.

3.2
Distinctive Characteristics of Mucosal Immunity

In mammals, immune responses at mucosal surfaces are provided by a defence system known as *mucosa-associated lymphoid tissue* (MALT). It comprises an integrated network of cells and molecules anatomically grouped and functionally divided into sites of antigen uptake and effector function. Mucosal immune responses are triggered in specialized zones that sample foreign material from the epithelial surface. The main inductive site of the upper respiratory tract is in the nasal-associated lymphoid tissue (NALT), and in the gastrointestinal tract it is the Peyer's patch (PP), the appendix, and the solitary lymphoid nodules, collectively called the *gut-associated lymphoid tissue* (GALT).

The respiratory epithelium contains four different cell types: alveolar macrophages, dendritic cells, M cells, and intraepithelial lymphocytes. The latter are relatively scarce in the respiratory tract. M cells are associated with lymphoid structures and differ in morphology from absorptive cells by their short microvilli, small cytoplasmic vesicles, and few lysosomes. The M cells selectively absorb antigen by endocytosis or pinocytosis and direct it to professional APC (macrophages, dendritic cells, B lymphocytes) or directly process and present antigen to T cells (Fig. 1). In the gut, specialized M cells

are thought to provide the main route by which complex antigens gain access to the immune system. Lymphocytes primed in the Peyer's patches express $\alpha_4\beta_7$ integrin, which binds to the mucosal adressin cell-adhesion molecule 1 (MAD-CAM1), which is expressed at high levels by the vascular endothelium in mucosal surfaces. Gut-derived T cells also express the chemokine receptor CCR9, which allows them to respond to the chemokine CCL25, which is expressed selectively by small bowel epithelial cells. By contrast, T cells that are primed peripherally typically display the $\alpha_4\beta_1$ integrin and CCR4 and so do not migrate or respond in mucosal sites. This selective expression explains why mucosal vaccination is often required to protect against mucosal infections, and why peripheral administration of vaccine antigens is often ineffective against mucosal infections. The establishment and maintenance of the immune response requires close cooperation between many cell types. Activated T cells help B cells to develop into plasma cells, a process that depends on the nature of the antigen and on the cytokines produced by T helper cells. TGF-β, IL-10, and IL-4 promote the switch from IgM to IgA production from stimulated B cells. Precursors of mucosal plasma cells derive from lymphoepithelial structures, mature in the regional lymph nodes, and enter the circulation via the thoracic duct. Then they can seed the lamina propria of distant mucosal sites (e.g. intestines, respiratory tract, genital tract, salivary gland, etc.), where they differentiate. In this way, antibodies can appear at different mucosal sites elsewhere within the common mucosal immune system. Natural killer cells may be also present in mucosal tissues and, together with

CTLs, are important for eliminating virus-infected cells.

There is strong evidence that various mucosal sites of the immune system can communicate. For instance, within 24 h of adoptive lymphocyte transfer, mesenteric lymph node (MLN) cells are found in recipient gut, cervix and vagina, uterus, and mammary glands. The ability of cells to migrate between mucosal zones has been extensively investigated. Thus, intranasal immunization may induce specific immunity in intestine, and boosted levels of IgA have been reported in the intestine after intranasal (rather than oral) administration. In these experiments, the relatively poor induction of gut immunity by oral administration could be due to degradation of antigen and dilution in the contents of the gastrointestinal tract.

Several examples show that intranasal immunization can result in efficient immunity at distant mucosal sites. For instance, intranasal administration of a recombinant HIV envelope protein formulated in CTB-associated GM1 lipid vesicles enhances mucosal IgA antibody responses in nasal and gut tissues. Administration of a DNA vaccine for herpes simplex induces antigen-specific cellular and secretory IgA responses in the gut, vagina, and oral cavity after intradermal, intraperitoneal, intravaginal, intranasal, or oral immunization. Taking advantage of the functional integrity of the common mucosal system, vaccines easily administrated into the nose may in the future include those against herpes simplex virus or HIV. For instance, recombinant adenovirus used to deliver antigens derived from HSV and HIV has been demonstrated to induce protective immunity in mice vaginal mucosa, particularly when cholera toxin was included in the vaccine preparation.

3.3
Multivalent Mucosal Vaccines

Delivery systems that allow simultaneous delivery of several antigens derived from one or more pathogens may be used to design a multivalent vaccine. Such a vaccine potentially may reduce the number of administrations needed and increase protection in the populations studied.

For example, RSV and human parainfluenza represent two of the most important viral agents of pediatric respiratory tract disease worldwide. Studies are under way that use recombinant bovine/human parainfluenza virus type 3 (rB/HPIV3), with bovine F and HN genes replaced with their HPIV3 counterparts and expressing the major surface antigens of respiratory syncytial virus. Both recombinant viruses were shown to replicate efficiently in the respiratory tract of hamsters and induced serum antibody titers similar to those induced by RSV or HPIV3 infection. Immunization of hamsters with rB/HPIV3-G1, rB/HPIV3-F1, or a combination of both viruses resulted in high level of resistance to challenge with RSV or HPIV3 28 days later. Schmidt et al. described a similar strategy for intranasal immunization against RSV subgroups A and B and human parainfluenza virus type 3 by using a live cDNA-derived vaccine in monkeys.

3.4
Edible Vaccines

Transgenic plant technology allows the production of large quantities of transgenic antigen from pathogenic microorganisms at low cost and would have particular benefits in underresourced areas. The first edible vaccine was obtained by inserting the gene encoding hepatitis B surface antigen into tobacco plants, resulting in an antigen very similar to that obtained from recombinant yeast. Similar methods have been used to make vaccines containing genes of *E. coli* heat-labile enterotoxin B (LT-B), rabies virus glycoprotein, F protein of RSV, Norwalk virus capsid protein, and *V. cholerae* (CT-B) in potatoes, tomatoes, and other plants. Materials from recombinant plants have been shown to be immunogenic in feeding experiments in animals, inducing systemic and mucosal responses, and promising clinical trials show that specific anti-LT-B responses can be demonstrated in humans given transgenic potatoes expressing LT-B. The possibility of using transgenic plants for the production of specific immunoglobulins for passive immunotherapy is also being explored, but the commercial viability of these technologies is unproven.

Oral administration could theoretically induce tolerance to pathogens, resulting in serious adverse consequences. Fear of this effect has limited the clinical trials of oral vaccination against hepatitis B with edible potatoes expressing hepatitis surface antigens to preimmune individuals.

3.5
Overcoming Preexisting Immunity or Tolerance

Immunity to vaccinia potentially limits the utility of vaccinia vectors in people previously immunized against smallpox. However, the studies of Belyakov et al. suggest that modified vaccinia Ankara (MVA) expressing HIV gp160 could successfully immunize via the mucosal route and induce potent specific systemic humoral and CTL responses in vaccinia-immune mice. It therefore seems possible that immune responses at mucosal sites may not be prevented by prior systemic immunization

and that this mucosal naiveté may be exploited in generating immune responses in previously vaccinated individuals.

Another challenge to immunization against RSV and measles is the inefficiency of vaccination in the presence of specific maternal antibody. It appears that mucosal immunization may overcome this difficulty. Mutwiri et al. concluded from studying immune responses in neonatal lambs that enteric immunization with a human adenovirus vector may be an effective approach for inducing both mucosal and systemic immune responses in neonates. Similarly, in the cotton rat model of measles mucosal vaccination with vesicular stomatitis virus expressing the hemagglutinin of measles virus induces seroconversion in the presence of maternal antibodies and leads to protection against measles challenge. On the basis of experience with highly effective live polio vaccine given at birth, it has been proposed that mucosal immunization may be safer and more effective than any other route of vaccination in young children.

4
Challenges for Mucosal Immunization

4.1
Mucosal Delivery Systems

4.1.1 Live Bacterial Vectors

Commensal flora as expression vectors
Genetic manipulation of normal surface bacteria is attractive in that no pathogen is deployed in vaccine production and the bacteria themselves can exhibit probiotic properties. However, the risk is that such bacteria will be tolerated and thus no immune response will be generated. Considerable work has been done

with *Lactococcus lactis*, some strains of *Lactobaccillus*, *Staphylococcus carnosus*, and *Streptococcus gordonii*. These bacteria are noninvasive or commensal organisms in healthy people. Internal adjuvanticity is provided by peptidoglycan, and lactobacilli tagged with green fluorescent protein are actively taken up by APC after intranasal administration, suggesting that antigens would be processed and presented. Recombinant antigens that have been expressed in commensal bacteria include the V3 domain of HIV-1 gp120, fusion and hemagglutinin from measles virus, fragment B of diphtheria toxin, peptides from *Plasmodium falciparum*, and epitopes from RSV.

Pathogens as expression vectors Many bacterial pathogens naturally invade via mucosal surfaces and induce strong mucosal immunity. Attenuated *Salmonella* has been extensively studied as a vector both for oral and intranasal delivery. Interestingly, i.n. vaccination induces better systemic and local immune responses to inserted antigens (e.g. hepatitis B core antigen) than oral or rectal immunization. Insertion of antigens from *Helicobacter pylori* into *Salmonella*, for instance, results in induction of specific CD4+ T cells producing IFN (interferon) and IL-10 in a mouse model, as well as good protection after just two i.n. doses of vaccine.

Attenuated strains of *Shigella* are also promising live bacterial vectors. In animal models, they were shown to induce specific serum antibody responses (IgG and IgA) to inserted antigens after i.n. immunization.

Bacille Calmette-Guérin (BCG) is the first attenuated vaccine introduced to humans and still is the most widely used vaccine in the world, since it is the only one available against tuberculosis. Since BCG induces cellular immunity and can accommodate foreign epitopes, it may work as a successful vector when specific cellular responses are needed. The safety of the vaccine given at birth suggests its relevance for immunizations is required very early in life. Candidates would therefore include measles and RSV vaccine. Indeed, recombinant BCG producing measles nucleoprotein provided protection against virus challenge in intranasally immunized infant rhesus macaques. Other examples of heterologous antigens expressed in BCG for mucosal vaccination include HIV, human papilloma virus, *Schistosoma haematobium*, *Plasmodium yoelii*, and *Toxoplasma gondi*.

Recently, attenuated *Bordetella pertussis* emerged as a bacterial vector for i.n. vaccination. Deletion of genes coding for pertussis toxin diminished the virulence of the bacteria but preserved their ability to colonize mucosal sites. Interestingly, the immunogenicity to the mayor antigen, filamentous hemagglutinin (FHA) was increased, as well as that to heterologous antigen fused to FHA. Additionally, significant protection was observed in a mouse model of *Schistosoma mansoni* infection following single i.n. immunization.

4.1.2 Virosomes

Cusi et al. investigated the efficacy of a vaccine composed of the RSV fusion protein associated with influenza virosomes (IRIV), which was administered intranasally together with *E. coli* heat-labile toxin (LT). After an intramuscular "priming" with influenza virus vaccine, mice were i.n. immunized with RSV-F/IRIV + LT or with RSV-F + LT or IRIV + LT. The results showed that mice immunized with RSV-F + LT developed Th2 type responses and that virosomal delivery greatly potentiated immune responses in animals. All mice immunized with RSV-F/IRIV + LT developed a balanced Th1/Th2 cytokine profile with mucosal IgA and high levels of serum IgG. More importantly, histological analysis of lung tissue of RSV-challenged mice did not reveal vaccine-enhanced pulmonary eosinophilia.

4.1.3 Mucosal DNA Vaccines

The immunogenic potential of DNA vaccination has been extensively tested in animal models. DNA vaccination has many theoretical advantages, including ease of production, stability of vaccine preparations, and the ability to induce both antibody and cell-mediated immune responses. The drawbacks are that success in animal models is poorly predictive of outcome in human studies and that long-term follow-up studies are necessary to assess the risk of vaccine involvement in autoimmune diseases or immune disregulation.

DNA is not usually taken up in non-degraded form from mucosal surfaces. However, DNA immunization in mice and chickens has been shown to induce antibodies to influenza nucleoprotein and to trigger both mucosal and systemic cellular protective immune responses. In a recent study, Sasaki et al. described immune responses induced by intranasally and intramuscularly delivered DNA encoding HIV-1 proteins in mice. Both routes produced similar levels of cell-mediated immunity, but intranasal immunization induced higher levels of intestinal S-IgA than intramuscular immunization, and the adjuvant QS21 enhanced both intranasal and intramuscular immune responses.

Encapsulation of DNA into microparticles prevents degradation of DNA and enhances immunogenicity; therefore, it is an attractive delivery system for a mucosal DNA vaccine. Such a vaccine, designed using rotavirus VP6 DNA encapsulated in PLG microparticles, was shown to be effective in inducing systemic and mucosal immunity after oral administration to mice.

4.2
Mucosal Adjuvants

Many substances are known to have adjuvant properties; however, the majority are used for parenteral immunization and only a few are suitable for the mucosal route. Alum, the only universally licensed adjuvant for use in humans, is not a suitable adjuvant for mucosal immunization. The choice of an appropriate adjuvant for mucosal vaccination is very important to its success. Many mucosal adjuvants are based on bacterial toxins and their derivatives, CpG-containing DNA, and various cytokines and chemokines. Since most antigens are poorly immunogenic when introduced via the mucosal route if no adjuvant is added, induction of tolerance is likely.

4.2.1 Biodegradable Polymeric Particles

Micro- and nanoparticles may be used to encapsulate vaccine antigens. Many such polymers have the advantages of being biodegradable and at the same time of protecting the antigen from premature degradation. Slow degradation of polymers controls the release of entrapped antigen; therefore, such vaccines can induce immune responses over prolonged periods of time. Poly(D,L-lactide-co-glycolide) (PLG) polymer is probably the most extensively studied polymer because of its safety record in humans. The hydrolysis of PLG polymer to release encapsulated antigen is controlled by the polymer composition and its molecular weight. PLG particles may be suitable mucosal vaccine carriers, as shown by using fimbriae from

B. pertussis encapsulated in PLG particles for oral delivery to mice. The technique was able to induce serum and mucosal antibody responses in mice as well as to protect against live bacterial challenges. Baras et al. demonstrated recently the feasibility of long-term *in vivo* release of antigens from PLG particles. They showed that a single nasal or oral immunization of glutathione S-transferase from *S. mansoni* in PLG microparticles could induce a long-lasting antigen-specific antibody response in mice, with a peak at 9 to 10 weeks after immunization.

4.2.2 Bacterial Toxins

Bacterial toxins have been used for a long time as adjuvants in experimental models, and some chemically detoxified toxins have been employed to prevent bacterial infectious diseases (e.g. formalin inactivation of *Corynebacterium diphtheriae* or *Clostridium tetani* exotoxins). Although bacterial toxins possess excellent adjuvant properties, they have been prohibited from wide use as adjuvants in humans because of their high toxicity. Two bacterial toxins have found particular attention for application as mucosal adjuvants: cholera toxin (CT) produced by *V. cholerae* and heat-labile enterotoxin (LT) produced by *E. coli* have similar structures, since both contain subunits A and B and the toxicity originates from the A subunit, which catalyzes ADP-ribosylation of the stimulatory GTP-binding proteins on the surface of the epithelial cells and raises the intracellular levels of cAMP. This then leads to secretion of water and electrolytes into the mucosal lumen. Today it is possible to obtain detoxified derivatives by mutagenesis of the toxin genes. With this technology, the genes are modified in such a way as to encode different amino acid(s) that are no longer able to function

in enzymatic activity. Such inactivated substances are safe and in the future could replace toxoids in existing vaccines as well as be used as mucosal adjuvants in new vaccination strategies.

Several mutant toxins have already been described; among them are mutations of the heat-labile toxin of *E. coli*: LTK63, LTR72, and LTG192. All of them have been studied in detail for their ability to induce systemic and local immune responses to coadministered antigen. LTR72 was the strongest adjuvant, as compared with the fully nontoxic LTK63 mutant, and showed only 0.6% of the enzymatic activity of the wild-type LT. LTK63 was shown to be 100 000-fold less toxic but 20 times less effective than LT. Moreover, CD4+ lymphocytes from animals immunized with ovalbumin together with LTR72 exhibited very strong proliferative responses, similar to those induced by wild type LT. LTK63 was also tested in the murine measles model, and mucosal coimmunization with the mutant and a synthetic peptide representing a CTL epitope from measles N protein was very effective in *in vivo* priming of peptide-specific and measles virus–specific CTL responses. Similarly, the addition of LTK63 to peptide RSV vaccine induced strong CTL responses and protection after intranasal administration into a mouse. Compared with administration of oral influenza vaccine alone, coadministration of vaccine with LTG192 provided enhanced protection from infection in the upper and lower respiratory tract, equivalent to and at similar doses as that obtained with wild-type LT. The mutant toxin augmented virus-specific IgG and IgA responses in serum, lung, and nasal washes and also the numbers of virus-specific antibody-forming cells in spleen, lung, and Peyer's patches in a manner comparable to that of wild-type LT.

Studies conducted in experimental animals have given promising results with respect to immune responses developed after intranasal vaccination by a number of delivery systems. Cholera toxin B given intranasally with measles virus stimulated systemic neutralizing antibodies, and intranasal immunization with a chimeric synthetic peptide containing two T-helper epitopes (MVF: aa 288–302) and one B-cell epitope (MVF: aa 404–414) produced both systemic and mucosal antibody responses that conferred protection against encephalitis after infection with neuroadapted measles virus.

A novel nontoxic form of chimeric mucosal adjuvant that combines the A subunit of mutant cholera toxin E112K with the pentameric B subunit of heat-labile enterotoxin from enterotoxigenic *E. coli* was constructed. Nasal immunization of mice with tetanus toxoid (TT) plus the new adjuvant elicited significant TT-specific immunoglobulin A responses in mucosal compartments and induced high serum immunoglobulin G and immunoglobulin A anti-TT antibody responses.

The suitability of CT and LT and their mutants in human vaccinology remains to be determined. Several clinical trials did not report side effects; however, at least one influenza vaccine coadministered with LT had to be withdrawn from study due to several instances of facial paresis.

4.2.3 CpG Oligodinucleotides

Synthetic oligodeoxynucleotides (ODN) that contain unmethylated CpG motifs (CpG ODN) are also novel candidates as adjuvants for mucosal immunization. Initially, it was reported that these motifs could induce *in vitro* production of IL-6 and IFNγ by CD4+ T cells, IL-6 and IL-12 by B cells, and IFNγ by NK cells. Such properties led to the use of CpG ODNs as an adjuvant in several experimental models and indeed, work published so far supports the view that Th1-type responses dominate after CpG coadministration with an immunogen. However, some authors indicate that these responses are very much dependent on the age of the primed animals, the route of antigen delivery, or the nature of the antigen. The potential of CpG motifs as adjuvants for delivery via mucosal surfaces is particularly promising, as can be judged from several recent publications.

4.2.4 Cytokines and Chemokines

Cytokines can be used as mucosal adjuvants, either added directly to a vaccine preparation or encoded in DNA. Recent advancements in cytokine applications are summarized in Table 2. As an example, combination of IL-1 and IL-12 or IL-18 and GM-CSF creates a potent stimulating environment for mixed Th1/Th2 immune responses with the effect of IFNγ, IgA, and induction of CTL responses. IL-2 and IL-6, coexpressed in recombinant bacteria, significantly increase specific antibody in serum as well as in mucosal IgA subsequent to i.n. inoculation in mice. Likewise, some chemokines are being explored for use to potentiate mucosal immunity, as reported by Lillard et al. and Eo et al. Particularly promising is RANTES, which is a chemoattractant for monocytes, T cells, and NK cells, and has strong ability to induce Th1 responses, particularly CTLs. Moreover, nasal coadministration of RANTES with a protein antigen was demonstrated to augment Th1 and Th2 local and systemic immune responses.

4.2.5 Saponins

Saponins purified from the bark of the tree *Quillaja saponaria molina* and their

Tab. 2 Examples of cytokines and chemokines coadministered with mucosal vaccines.

Cytokine/ chemokine	Delivery system	Model	Route of delivery	Effect
GM-CSF	Adenovector	Mouse	Intranasal	Increased IgG2a and IgG1, elevated levels of IFN-γ, TNF-α, and IL-10
	Plasmid DNA	Mouse	Rectal/ vaginal	Increased serum IgG, enhanced mucosal and fecal IgA
IL-2	Plasmid DNA- encapsulated in PLG microparticles	Mouse	Oral	Enhanced CTL responses
	Recombinant *L. lacti*	Mouse	Intranasal	Increased antibody titers
	Liposomes s	Mouse	Intranasal	Increased *P. aeruginosa* polysaccharide-specific pulmonary plasma cells, reduced mortality from pneumonia
IL-4	Plasmid DNA	Mouse	Rectal/ vaginal	Increased antibody levels; decreased CTL activity
IL-6	Recombinant *L. lactis*	Mouse	Intranasal	Increased antibody titers
IL-12	Liposomes	Mouse	Oral	Shift to IgG2a and IgG3, decreased IgE Abs, enhanced serum IFN-γ
	Plasmid DNA	Mouse	Rectal/ vaginal	Decreased antibody levels, enhanced CTL activity
IL-1alpha, IL-12, and IL-18	Peptide	Mouse	Intranasal	Specific anti-HIV IgA in saliva, fecal extracts, and vagina
RANTES	Peptide	Mouse	Intranasal	Augmented Th1 and Th2 responses
IL-10	Plasmid DNA	Mouse	Intranasal	Diminished Ag-induced delayed type hypersensitivity, production of Th1 cytokines
CCR7 ligands	Plasmid DNA	Mouse	Intranasal, intragastric	CD4+ T helper cell proliferation and CD8+ T cell-mediated CTL activity, serum IgG

derivatives are being used experimentally for mucosal immunization. Quil A can be incorporated into more potent adjuvant systems, such as ISCOMs (see below). Onjisaponins were shown in studies of Nagai et al. to be safe and potent adjuvants for intranasal inoculation together with influenza and DTP vaccines. Experiments in mice showed that the use of adjuvant significantly increased serum IgG and nasal IgA as well as inhibited proliferation

of mouse-adapted influenza virus in BAL of infected mice.

4.2.6 Immune Stimulating Complexes (ISCOMS)

ISCOMs (immune stimulating complexes) consist of cholesterol, phospholipids, viral proteins, and glycosides of the adjuvant Quil A. For several viruses, it has been shown that the incorporation of viral proteins into the ISCOM structure can dramatically enhance their immunogenicity. Strong B-cell and T-cell responses are usually observed together with induction of CTL responses, which are normally not evoked by nonreplicating vaccine preparations. ISCOM-based influenza vaccines are currently being evaluated in clinical trials, and initial studies have shown that individuals immunized with ISCOM preparations developed virus-specific CTL responses in addition to strong antibody responses.

Intranasal delivery of inactivated influenza vaccine plus the ISCOMATRIX$^{\varnothing}$ (IMX) adjuvant was able to induce serum hemagglutination inhibition (HAI) titers in mice better than those obtained with nonadjuvanted vaccine delivered subcutaneously. Furthermore, the IMX-adjuvanted vaccine delivered intranasally induced mucosal IgA responses in the lung, nasal passages, and large intestine, together with high levels of serum IgA.

4.2.7 **MF59**

MF59 is an adjuvant approved for use in humans and elicits higher antibody titers than alum when used in combination with a variety of recombinant and natural subunit antigens. MF59 is an oil-in-water emulsion that contains squalene (a metabolite of cholesterol), polysorbate 80 (a surfactant soluble in water), and sorbitan trioleate (a surfactant soluble in oil). Although the mechanisms responsible for the adjuvant action of MF59 are not fully understood, enhancement of humoral immune response to parenteral influenza vaccine has been shown in humans, and mucosal immune responses to intranasally administered influenza vaccine were evoked in mice. Thus, it appears that MF59 may be a promising formulation for future mucosal vaccines.

5
Vaccination via the Respiratory Tract

The first point of contact for inhaled pathogens is usually the nasal mucosa. Many infections are initiated at the mucosal surface and so it would appear desirable to induce neutralizing antibodies and specific cellular responses at the site of pathogen entry. The large, highly vascularized surface (150 cm^2) has the potential of very efficient absorption of delivered vaccine, and the presence of immune cells in this area enables the initiation of immune reactions. Nasal epithelium absorbs principally soluble antigens, so use of a suitable delivery system is likely to contribute to achieving protective responses after immunization via the nasal route.

The technique of intranasal immunization is very important if large doses are administered. Some vaccine may be swallowed, leading to oral delivery, or, in anesthetized animals, a proportion of the vaccine may reach the lung. The epithelial deposition of antigens in the respiratory mucosa depends on size of particles (aerodynamic diameter); therefore, in designing a vaccine for delivery via the respiratory route particle size is a critical issue.

5.1
Applications of Nasal Vaccination

Many preclinical studies have been conducted on vaccines administered via the respiratory tract. These are mostly delivered into nasal mucosa, but some trials involving deep lung deposition have also been described. As mentioned above, the functional integrity of mucosal system allows induction of immune responses in sites distant from the immunization site. It seems obvious, however, that intranasal immunization is particularly appropriate for prevention of those infectious diseases acquired by inhalation. Therefore, we focus here on recent developments of nasal vaccines against some respiratory viral diseases.

The protective efficacy of mucosal immunization with purified RSV fusion protein (F) or chimeric FG glycoprotein was shown in animal models when the vaccines were coadministered with CT adjuvant. Complete protection was also demonstrated in mice immunized with a synthetic peptide (residues 174–187) of the G protein mixed with CT, even though the peptide failed to induce a detectable level of secretory IgA. Effective mucosal immunization against RSV was reported after using F protein and a genetically detoxified toxin CT-E29H or CTL peptide from the M2 protein together with LTK63 as an adjuvant.

Recently, live, attenuated, cold-adapted viral vaccines have been developed as alternatives to inactivated vaccines. Thus, it is possible to produce an organism that replicates efficiently at 25 to 28 °C, that is, the temperature of the nasal passage, but not at 37 °C (the temperature of the lungs). Intranasal administration of cold-adapted vaccines usually induces good immune responses – including local IgA responses and secretory IgA antibodies, which can provide protection against pathogens that infect mucosal sites – although the magnitude of the serum antibody response depends on the extent of virus replication. Furthermore, live vaccines can induce CTL responses or can prime CTL responses induced during natural infection.

Several vaccines against influenza are licensed for use in humans. Among them, few are administered via the intranasal route. The trivalent Nasalflu Berna vaccine consists of influenza virosomes formulated from inactivated influenza surface glycoproteins, combined with lecithin and heat-labile toxin of *E. coli*. This vaccine has been reported to induce high levels of influenza-specific hemagglutination inhibition IgG and IgA in nasal mucosa and in the saliva. In clinical trials, 85% efficacy was reported in adults and nearly 90% efficacy in children. Although no significant adverse reactions were observed in initial studies, 43 instances of Bell's palsy were encountered among the first season's vaccine recipients (totaling more than 100 000 people). This resulted in suspension of sales and the launch of a detailed investigation of potential side effects.

Another human trial involving influenza vaccine compared intranasal and intramuscular trivalent whole virus vaccines in elderly people, and showed that the intranasal route was significantly more effective in inducing mucosal IgA response. Additionally, combined vaccination involving intramuscular inactivated and intranasal cold-adapted influenza vaccines had significantly increased efficacy in an elderly population. A different human study involving MF59-adjuvanted or nonadjuvanted subunit influenza vaccines indicated that in these trials immune responses, including mucosal IgA production, were not influenced by the presence of adjuvant.

Interesting results were obtained from a mouse study using a peptide mucosal influenza vaccine. A retro–inverso analog, encompassing the protective B-cell epitope sequence from hemagglutinin (HA) (91–108) conjugated to ovalbumin and coadministered with cholera toxin, produced strong systemic (serum IgG) and mucosal (lung IgA) antibody responses that protected against intranasal challenge with a lethal dose of influenza virus. The half-life of the retro–inverso analog in the presence of lung homogenate proteases was at least 700 times greater than that of the parent L-peptide. These results demonstrated that peptido-mimetic analogs with high resistance to proteolytic degradation are very effective immunogens for intranasal administration.

Existing measles vaccines have been successful in young children when administered as an aerosol or via the intranasal route. However, results from one study did not support these findings and implied that further investigations of the immunization protocol were required.

6
Oral Vaccines

Oral administration of infectious non-pathogenic agents is an ideal method of vaccination, a principal aim being to induce specific mucosal IgA immune responses. An outstanding example of a successful oral vaccine is the Sabin polio vaccine, which induces both local and systemic immune responses and provides good protection against poliovirus infection. Other current oral vaccines include killed whole-cell B sub-unit and live–attenuated cholera vaccines, live–attenuated typhoid vaccine, and live adenovirus vaccine. Oral immunization with live–attenuated vaccine against adenovirus successfully eliminated frequent epidemics at trainee camps and was in routine use for 25 years in US army recruits. Rotashield was a highly successful live oral vaccine giving good protection against infectious diarrhea, but was withdrawn because of possible links with intussusception. The decision to withdraw vaccination resulted in the reemergence of adenoviral infections and work is now in progress to bring back this effective and safe vaccine.

Oral administration of nonliving antigens can induce oral tolerance to systemic and local autoantigens or allergens, thereby protecting against DTH. This approach could potentially be used to reduce inflammatory reactions in chronic infections, autoimmune disorders, or allergies. However, oral tolerance could hinder successful oral vaccine development, and in practice oral vaccines have proved difficult to develop; therefore, relatively few have been licensed for human use.

7
Conclusions

An important key to the development of novel mucosal vaccines is to understand how mucosal adjuvants can lead to the induction of protective responses (particularly T-cell immunity and local antibody production), while avoiding oral tolerance, induction of regulatory T cells, or immunopathogenic immunity. Genetically engineered live vaccines designed to exploit newly understood immune mechanisms of tolerance and immunoregulation offer fresh hope in the race to develop new mucosal vaccines.

Acknowledgment

This work was supported by the Wellcome Trust, UK (programme grant 054797/Z/98/Z) and European Union grant 'Impressuvac' (QLRT-PL1999-01044).

See also Autoantibodies and Autoimmunity; Dendritic Cells; Human and Veterinary Classical Vaccines against Bacterial Diseases; Immune Defence, Cell Mediated; Immunologic Memory; Immunology.

Bibliography

Books and Reviews

Del Giudice, G., Pizza, M., Rappuoli, R. (1998) Molecular basis of vaccination, *Mol. Aspects Med.* **19**, 1–70.

Eriksson, K., Holmgren, J. (2002) Recent advances in mucosal vaccines and adjuvants, *Curr. Opin. Immunol.* **14**, 666–672.

McCluskie, M.J., Davis, H.L. (1999) Novel strategies using DNA for the induction of mucosal immunity, *Crit. Rev. Immunol.* **19**, 303–329.

Mielcarek, N., Alonso, S., Locht, C. (2001) Nasal vaccination using live bacterial vectors, *Adv. Drug Deliv. Rev.* **51**, 55–69.

Walker, R.I. (1994) New strategies for using mucosal vaccination to achieve more effective immunization, *Vaccine* **12**, 387–400.

Primary Literature

Barchfeld, G.L., Hessler, A.L., Chen, M., Pizza, M., Rappuoli, R., Van Nest, G.A. (1999) The adjuvants MF59 and LT-K63 enhance the mucosal and systemic immunogenicity of subunit influenza vaccine administered intranasally in mice, *Vaccine* **17**, 695–704.

Bastien, N., Trudel, M., Simard, C. (1999) Complete protection of mice from respiratory syncytial virus infection following mucosal delivery of synthetic peptide vaccines, *Vaccine* **17**, 832–836.

Beck, M., Smerdel, S., Dedic, I., Delimar, N., Rajninger-Miholic, M., Juzbasic, M., Manhalter, T., Vlatkovic, R., Borcic, B., Mihajic, Z. (1986) Immune response to Edmonston-Zagreb measles virus strain in monovalent and combined MMR vaccine, *Dev. Biol. Stand.* **65**, 95–100.

Belshe, R.B., Gruber, W.C. (2000) Prevention of otitis media in children with live attenuated influenza vaccine given intranasally, *Pediatr. Infect. Dis. J.* **19**, S66–S71.

Belshe, R.B., Mendelman, P.M., Treanor, J., King, J., Gruber, W.C., Piedra, P., Bernstein, D.I., Hayden, F.G., Kotloff, K., Zangwill, K., Iacuzio, D., Wolff, M. (1998) The efficacy of live attenuated, cold-adapted, trivalent, intranasal influenzavirus vaccine in children, *N. Engl. J. Med.* **338**, 1405–1412.

Belshe, R.B., Gruber, W.C., Mendelman, P.M., Mehta, H.B., Mahmood, K., Reisinger, K., Treanor, J., Zangwill, K., Hayden, F.G., Bernstein, D.I., Kotloff, K., King, J., Piedra, P.A., Block, S.L., Yan, L., Wolff, M. (2000) Correlates of immune protection induced by live, attenuated, cold-adapted, trivalent, intranasal influenza virus vaccine, *J. Infect. Dis.* **181**, 1133–1137.

Belyakov, I.M., Moss, B., Strober, W., Berzofsky, J.A. (1999) Mucosal vaccination overcomes the barrier to recombinant vaccinia immunization caused by preexisting poxvirus immunity, *Proc. Natl. Acad. Sci. U. S. A.* **96**, 4512–4517.

Bradney, C.P., Sempowski, G.D., Liao, H.X., Haynes, B.F., Staats, H.F. (2002) Cytokines as adjuvants for the induction of anti-human immunodeficiency virus peptide immunoglobulin G (IgG) and IgA antibodies in serum and mucosal secretions after nasal immunization, *J. Virol.* **76**, 517–524.

Brazolot Millan, C.L., Weeratna, R., Krieg, A.M., Siegrist, C.A., Davis, H.L. (1998) CpG DNA can induce strong Th1 humoral and cell-mediated immune responses against hepatitis B surface antigen in young mice, *Proc. Natl. Acad. Sci. U. S. A.* **95**, 15553–15558.

Broide, D., Raz, E. (1999) DNA-Based immunization for asthma, *Int. Arch. Allergy Immunol.* **118**, 453–456.

Chang, E.J., Zangwill, K.M., Lee, H., Ward, J.I. (2002) Lack of association between rotavirus infection and intussusception: implications for use of attenuated rotavirus vaccines, *Pediatr. Infect. Dis. J* **21**, 97–102.

Chun, S., Daheshia, M., Lee, S., Eo, S.K., Rouse, B.T. (1999) Distribution fate and mechanism of immune modulation following mucosal delivery of plasmid DNA encoding IL-10, *J. Immunol.* **163**, 2393–2402.

Coulter, A., Harris, R., Davis, R., Drane, D., Cox, J., Ryan, D., Sutton, P., Rockman, S., Pearse, M. (2003) Intranasal vaccination with ISCOMATRIX((R)) adjuvanted influenza vaccine, *Vaccine* **21**, 946–949.

Cusi, M.G., Zurbriggen, R., Correale, P., Valassina, M., Terrosi, C., Pergola, L., Valensin, P.E., Gluck, R. (2002) Influenza virosomes are an efficient delivery system for respiratory syncytial virus-F antigen inducing humoral and cell-mediated immunity, *Vaccine* **20**, 3436–3442.

Davis, H.L., Weeratna, R., Waldschmidt, T.J., Tygrett, L., Schorr, J., Krieg, A.M., Weeratna, R. (1998) CpG DNA is a potent enhancer of specific immunity in mice immunized with recombinant hepatitis B surface antigen, *J. Immunol.* **160**, 870–876.

Fromen-Romano, C., Drevet, P., Robert, A., Menez, A., Leonetti, M. (1999) Recombinant Staphylococcus strains as live vectors for the induction of neutralizing anti-diphtheria toxin antisera, *Infect. Immun.* **67**, 5007–5011.

Gallichan, W.S., Rosenthal, K.L. (1995) Specific secretory immune responses in the female genital tract following intranasal immunization with a recombinant adenovirus expressing glycoprotein B of herpes simplex virus, *Vaccine* **13**, 1589–1595.

Giuliani, M.M., Del Giudice, G., Giannelli, V., Dougan, G., Douce, G., Rappuoli, R., Pizza, M. (1998) Mucosal adjuvanticity and immunogenicity of LTR72, a novel mutant of Escherichia coli heat-labile enterotoxin with partial knockout of ADP-ribosyltransferase activity, *J. Exp. Med.* **187**, 1123–1132.

Glueck, R. (2001) Pre-clinical and clinical investigation of the safety of a novel adjuvant for intranasal immunization, *Vaccine* **20**, S42–S44.

Hathaway, L.J., Obeid, O.E., Steward, M.W. (1998) Protection against measles virus-induced encephalitis by antibodies from mice immunized intranasally with a synthetic peptide immunogen, *Vaccine* **16**, 135–141.

Hordnes, K., Tynning, T., Brown, T.A., Haneberg, B., Jonsson, R. (1997) Nasal immunization with group B streptococci can induce high levels of specific IgA antibodies in cervicovaginal secretions of mice, *Vaccine* **15**, 1244–1251.

Hvalbye, B.K., Aaberge, I.S., Lovik, M., Haneberg, B. (1999) Intranasal immunization with heat-inactivated Streptococcus pneumoniae protects mice against systemic pneumococcal infection, *Infect. Immun.* **67**, 4320–4325.

Jabbar, I.A., Fernando, G.J., Saunders, N., Aldovini, A., Young, R., Malcolm, K., Frazer, I.H. (2000) Immune responses induced by BCG recombinant for human papillomavirus L1 and E7 proteins, *Vaccine* **18**, 2444–2453.

Kato, H., Bukawa, H., Hagiwara, E., Xin, K.Q., Hamajima, K., Kawamoto, S., Sugiyama, M., Sugiyama, M., Noda, E., Nishizaki, M., Okuda, K. (2000) Rectal and vaginal immunization with a macromolecular multicomponent peptide vaccine candidate for HIV-1 infection induces HIV-specific protective immune responses, *Vaccine* **18**, 1151–1160.

Klinman, D.M., Yi, A.K., Beaucage, S.L., Conover, J., Krieg, A.M. (1996) CpG motifs present in bacteria DNA rapidly induce lymphocytes to secrete interleukin 6, interleukin 12, and interferon gamma, *Proc. Natl. Acad. Sci. U. S. A.* **93**, 2879–2883.

Kovarik, J., Bozzotti, P., Love-Homan, L., Pihlgren, M., Davis, H.L., Lambert, P.H., Krieg, A.M., Siegrist, C.A. (1999) CpG oligodeoxynucleotides can circumvent the Th2 polarization of neonatal responses to vaccines but may fail to fully redirect Th2 responses established by neonatal priming, *J. Immunol.* **162**, 1611–1617.

Kremer, L., Dupre, L., Riveau, G., Capron, A., Locht, C. (1998) Systemic and mucosal immune responses after intranasal administration of recombinant Mycobacterium bovis bacillus Calmette-Guerin expressing glutathione S-transferase from Schistosoma haematobium, *Infect. Immun.* **66**, 5669–5676.

Kweon, M.N., Yamamoto, M., Watanabe, F., Tamura, S., Van Ginkel, F.W., Miyauchi, A., Takagi, H., Takeda, Y., Hamabata, T., Fujihashi, K., McGhee, J.R., Kiyono, H. (2002) A nontoxic chimeric enterotoxin adjuvant induces protective immunity in both mucosal and systemic compartments with reduced IgE antibodies, *J. Infect. Dis.* **186**, 1261–1269.

Lee, S.W., Sung, Y.C. (1998) Immuno-stimulatory effects of bacterial-derived plasmids depend on the nature of the antigen in intramuscular DNA inoculations, *Immunology* **94**, 285–289.

Leung, N.J., Aldovini, A., Young, R., Jarvis, M.A., Smith, J.M., Meyer, D., Anderson, D.E., Carlos, M.P., Gardner, M.B., Torres, J.V. (2000) The kinetics of specific immune responses in rhesus monkeys inoculated with live recombinant BCG expressing SIV Gag, Pol, Env, and Nef proteins, *Virology* **268**, 94 103.

Lillard, J.W. Jr, Boyaka, P.N., Taub, D.D., McGhee, J.R. (2001) RANTES potentiates antigen-specific mucosal immune responses, *J. Immunol.* **166**, 162–169.

Lu, H., Xing, Z., Brunham, R.C. (2002) GM-CSF transgene-based adjuvant allows the establishment of protective mucosal immunity following vaccination with inactivated *Chlamydia trachomatis*, *J. Immunol.* **169**, 6324–6331.

Marinaro, M., Boyaka, P.N., Finkelman, F.D., Kiyono, H., Jackson, R.J., Jirillo, E., McGhee, J.R. (1997) Oral but not parenteral interleukin (IL)-12 redirects T helper 2 (Th2)-type responses to an oral vaccine without altering mucosal IgA responses, *J. Exp. Med.* **185**, 415–427.

Mason, H.S., Lam, D.M., Arntzen, C.J. (1992) Expression of hepatitis B surface antigen in transgenic plants, *Proc. Natl. Acad. Sci. U. S. A.* **89**, 11745–11749.

Mason, H.S., Ball, J.M., Shi, J.J., Jiang, X., Estes, M.K., Arntzen, C.J. (1996) Expression of Norwalk virus capsid protein in transgenic tobacco and potato and its oral immunogenicity in mice, *Proc. Natl. Acad. Sci. U. S. A.* **93**, 5335–5340.

Matsumoto, S., Yukitake, H., Kanbara, H., Yamada, T. (1999) Long-lasting protective immunity against rodent malaria parasite infection at the blood stage by recombinant BCG secreting merozoite surface protein-1, *Vaccine* **18**, 832–834.

McCluskie, M.J., Davis, H.L. (1998) CpG DNA is a potent enhancer of systemic and mucosal immune responses against hepatitis B surface antigen with intranasal administration to mice, *J. Immunol.* **161**, 4463–4466.

McDermott, M.R., Bienenstock, J. (1979) Evidence for a common mucosal immunologic system. I. Migration of B immunoblasts into intestinal, respiratory, and genital tissues, *J. Immunol.* **122**, 1892–1898.

Medaglini, D., Oggioni, M.R., Pozzi, G. (1998) Vaginal immunization with recombinant gram-positive bacteria, *Am. J. Reprod. Immunol.* **39**, 199–208.

Medaglini, D., Pozzi, G., King, T.P., Fischetti, V.A. (1995) Mucosal and systemic immune responses to a recombinant protein expressed on the surface of the oral commensal bacterium Streptococcus gordonii after oral colonization, *Proc. Natl. Acad. Sci. U. S. A.* **92**, 6868–6872.

Mielcarek, N., Riveau, G., Remoue, F., Antoine, R., Capron, A., Locht, C. (1998) Homologous and heterologous protection after single intranasal administration of live attenuated recombinant Bordetella pertussis, *Nat. Biotechnol.* **16**, 454–457.

Minutello, M., Senatore, F., Cecchinelli, G., Bianchi, M., Andreani, T., Podda, A., Crovari, P. (1999) Safety and immunogenicity of an inactivated subunit influenza virus vaccine combined with MF59 adjuvant emulsion in elderly subjects, immunized for three consecutive influenza seasons, *Vaccine* **17**, 99–104.

Morein, B., Sundquist, B., Hoglund, S., Dalsgaard, K., Osterhaus, A. (1984) Iscom, a novel structure for antigenic presentation of membrane proteins from enveloped viruses, *Nature (London)* **308**, 457–460.

Muller, C.P., Beauverger, P., Schneider, F., Jung, G., Brons, N.H. (1995) Cholera toxin B stimulates systemic neutralizing antibodies after intranasal co-immunization with measles virus, *J. Gen. Virol.* **76**, 1371–1380.

Niewiesk, S. (2001) Studying experimental measles virus vaccines in the presence of maternal antibodies in the cotton rat model (Sigmodon hispidus), *Vaccine* **19**, 2250–2253.

Olszewska, W., Partidos, C.D., Steward, M.W. (2000) Antipeptide antibody responses following intranasal immunization: effectiveness of mucosal adjuvants, *Infect. Immun.* **68**, 4923–4929.

Olszewska, W., Erume, J., Ripley, J., Steward, M.W., Partidos, C.D. (2001) Immune responses and protection induced by mucosal and systemic immunisation with recombinant measles nucleoprotein in a mouse model of measles virus-induced encephalitis, *Arch. Virol.* **146**, 293–302.

Partidos, C.D., Vohra, P., Steward, M.W. (1996) Induction of measles virus-specific cytotoxic T-cell responses after intranasal immunization with synthetic peptides, *Immunology* **87**, 179–185.

Pizza, M., Domenighini, M., Hol, W., Giannelli, V., Fontana, M.R., Giuliani, M.M., Magagnoli, C., Peppoloni, S., Manetti, R., Rappuoli, R. (1994) Probing the structure-activity relationship of Escherichia coli LT-A by site-directed mutagenesis, *Mol. Microbiol.* **14**, 51–60.

Pozzi, G., Oggioni, M.R., Manganelli, R., Medaglini, D., Fischetti, V.A., Fenoglio, D., Valle, M.T., Kunkl, A., Manca, F. (1994) Human T-helper cell recognition of an immunodominant epitope of HIV-1 gp120 expressed on the surface of Streptococcus gordonii, *Vaccine* **12**, 1071–1077.

Robinson, K., Chamberlain, L.M., Schofield, K.M., Wells, J.M., Le Page, R.W. (1997) Oral vaccination of mice against tetanus with recombinant Lactococcus lactis, *Nat. Biotechnol.* **15**, 653–657.

Sabin, A.B., Flores, A.A., Fernandez, d.C., Albrecht, P., Sever, J.L., Shekarchi, I. (1984) Successful immunization of infants with and without maternal antibody by aerosolized measles vaccine. II. Vaccine comparisons and evidence for multiple antibody response, *JAMA* **251**, 2363–2371.

Sabin, A.B., Flores, A.A., Fernandez, d.C., Sever, J.L., Madden, D.L., Shekarchi, I., Albrecht, P. (1983) Successful immunization of children with and without maternal antibody by aerosolized measles vaccine. I. Different results with undiluted human diploid cell and chick embryo fibroblast vaccines, *JAMA* **249**, 2651–2662.

Sasaki, S., Sumino, K., Hamajima, K., Fukushima, J., Ishii, N., Kawamoto, S., Mohri, H., Kensil, C.R., Okuda, K. (1998) Induction of systemic and mucosal immune responses to human immunodeficiency virus type 1 by a DNA vaccine formulated with QS-21 saponin adjuvant via intramuscular and intranasal routes, *J. Virol.* **72**, 4931–4939.

Schmidt, A.C., McAuliffe, J.M., Murphy, B.R., Collins, P.L. (2001) Recombinant bovine/human parainfluenza virus type 3 (B/HPIV3) expressing the respiratory syncytial virus (RSV) G and F proteins can be used to achieve simultaneous mucosal immunization against RSV and HPIV3, *J. Virol.* **75**, 4594–4603.

Schmidt, A.C., Wenzke, D.R., McAuliffe, J.M., St Claire, M., Elkins, W.R., Murphy, B.R., Collins, P.L. (2002) Mucosal Immunization of Rhesus Monkeys against Respiratory Syncytial Virus Subgroups A and B and Human Parainfluenza Virus Type 3 by Using a Live cDNA-Derived Vaccine Based on a Host Range-Attenuated Bovine Parainfluenza Virus Type 3 Vector Backbone. *J. Virol.* **76**, 1089–1099.

Shroff, K.E., Marcucci-Borges, L.A., de Bruin, S.J., Winter, L.A., Tiberio, L., Pachuk, C., Snyder, L.A., Satishchandran, C., Ciccarelli, R.B., Higgins, T.J. (1999) Induction of HSV-gD2 specific CD4(+) cells in Peyer's patches and mucosal antibody responses in mice following DNA immunization by both parenteral and mucosal administration, *Vaccine* **18**, 222–230.

Simasathien, S., Migasena, S., Bellini, W., Samakoses, R., Pitisuttitham, P., Bupodom, W., Heath, J., Anderson, L., Bennett, J. (1997) Measles vaccination of Thai infants by intranasal and subcutaneous routes: possible interference from respiratory infections, *Vaccine* **15**, 329–334.

Simmons, C.P., Hussell, T., Sparer, T., Walzl, G., Openshaw, P., Dougan, G. (2001) Mucosal delivery of a respiratory syncytial virus CTL peptide with enterotoxin-based adjuvants elicits protective, immunopathogenic, and immunoregulatory antiviral CD8+ T cell responses, *J. Immunol.* **166**, 1106–1113.

Simonsen, L., Morens, D., Elixhauser, A., Gerber, M., Van Raden, M., Blackwelder, W. (2001) Effect of rotavirus vaccination programme on trends in admission of infants to hospital for intussusception, *Lancet* **358**, 1224–1229.

Staats, H.F., Ennis, F.A. Jr (1999) IL-1 is an effective adjuvant for mucosal and systemic immune responses when coadministered with protein immunogens, *J. Immunol.* **162**, 6141–6147.

Staats, H.F., Bradney, C.P., Gwinn, W.M., Jackson, S.S., Sempowski, G.D., Liao, H.X., Letvin, N.L., Haynes, B.F. (2001) Cytokine requirements for induction of systemic and mucosal CTL after nasal immunization, *J. Immunol.* **167**, 5386–5394.

Steidler, L., Robinson, K., Chamberlain, L., Schofield, K.M., Remaut, E., Le Page, R.W., Wells, J.M. (1998) Mucosal delivery of murine interleukin-2 (IL-2) and IL-6 by recombinant strains of Lactococcus lactis coexpressing antigen and cytokine, *Infect. Immun.* **66**, 3183–3189.

Tacket, C.O., Mason, H.S., Losonsky, G., Clements, J.D., Levine, M.M., Arntzen, C.J. (1998) Immunogenicity in humans of a recombinant bacterial antigen delivered in a transgenic potato, *Nat. Med.* **4**, 607–609.

Tebbey, P.W., Scheuer, C.A., Peek, J.A., Zhu, D., LaPierre, N.A., Green, B.A., Phillips, E.D., Ibraghimov, A.R., Eldridge, J.H., Hancock, G.E. (2000) Effective mucosal immunization against respiratory syncytial virus using purified F protein and a genetically detoxified cholera holotoxin, CT-E29H, *Vaccine* **18**, 2723–2734.

Ulmer, J.B., Donnelly, J.J., Parker, S.E., Rhodes, G.H., Felgner, P.L., Dwarki, V.J., Gromkowski, S.H., Deck, R.R., DeWitt, C.M., Friedman, A. (1993) Heterologous protection against influenza by injection of DNA encoding a viral protein, *Science* **259**, 1745–1749.

Walsh, E.E. (1993) Mucosal immunization with a subunit respiratory syncytial virus vaccine in mice, *Vaccine* **11**, 1135–1138.

Walsh, E.E. (1994) Humoral, mucosal, and cellular immune response to topical immunization with a subunit respiratory syncytial virus vaccine, *J. Infect. Dis.* **170**, 345–350.

Welter, J., Taylor, J., Tartaglia, J., Paoletti, E., Stephensen, C.B. (1999) Mucosal vaccination with recombinant poxvirus vaccines protects ferrets against symptomatic CDV infection, *Vaccine* **17**, 308–318.

Wierzbicki, A., Kiszka, I., Kaneko, H., Kmieciak, D., Wasik, T.J., Gzyl, J., Kaneko, Y., Kozbor, D. (2002) Immunization strategies to augment oral vaccination with DNA and viral vectors expressing HIV envelope glycoprotein, *Vaccine* **20**, 1295–1307.

Wigdorovitz, A., Carrillo, C., Dus Santos, M.J., Trono, K., Peralta, A., Gomez, M.C., Rios, R.D., Franzone, P.M., Sadir, A.M., Escribano, J.M., Borca, M.V. (1999) Induction of a protective antibody response to foot and mouth disease virus in mice following oral or parenteral immunization with alfalfa transgenic plants expressing the viral structural protein VP1, *Virology* **255**, 347–353.

Yusibov, V., Hooper, D.C., Spitsin, S.V., Fleysh, N., Kean, R.B., Mikheeva, T., Deka, D., Karasev, A., Cox, S., Randall, J., Koprowski, H. (2002) Expression in plants and immunogenicity of plant virus-based experimental rabies vaccine, *Vaccine* **20**, 3155–3164.

Zhu, Y.D., Fennelly, G., Miller, C., Tarara, R., Saxe, I., Bloom, B., McChesney, M. (1997) Recombinant bacille Calmette-Guerin expressing the measles virus nucleoprotein protects infant rhesus macaques from measles virus pneumonia, *J. Infect. Dis.* **176**, 1445–1453.

Mucoviscidosis (Cystic Fibrosis), Molecular Cell Biology of

Gerd Döring[1] and Felix Ratjen[2]
[1]Universitätsklinikum Tübingen, Germany
[2]University of Essen, Germany

Encyclopedia of Molecular Cell Biology and Molecular Medicine, 2nd Edition. Volume 8.
Edited by Robert A. Meyers.
Copyright © 2005 Wiley-VCH Verlag GmbH & Co. KGaA, Weinheim
ISBN: 3-527-30550-5

Keywords

Airway Surface Liquid (ASL)
Fluid lining the apical surface of epithelial cells. ASL is depleted in the respiratory tract of CF patients. This increases mucus viscosity and impairs cilia function, resulting in defective mucociliary clearance of bacterial pathogens from CF airways.

Chaperones
Chaperones enable misfolded proteins to be maintained in a conformation that is competent for folding, its subsequent release from the quality control apparatus, and delivery to the cell surface. Certain chemical chaperones, such as phenylbutyrate, CPX, genisteine, and glycerol can increase F508del folding yield in cell culture systems.

Cystic Fibrosis Transmembrane Conductance Regulator (CFTR)
Chloride conducting channel in apical cell membranes. Mutations in CFTR cause the disease cystic fibrosis.

DF508 Mutation
Deletion of phenylalanine in position 508 (F508del) of the CFTR protein; the most common mutation causing CF worldwide. The mutation results in misfolding of the CFTR protein leading to premature intracellular degradation.

Pseudomonas aeruginosa
Most common pathogen in CF lung disease. *Pseudomonas aeruginosa* infection is associated with clinical deterioration and massive endobronchial inflammation in most patients with cystic fibrosis.

■ Cystic fibrosis (CF) is the most common autosomal recessive disorder in Caucasians, affecting ~1 : 2500 children, with a carrier frequency of 1 : 25. The causative gene, named *CF transmembrane conductance regulator* (*CFTR*), encodes a chloride channel in epithelial cells. Abnormal transport of chloride and sodium ions affects water movement across epithelia, leading to pathophysiological consequences in various organs including the respiratory, gastrointestinal and reproductive tract, the pancreas, and liver. The CF phenotype is rather heterogeneous due to many different mutations in CFTR and the influence of modifier genes. Chronic bacterial lung infections stimulate inflammatory defense mechanisms, leading to extensive tissue remodeling. The resulting emphysema and fibrosis mainly determine the reduced life expectancy in individuals with CF. Owing to improved symptomatic treatment strategies, including better nutrition and antibiotic therapy, the prognosis of CF individuals has considerably improved and many children now reach adult life. Research is focused on the development of pharmacological drugs correcting ion channel dysfunction, anti-inflammatory drugs and vaccines to prevent airway infections. The causative gene replacement therapy has not yet been successfully applied in CF patients.

1
Epidemiology and Diagnosis

1.1
Incidence and Prevalence

CF (cystic fibrosis) is the most common fatal inherited disease in the Caucasian population, affecting about $1:2500$ children, with a carrier frequency of $1:25$. Case-finding studies resulted in incidence rates between $1:7700$ in Sweden and $1:2500$ in the United Kingdom. In the United States, an incidence rate of $\sim1:3500$ live white births has been estimated and, based on registry data, the prevalence rate in the year 2000 was 8.9 CF patients in 100 000 individuals. Similar prevalence rates have been calculated in Germany (6.9/100 000), Italy (6.5/100 000), and United Kingdom (15/100 000). In other racial groups, CF is much less frequent. Estimates from non-Caucasian populations from the United States give incidence rates of $1:14 000$ for black Americans and $1:11 500$ for Hispanic births.

1.2
Diagnosis and Screening

CF is diagnosed on clinical symptoms including persistent cough and diarrhea caused by pancreatic insufficiency. The single most useful diagnostic procedure is the sweat test with chloride concentrations >60 mmol L^{-1} in typical cases of CF.

Generally, the diagnosis is confirmed by genotyping of the most common CFTR mutations, which vary between different geographic regions. Since CFTR mutations have been identified in persons with clinical conditions such as pancreatitis, infertility due to congenital bilateral absence of the vas deferens (CBAVD), allergic bronchopulmonary aspergillosis, disseminated bronchiectasis, and diffuse panbronchiolitis, it is generally thought that genotype alone is an insufficient basis for the diagnosis of CF.

Fifty percent of all CF patients in the United States are diagnosed by the age of 6 months and 90% by the age of 8 years. Neonatal screening has been recommended because of data revealing that screened patients have better growth and weight gain than controls and that early diagnosis decreases the rate of hospitalization and may lead to an improved lung function in the first 10 years of life. Neonatal screening programs are generally based on a two-step approach with immune reactive trypsinogen in dried blood spots and confirmation by mutation analysis, including multiple CFTR alleles in positive cases. Indeed, neonatal screening has now been established in several countries. It is important to note that clinical benefits from neonatal screening are dependent on adequate treatment, and the failure to provide an adequate standard of care after a diagnosis of CF has been made is probably a critical factor in outcome. Furthermore, early genetic counseling gives parents the choice of avoiding the birth of another child with CF.

Population screening for CF can be carried out to identify CF carriers or patients, whereas heterozygote screening allows carrier couples to improve family planning. Whether CF carrier screening reduces the number of CF births on a population level is still questionable. Recent guidelines recommend prenatal screening for CF in the United States.

1.3
Prognosis

Prognosis of CF has improved dramatically in some but not all countries as a result of better care and therapy and most children now reach adult life. In the United States, the median age has risen from 8.4 years in 1969 to 14.3 years in 1998 and the median survival rose from 14 years in 1969 to 32.3 years in 1998. Similar improvements in survival have occurred in Europe but significant differences persist because of differences in treatment strategies, access to specialized CF centers, and socioeconomic status. It is generally agreed that patients treated in a CF center with specialized medical care have a better survival. Since chronic lung disease contributes mostly to morbidity and mortality in CF, optimal antibiotic therapy may influence prognosis considerably. Patients with class IV or V mutations associated with residual CFTR function have better clinical outcomes compared to those with class II mutations (including F508del, see below). However, for most patients, prognosis cannot be determined on the basis of genotype analysis.

2
Genetics and Functions of CFTR

2.1
CFTR, Mutations, Genotype-phenotype Relations and Modifier Genes

CF is caused by mutations in a 230-kb gene on chromosome 7, encoding a 1480 amino acid polypeptide named *cystic fibrosis transmembrane conductance regulator* (*CFTR*). Over 1200 mutations and sequence variants have been described to date and reported to the Cystic Fibrosis Genetic Analysis Consortium. Most of these mutations are rare and only 4 mutations occur in a frequency of more than 1%. CFTR mutations are grouped into five classes: defective synthesis (I), defective processing (II), defective regulation (III), defective conductance (IV), and partially defective production or processing (V). Class I–III mutations are more common and are associated with pancreatic insufficiency. Patients with the rarer class IV–V mutations often are pancreatic sufficient. The most common mutation worldwide is a class II mutation caused by a deletion of phenylalanine in position 508 (F508del) of the CFTR protein, leading to misfolding. Of the 43 849 CF chromosomes tested, 66% are F508del. Linking mutations to the severity of lung disease has been unsuccessful, and patients who are homozygous for the F508del mutation exhibit a wide spectrum in the rate of development and severity of lung disease.

Because F508del homozygous patients differ largely in phenotype and in chloride conductance, it is believed that environmental factors and/or genes other than CFTR modify the development, progression, and disease severity of CF. At present, a number of candidate genes including genes that regulate aspects of innate lung defense and inflammatory cascades have been suggested, which await confirmation in large CF population studies. Modifier genes are now being identified using single nucleotide polymorphisms in genome-wide searches and complemented by proteomic approaches. Modifying genes are likely to affect many different aspects of the CF phenotype, which therefore remains difficult to predict from CFTR mutation data alone.

2.2
CFTR Function and Structure

CFTR functions as a chloride channel in apical membranes. The primary structure of CFTR indicated that it belongs to a family of transmembrane proteins called *ATP-binding cassette (ABC) transporters*. ABC transporters (or traffic ATPases) form a large family of proteins responsible for the translocation of a variety of compounds across membranes of both prokaryotes and eukaryotes. CFTR is composed of five domains: two membrane-spanning domains (MSDs), two nucleotide binding domains (NBDs), and a regulatory (R) domain (Fig. 1). The F508del

mutation occurs in the DNA sequence that codes for the NBD1. In wild-type CFTR, an extracellular glycosylation site is present and the NBDs and R domain are located on the intracellular side of the membrane.

CFTR channel activity is controlled by the balance of kinase and phosphatase activity within the cell and by cellular ATP levels. Activation of a cAMP-dependent protein kinase (PK) (such as PKA and PKC) causes the phosphorylation of multiple serine residues within the R domain. Once the R domain is phosphorylated, channel gating is regulated by a cycle of ATP binding and hydrolysis at the NBDs, which form a head-to-tail dimer. The two NBDs are nonequivalent in their interactions with nucleotides. Whereas NBD1 acts as a site of stable nucleotide interaction, NBD2 constitutes a site of fast turnover. Finally, protein phosphatases dephosphorylate the regulatory domain and return the channel to its quiescent state. It has been proposed that energy from ATP hydrolysis by the conserved NBDs causes conformational changes in the MSDs, which are coupled with the opening and closing of the pore. The

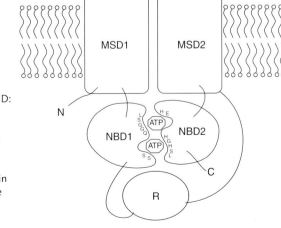

Fig. 1 Schematic model of CFTR. MSD: membrane-spanning domain, NBD: nucleotide binding domain; R: regulatory domain. Also indicated is a putative nucleotide binding domain. (From Lewis, H.A. et al. (2004) Structure of nucleotide-binding domain 1 of the cystic fibrosis transmembrane conductance regulator, *EMBO J.* **23**, 282–93).

sixth transmembrane segment seems to play a dominant role in determining the pore properties of CFTR. The CFTR pore has a deep wide intracellular vestibule but a shallow wide extracellular vestibule. The resolution of the crystal structure of mouse NBD1 confirms some of these notions (Fig. 2). This NBD1 differs from typical ABC domains in having added regulatory segments, a foreshortened subdomain interconnection, and unusual nucleotide conformation and undetectable ATPase activity.

The complex secondary structure of the protein suggests that CFTR possesses other functions in addition to being a chloride channel. Amongst others, CFTR has been described as a regulator of other apical membrane conductance pathways through interactions with the amiloride-sensitive epithelial sodium channel and the outwardly rectifying chloride channel.

Whether the activity of the epithelial sodium channel (ENaC) is inversely related to the activity of CFTR or whether its activity increases with CFTR activity is not clear at present. ENaC is also regulated by the serum and glucocorticoid-dependent kinase (SGK1), which is increased in CF lung tissue because of inflammatory processes. F508del CFTR has defects in both channel gating and endoplasmic reticulum-to-plasma membrane processing. Recent data suggest that CFTR also transports HCO_3^- as well as glutathione or regulates HCO_3^- transport through epithelial cells.

2.3
CFTR Trafficking and Tissue Location

F508del is retained in the endoplasmic reticulum (ER) because of improper folding and, therefore, will not reach the Golgi

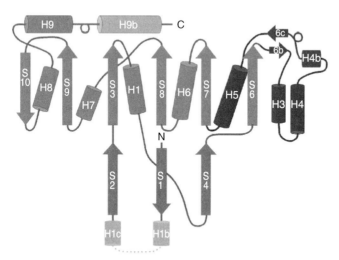

Fig. 2 Topology diagram of mNBD1. The F1-type ATP-binding core subdomain is shown in gold, the ABC α-subdomain in cyan, and the ABC ß-subdomain in green. Regions of mNBD1 that are different from previous ABC structures are shown in gray. Circles indicate the positions of 3_{10} helices. (From Lewis, H.A. et al. (2004) Structure of nucleotide-binding domain 1 of the cystic fibrosis transmembrane conductance regulator. *EMBO J.*, **23**, 282–293) (see color plate p. xxiv).

complex for the acquisition of complex N-linked oligosaccharide chains, important in the quality control pathway to prevent the development of improperly folded glycoproteins. Thus, the F508del mutation is primarily a protein-folding defect with altered protein trafficking or kinesis secondary to altered folding. Although some F508del can be delivered to the plasma membrane, channel gating is profoundly defective, because PKA-dependent phosphorylation of the R domain activates F508del CFTR to a much lesser degree than wild-type CFTR. Like other membrane glycoproteins, CFTR interacts with chaperone proteins, which prevent premature and inappropriate folding interactions during biosynthesis and increase the overall yield of correctly folded product. In this context, hsp70, hsc70, and calnexin have been identified. CFTR has also been shown to form a macromolecular complex in apical membranes with the ß2 adrenergic receptor and the ezrin-/radixin-/moesin-binding phosphoprotein 50. The assembly of the complex is regulated by PKA-dependent phosphorylation and deletion of the regulatory domain of CFTR abolishes PKA regulation of complex assembly.

In general, CFTR is found in tissues that are clinically affected by CF although low levels also occur elsewhere. The most common site is in the apical membrane of epithelial cells that line exocrine ducts or airways, and this is consistent with the proposed chloride channel function. By immunohistochemistry, wild type but not F508del CFTR was detected at the luminal membrane of crypt colonocytes, sweat glands, and respiratory epithelial cells. Both B and T lymphocytes express CFTR, reveal abnormal chloride transport, although this seems to have little functional importance. No important functional abnormalities have been shown in the heart or the placenta, both sites of CFTR expression, and the electrolytes of ocular humor, breast milk, and seminal fluid are not significantly altered in CF.

3
Basic Defect-pathology Relations

CF leads to pathologic changes in organs that express CFTR; therefore, secretory cells, sinuses, lungs, pancreas, liver, and reproductive tract are involved. The most dramatic changes are observed in CF airways where the basic defect causes mucus retention, chronic bacterial infection, and inflammation.

3.1
Respiratory Tract

Lung infections with *Pseudomonas aeruginosa* constitute a predominant disease phenotype in CF patients. Infections with *Staphylococcus aureus* and *Haemophilus influenzae* are also frequent. Several hypotheses have been offered to explain the failure of mucosal defense and the high prevalence of *P. aeruginosa* in the CF lung. It has been proposed that *P. aeruginosa* binds to CF airway epithelial cell membranes in higher density than to respective cells from normal individuals owing to an increased *P. aeruginosa* asialo-GM1 receptor density. The higher bacterial number would then lead to infection in CF airways. Alternatively, wild-type CFTR (but not mutated CFTR) has also been shown to be a receptor for *P. aeruginosa*, which mediates bacterial cell internalization and *P. aeruginosa* killing. In CF airways, therefore, *P. aeruginosa* would not be eradicated intracellularly and could multiply and cause infection. Other studies, however, reveal that both *P. aeruginosa* and *S. aureus* are located in the mucus layer on respiratory epithelial cells rather than directly on cell membranes and that no difference in location and number of adhering bacteria is visible regardless of whether normal or CF primary respiratory cells are used or infected CF lung tissue is investigated for *P. aeruginosa* or *S. aureus* adhesion. Additionally, on the basis of the assumption of an increased sodium chloride concentration due to a defective CFTR channel on the luminal side of the respiratory epithelium, it has furthermore been suggested that salt-sensitive cationic antimicrobial peptides (defensins) are inactivated in the airway surface liquid (ASL) of CF patients, which would lead to bacterial multiplication and subsequent infection. However, not all defensins are salt sensitive and it has been difficult to prove that the ASL in CF is indeed hypertonic.

In contrast, most *in vivo* data reveal that the ASL from normal and CF individuals is isotonic. This notion has been explained by a mechanism whereby the airway epithelium reduces its volume in order to reach isotonicity (Fig. 3).

Owing to defective chloride secretion of the CF respiratory epithelial cells, sodium and therefore also water are excessively reabsorbed from the periciliary liquid, which reduced the volume of the ASL from ∼20 to 6 μm. Consequently, mucus viscosity is increased and cilia function is impaired, which leads to a defect in mucociliary clearance of bacterial pathogens from CF airways. Bacteria invading the CF lung are trapped in this viscous mucus layer on top of respiratory epithelial cells, where they encounter microaerophilic or anaerobic growth conditions due to abnormal oxygen consumption of the CF cell. The physical conditions in the viscous matrix trigger a switch of *S. aureus* and *P. aeruginosa* from nonmucoid to mucoid cell types, the predominant bacterial phenotype in CF lungs. Apparently, the size of the mucoid bacterial microcolonies impairs phagocytosis and eradication of the pathogens by phagocytis cells and thus is thought to be one important factor in the pathogenicity of lung disease. The anaerobic viscous environment on the respiratory epithelial mucus layer, which is also present in large mucus (sputum) plugs later in the course of the infection, is the main reason for the chronicity of the bacterial lung infection and inflammation in CF lungs (see below).

3.2
Other Organs

Water retention as a consequence of defective chloride secretion and excessive sodium absorption is also the explanation for the pathophysiologic events in the pancreas of CF patients, leading to exocrine pancreatic insufficiency in approximately 90% of CF patients. Without sufficient fluid and HCO_3^-, digestive proenzymes are retained in pancreatic ducts and prematurely activated, which ultimately leads to tissue destruction and fibrosis. As a consequence of increasing Langerhans cell destruction, insulin secretion is decreased and delayed in older CF patients, leading to increasing prevalence of both type I and II diabetes. The consequences of CFTR mutations also lead to abnormal liver function in at least one-third of patients and biliary cirrhosis in less than 10% of the patients. Ninety-eight percent of males with CF are infertile, with aspermia secondary to atretic or absent vasa deferentia and dilated or absent seminal vesicles. Sexual potency

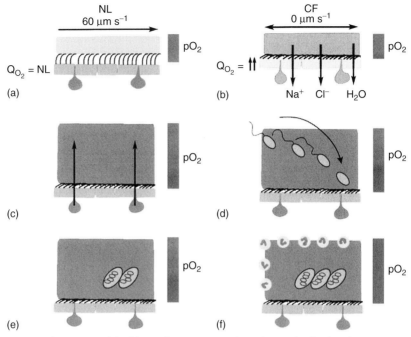

Fig. 3 Schematic model of the pathogenic events hypothesized to lead to chronic *P. aeruginosa* infection in airways of CF patients. (a) On normal airway epithelia, a thin mucus layer (light green) resides atop the PCL (clear). The presence of the low-viscosity PCL facilitates efficient mucociliary clearance (denoted by vector). A normal rate of epithelial O_2 consumption (Q_{O2}; left) produces no O_2 gradients within this thin ASL (denoted by red bar). (b–f) CF airway epithelia. (b) Excessive CF volume depletion (denoted by vertical arrows) removes the PCL, mucus becomes adherent to epithelial surfaces, and mucus transport slows/stops (bidirectional vector). The raised O_2 consumption (left) associated with accelerated CF ion transport does not generate gradients in thin films of ASL. (c) Persistent mucus hypersecretion (denoted as mucus secretory gland/goblet cell units; dark green) with time increases the height of luminal mucus masses/plugs. The raised CF epithelial Q_{O2} generates steep hypoxic gradients (blue color in bar) in thickened mucus masses. (d) *P. aeruginosa* bacteria deposited on mucus surfaces penetrate actively and/or passively (due to mucus turbulence) into hypoxic zones within the mucus masses. (e) *P. aeruginosa* adapts to hypoxic niches within mucus masses with increased alginate formation and the creation of macrocolonies. (f) Macrocolonies resist secondary defenses, including neutrophils, setting the stage for chronic infection. The presence of increased macrocolony density and, to a lesser extent neutrophils, render the now mucopurulent mass hypoxic (blue bar) (From Worlitzsch, D., Tarran, R., Ulrich, M. et al(2002) Effects of reduced mucus oxygen concentration in airway Pseudomonas infections of cystic fibrosis patients, *J. Clin. Invest.* **109**, 317–325) (see color plate p. xxiv).

as well as spermatogenesis are normal, and CF males have become fathers with techniques such as microscopic epididymal sperm aspiration and intracytoplasmatic sperm injection. Female reproductive function is normal, although cervical mucus has been reported to be dehydrated, which may impair fertility.

4
Immunology of Chronic Respiratory Infections

The inability of the immune defense system to keep the airways free of bacteria led to speculations that the then unknown genetic abnormality causes an immunological defect. With the knowledge that the CF gene codes for an epithelial ion-transport protein and data showing essentially normal systemic immune function in CF, the immunology of CF was then regarded no differently from the immunology of any other chronic infection. However, currently, CF lung disease is thought to result from a defect in innate immune response that leads to chronic lung infection (see above) and chronic inflammation. The question of whether the consequences of the basic defect provoke abnormal respiratory inflammation without prior bacterial stimuli has not been resolved as yet.

4.1
Type III Hypersensitivity Reaction

CF patients respond with the production of antibodies to a large number of *P. aeruginosa* antigens and antigens of other microbial pathogens. Antibody production provokes hypergammaglobulinaemia in infected CF patients, and leads to immune complexes, which are detectable in patients' sputa, bronchial secretions, or serum samples. Immune complexes stimulate the major phagocytic cell entering the endobronchial airways, the polymorphonuclear leukocytes (neutrophils). High immune complex levels (and antibody titers) correlate with poor clinical status of the patients, suggesting that neutrophil stimulation is responsible for tissue damage or remodeling in CF airways. Nevertheless, some of the antipseudomonal antibodies are nonopsonic or even blocking phagocytosis, particularly antibodies directed against *P. aeruginosa* polysaccharide components. This phenomenon has not yet been fully explained and might be related to a subsequent loss of avidity in antigen binding during the chronic course of the infection. During this type III hypersensitivity reaction, neutrophils are attracted from the blood vessels to the site of infection by chemoattractants, includingleukotriene B_4 (LTB$_4$) and interleukin (IL)-8.

As mentioned above, bacterial microcolony formation triggered by the anaerobic viscous environment on the respiratory epithelial mucus layer and in the large mucus (sputum) plugs (Fig. 4) is one reason for the chronicity of the bacterial lung infection. In addition, other factors may be involved. The highly viscous mucus may affect the motility of neutrophils, resulting in retarded neutrophil transmigration of the mucus layer overlaying the epithelial cells in CF airways, which could give bacteria a temporal advantage to form their protecting microcolony polysaccharide coat. Owing to the anaerobic environment in the viscous mucus, antimicrobial activity of neutrophils related to the production of reactive oxygen species is impaired and bacteria with intrinsic or acquired resistance to antimicrobial peptides (defensins) may survive phagocytosis under these circumstances. All three conditions may be present at the same time in infected CF mucus and lead to chronicity of the infection characterized by large volumes of necrotic neutrophils and accumulation of polymerized DNA strands that form highly viscous sputum plugs (Fig. 4).

Fig. 4 Mucus plugs obstructing CF airways consisting of mostly necrotic neutrophils and bacterial microcolonies. By courtesy of Dr. Uschi Weber, Wangen im Allgäu, Germany.

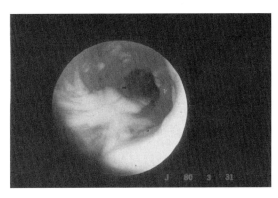

The anaerobic nature of the mucus plugs obstructing airways of CF patients implies that lysosomal enzymes for neutrophils such as elastase rather than reactive oxygen species play a major role in the pathophysiology of endobronchial inflammation. Neutrophil elastase cleaves a variety of substrates including the endogenous serine proteinase inhibitors α_1-proteinase inhibitor (α_1-PI) and secretory leukocyte proteinase inhibitor (SLPI), the structural components or airways and alveoli, elastin, and proteins of the innate and specific immune systems such as immunoglobulines, immune complexes, complement components, and cell surface receptors on neutrophils or lymphocytes. Since neutrophil-mediated phagocytosis is dependent on intact complement receptors (CR1, CR3), complement components (C3b, C3bi) and intact, opsonic antibodies, opsonophagocytosis is heavily impaired in plugs in the CF lung. On the other hand, neutrophil elastase may act as a regulatory enzyme in chronic inflammation by lowering signal transduction between inflammatory molecules and effector cells. Thus, the impairment of opsonophagocytosis and T cell function by neutrophil serine proteases may have beneficial and deleterious consequences for the host: temporal down-regulation of inflammation on the one hand and allowance of bacterial survival on the other hand. Similarly, stimulation of airway gland secretion by neutrophil serine proteases may be seen as beneficial and deleterious for the host. Mucus hypersecretion may be regarded as beneficial, because it removes bacterial pathogens from close contact with airway epithelial cells into the airway lumen, and deleterious because airway obstruction leads to lung function impairment.

Besides neutrophil serine proteases, macrophage-derived metalloproteinase and metalloproteinases from neutrophils including neutrophil gelatinase (type IV collagenase) may contribute to tissue remodeling in CF airways, particularly in the alveoli.

Despite the wealth of information on the neutrophil-dominated pathophysiology of CF lung disease, little is known about how endobronchial infection/inflammation impacts on fibrosis and emphysema in the alveolar space of CF patients.

4.2
Cytokines, Growth Hormones, and Nitric Oxide (NO)

The inflammatory system communicates by a large repertoire of chemical messengers called *cytokines* produced by epithelial cells, T lymphocytes and phagocytic cells. In bronchoalveolar lavage (BAL) samples of CF patients, increased proinflammatory cytokine levels including TNFα, IL-6, and IL-8 have been measured. Persistent high cytokine levels may have deleterious consequences. For example, TNFα may cause cachexia and osteoporosis, and IL-8, a chemotactic attractant for neutrophils, may lead to a "self-perpetuating" inflammatory process. Furthermore, BAL fluids from CF patients contain significantly less IL-10 than BAL from normal control individuals, suggesting an imbalance between upregulating and downregulating cytokines during chronic inflammation. Besides cytokines, growth factors such as insulinlike growth factor 1 (IGF-1), which can be produced by many cell types including lymphocytes and alveolar macrophages may contribute to lung pathology in CF patients. One of the consequences of enhanced IGF-1 production is fibrosis.

NO originates from the biotransformation of L-arginine to L-citrulline by enzymes called *NO synthases* (*NOS*). Although one would expect that NO production is increased in CF due to chronic lung inflammation, this is not the case and NO levels in exhalations of CF patients are low. This phenomenon has not been completely resolved and attributed in part to low levels of the inducible isoform of NOS (iNOS) in the respiratory epithelium of CF patients and consumption of NO in airway secretions in CF airways during inflammation. Since NO has multiple functions (e.g. bronchodilation, phagocytic killing of bacteria, stimulation of ciliary function), low NO concentrations in airways of CF patients may be of pathophysiologic relevance.

5
Causative and Symptomatic Therapy Strategies

5.1
Gene Replacement Therapy

The ultimate curative treatment for CF is to restore CFTR function by transfecting CF cells with wild-type CFTR. Since the initiation of the first clinical trials for CF lung disease using recombinant adenovirus in the early 1990s, the field has encountered numerous obstacles including (1) difficulties with vector production, (2) inefficient delivery of vector, and (3) vector-induced inflammation. Over 30 trials on CFTR gene therapy, mainly using adenoviral vectors or liposomes have been published and have established the proof of principle for gene transfer to the lung. However, efficiency is generally low. Viral vectors have encountered the problem of host response to the vector that either results in mucosal inflammation or inefficiency of gene transfer on repeat dosing. Despite these obstacles, several research groups are currently involved in preclinical and clinical studies to enhance the efficiency and selectivity of gene delivery to the lung.

5.2
Pharmacological Treatment of CFTR

The intracellular production, trafficking or activation of CFTR are possible targets of therapeutic interventions. Class I mutations lead to decreased production of CFTR mRNA and gentamicin has been shown to partially overcome this problem. The frequency of the F508del allele among CF patients and the possibility that chloride channel activity can be restored in these patients suggest strategies for CF treatment based on increasing the efficiency of folding and intracellular processing of this mutant. Loss of this amino acid decreases the efficiency of folding of NBF1 *in vitro* and appears to increase the probability that nascent CFTR chains become committed to degradation by a pathway that involves at least in part the ubiquitin/proteosome pathway. Certain chemical chaperones, such as phenylbutyrate, CPX, genisteine, and glycerol can increase F508del folding yield, possibly by stabilizing an early intermediate in CFTR folding, maintaining it in a conformation that is competent for folding and its subsequent release from the quality control apparatus and delivery to the cell surface. Moreover, 4-phenylbutyric acid has also been shown *in vitro* to increase production of mature CFTR and chloride transport at the cell surface, most likely by a mechanism involving upregulation at the transcriptional level and modulation of protein-folding steps. Other agents enhance the activity of F508del CFTR chloride channels at the cell surface. The best studied of these agents is the flavonoid genistein, which interacts directly with NBD2, to stabilize the open-channel configuration. Agents termed *pharmacological chaperones* (e.g. MPB compounds) potentiate the activity of mutant chloride channels present at the cell surface and rescue the cell surface expression of the same mutants. High-throughput screening technology is currently being used in the search for further compounds that either rescue the cell surface expression of mutant CFTR (termed *CFTR correctors*) or enhance the activity of mutant chloride channels present at the cell surface (termed *CFTR potentiators*). Vitamin C has been shown to regulate CFTR-mediated Cl secretion in epithelia and, therefore, represents a potential

drug in CF. Furthermore, curcumin, a Ca-adenosine triphosphatase pump inhibitor, corrected defective nasal potential difference in F508del CFTR mice.

Whether treatment with chaperones will result in sufficient concentrations of CFTR on the cell surface of epithelial cells to improve CF related pathology is unclear at present. Furthermore, mutated CFTR that reaches the cell membrane may fail to function normally. Additional targets of CFTR pharmacotherapy include activators of chloride secretion such as UTP or inhibitors of sodium absorption such as amiloride. Both drugs have a very short half-life, limiting their efficacy, but newer compounds with an improved pharmacokinetic profile are being tested in CF trials. A double-blind phase II inhalation study of the UTP analog INS37217 in patients with mild CF lung disease was safe and resulted in significantly better FEV1, FEF 25 – to 75%, and FVC in the treatment group compared to patients receiving placebo. INS37217 is believed to enhance the lung's mucosal hydration and mucociliary clearance mechanisms by activating an alternative ion channel that acts in the same way as the defective ion channel in moving salt and water to the surface of the airways. Similarly, hypertonic saline is mechanistically active and early trials show beneficial results. Thus, the notion that getting water to the surface of the respiratory epithelium of CF patients will correct the pathological consequences of the basic defect becomes more and more realistic.

5.3
Prevention of Infection

Prevention of bacterial lung infection is regarded as the primary aim for

CF. Epidemiological studies suggest that transmission of *P. aeruginosa* and other pathogens occurs either by direct patient-to-patient contacts or from various environmental bacterial reservoirs. Improved hygienic measures and separation regimes have therefore been implemented in several CF centers to limit cross-infection. Prevention of *P. aeruginosa* lung infection by active vaccination may also represent a suitable alternative strategy in CF. Whereas studies using *P. aeruginosa* lipopolysaccharide preparations were unsuccessful, a small open vaccine trial using a polysaccharide-exotoxin A conjugate as antigen showed promising results. Currently, phase III multicenter vaccine trials using a *P. aeruginosa* flagella vaccine and the polysaccharide-exotoxin A conjugate vaccine are underway in Europe.

5.4
Treatment of Infection

Improved antibiotic therapy strategies against respiratory tract infections are considered as the main reason for the increased life expectancy of CF patients that has been achieved over the last decades. Antibiotics are administered by the intravenous, aerosol, or oral route, and treatment strategies include regular courses independent of the clinical status of the patient or restricted to episodes of acute exacerbations. Decisions about optimal antibiotic therapy regimes are difficult to take since eradication of the bacteria is not normally achieved. Owing to poor penetration of antibiotics into the anaerobic sputum plugs and the rapid development of mutator strains that display increased resistance against antimicrobial drugs, even with intensive antibiotic regimens, mucoid *P. aeruginosa* cannot be eradicated. A major improvement in the strategy to fight pulmonary *P. aeruginosa* infection in CF patients is early antibiotic therapy. Anti-staphylococcal treatment for at least two weeks results in an eradication rate of ~75% and only a low rate of patients harbor *S. aureus* for more than six months thereafter. In the early phase of *P. aeruginosa* colonization, antibiotic therapy may avoid the shift to chronic mucoid *P. aeruginosa* infection. The combination of inhaled colistin with oral ciprofloxacin, inhaled colistin or inhaled tobramycin alone has been used to treat early *P. aeruginosa* colonization. Long-term follow-up of patients treated with inhaled tobramycin following initial colonization with *P. aeruginosa* demonstrated that treatment not only postponed chronic infection but also led to eradication of the bacteria in treated patients. This treatment regime will most probably reduce the incidence of chronic airway infection with *P. aeruginosa* in CF patients.

5.5
Anti-inflammatory and Mucolytic Drugs

Since chronic inflammation is the cause of tissue remodeling and destruction in CF airways, the administration of anti-inflammatory drugs to other treatment arms is a logical strategy. High-dose prednisone was found to reduce the decline of lung function for the first two years in *P. aeruginosa* positive patients, but serious side effects such as growth retardation, glucose intolerance, and cataracts were found in treated patients. Therefore, the risk/benefit ratio does not favor long-term prednisone therapy in CF patients. Treatment with inhaled steroids did not result in clear benefits for pulmonary function in CF patients. High-dose ibuprofen slowed the lung function decline in CF patients, but the beneficial effect was concentrated to a subgroup of patients aged 5 to 13 years. Increasing antiproteinase levels in the lung by supplementation with suitable inhibitors has been proposed. However, neither aerosolized recombinant human secretory leukocyte proteinase inhibitor (rSLPI) nor aerosolized α_1-PI has been tested in large numbers of CF patients. A multicenter trail with a small neutrophil elastase inhibitor, EPI-hNE4 is currently carried out in CF patients. Recombinant human DNase has been found to reduce sputum viscosity, improve pulmonary function, and reduce the number of pulmonary exacerbation in CF patients with moderate and mild lung disease. Lipoxins are anti-inflammatory lipid mediators that modulate neutrophilic inflammation. Furthermore, administration of a metabolically stable lipoxin analog to infected and inflamed murine lungs suppressed neutrophilic inflammation, decreased pulmonary bacterial burden, and attenuated disease severity, suggesting that lipoxins have therapeutic potential in CF.

5.6
Transplantation

Double lung or heart lung transplantation is a therapeutic option for CF patients with end-stage lung disease. Overall survival of lung transplant patients is poorer compared to other organ transplantation with a three-year survival of approximately 60% in CF patients. Generally, survival is better for adults than for children, but individual centers have reported a survival benefit through lung transplantation in children.

5.7
Treatment of Pancreatic Insufficiency, Nutrition, and Liver Disease

CF patients with poor nutritional status are more prone to chest infections and patients with normal fat absorption have a better pulmonary prognosis. Poor nutritional status has been linked to worse prognosis of CF patients, a finding that has prompted an aggressive approach of maintaining normal weight in CF patients. The introduction of acid-resistant microspheric pancreatic enzyme preparations in the 1970s has greatly improved but not normalized the weight of CF patients, and a European consensus was reached on optimized nutrition in CF. Unfortunately, both oral caloric supplements and enteral feeding by either nasogastric tube or gastrostomy tube have been introduced clinically prior to performing adequate scientific studies. Ursodeoxycholic acid has been shown to normalize elevated liver enzyme levels, but its long-term effect on the evolution of liver disease remains largely unproven. Liver disease is a life-limiting factor in only a few patients, but liver transplantation has been successfully performed in individual CF patients with advanced liver and limited pulmonary disease.

5.8
Animal Models

The possibility to directly modify CFTR in experimental animals by molecular techniques allows insight into the physiologic functions of CFTR and its malfunction in the course of CF. Several CFTR mouse models have been constructed. However, owing to differences in gland distribution between man and mouse, the most important trait of CF, chronic pulmonary infection, has not been reproduced in these animal models. One mouse strain has recently been shown to exhibit pathologic alterations in all organs typically affected in CF, but the pulmonary phenotype lacks chronic infection and inflammation. However, on the basis of the hypothesis that mutations in the CFTR gene result in increased airway sodium absorption, mice were generated with airway-specific overexpression of ENaC. Indeed, these mice revealed ASL volume depletion, increased mucus production, delayed mucus transport, and mucus adhesion to airway surfaces. Furthermore, defective mucus transport caused a severe spontaneous lung disease, sharing features with CF, including mucus obstruction, goblet cell metaplasia, neutrophilic inflammation, and poorer bacterial clearance. However, the chronic type of infection as seen in CF patients was not observed. Other approaches to study particularly CF lung pathophysiology are based on human tracheal xenografts in immunodeficient severe combined immunodeficiency (SCID) mice and nude mice.

See also Calcium Biochemistry; Endocrinology, Molecular; Neuron Chemistry.

Bibliography

Books and Reviews

Boucher, R.C. (2004) New concepts of the pathogenesis of cystic fibrosis lung disease, *Eur. Respir. J.* **23**, 146–158.

Davis, P.B., Drumm, M., Konstan, M.W. (1996) Cystic fibrosis, *Am. J. Respir. Crit. Care Med.* **154**, 1229–1256.

Döring, G., Hoiby, N. for the Consensus Study Group. (2004) Early intervention and prevention of lung disease in cystic fibrosis: a European consensus, *J Cystic Fibrosis* **3**, 67–91.

Döring, G., Knight, R., Bellon, G. (2000) Immunology of Cystic Fibrosis, in: Hodson, M.E., Geddes, D. (Eds.) *Cystic Fibrosis*, Arnold, London, UK, pp. 109–141.

Driskell, R.A., Engelhardt, J.F. (2003) Current status of gene therapy for inherited lung diseases, *Annu. Rev. Physiol.* **65**, 585–612.

Gibson, R.L., Burns, J.L., Ramsey, B.W. (2003) Pathophysiology and management of pulmonary infections in cystic fibrosis, *Am. J. Respir. Crit. Care Med.* **168**, 918–951.

Pilewski, J.M., Frizzell, R.A. (1999) Role of CFTR in airway disease, *Physiol. Rev.* **79**(Suppl. 1), S215–S255.

Quinton, P.M. (1999) Physiological basis of cystic fibrosis: a historical perspective, *Physiol. Rev.* **79**(Suppl. 1), S3–S22.

Sinaasappel, M., Stern, M., Littlewood, J. et al (2002) Nutrition in patients with cystic fibrosis: a European consensus, *J. Cystic Fibrosis* **2**, 51–75.

Verkman, A.S. (2004) Drug discovery in academia, *Am. J. Physiol.* **286**, C465–C474.

Primary Literature

Boucher, R., Cotton, C.U., Gatzy, J.T. et al (1988) Evidence for reduced Cl− and increased Na+ permeability in cystic fibrosis human primary cell cultures, *J. Physiol.* **404**, 77–103.

Cai, Z., Scott-Ward, T.S., Sheppard, D.N. (2003) Voltage-dependent gating of the cystic fibrosis transmembrane conductance regulator Cl− channel, *J. Gen. Physiol.* **122**, 605–620.

Chappe, V., Hinkson, D.A., Howell, L.D., Evagelidis, A., Liao, J., Chang, X.B., Riordan, J.R., Hanrahan, J.W. (2004) Stimulatory and inhibitory protein kinase C consensus sequences regulate the cystic fibrosis transmembrane conductance regulator, *Proc. Natl. Acad. Sci. U. S. A.* **101**, 390–395.

Cheng, S.H., Gregory, R.J., Marshall, J. et al (1990) Defective intracellular traffic and processing of CFTR is the molecular basis of most cystic fibrosis, *Cell* **63**, 827–834.

Clarke, L.L., Grubb, B.R., Yankaskas, J.R., Cotton, C.U., McKenzie, A., Boucher, R.C. (1994) Relationship of a non-cystic fibrosis transmembrane conductance regulator-mediated chloride conductance to organ-level disease in Cftr(−/−) mice, *Proc. Natl. Acad. Sci. U. S. A.* **91**, 479–483.

Dormer, R.L., Derand, R., McNeilly, C.M., Mettey, Y., Bulteau-Pignoux, L., Metaye, T., Vierfond, J.M., Gray, M.A., Galietta, L.J., Morris, M.R., Pereira, M.M., Doull, I.J., Becq, F., McPherson, M.A. (2001) Correction of delF508-CFTR activity with benzo(c)quinolizinium compounds through facilitation of its processing in cystic fibrosis airway cells. et al, *J. Cell Sci.* **114**, 4073–4081.

Durie, P.R., Kent, G., Phillips, M.J., Ackerley, C.A. (2004) Characteristic multiorgan pathology of cystic fibrosis in a long-living cystic fibrosis transmembrane regulator knockout murine model, *Am. J. Pathol.* **164**, 1481–1493.

Egan, M.E., Pearson, M., Weiner, S.A., Rajendran, V., Rubin, D., Glockner-Pagel, J., Canny, S., Du, K., Lukacs, G.L., Caplan, M.J. (2004) Curcumin, a major constituent of turmeric, corrects cystic fibrosis defects, *Science* **304**, 600–602.

Farrell, P.M., Kosorok, M.R., Laxova, A. et al (1997) Nutritional benefits of neonatal screening for cystic fibrosis, *N. Engl. J. Med.* **337**, 963–969.

Fischer, H., Schwarzer, C., Illek, B. (2004) Vitamin C controls the cystic fibrosis transmembrane conductance regulator chloride channel, *Proc. Natl. Acad. Sci. U.S.A.* **101**, 3691–3696.

Galietta, L.V., Jayaraman, S., Verkman, A.S. (2001) Cell-based assay for high-throughput quantitative screening of CFTR chloride transport agonists, *Am. J. Physiol. Cell Physiol.* **281**, C1734–C1742.

Hung, L.W., Wang, I.X., Nikaido, K., Liu, P.Q., Ames, G.F., Kim, S.H. (1998) Crystal structure of the ATP-binding subunit of an ABC transporter, *Nature* **396**, 703–707.

Huang, P., Lazarowski, E.R., Tarran, R., Milgram, S.L., Boucher, R.C., Stutts, M.J. (2001) Compartmentalized autocrine signaling to cystic fibrosis transmembrane conductance regulator at the apical membrane of airway epithelial cells, *Proc. Natl. Acad. Sci. U. S. A.* **98**, 14120–14125.

Hwang, T.C., Sheppard, D.N. (1999) Molecular pharmacology of the CFTR Cl− channel, *Trends Pharmacol. Sci.* **20**, 448–453.

Karp, C.L., Flick, L.M., Park, K.W., Softic, S., Greer, T.M., Keledjian, R., Yang, R., Uddin, J., Guggino, W.B., Atabani, S.F., Belkaid, Y., Xu, Y., Whitsett, J.A., Accurso, F.J., Wills-Karp, M., Petasis, N.A. (2004) Defective lipoxin-mediated anti-inflammatory activity in the cystic fibrosis airway, *Nat. Immunol.* **5**, 388–392.

Knowles, M.R., Clarke, L.L., Boucher, R.C. (1991) Activation by extracellular nucleotides of chloride secretion in the airway epithelia of

patients with cystic fibrosis, *N. Engl. J. Med.* **325**, 533–538.

Ko, S.B., Zeng, W., Dorwart, M.R., Luo, X., Kim, K.H., Millen, L., Goto, H., Naruse, S., Soyombo, A., Thomas, P.J., Muallem, S. (2004) Gating of CFTR by the STAS domain of SLC26 transporters, *Nat. Cell Biol.* **6**, 343–350.

Kogan, I., Ramjeesingh, M., Li, C., Kidd, J.F., Wang, Y., Leslie, E.M., Cole, S.P., Bear, C.E. (2003) CFTR directly mediates nucleotide-regulated glutathione flux, *EMBO J.* **22**, 1981–1989.

Konstan, M.W., Byard, P.J., Hoppel, C.L. et al (1995) Effect of high-dose ibuprofen in patients with cystic fibrosis, *N. Engl. J. Med.* **332**, 848–854.

Kopelman, H., Durie, P., Gaskin, K. et al (1985) Pancreatic fluid secretion and protein hyperconcentration in cystic fibrosis, *N. Engl. J. Med.* **312**, 329–334.

Lewis, H.A. et al. (2004) Structure of nucleotide-binding domain 1 of the cystic fibrosis transmembrane conductance regulator, *EMBO J.* **23**, 282–293.

Locher, K.P., Lee, A.T., Rees, D.C. (2002) The E. coli BtuCD structure: a framework for ABC transporter architecture and mechanism, *Science* **296**, 1091–1098.

Mall, M., Grubb, B.R., Harkema, J.R., O'Neal, W.K., Boucher, R.C. (2004) Increased airway epithelial Na+ absorption produces cystic fibrosis-like lung disease in mice, *Nat. Med.* **10**, 487–493.

Mall, M., Kreda, S.M., Mengos, A., Jensen, T.J., Hirtz, S., Seydewitz, H.H., Yankaskas, J., Kunzelmann, K., Riordan, J.R., Boucher, R.C. (2004) The DeltaF508 mutation results in loss of CFTR function and mature protein in native human colon, *Gastroenterology* **126**, 32–41.

Matsui, H., Grubb, B.R., Tarran, R. et al (1998) Evidence for periciliary liquid layer depletion, not abnormal ion composition, in the pathogenesis of cystic fibrosis airways disease, *Cell* **95**, 1005–1015.

Morral, N., Bertranpetit, J., Estivill, X., Nunes, V., Casals, T., Gimenez, J., Reis, A., Varon-Mateeva, R., Macek, M. Jr., Kalaydjieva, L. et al (1994) The origin of the major cystic fibrosis mutation (delta F508) in European populations, *Nat. Genet.* **7**, 169–175.

Naren, A.P., Cobb, B., Li, C., Roy, K., Nelson, D., Heda, G.D., Liao, J., Kirk, K.L., Sorscher, E.J.,

Hanrahan, J., Clancy, J.P. (2003) A macromolecular complex of beta 2 adrenergic receptor, CFTR, and ezrin/radixin/moesin-binding phosphoprotein 50 is regulated by PKA, *Proc. Natl. Acad. Sci. U.S.A.* **100**, 342–346.

Quinton, P.M. (2001) The neglected ion: HCO3−, *Nat. Med.* **7**, 292–293.

Ramsey, B.W., Pepe, M.S., Quan, J.M. et al (1999) Intermittent administration of inhaled tobramycin in patients with cystic fibrosis, *N. Engl. J. Med.* **340**, 23–30.

Ratjen, F., Döring, G., Nikolaizik, W. (2001) Eradication of *Pseudomonas aeruginosa* with inhaled tobramycin in patients with cystic fibrosis, *Lancet* **358**, 983–984.

Reddy, M.M., Light, M.J., Quinton, P.M. (1999) Activation of the epithelial Na+ channel (ENaC) requires CFTR Cl− channel function, *Nature* **402**, 301–304.

Riordan, J.R., Rommens, J.M., Kerem, B.S. et al (1989) Identification of the Cystic Fibrosis Gene: Cloning and characterization of complementary DNA, *Science* **245**, 1066–1073.

Rommens, J.M., Iannuzzi, M.C., Kerem, B.S. et al (1989) Identification of the Cystic Fibrosis Gene: Chromosome walking and jumping, *Science* **245**, 1059–1065.

Rosenberg, M.F., Kamis, A.B., Callaghan, R., Higgins, C.F., Ford, R.C. (2003) Three-dimensional structures of the mammalian multidrug resistance P-glycoprotein demonstrate major conformational changes in the transmembrane domains upon nucleotide binding, *J. Biol. Chem.* **278**, 8294–8299.

Rosenstein, B.J., Cutting, G.R. (1998) The diagnosis of cystic fibrosis: a consensus statement, *J. Pediatr.* **132**, 589–595.

Schwiebert, E.M., Morales, M.M., Devidas, S. et al (1998) Chloride channel and chloride conductance regulator domains of CFTR, the cystic fibrosis transmembrane conductance regulator, *Proc. Natl. Acad. Sci. U.S.A.* **95**, 2674–2689.

Shcheynikov, N., Kim, K.H., Kim, K.M., Dorwart, M.R., Ko, S.B., Goto, H., Naruse, S., Thomas, P.J., Muallem, S. (2004) Dynamic control of cystic fibrosis transmembrane conductance regulator Cl(−)/HCO3(−) selectivity by external Cl(−), *J. Biol. Chem.* **279**, 21857–21865.

Teem, J.L., Berger, H.A., Ostedgaard, L.S., Rich, D.P., Tsui, L.C., Welsh, M.J. (1993) Identification of revertants for the cystic fibrosis delta F508 mutation using STE6-CFTR chimeras in yeast, *Cell* **73**, 335–346.

Ulrich, M., Herbert, S., Berger, J., Bellon, G., Louis, D., Münker, G., Döring, G. (1998) Localization of *Staphylococcus aureus* in infected airways of patients with cystic fibrosis and in a cell culture model of *S. aureus* adherence, *Am. J. Respir. Cell Mol. Biol.* **19**, 83–91.

Varga, K., Jurkuvenaite, A., Wakefield, J., Hong, J.S., Guimbellot, J.S., Venglarik, C.J., Niraj, A., Mazur, M., Sorscher, E.J., Collawn, J.F., Bebok, Z. (2004) Efficient intracellular processing of the endogenous cystic fibrosis transmembrane conductance regulator in epithelial cell lines, *J. Biol. Chem.* **279**, 22578–22584.

Wagner, C.A., Ott, M., Klingel, K., Beck, S., Melzig, J., Friedrich, B., Wild, K.N., Broer, S., Moschen, I., Albers, A., Waldegger, S., Tummler, B., Egan, M.E., Geibel, J.P., Kandolf, R., Lang, F. (2001) Effects of the serine/threonine kinase SGK1 on the epithelial Na(+) channel (ENaC) and CFTR: implications for cystic fibrosis, *Cell. Physiol. Biochem.* **11**, 209–218.

Wang, F., Zeltwanger, S., Hu, S., Hwang, T.C. (2000) Deletion of phenylalanine 508 causes attenuated phosphorylation-dependent activation of CFTR chloride channels, *J. Physiol.* **524**, 637–648.

Wooldridge, J.L., Deutsch, G.H., Sontag, M.K., Osberg, I., Chase, D.R., Silkoff, P.E., Wagener, J.S., Abman, S.H., Accurso, F.J. (2004) NO pathway in CF and non-CF children, *Pediatr. Pulmonol.* **37**, 338–350.

Worlitzsch, D., Tarran, R., Ulrich, M. et al (2002) Effects of reduced mucus oxygen concentration in airway Pseudomonas infections of cystic fibrosis patients, *J. Clin. Invest.* **109**, 317–325.

Zhang, X.M., Wang, X.T., Yue, H., Leung, S.W., Thibodeau, P.H., Thomas, P.J., Guggino, S.E. (2003) Organic solutes rescue the functional defect in delta F508 cystic fibrosis transmembrane conductance regulator, *J. Biol. Chem.* **278**, 51232–51242.

Zielinski, J., Rozmahel, R., Bozon, D. Kerem, B., Grzelczak, Z., Riordan, J.R., Rommens, J., Tsui, L.C. (1991) Genomic DNA sequence of the cystic fibrosis transmembrane conductance regulator, *Genomics* **10**, 241–248.

Glossary of Basic Terms

The most basic terms in molecular cell biology are defined below. These, in combination with the key words listed at the head of each article, provide definitions of all essential terms in this Encyclopedia.

Alleles

Alternative forms of a given gene, inherited separately from each parent, differing in nucleotide base sequence and located in a specific position on each homologous chromosome, affecting the functioning of a single product (RNA and/or protein).

Amino Acid

An organic compound containing at least one amino group and one carboxyl group. In the 20 different amino acids that compose proteins, an amino group and carboxyl group are linked to a central carbon atom, the α-carbon, to which a variable side chain is bound (see pages at the back of each volume).

Amplification

The process of replication of specific DNA sequences in disproportionately greater amounts than are present in the parent genetic material, for example, PCR is an *in vitro* amplification technique.

Apoptosis

Regulated process leading to nonpathological animal cell death via a series of well-defined morphological changes; also called *programmed cell death*.

Bacteriophage (phage)

Any virus (containing DNA or RNA) that infects bacterial cells. Some bacteriophages are widely used as cloning vectors.

Base Pair

Association of two complementary nucleotides in a DNA or RNA molecule stabilized by hydrogen bonding between their base components. Adenine pairs with thymine or uracil (A–T; A–U) and guanine pairs with cytosine (G–C) (see pages at the back of each volume).

Bioinformatics

Computational approaches to answer biological questions and enhance the ability of researchers to manipulate, collect, and analyze data more quickly and in new ways. Experts predict that more biologists will do their work *in silico*, using the computer to synthesize, analyze, and interpret the many terabytes of data now being generated.

Encyclopedia of Molecular Cell Biology and Molecular Medicine, 2nd Edition. Volume 8
Edited by Robert A. Meyers.
Copyright © 2005 Wiley-VCH Verlag GmbH & Co. KGaA, Weinheim
ISBN: 3-527-30550-5

cDNA (complementary DNA)

A DNA copy of an RNA molecule synthesized from an mRNA template *in vitro* using an enzyme called *reverse transcriptase*; often used as a probe.

Cell Cycle

Ordered sequence of events in which a cell duplicates its chromosomes and divides itself into two. Most eukaryotic cell cycles can be commonly divided into four phases: G_1 (G1) period after mitosis but before DNA synthesis occurs; S-phase when most DNA replication occurs; G_2 (G2) phase period of cell cycle when cells contain twice the G1 complement of DNA; and M-phase when cell division occurs, yielding two daughter cells (mitosis) each with one complete genome.

Cell Differentiation

Progressive restriction of the developmental potential and increasing specialization of function that takes place during the development of the embryo and leads to the formation of specialized cells, tissues, and organs.

Cell Division

Separation of a cell into two daughter cells. In higher eukaryotes, it involves division of the nucleus (mitosis) and of the cytoplasm (cytokinesis); mitosis often is used to refer to both nuclear and cytoplasmic division.

Cell Line

A defined unique population of cells obtained by culture from a primary implant through numerous generations.

Chromatin

The complex of nucleic acids (DNA and RNA) and proteins (histones) comprising eukaryotic chromosomes.

Chromosome

In prokaryotes, the usually circular duplex DNA molecule constituting the genome; in eukaryotes, a threadlike structure consisting of chromatin and carrying genomic information on a DNA double helix molecule. A viral chromosome may be composed of DNA or RNA.

Cloning

Asexual reproduction of cells, organisms, genes, or segments of DNA identical to the original.

Cloning Vector *see* Vector

Codon

Sequence of three nucleotides in DNA or mRNA that specifies a particular amino acid during protein synthesis; also called *triplet*. Of the 64 possible codons, three are stop codons, which do not specify amino acids (see pages at the back of each volume).

Complementary Base Pairing

Nucleic acid sequences on paired polymers with opposing hydrogen-bonded bases adenine (designated A) bonded to thymine (T), guanine (G) to cytosine (C) in DNA and adenine to uracil (U) replacing adenine to thymine in RNA (see pages at the back of each volume).

Complementary DNA see cDNA

Dalton

Unit of molecular mass approximately equal to the mass of a hydrogen atom $(1.66 \times 10^{-24}$ g$)$.

Deoxyribonucleic Acid *see* **DNA**

Diploid
The number of chromosomes in most cells except the gametes. In humans, the diploid number is 46.

DNA (Deoxyribonucleic Acid)
The molecular basis of the genetic code consisting of a poly-sugar phosphate backbone from which thymine, adenine, guanine, and cytosine bases project. Usually found as two complementary chains (duplex) forming a double helix associated by hydrogen bonds between complementary bases.

DNA Cloning (Gene Cloning)
Recombinant DNA technique in which specific cDNAs or fragments of genomic DNA are inserted into a cloning vector, which then is incorporated into cultured host cells (e.g., *E. coli* cells) and maintained during growth of the host cells.

DNA Library
Collection of cloned DNA molecules consisting of fragments of the entire genome (genomic library) or of DNA copies of all the mRNAs produced by a cell type (cDNA library) inserted into a suitable cloning vector.

DNA Polymerase
Enzymes that catalyze the replication of DNA from the deoxyribonucleotide triphosphates using single- or double-stranded DNA as a template.

DNA Transcription *see* **Transcription**

E. coli (Escherichia coli)
A colon bacillus, which is the most studied of all forms of life.

Embryonic Stem Cells (ES)
Cultured cells derived from the pluripotent inner cell mass of blastocyst-stage embryos.

Epigenetics
Mechanisms of storing and transmitting cellular information additional to those based on DNA sequences.

Escherichia coli see E. coli

Eukaryotes
Organisms whose cells have their genetic material packed in a membrane-surrounded, structurally discrete nucleus and with well-developed cell organelles. Eukaryotes include all organisms except *archaebacteria* and *eubacteria*.

Expression
The process of making the product of a gene, which is either a specific protein giving rise to a specific trait or RNA forms not translated into proteins (e.g. transfer ribosomal RNAs).

Functional Genomics
A discipline that aims to understand how genes are regulated and what they do, largely through massive parallel studies of gene expression over time and in a variety of tissues.

Gamete
Specialized haploid cell (in animals either a sperm or an egg) produced by meiosis of germ cells; in sexual reproduction, the union of a sperm and an egg initiates the development of a new diploid individual.

Gene Cloning *see* **DNA Cloning**

Gene
A DNA sequence, located in a particular position on a particular chromosome,

which encodes a specific protein or RNA molecule.

Genomics
Comparative analysis of the complete genomic sequences from different organisms; used to assess evolutionary relations between species and to predict the number and general types of proteins produced by an organism.

Genotype
Entire genetic constitution of an individual cell or organism; also, the alleles at one or more specific loci.

Haploid
The number of chromosomes in a sperm or egg cell, half the diploid number.

Heterozygous
Having two different alleles for a given trait in the homologous chromosomes.

Homologies
Similarities in DNA or protein sequences between individuals of the same species or among different species.

Homologous Chromosomes
Chromosome pairs, each derived from one parent, containing the same linear sequence of genes, and as a consequence, each gene is present in duplicate (e.g., humans have 23 homologous chromosome pairs, but the toad has 11 pairs, the mosquito has three pairs, and so on).

Homozygous
Having two identical alleles for a given trait in the homologous chromosomes.

Hybridization
The formation of a double-stranded polynucleotide molecule when two complementary strands are brought together at moderate temperature. The strands can be DNA or RNA or one of each; a technique for assessing the extent of sequence homology between single strands of nucleic acids.

Ligation
The formation of a phosphodiester bond to join adjacent terminal nucleotides (nicks) to form a longer nucleic acid chain (DNA of RNA); catalyzed by ligase.

Marker
A gene or a restriction enzyme cutting site with a known location on a chromosome and a clear-cut phenotype (expression), or pattern of inheritance, used as a point of reference when mapping a new mutant.

Meiosis
In eukaryotes, a special type of cell division that occurs during maturation of germ cells; comprises two successive nuclear and cellular divisions, with only one round of DNA replication resulting in production of four genetically nonequivalent haploid cells (gametes) from an initial diploid cell.

Messenger RNA *see* mRNA

Mitosis
In eukaryotic cells, the process whereby the nucleus is divided, involving condensation of the DNA into visible chromosomes, to produce two genetically equivalent daughter nuclei with the diploid number of chromosomes.

mRNA (messenger RNA)
RNA used to translate information from DNA to ribosome where the information is used to make one or several proteins.

Mutation

The heritable change in the nucleotide sequence of a chromosome.

Nucleotide

The monomer which, when polymerized, forms DNA or RNA. It is composed of a nitrogenous base bonded to a sugar (ribose or deoxyribose), bonded to a phosphate.

Oligonucleotide

A polynucleotide 2 to 20 nucleotide units in length.

Operon

A series of prokaryote genes encoding enzymes of a specific biosynthesis pathway and transcribed into a single RNA molecule.

Organelle

Any membrane-limited structure found in the cytoplasm of eukaryotic cells.

Phage *see* **Bacteriophage**

Phenotype

The observable characteristics of a cell or organism as distinct from it's genotype.

Plasmid

An extrachromosomal circular DNA molecule found in a variety of bacteria encoding "dispensable functions," such as resistance to antibiotics. Often found in multiple copies per cell and reproduces every time the bacterial cell reproduces. May be used as a cloning vector.

Polymorphism

Difference in DNA sequence among individuals expressed as different forms of a protein in individuals of the same interbreeding population.

Polynucleotide

The polymer formed by condensation of nucleotides.

Probe

A radioactively fluorescent or immunologically labeled oligonucleotide (RNA or DNA) used to detect complementary sequences in a hybridization experiment, for example, identify bacterial colonies that contain cloned genes or detect specific nucleic acids following separation by gel electrophoresis.

Procaryotes (Prokaryotes)

Typically unicellular or filamentous with DNA not located within a nuclear envelope. Prokaryotes include archaebacteria, eubacteria, cyanobacteria, prochlorophytes and mycoplasmas.

Programmed Cell Death *see* **Apoptosis**

Prokaryotes *see* **Procaryotes**

Protein

A linear polymer of amino acids linked together in a specific sequence and usually containing more than 50 residues. Proteins form the key structure elements in cells and participate in nearly all cellular activities.

Proteomics

A discipline that promises to determine the identity, function, and structure of each protein in an organelle or cell and to elucidate protein–protein interactions.

Replication

The copying of a DNA molecule duplex yielding two new DNA duplex molecules, each with one strand from the original DNA duplex. Single-stranded DNA replication results in a single-stranded DNA molecule.

Repressor
A protein that binds to a specific location (operator) on DNA and prevents RNA transcription from a specific gene or operon.

Restriction Fragment Length Polymorphism *see* RFLP

Restriction Mapping
Uses restriction endonuclease enzymes to produce specific cuts (cleavage) in DNA, allowing preparation of a genome map describing the order and distance between cleavage sites.

Reverse Transcription
The synthesis of cDNA from an RNA template as catalyzed by reverse transcriptase.

RFLP (Restriction Fragment Length Polymorphism)
DNA fragment cut by enzymes specific to a base sequence (restriction endonuclease) generating a DNA fragment whose size varies from one individual to another. Used as markers on genome maps and for screening for mutations and genetic diseases.

Ribonucleic Acid *see* RNA

Ribosomes
Small cellular components composed of proteins plus ribosomal RNA that translate the genetic code into synthesis of specific proteins.

RNA (Ribonucleic Acid)
A single-stranded polynucleotide with a phosphate oxyribose backbone and four bases that are identical to those in DNA,

with the exception that the base uracil is substituted for thymine.

RNA Interference (RNAi)
Intracellular degradation of RNA that removes foreign RNAs such as those from viruses. These fragments (small, micro, or mini RNA) cleaved from free double-stranded RNA (dsRNA) direct the degradative mechanism to other similar RNA sequences. Used as a technique to silence the expression of targeted genes in a sequence-dependent mode.

RNA Polymerase
The enzyme (peptide) that binds at specific nucleotide sequences, called promoters, in front of genes in DNA, which catalyze transcription of DNA to RNA.

RNA Translation *see* Translation

Stem Cell
A self-renewing cell that divides to give rise to a cell with an identical developmental potential and/or one with a more restricted developmental potential.

Structural Biology
The discovery, analysis and dissemination of three-dimensional structures of protein, DNA, RNA, and other biological macromolecules representing the entire range of structural diversity found in nature.

Transcription (DNA transcription)
Synthesis of an RNA molecule from a DNA template (gene) catalyzed by RNA polymerase.

Transfer RNA *see* tRNA

Translation (RNA translation)
The process on a ribosome by which the sequence of nucleotides in a mRNA

molecule directs the incorporation of amino acids into protein.

tRNA (transfer RNA)

RNA molecules that transport specific amino acids to ribosomes into position in the correct order during protein synthesis.

Vector

A DNA molecule originating from a virus, a plasmid, or a cell of a higher organism into which another DNA fragment can be integrated without loss of the vector's capacity for self-replication. Vectors introduce foreign DNA into host cells where it can be reproduced in large quantities.

Virus

A small parasite consisting of nucleic acid (RNA or DNA) enclosed in a protein coat that can replicate only in a susceptible host cell; widely used in cell biology research.

Wild type

Normal, nonmutant form of a macromolecule, cell or organism.

Zygote

A fertilized egg; a diploid cell resulting from fusion of a male and female gamete.

The Twenty Amino Acids that are Combined to Form Proteins in Living Things

Amino acids with nonpolar side chains

Glycine
Gly
G

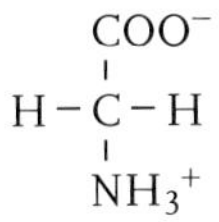

$$H-\underset{\underset{NH_3^+}{|}}{\overset{\overset{COO^-}{|}}{C}}-H$$

Alanine
Ala
A

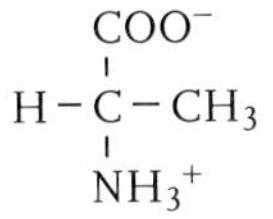

$$H-\underset{\underset{NH_3^+}{|}}{\overset{\overset{COO^-}{|}}{C}}-CH_3$$

Valine
Val
V

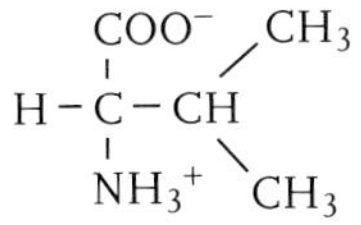

$$H-\underset{\underset{NH_3^+}{|}}{\overset{\overset{COO^-}{|}}{C}}-CH\overset{CH_3}{\underset{CH_3}{}}$$

Leucine
Leu
L

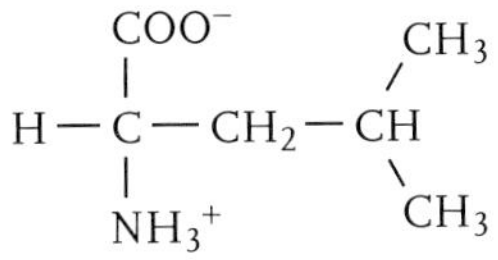

$$H-\underset{\underset{NH_3^+}{|}}{\overset{\overset{COO^-}{|}}{C}}-CH_2-CH\overset{CH_3}{\underset{CH_3}{}}$$

Isoleucine
Ile
I

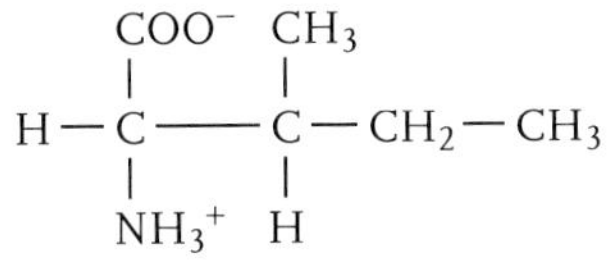

$$H-\underset{\underset{NH_3^+}{|}}{\overset{\overset{COO^-}{|}}{C}}-\underset{\underset{H}{|}}{\overset{\overset{CH_3}{|}}{C}}-CH_2-CH_3$$

Methionine
Met
M

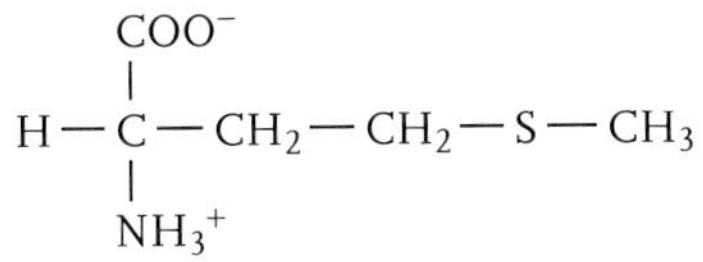

$$H-\underset{\underset{NH_3^+}{|}}{\overset{\overset{COO^-}{|}}{C}}-CH_2-CH_2-S-CH_3$$

Proline
Pro
P

$$\overset{H_2}{\underset{N}{C}}$$

COO⁻ C² C³—CH₂(4) CH₂(5) H N⁺(1) H₂

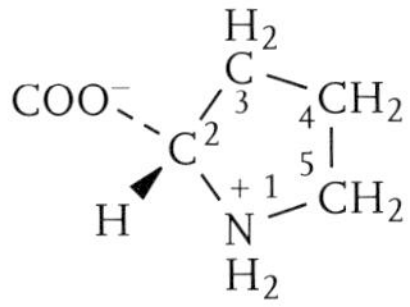

Encyclopedia of Molecular Cell Biology and Molecular Medicine, 2nd Edition. Volume 8
Edited by Robert A. Meyers.
Copyright © 2005 Wiley-VCH Verlag GmbH & Co. KGaA, Weinheim
ISBN: 3-527-30550-5

Amino acids with nonpolar side chains (continued)

Phenylalanine
Phe
F

$$H-\underset{\underset{NH_3^+}{|}}{\overset{\overset{COO^-}{|}}{C}}-CH_2-\bigcirc$$

Tryptophan
Trp
W

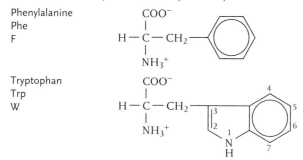

Amino acids with uncharged polar side chains

Serine
Ser
S

$$H-\underset{\underset{NH_3^+}{|}}{\overset{\overset{COO^-}{|}}{C}}-CH_2-OH$$

Threonine
Thr
T

$$H-\underset{\underset{NH_3^+}{|}}{\overset{\overset{COO^-}{|}}{C}}-\underset{\underset{OH}{|}}{\overset{\overset{H}{|}}{C}}-CH_3$$

Asparagine
Asn
N

$$H-\underset{\underset{NH_3^+}{|}}{\overset{\overset{COO^-}{|}}{C}}-CH_2-C\overset{\diagup\diagup O}{\underset{\diagdown NH_2}{}}$$

Glutamine
Gln
Q

$$H-\underset{\underset{NH_3^+}{|}}{\overset{\overset{COO^-}{|}}{C}}-CH_2-CH_2-C\overset{\diagup\diagup O}{\underset{\diagdown NH_2}{}}$$

Tyrosine
Tyr
Y

$$H-\underset{\underset{NH_3^+}{|}}{\overset{\overset{COO^-}{|}}{C}}-CH_2-\bigcirc-OH$$

Cysteine
Cys
C

$$H-\underset{\underset{NH_3^+}{|}}{\overset{\overset{COO^-}{|}}{C}}-CH_2-SH$$

Amino acids with charged polar side chains

Lysine
Lys
K

$$H-\underset{\underset{NH_3^+}{|}}{\overset{\overset{COO^-}{|}}{C}}-CH_2-CH_2-CH_2-CH_2-NH_3^+$$

Amino acids with charged polar side chains (continued)

Arginine
Arg
R

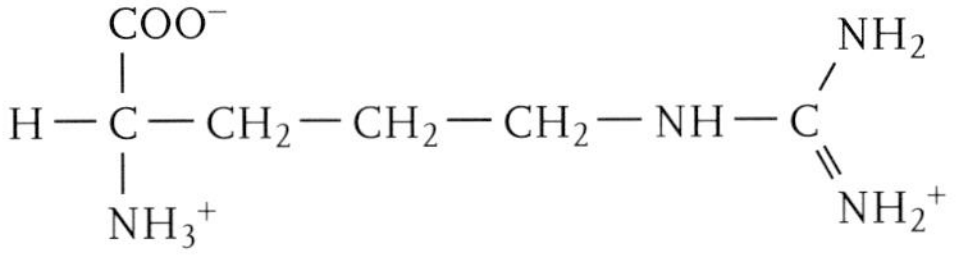

$$H-\underset{\underset{NH_3^+}{|}}{\overset{\overset{COO^-}{|}}{C}}-CH_2-CH_2-CH_2-NH-C\overset{\diagup NH_2}{\underset{\diagdown NH_2^+}{}}$$

Histidine
His
H

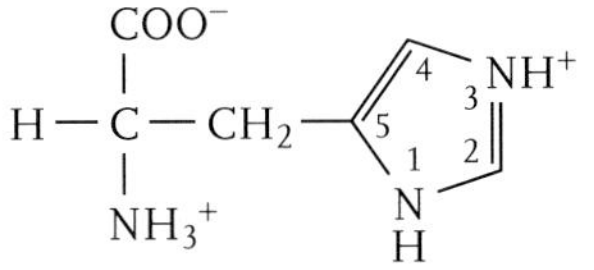

$$H-\underset{\underset{NH_3^+}{|}}{\overset{\overset{COO^-}{|}}{C}}-CH_2-\overset{4\quad 3}{\underset{\underset{\overset{N}{H}}{1\quad 2}}{5}}NH^+$$

Aspartic acid
Asp
D

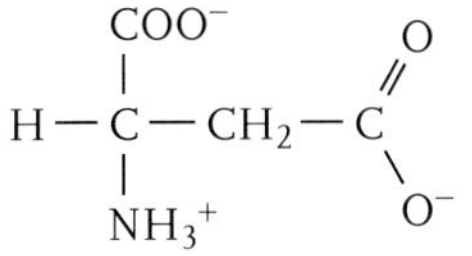

$$H-\underset{\underset{NH_3^+}{|}}{\overset{\overset{COO^-}{|}}{C}}-CH_2-C\overset{\overset{O}{\diagup\diagup}}{\underset{\diagdown O^-}{}}$$

Glutamic acid
Glu
E

$$H-\underset{\underset{NH_3^+}{|}}{\overset{\overset{COO^-}{|}}{C}}-CH_2-CH_2-C\overset{\overset{O}{\diagup\diagup}}{\underset{\diagdown O^-}{}}$$

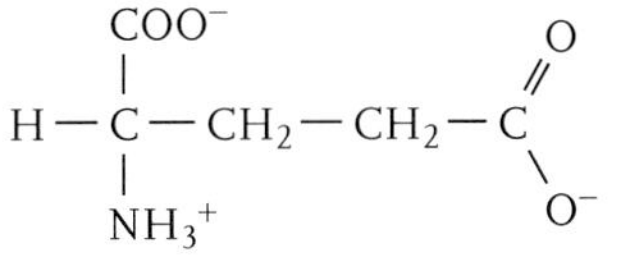

(Figures with kind permission from Voet, D., Voet, J.G., Pratt, C.W. (2001) *Fundamentals of Biochemistry*, Wiley, New York)

The Twenty Amino Acids with Abbreviations and Messenger RNA Code Designations

Amino acid	One letter symbol	Three letter symbol	mRNA code designation
alanine	A	ala	GCU, GCC, GCA, GCG
arginine	R	arg	CGU, CGC, CGA, CGG, AGA, AGG
asparagine	P	asn	AAU, AAC
aspartic acid	D	asp	GAU, GAC
cysteine	C	cys	UGU, UGC
glutamic acid	E	glu	GAA, GAG
glutamine	Q	gln	CAA, CAG
glycine	G	gly	GGU, GGC, GGA, GGG
histidine	H	his	CAU, CAC
isoleucine	I	ile	AUU, AUC, AUA
leucine	L	leu	UUA, UUG, CUU, CUC, CUA, CUG
lysine	K	lys	AAA, AAG
methionine	M	met	AUG
phenylalanine	F	phe	UUU, UUC
proline	P	pro	CCU, CCC, CCA, CCG
serine	S	ser	UCU, UCC, UCA, UCG, AGU, AGC
threonine	T	thr	ACU, ACC, ACA, ACG
tryptophan	W	trp	UGG
tyrosine	Y	tyr	UAU, UAC
valine	V	val	GUU, GUC, GUA, GUG

Complementary Strands of DNA with Base Pairing

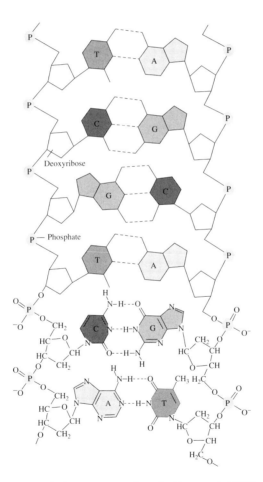

Two nucleotide chains associate by base pairing to form double-stranded DNA. A (Adenine) pairs with T (Thymine), and G (Guanine) pairs with C (Cytosine) by forming specific hydrogen bonds. (Figure with kind permission from Voet, D., Voet, J.G., Pratt, C.W. [2001]: Fundamentals of Biochemistry. Wiley: New York.)